BIOLOGY

William A. Jensen
Bernd Heinrich
David B. Wake
Marvalee H. Wake
University of California, Berkeley

Stephen L. Wolfe
University of California, Davis

The cell biology, genetics, and molecular
evolution chapters were adapted from
Stephen L. Wolfe's *Biology: The Foundations*

Wadsworth Publishing Company
A Division of Wadsworth, Inc.
Belmont, California

Biology Editor: Jack C. Carey

Production Editor: Mary Arbogast

Art Director: Nancy Benedict

Copy Editor: Beth Luey

Art Editor: Hedy McAdams

Illustrators: Darwen Hennings, Vally Hennings,
    Catherine Brandel, Victor Royer, Judy Waller,
    Judy Morley, John Foster

Cover photo: Steve Wilson / Entheos

Acknowledgments

The following essays are from previously published sources.

Essay 4-2 is from Hans Krebs, "The Tricarboxylic Acid Cycle," *Perspectives in Biology and Medicine,* vol. 14, 1973.

Essay 5-1 is from James D. Watson, *The Double Helix,* Atheneum, 1968.

Essay 7-1 is from Garrett Hardin, *Nature and Man's Fate,* Mentor, 1959.

Printed in the United States of America

1 2 3 4 5 6 7 8 9 10—83 82 81 80 79

L.C. Cat. Card No.: 79-3940
ISBN 0-534-00621-3

Biology books under the editorship of William A. Jensen, University of California, Berkeley

Botany: An Ecological Approach, Jensen and Salisbury

Biology: The Foundations, Wolfe

Biology of the Cell, Wolfe

Plant Physiology, Salisbury and Ross

Plant Physiology Laboratory Manual, Ross

Plant Diversity: An Evolutionary Approach, Scagel *et al.*

An Evolutionary Survey of the Plant Kingdom, Scagel *et al.*

Plants and the Ecosystem, 3rd, Billings

Plants and Civilization, 3rd, Baker

Other Books in the Wadsworth biology series

Biology: The Unity and Diversity of Life, Kirk, Taggert, Starr

Living in the Environment, 2nd, Miller

Energy and Environment: Four Energy Crises, Miller

Replenish the Earth: A Primer in Human Ecology, Miller

Oceanography: An Introduction, 2nd, Ingmanson and Wallace

Preface

Dealing as it does with the living world that surrounds us—and with our own bodies—biology has an immediacy not shared by other sciences. Because it extends from the atom to the ecosystem, it is also more complex than other sciences. In fact, biology seems at times to border on the incomprehensible. Yet, comprehend it we must if human beings are to survive. We have written this book in an effort to make biology both comprehensible and interesting to the college reader.

A Team of Specialists and a Unified Book Because it is such a broad subject, we believe that biology can best be treated by a group of specialists rather than by any one author. The benefits of this plan are obvious. Each section of the book will contain the newest work, the latest theories, and the most accurate information available. The dangers are equally obvious: Researchers may lose touch with students, specialists may lose sight of the larger picture, and too many authors may spoil the book (or write what sounds like several books instead of one). Against these dangers we have several safeguards: All of us are both teachers and researchers, teaching general biology as well as our own field. We are all at the same university and in close contact. And when the manuscript was finished, senior author William Jensen and Autumn Stanley, a professional writer, rewrote it, guided in part by more than sixty reviewers who watched, among other things, for variations in level.

The book presents material in a style and at a level that is easy to read by those approaching biology for the first time. The book is aimed at the general biology course, whether composed of students intending to major in biology or of both majors and nonmajors. We have kept scientific terms to a necessary minimum and have avoided scientific jargon.

We stress an understanding of basic concepts by stating them clearly in the chapters and illustrating them with many examples chosen from a wide range of organisms. Where appropriate, we have included human examples, but the emphasis is on comparative, rather than human, biology.

Core-Supplement Format The supplemental material presented here seems to be of two kinds (short and long) but has the same function—to kindle and rekindle interest by providing: (1) more information on a topic mentioned in the text, (2) another viewpoint from that taken in the text, or (3) slightly more advanced material (usually not more difficult, but rather closer to the frontiers of science). Varying from thoughts on the flow of evolution to a recipe for yogurt, the shorter pieces appear within the chapters themselves. The longer pieces appear at the ends of chapters. Often they relate the information in the chapter to human concerns. For example, the chapter on movement has a supplement on the physiology of long-distance running, the chapter on nutrition has a supplement on human nutrition, and the chapter on bacteria has a supplement on the preservation of food.

The Scientific Method in Real Life Biology is more than a compilation of concepts and facts. It is a living science constantly being extended by men and women working in the laboratory and in the field. To understand biology is to appreciate how it grows. We start with an explanation of the scientific method in Chapter 1 and present the historical development of various areas of biology within the chapters and supplements throughout the book. In addition, we have included personal accounts of how various scientists go about the business of research. These vivid and very human tales should counteract the popular image of the scientist as inhuman white-coated automaton. They should also destroy the image of the scientific method as a rigid set of rules that scientists slavishly follow. In short, they should bring science to life.

Organization The possible approaches to organizing an introductory biology text are many. We chose to start at the cellular level, after a review of the chemistry that underlies all modern biology. In this section we present a unified picture of cell structure and function. Through essays and supplements we present some of the recent developments that have affected and will affect our daily lives. Next comes the genetics section, starting with classical genetics, moving to molecular genetics, and ending with current concepts of the gene. Again the supplements bring out the implications of recent findings in this important field. In the following section on development we discuss plant and animal reproduction and development in separate chapters. We do this because the principles are so different for plant and animal development that this seems the most logical way to proceed. We next cover organismic form and function, using a comparative approach centered on a particular process. Thus, we have chapters on nutrition, gas exchange, osmoregulation, and so on, with the examples drawn from all types of animals and plants. We cover ecology and evolution together in the next section because we feel this treatment is in keeping with current trends. Our treatment of evolution reflects the most modern thinking in this important field. Finally, we consider the origin of life and the diversity of life on earth today. Here we use a five-kingdom approach because we feel this most accurately reflects the current state of our knowledge.

We know that this organization will not fit every biology course. We have, therefore, made the sections independent of one another, both by organizing them so and by repeating key definitions and concepts. By assigning or not assigning the optional supplements instructors can also change the nature of the course. Every effort has been made to make the text flexible.

Supplements Text books such as this one are only part of a biology course. We have expanded the usefulness of the book by designing a series of supplemental materials to be used with it. These include the traditional laboratory manual, student study guide, and instructor's manual. We have tried to introduce some new and different ideas in all three of these essential aids. In addition, we have been fortunate to be able to work with BIOLOGY MEDIA, a company that produces media aids for biology teaching. They have made available a series of multi-image lectures and slide-tape modules that can be used in conjunction with this text. Such materials can be used either in conventional lecture-laboratory courses or in self-paced individualized courses.

Acknowledgments We would like to thank the many people who have helped us in preparing this book. First among these is Stephen Wolfe of the University of California, Davis, who wrote the initial drafts of the cell biology and genetics sections and the chapter on the origin of life. We owe particular thanks also to Autumn Stanley and Robert McNally who helped in the rewriting and to Mary Arbogast who so skillfully oversaw the transformation of a manuscript into a book. We also want to thank the more than sixty reviewers who helped us get the facts and the level right.

Reviewers

Henry C. Aldrich University of Florida
Glenn D. Aumann University of Houston
James M. Barrett Marquette University
Marjorie P. Behringer University of North Dakota
Albert F. Bennett University of California, Berkeley
James Bonner California Institute of Technology
Richard Boohar University of Nebraska
David W. Borst University of California, Berkeley
George M. Briggs University of California, Berkeley
David D. Cass University of Alberta
Robert Catlett University of Colorado
Chris J. Chetsanga University of Michigan
Roger W. Comeau Middle Georgia College
David W. Deamer University of California, Davis
Alfred G. DeBall Macon Junior College
Warren D. Dolphin Iowa State University
Ralph P. Eckerlin Northern Virginia Community College
Gordon G. Evans Tufts University
Arlene F. Foley Wright State University
Douglas Fratianne Ohio State University
B. E. Frye University of Michigan
Ernest M. Gifford, Jr. University of California, Davis
Hal T. Gordon University of California, Berkeley
George C. Gorman University of California, Los Angeles
George E. Goslow, Jr. Northern Arizona University
Anthony J. Griffiths University of British Columbia
Harlo H. Hadow Coe College
William M. Harris University of Arkansas
Rachel Hays Ecology Consultants, Inc., Fort Collins, Colorado
S. Robert Hilfer Temple University

James A. Hopson University of Chicago
Charles E. Keys University of North Alabama
George H. Krasilovsky Rockland Community College
Fred Landa Virginia Commonwealth University
Omer R. Larson University of North Dakota
Harvey B. Lillywhite University of Kansas
Lon L. McClanahan California State University, Fullerton
Lawrence B. Mish Bridgewater State College
Robert E. Moore Montana State University
Bruce C. Parker Virginia Polytechnic Institute
Eric T. Pengelley University of California, Riverside
Ralph S. Quantrano Oregon State University
Alvin F. Reeves University of Maine
Fred R. Rickson Oregon State University
Kenneth D. Roeder Tufts University
Cleon W. Ross Colorado State University
Frank B. Salisbury Utah State University
Jan Savitz Loyola University
Valerie Seeley Queensborough Community College
David Shappirio University of Michigan
Kemet D. Spence Washington State University
Herbert Stern University of California, San Diego
Daryl Sweeney University of Illinois
Rosalie Talbert Nassau Community College
Norman S. Tweed Fort Steilacoom Community College
Jon Weil University of California, San Francisco
Russell Wells St. Lawrence University
Harry S. Wessenberg San Francisco State University
Barbara L. Winternitz The Colorado College
Robert Woollacott Harvard University

Brief Contents

Contents

ONE
_____

Introduction

1

The Science of Biology

The man checks his camera once more and shifts his body into a more comfortable position in the cramped blind. His mind becomes alert as he hears a noise in the bushes ahead of him and the olive green female bowerbird appears. The male in his elaborately constructed bower begins his courting rituals. The man in the blind happily photographs and takes notes, the hours of waiting over.

The elderly woman sits down before the electron microscope. With the ease of years of practice, she inserts the specimen and turns on the beam of electrons. Looking for the presence of virus particles in the transport tissues of the plants she is studying, she scans the section of tissue. Today she may get the evidence that will further her work—or she may not, and the hours of work will show no results. Yet her years of experience tell her she will ultimately succeed.

A group of young men and women sit around a crowded table drinking coffee. They look once more at the number-covered papers before them. A feeling of excitement runs through the group. They have taken the first step in manipulating the gene, the basic unit of heredity, but can they make the next step and transfer it to a new host? The excitement would be even higher if all were not remembering the days and nights of laboratory work required to get this far. There is also the nagging question of just how far a second group of researchers has progressed. If the rival group were to publish first, all the work they had done, though important, would be secondary.

Each of the individuals in these sketches is a biologist. With thousands of other biologists, they are involved in the study of the living world that surrounds us. Using research tools as unlike as dugout canoes and electron microscopes, biologists have probed the life around us and within us. This book is a survey of some of the data and concepts that people like these have collected and created over the 300 years or so that modern biology has been in existence.

Biology is a human endeavor. It is a science, an attempt by humans to understand the living world. This world existed long before any human thought to ask questions about it, but biology did not exist until that moment. Like other sciences, biology has developed a method for its work.

The Methods of Science

Scientists go about the business of investigation in a variety of ways. Their methods of study in part reflect the areas they work in, and in part reflect their personalities. It is easy to portray "the scientist" using "the scientific method" to solve an "important scientific problem," but this is a gross oversimplification of what actually takes place. There is, however, a generally accepted method of investigating problems in science that has been called the *scientific method.*

The generalized scientific method can be stated as follows: (1) Select a problem related to some phenomenon of manageable size and collect as much information about it as possible (this may be from observation or from the published reports of others); (2) formulate a testable statement, known as a *hypothesis,* about the phenomenon; (3) experiment or make further observations to test the hypothesis; (4) if the hypothesis fits the observations, keep it as a useful explanation of the phenomenon. Its usefulness lasts until more information disproves it or a new tested hypothesis explains the data better.

We can look again at the courting bowerbirds to see how the scientific method operates. The man in the blind *observes* that the male bowerbird builds an elaborate nest or bower and puts a collection of found objects in front

Figure 1–1 A male bowerbird and his bower.

of it (Figure 1–1). Often stolen from nearby houses, the objects are usually brightly colored, shiny, or both. The female bowerbird seems to notice first the bright objects and then the male bird; then she enters the bower. From these observations, the watching biologist guesses that the male bird puts the objects before his bower to attract the female. However, this guess, or hypothesis, is useful only *if it can be tested*. To test it, the biologist takes away the objects in front of a nest. Now the female will have nothing to do with the male and will not enter the bower. This observation does not prove the hypothesis is right, but it does indicate that the explanation is reasonable and one on which the biologist can base many similar experiments.

We can look at one other example of the scientific method. The observation had been made before the turn of the century that the shoot tips of grass seedlings newly emerged from the soil would bend toward the light when illuminated from one direction. This is but one example of the general phenomenon of plants growing toward light, but it provided a manageable system with which to experiment. The question became: What caused the seedling to bend toward the light? In 1927, a young Dutch botanist, Frits Went, decided to investigate this question.

Searching published accounts of other scientists' experiments, he found that various explanations could be discarded. For example, previous work suggested it was unlikely that electrical impulses, such as those found in nerves of animals, were involved. Went hypothesized that a growth-promoting chemical produced in the shoot might move down from the tip along the dark side of the seedling and cause growth of cells on one side of the shoot, thus accounting for the bending. To test this hypothesis, he designed the following experiment: The tips of seedlings were cut off and placed on tiny blocks of a gelatinlike substance known as agar. Went reasoned that if a chemical produced in the tip normally moved down the shoot from the tip, this substance might move into the agar instead. Next, Went cut the tips from a second set of seedlings grown and kept in the dark (so that they would grow straight). On each shoot stump from this second set of seedlings, he placed one of the agar blocks he had obtained from the previous step, but he placed it to one side of the shoot stump. If a growth-promoting chemical was present in the agar, it should move down the shoot only on the side below the block, and the shoot should therefore bend away from the side with the agar on it. Went also placed untreated agar blocks on some of the shoots in an identical manner to the treated agar blocks. He performed this *control test* to see whether the mere presence of plain agar pieces could somehow cause bending.

At 3:00 in the morning Went got the first results of his experiment: The shoots with the treated agar blocks were bending, but the ones with the untreated control blocks were still growing straight. He ran to his father, a famous Dutch botanist, and urged him to get out of bed and come to the laboratory to see the results himself. The elder Went told his son that either the experiment would work again the next day, or there would be no point in observing it now. With this he turned over and went back to sleep! The point the elder Went made is a key feature of science—experiments or observations must be repeatable, not only by the original investigator but by others as well.

Went was able to repeat his experiment with the same results the next day, and many other experimenters around the world were able to repeat it in the months and years that followed. But what did Went's results mean? They showed that his explanation was a reasonable one, but not that it was the only one or even necessarily the best one. That is, they did not prove that Went's hypothesis was right, only that it was not wrong. However, Went's hypothesis also offered the possibility of additional experiments. When several types of experimentation lead scientists to the same conclusion, that conclusion is generally accepted as correct, but it may later be modified or even rejected if another hypothesis proves to describe the phenomenon more accurately.

Characteristics of Science

By this roundabout and tedious method, scientists slowly construct explanations of natural phenomena. These explanations are subject to constant testing through observations and experimentation: They are accepted only as long as they are shown not to be wrong.

The more general the area explained by the results—the more organisms they apply to—the more important the work. For example, the discovery of the molecular structure of DNA (a molecule contained within chromosomes) was a scientific bombshell because it applied to all living organisms. Went's research laid the foundation for the field of growth-hormone research in plants. Occasionally, scientists can tie together seemingly unrelated observations through a single great theory. Darwin's theory of natural selection to explain evolution is an example. Such truly great concepts are rare, but hypotheses and theories tying smaller areas together are also important, and far more frequent.

In the pursuit of scientific truth, certain pitfalls must be avoided. One such pitfall is being more interested in being right than in being objective. In the enthusiasm of following a line of thought, it is easy to overlook pieces of data or to make observations that conform more to what the investigator wants to prove than to what is actually taking place (see Essay 1–1). In rare cases, data may actually be fabricated and deliberate fraud committed. Such deceptions have been very few in the history of biology, considering the number of individuals involved in research.

Before we leave the topic of the characteristics of science, we should consider the role of luck. Much has been said concerning this elusive factor in discoveries in science. One of the most famous examples is the discovery of antibiotics by Alexander Fleming in 1928. While working with the staphylococcus bacterium, Fleming noticed a bacteria-free circle around a mold growth that was contaminating a culture of the bacteria. This observation could have been made by many, because such mold contaminations are frequent in laboratories. Yet only Fleming was lucky enough to understand the significance of the observation—that the mold was producing a substance (penicillin) that killed the bacteria surrounding it. Fleming had for years been looking for antibacterial substances and had already discovered an antibacterial agent in tears and saliva. He had not planned the experiment with the bacteria and the mold, but he capitalized on a chance observation—an observation for which he was prepared by years of training and experience. When a famous baseball player was asked why he was so lucky in making impossible catches, he said his luck consisted of eight hours of hard practice every day! This is the same kind of luck that operates in science.

Science and Scientists

What are science and scientists really like? What is research in science, especially biology, like? As we tried to show at the beginning of this chapter, biologists come in all shapes, ages, and sexes. Few actually wear white coats or work in shining laboratories, but some do. Many work in the field, while others would not know what to do if they found themselves outside their laboratories. We said that biology is a human endeavor, and this is reflected in the people who are biologists. Some are retiring and prefer to work by themselves; others are as aggressive as any business tycoon and run laboratories with personnel and budgets as big as many a business. Although women in biology are still too few, they have played an important role in the development of biology, and their number increases each year.

How do these people go about doing research in biology? Is the progress of science as cut and dried as the scientific method implies? Research is anything but cut and dried. It isn't a formalized routine that invariably produces exciting data. Research is a highly individual business. Some biologists may work a lifetime making observations and never formally present a hypothesis that they test, yet they work within the general framework of the scientific method. Others may spend months devising a specific set of experiments, which may take only a few days to run, to test a formal hypothesis. The area of study and the method of attack of most biologists reflect their personalities, as modified by the training they have received.

Research usually goes in fits and starts. As new ideas and equipment become available, the research moves forward, and as ideas and techniques become inadequate, the work slows. One part of research that never changes is that so much of it is repetitive. Repeating a procedure or an observation over and over again to verify data is anything but exciting. The seemingly endless hours spent getting a procedure to work perfectly, and then the additional hours spent doing the procedure under the right conditions to obtain valid results, may be discouraging and time-consuming.

Yet the excitement is there, and it is real. Knowing something no one else knows is an exhilarating experience. The fact that your work can lead to the cure of a disease or solve a problem that has puzzled others for years is exciting. To see the fruits of one's own intellect and ingenuity recognized is a very fine feeling indeed.

But science is humbling as well as exciting. Because the object of all good science is to continually challenge existing hypotheses and question all facts and concepts, your conclusions may not be valid for long. Yesterday's great discovery may be tomorrow's footnote. However, science is a progressive, cumulative enterprise. Each observation, each piece of data, adds to the whole and contributes to the progress of the entire field. Supplement 1–1 explains how data and conclusions are communicated among biologists.

Science, including biology, is a never-ending quest. It is easy when reading a book such as this to get the impression that most of the facts are known, that most of the great thoughts have been expressed, that biology is complete, with only the details left to be filled in. Nothing could be further from the truth. Vast areas remain to be explored, and important problems remain unsolved.

In this introductory chapter, we will present some examples of progress in biological knowledge. We will pick a few major points from the various sections, to give you a preview of the book's organization and to show how major ideas and discoveries have shaped the different areas of biology.

Cell Biology The cell is the basic unit of life. All organisms are composed of either one cell or collections of cells. Yet the cell was not known until the invention of suitable microscopes in the late 1600s. The sight of actual cells sparked a burst of activity that extended into the early 1700s and showed that cells were important in understanding both plants and animals. This work came to a halt, however, as scientists reached the limits of these early microscopes.

The early 1800s brought new and better microscopes, and a whole new era of cell exploration. The cell theory, which stated that the cell was the structural and functional unit of life, spurred on these studies. The development of organic chemistry (chemistry of carbon-containing compounds like those found in living organisms) also encouraged the exploration of the cell.

By the early part of this century, biologists knew the various parts of the cell and how cells reproduced because they could see some cell structures with better light microscopes. They also knew they still had much to learn about the function and composition of the various cell parts. Before they could go further, they would need not only better microscopes, but a better understanding of cell chemistry.

After World War II came several developments, including the electron microscope with its astonishing powers of magnification; radioactive atoms that could be followed in reactions within the cell; and tremendous advances in biochemistry. Cell research blossomed and is still blooming to this day. We have found parts of the cell that we could not see before, we have learned much about the workings of the cell parts, and we understand more clearly why cells are so important in the growth and functioning of organisms. Great unifying concepts played their role in this history of the cell (discussed in Section 2 of this text).

Essay 1–1

How We Become Biased by Our Hypotheses

The testable hypothesis has a central role in the scientific method. How one goes about formulating a useful hypothesis and what one does with it is a highly personal affair. To gain an insight into the scientific method as it is actually practiced, we have asked Kenneth D. Roeder of Tufts University to tell of his experiences.

Much of experimental research in biology asks how the mechanisms of living things work and how these mechanisms have evolved. In putting such questions to insects regarding their sense organs and nervous systems, I have learned that a major stumbling block in the way of obtaining valid answers is bias on the part of the experimenter. The bias begins with the formulation of a hypothesis, but without a hypothesis it is impossible to plan meaningful experiments. For instance, one cannot observe and become curious about certain actions of an animal without forming some sort of tentative theory. This may be to the effect that the animal uses a particular set of muscles in the action, or that the action is steered through a particular sense, or that the survival value of the action lies in enabling the animal to find food or a mate or to avoid being eaten by a predator. The working hypothesis suggests the research plan, which is merely a set of tests whose results logically relate to one another if the hypothesis is correct.

The conception of a hypothesis is an exciting event, rather like the conception of a new living creature. But either type of conception will become valid and viable only if it is followed by research in the first case and embryonic development in the second. Both of these processes require time, hard work, and the expenditure of energy. Indeed, experimenters become biased in favor of their hypotheses much as parents become biased in favor of their child. This must be so if the hypothesis is to become bolstered by a logical structure of facts and the child is to be nurtured to maturity. But both experimenter and parent pay a price for developing "tunnel vision." Researchers are likely unconsciously to ignore or simply not "see" subtle indications that run counter to their hypotheses, just as loving parents are apt to discount deficiencies in their child.

But the comparison ends here, for we cannot discard a child who fails to come up to our expectations, while the researcher must always be ready to discard an untenable hypothesis. This can be a most painful and difficult experience, particularly if it becomes necessary after years of research and there is no new hypothesis to take its place. But if one admits that one's brain is more than a logic machine and subject to this very human tendency to become biased, then the "moment of truth" can be one of the most joyful and exciting experiences a scientist can have. For the hypothesis may become untenable because of mounting evidence favoring another explanation, evidence that has been unconsciously disregarded in the past but which has been figuratively crying for attention. One feels that one is suddenly permitted in spite of one-

---

Genetics Genetics has advanced by a combination of great ideas and the selection of the right organisms to study them in. The story of modern genetics starts with Gregor Mendel, an Austrian monk who discovered the fundamental rules of heredity and proposed the existence of the heredity unit we now call the gene. Working with the garden pea, an ideal plant for his experiments, he published his data in 1864. It is one of the ironies of modern biology that the importance of this work was not recognized until 1900, years after Mendel's death.

The rediscovery of Mendel's paper stimulated the study of the gene, and many great biologists attacked the problem. One of the greatest was Thomas Hunt Mor-

gan, whose work helped establish that the genes are located on the chromosomes, structures present in the nucleus of the cell. Morgan picked the common fruit fly as his experimental organism. Easy to raise and remarkably short in generation time, the fruit fly proved a boon to the study of the gene.

By the 1940s geneticists were asking how genes could control function and development within cells and organisms. A major breakthrough came when Beadle and Tatum showed a one-to-one correlation between genes and enzymes. Enzymes are proteins that are important in chemical reactions occurring in the cell. Beadle and Tatum used the pink bread mold as their experimental organism.

self to take a completely new view of some aspect of nature.

Three or four times in my career the weight of evidence has forced me to discard a favorite but outworn hypothesis. The most recent case involved the mechanism of hearing in a small group of hawkmoths. Behavioral evidence showed that these moths could detect high-pitched sounds despite the absence of a thoracic tympanic organ, or ear, like that found in other moth groups. I soon located the hawkmoths' acoustic detector on the head and found that the vibration of one of the mouthparts—the palp—played a central part in its operation. In considering how the palp might serve as an ear I arrived at a hypothesis that still today seems to be logical and to have good physical justification but that was nevertheless completely wrong.

The palp is very light and small and is delicately articulated with the moth's cranium. It is shaped like a small handle, and I reasoned that the palp must be rotated slightly by each arriving sound wave so that it vibrates in its articulating joint. All that is needed, then, is a sense organ at the palp's articulation to respond to the slight rotation and generate nerve impulses.

My colleagues and I spent two years in a fruitless search for the sense organ postulated to lie near the palp base and respond to palp deflection. Numerous experiments were carried out, all but one of which supported the palp-deflection hypothesis, but no sense organ could be found. However, I still clung to my hypothesis in spite of minor inconsistencies in the experimental facts.

Then one day the hypothesis was abolished in a flash.

I was idly touching various structures on a moth's head with a fine bristle while listening to acoustic nerve impulses in its nervous system. When I touched an insignificant projection on the ventral surface of the head at some distance from the palp articulation, a burst of impulses came from the acoustic nerve fibers. This projection, which I later learned was called the pilifer, is normally covered by the palp and therefore not visible from outside. Vibrations in the palp caused by the sound waves are mechanically transferred to the pilifer, the two organs being in contact when the palp is held in its natural position. When I moved the palp away so that I could see the pilifer, the system would not work as an ear, and it was mere chance that the mechanical disturbance caused by my bristle brought the pilifer to my attention and led to an explanation of how this ear works. Without the palp-deflection hypothesis, I would probably never have begun the search in the first place, but the fact that it was a false hypothesis greatly lengthened the hunt.

The technical aspects of this story are to be found in the two papers listed below. The first was published before discovering the palp-pilifer mechanism.

Roeder, K. D., and A. E. Treat. 1970. An acoustic sense in some hawkmoths (Choerocampinae). *Journal of Insect Physiology* 16:1069–1086.

Roeder, K. D. 1972. Acoustic and mechanical sensitivity of the distal lobe of the pilifer in choerocampine hawkmoths. *Journal of Insect Physiology* 18:1249–1264.

This common fungus could be manipulated easily in the laboratory and provided a means of investigating the chemical basis of genetics.

The next great step in learning how genes control cell development and function came with the realization that the nucleic acids, a special group of compounds in the cell, were involved in protein synthesis and hence in enzyme synthesis, since enzymes are composed of proteins. The work of Watson and Crick presented a basis for understanding how one nucleic acid, deoxyribonucleic acid, or DNA, located in the cell nucleus as part of the chromosomes, could provide the means for containing and transferring the information necessary to produce an enzyme. Work with viruses and a bacterium known as *Escherichia coli,* or *E. coli,* has greatly expanded our knowledge of gene action in recent years.

Genetics is one of the most active and controversial areas of biological research today. We will examine some of the questions of current interest in Section 3.

Development Our knowledge of plant and animal development is closely related to our understanding of the cell. The improvement of the microscope in the last half of the 1800s opened the modern era of plant and animal embryology. Although animal sperm had been seen in some of the first microscopes, biologists did not understand fertilization in either animals or plants until microscopes and microscopic staining and preparing techniques were improved.

The great period of experimental animal embryology came in the first quarter of this century, as two great embryologists, the American Harrison and the German Spemann, opened the way to understanding the forces involved in early animal development. Using simple techniques in highly ingenious experiments, they worked out a broad outline of the events and chemical interactions of embryogenesis.

Progress in both animal and plant development slowed as the problems outdistanced the means available to solve them. As in cell biology, several technical advances, as well as advances in other fields, came to the rescue. The electron microscope, cell and tissue culture (whereby cells and small pieces of tissue can be kept alive and grown outside the animal or plant), and increases in our knowledge of cell chemistry all contributed to the current resurgence of interest in development.

Anatomy and Physiology Anatomy (study of structure) and physiology (study of the chemical and physical processes occurring in plants and animals) touch areas of immediate concern to us. We all care how our bodies function and what they look like. Often we can do little about either, but this does not dampen our enthusiasm for knowing about them. Since we all directly or indirectly depend on plants for our food, we also have a natural interest in them and in how they function.

These concerns are mirrored in the historical development of these areas of biology. Because of their application to health, anatomy and physiology were among the first fields to be intensively explored, and they continue to be in the forefront of current research.

The way animals move, the circulation of their blood, the transmission of nerve impulses, the contraction of muscles, the chemical coordination of hormones, the digestion of food, the exchange of gases within the body, and many, many more problems have been and are being attacked. Our ideas become ever more sophisticated and our data more detailed, but we really do not know how many of these basic phenomena work. Research in these and other areas of animal physiology continues, using radioactive isotopes, electrical measuring devices, the electron microscope, tissue cultures, X-ray analysis, and hundreds of other instruments and methods. Because of the medical implications of so much of this research, funds are readily available—no small consideration in modern biology.

Plant physiological research has not moved quite so fast, but the advances have been steady. Studies of photosynthesis, water movement up a tree, dissolved sugar movement within a plant, adjustment of plants to their environments, plant responses to changes in their surroundings, such as changes in the length of day—all these are topics of current research. Again, as in animal physiology, the techniques used are diverse and becoming more and more sophisticated each year. Studies in physiology have indirectly led to improved food production and crop yield, improved lumber production, and many other advances of great importance to humanity.

Evolution and Ecology Charles Darwin's theory of natural selection as the basis of evolution has dominated the study of evolution since the 1860s. Evolution is the concept that organisms arise through gradual change in preexisting organisms. This idea was not new with Darwin, but he provided a hypothesis to explain how it worked. In so doing, he unified a large body of data and gave us a way of thinking about a complex phenomenon. The concepts of evolution and natural selection have dominated large sectors of biology for generations.

We are in what many biologists consider the second great period of research in evolution. This time the stimulus has come from genetics. The modern evolutionary biologist seeks to understand the genetic basis for change in populations. Active research in this area is giving us exciting insights into the process of evolution.

Related to evolution is ecology. This is the study of the interaction between the organism and its environment. Because of pollution and expanding human population, and concern about their combined impact on the environment, ecology has become a watchword in recent years. Yet the science of ecology is still an infant. We are only beginning to learn how organisms and the environment interact.

In both evolution and ecology, it is difficult to pose meaningful hypotheses and do experiments. Because of the many factors involved in any problem in these fields, the research moves more slowly than in some other areas, and produces less clear-cut results. But the importance of ecology to our future makes this a field with great potential and an exciting future.

Survey of Living Organisms The final section of the book surveys the organisms that make up the living world. The sheer diversity of plants, animals, and microbes that share this planet is hard to comprehend. The great impetus to organizing this mass of living creatures was given by the Swedish biologist Carl Linnaeus, who in the 1700s began systematically cataloging plants and animals. He popularized the idea of giving all organisms names of two words: one that identified the large group of organisms to which it was related (genus), and a second that identified the specific kind of organism (species).

In the 300 years of the modern study of organisms, we have learned a great deal about our fellow inhabitants of the earth. The final chapters of the book will introduce at least some of them to you. We also treat briefly the question of how life may have originated on earth. This is an area of great speculation and relatively little data, but it is one of the most fascinating in modern biology.

Biology as a Human Endeavor

Throughout this introductory chapter we have tried to make the point that biology is not a disembodied group of facts collected by robots in white coats. It is a very human activity conducted by men and women who may have started out taking a biology course and using a biology text much like this one. Many of you who are using this book now will be the source of new information in the future. Many of you will have the equally important task of applying the biology you will learn to a host of practical problems, including the healing of your fellow humans. To others, biology is but an interest in the world and organisms around us. This concern is no less important, for to be human is to be curious, and to satisfy this curiosity through study is to be educated.

Summary

In this chapter we have explored the way scientific knowledge is gathered. We have discussed the scientific method and the need for testable hypotheses. Stressing that biology is a human endeavor, we have shown that progress comes when many individuals with diverse personalities and methods of approach all work toward the goal of increasing our knowledge of the living world. We also previewed the organization of the book with glimpses of ideas and techniques that helped shape the various fields. Finally, we emphasized that biology is an active field of research, that not all the questions are answered, and that not all the data have been collected. Biology is the science of life, and there is a lot of life in biology!

Supplement 1–1
The Art of the Scientific Paper

No matter how brilliant or important research is, if it never gets published, it is as if it had never been done. We have tried to show how ideas in science develop and how ideas and data interact to produce an understanding of the way in which nature works. But this growth depends on the communication of the data and ideas.

The major medium for this exchange is the scientific journal. First started in the 1600s, these journals now number in the thousands for the biological sciences alone. Most biologists examine only a few dozen of them and actually read only a small number of papers per issue. Yet these journals constitute the major means of advancing biological knowledge.

How does one publish in a journal? Some journals are managed by scientific societies, and you must be a member to submit manuscripts. Other journals will publish acceptable papers submitted by anyone. This does not mean that all manuscripts are published. In most journals, the editor, assisted by an editorial board and technical reviewers, has the final say on whether a paper will be published. To see how the process works, let's follow a manuscript from submission to publication.

Say you have done a piece of research that you think is worthy of being shared with your fellow biologists. First, you write the paper and document your findings as fully and clearly as possible. You then send the manuscript to the editor of the journal that you decide is most appropriate for the subject being discussed. The editor reads the manuscript and sends it to two reviewers, people the editor feels are experts in the field of your research. In most biological journals these reviewers are anonymous. They read the manuscript, looking carefully to see whether you have presented enough original, carefully collected data to justify the conclusions you have drawn. They also check to see whether the writing is clear and the data are adequately illustrated. Finally, they evaluate the importance of the work. If they are happy with the paper, they will advise the editor to publish it as it stands, and usually he or she will do so.

Normally, however, the reviewers will recommend publication only after certain changes they have suggested are made. Again, the editor will probably agree, and may even make some further suggestions. In this case, the manuscript comes back to you, so you can make the suggested changes and return the paper to the editor.

On the other hand, the reviewers and the editor may feel that the work does not warrant publication for any number of reasons, and the paper is then rejected. There are few scientists who have not had a paper rejected—no matter how famous they may be!

After the revised manuscript is returned to the journal, the editor sends it to the printer, who sets it in type. A proof is made of this type and sent to you. You read it for typesetting errors and return it to the editor, who returns it to the printer with the corrections indicated. The journal is now assembled in its final form, printed, bound, and mailed out. At this time you can order a special printing of only your paper. These "reprints" are important in the communication of data, because few people subscribe to more than a few journals, but they like to have copies of specific papers from a wide range of journals. They will therefore send you a postcard or letter asking for a reprint of your paper. You may also decide to send a reprint to a person you think will be interested in your work. Thus, most scientists have large reprint files of papers in their fields.

Until now there has been no mention of money. But you will have to pay for the reprints, sometimes with an additional charge per page or per plate of figures. These charges vary greatly from journal to journal, but they are usually stated clearly in the journal itself, and you learn about them early in your dealings with the editor. If you cannot pay them, the paper is usually published anyway. In no case are you paid for the publication of a research paper.

Research papers have an almost formal organization, following the same general plan with many variations in detail. The paper starts with an introduction in which the author states the problem or question to be answered in the paper and then provides the necessary background for it. Next comes a section on materials and methods. Here the author tells you what organisms are used, how they were grown, or gives any other information on materials or methods necessary to understand the research being presented.

Then comes a description of the results the author got, using these organisms and methods. The results are presented with great care and may be illustrated by graphs, tables, drawings, or photographs. The results section is the most important part of the paper, since the data here will ultimately be the most useful to other researchers. But the discussion section that follows it is often the most interesting, since the author will tell you what she or he thinks the data mean. You may accept those conclusions, or you may draw your own from the data.

Another important part of the paper is the summary, which presents a brief report of the work and the author's conclusions. The summary is also called an abstract and is often printed at the front of the paper, before the introduction. This is because the summary or abstract is usually the first (and often the only) part of a paper that busy scientists read. Throughout the paper you will see numbers or names followed by a year (Jensen and Salisbury, 1972). These are references to other papers, and a list of these can be found at the end of the paper.

Research papers are usually not easy to read. There are several reasons for this. One is that you are coming in on the middle of a continuing story that may have many different authors in many parts of the world. Without enough background, it is hard to see how this work fits into the overall picture. There is also a formidable vocabulary that has been introduced to ensure precision of statement and communication. Furthermore, the style used by many scientists is not the most readable, but it is tolerated if the meaning is clear. Finally, the research is usually not presented in the way it was actually conducted; rather it is logically organized for the purpose of the paper.

How many people read a given scientific paper? That depends on the paper. Some, like the famous Watson and Crick paper on DNA, have been read by thousands, but these are the exception rather than the rule. Important papers may be read by hundreds, but a run-of-the-mill paper will probably be read by only a dozen people around the world. Yet the data are there, and slowly and surely they will accumulate, changing our ideas and concepts. The art of the scientific paper is an essential part of the advancement of science.

TWO

The Cell

2

Matter, Energy, and Life

Look for a moment at an ordinary lead pencil. Where the paint is sharpened away, you may be able to see a little of the grain of the wood (Figure 2–1). If you put a section of the wood under a light microscope, you can see that it consists of the remains of its basic units, called **cells.** The thick walls of these cells are clearly visible with only slight magnification. Using the much greater magnification of an electron microscope, you could resolve the wall into a series of subunits known as microfibrils. At this point we run out of ways to analyze cell-wall material visually and must turn to chemical means.

Using chemical methods, we can determine that the cell-wall material is composed of carbon, hydrogen, and oxygen atoms. **Atoms** are the basic structural units on which the chemical characteristics of matter are based. The notion of an indivisible as the basic unit of matter was introduced by the ancient Greeks, although their understanding of the concept was quite different from ours (see Essay 2–1).

Elements are pure substances consisting of a single kind of atom; they cannot be broken down into simpler substances by ordinary chemical or physical means. The carbon, hydrogen, and oxygen in the cell walls of wood are 3 of the 106 known elements and are among the most common found in living matter.

The atoms we have uncovered in the cell-wall material are not found in random amounts, but in fixed proportions to one another. A pure substance composed of two or more elements held together in fixed proportions by **chemical bonds** is called a **compound.** In compounds, such as sugar and water, the molecule is the smallest unit into which the substance can be subdivided without destroying its characteristic properties. A **molecule** is a structural unit consisting of two or more like or unlike atoms, held together by chemical bonds.

In the materials making up the cell wall, such as cellulose, we find compounds that consist of huge molecules composed of repeating units. Such large molecules are termed **polymers,** and the repeating units in them are themselves formed from molecules. In cellulose, the basic unit is the simple sugar glucose. Polymers are particularly

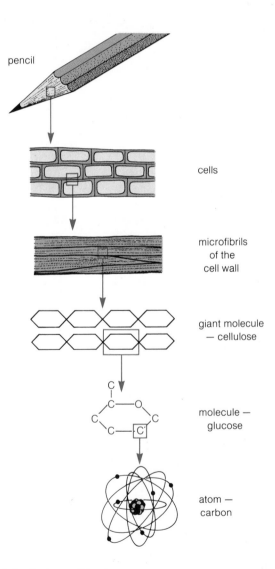

Figure 2–1 From pencil to atom.

Greeks, Romans, and Atoms

The concept of the atom and the word itself go back to the ancient Greek philosophers. The Greeks, however, did not have experimental skills or equipment, to say nothing of the concept of experimental proof. How, then, did they arrive at the idea of the atom? The story goes that the Greek philosopher Leucippus was strolling along a beach with his student Democritus when he began to wonder whether the water of the sea was continuous. Leucippus reasoned that from a distance the grains of sand on the beach looked continuous, but when viewed more closely they could be seen to be individual grains. Could the same hold true for water? Could water be divided into smaller and smaller drops? Could the division process be continued indefinitely, or was there a limit beyond which further division was impossible? Leucippus reasoned solely on intuitive grounds that there was a limit to which the division could be carried and that a particle would be found that could not be subdivided.

Democritus named this ultimate particle the *atomos,* meaning "indivisible" *(a* = without, *tomos* = cut), from which we derive the word *atom.* Democritus expanded the atomic theory of matter to include the concept that the atoms of each element (and to the Greeks there were only four: fire, water, air, and earth) had distinctive sizes and shapes. The theories of Democritus were taken over by Epicurus, a philosopher of the third century B.C., who incorporated them into a materialistic philosophy of life. The fullest description of this philosophy comes from Lucretius in his long poem "On the Nature of Things," written in the first century B.C. Some of the essential statements of this impressive work are: (1) all matter is composed of atoms, and all changes in matter are merely changes in the groupings of atoms; (2) there are many kinds of atoms, of different sizes and shapes, and the properties of bulk matter reflect in some way the properties of the atoms of which it is composed; (3) the only forces are those of the collision between atoms; (4) apart from atoms, the only reality is infinite empty space through which the atoms move.

Many of these points seem remarkably modern in concept, but it must be stressed that the Greek idea of the atom was based on intuitive rather than experimental grounds, and is actually quite different from ours. We know now that the atom is divisible into quite a number of smaller units and that, although the atoms of the various elements are different, they are composed of the same basic units. Our modern concept of the atom dates from the chemists of the seventeenth century, not the Greeks of the fifth century B.C. Yet it is interesting that solely on intuitive evidence the Greeks could come to surprisingly modern ideas. The problem with such evidence is that it does not establish what is reality, but merely gives a personal view of reality that may logically be rejected by others on intuitive grounds. Plato and Aristotle rejected the atomic concept and refused to reduce the whole of reality, including humanity, to a system of moving atoms. Many people today are willing to accept the concept of atoms but hold a curious reservation concerning parts of human activities. They find it difficult to conceive of thought, memory, consciousness, and similar phenomena as systems of atoms. Yet biologists expect and work toward an understanding of even such difficult topics in chemical terms.

characteristic of living organisms, and we will encounter them again and again throughout our study of biology.

The atoms in the molecule are held together by chemical bonds that require energy to be formed and release energy when broken. You know that the wood in the pencil will burn and give off considerable heat and light in the process. The heat and light are the energy released by breaking the chemical bonds that were synthesized when the cell originally made its wall. Making the wall involved a large number of chemical reactions that also conform to a fixed set of physical laws.

This chapter is essentially about the wood in a pencil. We will cover some of the basic physical and chemical properties of matter and energy. We do so now, early in the book, because these topics are basic to all aspects of biology: You will need to know something about them whether you are studying molecules, cells, organisms, or biological communities. In this chapter we will consider the nature of matter, what chemical bonds are and how they are formed, some general properties of energy and how the cell handles it, and some of the major types of molecules found in organisms.

Elements, Atomic Nuclei, and Isotopes

C HOPKINS CaFe is a memory device used by generations of biology students to recall the major elements found in organisms (Table 2–1). The letters are standard symbols used to designate the elements: C stands for carbon, H for hydrogen, O for oxygen, P for phosphorus, K for potassium, and so forth. These are by no means the only elements found in living systems, but they are among the most common. From some two dozen elements are constructed the thousands of compounds necessary for life. The properties of elements are determined by the structure of the atoms of which they are composed. To understand the chemistry of life, it is necessary to appreciate the structure and properties of atoms.

An atom consists of an **atomic nucleus** surrounded by smaller, fast-moving particles called **electrons** (Figure 2–2). Most of the space taken up by an atom contains the electrons; the nucleus occupies only about 1/10,000 of the total volume of the atom. However, the nucleus is much heavier than the entire complement of electrons and makes up more than 99 percent of the total mass of an atom.

Table 2–1 Some Important Elements in Living Organisms

Atom	Symbol	Atomic Number	Electrons in Outer Shell	Biological Occurrence
Carbon	C	6	4	Basic atom of all organic compounds
Oxygen	O	8	6	Component of most biological molecules; final electron acceptor in energy-yielding reactions
Hydrogen	H	1	1	Component of most biological molecules; H^+ ion important component of solutions
Nitrogen	N	7	5	Component of proteins, nucleic acids, and many other biological molecules
Sulfur	S	16	6	Component of many proteins
Phosphorus	P	15	5	Component of nucleic acids and molecules carrying chemical energy; found in many lipid molecules
Iron	Fe	26	2	Important in energy-yielding reactions; component of oxygen carriers in blood
Calcium	Ca	20	2	Found in bones and teeth
Potassium	K	19	1	Important in conduction of nerve impulses
Sodium	Na	11	1	Ion in solution in living matter
Chlorine	Cl	17	7	Ion in solution in living matter
Magnesium	Mg	12	2	Part of molecules important in photosynthesis
Copper	Cu	29	1	Important in photosynthesis and energy-yielding reactions
Iodine	I	53	7	Component of hormone produced by thyroid gland
Fluorine	F	9	7	Found in trace amounts
Manganese	Mn	25	2	Found in trace amounts
Zinc	Zn	30	2	Found in trace amounts
Selenium	Se	34	6	Found in trace amounts
Molybdenum	Mo	42	1	Found in trace amounts

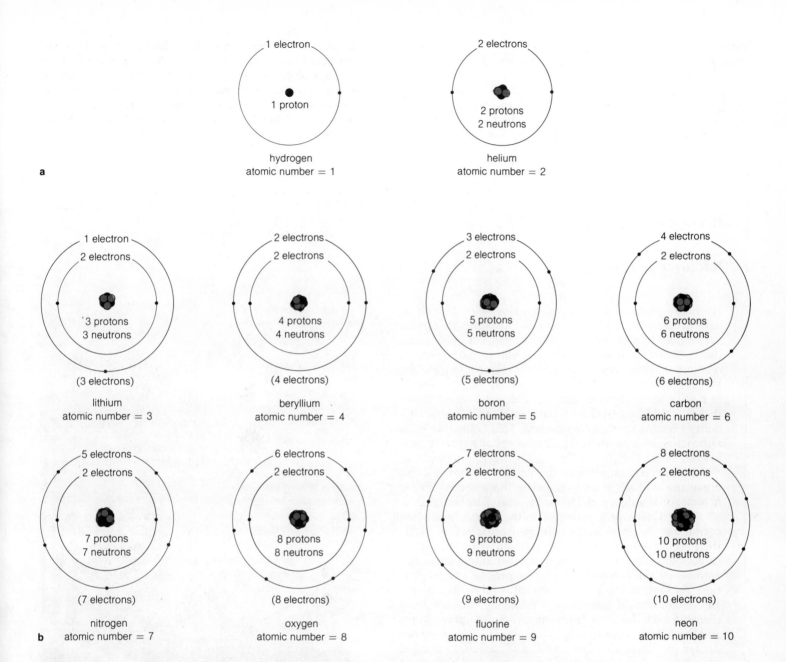

Figure 2–2 The arrangement and number of electrons, protons *(p)*, and neutrons *(n)* in ten common elements. **(a)** Hydrogen and helium, with one and two protons respectively, have atomic numbers of 1 and 2; the atomic number is equal to the number of protons in the nucleus of the atoms. **(b)** Atomic numbers of more complex atoms.

The atomic nucleus is this heavy because it contains a number of dense particles packed closely together. Present in all atomic nuclei are one or more positively charged particles called **protons.** The number of protons in an atom of each element is constant and ranges from a single proton in the nucleus of hydrogen to 92 protons in the nucleus of uranium, the heaviest naturally occurring atom. Since the number of protons in a given type of atom is constant, this number, called the **atomic number,** specifically identifies an atom. Hydrogen has an atomic number of 1, carbon an atomic number of 6, oxygen an atomic number of 16; each number tells you the number of protons present.

The atoms of every element except hydrogen also have a number of uncharged particles in their nuclei. These particles, called **neutrons,** occur in numbers approximately equal to the numbers of protons. For example, in its most common form, carbon has 6 protons and 6 neutrons in its nucleus (Figure 2–3). A rarer form of carbon, making up about 1 percent of naturally occurring carbon, has 6 protons and 7 neutrons. A still rarer form has 6 protons and 8 neutrons (Figure 2–3). These different forms of an element, all with the same number of protons but varying numbers of neutrons, are called **isotopes.**

Essay 2–2

Radioactive Dating

For many years, archaeological remains were difficult to date in absolute terms. It was possible, through careful excavation, to determine the relative dates of objects and the civilizations they represented, but to date them in relation to years from the present seemed impossible. When the radioactive isotope of carbon, ^{14}C, was found in nature, it occurred to the U.S. chemist Willard Libby that it would be possible to date with considerable certainty remains that contain carbon. This is because each radioactive isotope decays, or breaks up, at a specific rate, known as its half-life. Thus, comparing the amount of ^{14}C in an ancient object with the amount present in living organisms will allow us to calculate the amount of time it took for the decay of ^{14}C in the object to be dated.

The ^{14}C isotope is produced in the atmosphere when cosmic-ray neutrons strike nitrogen nuclei. The ^{14}C is then converted chemically by reaction with oxygen to carbon dioxide, which is incorporated into plants and animals by the processes of photosynthesis and respiration. As long as the plant or animal is alive, its carbon-14 content should remain on a relatively constant level (Figure E2–1). But after death the carbon-14 uptake stops, and the carbon-14 originally incorporated in the organism decays and decreases. The half-life of carbon-14 is 5,730 years; after that length of time, the original carbon-14 content has decreased to one-half. After two half-lives (11,460 years), one-fourth of the original carbon-14 is left, and so on. Thus, measuring the amount of radioactive carbon that remains provides an estimate of the age of an object (Figure E2–2).

The carbon-14 dating method is limited to objects less than approximately 50,000 years old, because it is difficult to measure the small amounts of carbon-14 left in older objects. A number of other methods based on different isotopes are used to analyze these older items.

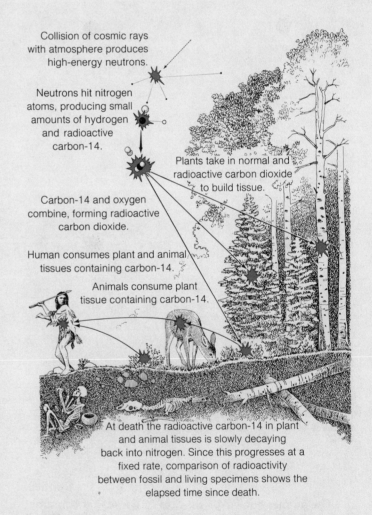

Collision of cosmic rays with atmosphere produces high-energy neutrons.

Neutrons hit nitrogen atoms, producing small amounts of hydrogen and radioactive carbon-14.

Plants take in normal and radioactive carbon dioxide to build tissue.

Carbon-14 and oxygen combine, forming radioactive carbon dioxide.

Human consumes plant and animal tissues containing carbon-14.

Animals consume plant tissue containing carbon-14.

At death the radioactive carbon-14 in plant and animal tissues is slowly decaying back into nitrogen. Since this progresses at a fixed rate, comparison of radioactivity between fossil and living specimens shows the elapsed time since death.

Figure E2–1 How carbon-14 can be used to date archaeological artifacts that were once living.

The isotopes of an element have essentially the same chemical properties, but they differ in mass and other physical characteristics. The *mass number* is used to distinguish the various isotopes of the same element and is based on the total number of protons and neutrons, each assigned the arbitrary mass of 1, present in the atomic nucleus. The three isotopes of carbon found in nature have mass numbers of 12, 13, and 14 and are designated as ^{12}C, ^{13}C, and ^{14}C.

Some of the isotopes of an atom are unstable and tend to undergo *radioactive decay*. For example, the carbon isotope ^{14}C is unstable, and one neutron in the nucleus splits into a proton and an electron. The proton is retained in the nucleus, giving a new total of 7 protons and 7 neutrons (which means that the carbon has been transformed into nitrogen), and the electron is ejected as a radioactive particle.

The radioactive isotopes of carbon, hydrogen, nitrogen, sulfur, and phosphorus, among others, have played an important role in biological research. Because of their radioactivity, such atoms can be traced as they go through chemical reactions in living organisms. Radioactive isotopes occur naturally in extremely small amounts, but they can be manufactured in nuclear reactors. Thus, the great burst of biochemical and molecular research since World War II is in part a by-product of the development of the atomic bomb. Naturally occurring radioactive ^{14}C is important in the dating of archaeological artifacts (see Essay 2–2).

The Electrons

The electrons surrounding the nucleus carry a negative charge. Normally there are as many electrons as there are positively charged protons. Electrons are in constant rapid motion around the nucleus. A given electron may be found in nearly any location, ranging from the immediate vicinity of the nucleus to the very outer limits of the atom, moving so rapidly that it can almost be regarded as being in all locations at the same time. However, if an electron could be tracked in its movements around the nucleus, it would be found to pass through some locations much more often than others. These most probable locations surround the atomic nucleus in layers of different shapes, called **orbitals**. Figure 2–3a shows the orbital of the single electron moving around the hydrogen nucleus as a shaded region, with the most probable locations for the electron at any given time shaded most densely. Although either one or two electrons may occupy a given orbital, the most stable and balanced conditions occur in orbitals occupied by a pair of electrons.

The most probable locations for electrons in orbitals lie in successive layers, or shells, around the nucleus. The innermost shell may contain a maximum of 2 electrons, and the second shell may contain a maximum of 8. Larger atoms have successive layers of orbitals that may contain more than 8 electrons, up to a maximum of 32 for any single shell. Although the outermost shell may contain more than 8 electrons in atoms with three or more orbitals, the most stable conditions are reached when the outer-

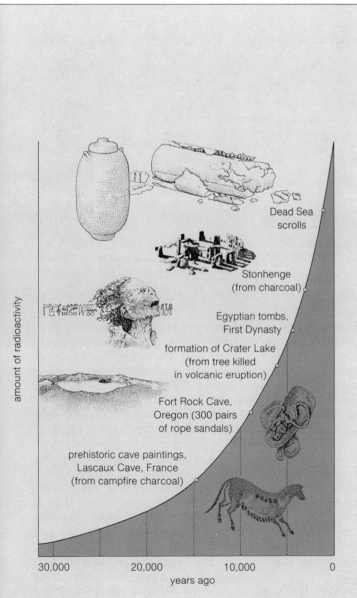

Dead Sea scrolls

Stonhenge (from charcoal)

Egyptian tombs, First Dynasty

formation of Crater Lake (from tree killed in volcanic eruption)

Fort Rock Cave, Oregon (300 pairs of rope sandals)

prehistoric cave paintings, Lascaux Cave, France (from campfire charcoal)

amount of radioactivity

30,000   20,000   10,000   0

years ago

Figure E2–2   Carbon-14 content has been used to estimate the age of artifacts.

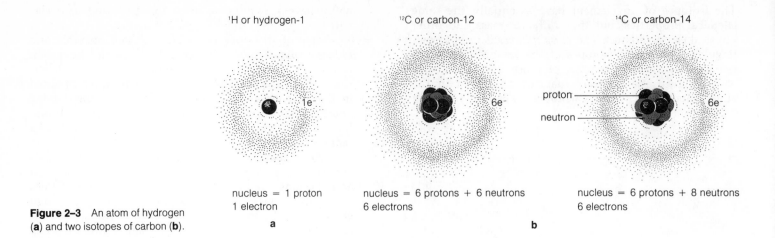

¹H or hydrogen-1   ¹²C or carbon-12   ¹⁴C or carbon-14

proton

neutron

1e⁻   6e⁻   6e⁻

nucleus = 1 proton   nucleus = 6 protons + 6 neutrons   nucleus = 6 protons + 8 neutrons
1 electron   6 electrons   6 electrons
a   b

Figure 2–3 An atom of hydrogen (**a**) and two isotopes of carbon (**b**).

most shell contains a total of 8 electrons (see Figure 2–2).

The chemical properties of an atom are determined largely by the number of electrons in the outermost shell. Helium, with two electrons in its single shell, and atoms such as argon and neon, with eight electrons in their outer shells, are stable and essentially inert chemically. Atoms with electrons near these numbers tend to share or exchange electrons to approximate these stable configurations; in other words, they tend to achieve full outer shells (Figure 2–4). This tendency to share or exchange electrons is essential to the formation of chemical bonds between atoms.

Molecules and Chemical Bonds

The cellulose molecules (and the glucose subunits of which they are composed) in the wood of a pencil consist of atoms of carbon, hydrogen, and oxygen arranged in precise ways. In glucose, there are always 6 atoms of carbon, 12 of hydrogen, and 6 of oxygen; in the shorthand of chemistry, glucose is $C_6H_{12}O_6$. Since cellulose is a polymer composed of an undefined number of glucose subunits, its chemical formula is written $(C_6H_{12}O_6)_n$. Clearly, if atoms are present in precise amounts, they must be held in specific relations to one another within the molecule. This is accomplished through **chemical bonds.** There are several kinds of chemical bonds, which differ in character and strength. We will first examine the various types of chemical bonds and then turn to the important relationship between energy and chemical bonds.

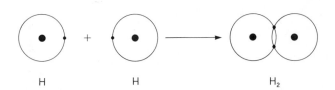

H   H   H₂

Figure 2–4 Electron-sharing between two atoms of hydrogen to form H_2; the most probable locations of the electrons are shown as thin lines.

Electrostatic or Ionic Bonds If two atoms that differ by only a single electron—one of them having a single electron in its outer shell and the other having seven electrons in its outer shell—move close together, an electron may pass from one to the other, achieving stable configurations in both. To illustrate this, let's look at one of the most familiar of all chemicals—common table salt, or sodium chloride (NaCl). If sodium, which tends to lose an electron, is brought together under the right conditions with chlorine, which tends to gain an electron, the sodium atom readily loses an electron to chlorine (Figure 2–5). After the transfer, the sodium atom carries a single positive charge, because it has one more proton than it has electrons, and the chlorine carries a single negative charge, because it has one excess electron. In this charged condition, the sodium and chlorine atoms are called **ions,** and are identified as Na⁺ and Cl⁻. Both sodium and chloride ions are made from single atoms and carry single charges. It is possible, however, to have ions that consist of two or more atoms each; these may or may not have more than one charge. For example, the ammonium ion is composed of one nitrogen atom and four hydrogen atoms and has a single positive charge—NH_4^+—while

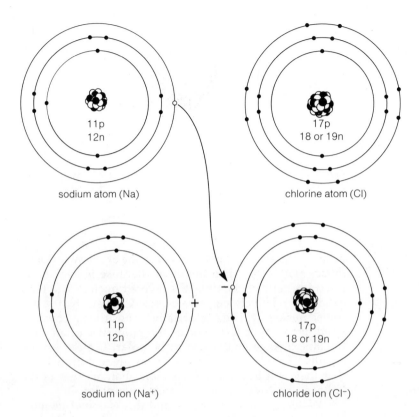

sodium atom (Na)

chlorine atom (Cl)

sodium ion (Na⁺)

chloride ion (Cl⁻)

Figure 2–5 Charge differences in the sodium and chloride ions set up the attraction that brings the atoms close together and holds them there.

the sulfate ion has one atom of sulfur and four of hydrogen and bears two negative charges—SO_4^{2-}.

The charge differences in the ions set up an attraction that brings the atoms close together in space and holds them there (Figure 2–5). This attraction is an **electrostatic** or **ionic bond.** Electrostatic bonds are readily broken and rejoined. If crystals of sodium chloride are placed in water, molecules of water take up positions between the sodium and chloride ions, greatly reducing their attraction for one another. The ions, each surrounded by a layer of water molecules, may then separate and diffuse through the solution as free ions. If water molecules are removed by evaporation, the Na^+ and Cl^- ions will reassociate into solid crystals. This reaction with water is related to still other properties of molecules and bonds that we will look at shortly.

Covalent Bonds   In ionic bonds, electrons are given and taken; but in another extremely important type of bonding, the electrons are actually shared. Chemical bonds based on shared electrons are called **covalent.** Covalent bonds are more stable than ionic bonds and considerable

energy is required to break them. These bonds are formed when the atoms involved have little tendency to gain or lose electrons completely. An example is carbon, with four electrons in its outer shell. In molecular diagrams, a covalent bond is designated by a pair of dots to represent the shared electrons, or by a single line:

$$\text{H} \cdot + \cdot \text{H} \longrightarrow \text{H} \colon \text{H} \text{ or } \text{H–H}$$

Carbon, with four unpaired outer electrons, can form four separate covalent bonds to complete its outermost shell. An example of this process is shown in Figure 2–6, in the formation of methane (CH_4) by electron-sharing between one carbon atom and four hydrogen atoms. The ability of carbon to form covalent bonds so readily is one of the characteristics that make it so important in the chemistry of living things. The whole science of organic chemistry is essentially the chemistry of carbon compounds. The other atoms most commonly found in organic molecules—nitrogen (N), oxygen (O), and hydrogen (H)—also form covalent linkages readily and are commonly found with carbon.

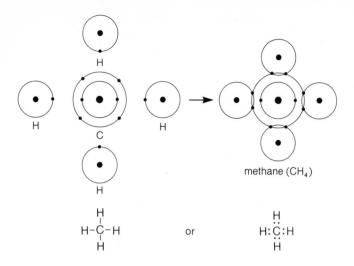

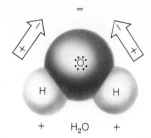

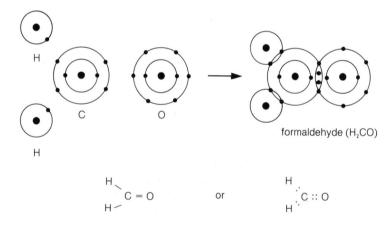

Figure 2–6 Above: Electron-sharing between four hydrogen atoms and a carbon atom to form methane. Below: Two ways of diagramming covalent bonds.

Figure 2–7 Above: Formation of a double bond in formaldehyde. Below: Two ways of diagramming double bonds.

Figure 2–8 Diagram of a water molecule showing the asymmetrical distribution of the two hydrogen atoms that results in the molecule having positive (+) and negative (−) ends.

In some cases, two pairs of electrons are shared between the atoms involved in the linkage. This results in the formation of a **double bond** (Figure 2–7). Such double bonds frequently represent places where additional atoms or groups of atoms can be more easily added to the compounds. Double bonds also are involved in light absorption, and pigment molecules in both plants and animals are rich in them.

Polar Bonds   Sharing of electrons between two covalently bound atoms is not always equal. In traveling their orbitals, the electrons may pass near one of the two atomic nuclei more often than the other. Because it retains the electrons more of the time, the favored nucleus will tend to carry a relatively negative charge. Conversely, the deprived nucleus will be made relatively positive. A covalent bond of this type has some of the properties of an electrostatic linkage and is said to be **polar.** If electrons are equally shared, so that there is no greater probability that they will be closer to one atom than another in the linkage, the covalent bond is called **nonpolar.**

Water is a polar molecule, and this is one of its most important characteristics. In the water molecule, the two hydrogen atoms are asymmetrically located at one side (Figure 2–8), giving this end of the molecule a relatively positive charge with respect to the relatively negative oxygen end. It is this feature that makes water such an effective solvent. Let's reconsider the solution of sodium chloride in water (Figure 2–9). When we drop a crystal of sodium chloride into water, the strong electrostatic forces between the Na^+ and the Cl^- ions must be overcome before the ions can separate. This occurs when the Na^+ and Cl^- ions on the surface of the crystal strongly attract and are attracted by the positive and negative ends of the polar water molecules. The forces are strong enough to pull the ions away from the corners, edges, and surfaces of the NaCl crystal. Once the ions are free, they are surrounded by water molecules, with the negative ends of the water molecules oriented toward the Na^+ and the positive ends toward the Cl^-. This *hydration* of the ions hinders them from recombining (or recrystallizing) to return to the NaCl crystal and helps spread them uniformly throughout the solvent.

Polar molecules attract and align themselves with other charged ions and molecules. These polar environments tend to exclude nonpolar substances, which also tend to be drawn together. This can be easily illustrated by mixing a nonpolar substance, such as oil, with a polar substance, such as water. No matter how vigorously the oil and water are shaken or mixed, they will quickly sepa-

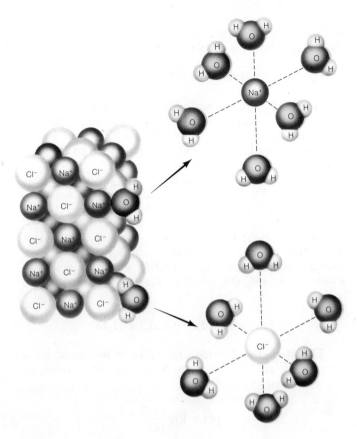

Figure 2-9 Dissolving an ionic solid (sodium chloride) in a polar solvent (water).

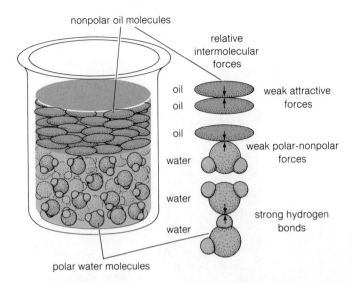

nonpolar oil molecules

relative intermolecular forces

oil
oil — weak attractive forces

oil
water — weak polar-nonpolar forces

water

water — strong hydrogen bonds

polar water molecules

Figure 2-10 Oil and water don't mix because the hydrogen-bond forces between water molecules are stronger than the weak attractive forces between nonpolar oil molecules and polar water molecules.

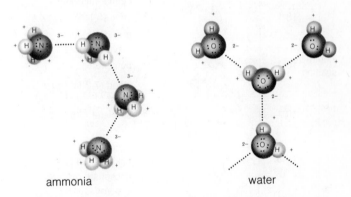

ammonia

water

Figure 2-11 Hydrogen forces or bonding (indicated by dotted lines).

rate into their respective nonpolar and polar environments (Figure 2–10). Part of this effect is the result of still another type of bond, one particularly important in reactions between molecules, called the hydrogen bond.

Hydrogen Bonds Electrostatic and covalent bonds are forces set up between *atoms,* but chemical bonds can also exist between *molecules.* The strongest of the intermolecular forces, the **hydrogen bond,** is still much weaker than the covalent bonds within molecules. The basis of the hydrogen bond is the unequal sharing of electrons in polar covalent linkages. In the example of water we discussed earlier, the hydrogen nuclei are relatively positive because the electrons shared with oxygen tend at any instant to be nearer the oxygen nucleus. These hydrogen nuclei may be attracted to the relatively negative oxygen atom of an *adjacent* water molecule (Figure 2–11). Hydrogen bonds are often depicted by a dotted line between the hydrogen and the adjacent atom involved

in the linkage. Hydrogen bonds exist between polar covalent molecules that contain at least one hydrogen atom bonded to an oxygen, nitrogen, or fluorine atom. Both oxygen and nitrogen with attached hydrogen atoms are common in biological compounds, so hydrogen bonding is frequently encountered in living systems.

Although individually weak, hydrogen bonds are collectively strong and lend stability to the three-dimensional structure of complex molecules. In cellulose, for example, they are important in maintaining the continuity and positioning of the long strands of molecules.

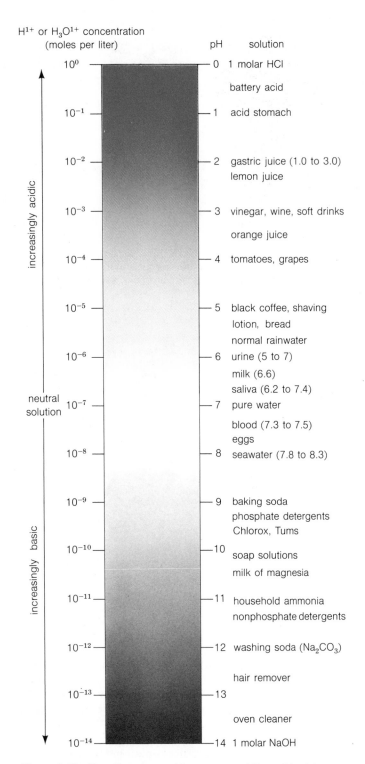

H^{1+} or H$_3$O^{1+} concentration
(moles per liter)   pH   solution

H$^+$ conc.	pH	solution
10^0	0	1 molar HCl
		battery acid
10^{-1}	1	acid stomach
10^{-2}	2	gastric juice (1.0 to 3.0)
		lemon juice
10^{-3}	3	vinegar, wine, soft drinks
		orange juice
10^{-4}	4	tomatoes, grapes
10^{-5}	5	black coffee, shaving
		lotion, bread
		normal rainwater
10^{-6}	6	urine (5 to 7)
		milk (6.6)
		saliva (6.2 to 7.4)
10^{-7}	7	pure water
		blood (7.3 to 7.5)
		eggs
10^{-8}	8	seawater (7.8 to 8.3)
10^{-9}	9	baking soda
		phosphate detergents
		Chlorox, Tums
10^{-10}	10	soap solutions
		milk of magnesia
10^{-11}	11	household ammonia
		nonphosphate detergents
10^{-12}	12	washing soda (Na$_2$CO$_3$)
		hair remover
10^{-13}	13	
		oven cleaner
10^{-14}	14	1 molar NaOH

increasingly acidic — neutral solution — increasingly basic

Figure 2–12 The pH scale, used to measure acidity and basicity.

Acids and Bases   Water has a tendency to ionize according to the following equation:

$$HOH \longrightarrow H^+ + OH^-$$

(Hydrogen does not actually exist as an ion, as shown, but as H$_3$O$^+$. This makes no difference in our discussion, and it is easier to deal with H$^+$ ions.) In pure water, the numbers of H$^+$ and OH$^-$ ions are equal, and a relatively small number are present.

In an acidic solution, the number of H$^+$ ions exceeds that of the OH$^-$ ions; in an alkaline or basic solution, the number of OH$^-$ ions exceeds the number of H$^+$ ions. The more ionized an acid is, the stronger it is. Hydrochloric acid (HCl) is a strong acid because it is almost completely ionized. The same is true for a strong base such as sodium hydroxide (NaOH). When an acid and a base are mixed, the result is always a salt and water; for example:

$$HCl + NaOH \longrightarrow NaCl + H_2O$$

The strength of acids and bases is measured by a term called *pH.* The term is based on the number of hydrogen ions present:

$$pH = \log \frac{1}{H^+}$$

If the concentrations of H$^+$ and OH$^-$ are equal, a solution is said to be *neutral,* with a pH value of 7. Acid solutions have greater concentrations of H$^+$ than OH$^-$ ions and pH values lower than 7; basic solutions, with their higher concentrations of OH$^-$ ions, have values greater than 7. The values in the pH scale (Figure 2–12) represent logs to the base 10 of concentration; thus, one pH unit represents a multiplication by 10. For example, a solution at pH 6 has 10 times as many H$^+$ ions as a solution at pH 7.

Essentially all biological reactions occur in the pH range of 6 to 8 (Figure 2–12). Blood, for example, has a pH of 7.4. Biological systems rarely involve strong acids or bases, although the human digestive system is an exception, with stomach cells actually secreting HCl. Weak acids and bases, however, are common in living systems. The organic acids, such as acetic acid and lactic acid, are important in cell metabolism. They all contain the carboxyl group (–COOH), which yields H$^+$ ions in water. Carboxyl groups are present in the amino acids and give these important compounds many of their distinguishing features. Also present in amino acids is the amino group (–NH$_2$), which accepts H$^+$ ions and acts as a weak base. Thus, an amino acid molecule can act either as an acid or as a base, depending on the pH of the solution it is in.

Despite the variety of compounds found within them, cells are able to maintain an essentially constant pH, usually close to 7. This is also true of our blood, which maintains a pH of 7.4 in the face of a vast range of compounds that routinely enter it. This is because cells and blood, as well as other living systems, have *buffers*. A buffer is a system that will contribute either H^+ ions or OH^- ions, depending upon the concentrations of these ions in solution. One buffer system found in cells is based on carbonic acid (H_2CO_3) and bicarbonate ion (HCO_3^-):

$$H_2CO_3 \rightleftharpoons H^+ + HCO_3^-$$

When an acid is added to the system, it combines with the HCO_3^- to form H_2CO_3. When a base is added, it combines with H_2CO_3 to form H^+ and HCO_3^-. Thus, the cell can compensate for relatively small changes in the acid or base level and maintain a relatively constant pH.

Energy and Life

Life is based on energy. A human can survive only a few days without food, and a green plant will die in an equally short time without light. The energy all earthly organisms require for life comes ultimately from the sun through the mediation of photosynthesis. Understanding the flow of energy in cells and organisms is one of the underlying goals of biologists and is fundamental to any real understanding of biology.

The first step to such an understanding is to ask the deceptively simple question: What is energy? Energy is the capacity to do work. We all have intuitive answers to the question that somehow tie energy to the idea of work. We eat a good breakfast to get the energy to take a long hike or have a snack to get the energy to start a term paper. Our intuition is right: There is a real correlation between energy and work. But, as with so many things we will cover in this book, that is only part of the picture.

Energy can be either *potential*, which means that it is in a stored form, or *kinetic*, which means that it is involved in work, which in turn means that motion is involved (Figure 2–13). The energy in the muscles of your arm and hand holding a pencil is potential as long as your arm is at rest. But when you move your arm to write, the energy becomes kinetic. The energy in the chemical bonds of the cellulose of the wood in the pencil is potential energy, but when the wood burns, it is released as kinetic energy in the form of heat. There are many ways in which energy is released and transmitted. Life exists because organisms have mechanisms for making use of energy and changing it from one form to an-

other. Cells are essentially machines for the transformation of energy.

The Study of Changes in Energy Since energy is so important not only to life itself, but to physics, chemistry, and biology, it is hardly surprising that a whole science has developed around the study of energy and its changes. This is the science of thermodynamics, which we will consider only at a general level. The central knowledge of thermodynamics consists of a number of laws that sound like truisms but in fact provide the bases for profound insights into nature.

The *first law of thermodynamics,* for example, is a formal statement of the common observation that you can't get something for nothing. More precisely, the first law states that energy can be neither created nor destroyed. Although it may be transformed from one form to another, the total quantity of energy in the universe remains the same.

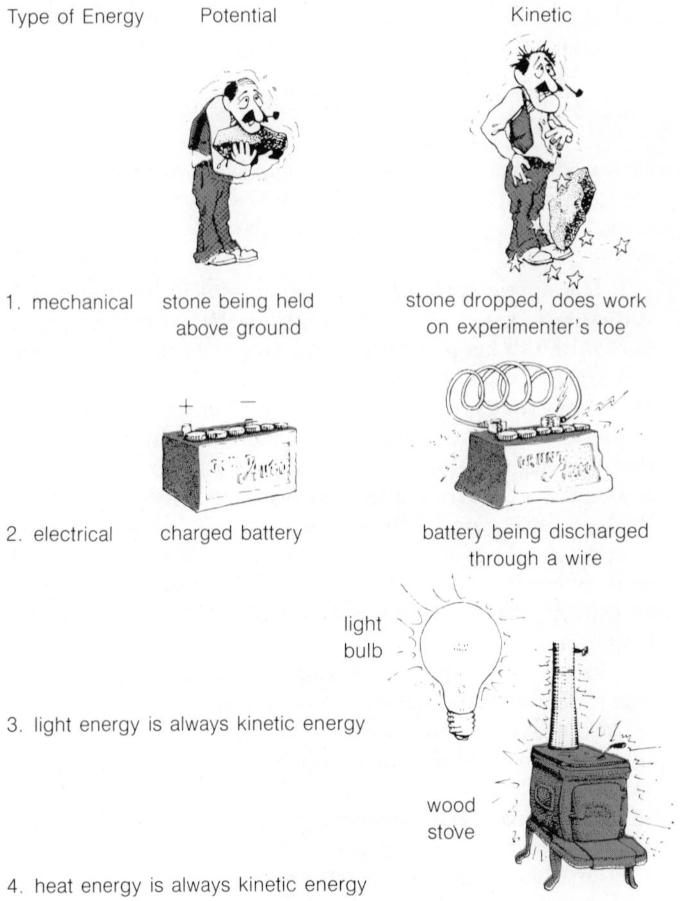

Type of Energy	Potential	Kinetic
1. mechanical	stone being held above ground	stone dropped, does work on experimenter's toe
2. electrical	charged battery	battery being discharged through a wire
3. light energy is always kinetic energy		
4. heat energy is always kinetic energy		

Figure 2–13 Potential and kinetic energy are the two major types of energy.

The spontaneous tendency toward disorder......

high order
low entropy

ONE WAY

less order
more entropy

Figure 2–14 The second law of thermodynamics. Any spontaneous change is toward a less ordered state—toward greater entropy.

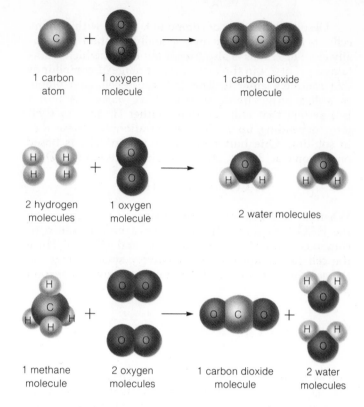

1 carbon atom   1 oxygen molecule   1 carbon dioxide molecule

2 hydrogen molecules   1 oxygen molecule   2 water molecules

1 methane molecule   2 oxygen molecules   1 carbon dioxide molecule   2 water molecules

Figure 2–15 In a chemical reaction, atoms, ions, or molecules are rearranged by the breaking of old bonds and the formation of new bonds.

The *second law of thermodynamics* is more subtle and difficult to grasp. The observation that constant effort is required to keep anything in order is its common-sense basis (Figure 2–14). In more precise terms, the second law states that spontaneous change of any kind occurring anywhere in the universe always increases the total disorder, or randomness, of the universe. If order is increased in one place, as when a tree grows from a seed, assimilating and arranging additional matter as it grows, this increase can take place only at the expense of a larger reduction of order somewhere else. This losing trade-off, amounting to an increase in randomness, is called **entropy.**

In its most applicable and general form, then, the second law affirms that as change takes place, the entropy of the universe always increases. Since the sun is the ultimate source of the earth's energy, the order established in the growth of the tree—and all living things on earth—is established at the expense of energy derived from a decrease of order in the sun.

Finally, before we leave the subject of energy changes and thermodynamics, we should consider the concept of **free energy.** Free energy is essentially potential energy, in that it represents energy potential for producing change. Free energy is related to entropy in that it is a measure of the orderliness of a system. The more orderly a system is, the higher its free energy will be because of its greater potential to move to a disorganized state.

Chemical Reactions   A *chemical reaction* involves a rearrangement of molecules, atoms, ions, or electrons so that other atoms, ions, or molecules with distinctive physical and chemical properties are formed (Figure 2–15). Thus, in a chemical reaction the molecules originally present disappear, and molecules that were not initially present appear. If we mix sodium hydroxide (a strong base) and hydrochloric acid (a strong acid), we cause a chemical reaction that results in the formation of sodium chloride (a salt) and water. Sodium chloride is neutral (neither acid nor alkaline) and has completely different properties than either sodium hydroxide or hydrochloric acid. Using standard chemical symbols, this reaction can be written as:

$$NaOH + HCl \rightleftharpoons NaCl + H_2O$$

sodium hydroxide   hydrochloric acid   sodium chloride   water

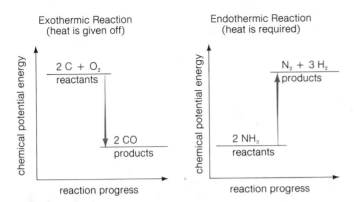

Figure 2–16 Exothermic and endothermic reactions. The partial combustion of coal (C) to form carbon monoxide (CO) gives off heat as the chemical potential energy falls to a state of lower chemical potential energy (the reaction is exothermic). The decomposition of ammonia gas (NH_3) into N_2 and H_2 gas requires an outside input of energy (usually as heat) to make the reaction occur (the reaction is endothermic).

This reaction illustrates several points that are universal to all chemical reactions. Note, for example, that the same numbers and kinds of atoms are present on both sides of the chemical equation; the number of sodium, oxygen, hydrogen, and chlorine atoms is balanced on the left and right sides. The small arrow pointing to the left indicates that the reaction is reversible. This is a universal feature of chemical reactions, although the conditions may be such that the reaction appears to be going in only one direction. As written here, the equation tells you nothing about the rate or the energy requirements of the reaction.

Some chemical reactions are spontaneous, but others will occur only when energy is added. The first and second laws of thermodynamics allow us to predict whether a reaction is possible and, if so, whether it will take place spontaneously and release energy (be an *exothermic* reaction) or will require energy in order to take place (be an *endothermic* reaction) (Figure 2–16). If the reaction does take place spontaneously, it always tends toward a final condition of minimum total energy and maximum disorder or entropy. In proceeding to its final state, a spontaneously reacting or "downhill" system gives off energy to its surroundings. Energy given off in this way is available to do work (Figure 2–16). Some reacting systems will not proceed to completion spontaneously because they are moving "uphill" in the direction of greater order and energy (Figure 2–16). If such reactions are possible at all, they can proceed only with energy added from the outside, at the expense of other spontaneously occurring reactions that release free energy. This is an important principle, because it is the basis for all the reactions of

living organisms, including growth, reproduction, movement, and response to the environment. In these life processes, living organisms *couple* systems giving off free energy to those requiring an input of energy. If the entropy changes for all of these coupled systems are summed together, the total randomness always increases.

In the case of a cellulose molecule in your pencil, solar energy is used to produce the bonds between the carbon, hydrogen, and oxygen atoms making up the glucose molecules. This is not a direct incorporation of solar energy, but involves a complex series of reactions known as photosynthesis. In the final synthesis of the polymer cellulose, additional energy is needed to bond the glucose molecules together. Now, if you hold a match to a pencil and the wood begins to burn, the energy from the chemical bonds is released and carbon dioxide and water are re-formed. The energy, in the form of heat, is not immediately useful to us as organisms, but it can be used to heat water and produce steam to run a machine. You can also bury a pencil in the ground, where a variety of molds will begin to grow on and in it. In this case, too, the chemical bonds of the cellulose are broken apart, but in a controlled manner that transfers some of the released energy to other compounds useful to the organisms involved. Some of the energy is also released as heat.

Oxidation and Reduction Much of the energy passed from one substance to another is transferred in the form of electrons, because many substances have the ability to accept and donate electrons. In such reactions, electrons are removed from a donor molecule and accepted by an acceptor molecule (Figure 2–17). Removal of electrons is called **oxidation** and acceptance of electrons is called **reduction;** a substance from which electrons are removed is said to be *oxidized,* and the accepting substance is *reduced.* In general, a given substance contains more energy in the reduced than in the oxidized state. Since the electrons the acceptor receives in reduction must come

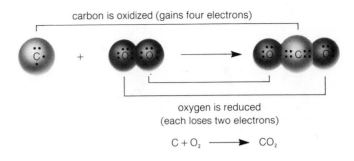

Figure 2–17 Oxidation-reduction reaction.

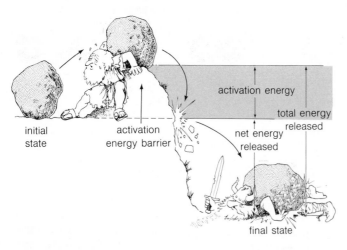

Figure 2–18 An analogy of the activation energy for a chemical reaction. If enough energy is supplied to the boulder to push it over the hill (activation energy barrier), it will spontaneously roll down the mountain, releasing energy as it moves to a lower state of potential energy. The rate of boulders being pushed over the cliff will depend on the height of the activation energy barrier.

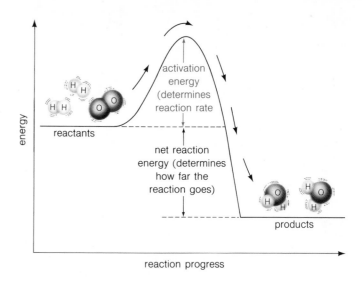

Figure 2–19 The activation energy of the net reaction between 2 H_2 and O_2 to form 2 H_2O is so high that the rate of reaction is extremely slow.

from somewhere, there is always a paired reaction of oxidation and reduction.

The electrons removed from an oxidized substance have a characteristic energy that depends on the orbitals they occupied before their removal. This energy, often called the *redox potential,* can be measured and used as a guide to the direction in which the reactions and energy transfers will take place. Carbohydrates, as the end product of photosynthesis, are highly reduced compounds with a high redox potential. At the opposite end of the scale is water, highly oxidized, with low redox potential. During cellular respiration, which we will discuss at greater length in Chapter 4, a series of oxidation-reduction reactions occurs going from glucose to water and carbon dioxide—from high-energy compounds to low. In each of these paired reactions, electrons are moved to lower and lower states, with energy released at each step.

The Rates of Chemical Reactions and Enzymes All chemical reactions do not occur at the same rate. If we look at the amount of time it takes for a quantity of reactant to be converted into product, we find that some reactions take less time to form a given amount of product than do others. What makes some reactions slow and others very rapid? First, all atoms, ions, and molecules are in motion. Second, they must collide before a reaction can take place. The more frequently they collide, the faster the rate of reaction. The problem, however, is not so simple. Just because two molecules collide does not mean that they will necessarily react. The rate of reaction depends not only on the collision frequency, but also on the collision energy and collision geometry. The collision must be effective—the molecules must collide with sufficient energy and with the right geometry, or orientation, to bring about a reaction. The minimum energy required to bring about a reaction is known as the **activation energy.** There is an energy barrier, or "hill," that must be surmounted before the reaction can proceed, just as a boulder may have to be rolled over a small hill before it can spontaneously roll down a mountain (Figure 2–18). The height of the initial energy barrier determines how fast the change will occur, and the difference in energy between the initial and final states determines how far the boulder goes and how much net energy is released in the process.

We can use a similar diagram (Figure 2–19) to explain why H_2 and O_2 (gaseous hydrogen and gaseous oxygen) don't react at room temperature, even though this is a downhill, or spontaneous, reaction from a net energy standpoint. The initial activation energy barrier is high enough that at room temperature most of the colliding molecules do not have sufficient energy to bring about a fruitful collision. The height of the activation energy barrier (top part of the diagram) determines the rate of the reaction. The bottom part of the diagram illustrates the net energy released and how far the reaction goes.

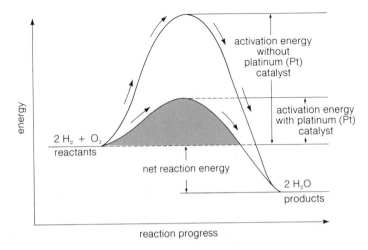

Figure 2–20 A catalyst increases the rate of a reaction by lowering the activation energy, so that more reactant molecules collide with enough energy to surmount the lower energy barrier.

Many reactions that do not proceed at all or have slow rates when the reactants are mixed can be made to take place much more rapidly by introducing an outside substance. Substances that increase the rate of a reaction without themselves appearing in the overall net equation are called **catalysts.**

The following equation represents a catalyzed reaction, where platinum metal (Pt) is the catalyst added to speed up the rate of reaction:

$$2 H_2 + O_2 + Pt \longrightarrow 2 H_2O + Pt$$

Although the catalyst somehow speeds up the reaction rate, it is not used up. This means that a tiny amount of catalyst can be used over and over again. Thus, the net reaction may be written as follows:

$$2 H_2 + O_2 \overset{Pt}{\longrightarrow} 2 H_2O$$

How does a catalyst speed up the rate? It is generally agreed that a catalyst increases the rate of reaction by lowering the activation energy for the reaction. For example, although the rate of reaction between H_2 and O_2 is extremely slow at room temperature, the addition of finely divided platinum metal as a catalyst causes the reaction to go very rapidly because the activation energy is lowered (Figure 2–20).

Many reactions, although spontaneous, proceed so slowly without a catalyst that their rate at room temperature cannot be measured. A good example is the break-down of glucose to carbon dioxide and water. A mound of dry glucose could be allowed to lie on a table exposed to oxygen for hundreds of years, and you could barely detect any change. Introduce a fly that eats the sugar, and the reaction will proceed to completion inside the fly in a matter of minutes because the right catalysts are present within the fly's cells. The catalysts found in living systems are called **enzymes.** All enzymes are proteins (see page 35).

There are a number of things that enzymes are not and things they do not do. Enzymes are not used as building blocks by cells or burnt for energy. They cannot initiate reactions. Instead, they do what a catalyst does: lower the activation energy required for the reaction and thus increase its rate. Only small amounts of an enzyme are needed to speed the interaction of a large number of reactants, because each enzyme molecule, after release, may immediately enter into another sequence. This alternate entry and release may take place as rapidly as 3 million times a minute.

The question now is, how do enzymes lower the activation energy of a reaction? The full answer to this question is not known, but it is certain that the geometry of the reaction is involved. It is known that enzymes provide a site to which the reacting atoms, molecules, or ions adhere and on which the reaction takes place. The place on the enzyme where this takes place is known as the *active site.* There is a precise spatial relationship between the reacting atoms, molecules, or ions—known as the *substrate*—and the active site (Figure 2–21). This arrangement of the reactants dramatically increases the chances that a reaction will occur and consequently lowers the activation energy. In addition, the juxtaposition of the substrate molecules and the enzyme may cause stresses in the bonds of substrate molecules that aid in breaking or forming bonds. Scientists do not know all the interactions between the substrate and enzyme molecules, and research continues on these questions.

One of the most studied enzymes is lysozyme, which breaks down certain components in the cell walls of bacteria. Scientists can now completely enumerate the structure of lysozyme and describe how a part of its surface configuration (its active site) matches that of its substrate. Other enzymes are nearly as well understood as lysozyme, and more are being analyzed every day.

We said earlier that, when a reaction is complete, enzymes are released unchanged. This is true for enzymes in general, but as particular enzymes cycle thousands or millions of times through a reaction sequence, they slowly wear out and are themselves broken down by other enzymes.

Enzyme Specificity Each of the myriad reactions taking place in living cells is catalyzed by an enzyme specific

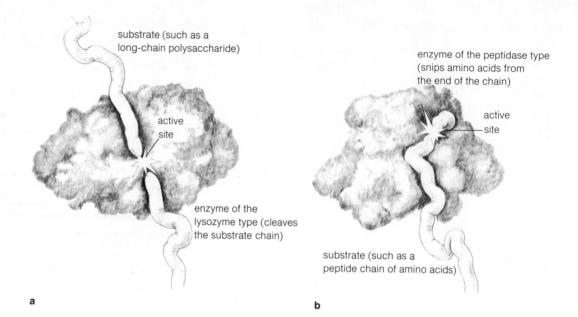

substrate (such as a
long-chain polysaccharide)

active
site

enzyme of the
lysozyme type (cleaves
the substrate chain)

enzyme of the peptidase type
(snips amino acids from
the end of the chain)

active
site

substrate (such as a
peptide chain of amino acids)

Figure 2–21 Enzymes in action. In (**a**) the substrate lies in a cleft in the side of the enzyme; in (**b**) the substrate penetrates to the interior of the enzyme.

a

b

for that reaction. Moreover, enzymes are often part of complex sequences of reactions where the end product of one enzyme is the substrate for the next. At any time, a cell contains many thousands of different enzymes, each capable of increasing the rate of a single reaction. This specificity depends on the structure of the active site. These three-dimensional sites contain patterns of charged, polar and nonpolar groups that exactly match and bind to parts of the reactant molecules. Several poisons act by competing with the substrate for the active site of the enzyme molecule. If the active site is filled by the poison, the necessary reaction cannot occur, and the result may be fatal to cell and organism.

Some enzymes require the presence of another organic molecule (known as a coenzyme) or a mineral ion. Coenzymes are derived from vitamins, so when you eat a breakfast cereal that promises to fill your daily requirement for vitamins and minerals, you are actually supplying the needs of your enzymes. Enzymes are known to be easily inactivated by such factors as high temperatures, the presence of organic solvents, and certain heavy-metal ions, such as those of lead and mercury. All these properties are related to the fact that enzymes are proteins. Later in this chapter, we will consider the complex nature of the protein molecule.

Finally, the names of most enzymes are related to their specificity. Traditionally, enzymes are named after their substrate plus the ending *ase.* Thus, the enzyme that breaks down cellulose is called cellulase. We also speak of enzymes collectively by the type of reaction

they carry out. Many enzymes remove hydrogen from substrate molecules and are termed *dehydrogenases.* But there are specific dehydrogenases for the removal of hydrogen from particular molecules, and so the specific names include the substrate names—for example, succinic acid dehydrogenase.

The Molecules of Life

We started our discussion of matter and energy by considering a common wooden pencil. We noted that with proper examination it is possible to demonstrate the presence of large molecules of cellulose in the wood (Figure 2–1). Moreover, the cellulose in turn was shown to consist of chains of smaller molecules joined together. Large molecules like cellulose are called polymers and are extremely important in biological systems. The enzymes we have just been discussing are also composed of smaller repeating units, in this case amino acids. We can continue our examination of some of the basic chemistry of living systems by looking at the major groups of organic compounds present in such systems and at some of the important subunits present in each.

All organic molecules are based on carbon and its interaction with other atoms, usually hydrogen, oxygen, nitrogen, and, in lesser quantities, phosphorus and sulfur. From these atoms are formed the four major classes of organic molecules present in all living systems: carbohy-

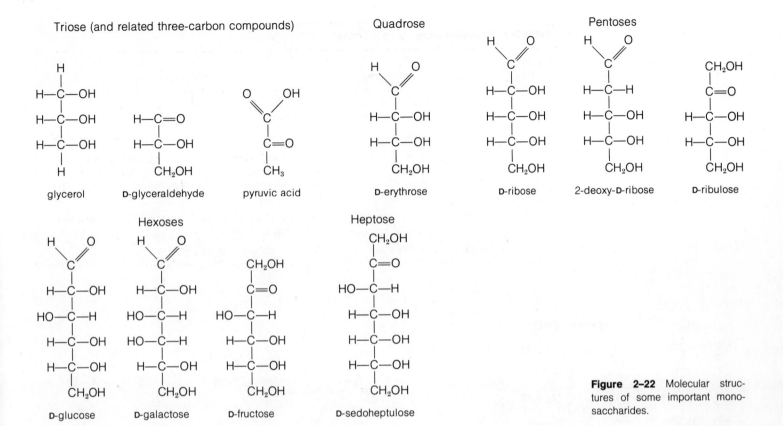

Triose (and related three-carbon compounds)

glycerol D-glyceraldehyde pyruvic acid

Quadrose

D-erythrose

Pentoses

D-ribose 2-deoxy-D-ribose D-ribulose

Hexoses

D-glucose D-galactose D-fructose

Heptose

D-sedoheptulose

Figure 2–22 Molecular structures of some important monosaccharides.

drates, lipids, proteins, and nucleic acids. All of these molecules in turn are based on a relatively few organic subunits that are repeatedly encountered in biological chemistry, primarily organic acids, alcohols, and aldehydes. Amino acids and fatty acids are examples of organic acids. These smaller subunit molecules owe their acidic properties to the –COOH (carboxyl or acid) group that ionizes in solution to release a H^+ ion:

Acid

Alcohols, important building blocks of fats and many carbohydrates, have in common the reactive group:

Alcohol

Linkages formed by the alcohols in building more complex substances are based primarily on the reactivity of the –OH segment of this group of atoms. Aldehydes, which with the alcohols compose the basic building

blocks of carbohydrates, are characterized by the reactive group:

Aldehyde

Now let's look at the major classes of molecules found in biological systems—carbohydrates, lipids, proteins, and nucleic acids—and see how the chemistry we have been discussing relates at the molecular level to organisms.

Carbohydrates Much of the organic matter on earth, perhaps as much as 50 percent, consists of **carbohydrates.** We eat them, drink them, and clothe our bodies in them. They are the source of energy to the cells of both plants and animals and are the major structural element in the cell walls of plants. Carbohydrates may take the form of simple sugars or *monosaccharides,* such as glucose, or of large molecules made of several simple sugars and known as *polysaccharides,* such as starch, cellulose, and glycogen.

The simplest carbohydrates found in organisms are the monosaccharides (Figure 2–22). They are the most

Table 2–2 Monosaccharides Found in Nature

Number of Carbons	Type	Examples	Importance
3	Triose	Glyceraldehyde, dihydroxyacetone	Intermediates in energy-yielding reactions and photosynthesis
4	Tetrose	Erythrose	Rare in nature; an intermediate in photosynthesis
5	Pentose	Ribose, deoxyribose, ribulose	Intermediates in photosynthesis; components of molecules carrying energy; components of the informational nucleic acids DNA and RNA; structural molecules in cell walls of plants
6	Hexose	Glucose, fructose, galactose, mannose	Fuel substances; products of photosynthesis; building blocks of starches and cellulose
7	Heptose	Sedoheptulose	Rare; intermediate in photosynthesis

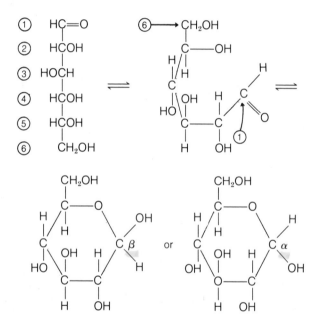

Figure 2–23 Formation of the ring structures of sugars, specifically D-glucose.

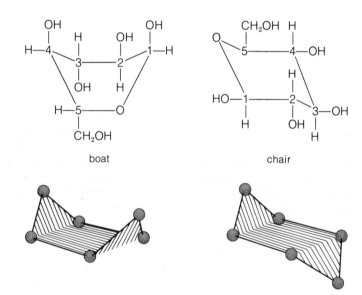

boat

chair

Figure 2–24 The boat and chair forms of a hexose-ring sugar molecule like glucose.

common energy source for the cell, because they are readily broken down and their bond energy released. Monosaccharides, like all carbohydrates, contain carbon, hydrogen, and oxygen in the approximate ratio 1:2:1, or CH_2O. The basic subunits are short chains of carbon, usually from three to seven carbons long. Monosaccharides are named according to the number of carbons they contain:

those with three carbons are triose, with four are tetrose, with five are pentose, with six are hexose, and with seven are heptose. Only the three-, five-, and six-carbon sugars are common. Glucose, for example, is a six-carbon sugar, or hexose. Table 2–2 lists a number of monosaccharides and some of their functions in organisms.

Many sugars in solution form rings rather than

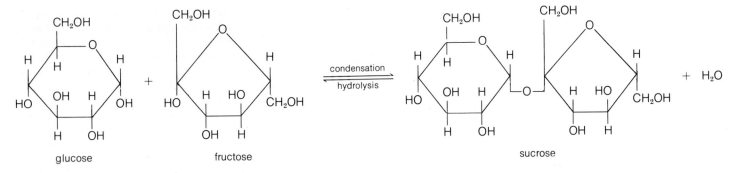

Figure 2-25 A disaccharide is formed by linking two monosaccharide, or simple sugar, molecules.

straight chains. In glucose, the ring forms by the carbon numbered 1 reacting through oxygen with the 5-carbon to form a six-membered ring, as shown in Figure 2–23. When this occurs, the 1-carbon has four atoms attached to it (instead of three); thus it becomes asymmetric. This means that there are two possible ways for the ring to form. As indicated in the figure, the –OH group can point down or up. The two forms are called alpha (α) and beta (β). These configurations become important when we consider polysaccharides.

There is an additional complication in the structure of monosaccharides that is of great importance in the organization and properties of polysaccharides. The carbon molecules in the chain are not really straight when the ring forms, but are zigzagged. Thus, the ring, instead of being flat, is distorted in the form of either a "boat" or a "chair" (Figure 2–24).

Two monosaccharides may be linked together to form *disaccharides,* or a greater number may be linked to form *polysaccharides.* The basic mechanism in the formation of these compounds is shown in Figure 2–25. Here glucose and fructose are shown in a *condensation* reaction, in which the elements of water are removed as the subunits are linked together. The result is sucrose (common table sugar), one of the most important sugars in the plant kingdom. Sucrose is the form in which sugars are moved through plants, so we can easily refine sucrose from sugar cane and sugar beet. For this reason, sucrose is the sugar most commonly used in cooking. Sucrose can easily be split into its two component monosaccharides by adding the elements of water in what is known as a *hydrolytic* reaction. As Figure 2–25 indicates, the reaction between glucose and fructose, on the left, and sucrose and water, on the right, is reversible; condensation and hydrolysis are thus reverse forms of the same reaction. Specific enzymes mediate the condensation and hydrolytic reactions in cells.

The same basic condensation reactions are involved in the formation of polysaccharides. Cellulose, starch, and glycogen all consist of chains of glucose molecules. Each is an important molecule and each differs physically and chemically from the others. Cellulose is insoluble in water and organic solvents. Wood is about 50 percent cellulose, and cotton is almost pure cellulose. Starch is the principal food-storage form in plants, and humans the world over use starchy foods, such as potatoes and cereal grains, as diet staples. Starch, in contrast to cellulose, is water soluble and easily degraded into its monosaccharide components. Glycogen is the major carbohydrate-storage form in animals. We accumulate it in the muscles and liver, breaking it down into glucose in response to exercise and other situations where we need more energy.

The differences among these three large molecules are related to the structure of the units composing them and the way in which they are linked together (Figure 2–26). Cellulose molecules consist of glucose molecules in the β chair form. Since the molecules hook together through the groups that were formed by closing the ring in the β configuration, the glucose molecules are said to be joined by β linkages. A single molecule of cellulose contains 3,000 to 10,000 glucose molecules in an unbranched chain.

The relevance of all this chemistry to cellular structure and function is that when β chairs of glucose are hooked together through β linkages, the chain proves to be almost straight. Furthermore, several of these straight chains can fit nicely alongside one another, bonding to each other through hydrogen bonds. This configuration makes cellulose insoluble and gives it high tensile strength (exceeding that of steel).

Starch also consists of glucose, but in the boat form and with the α linkage. This produces coiled instead of straight chains (Figure 2–26). (Incidentally, the coil is of just the right dimensions to accommodate iodine molecules and produce the characteristic blue color used in the iodine test for starch. For a novel use of the same reaction see Essay 2–3). One form of starch (amylose) consists of unbranched glucose chains and is fairly soluble

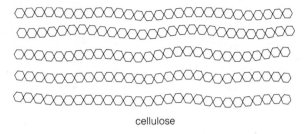

cellulose

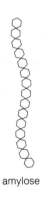

amylose

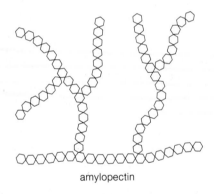

amylopectin

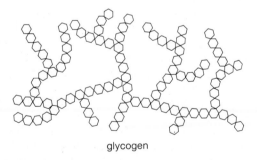

glycogen

Figure 2–26 A schematic representation of the arrangement of glucose in cellulose, in two types of starch (amylose and amylopectin), and in glycogen.

in water. The other form (amylopectin) has highly branched chains and is somewhat less soluble, although it too can be dissolved in hot water. The relative insolubility of starch, even though it consists of glucose molecules highly soluble in water, comes from the large size of the molecules, especially those of amylopectin.

Glycogen, sometimes called "animal starch," is quite similar to amylopectin except that it is more highly branched, with branches occurring every 8 to 10 glucose units (Figure 2–26). This highly branched structure aids in the rapid breakdown of the molecules.

Lipids The **lipids** are a diverse collection of organic compounds found in plants, animals, and microorganisms. Characteristically, they are greasy to the touch and insoluble in water but soluble in alcohol, ether, and other organic solvents. The lipids include fats, oils, waxes, and steroids such as cholesterol and sex hormones. Fats and oils are the most abundant lipids in nature. Fats are the most concentrated source of energy in our food, providing more than twice as much energy per gram as carbohydrates and three times as much energy per gram as protein. They are stored as reserve energy rather than being used for quick energy like sugars. At room temperature, animal fats tend to be solids, and vegetable fats (such as olive oil, linseed oil, soybean oil, and corn oil) tend to be liquid. Thus, oils are simply fats that exist as liquids at room temperature.

Essay 2–3

Starch Chemistry and John Dillinger

Knowledge of a little chemistry can be useful to anyone, even criminals. Plant starch can be detected chemically by the blue color it forms when mixed or painted with an iodine (I_2) solution. This chemical knowledge was used in the 1930s by the notorious American gangster John Dillinger. While serving a jail sentence, he asked for a potato. He also deliberately cut his finger and asked for some iodine to put on the cut. After carving the potato into a pistol, he painted it with iodine until it resembled blue steel. Then he called a guard and, using his potato pistol, made one of the most famous "chemical escapes."

The neutral lipids, commonly found in living organisms as storage fats, are so named because they have no charged groups at a neutral pH. They can be used to illustrate the basic molecular structure of cell lipids. A neutral lipid consists of the alcohol glycerol combined with long-chain fatty acids. Glycerol has three sites avail-able for forming linkages with fatty acids and, if all three sites are occupied, a triglyceride is formed (Figure 2–27). The neutral lipids in living systems are primarily triglycerides.

Fatty acids are long molecules consisting of a chain of carbon atoms with attached hydrogens (Figure 2–28).

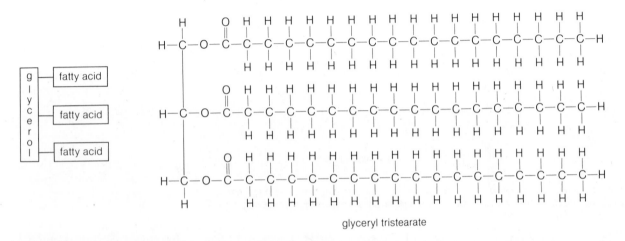

glyceryl tristearate

Figure 2–27 The basic structure of triglycerides (fats, oils, and waxes), formed by interaction between glycerol and three fatty acids. The three fatty acids in the interaction may be all the same, two the same and one different, or all different.

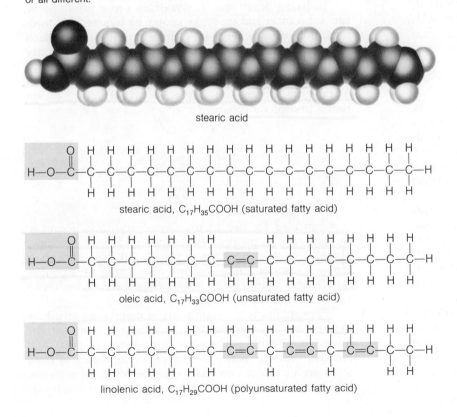

stearic acid

stearic acid, $C_{17}H_{35}COOH$ (saturated fatty acid)

oleic acid, $C_{17}H_{33}COOH$ (unsaturated fatty acid)

linolenic acid, $C_{17}H_{29}COOH$ (polyunsaturated fatty acid)

Figure 2–28 A comparison of saturated, unsaturated, and poly-unsaturated fatty acids. Stearic acid is also shown by a space-filling model.

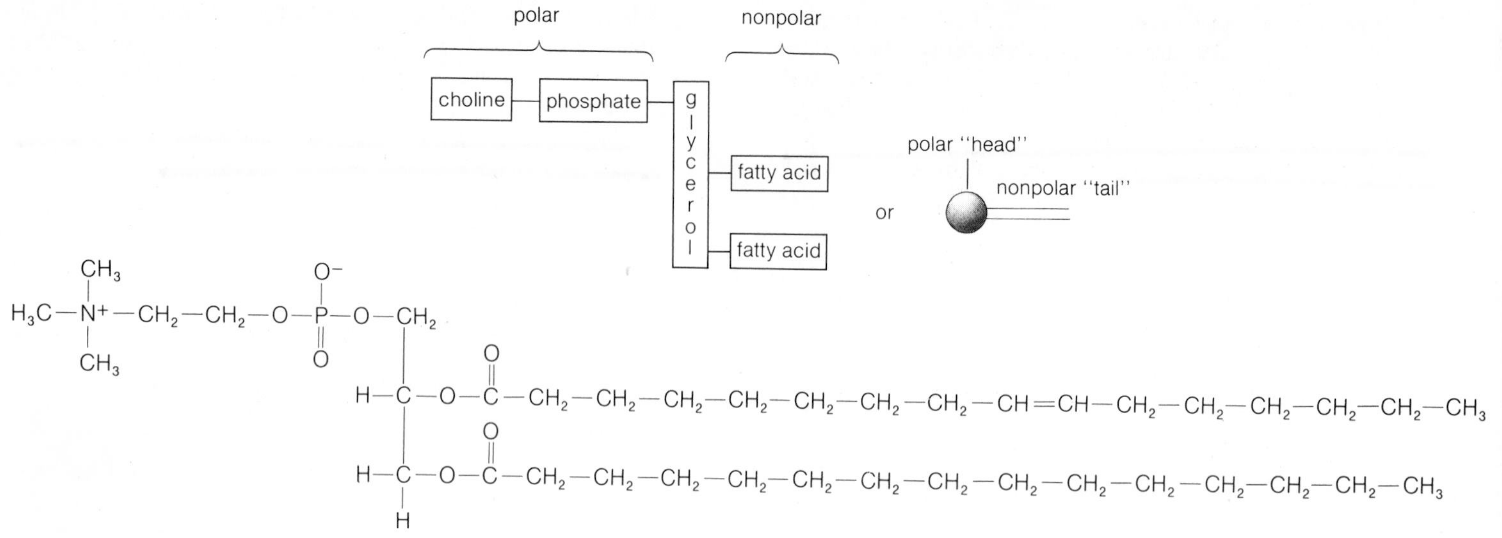

Figure 2–29 An important membrane phospholipid, phosphatidyl choline. Phospholipids are often diagrammed as a circle with two legs (at the upper right). The circle stands for the polar phosphate group and the legs for the two long, nonpolar fatty acid chains of the molecule.

A terminal carboxyl (–COOH) group gives the molecule its acidic properties and is the site where the fatty acids bind to the glycerol. Most fatty acids in natural lipids have chains with 14 to 22 carbons, usually either 16 or 18.

If the carbons have as many hydrogens attached to them as they can hold, then the fatty acid (and thus the fat containing it) is said to be *saturated* (Figure 2–28). In such a molecule, since all the carbon bonding sites are occupied by hydrogens, there will be no double bonds between the carbons. Saturated fats have been implicated, along with cholesterol, in one type of arteriosclerosis (hardening of the arteries), although the cause and effect have not been definitely established.

Unsaturated fatty acids (Figure 2–28) have some carbon bonding sites that are not occupied by hydrogens, and thus will contain some double bonds between carbons. Fats with unsaturated fatty acids have lower melting points than saturated fats. Certain natural vegetable oils, for example, are usually liquid at room temperature; they can be "hydrogenated," or saturated, to transform them into the solid shortenings preferred for cakes and pastries. Unsaturated fats are more abundant than saturated fats in living organisms. *Polyunsaturated,* a term much used in advertising, means simply that several sites are unoccupied, and thus several double bonds between carbons will be present.

In living organisms, triglycerides serve primarily as fuel substances and are broken down by biochemical reactions that release energy for cellular activities. Seeds such as peanuts and corn are rich sources of unsaturated fatty acids.

Phospholipids, important components of cellular membranes, are similar to triglycerides in structure, except that a phosphate group,

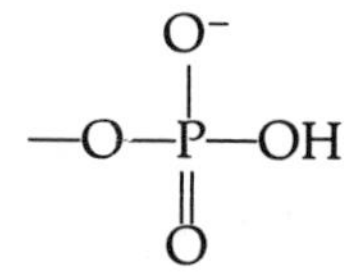

$$-O-\overset{\overset{\displaystyle O^-}{\displaystyle |}}{\underset{\underset{\displaystyle O}{\displaystyle \|}}{P}}-OH$$

is substituted for one of the three fatty acids bound to the glycerol. In membrane phospholipids such as phosphatidyl choline (lecithin, Figure 2–29), the phosphate group is usually linked to one of a group of alcohols, usually containing one or more nitrogen atoms. Lecithin's particular alcohol is choline.

Phospholipids are significant in membrane structure because of their relationship to water. The end of the molecule containing the phosphate group is polar and has an affinity for water, whereas the long chains of fatty acids are nonpolar and have affinities for lipid solvents. When a phospholipid is placed in water, a layer one mole-

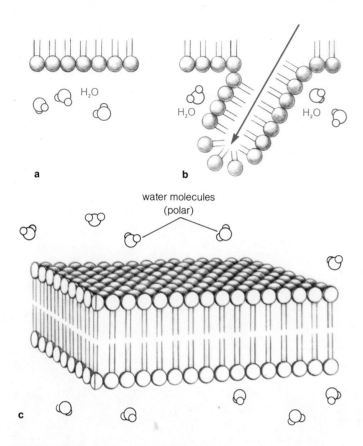

Figure 2–30 (**a**) On the surface of water, lipids form a monolayer with the polar "head" down and the nonpolar "tail" up. (**b**) When pressed into water, lipids form a bilayer with heads out and tails in. (**c**) A lipid bilayer surrounded by water.

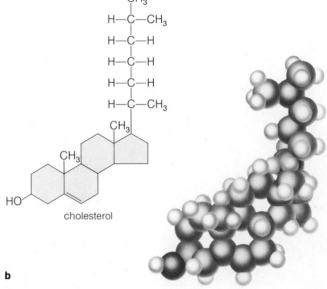

Figure 2–31 The steroid skeleton and the molecular structure and model of cholesterol.

cule thick (a monolayer) will form. If this monolayer is drawn down into the water, the phospholipid will form a bilayer two molecules thick in which the polar phosphate groups face toward the top and bottom surfaces of the layer in contact with the water (Figure 2–30). The long, nonpolar fatty acid chains face each other in the interior of the bilayer in a nonpolar environment that excludes water. Chapter 3 will discuss bilayers and membrane structure in more detail.

Although classed as lipids because of their solubility properties, steroids are structurally different from other lipids, being derived from a system of four interlocking rings of carbon atoms (Figure 2–31). One of the most widely distributed steroids is cholesterol (Figure 2–31b). Found in the surface membrane of all animal cells, cholesterol also makes up about one-tenth of the brain and is a major component of certain types of gallstones, from which it derives its name—"bile solid" in Greek. It is also found in deposits in hardened arteries, although it

has not been established whether it is there as a cause or as an effect of some other process. Other steroids occur as part of the surface membranes in plant cells and as **hormones,** substances that regulate the complex chemistry of both higher plants and animals.

Proteins **Proteins** are large complex molecules that serve two vital functions in living organisms: They act as structural molecules, providing much of the framework of cells, and as enzymes. All protein molecules are basically similar in structure, consisting of one or more long chains of **amino acids.** Although only 20 different kinds of amino acids occur naturally in proteins, the total number of molecules of these different amino acids in a protein may vary from as few as 50 to more than 20,000. Each different sequence of amino acids produces a different protein. Thus, an almost infinite variety of proteins can be imagined. Even in the shortest protein, consisting of

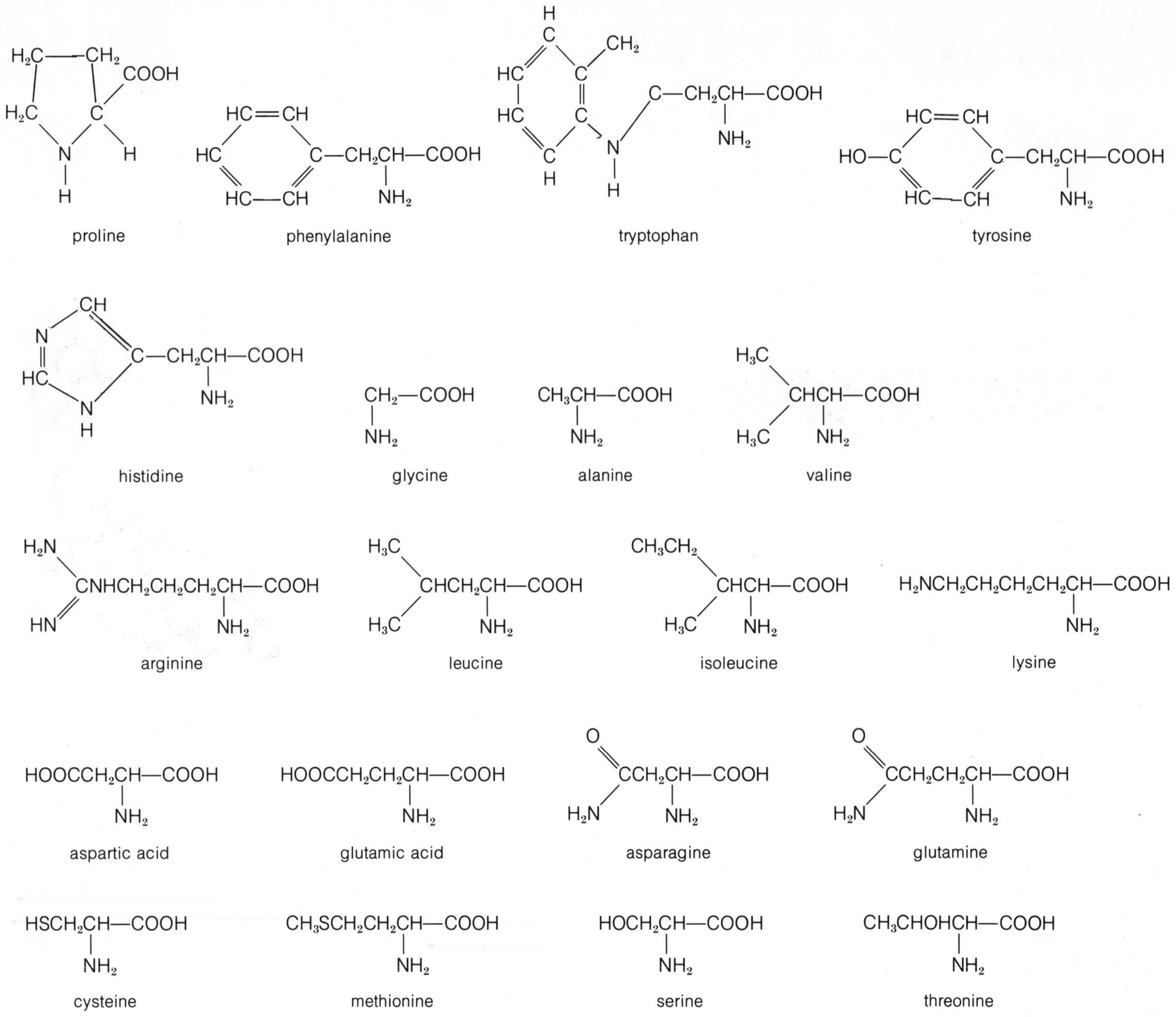

Figure 2-32 Molecular structures of the 20 amino acids.

50 amino acid molecules, 20^{50} different possible sequences could occur. This number is roughly equivalent to one unique protein for every gram of matter in the universe!

Nineteen of the 20 amino acids are based on the same structural plan (Figure 2-32)—a central carbon atom with an amino group ($-NH_2$) attached on one side and a carboxyl group ($-COOH$) on the other. One of the remaining bonds on the central carbon is linked to a hydrogen atom, giving this basic arrangement:

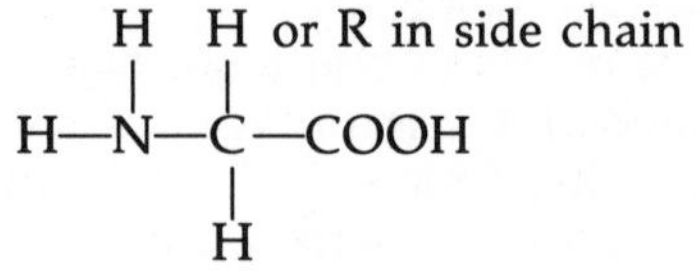

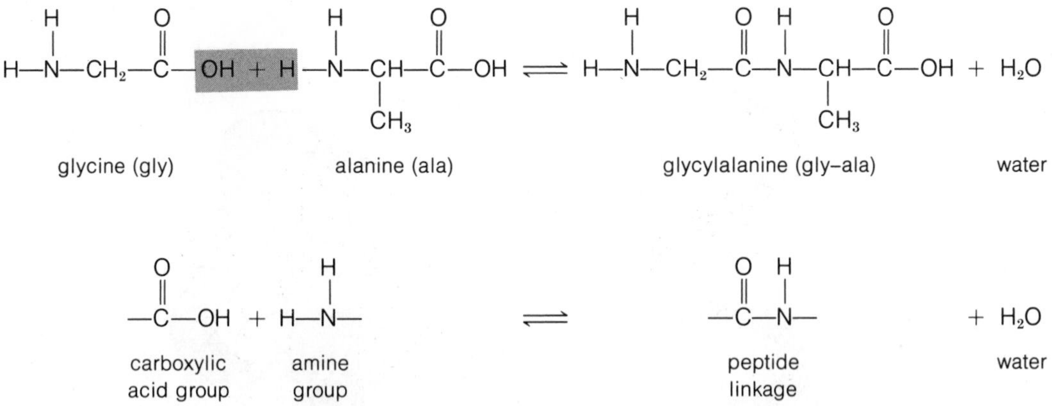

glycine (gly) alanine (ala) glycylalanine (gly–ala) water

carboxylic acid group amine group peptide linkage water

Figure 2–33 When two amino acids join together, they form a peptide linkage by splitting out a water molecule. An additional amino acid can be attached through a second peptide linkage, and the process can be repeated until a long-chain polymer, or protein, results.

The fourth bond of the central carbon may be attached to any one of 19 different side chains (indicated by H or R), ranging in complexity from a single hydrogen atom in the simplest amino acid, glycine, to long chains or rings of carbons in the more complex amino acids. Some of the more complex side chains contain oxygen, nitrogen, or sulfur atoms—often in the form of amide ($-NH_2$), hydroxyl ($-OH$), and sulfhydryl ($-SH$) groups—in addition to carbon and hydrogen. The exception to this general pattern is proline, in which the structure is a ring without a central carbon atom (Figure 2–32).

The covalent bonds linking the individual amino acid subunits into a protein chain, called *peptide linkages,* are produced by a reaction between the amino group of one amino acid and the carboxyl group of a second amino acid (Figure 2–33) and are formed between the carbon and nitrogen atoms. The short chain yielded by this reaction, consisting of the two amino acids linked side by side, has an amino group at one end and a carboxyl group at the other. As a result, the chain has the same reactive groups available at its two ends as any single amino acid. Although either end of this two-unit chain (called a *dipeptide*) could conceivably enter into a reaction with a third amino acid, in nature, new links in the chain are added only at the carboxyl end. These are added repeatedly, producing long chains of amino acids.

The reaction shown in Figure 2–33 is a typical biochemical pathway for assembling the building blocks of larger molecules. Note that as the two amino acids link together to yield a dipeptide, a molecule of water is also produced.

The sequence of amino acids tell us only part of what we need to know about the structure of a protein molecule. The chain of amino acids is folded and twisted about itself into a three-dimensional arrangement that is significant for the function of the protein (Figure 2–34). For example, even small changes in the three-dimensional shape of enzymes can completely inactivate them.

On the basis of this folding, proteins can be roughly grouped into two classes: *fibrous proteins* that are long linear molecules, and *globular proteins* that are more highly coiled and folded into nearly spherical configurations. Fibrous proteins are poorly soluble in water and are found in such structural elements of the animal body as muscle, skin, hair, and ligaments. Globular proteins are often soluble in water and make up most enzymes, some hormones, storage proteins (as in egg whites), and blood proteins.

Much of our information on the three-dimensional structure of proteins comes from the research of Linus Pauling and Robert B. Corey of the California Institute of Technology. Pauling and Corey worked on proteins from the late 1930s into the 1950s. Their first step was to deduce the structure of the peptide bond. Later they discovered that the amino acid chain of proteins is often twisted into a regular spiral. This spiral, which they called the alpha helix (Figure 2–34), evidently occurs as a part of the folded structure of all proteins. Much of the work that Pauling did was on fibrous protein, such as that in hair.

The alpha-helical portions of protein molecules resist bending and deformation; where sharp bends are found in folded backbone chains, the alpha helix gives way to a much less regular arrangement known as a random coil.

How do proteins attain their unique shape? Proteins are held in their globular or fibrous shape by a number of bonds, the most important being hydrogen bonds and disulfide linkages. Each peptide linkage has hydrogen and oxygen atoms capable of forming hydrogen bonds (Figure 2–33). Such bonds, as already mentioned, are individually weak, but in this case many are formed and they restrict the twisting of the amino acid backbone in a way that

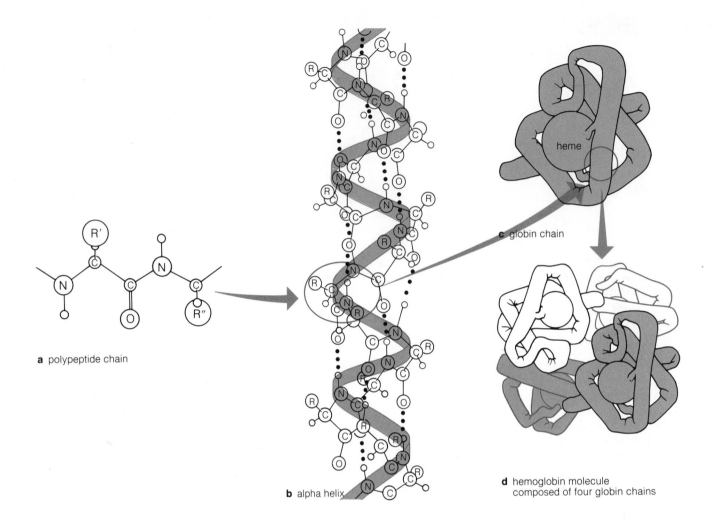

a polypeptide chain

b alpha helix

c globin chain

heme

d hemoglobin molecule composed of four globin chains

Figure 2–34 Organization of protein molecules. Hemoglobin, the oxygen-containing protein in your blood, is shown because it is so well understood. Note that the structure utlimately rests on the polypeptide chain of amino acids and that the final molecule may contain extra groups (in this case, heme, which contains iron).

limits the number of folding possibilities for a protein of a given sequence. The greater the number of internal hydrogen bonds formed, the less flexible the three-dimensional structure of the chain becomes.

Disulfide linkages (Figure 2–35) are produced by covalent bonding between the –SH (sulfhydryl) groups of two cysteine amino acids located at different points in the backbone chain. In regions where they form, these links anchor the protein in a permanently folded position. The fibrous protein of human hair is rich in –SH groups, and disulfide linkages can easily be made. This fact is

taken advantage of when hair stylists give "permanents." Heat and chemicals are used to facilitate the formation of disulfide bonds in ways that keep the hairs twisted in the directions dictated by fashion.

Also important in producing the three-dimensional shape of proteins are the forces produced by attraction and repulsion between charged side groups of different amino acids. Still another factor determining shape is the tendency of polar and nonpolar side groups of different amino acids to associate in distinct subregions within the protein. In addition, some proteins (such as hemoglobin)

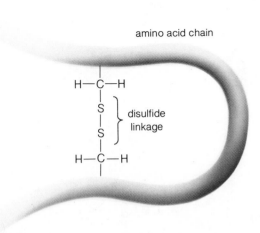

amino acid chain

H—C—H
|
S
| } disulfide
S   linkage
|
H—C—H

Figure 2-35 A disulfide linkage, formed by an interaction between sulfhydryl (–SH) groups on cysteine amino acids located in different parts of an amino acid chain. Disulfide linkages are one of the forces holding protein molecules in a three-dimensional configuration.

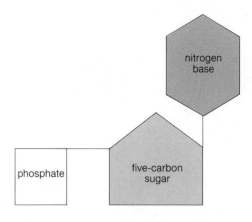

nitrogen base

phosphate

five-carbon sugar

Figure 2-36 A symbolic representation of a nucleotide unit.

are built up from several amino acid chains, folded and twisted together into the three-dimensional structure.

The three-dimensional folding of the amino acid chains in hemoglobin is shown in Figure 2–34. This folding configuration is assumed by the protein at temperatures and pHs approximating the conditions in the living organism. All of the factors just discussed are involved in the molecule's shape. These include the number of chains in the protein, the relative amounts and distribution of alpha helix and random coil, the position and number of hydrogen bonds and disulfide linkages, and the electrostatic, polar, and nonpolar forces. These factors depend in turn on the sequence of amino acids in the component chain. Finally, all of these factors are modified by the environment surrounding the protein molecule— factors such as temperature, number and location of ions in solution, and pH. It is the combination of complex structure and relative plasticity that makes proteins so important to life.

The disruption of any of the bonds in a protein will change its properties. The hydrogen bonds are the most easily disrupted and can be broken by elevated temperatures, excessive acidity, chemicals, radiation, and so on. When the hydrogen bonds are broken, the molecule unfolds into a random, extended shape and loses its functional activity. This type of disorganization, called **denaturation,** may in some cases be reversible. Temperatures above 50°C will break most of the hydrogen bonds, causing protein denaturation, and this is one of the primary reasons why few living organisms can tolerate temperatures above 45 or 55°C. In extreme denaturation, a protein loses its solubility in water and undergoes other massive changes in structure. The protein is then said to be *coagulated,* a condition that is irreversible. The protein albumin in the white of a hard-boiled egg is coagulated.

Nucleic Acids **Nucleic acids** are not used as sources of energy, as structural elements, or as enzymes, as are the carbohydrates, lipids, and proteins. Instead, the role of these molecules is to store and transmit information. How molecules can achieve these feats has been the primary question asked by molecular biologists and geneticists from the 1950s to the present. The answer to the question also represents the triumph of those fields and will be discussed in later chapters. Here we will look only at the structure and composition of nucleic acids.

There are two classes of nucleic acids, *deoxyribonucleic acid* (DNA) and *ribonucleic acid* (RNA). Both are long structures built up from chains of repeating building blocks, the **nucleotides.** Only five different nucleotides are commonly found in the two nucleic acids; the sequence of these nucleotides forms the basis of a code that stores and carries all the information required for the growth and reproduction of living organisms. Some nucleotides transfer energy (such as ATP) or regulate cell functions (cAMP).

Nucleotides are the structural units of all nucleic acids. Each nucleotide (Figure 2–36) consists of (1) a nitrogen-containing base, (2) a five-carbon sugar, and (3) a phosphate group, all linked together by covalent bonds.

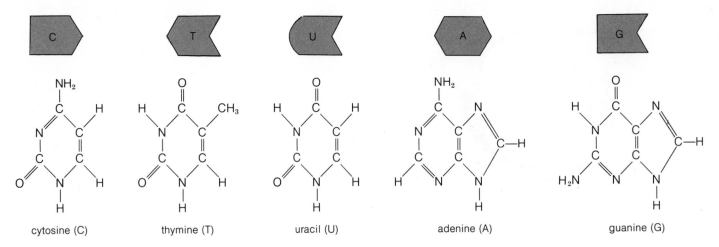

Pyrimidines: Single-Ring Bases

cytosine (C)

thymine (T)

uracil (U)

Purines: Double-Ring Bases

adenine (A)

guanine (G)

Figure 2–37 The five major nitrogen bases found in DNA and RNA. The nitrogen ring is common to all these bases. Thymine is found only in DNA, and uracil is found only in RNA.

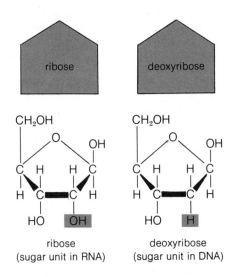

ribose
(sugar unit in RNA)

deoxyribose
(sugar unit in DNA)

Figure 2–38 RNA and DNA are named after the pentose sugars ribose or deoxyribose found in their nucleotide units. The only difference between the sugar molecules is the shaded group.

The nitrogen base will be either a pyrimidine or a purine (Figure 2–37). Both are ring-shaped molecules containing carbon and nitrogen atoms. Three pyrimidine bases, cytosine (C), thymine (T), and uracil (U), and two purine bases, adenine (A) and guanine (G), are the ones most often found in natural nucleotides. Other pyrimidines

and purines are found in nature; for example, caffeine is a purine.

The five-carbon sugar of a nucleotide will be either ribose (in RNA) or deoxyribose (in DNA). By convention, the carbons of these pentose sugars are numbered from 1 to 5 as in Figure 2–38; the covalent bond between ribose or deoxyribose and the nitrogen base is always at the 1-carbon of the sugar. The ribose or deoxyribose also carries one or more phosphate groups in a chain that links to the 5-carbon of the sugar.

As we have already said, nucleotides are linked in long chains to form the two nucleic acids DNA and RNA (Figure 2–39). The backbone of the nucleic acid chain is formed by alternating sugar and phosphate groups. The phosphate groups, joined to their "parent" nucleotides by a bond to the 5-carbon of the sugar, are also linked to the 3-carbon of the sugar next in the chain. Thus 3- and 5-carbon linkages alternate along the backbone of the molecules. DNA consists of two long chains of nucleotides linked together in this way, each nucleotide containing the sugar deoxyribose and one of the four bases adenine, thymine, guanine, or cytosine (A, T, G, C) (Figure 2–40). Each nucleotide in the single chain making up RNA contains instead the sugar ribose and one of the four bases adenine, uracil, guanine, or cytosine (A, U, G, C). Thus, RNA and DNA differ chemically in the sugar present (ribose or deoxyribose), in one component base (uracil or thymine), and in the number of nucleotide chains. As you would expect, the two nucleic acids have similar chemical properties.

DNA

RNA

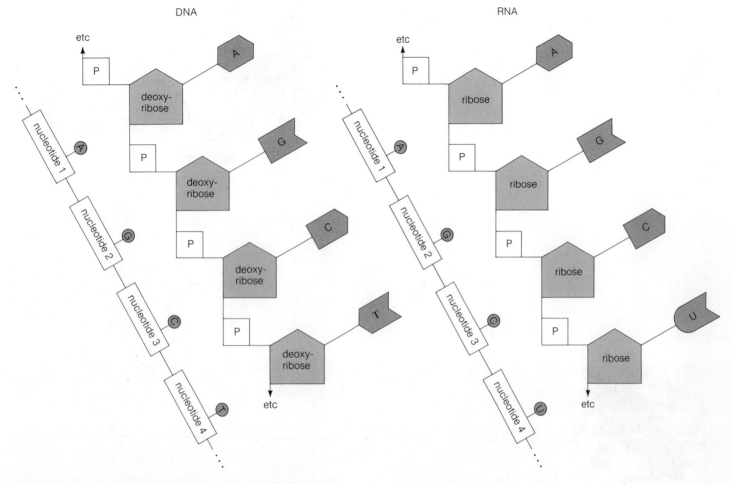

Figure 2–39 Four-nucleotide segments of the DNA and RNA polymers. In DNA, thymine (T) is present instead of the uracil (U) of RNA, and deoxyribose is in the place of the ribose.

ATP: The Coupling Agent One of the most important of all the molecules found in organisms is the nucleotide **adenosine triphosphate,** or ATP. All living organisms, from bacteria to humans, use ATP for the short-term storage of energy and as a coupling agent between uphill and downhill reactions. In the cells of the tree that produced the wood in your pencil, ATP was used as the source of energy for the immediate reactions that synthesized cellulose and the other cell-wall materials. In your hand, the muscles can contract because ATP is available. The amount of ATP used by a cell is enormous—several million molecules per second. For long-term energy storage, organisms use compounds like starch and fats. Such compounds are like money in a stock or bond that can only be used after some work has been done to get it out. In contrast, ATPs are like money orders or traveler's checks that can be converted instantly into cash. Because

of analogies like this, ATP has been called the universal currency of the cell.

Structurally, ATP is a nucleotide. It is composed of adenine, ribose, and three phosphate groups that form a tail (Figure 2–41). If these phosphate groups are removed one by one, the molecules adenosine diphosphate (ADP) and adenosine monophosphate (AMP) are produced.

Basically, ATP is a high-energy compound that requires a large input of energy to be formed and releases a large amount of energy when one of the phosphate groups is split off. The conversion between ATP and ADP, by the addition or removal of the terminal phosphate group and water, can be written:

$$ATP + H_2O \rightleftharpoons ADP + HPO_4{}^{2-} + \text{energy (7,000 cal)}$$

This reaction is readily and fully reversible. If it is moving

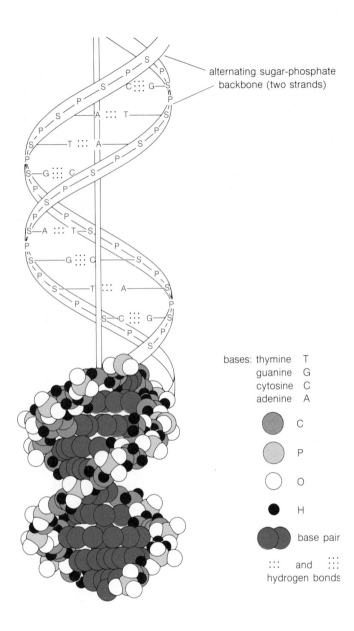

Figure 2–40 The Watson-Crick model of DNA as a double-stranded helix. The two strands are held together by hydrogen bonds (dotted lines) between nitrogen bases in each strand.

alternating sugar-phosphate backbone (two strands)

bases: thymine T
guanine G
cytosine C
adenine A

C
P
O
H
base pair

::: and :::
hydrogen bonds

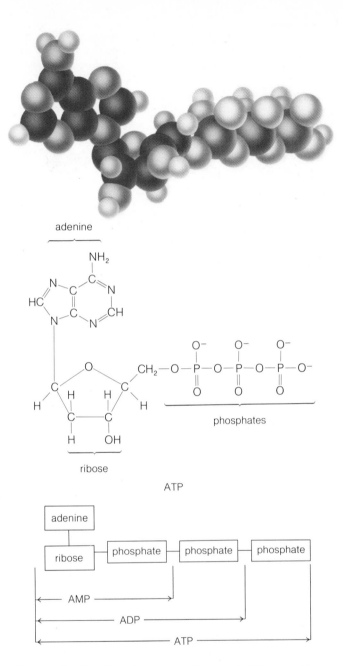

Figure 2–41 Simplifying the complex structure of ATP to show its major reaction site.

adenine

phosphates

ribose

ATP

AMP

ADP

ATP

toward the right in the equation, in which the terminal phosphate is removed from ATP, the reaction releases free energy to the environment, in the amount of 7,000

calories for each molecular weight of ATP converted to ADP. Running the reaction toward the left, in the uphill direction, requires an input of 7,000 calories for each molecular weight of ATP produced. (A calorie is an amount of energy equivalent to the heat required to raise the temperature of 1 g of water 1°C.)

The bonds holding the terminal phosphates of ATP are frequently called *high-energy bonds* because such a large amount of energy is released when they are broken. They are usually designated by a wavy line to show their unusual nature: $\sim P \sim P$. It should be realized, however, that the energy is not actually confined to one bond but is distributed over several. The entire molecule is under stress as a result of several structural features and requires sufficient chemical energy to maintain its stability. When ATP is converted to ADP, the stresses are minimized and energy is released.

During photosynthesis, the reaction is driven uphill, from ADP to ATP, by energy derived from light. Similarly, in the breakdown of sugar during cellular respiration, ADP is converted to ATP using the energy released during this breakdown. The energy captured in ATP is released elsewhere in the cell, at the site of other activities requiring energy. These may be the synthesis of proteins, the contraction of muscle fibers, bioluminescence, or the synthesis of the thousands of compounds needed by the cell.

Summary

In this chapter we have summarized some of the chemical background needed for an appreciation of biology. We have explored, however briefly, the structure of the atom and the forces that are at work between atoms and molecules in the form of chemical bonds. This led to a discussion of the role of energy in biological systems, particularly the energy considerations in chemical reactions. Chemical reactions result in new compounds, and we next covered the major types of compounds found in organisms. Carbohydrates, lipids, proteins, and nucleic acids all have their various and unique roles in the cell. In each of these cases we have tried to show how the fundamental interactions of forces at the atomic and ionic levels result in particular structures of the molecules that in turn affect their role in the organism.

This introduction to the chemistry of life is in no way complete, and other topics will appear in later chapters. Students who are serious in their study of biology—any field of biology—are urged to obtain as thorough a background in chemistry as possible. It is clear that topics such as those covered in this chapter are becoming more important to biologists each year.

3

~ The Cell: Unit of Life ~

Cells are the basic functional units of life. In Chapter 2 we discussed ions, atoms, and molecules found in living systems. Now we will see how these are packaged so that they can exist and be reproduced on earth. The package is, of course, the cell. Humans and all other living things exist because of the activities of the individual cells of which they are composed. Some organisms, such as the protozoa, many algae, and all bacteria, have only one cell. Other organisms, both plant and animal, have many cells—in most cases, many millions of cells. The cells of such organisms are interdependent, but single cells removed from a multicellular organism, if kept under proper conditions, may retain the quality of life indefinitely, and can grow and reproduce. We speak of such cells as being *cultured,* and several types of cells from mammals, such as mice and humans, have been kept growing for decades. But once a cell is damaged or broken open, it dies. Thus, the organization of matter that we recognize as life does not exist in units smaller than the cell.

This point is well made by considering the viruses. A **virus** is a particle that contains the information necessary to tell a cell how to make more virus particles. Consisting of a piece of nucleic acid surrounded by a protective protein coat, viruses can be isolated and crystallized. They can be stored in this form for years in a bottle on the shelf and nothing will happen to them. But, placed in a solution with cells that can act as their hosts, they will infect the cells and convert the machinery of the cells they have invaded to produce more viruses. Thus, having only part of the system is not enough: An integrated, functional whole is necessary to sustain life, and that whole is present only in the cell and in organisms composed of cells.

This concept of the cell as the basic unit of life is one of the central themes of biology. The study of the cell, called **cytology** or **cell biology,** has been going on for slightly over 300 years and is currently in one of its most exciting periods. When we reach the level of the cell, we are in a realm where we can see the structures we are talking about. In Chapter 2 we were dealing with units so small—ions, atoms, and molecules—that we must deduce their properties and form. Only a few of the largest protein molecules, for example, have been seen with the most powerful electron microscopes. In contrast, cells and many of their parts can be seen with the ordinary light microscope, and smaller parts and details of cells are revealed by the electron microscope. In the remaining chapters of this section, we will see how molecules and chemical reactions are integrated to form the living cell. In this chapter, after looking briefly at the history of research on the cell, we will take a quick overview of the two basic types of cells and then examine in some detail the structure and function of membranes, particularly the membrane surrounding the cell.

Background of Cell Biology

A few cells are large enough to be seen with the unaided eye, such as the eggs of birds, frogs, and other amphibians, the nerve cells of the squid, the juice cells in an orange, the fiber cells in flax, a few single-celled algae, and some others. However, the usual size range of cells is 0.5 to 40 micrometers in diameter (see Essay 3–1, p. 52, for the units of measurement used in cell biology). This means that the discovery and study of the cell are dependent on the microscope and related equipment that allow us to study things too small for us to see otherwise.

Using one of the first compound microscopes (that is, a microscope with more than one lens), the great English scientist Robert Hooke traced out compartments in cork and woody plant tissues in 1665. He called the compartments *cells,* because they reminded him of monastery cells. These are similar to the cells we saw in the wood of the pencil in Chapter 2. The impression is often given that Hooke saw only empty or dead cells, but it is clear from his writings that he actually saw substances within the cells, probably in the living state. Hooke's descriptions of cells were extended by the observations of a remarkable Dutch amateur microscopist, Anton van Leeuwenhoek. Over the period 1673–1723, Leeuwenhoek wrote some 375 letters to the British Royal Society, of which Hooke was secretary, and he became a fellow. In these letters, Leeuwenhoek outlined his remarkably accurate observations of the microscopic structure of protozoa, sperm cells, blood cells, and many other "animalcules," as he called them.

Other scientists of the early 1700s also used microscopes to probe the structure of plants and animals. This research revealed that organisms had an elaborate internal structure, but the true nature of this structure could not be pursued because of limitations in the microscopes being used. Serious study of the cell was therefore frustrated and had ceased by the mid-1700s.

Not until the early nineteenth century was it possible to make lenses for microscopes that compensated for the errors found in the earlier lenses. This improvement was coupled with design modifications in the body of the microscope to produce the forerunner of the modern light microscope. Almost immediately there arose a renewed interest in studying the cell.

The Cell Theory In the early 1830s, the Scottish botanist Robert Brown began investigating the microscopic structure of plants. In an influential paper published in 1833, Brown pointed out that all the cells he had examined contained a relatively large spherical structure, which he called the nucleus. He concluded that the nucleus was a universal feature of plant cells.

Brown's findings were broadened and extended by the German scientist Theodor Schwann, who first described the cellular nature of animal tissue in his study of cartilage published in 1839. Soon thereafter Schwann came under the influence of his countryman M. J. Schleiden, a controversial scientist whose greatest contribution seems to have been arousing the interest (and often the antagonism) of other investigators (see Essay 6–1). Schleiden was convinced that studying development was the best way to understand cell structure and hypothesized that the nucleus was important in cellular development.

Schleiden's views so impressed Schwann that he reexamined his own evidence in light of them. From this examination, and a review of other investigators' results, Schwann concluded that *all living organisms are composed of one or more nucleated cells,* and that *the nucleated cell is the unit of function in living organisms.*

Other cell researchers were prompted in their work, as Schwann had been, by Schleiden's controversial views.

While Schwann was developing his theories on the cellular nature of living organisms, other scientists observed that both plant and animal cells multiply by the division of parent cells into daughter cells. By 1855 this work had progressed far enough for the prominent German scientist Rudolf Virchow to state with assurance that *all cells arise from preexisting cells by a process of division.*

In summary, the cell theory, which has remained virtually unchanged and has withstood all experimental tests to this day, concludes that (1) all living organisms are composed of cells, (2) the cell is the unit of function of living organisms, and (3) cells arise only by the division of preexisting cells.

Development of Knowledge of the Cell Improvements in the microscope and in microscopic techniques for examining tissue led to an intensive period of cell investigation in the second half of the 1800s. This was an exciting period, in which the major features of the cell were discovered. During this time and the early 1900s, all the cell parts that could be seen with the light microscope were observed. Still, the exact nature and functions of many of these parts were not clearly understood. The reasons for these uncertainties were twofold. First, the limit of the light microscope had been reached. Things could be usefully magnified about 1,000 to 2,000 times, but no more, allowing researchers to see the larger cell parts but not their interior structure or smaller particles. Since this limitation was based on the wavelength of light, it could not be overcome by designing finer lenses or better microscope bodies. Second, chemistry—and particularly organic chemistry—had not reached a state where it could be used to understand the complex composition and reactions of the various cell parts. And when, during the first half of this century, organic chemistry and biochemistry did reach the necessary levels of sophistication, technical problems in dealing with cells prevented discoveries in these fields from being applied to cell biology right away.

During the 1940s and 1950s, several dramatic developments were to change the nature of cell biology. One was the coming of age of the electron microscope as a biological research tool. In this kind of microscope, a beam of electrons replaces light, and magnetic lenses replace glass ones (Figure 3–1). Magnifications simply not possible with the light microscope could now be achieved. At last the limitations imposed by the properties of light could be overcome, and a new world of cell structure opened before the cell biologist.

Coupled with the invention of the electron microscope were advances in handling individual cell parts. Procedures were discovered that allowed a cell to be taken apart and each of its components separated from all the others (Figure 3–2). Despite all of these manipulations, after separation the individual parts maintained the function they had in the intact cells. Thus it was possible not simply to study the cell as a bag of chemicals, but to look at what each part was made of, what it did, and how it related to the whole. Radioactive isotopes, which became available in the 1950s, greatly aided in this research, because individual elements such as carbon could now be traced through all the cell activities. Likewise, the development of chemical methods that allowed researchers to localize compounds and activities directly within the cell also broadened our understanding of cell function.

Finally, the discovery of the structure of DNA in 1953 provided a tremendous stimulus to our appreciation of the cell as the place where genetic information is translated into biological action. Advances in genetics, biochemistry, and cell biology all crystallized in the 1960s and 1970s to give us an understanding of protein synthesis, nucleic acid metabolism, and gene function. Cell biology is still growing, and research is daily contributing to our knowledge of the cell.

The Organization of Living Cells

As a result of intensive investigation, cells have been classified into two groups according to the arrangement of their cell membranes and the locations of cellular functions on these membranes. As we shall see shortly, membranes are vital to the structure and function of the cell. Indeed, membranes determine cellular structure. There are only two basic plans for this structure, and all organisms on earth can be divided into two groups according to which plan they follow. In the simpler plan of the two (Figure 3–3a), found only in the bacteria and the blue-green algae, the cell has only one membrane, the surface membrane that completely surrounds it. Any membranous tissue found inside such cells will be inward extensions of the surface membrane. The **nucleus** by definition is enclosed in its own membrane or membranes, so these cells obviously have no nuclei *as such*. However, the name chosen for this kind of organization, **prokaryotic,** does not mean "without a nucleus," but "before a nucleus," or "prenucleic." This choice of Greek prefix indicates that scientists consider bacteria and blue-green algae primitive forms of life, and the fossil record supports the idea. The prokaryotes were apparently the first to evolve of all the organisms that have survived to the present. Fossils believed to be the remains of prokaryotes date back 3.3 billion years.

The second plan is much more elaborate (Figure 3–3b). Here, too, there is an external membrane, but there are also internal membranes that separate certain parts

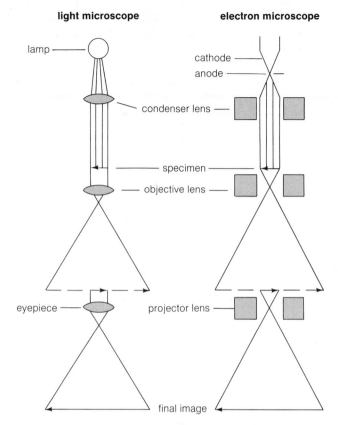

Figure 3–1 Comparison between the components of the light and electron microscopes.

of the cell. These membrane-bound interior structures, called **organelles,** are specialized to carry out the various cellular functions with a high degree of efficiency. All plants and animals on earth have cells of this sort and are called **eukaryotes,** the term being based on the presence of a true *(eu-)* nucleus *(karyon)*. Organisms with eukaryotic cells are believed to have evolved later than prokaryotes, and again the fossil record appears to support this theory.

Prokaryotic Cells Bacteria and blue-green algae are the principal groups of organisms composed of prokaryotic cells. We are all familiar with the bacteria and have, at one time or another, suffered from a bacterial disease. Bacteria are important to humans, as well as being a bother, and contribute much to the environment. The blue-green algae are much less familiar, but they too are important to us and the environment. We will discuss both the bacteria and blue-green algae as organisms in Chapter 33; here it is necessary to point out only that

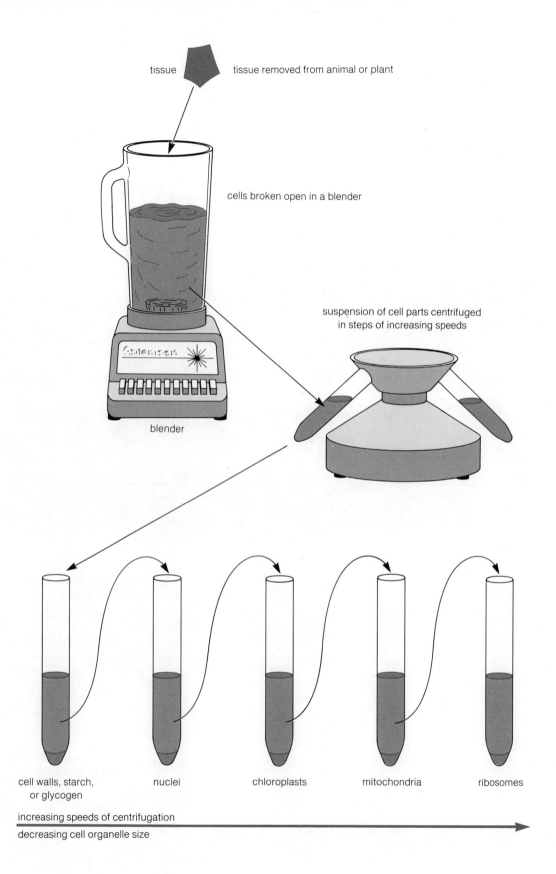

tissue — tissue removed from animal or plant

cells broken open in a blender

blender

suspension of cell parts centrifuged
in steps of increasing speeds

cell walls, starch,
    or glycogen

nuclei

chloroplasts

mitochondria

ribosomes

increasing speeds of centrifugation

decreasing cell organelle size

Figure 3–2 Outline of procedures used to separate the various
parts of the cell. At different centri-
fuge speeds, different-sized particles separate out.

eukaryotic (animal) cell

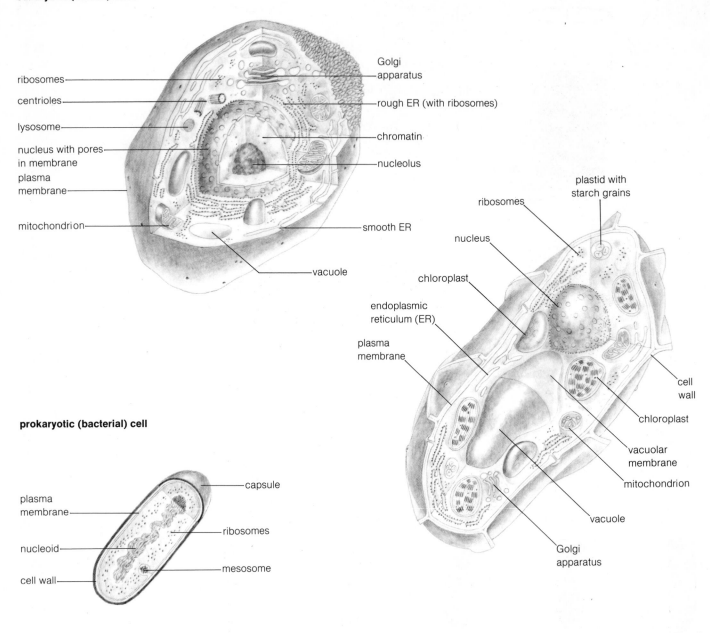

ribosomes

centrioles

lysosome

nucleus with pores
in membrane

plasma
membrane

mitochondrion

Golgi
apparatus

rough ER (with ribosomes)

chromatin

nucleolus

smooth ER

vacuole

plastid with
starch grains

ribosomes

nucleus

chloroplast

endoplasmic
reticulum (ER)

plasma
membrane

cell
wall

chloroplast

vacuolar
membrane

mitochondrion

vacuole

Golgi
apparatus

eukaryotic (plant) cell

prokaryotic (bacterial) cell

capsule

plasma
membrane

ribosomes

nucleoid

mesosome

cell wall

a

b

Figure 3–3 (**a**) A generalized prokaryotic cell. (**b**) Generalized eukaryotic animal and plant cells. The drawings are based on the appearance of cellular organelles in electron micrographs.

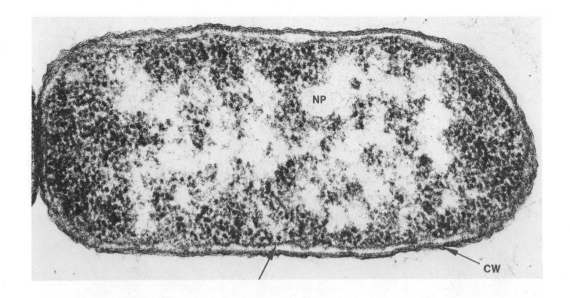

Figure 3–4 A prokaryotic cell, the bacterium *Escherichia coli*, magnified 28,500 times. The nucleoid (NP), the prokaryote equivalent of a nucleus, occupies the center of the cell. The cytoplasm surrounding the nucleus is packed with ribosomes. The cell is surrounded by the cell wall (CW); the plasma membrane (arrow) lies just beneath the cell wall. ×21,500.

they are important members of the living world, occupying perhaps as much as 50 percent of the total mass of living matter on earth.

Compared to eukaryotic cells, with their wealth of membrane systems and organelles, prokaryotic cells look simple (Figure 3–3a). They are comparatively small, seldom more than a few micrometers long and one micrometer wide. The outer membrane of the cell, called the **plasma membrane,** is surrounded by a rigid external cell wall, which in turn may be coated with a thick, jellylike capsule or slime. The plasma membrane may be smooth or marked by infoldings that extend into the cell, forming structures called **mesosomes.** Most of the various energy reactions of the cell take place in the plasma membrane. In addition, the plasma membrane of bacteria is thought to play a role in cell replication and division.

The interior of prokaryotic cells (Figures 3–3a and 3–4) contains a central nucleuslike body called the **nucleoid,** surrounded by **cytoplasm** (all of the substance enclosed by the plasma membrane but outside the nucleoid). No membranes separate the nucleoid from the cytoplasm. The nucleoid contains the genetic information of the cell, which in bacteria consists of a single piece of DNA about one millimeter long, closed into a circle. The cytoplasm is crowded with small spherical bodies about 20 nanometers in diameter, the **ribosomes.** They are the site of protein synthesis. The cytoplasm of the more complex bacteria and blue-green algae may also contain vacuoles (saclike structures), vesicles (small vacuoles), and reserve deposits of polysaccharide or inorganic phosphates.

Many bacteria are capable of rapid movement, gener-

ated by the action of threadlike structures, the flagella, covering the cell surface.

The seeming simplicity of prokaryotic cells is deceptive, because these microorganisms can assemble all the complex molecules of life from the simplest substances, such as elemental carbon or sulfur or inorganic salts and water. For example, they can synthesize vitamins that humans must get from their food or produce in complicated laboratories. They have survived for billions of years, enduring changes that destroyed dinosaurs, mastodons, and earlier versions of human beings. Yet the prokaryotic organization seems to have lacked the potential to give rise to true multicellular organisms composed of prokaryotic cells.

Eukaryotic Cells In contrast to prokaryotic cells, eukaryotic cells (Figures 3–3b and 3–5) contain systems of internal membranes forming distinct organelles that are completely separate from the plasma membrane. The plasma membrane, of course, is present, and it functions as a gatekeeper to all substances entering and leaving the cell, just as it does in prokaryotic cells.

Often the most prominent organelle in the eukaryotic cell is the nucleus (Figures 3–3b and 3–5), surrounded by two concentric membranes known collectively as the **nuclear envelope.** Within the nucleus is the **chromatin,** which consists of DNA and proteins. Also within the nucleus are one or more irregularly shaped bodies, called **nucleoli** (singular **nucleolus**), which function in the synthesis of ribosomes.

The cytoplasm of the eukaryotic cell contains many

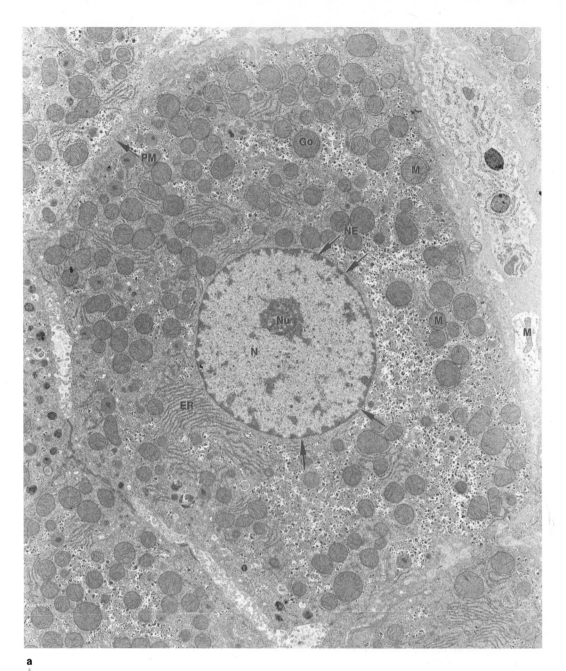

Figure 3–5 (**a**) Electron micrograph of a pancreas cell of an animal. ×15,000. (**b**) Electron micrograph of a plant cell. ×10,000. Both are good examples of eukaryotic cells. Identified are the nucleus (N), nucleolus (Nu), mitochondrion (M), endoplasmic reticulum (ER), Golgi apparatus (Go), plasma membrane (PM), nuclear envelope (NE), plastids (Pl), vacuoles (V), and cell wall (CW). Arrows point to pores in nuclear envelope.

ribosomes and membrane-bound organelles. Among the most common of these organelles are the **mitochondria** (Figures 3–3b and 3–5). These small, usually rod-shaped bodies carry out chemical reactions that release energy to drive all the activities of the cell.

Conspicuous in plant cells are the various types of **plastids,** which are involved in food synthesis and storage in the cell. The most familiar of these are the green

chloroplasts, in which phostosynthesis takes place (see Figure 4–1).

The many ribosomes in the eukaryotic cytoplasm may be freely suspended in clusters or attached to a membranous system known as the **endoplasmic reticulum** or ER (Figures 3–3b and 3–5). Ribosomes are also found in mitochondria and plastids. It is a curious fact that the ribosomes found in prokaryotic cells and those found

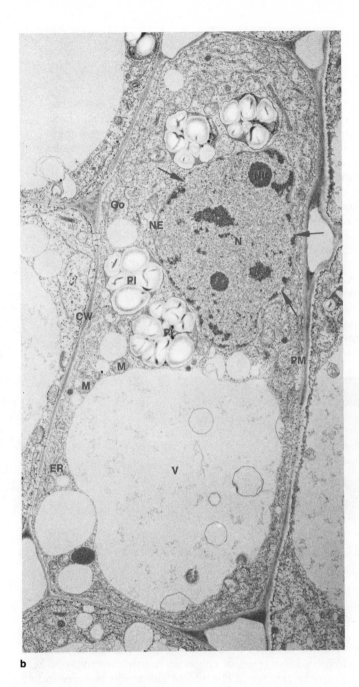

b

Camillo Golgi, who discovered them in 1898. In animal cells, they appear to function as the packaging sites for protein secreted from the cell, as in the case of cells that secrete digestive enzymes into our stomachs. In plant cells, they are involved in the packaging and secretion of certain types of cell-wall materials.

Lysosomes are vesicles that contain many different enzymes, together capable of breaking down most biological molecules. Lysosomes may fuse with and digest matter absorbed by the cell, or they may help dispose of other cell organelles that are worn out or no longer needed.

Vacuoles (Figures 3–3b and 3–5) are found in most plant and some animal cells, but they are usually much larger and far more conspicuous in plant cells, where nine-tenths or more of the cell may be occupied by one or more of them. Vacuoles are saclike structures, bounded by membranes. They contain dilute solutions of a variety of substances, including organic and inorganic salts, organic acids, sugars, and various pigments. The colors of flowers are often due to pigments concentrated in vacuoles.

The cytoplasm of both plant and animal cells also contains **microtubules.** These structures are long, hollow cylinders, ranging from 18 to 27 nanometers in diameter, and varying in length from a few nanometers to many micrometers. Microtubules in the form of spindle fibers are involved in cell division and, in some cells, seem to be involved in support, transport, and motility. In regions of the cell with a motile function, microtubules are often found closely associated with **microfilaments,** thin fibers 5 to 7 nanometers in diameter.

In all animal cells and some plant cells, such as algae, a pair of **centrioles** are found near the nucleus (Figure 3–3b). The centrioles are apparently important in cell division in animal cells, and in both animal and plant cells they give rise to the eukaryotic **flagella** (singular, **flagellum**). These are hairlike extensions of the cell surface of motile plant and animal cells. Flagella consist of microtubules and are found in a wide range of cells either singly or in groups. Some organisms, such as the protozoa *Paramecium,* are covered with numerous short flagella called **cilia.** Sperm swim because of flagella, and the human respiratory tract is lined with ciliated cells. These cells help keep our air passages clear of dirt and other particles, their whiplike motion forcing the particles back up the respiratory tract. One of the problems associated with smoking is that it inhibits cilia motion.

Most plant cells, in contrast to animal cells, are surrounded by **cell walls.** These are layered structures outside the plasma membrane and thus outside the cell (Figure 3–5b). The properties and composition of the wall change as the cell grows and develops.

In this rapid and superficial survey of the eukaryotic cell, we have not tried to discuss the function or structure

in plastids and mitochondria are similar to one another in size, and smaller than those found in the cytoplasm of the eukaryotic cell. In all cases, ribosomes appear to be the site of protein synthesis.

Also found in the cytoplasm are collections of membranous vesicles that often take the form of closely stacked, flattened sacs (Figures 3–3b and 3–5) known as the **Golgi complex** or Golgi apparatus after the Italian

Essay 3–1

Units of Measurement Used in Cell Biology

Much information about cell structure is given in terms of physical dimensions. The units most frequently used are the micrometer, μm (formerly called the micron), the nanometer, nm, and the angstrom (A). These units are compared in Table E3–1. The micrometer, equivalent to 0.001 millimeter, is convenient for describing dimensions of whole cells or larger cell structures such as the nucleus, and is much used in this book. Most cells are between 5 and 200 micrometers in diameter, although some animal eggs may reach much larger sizes, ranging up to several centimeters. Objects at about 200 micrometers in (0.2 millimeters) in diameter are just visible to the unaided eye.

Dimensions ranging downward from particles roughly the size of mitochondria and chloroplasts (objects in this smallest range of dimensions are called the "ultrastructures" of the cell) are measured by the nanometer or the angstrom, which are useful for describing dimensions from this level down to particles as small as molecules and atoms. The nanometer, equivalent to 0.001 micrometer, is difficult to visualize, but with experience a relative appreciation can be acquired of the size of objects given in this unit. Lipid molecules, for example, are about 2 nanometers in length; amino acids are about 1 nanometer long; protein molecules may be 10 nanometers or so in diameter. On the level of cell organelles, membranes are 7 to 10 nanometers thick, and ribosomes are about 25 nanometers in diameter. The electron microscope, incidentally, can "see" objects with diameters as small as 1 nanometer or less. Angstrom units (10 angstroms equals 1 nanometer) are also frequently used for the smallest dimensions, those below 1 nanometer.

Table E3–1 Units of Measurement Used in Cell Biology

Unit	Equivalence in Millimeters	Equivalence in Micrometers	Equivalence in Nanometers	Equivalence in Angstroms
Millimeter (mm)	1	1,000	1,000,000	10,000,000
Micrometer (μm)	0.001	1	1,000	10,000
Nanometer (nm)	0.000001	0.001	1	10
Angstrom (A)	0.0000001	0.0001	0.1	1

of the various cell organelles in detail. We will do this in later chapters of this section. It is clear, however, even from this sketchy overview, that we are dealing with a much more complex structure than is present in the prokaryotic cell.

Membranes and the Cell

Both prokaryotic and eukaryotic cells exist because they are surrounded by membranes. The plasma membrane is vital for the survival of all cells. Why? Because the molecules and systems necessary for life—in thermodynamic terms, the processes needed to maintain order in the face of disorder—must be kept together and maintained in the proper environment. Enzymes must be kept close to supplies of substrates, and groups of enzymes involved in a series of consecutive reactions must be kept properly arranged. The pH of the medium surrounding the enzymes must be correct for them to function, and various salt concentrations must be kept within narrow ranges. If the plasma membrane of a cell is broken, the contents spill out; the individual enzymes may continue to function, but the cell is dead.

The plasma membrane does more than keep things in the cell. It also keeps things out of the cell that would be harmful, while admitting substances necessary for cell function. Water, oxygen, carbon dioxide, minerals, and similar substances must be able to enter the cell. At the same time, foreign protein molecules and other compounds must be kept out. In short, the plasma membrane must be *differentially permeable,* allowing some substances to pass through it while restricting or preventing the passage of others.

The internal membranes of the eukaryotic cell compartmentalize the various activities of the cell. Thus, for example, the membranes that surround the chloroplast enclose the enzymes and materials necessary for photosynthesis. It is also clear that, in the case of such organelles as chloroplasts and mitochondria, the membrane does more than hold things together. The enzymes themselves are arranged in the membranes, and the spatial alignment of the various systems is a function of the membranes.

With membranes playing such important roles in the life of the cell, it is hardly surprising to find that they have been intensively studied by cell biologists. Let us look at some of that research and the current status of our concept of membrane structure.

Membrane Structure Much of the early research on membranes used plasma membranes, particularly those of the red blood cells, because they were easily obtained. Some of the earliest data showed that molecules soluble in lipids enter the cell faster than insoluble molecules (an exception being water, which readily enters and leaves the cells). This suggested that the surface layer of the cell is composed of lipids. Later analysis showed that lipids are indeed present, as are proteins.

Putting together all the available data, the English researchers Hugh Davson and J. P. Danielli created the model of membrane structure shown in Figure 3–6. In this model, the lipid molecules form a double layer (or bilayer), with their polar "heads" outward and their nonpolar "tails" inward. This lipid bilayer was postulated to be coated on both sides with a layer of protein.

For some years, this was accepted as the universal model for cell membranes. It was supported by a considerable body of data, including some from the electron microscope. But conflicting data also began to accumulate. Many membranes were found to contain carbohydrates as well as lipids and proteins, and some seemed to have different structures from the one Davson and Danielli had visualized.

A key method in reevaluating the model and learning more about membrane structure has been the freeze-fracture technique. In this technique, cells are quick-frozen in liquid nitrogen (−196°C). The frozen cells are split by a knife edge, and the fracture travels through the frozen specimen, primarily following membrane surfaces, but sometimes splitting through the interior of membranes, separating the membrane into top and bottom halves. Figure 3–7 shows both sides and the interior of a membrane exposed by the freeze-fracture technique.

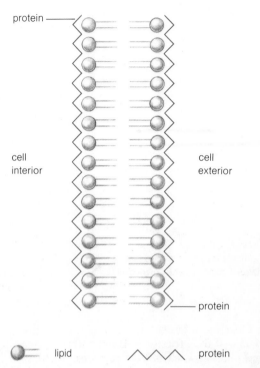

protein

cell interior

cell exterior

protein

○━ lipid   ∧∧∧ protein

Figure 3–6 Diagram of a membrane, according to the Davson-Danielli model.

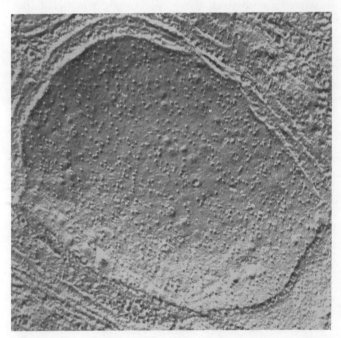

D. W. Deamer

Figure 3–7 A surface replica of a membrane system prepared for electron microscopy by the freeze-fracture method. In the entire central region of the micrograph, the path followed by the fracture has exposed the membrane interior. The particles exposed by the fracture are proteins that extend into the membrane interior. ×66,000.

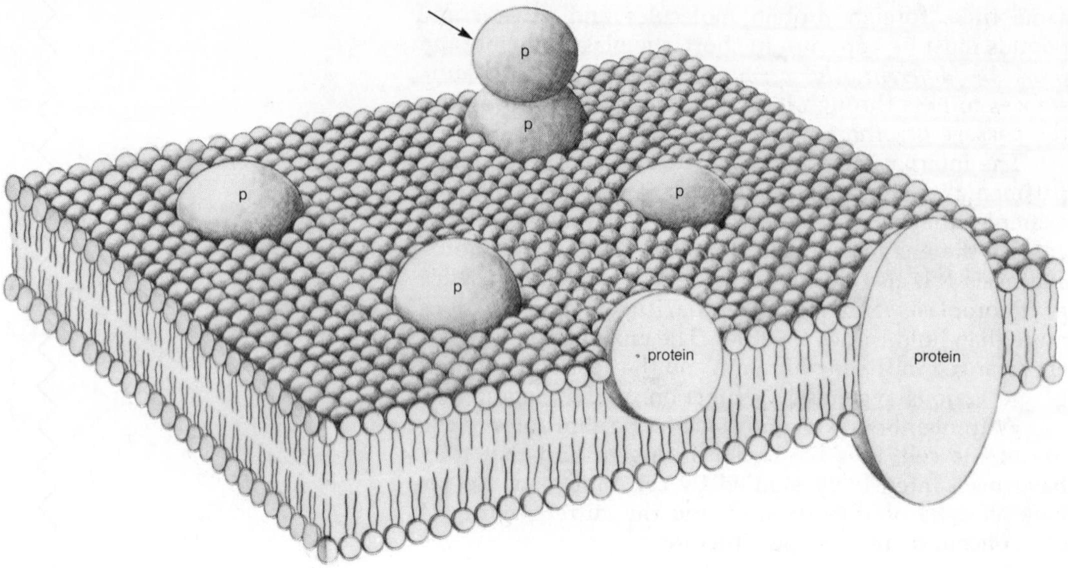

Figure 3–8 The fluid mosaic model for membrane structure. In this model, the membrane is basically a lipid bilayer with proteins embedded in it. Some proteins extend through the structure, others only halfway, while still others are more loosely bound to the membrane surface (arrow). Redrawn; courtesy S. J. Singer. Copyright 1972 by the American Association for the Advancement of Science.

Electron microscopy of freeze-fractured specimens shows that the proteins of the membrane, instead of being evenly spread on the two surfaces, are spaced intermittently, both on and within the membrane (Figure 3–7). Some proteins are attached here and there on both surfaces of the membrane while others are embedded within it, some extending all the way through, somewhat like a group of corks bobbing in a sea of lipid. These electron microscope findings and other evidence have led to a new model for membrane structure, called the fluid mosaic model, shown in Figure 3–8.

Furthermore, the relative amounts of protein and lipid vary with the type of membrane. So far we have been discussing plasma membranes. These outer cell-surface membranes, with fewer enzymatic functions than some of the interior structures of the cell, have rather sparsely distributed proteins, and much of the membrane consists of uninterrupted lipid bilayers. In some of the interior membranes, however, such as those of chloroplasts and mitochondria, so much enzymatic protein is present that much of the bilayer is replaced by it.

It is important to realize that, despite the new data and the new models, we still do not fully understand the structure of membranes or all their many functions. For instance, we are just beginning to learn how cell-surface membranes may be involved in cells' recognizing each other. Cells that an organism's defense cells do not recognize as belonging to that organism will be destroyed. This is what happens when a heart transplant is rejected. On the other hand, cancer cells are somehow able to get themselves accepted even though they are different from surrounding cells (see Essay 3–2). Naturally, we want to know how. These are currently areas of intense research, and we should expect major developments in

the next few years to change our model of the membrane once again.

Membrane Functions How does the current model of membrane structure fit with what we know of membrane function? Earlier we stated that the plasma membrane was differentially permeable; that is, some ions and compounds moved across the membrane freely in both directions, while others could not. Compounds soluble in lipids move readily through the membrane, and this is easily explained on the basis of the fluid mosaic model. But what of water and ions? In the case of ions, negatively charged ones move through membranes much faster than positively charged ones. Also, in general, the smaller its molecule, the faster a substance passes through. Pores in the lipid bilayer were first postulated to account for these results, but no pores have been seen with any of the many techniques used to look for them. Instead, recent results suggest that the columns of proteins that run through the membrane from one surface to the other may be the pathway for water and other polar substances.

But why should any substance pass into or out of a cell? Many substances, including water, nonpolar molecules, and negatively charged ions, pass through membranes in a direction and at a rate that reflect their relative concentrations inside and outside the cell. In other words, if Cl⁻ ions are more highly concentrated inside the cell than in the fluids bathing the cell, the ions should pass outward into the surrounding fluid; but if the cytoplasm is low in Cl⁻ ions, then these ions will begin to pass into the cell from the surrounding fluid. To understand the basis of this movement, we must first discuss the phenomenon of diffusion.

Essay 3–2

Cell Membranes and Cancer

Cancer cells are cells gone wild. They are groups of cells that begin a series of rapid cell divisions and then continue to divide when other cells would have ceased, forming a mass called a tumor. Then individual cells break away from the mass of dividing cells (they *metastasize*) and migrate to other parts of the body where they may form new tumors. These features are associated with the plasma membrane of the cell.

In culture, most normal cells stop dividing when they become crowded together. In other words, their growth is density-dependent, and they stop growing when a single layer of cells is formed. Cells that have been transformed into the rapidly dividing stage, however, continue to divide and pile up layer after layer. Normal cells also adhere quite strongly to one another, but transformed cancer cells appear to be less sticky. This allows them to metastasize. Since the plasma membrane is the structure through which cells communicate and interact, attention has been focused on alterations in the cell membrane that could be correlated with this transformation.

Data have accumulated indicating that changes in the glycoproteins (carbohydrates attached to proteins) and glycolipids (carbohydrate-lipid combinations) of the plasma membrane are altered in various ways with transformation. By using a plant protein called lectin concanavalin A, or Con A, which binds to glycoproteins and glycolipids, changes in the cell surface have been found.

For instance, transformed cells aggregate (agglutinate) in the presence of Con A, while normal cells do not. One interpretation of this result is that the glycoproteins and glycolipids within the membrane become more mobile in the transformed cells than in the normal cells. According to this hypothesis, the surface membranes of the transformed cells are dynamic fluid structures in which membrane components may migrate laterally in the membrane. Such migration would result in the clustering in the membrane of certain compounds that increase the cells' tendencies to agglutinate.

Involved in the whole question of changes in plasma membranes and cell transformation are recent observations by Mark Willingham and Ira Pastan of the National Cancer Institute. These investigators and others have been studying the tiny hairlike structures called *microvilli* on the surface of some animal cells. Microvilli have been observed to be much more numerous in tumor cells than in normal cells that were not dividing. The function of microvilli is unknown, but it is clear that a cell covered with microvilli has a much greater surface area than one without them. This increased surface area may be related to an increased ability of cells to bind together in the presence of Con A. It may also be related to the increased rate of cell division and the increased motility of the transformed cells.

Clearly, not all the data are in, but those available suggest that one important aspect in an understanding of cancer is locked in the structure and composition of the plasma membrane.

Diffusion is the movement of suspended or dissolved particles from a more concentrated to a less concentrated region as a result of the random movement of the individual particles. Diffusion results in the uniform distribution of the particles throughout the available space. Small particles, including atoms, ions, and molecules, are constantly moving as a consequence of heat energy. But how can random movement account for movement from more to less concentrated regions? Let's consider a sugar cube carefully dropped to the bottom of a mug of tea (Figure 3–9). The sugar cube dissolves, and we can see a thick layer of concentrated sugar in the bottom of the mug. With time, this layer will slowly disappear and the sugar molecules will be found uniformly distributed throughout the tea. The sugar molecules, if we could see them, would be seen to be moving independently and at random. They do, however, tend to move away from the center of concentration even though their movement is random. For one thing, there is simply more room to move in the upper part of the glass. Thus, the probability of hitting another sugar molecule, which results in loss of forward motion, is lower in the part of the glass with fewer sugar molecules. Because the movement of the molecules is random, some are always moving toward the concentrated sugar, but the net result of their movement is away from the center of concentration.

In our discussion of thermodynamics, we said that free energy is essentially potential energy in that it represents energy potential for producing change. The more orderly the system, the higher will be its free energy because of its greater potential to move to a disorganized state. This can be directly applied to the phenomenon

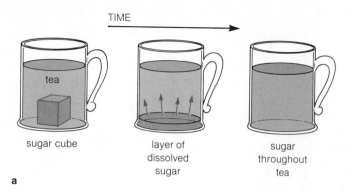

TIME →

tea

sugar cube

layer of
dissolved
sugar

sugar
throughout
tea

a

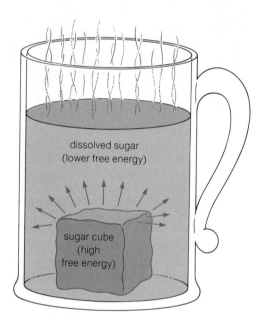

dissolved sugar
(lower free energy)

sugar cube
(high
free energy)

b

Figure 3–9 (**a**) Diffusion of sugar in tea. (**b**) Diffusion is the movement of particles from regions of higher free energy to regions of lower free energy of the same substance.

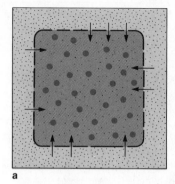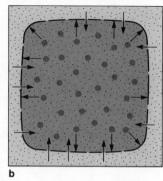

a   b

Figure 3–10 (**a**) A differentially permeable membrane sac containing sugar (●) and water (·) molecules is placed in water. Water enters the membrane, but the sugar molecules cannot pass through it because of the nature of the membrane. (**b**) Pressure thus builds up in the membrane and the sac swells.

of diffusion. The sugar cube has a higher free energy than its surroundings because it is more highly organized (Figure 3–9). Thus we can say that, in diffusion, particles move from regions of greater free energy to those of lesser free energy. It is important to recognize that in considering free energy we are talking only of the differences in free energy of one type of particle. In the example of the glass of tea, only the free energy of the sugar cube in relation to other sugar molecules in the tea is important, not the free energy of the tea or water molecules.

We can continue our discussion of diffusion and free energy by considering an important phenomenon connected with water movement in both plants and animals,

namely, osmosis. **Osmosis** can be defined as the movement of water (or some other solute) across a semipermeable membrane. Osmosis is important because of the basic role water plays in the cell. Water is the universal solvent in biological systems, and organisms have evolved elaborate systems to ensure an adequate supply of water to their cells.

Now let's look at a cell complete with plasma membrane (Figure 3–10). This membrane, as we have noted, is differentially permeable, or semipermeable. Water can pass in both directions through the membrane, but sugar molecules and proteins, for example, cannot pass out. Under these conditions, the free energy of the water molecules outside the cell will be higher than the free energy of the water molecules within the cell. This difference is based on the relative disorder of the water molecules inside and outside the cell: The presence of nonwater molecules, such as the dissolved sugar and proteins, within the cell makes the water there less orderly than the purer water outside the cell. Hence, water will move into the cell. When the free energy of the water molecules inside the cell equals that of the molecules outside the cell, the movement stops. Thus, osmosis results in the movement of water into the cell and the build-up of pressure within the cell.

If, on the other hand, the cell is placed in a solution containing large numbers of dissolved molecules—more than in the cytoplasmic fluid—the free-energy level of the water will be lower outside the cell than inside. Under these conditions, water will leave the cell.

When the number of dissolved particles is the same on both sides of the membrane, there is no net movement of water across it and the solution is said to be *isotonic*. If, on the other hand, the number of dissolved molecules

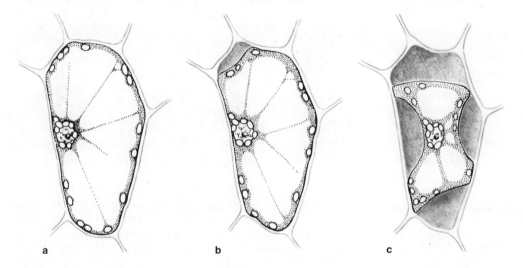

Figure 3–11 Rapid wilting in response to salt solution. (**a**) At time 0 (on the 24-hour clock), 10 grams of salt in about 60 milliliters of water are added to a pot containing four tomato plants growing in vermiculite (an inert insulating material in which plants may be grown if nutrient is added—as in this case). (**b**) Noticeable wilting has occurred after about 11 minutes—indeed, it was possible to watch the plants collapse. (**c**) Plants have wilted severely after about 27 minutes. The cell drawings show progressive plasmolysis, the shrinking of the protoplast (cell contents) away from the cell wall. This loss of turgor in the cells is what causes the plants to wilt.

is different on the two sides of the membrane, the side with the greater concentration is termed *hypertonic* and the side with the smaller number of molecules is called *hypotonic.*

Plant cells are usually hypertonic in relation to the medium that surrounds them, so there is a net movement of water into them. This movement of water into the cell creates pressure within the cell because of the rigid cell wall that surrounds plant cells. This pressure is known as the **turgor pressure.** Turgor pressure is important in the growth of plant cells and in maintaining the rigidity of the cells when mature. If the solution around a plant

cell is made hypertonic, water will move from the cell, and the cytoplasm and vacuole will decrease in volume and pull away from the wall. Such a cell is said to be **plasmolyzed** (Figure 3–11). A plant wilts when the water available to the cells is no longer adequate to maintain a high enough turgor pressure.

The movement of water—and of some salts—in and out of the cell does not involve the expenditure of energy by the cell. In some cases, however, energy supplied by the cell is necessary to move substances across the plasma membrane. This phenomenon is called **active transport.**

One of the best-known examples of active transport

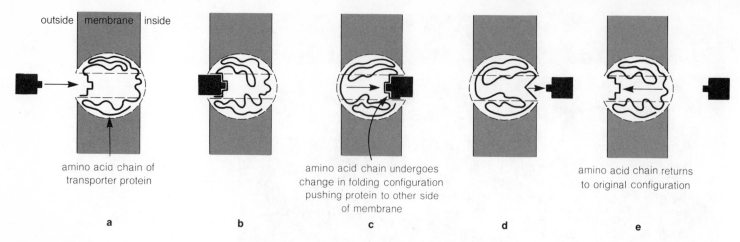

outside | membrane | inside

amino acid chain of
transporter protein

amino acid chain undergoes
change in folding configuration
pushing protein to other side
of membrane

amino acid chain returns
to original configuration

a   b   c   d   e

Figure 3–12 The carrier system in the plasma membrane postulated to account for active transport.

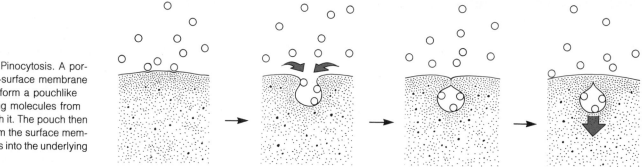

Figure 3–13 Pinocytosis. A portion of the cell-surface membrane invaginates to form a pouchlike pocket, carrying molecules from the exterior with it. The pouch then pinches off from the surface membrane and sinks into the underlying cytoplasm.

involves Na⁺. Many cells are surrounded by seawater (or blood that has a salt content almost identical with that of seawater), so that the concentration of Na⁺ is greater outside the cell than inside. (That is, there is a concentration gradient from greater to lesser concentration.) As a result, Na⁺ is constantly diffusing into the cell and would soon reach amounts that would injure the cell if there were not a mechanism to "pump" the excess Na⁺ out of the cell. Such a mechanism is operating in the face of a concentration gradient, and energy is necessary for this to take place. This energy is supplied in the form of ATP.

Active transport can also be used to accumulate a substance within the cell—the opposite of "pumping out." A classic example of accumulation against a concentration gradient is found in certain marine algae, which maintain an internal concentration of iodine a million times greater than its concentration in the seawater that surrounds them. The positive ions sodium and potassium

are often moved simultaneously in opposite directions across the plasma membrane by active transport, both against a concentration gradient. Nerve impulses, the basis for all coordination of the complex activities of animals, are also conducted partly by active transport of ions.

The actual mechanism in active transport is not clearly understood. The best model currently available to explain the evidence postulates a specific carrier located in the plasma membrane. This model is illustrated in Figure 3–12. The substance to be transported is linked to the carrier at the surface of the membrane. The carrier then changes shape, pushing the substance to the opposite membrane surface, where the substance is released.

Large molecules, aggregates of molecules, or even whole organelles or other cells, may enter the cell by mechanisms unrelated to active transport. These mechanisms involve membrane invaginations (indentations in the membrane) and membrane fusion. When materials

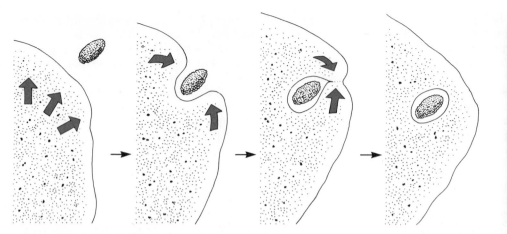
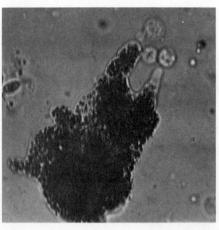

Figure 3–14 Phagocytosis. Part of the cell flows around a large particle or even a whole cell to form a large pocket in the cell membrane. The pocket pinches off as a large vesicle, which sinks into the cytoplasm. The enclosed material may be released directly or digested by enzymes secreted into the visicle. The light micrograph shows an amoeba engulfing a food particle by phagocytosis. ×240.

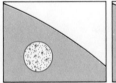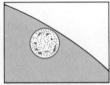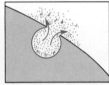

Figure 3–15 Exocytosis in cells. A vesicle moves toward the cell's plasma membrane. The vesicle membrane then fuses with the plasma membrane, and its contents are secreted.

move into the cell by these means, we call it **endocytosis.** In contrast, the cell can secrete material by what appears to be the reverse of these processes, in **exocytosis.**

There are two forms of endocytosis: pinocytosis and phagocytosis. In **pinocytosis** (Figure 3–13), large molecules or molecular aggregates become attached to the outside of the membrane. The membrane then invaginates until the material is enclosed and protruding into the cytoplasm. The inpocketing is sealed at the outside and pinches off to form a membrane-enclosed bundle called a vesicle, suspended in the cytoplasm of the cell. The membrane may then break down, introducing the contained material directly into the cell interior. Alternatively, lysosomes may fuse with the vesicles and release enzymes into them that break down the enclosed substances into molecules small enough to pass through the vesicular membrane and out into the rest of the cell.

Phagocytosis is a basically similar process by which parts of cells or even whole cells such as bacteria may be engulfed by another cell. In this process, parts of the cell surface extend outward and flow around the matter to be engulfed (Figure 3–14). Flow and extension continue until the material is completely enclosed in a large vesicle, which then pinches off from the plasma membrane and moves into the cell. As in pinocytosis, the material in the vesicle may either be released directly into the cell interior by disintegration of the vesicular membrane, or may first be acted upon by digestive enzymes secreted into the vesicle by lysosomes. Amoebae obtain their food in this way, ingesting bacteria and other material by phagocytosis. Our white blood cells also phagocytose bacteria, thereby forming one of our main lines of defense against disease.

Exocytosis occurs when cells secrete material. In animals, this material is frequently enzymes necessary for the digestion of food; in plants, it is usually substances used in the cell wall. In either case, the material to be secreted is formed within a membrane-bound vesicle that moves to the plasma membrane of the cell. There the membrane of the vesicle fuses with the plasma membrane, and its contents are released to the exterior of the cell (Figure 3–15).

Summary

Our knowledge of the cell is slightly over 300 years old. During that time we have learned that all organisms are composed of cells and that all cells arise from the division of existing cells. We have also discovered the major parts of the cell and the two major plans of cell organization.

These are the prokaryotic cells, which lack extensive internal membrane structure, and the eukaryotic cells, with membrane-bound cell parts. In both types of cells, membranes are important, and the presence of an intact plasma membrane surrounding the cell is a condition of life. The structure of membranes has been intensively investigated for some 50 years, and our concepts of that structure have changed. Currently there is great interest in membrane form and function. The model of membrane structure presently accepted is of a lipid bilayer with proteins distributed throughout. Membrane function is critical to the cell, and the semipermeable nature of the plasma membrane is the basis of the important phenomenon of osmosis. The plasma membrane is also involved in the active transport of substances and in the endocytosis and exocytosis of materials. Although we have had 300 years to explore the cell, much remains to be learned and much research is currently being undertaken.

4

The Flow of Energy in Living Organisms

Across the midwestern United States stretch millions of acres planted with corn. Growing in some of the richest soil in the world, this corn represents an increasingly precious resource that feeds not only Americans but a significant number of people in other countries. Under the hot summer sun, corn is a highly efficient solar-energy converter. During **photosynthesis,** light energy is trapped and used to form chemical bonds. From the raw materials carbon dioxide and water, corn—like other green plants—makes carbohydrates and other organic molecules. The importance of the process of photosynthesis cannot be overemphasized: It is the *only* means available in the living world to convert solar energy into chemical energy.

But why is this chemical energy so important? Because of the second law of thermodynamics: Living organisms must expend energy to maintain their highly organized state in the face of the natural tendency toward disorganization or randomness. Cells use energy to synthesize the molecules of which they are composed and to carry on the processes that make life possible.

This energy must be in a form that can be used to make chemical bonds, and that usually means that it is stored in the energy-rich molecule ATP. Photosynthesis is only one part, although the most fundamental one, of the total picture of energy flow in living organisms. The other part is cellular respiration, in which carbohydrates and other organic molecules are broken down and the energy of their chemical bonds converted to ATP. The energy in ATP can be used in thousands of different reactions, and, as we noted in Chapter 2, forms the energy currency of the organism.

To see the importance of photosynthesis and cellular respiration, let's return to the corn plants. During the day, they carry on both photosynthesis and respiration, although the amount of photosynthesis is much greater than the amount of respiration. During the warm summer night, however, they only respire, using the sugars produced by photosynthesis during the day as sources of energy for the activities of the night. Not all of the carbo-

hydrates produced during the day are used at night. The excess is transported to the developing ears of corn and stored in the kernels as starch or converted to protein and oils. It is this excess starch that farmers harvest and feed to their pigs, cattle, and chickens. These animals survive by using the energy in the food, made available to them through their own cellular respiration. They use this energy for growth, by synthesizing new molecules, and for their many other life processes that require energy. Eventually the farmers sell these pigs, cattle, and chickens, and they are killed and prepared for human consumption. We eat the meat of these animals and, in turn, through the process of cellular respiration, use the energy in the chemical bonds of the molecules of flesh to make the molecules of our own bodies. Ultimately all this energy can be traced back to the solar energy trapped by the corn plants during photosynthesis.

Only plants containing the green pigment **chlorophyll** can carry on photosynthesis. Since they can convert solar energy into chemical-bond energy, and thus make their own food, such plants are self-sustaining and are called **autotrophs.** All other plants and all animals must rely on green plants for energy—either directly, by eating them, or indirectly, by eating animals that have eaten green plants. They are called **heterotrophs.** Energy is lost at each transition from organism to organism, so it would be much more efficient from the point of view of energy if humans simply ate the corn rather than feeding it to the pigs and chickens and then eating them.

Photosynthesis and cellular respiration are the two sides of the energy coin, as can be seen if we compare the two overall reactions.

In photosynthesis:

$$CO_2 + H_2O \rightarrow O_2 + sugar$$

In respiration:

$$O_2 + sugar \rightarrow H_2O + CO_2$$

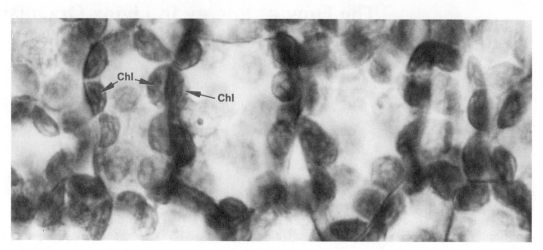

a.

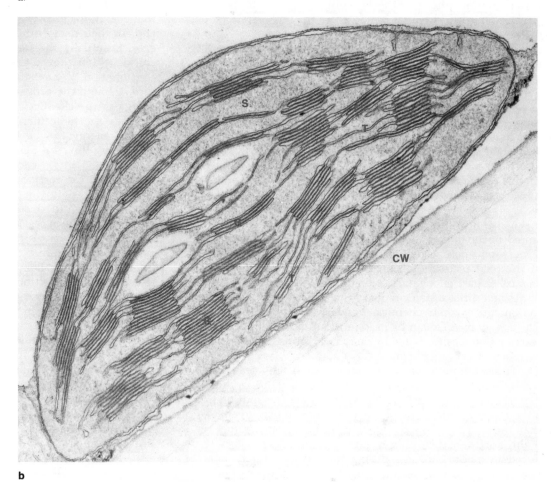

Figure 4–1 Chloroplasts viewed
with the light (**a**) and electron (**b**)
microscopes. Grana (G), stroma
(S), and thylakoids (T) are shown.

b

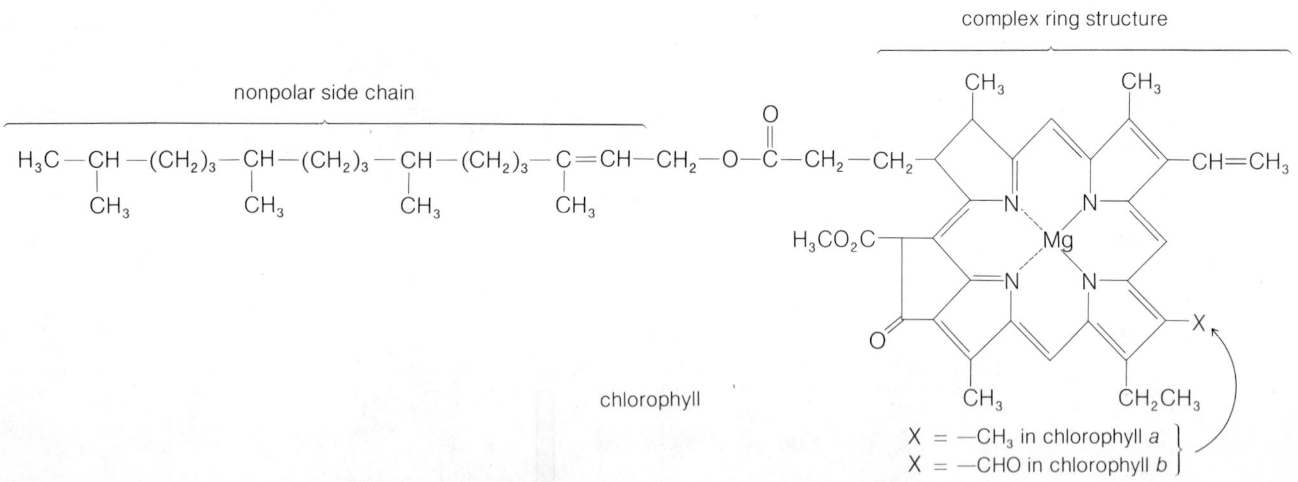

Figure 4–2 The arrangement of the atoms in a molecule of chlorophyll. Note that the difference between chlorophyll *a* and chlorophyll *b* is only a minor modification of atoms at the site marked X.

These are extremely generalized versions of the reactions involved—much as if we described life by saying birth plus 75 years equals death. Such generalizations are true, but they conceal as much as they reveal. The important thing these particular simplifications point out is that photosynthesis and respiration are essentially reversals of each other. This one fact can help clarify much of what follows, but we must examine the details of the reactions to see where they are similar and where they are different. We can start by noting that, in the eukaryotic cell, different organelles are involved in the two processes. Photosynthesis occurs only in the chloroplasts, whereas part of the process of respiration takes place in the cytoplasm and the rest in the mitochondria.

Photosynthesis

Chloroplasts: Site of Photosynthesis In eukaryotes, photosynthesis occurs only in chloroplasts. Chloroplasts had been known to be the site of photosynthesis for over 100 years, and their structure had been carefully examined with the light microscope. Nevertheless, few researchers were prepared for the elaborate and beautiful ultrastructure revealed in the late 1950s by the electron microscope (Figure 4–1). The chloroplast consists of an interior cavity, called the *stroma,* surrounded by a double membrane. The inner membrane is thrown into folds and saclike pockets known as *thylakoids.* Particularly in the higher plants, there are stacks of disk-shaped thylakoids known as *grana.* Such arrangements increase the total surface area of membrane within a chloroplast. Deposits of DNA, ribosomes, and starch granules are also present inside chloroplasts in the stroma.

Chlorophyll and the Chloroplast Chloroplasts can be separated from the leaf and their pigments analyzed chemically. When this is done, two major classes of pigments are always found—the green chlorophylls and the yellow carotenes. There are several different types of chlorophyll molecules (of which chlorophyll *a* and chlorophyll *b* are the most common), but they all have the same general structure—namely, a flat head and a long tail at right angles to the head (Figure 4–2). The head consists of four repeating units in which carbon, hydrogen, and nitrogen atoms surround a magnesium (Mg) atom. The tail gives the molecule many of its lipidlike characteristics. The general form of chlorophyll is similar to the heme portion of hemoglobin, the molecule that carries oxygen in our blood. In hemoglobin, iron is found in place of magnesium. When plants grow in conditions lacking magnesium, they soon become yellow because of inadequate chlorophyll production. The same thing will happen if they are grown without iron. Iron is not present in the chlorophyll molecule but is necessary for its production. In either case, the leaves are said to be *chlorotic.*

The **carotenes** are another family of pigmented lipid molecules closely related to one another in structure. The various carotenes, such as the β-carotene shown in Figure 4–3, are built up from a single chain 40 carbon atoms long. Carotenes are always found in chloroplasts, but they

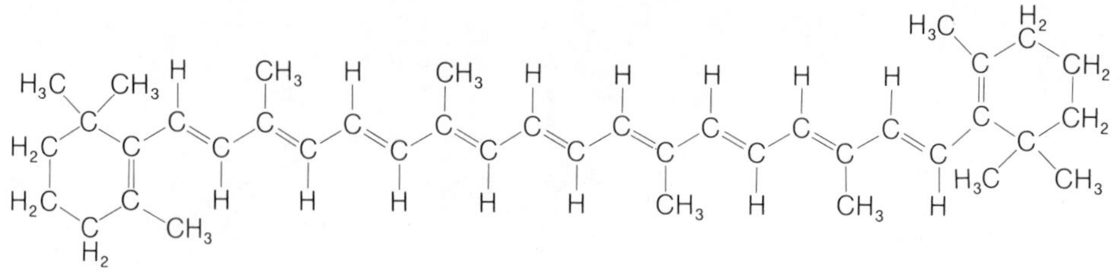

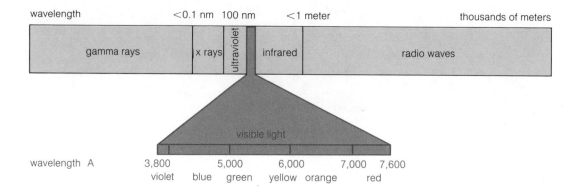

Figure 4–3 The arrangement of atoms in β-carotene.

β-carotene

wavelength		<0.1 nm	100 nm		<1 meter		thousands of meters

gamma rays	x rays	ultraviolet	infrared	radio waves

visible light

Figure 4–4 The electromagnetic spectrum. Visible light makes up only a small portion of the entire spectrum.

wavelength Å

3,800	5,000	6,000	7,000 7,600
violet	blue green	yellow orange	red

may also be found in high concentrations in other types of plastids, called **chromoplasts,** in carrots and other plants. Animals produce vitamin A by splitting carotene molecules in half.

Light   It is important to recognize that we do not really understand the physical nature of light. What we have are two models of light that allow us to predict its behavior with a high degree of accuracy. The problem is that the two models at times appear so different that they are incompatible. Scientists have learned to live with this schizophrenic condition, using each model where it works best.

The first model states that light is a wave phenomenon. In this model, light is treated as a form of radiant energy traveling in waves at a constant velocity. The wavelength—that is, the distance from the peak of one wave to the peak of the next—is different for different types of light. White light, for example, can be shown to consist of a complete spectrum of colors from purple to red, with each color having a specific wavelength. It can also be shown that these visible colors are only part of the total electromagnetic spectrum of radiation that spans wavelengths from X rays measured in nanometers

to radio waves measured in meters (Figure 4–4). One problem with the wave model of light is that there is no apparent medium through which the waves travel.

The second model treats light as a number of particles, called **photons** or **quanta,** having no mass. The energy in a photon is proportional to the wavelength: the shorter the wavelength, the more energy per photon. For example, a photon of blue light has almost twice the energy of a photon of red light. An important aspect of the particle model is that it helps us to understand what happens when a molecule absorbs energy. We know that molecules absorb photons individually. When this occurs, an electron in an atom is moved to a higher level; that is, it is moved farther from the nucleus. Since the electron and the nucleus attract each other (one is positive, the other negative), moving them apart requires an input of energy—the energy of a photon.

When a molecule absorbs a photon, its energy level changes, causing the molecule to be in the so-called excited state. The molecule then falls back to the ground state (for example, an electron falls to a lower orbit), and any one of three things can happen:

1. The energy between the two states can be contributed to the overall kinetic energy of the system. In other words,

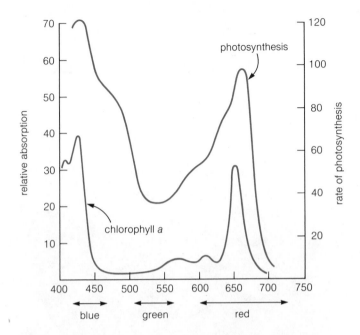

Figure 4–5 The absorption spectrum of chlorophyll *a* and the action spectrum for photosynthesis.

the energy absorbed as a photon may be converted to heat. This is what goes on when something warms in the sunlight.

2. Another photon can be emitted; that is, it can give off light (or any radiant energy). The emitted photon is always less energetic (has a longer wavelength) than the original one absorbed. This process is known as *fluorescence* when the light is given off in an extremely short time (10^{-8} sec). If the light is given off over a longer interval of time (up to minutes), the phenomenon is called *phosphorescence*.

3. The energy can be incorporated into a chemical bond. This is the fundamental reaction of photosynthesis.

Chlorophyll and Light Absorption   Chlorophyll is a green pigment. Any pigmented object appears colored to our eye because it selectively absorbs certain colors of light and reflects others. Our eyes see only the reflected light, not the light that is absorbed. In chlorophyll, when a beam of white light (which consists of all colors) illuminates the pigment, all the colors except green are absorbed while green is reflected. It is possible to determine accurately the color of light that is absorbed by chlorophyll

and to express the results in a graph in which the amount of light absorbed is plotted against the wavelength of that light (Figure 4–5). Such a graph portrays the **absorption spectrum** of chlorophyll.

We can next ask what colors of light are most effective in photosynthesis; that is, we can determine the **action spectrum** for the process. This can be done in a number of ways. One of the most precise is to shine equal amounts of light of known wavelengths on a suspension of algae and to measure the amount of oxygen released. From the overall equation for photosynthesis, it is clear that the amount of oxygen released is a direct measure of the amount of photosynthesis being carried on. The amount of oxygen liberated can be measured accurately, and the data collected can be plotted graphically (Figure 4–5). This time, the amount of photosynthesis is plotted against the wavelength of the light used.

The two spectra are similar. When the absorption spectra of other pigments, such as the carotenoids, are plotted, they show little or no similarity to the action spectrum of photosynthesis. Both the absorption and the action spectra illustrate the same important point: Not all wavelengths of light are used with equal effectiveness in photosynthesis. Blue and red light are the most effective and important.

All available evidence indicates that chlorophyll is the center of light absorption in photosynthesis. Light absorbed by the carotenes is passed on, probably in the form of high-energy electrons, to chlorophyll for transfer to electron acceptors. There are several interesting points to consider about the involvement of chlorophyll molecules in light absorption. First, groups of molecules operate together in what have been termed **photosynthetic units,** probably made up of several hundred molecules. Second, chlorophyll and the carotenes are located in a membrane within the elaborate structure of the chloroplast. The structure of the membrane keeps all the various molecules of the system in proper spatial relationship with one another. Finally, the various chlorophyll and carotene molecules act together as a kind of funnel in which light energy enters at the top and is eventually concentrated in one highly active chlorophyll molecule at the bottom.

The Light and Dark Reactions of Photosynthesis
Photosynthesis can be divided into two sets of reactions (Figure 4–6): the **light reactions,** in which chlorophyll absorbs light and energy-rich compounds (including ATP) are produced together with oxygen as a by-product, and the **dark reactions,** or carbon dioxide fixation reactions, in which the energy-rich compounds produced in the light reactions are used to synthesize (or "fix") carbon dioxide into organic molecules. Light is necessary for the light reactions to take place, and the name is easy to

understand. In the case of the dark reactions, however, the name does not mean that the reactions take place *only* in the dark, but that light is not directly involved. Both the light and dark reactions normally take place in the light, but the dark reactions can and do take place in the dark if the proper energy-rich compounds are available. The dark reactions can take place in animals; our livers contain the enzymes of the dark reactions and are capable of fixing carbon dioxide. Both reactions take place in the chloroplasts in the leaves of the plant, but in different parts. The light reactions occur in the membranes of the thylakoids; the dark reactions in the stroma.

In energy terms, photosynthesis entails the use of light energy to convert compounds of low potential energy, namely water and carbon dioxide, into compounds with high chemical potential energy, such as carbohydrates. A useful analogy that helps make this point is that if you simply combine carbon dioxide and water without photosynthesis, you get carbonated water, not carbohydrates. The difference in energy between these two compounds is well known to anyone on a diet.

The light reactions basically involve the absorption of light by chlorophyll and the loss of a pair of energetic electrons from the chlorophyll to an acceptor molecule. Next, the electrons are passed through a series of compounds, and the energy of the electrons is transferred to energy-rich compounds. There are two major patterns of the light reactions: *cyclic photosynthesis*, in which the high-energy electrons are eventually returned to the chlorophyll molecule after most of their energy has been tapped, and *noncyclic photosynthesis*, in which the high-energy electrons are replaced with lower-energy electrons from some other source, most commonly water. Let's look first at noncyclic photosynthesis.

In noncyclic photosynthesis (Figure 4–7), there are two photochemical systems, labeled in the figure chlorophyll II and chlorophyll I. Light energy is absorbed by the chlorophyll molecules of the photosynthetic unit and funneled to one special chlorophyll molecule known as chlorophyll II. Let's assume that this molecule has just filled its complement of electrons by taking two from water to replace two it had lost earlier in photosynthesis. Now two of the electrons of chlorophyll II become highly energetic because of the light energy that has been absorbed, and they leave the chlorophyll II molecule and are accepted by the first of a series of electron-transfer compounds. Several of these compounds are **cytochromes,** which contain iron in a ring much like the head of a chlorophyll molecule. The iron-containing ring portion of the molecule is surrounded by a protein containing some 100 amino acids. We will meet the cytochromes again when we consider cellular respiration.

The electron pair is transferred from one compound to the next in the chain. At each step an oxidation-reduction reaction occurs and energy is lost. This energy is

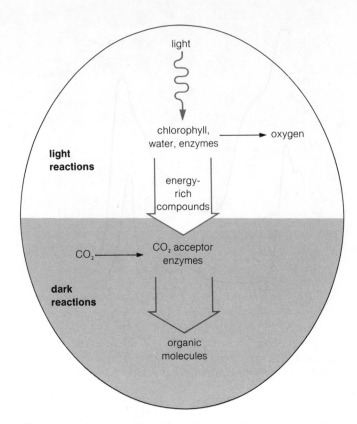

Figure 4–6 Summary of the light and dark reactions of photosynthesis.

used in the synthesis of ATP from ADP. Thus, in this first series of reactions, the cell has gained energy in the form of ATP molecules that can be used in other chemical reactions. But the reaction does not stop here. The electrons are next transferred to a second photochemical system, complete with a photosynthetic unit and a special chlorophyll molecule, chlorophyll I. Light energy is again used to boost the energy of a pair of electrons, which in turn are passed to another set of electron receptors. The end result is the formation of reduced **nicotinamide adenine dinucleotide phosphate,** or reduced NADP. This high-energy compound is used in the synthesis of many other substances in the cell. In this second set of reactions, the electron pair starts at a higher energy level than in unexcited chlorophyll II and is excited to still higher energy levels.

The end products of noncyclic photosynthesis are ATP, reduced NADP, oxygen, and protons (or hydrogen ions). The first two compounds are very useful to the cell and are vital to the dark reactions of photosynthesis. The last two products result from the dissociation (breakup) of the water molecules from which electrons have been removed.

Cyclic photosynthesis is essentially the second half

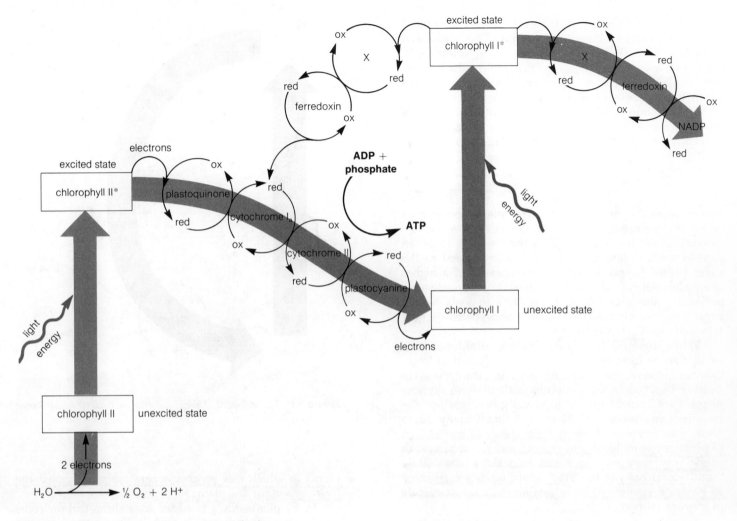

Figure 4–7 Electron flow in noncyclic photosynthesis.

of noncyclic photosynthesis (Figure 4–8). Starting with chlorophyll I, the electrons are passed through a series of electron carriers and returned to the chlorophyll molecule. During the electron-transfer reactions, ATP is synthesized from ADP. No reduced NADP is formed, and, since the original electrons are returned to the chlorophyll molecule, no water is broken down and no oxygen released. As a result, the only end product of cyclic photosynthesis is ATP. This form of photosynthesis is believed to be the more primitive of the two types of light reactions.

During the carbon dioxide fixation reactions, or dark reactions, the ATP and reduced NADP formed in the light reactions are used to convert carbon dioxide and water to carbohydrates. It might be thought that this would occur by one CO_2 molecule building on another until a six-carbon molecule is produced. This is not the

case. Instead, a more indirect process occurs—a cycle in which one unit of CO_2 is handled at a time and the compounds involved in the cycle are regenerated during the course of the cycle.

Two pathways of CO_2 fixation are known to be present in plants. The first to be worked out and the best understood is the **three-carbon cycle,** or **Calvin cycle,** worked out by Melvin Calvin, Andrew A. Benson, and their colleagues at the University of California, Berkeley. Many plants, however, including corn, sugar cane, rice, and a number of desert plants, have a **four-carbon pathway,** also known as the **Hatch-Slack pathway.**

The major reactions of the three-carbon cycle are shown in Figure 4–9. Carbon dioxide enters the cycle (upper right) and attaches itself to ribulose diphosphate, a molecule with five carbons and two phosphate groups. (A phosphate group, PO_3^-, is represented by (P).) With the addition of CO_2, an unstable six-carbon compound (shown in braces in Figure 4–9) forms. This compound immediately breaks into two three-carbon compounds, called **phosphoglyceric acid,** or PGA. Unlike the six-carbon compound, PGA is stable and may be considered the first product of CO_2 fixation. (Because PGA is a three-carbon compound and the first stable product, the cycle has been named the three-carbon, or C_3, cycle.)

Thus far, the energy made available from the light reaction has not been used. This happens in the next couple of reactions, where the energy of the ATP and the reduced NADP is incorporated through the reactions shown in the lower left of Figure 4–9. Through the addition of a phosphate (which requires the investment of one ATP), the subsequent loss of the phosphate, and the conversion of reduced NADP to the oxidized form, an oxygen atom is removed and PGA is converted to the sugar triose phosphate. Triose phosphate is at a higher energy level than PGA. Finally, two triose phosphate molecules are joined end to end, resulting in a six-carbon sugar with two phosphates. The two phosphates are next removed and $C_6H_{12}O_6$ results.

While these reactions are occurring, additional reactions, starting from triose phosphate, result in the formation of ribulose diphosphate. At this point, the cycle takes another turn, when CO_2 couples with ribulose diphosphate. Each turn of the cycle actually involves not the few reactions shown in Figure 4–9, but usually 10 to 15 separate reactions. Three turns of the cycle are required to produce one molecule of surplus triose phosphate; each single turn may be considered to yield a one-carbon "unit" of carbohydrate (CH_2O). Six turns are required to generate enough units to manufacture one molecule of glucose with six carbons.

Tracing one complete turn of the cycle reveals that, for each unit of carbohydrate generated, three molecules of ATP and two molecules of reduced NADP are converted to ADP and oxidized NADP. These cycle back to the light reactions of photosynthesis, to be converted back to ATP and reduced NADP.

Although carbohydrates may be considered the major products of photosynthesis, many other compounds are produced from one or another of the intermediate compounds. Amino acids, fatty acids, and glycerol are also products of photosynthesis.

Recently, two Australian investigators, M. D. Hatch and C. R. Slack, confirmed an alternative pathway to the C_3 cycle. Known as the four-carbon pathway, or Hatch-Slack pathway, it is important in the initial CO_2 fixation reactions in many plants, particularly the grasses. When these plants are exposed to radioactive CO_2, the radioactive carbon first appears in three four-carbon acids, oxaloacetic acid, malic acid, and aspartic acid, instead of the three-carbon PGA as in the C_3 cycle (Figure 4–10). Because these four-carbon acids are produced first, the

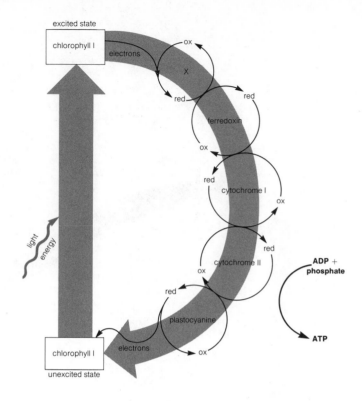

Figure 4–8 The sequence of electron carriers in cyclic photosynthesis.

plants in which this process occurs are often called the four-carbon, or C_4, plants.

In C_4 plants, CO_2 is added to a three-carbon compound phosphoenolpyruvic acid (PEP) and the four-carbon acid, oxaloacetic acid, is formed. Oxaloacetic acid, by means of side reactions, is readily converted into either malic or aspartic acid. These acids do not substitute for PGA in the C_3 cycle but are broken down enzymatically to yield CO_2 and pyruvic acid. The CO_2 is transferred to ribulose diphosphate of the C_3 cycle, and the pyruvic acid is used to form more PEP molecules. ATP is necessary for the reaction.

It is only in the initial steps of carbon fixation that C_4 photosynthesis differs from C_3 photosynthesis. Essentially, the C_4 reactions provide a method for supplying the three-carbon cycle with CO_2. Why should such a system arise? The answer appears to lie in the relative efficiency of the capture of CO_2 by PEP (in C_4 plants) and ribulose diphosphate (in C_3 plants). The enzyme involved in adding CO_2 to PEP is particularly efficient at low concentrations of CO_2 and at higher temperatures. Under conditions such as those found in the tropics, C_4 plants are better equipped to fix CO_2 than are the C_3 plants. Corn, sugar cane, sorghum, and rice—all C_4

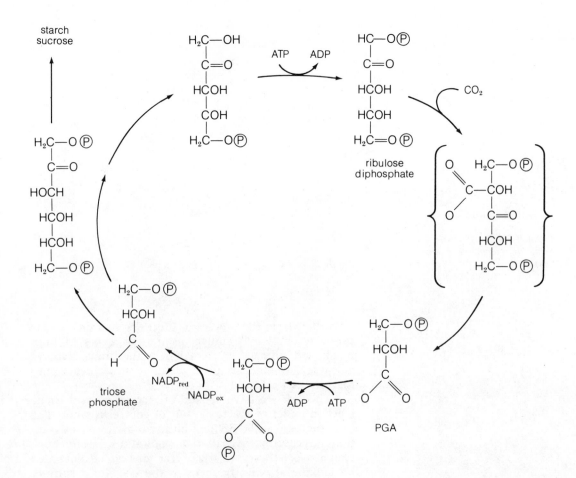

Figure 4–9 The three-carbon CO_2 fixation cycle of photosynthesis (Calvin cycle).

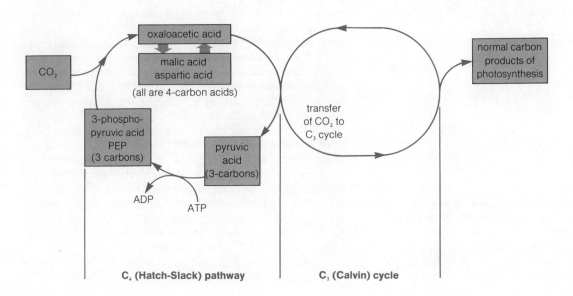

Figure 4–10 Summary of the C_4 and C_3 cycles.

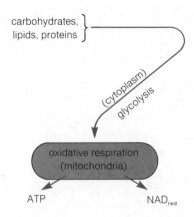

Figure 4–11 Summary of cellular respiration. Note that glycolysis generally occurs in the cytoplasm, while oxidative respiration occurs within a specific organelle—the mitochondrion.

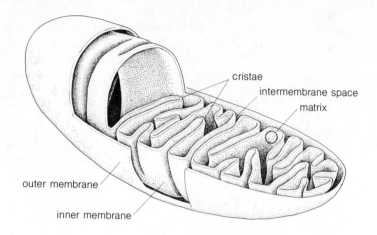

Figure 4–13 A diagram of a mitochondrion.

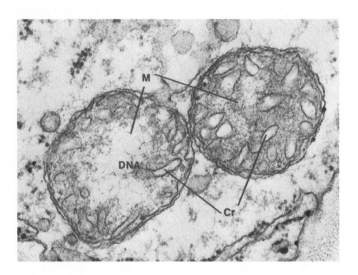

Figure 4–12 Mitochondria (M) as seen with the electron microscope.

plants—originated in the tropics; wheat, rye, and oats are all temperate C_3 plants.

Cellular Respiration

As we said earlier, photosynthesis is one side of the energy coin in living organisms; the other is **cellular respiration.** During cellular respiration, the molecules made either directly or indirectly from photosynthesis are broken down and the potential energy of the chemical bonds extracted. We speak of this as cellular respiration to dis-

tinguish it from the process of breathing in animals, also called respiration. Cellular respiration occurs in both plants and animals. In animals it is the primary source of energy, whereas in green plants it supplements photosynthesis.

Thus far we have spoken of respiration as a single reaction or set of reactions, but of course it is not. The total process can be divided into two basic sets of reactions: **glycolysis** and **oxidative respiration.** The relationship between these two sets of reactions can be illustrated by looking at what happens in the body of a sprinter during a race. When the gun sounds, the sprinter's muscles are contracted at a furious pace and the demand for energy in the muscles is tremendous. First, available ATP is used, and then sugars present in the cells are broken down by glycolysis. The sugars are only partially consumed in these reactions, but energy is rapidly made available to the muscles. The reason the sugars are not completely used up is that the second set of reactions, oxidative respiration, requires oxygen, and the runner's blood flow cannot provide enough oxygen soon enough to have oxidative respiration supply the energy needed in a sprint. The race over, the runner, gasping for air, heart pounding and blood coursing, makes up the oxygen deficit in his or her cells. Now oxidative respiration takes over, and the end products of glycolysis, still rich in energy, are broken down to water and carbon dioxide. Much more energy is released through oxidative respiration than through glycolysis, but, because of the need for oxygen, it is often released more slowly. In either case, the energy is released in the form of the old familiar ATP and reduced NAD (**nicotinamide adenine dinucleotide,** closely related to the NADP of photosynthesis).

The reactions of glycolysis take place in the cytoplasm of the eukaryotic cell outside any organelle (Figure 4–

Lysosomes

When mitochondria were first isolated from cells and studied enzymatically, a variety of enzymes were found that didn't seem to belong in mitochondria. These were enzymes whose function is to break down proteins, nucleic acids, and polysaccharides into smaller units; in other words, they are hydrolytic enzymes. Further work showed that, in fact, the mitochondrial preparations contained a second cell organelle with roughly the same size and shape as the mitochondrion. These organelles were found to be limited by a single membrane and to contain the hydrolytic enzymes. They were given the name *lysosomes,* derived from the Greek *lys,* meaning "to dissolve," and *soma,* meaning "body." As mentioned in Chapter 3, it was found that, when a particle of food is taken into the cell through phagocytosis, the membrane of the lysosome fuses with the membrane of the food vacuole, and the hydrolytic enzymes in the lysosome are emptied into the food vacuole. There they broke down the food into units small enough to enter the cytoplasm. Lysosomes were also found to be involved in the breakdown of specific cell parts. In many cells, there are striking changes in the number and type of cell organelles during development. In cases where cell organelles are to be eliminated, they are frequently seen to be isolated in a membrane. Then lysosomes fuse with this membrane in the same fashion as with the food vacuole. Lysosomes were once described as the "suicide capsules" of the cell that broke open when the cell was injured, releasing enzymes that would dissolve it. This is no longer thought to be true. Instead, lysosomes are believed to be important in the normal development and function of the cell.

11). During glycolysis, the "food" molecules, whether carbohydrates, lipids, proteins, or nucleic acids, are converted into short, two- or three-carbon segments that are the immediate fuels of oxidative respiration. Oxidative respiration takes place in the mitochondria of the cell (Figure 4–11). These cell organelles are much smaller than chloroplasts and much less elaborate in structure. Most ATP and reduced NAD are generated inside the mitochondria, with only limited amounts produced during glycolysis.

A few organisms are adapted to living exclusively by glycolysis and closely related reactions in the absence of oxygen, but most plants and animals use both sets of reactions.

The Mitochondria   Almost all eukaryotic cells contain mitochondria. The exceptions are a few metabolically inert types, such as the red blood cells of higher animals, which contain no cytoplasmic organelles of any kind when fully mature.

Mitochondria are apparently nonrigid and exhibit a variety of shapes in the cytoplasm. They can be observed to change slowly from spherical forms to long filamentous structures, and to fuse and divide. In most cells, mitochondria are spherical or slightly elongated, about 2 micrometers in length and 0.5 micrometers in diameter, approximating the dimensions of a bacterial cell.

Observations made since the 1800s reveal that mitochondria are often found in the cytoplasm near activities requiring an energy input, or near deposits of fuel substances such as glycogen (animal starch) or fat droplets. In muscle tissue, mitochondria are interspersed in tightly packed rows between the contractile fibers in the muscle. Thus, as we saw with the sprinter, ATP synthesized in the mitochondria reaches the energy-requiring contractile mechanisms almost instantly.

In prokaryotic forms, the energy-yielding mechanisms are distributed between the cytoplasm and the plasma membrane; no organelles equivalent to mitochondria or chloroplasts are present. To understand how energy functions are carried out in prokaryotic cells, see Supplement 4–1.

Mitochondria are surrounded by a smooth, continuous *outer membrane* and a second *inner membrane* thrown into folds called *cristae* (Figures 4–12 and 4–13). The innermost space of the mitochondrion, into which the cristae extend, is called the *matrix* and includes deposits of DNA and ribosomes.

The Initial Breakdown of Carbohydrates: Glycolysis
Glycolysis is the first set of reactions in the breakdown of carbohydrates. In these reactions six-carbon sugars, such as glucose, are split into three-carbon compounds. The cell gains ATPs from these reactions, but for this

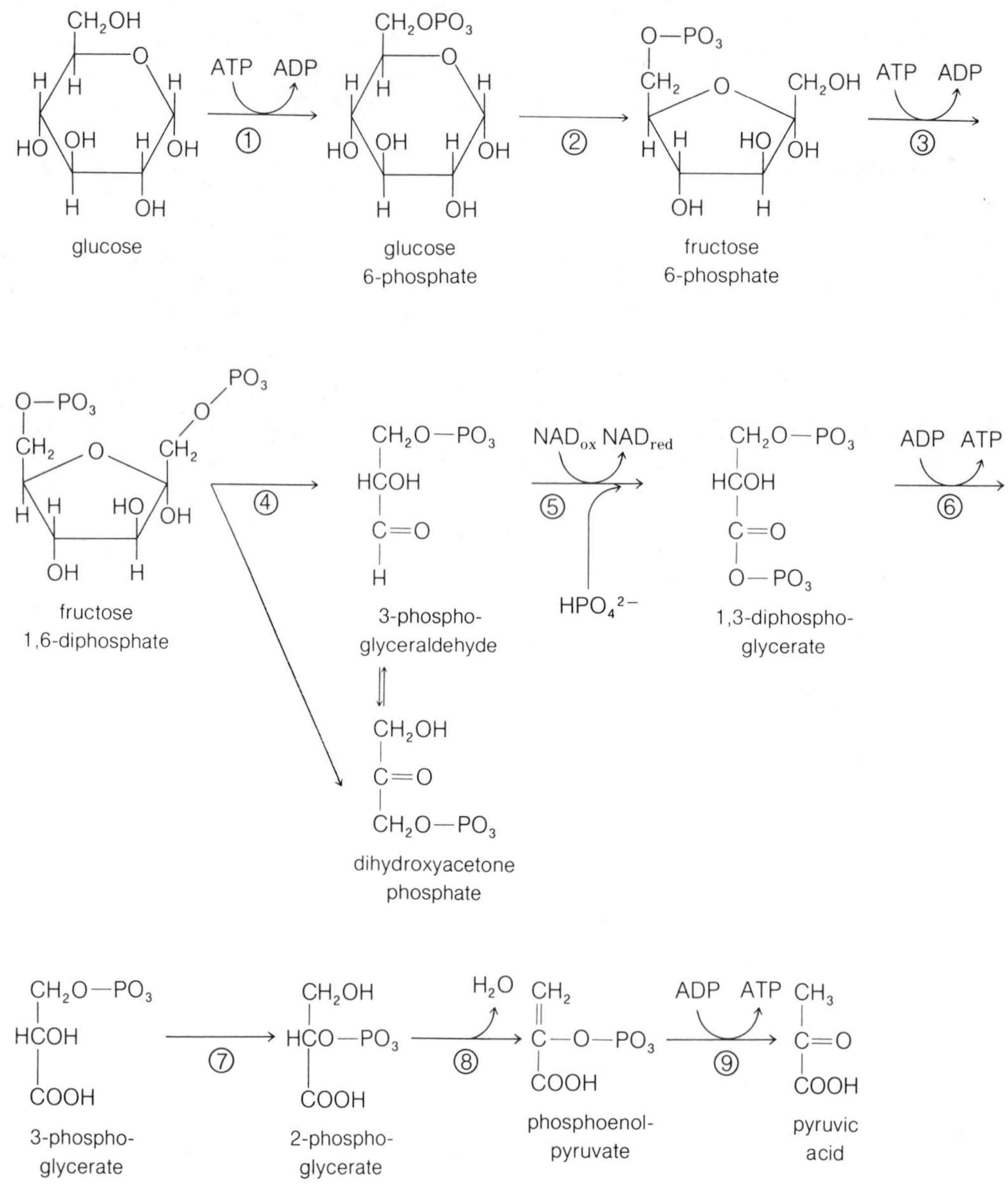

Figure 4–14 Reactions of glycolysis.

to happen ATP must also be invested. This means that the net yield (ATPs produced minus ATPs used) is small. In cells that oxidize carbohydrates through the entire pathway from glucose to H_2O and CO_2, glycolysis serves primarily as a "make-ready" sequence preparing the three-carbon segments that then enter the mitochondria for further oxidation. However, in cells existing either temporarily or permanently without oxygen, glycolysis becomes the central pathway for obtaining ATP.

The first question concerning glycolysis is, Where is the energy put into the reactions? A look at Figure 4–14, which presents some of the details of the glycolytic sequence, shows that the investment of energy occurs in the first three reactions where ATP becomes ADP. This series of reactions ends with the formation of fructose 1,6-diphosphate. The resulting breakdown of this compound into two three-carbon compounds (reaction 4, Figure 4–14) and the changes that occur in these compounds result in the formation of ATP from ADP. The energy changes involved in these two sets of reactions are summarized in Figure 4–15.

The second question is, Why is energy necessary to

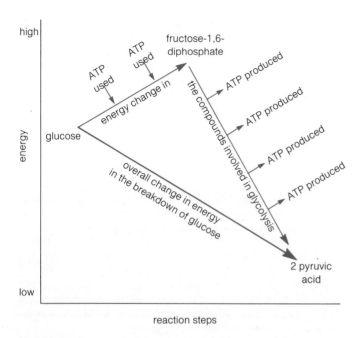

ATP used

ATP used

fructose-1,6-diphosphate

energy change in

energy

glucose

ATP produced

the compounds involved in glycolysis

ATP produced

ATP produced

overall change in energy in the breakdown of glucose

ATP produced

2 pyruvic acid

low

reaction steps

Figure 4–15 Energy changes occurring during glycolysis. Energy is released during the breakdown of the sugar, but to make this possible energy must first be invested.

make the reactions proceed? The answer lies in the nature of glucose and similar six-carbon sugars. Chemically they are stable compounds and, as we noted in Chapter 2, a pile of sugar kept in a cool, dry place protected from organisms will remain unchanged for years. As a consequence of this stability, the first series of reactions of glycolysis results in the formation of a more energetic, less stable sugar. But this can only occur with the investment of ATPs. These first reactions (1–3, Figure 4–14) can be summarized as:

$$glucose + 2\ ATP \rightarrow fructose\ 1,6\text{-diphosphate} + 2\ ADP$$

The second series of reactions breaks this highly reactive sugar into shorter segments and oxidizes them, meaning that glycolysis as a whole produces more ATP than it uses. These reactions start with the splitting of the fructose 1,6-diphosphate into two three-carbon compounds, each with one phosphate group (reaction 4). These two three-carbon compounds differ in where the phosphate group is attached, but since they are interconvertible, the only real product that moves to the next reaction is 3-phosphoglyceraldehyde.

At the next step (reaction 5), two electrons and two hydrogen ions are removed from 3-phosphoglyceraldehyde, and part of the energy released is used to replace

them with another phosphate group (derived from inorganic phosphate ions in the medium, not from ATP), yielding 1,3-diphosphoglycerate. The detached electrons and one of the two hydrogen ions are accepted by NAD, or nicotinamide adenine dinucleotide, and the second hydrogen ion enters the pool of H^+ ions in the cytoplasm. Reduced NAD produced by this reaction is an energy-rich compound. NAD is also known as niacin, or nicotinic acid, and is a vitamin of the B complex. Its function is analogous to that of NADP in photosynthesis. The overall reaction we have been discussing releases free energy and proceeds spontaneously to completion:

(5) $$3\text{-phosphoglyceraldehyde} + NAD_{oxidized} + \textcircled{P} \rightarrow$$
$$1,3\text{-diphosphoglycerate} + NAD_{reduced}$$

Removal of one of the two phosphate groups from 1,3-diphosphoglycerate at the next step (reaction 6) produces 3-phosphoglycerate and a second large increment of free energy. Some of this free energy is captured when the phosphate group removed from 1,3-diphosphoglycerate is attached to ADP, yielding one molecule of ATP for each molecule of the three-carbon sugar entering the reaction:

$$1,3\text{-diphosphoglycerate} + ADP \rightarrow$$
$$3\text{-phosphoglycerate} + ATP$$

The 3-phosphoglycerate found in reaction 6 is rearranged in two enzymatically catalyzed steps (reactions 7 and 8) into phosphoenolpyruvate. The remaining phosphate group is removed in the last step (reaction 9), yielding pyruvic acid. Part of the free energy thus released is used to attach this phosphate to ADP, forming one molecule of ATP for each molecule of phosphoenolpyruvate entering the reaction.

For every molecule of glucose that enters the pathway to start with, two molecules of reduced NAD are formed at reaction 5, and two molecules of ATP at reactions 6 and 9. Thus, the second half of the glycolytic sequence produces a total of two reduced NAD and four ATP. Subtracting the two molecules of ATP used to phosphorylate (add phosphate groups to) intermediate compounds at reactions 1 and 3, we find a net gain of two molecules of reduced NAD and two of ATP for the entire glycolytic sequence.

All of the reactions of glycolysis are essentially reversible. That is, the sequence can run in either direction, depending on the concentration of reactants, the balance between ADP and ATP, and the relative quantities of the reduced and oxidized forms of NAD.

Note that many of the intermediates and products of the final series of reactions in glycolysis, including 3-phosphoglyceraldehyde, 1,3-diphosphoglycerate, 3-phosphoglycerate, phosphoenolpyruvate, and pyruvic acid, are also a part of the dark reactions of photosynthesis (compare Figure 4–10 with Figure 4–14).

Any of these intermediates in the plants may enter glycolysis at the points where the dark reactions interact with it, and the sequence will then run in either direction, depending on the relative concentration of reactants. The first half of the glycolytic sequence is also important in photosynthesis; glucose is synthesized from 3-phospho-glyceraldehyde, the primary product of the Calvin cycle, essentially by a reversal of glycolysis reactions 1–4.

We must also consider some important variations of the basic glycolytic pathway. The reactions of glycolysis form a central pathway in the metabolism of carbohydrates in both plant and animal cells. Starch (in plants) and glycogen (in animals), both long-chain polysaccharides made up of glucose links (see Figure 2–28), enter the glycolytic pathway after being broken sequentially by hydrolytic enzymes into individual glucose molecules. Other sugars, including a wide variety of mono- and disaccharides, enter glycolysis after being converted by enzymes into one of the initial molecules of the pathway.

At the opposite end of the sequence, there may be other final products besides pyruvic acid. In our example of the sprinter, pyruvic acid may be converted into lactic acid after accepting electrons from reduced NAD. A significant feature of this pathway is that the oxidized NAD produced is free to cycle back and accept electrons in reaction 5 of glycolysis (Figure 4–14):

$$\text{(glucose)} \rightarrow 2 \text{ pyruvic acid} + 2 \text{ NAD}_{reduced} \rightarrow$$
$$2 \text{ lactic acid} + 2 \text{ NAD}_{oxidized}$$

Because oxidized NAD is continually regenerated by this alternative pathway, glycolysis can continue to "run" on its net production of ATP, and cells temporarily deprived of oxygen can survive. At the end of the race, the lactic acid level in the runner's muscles has risen significantly, but within seconds it will begin to fall and, after a few minutes (in a person in good physical condition), the acid will be completely oxidized. If the level continues to rise, however, or if the runner is unable to respond with increased oxygen soon enough, muscles will cramp painfully.

Another important glycolytic variation occurs in organisms such as yeasts. Here the pyruvic acid formed by the final steps in glycolysis accepts electrons from reduced NAD and is converted by additional reactions into ethyl alcohol (two carbons) and CO_2. This variation of the glycolytic sequence, called *alcoholic fermentation,* is of great commercial importance, since it forms the biochemical basis of the brewing and baking industries.

$$\text{(glucose)} \rightarrow 2 \text{ pyruvic acid} + 2 \text{ NAD}_{reduced} \rightarrow$$
$$2 \text{ ethyl alcohol} + 2 \text{ CO}_2 + 2 \text{ NAD}_{oxidized}$$

Bread rises because the yeast in the dough converts sugar to CO_2 and alcohol. The gaseous CO_2 is trapped in the dough, and the whole mass increases in volume. The heat of baking further increases the volume of the trapped CO_2 while evaporating the alcohol. In beer and sparkling wines, the bubbles are CO_2 from fermentation trapped inside the bottle, which was closed while fermentation was still occurring; in still wines, the CO_2 is allowed to escape before bottling, and no fermentation takes place in the bottle.

Oxidative Respiration: The Citric Acid Cycle   The second major stage of oxidative reactions in cells is the **citric acid cycle,** also called the **Krebs cycle** after the German-born English biochemist, Hans Krebs, who first understood that a cycle was involved (see Essay 4–2). In this cycle, pyruvic acid, the three-carbon molecule produced by the final steps of the glycolytic sequence, is completely oxidized to carbon dioxide and water, releasing large quantities of energy; the net yield of ATP here far exceeds the amount obtained through glycolysis.

In the presence of oxygen, a carbon atom is removed from the pyruvic acid, and an activated two-carbon compound is formed. This compound unites with a four-carbon compound, oxaloacetic acid, to form the six-carbon compound citric acid. Then, through a series of reactions shown in Figure 4–16, two CO_2 molecules are released and oxaloacetic acid is formed. This means that a cycle is operating, for each turn of the cycle results in the reforming of the oxaloacetic acid. For each CO_2 molecule produced—and there are 2 released per turn of the cycle—6 ATP molecules are ultimately produced, for a total of 12 ATPs.

As we have said, the citric acid cycle and the electron-transport systems associated with it are located in the mitochondria. Because of the large amount of ATP generated in conjunction with the citric acid cycle, the mitochondria are frequently called the powerhouses of the cell. To understand the citric acid cycle, one of the most important metabolic pathways in the cell, let us look at it in more detail.

Reactions and Products of the Citric Acid Cycle   Now let us see what chemical reactions take place in the mitochondria during the citric acid cycle. The pyruvic acid produced by glycolysis enters the mitochondria and is broken into two-carbon acetate units. This reaction produces one molecule of CO_2 and one reduced NAD for every molecule of pyruvic acid entering the sequence. The resulting acetate units are attached to coenzyme A (CoA), forming a compound known as acetyl-CoA. Coenzyme A is important in the metabolism of the cell and is the active form of pantothenic acid, a vitamin of the

Essay 4–2

On Discovery in Biology

One of the outstanding figures in modern biochemistry is Hans Krebs. Born and trained in Germany, as a student of the great German biochemist Otto Warburg, he did most of his research in England. He was the first to recognize the citric acid cycle, which is frequently called the Krebs cycle in his honor. For his work in cellular metabolism, Krebs was awarded the Nobel Prize. In this excerpt from his paper "The Tricarboxylic Acid Cycle," Krebs gives us a valuable insight into the way he did research and the basis for his discovery of the citric acid cycle.

The history of the tricarboxylic acid (citric acid) cycle makes it obvious that many biochemists contributed to the development of the concept, each standing on the shoulders of his predecessors. The names of Szent-Györgyi, Martius, and Knoop are prominent among those who built up the body of knowledge from which the concept finally emerged. I cannot help feeling the aptitude of the remarks of Kekulé, the discoverer of the ring structure of benzene, on the progress of scientific research. To the detached onlooker, he pointed out, the comparatively simple story which represents the final result is quite deceptive in respect to the mental processes which have led to the final picture. If people believed that the theory of the benzene structure had, complete and finished, sprung like Pallas Athena from the head of a chemical Zeus, they were mistaken. The idea had evolved slowly in the course of many years, though it is true that when one's mind hovers over a problem week after week and year after year, lightning flashes of vision may occur which help the thought along. He relates two such sudden visions, one of which occurred to him on the top of a London bus, but is quick to emphasize that such visions and flashes are no more than a very small part of a long-drawn-out mental process.

Retrospectively, one may well ask why Martius did not arrive at the concept of the tricarboxylic acid cycle before me. Why had it not occurred to him that the reactions which he had discovered and studied may be components of the main energy-yielding process in living matter? My guess is that this was a matter of scientific outlook—of "philosophical" attitude. Influenced by his teacher, F. Arndt, Martius regarded himself at that time (so he once told me) as a "theoretical organic chemist," interested in reaction mechanisms. The oxidative degradation of citrate was for him in the first instance a chemical and not a biological problem. He was therefore satisfied when he had clarified, with great ingenuity, the pathway citrate $\rightarrow$ *cis*-aconitate $\rightarrow$ isocitrate $\rightarrow$ α-oxoglutarate. He did not concern himself with the question of the physiological role of this pathway. Therefore he did not explore the quantitative aspects of the activity of the enzymes of the pathway or compare them with overall metabolic rates, nor did he measure the occurrence of the enzymes in tissues generally.

My outlook was that of a biologist trying to elucidate chemical events in living cells. I was thus accustomed to correlating chemical reactions in living matter with the activities of the cell as a whole. By putting together pieces of information in jigsaw-puzzle manner, and by attempting to discover missing links, I tried to arrive at a coherent picture of metabolic processes. So my mind was prepared to make use of any pieces of information which might have a bearing on the intermediary stages of the combustion of foodstuffs. This difference in outlook was, I believe, an important factor in determining who first stumbled on the concept of the tricarboxylic acid cycle. . . .

As for my motives, a major one is simply an insatiable curiosity and the thrill one gets from satisfying this curiosity. As Konrad Lorenz has said, scientific research is in fact equivalent to the playful curiosity of young animals. The thrill of finding the solution to a puzzle is perhaps related to the pleasure which people derive from solving crossword puzzles, though manmade puzzles have never interested me when there are so many natural puzzles awaiting solution. But what is the nature of the pleasure which the solving of a puzzle provides? Is it the feeling of satisfaction to have been clever enough? Gratification derived from expressions of appreciation, either by one's peers or by those who offer posts, is certainly an inspiring factor.

Another driving force, especially in my earlier days, was my ambition to justify my choice of career as a scientist vis-à-vis those who were doubtful about my ability to make a success in this field. These included my father, a surgeon who had inspired me to take up medicine, my teacher Otto Warburg (I believe)—and myself.

A third force, which has not left me to this day, is the justification vis-à-vis those who support me by putting financial resources and facilities at my disposal. This desire to justify a trust is, I suppose, the cause of the same sense of commitment and gratification which every worker experiences from knowledge of a well-done job.

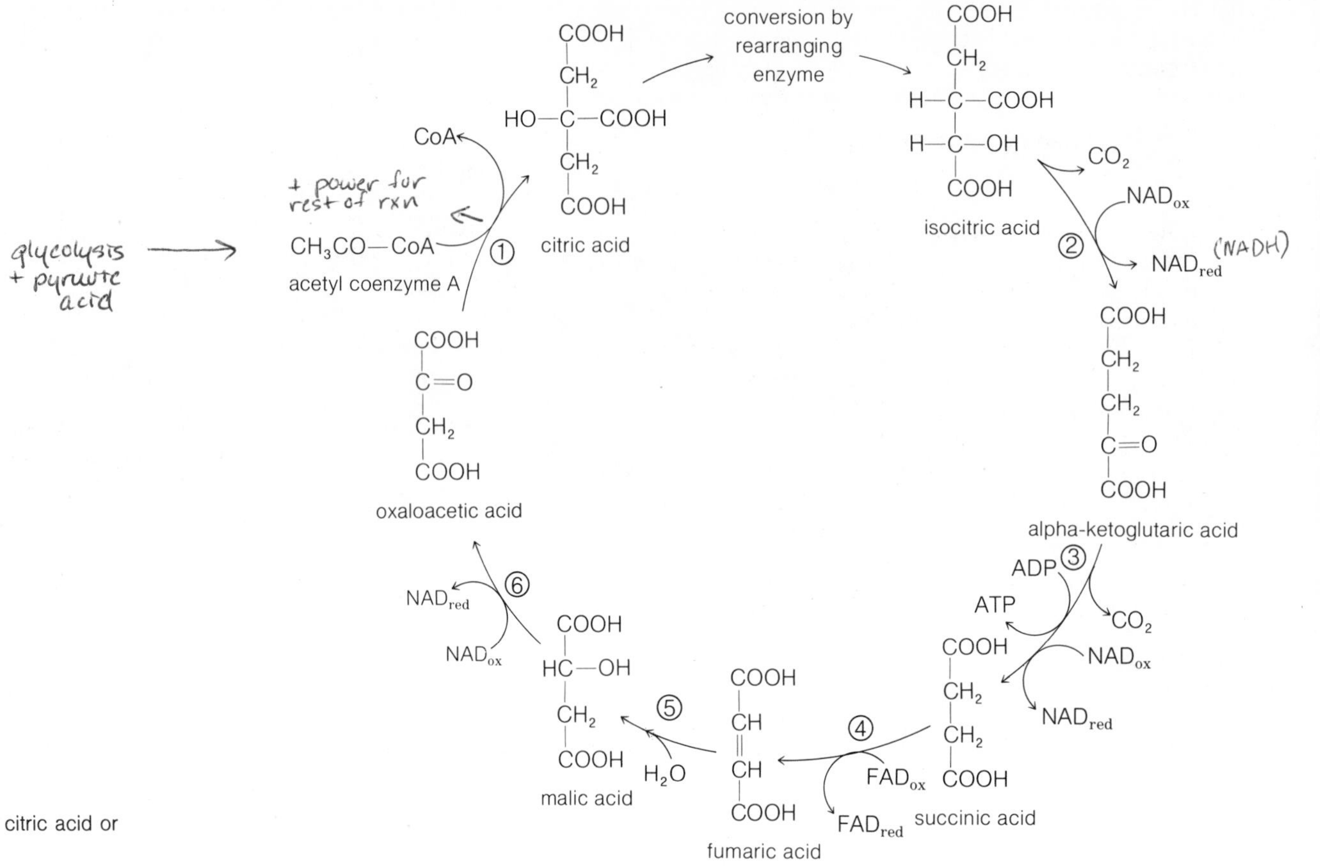

COOH
CH₂
HO—C—COOH
CH₂
COOH
citric acid

conversion by rearranging enzyme →

COOH
CH₂
H—C—COOH
H—C—OH
COOH
isocitric acid

CoA

+ power for rest of rxn

CH₃CO—CoA
acetyl coenzyme A

glycolysis + pyruvic acid →

① citric acid

② CO_2, NAD_{ox} → NAD_{red} *(NADH)*

COOH
CH₂
CH₂
C=O
COOH
alpha-ketoglutaric acid

COOH
C=O
CH₂
COOH
oxaloacetic acid

⑥ NAD_{red}, NAD_{ox}

③ ADP, ATP, CO_2, NAD_{ox}, NAD_{red}

COOH
CH₂
CH₂
COOH
succinic acid

COOH
HC—OH
CH₂
COOH
malic acid

⑤ H_2O

COOH
CH
CH
COOH
fumaric acid

④ FAD_{ox}, FAD_{red} succinic acid

Figure 4–16 The citric acid or Krebs cycle.

B complex. These initial reactions linking glycolysis to the citric acid cycle can be summarized:

$$2 \text{ pyruvic acid} + 2 \text{ CoA} + 2 \text{ NAD}_{oxidized} \rightarrow$$
$$2 \text{ acetyl-CoA} + 2 \text{ CO}_2 + 2 \text{ NAD}_{reduced}$$

"before" Kreb's cycle begins, this occurs.

In the first reaction of the citric acid cycle (Figure 4–16), an acetate unit is transferred from the CoA to the four-carbon oxaloacetic acid, yielding the six-carbon substance citric acid. The free energy released when the acetate unit is removed from the CoA powers the rest of the reaction.

In the second step of the cycle (reaction 2), two electrons, two hydrogen ions, and one carbon atom are removed from citric acid, yielding the five-carbon α-ketoglutaric acid. The lost carbon is released as CO_2. Wherever CO_2 is removed, thiamine, another B vitamin, is involved. The electron acceptor for this reaction, which also combines with one of the two hydrogen ions removed, is NAD.

The complex third reaction is shown only in skeletal form in Figure 4–16. In this step, α-ketoglutaric acid is oxidized to succinic acid; that is, two electrons and two hydrogen ions are removed and accepted by NAD. The carbon chain is again shortened, producing more CO_2, and one molecule of ATP is synthesized for every molecule of α-ketoglutaric acid oxidized:

③ $$\alpha\text{-ketoglutaric acid} + \text{NAD}_{oxidized} + \text{ADP} + \textcircled{P} \rightarrow$$
$$\text{succinic acid} + \text{NAD}_{reduced} + \text{ATP} + \text{CO}_2$$

The next oxidation of the citric acid cycle occurs at reaction 4, catalyzed by the enzyme succinic acid dehydrogenase. This enzyme carries its own electron acceptor, FAD (flavin adenine dinucleotide), and is the only molecule of the citric acid cycle that is attached to the internal membranes of the mitochondrion. The product of reaction 4, fumaric acid, is then rearranged at reaction 5, with the addition of a molecule of water, to malic acid. In the final step of the citric acid cycle (reaction 6), malic acid is oxidized to oxaloacetic acid, completing the cycle. NAD is the electron acceptor for this final oxidation.

We are now in a position to add up the products of the citric acid cycle and to summarize the cellular oxidation of glucose up to this point. As the Krebs cycle proceeds through one complete turn, one two-carbon acetate unit is consumed, and two molecules of CO_2 are released. At each of four reactions in the cycle, electrons are removed. At three of these steps, NAD is the acceptor, producing three molecules of reduced NAD; one step produces one molecule of reduced FAD instead. As a part of the complex series of interactions involving coenzyme A, one molecule of ATP is generated (in reaction 3). Oxaloacetic acid, used in the initial reaction of the citric acid cycle, is regenerated in the final reaction. Thus, as overall reactants and products, the citric acid cycle includes:

$$\text{an acetate unit} + 3 \text{ NAD}_{\text{oxidized}} + \text{FAD}_{\text{oxidized}} + \text{ADP} +$$

(2 carbons)

$$+ \textcircled{P} \rightarrow 2 \text{ CO}_2 + 3 \text{ NAD}_{\text{reduced}} + \text{FAD}_{\text{reduced}} + \text{ATP}$$

With this information, we can analyze the complete oxidation of glucose to carbon dioxide from glycolysis through the citric acid cycle. For each molecule of glucose entering the series, the citric acid cycle will turn twice. Glycolysis and the formation of acetyl coenzyme A yield 2 reduced NAD, 2 ATP, and 2 CO_2. Thus we have:

$$\text{glucose} + 4 \text{ ADP} + 4 \textcircled{P} + 10 \text{ NAD}_{\text{oxidized}}$$
$$+ 2 \text{ FAD}_{\text{oxidized}} \rightarrow$$
$$4 \text{ ATP} + 10 \text{ NAD}_{\text{reduced}} + 2 \text{ FAD}_{\text{reduced}} + 6 \text{ CO}_2$$

Note that little ATP has yet been produced. However, the electrons carried by the 10 molecules of reduced NAD and 2 molecules of reduced FAD exist at a high energy level and contain most of the energy obtained from oxidation of glucose to 6 CO_2. This energy is released in the cell through a series of electron carriers and is used to drive the synthesis of large quantities of ATP. Most of the ATP generated in eukaryotic cells is produced during this electron transfer, which uses oxygen as the final electron acceptor.

Direct Oxidation Pathway   Before we look at the electron-transport system, let's briefly consider another pathway for the breakdown of carbohydrates. This is the *direct oxidation pathway,* or the *pentose shunt mechanism,* known to exist in bacteria, plants, and animals. More involved in some ways than the glycolysis–citric acid cycle reactions, it has the same overall equation:

$$C_6H_{12}O_6 + 6 \text{ O}_2 \rightarrow 6 \text{ CO}_2 + 6 \text{ H}_2\text{O}$$

The first steps are the same as in glycolysis, but 1 CO_2 and a five-carbon sugar are then formed. This sugar is broken down through a complex set of reactions involving a number of three-, six-, and seven-carbon compounds. Ultimately, for every 6 glucose molecules entering the cycle, 6 five-carbon sugars, 6 CO_2 molecules, and 12 reduced NAD molecules are produced. The energy of the reduced NAD can be converted to 36 ATP molecules. The reaction is thus nearly as efficient as the glycolysis–citric acid cycle pathway. The five-carbon sugars produced are important in the synthesis of ATP and various other cellular constituents. Moreover, these same reactions play an important role in the CO_2 fixation cycle in photosynthesis.

The Electron-Transport System   The major result of the glycolysis–citric acid cycle or the direct oxidation pathway is the production of reduced NAD. This compound contains electrons at a high energy level, and their energy can be used by the cell directly. However, in the presence of oxygen, the electrons are transported from reduced NAD through a series of compounds to hydrogen ions that combine with oxygen to form water. As the electrons flow through the system, they lose part of their energy at each step, sometimes enough to drive the uphill synthesis of ATP from ADP. Three ATPs are formed for each pair of electrons passing through the system. The system transporting electrons from reduced NAD to oxygen in mitochondria is basically similar to the transport mechanism of the light reactions of photosynthesis.

These electron carriers are believed to be organized in mitochondria in the sequence shown in Figure 4–17. Reduced NAD from glucose oxidation transfers its electrons and hydrogen to FAD, the first carrier in the chain. In this transfer, NAD is oxidized and FAD is reduced; the oxidized NAD is then free to cycle back to the reactions of pyruvic acid oxidation and the citric acid cycle. The electrons flow, in turn, through a series of a quinone, at least three cytochromes, and finally to oxygen. The carrier delivering electrons to oxygen (cytochrome III in Figure 4–17) is responsible for the reduction of oxygen to water in the reaction:

$$\frac{1}{2} \text{ O}_2 + 2 \text{ H}^+ + 2 \text{ electrons} \rightarrow \text{H}_2\text{O}$$

The hydrogens (H^+) entering this reaction are derived from the pool of hydrogen ions in the surrounding medium. This pool is constantly replenished by hydrogens removed from intermediates during glucose oxidation.

Total ATP Production from Glucose Oxidation   We can now calculate the total ATP production of mitochondrial electron transport and derive a grand total for the entire sequence of oxidations from glucose to CO_2 and H_2O (Figure 4–18). Oxidation of glucose to CO_2, from glycolysis through the citric acid cycle, yields a total of 4 ATP, 10 reduced NAD, and 2 reduced FAD. The elec-

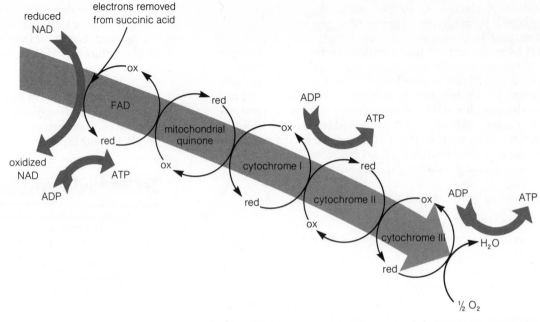

Figure 4–17 The electron-transport system of mitochondria. ATP is synthesized at three points as electrons flow along the chain from NAD to oxygen.

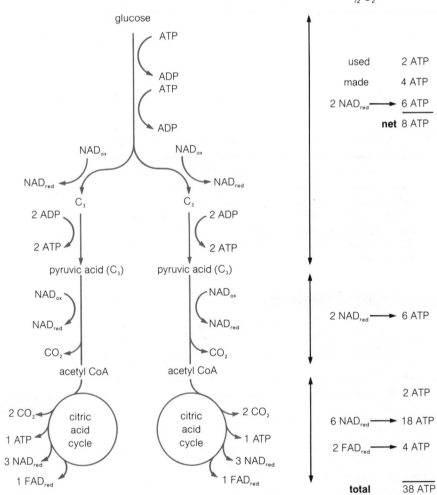

Figure 4–18 Summary of glycolysis and the citric acid cycle showing generation of ATPs.

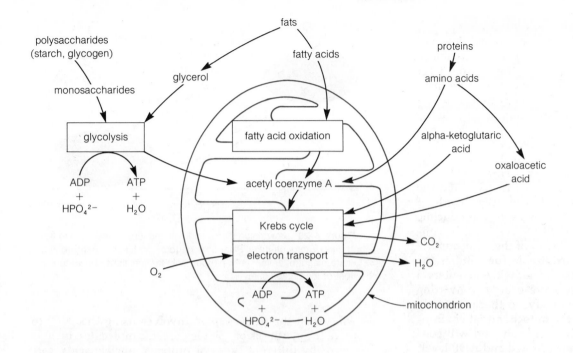

Figure 4–19 Summary of the way various food molecules enter cellular respiration.

trons carried by the 10 molecules of reduced NAD will drive the synthesis of 30 ATP as they pass through the electron-transport system. The electrons carried by the 2 molecules of reduced FAD induce the synthesis of only 4 ATP because they enter the sequence of carriers at a point farther along in the chain. This gives a total for electron transport of 30 ATP (from 10 NAD) + 4 ATP (from 2 FAD) = 34 ATP. Adding this total for electron transport to the net ATP production of glycolysis, acetyl coenzyme A formation, and the citric acid cycle yields a grand total of 38 ATP. In summary:

$$\text{glucose} + 38 \text{ ADP} + 38 \text{P} + 6 O_2 \rightarrow$$
$$6 CO_2 + 38 \text{ ATP} + 44 H_2O$$

Cellular Oxidation of Fats and Proteins Both animals and plants use carbohydrates as sources of immediate energy and fats as a means of storing energy for later use. As many of us are painfully aware, our bodies are capable of converting the food we eat into fat, which they often seem reluctant to use for anything.

When used as an energy source, both fats and proteins are split by preliminary reactions into shorter segments that enter the cellular respiratory system at various levels. In many cases, the shorter segments derived from these molecules are acetate units, subsequently carried to the citric acid cycle by coenzyme A.

In the cytoplasm outside mitochondria, fats are first split into glycerol and their component fatty acids. A phosphate group is then added to glycerol at the expense of ATP to produce a three-carbon, one-phosphate sugar. This sugar, which is also a product of reaction 4 in glycolysis (see Figure 4–14), enters the glycolytic pathway directly at this point. Next, the fatty acid chains enter mitochondria, where they are broken into two-carbon acetate units linked to coenzyme A. From this point onward, oxidation is the same as for two-carbon units derived from glucose. Because the fatty acid chains are long, containing from 14 to 22 carbons in almost all naturally occurring fats, many acetate groups are supplied to the citric acid cycle as the chains split into two-carbon segments. In the overall reaction from fats to CO_2 and H_2O, approximately 18 ATP are produced for each two-carbon segment in the fatty acid residues. This represents an ATP yield about double that of carbohydrates (by weight) and explains why fats are such an excellent energy source.

Segments split off from proteins are also oxidized through the citric acid cycle. The amino groups are first removed and then the molecules split into individual amino acids by reactions occurring in the cytoplasm outside the mitochondria (Figure 4–19). Some of these amino acids are further broken into two-carbon acetate units linked to coenzyme A. These acetate units then enter the citric acid cycle. Other amino acids are converted into intermediates of the citric acid cycle, some into oxaloacetic acid, and enter the cycle directly in this form.

The cellular oxidation of carbohydrates, fats, and proteins is summarized in Figure 4–19. Notice the central position of coenzyme A in linking the oxidation of these different substances in the cell.

Mitochondrial Oxidation and Cellular Activity The mitochondrial oxidation we have just detailed does not take place in a vacuum; it occurs in a living organism. It is finely tuned to the demands of the organism for chemical energy—turned on when these needs are high, and turned off when they are low. The key to this control mechanism seems to be the concentration of ADP in the organism's cells. When ADP concentrations are low, mitochondrial oxidation slows down, even when plenty of oxygen and raw materials are present. If ADP concentration increases, mitochondrial activity is stimulated, continuing rapidly until ADP is converted to ATP, and the ADP concentration is lowered again. In other words, if the organism is not growing or moving or reproducing, it will be using little energy, and little ATP will be converted to ADP. The oxidative activity of the mitochondria will remain low, conserving oxidizable fuel. But if the organism starts moving around or growing or enters a reproductive cycle, it will use more energy, converting ATP to ADP somewhere in its body, so that eventually the concentration of ADP in the mitochondrial environment will rise. Oxidation will then begin and will continue until ATP concentration is restored and ADP levels are once again low.

In plant cells with both mitochondria and chloroplasts the activity of the two organelles in combination provides all the ATP required for life. The interactions of these organelles in plant cells are diagrammed in Figure 4–20. Carbohydrates, fats, and some proteins are synthesized in the chloroplasts (to the left of the broken line in Figure 4–20) from the raw materials CO_2, H_2O, and various minerals by reactions using energy derived from light. These carbohydrates, fats, and proteins can be oxidized for energy in mitochondria (to the right of the broken line), yielding ATP, CO_2 and H_2O. Thus, CO_2 and H_2O cycle between the two systems. Plant cells may contain all of the machinery shown in Figure 4–20; animals have only the functions to the right of the broken line. The flow of energy in the diagram is from left to right, in the form of electrons that are excited to high energy by absorbing the energy of sunlight. These electrons, carried in various molecules, gradually release the captured energy as they flow through the energy-requiring activities of plant and animal cells.

Summary

The energy derived from oxidations in the cytoplasm of both prokaryotes and eukaryotes is used to drive the "uphill" reactions of the activities of life. In these oxidative reactions, as we have seen, energy generated when high-energy electrons are removed from carbohydrates (and other molecules) is eventually captured in the synthesis of ATP. Much of the remaining activity of life, as we

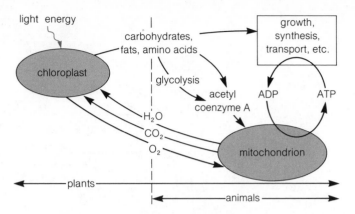

Figure 4–20 How chloroplasts and mitochondria interact in the flow of energy through cells. Plant cells perform the functions shown in the entire diagram; animal cells perform only the functions shown to the right of the broken line.

will see in the next chapter, involves using this ATP to drive the synthesis of complex organic molecules, including many different types of proteins, nucleic acids, carbohydrates, and lipids.

Supplement 4–1
Respiration and Photosynthesis in Prokaryotes

Bacteria and blue-green algae are prokaryotes, lacking mitochondria and chloroplasts. How do such organisms carry out their basic energy functions? Are they the same or different from their eukaryotic relatives? These questions become important when we consider the tremendous significance of bacteria in diseases of both animals and plants. Bacteria are also the source of many chemicals used by humans in a wide range of commercial activities. The blue-green algae are not important in either disease or commerce, but they are a major contributor to water pollution and thus become important to us. For these reasons and many more, let us look at some of the mechanisms of respiration and photosynthesis found in the bacteria and blue-green algae.

Bacteria

Oxidative Mechanisms in Bacteria Almost all bacteria use carbohydrates in some form as an energy source. Or-

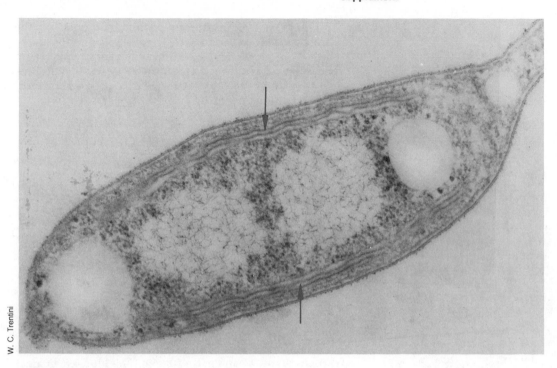

W. C. Trentini

Figure 4–21 Thylakoid membranes (arrows) in the cytoplasm of a photosynthetic bacterium. X59,000.

ganic acids and amino acids are also commonly oxidized by these organisms. Some bacteria can oxidize a wide variety of organic molecules; others are so restricted that their preferences in fuel substances can be used as a key to their identification. However, as in the eukaryotes, carbohydrates of different kinds are the primary energy source for most bacteria.

Aerobes and Anaerobes   At a minimum, most bacteria can oxidize carbohydrates through sequences closely resembling the glycolytic pathway. Because oxygen itself is not the final electron acceptor in these oxidations, this type of metabolism, if it is the only type used by the bacterium, is known as *anaerobic oxidation.* In anaerobic oxidation, the electrons carried by reduced NAD are accepted by an organic molecule instead of oxygen. Often pyruvic acid acts as the final acceptor, in a reaction similar to lactic acid generation in muscle tissue:

$$\text{pyruvic acid} + \text{NAD}_\text{reduced} \rightarrow \text{lactic acid} + \text{NAD}_\text{oxidized}$$

Anaerobic oxidation of this type, in which the final electon acceptor is an organic molecule, is also called *fermentation* in bacteria. By different pathways, bacterial fermentation may yield a wide variety of final products, including most common acids, such as the lactic acid of yogurt or the acetic acid of vinegar, and alcohols, such as ethyl or butyl alcohol.

Bacteria limited to fermentation are called *strict* or *obligate anaerobes.* Many bacteria, however, can also carry out reactions that use oxygen as a final electron acceptor if air is present. This is the case for the *Clostridium botulinum* bacteria, which grow in improperly sealed cans. Such bacteria, termed *aerobes,* have all the enzymes and mechanisms for acetyl coenzyme A formation, oxidation via the citric acid cycle, and electron transport. Some can "switch" between anaerobic fermentation and aerobic respiration, depending on the supply of oxygen in their environment; cells of this type are called *facultative aerobes.* If no oxygen is present, carbohydrates in these bacteria are oxidized through glycolysis, and an organic molecule such as pyruvic acid serves as the final electron acceptor. If oxygen becomes available, the facultative aerobe switches to the full pathway, including acetyl coenzyme A formation, the citric acid cycle, and electron transport. Some bacteria are *strict* or *obligate aerobes* and can live only if an adequate supply of oxygen is present.

Bacterial Photosynthesis   Bacterial photosynthesis is limited to a relatively few groups of bacteria named by the color of their cells, the purple and green bacteria (Figure 4–21). In addition to carotenes, the purple and green bacteria contain a distinct family of chlorophylls called *bacteriochlorophylls,* which differ slightly from the chlorophylls of eukaryotic plants. These bacterial chlorophylls absorb light most strongly in the far red wavelengths,

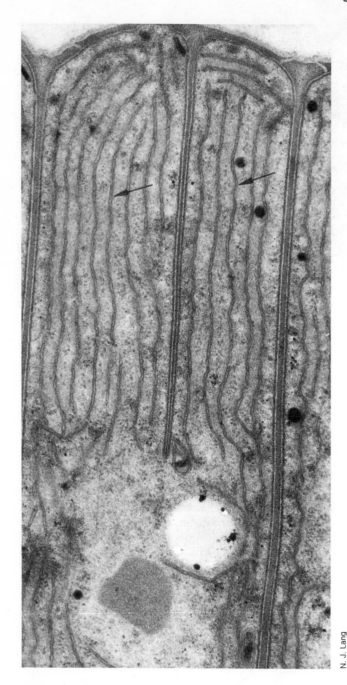

Figure 4–22 Thylakoid membranes suspended directly in the cytoplasm of a row of blue-green algae cells. X44,500.

N. J. Lang

These bacteria apparently possess only a part of the photosynthetic apparatus occurring in eukaryotic chloroplasts. The purple and green bacteria cannot split water as a source of electrons and hydrogen for photosynthesis, and they do not release oxygen as a by-product. Instead, in the reactions water is replaced by hydrogen, H_2:

$$CO_2 + 2\ H_2 \xrightarrow{\text{light}} (CH_2O) + H_2O$$

or hydrogen sulfide, H_2S:

$$CO_2 + 2\ H_2S \xrightarrow{\text{light}} (CH_2O) + H_2O + 2\ S$$

Blue-Green Algae

Respiration Research into the oxidative mechanisms of blue-green algae has been limited, and comparatively little is known of the abilities of these organisms to use carbohydrates as an energy source. The blue-green algae appear to live primarily by photosynthesis and evidently obtain most of the ATP required for their activities through electron transport in the light reactions of photosynthesis. If grown in the dark, some blue-green algae are able to take up a carbohydrate, such as glucose, and slowly oxidize it to CO_2. The pathways of this oxidation are not clear as yet, and it is not certain whether blue-green algae contain all the components of glycolysis and the citric acid cycle. Although at least some of the enzymes of each of these oxidative sequences have been isolated from blue-green algae, others have never been detected or identified.

Photosynthesis The blue-green algae, although typically prokaryotic in cell structure, can use water as a source of electrons and hydrogen for photosynthesis, and they evolve oxygen as a by-product. As far as can be determined, photosynthesis in these prokaryotes resembles the system of eukaryotic plants rather than bacteria, and all or most of the pathways shown in Figure 4–8 also occur in blue-green algae. As you would expect from this capability, the chlorophylls typical of eukaryotes, chlorophylls *a* and *b*, are present in the blue-green algae instead of bacteriochlorophylls, along with a variety of carotenes.

The light reactions of prokaryotes are associated with thylakoid membranes suspended directly in the cytoplasm (Figure 4–22). No discrete, separate organelles equivalent to chloroplasts are present in either the bacteria or the blue-green algae.

which are almost invisible to humans, and transmit almost all wavelengths of visible light. As a result, the bacteriochlorophylls contribute no distinctive color to either bacterial group, and the color of these bacteria results from the light transmitted by the carotenes present.

5

Protein Synthesis

Consider the puzzle that arises when you eat a roast beef sandwich. You consume the bread, lettuce, and meat, but when you have finished and your body has broken down and used them, you do not turn green like the lettuce, your flesh is still human flesh and not beef—you remain yourself. The mystery deepens when you consider your development. Nourished by the enormous variety of foodstuffs you eat, your body grows from child to adult. Your growth is not random, but takes place along clearly defined pathways that result in your unique variation on the general characteristics of your family and your species.

This single mystery forms the basis for some of the most profound questions in biology. How do organisms synthesize their component molecules from the basic materials to which they have access? What parts of the cell are involved? How is synthesis directed and controlled? In other words, how do cells "know" when to make a specific compound? How is this information passed on from one generation to the next?

Proteins are the key to cellular synthesis because all enzymes are proteins and, as we saw in Chapter 4, enzymes are critical to cell metabolism. The basic question then becomes: How are proteins made in the cell? We cannot give a detailed answer to this question, but 30 years of intensive work by cell biologists, biochemists, and molecular biologists have given us an excellent outline of the process. Here we will concentrate on what we know of how proteins are made by cells.

The Central Dogma

We know that proteins are made up of precise sequences of amino acids. The key to making a protein, therefore, must lie in knowing the particular sequence for that pro-

tein and then figuring out how the amino acids are joined together. Involved in both of these problems are the nucleic acids, DNA and RNA. The information that determines the sequence is contained in DNA, for the arrangement of the nucleotides that make up the pertinent segment of DNA will determine the arrangement of the amino acids in the protein. However, DNA is not immediately involved in protein production. Instead, the information in the DNA is used to make a special RNA, aptly called **messenger RNA** (mRNA), that moves from the DNA to a group of ribosomes made up of RNA and protein. There, amino acids temporarily associated with still another type of RNA (transfer RNA) are arranged in the proper order for the protein being made and are joined together to form that protein (Figure 5–1).

Thus, DNA directs the making of an RNA that directs the making of a protein. This is often called the central dogma in molecular biology. Although grossly oversimplified in this form, it gives the basic idea of how proteins are formed in living cells. To get a better idea of what is actually involved, let us look at some of the details.

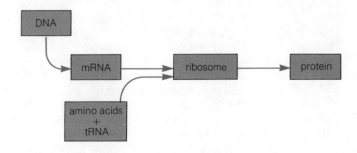

Figure 5–1 An overview of protein synthesis in the cell.

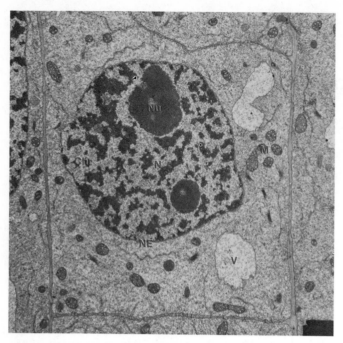

Figure 5–2 An electron micrograph of an onion root-tip cell, showing the nucleus and associated features.

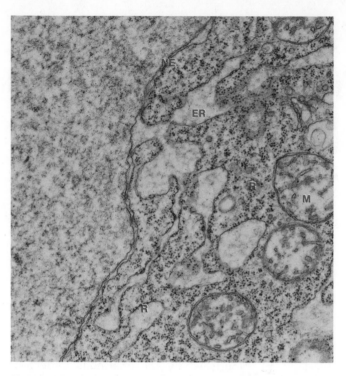

Figure 5–3 An electron micrograph of the nuclear envelope of a plant cell, showing the continuity of the endoplasmic reticulum and the nuclear membrane.

The Nucleus, Chromosomes, and Ribosomes

First let's try to place the central dogma in the context of the cell by asking two questions. Where are DNA and RNA located in the cell? Where does protein synthesis take place, and what cell parts are involved?

In the eukaryotic cell, the great bulk of the DNA is located within the nucleus (Figure 5–2), but DNA is also found in the plastids and mitochondria. Within the nucleus, the DNA is found associated with protein in the chromatin. Under the microscope, chromatin looks like threads distributed throughout the nucleus. When cell division takes place, the chromatin condenses into discrete bodies called **chromosomes.** Each chromosome contains thousands of **genes** that control all aspects of cell and organism development and function.

The nucleus also contains one or more roughly spherical bodies, the nucleoli (singular = nucleolus). Nucleoli are composed of dense, irregularly shaped masses of fibers and granules suspended among the chromatin. The granules are the most conspicuous feature of the nucleolus. The resemblance of these granules to cytoplasmic ribo-

somes is more than a coincidence, for research has shown that the subunits of ribosomes are made in the nucleolus. The nucleolus is composed primarily of RNA and protein.

Surrounding the nucleus are two surface membranes, together called the nuclear envelope (Figure 5–3). The outer membrane covers the entire surface of the nucleus and faces the surrounding cytoplasm. The inner membrane lies 20 to 30 nanometers inside the outer one and is concentric with it. The nuclear envelope is perforated by many pores that allow communication between the nucleus and the cytoplasm. In surface view, the pores are octagonal and some 70 nanometers in diameter. The pores appear to be filled with a plug of material known as the *annulus,* which may be perforated by a narrow channel. Often, these channels seem to be filled by particles about the right size to be ribosomal subunits in transit from the nucleus to the cytoplasm.

The principal sites of protein synthesis in the cytoplasm, as mentioned earlier, are the ribosomes. These may be found either loose in the cytoplasm (usually clustered in small groups) or attached to the membranes of the endoplasmic reticulum, or ER (Figure 5–4). Ribosomes are also found inside plastids and mitochondria.

Ribosomes are composed of roughly equal amounts

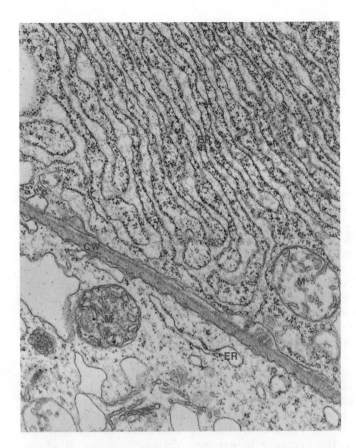

Figure 5–4 A section through two adjacent plant cells showing extensive rough endoplasmic reticulum (ER) in the upper cell. The small dark dots are ribosomes, some of which are attached to the ER and some of which are free.

of RNA and protein. Physically, they are made up of two subunits, one slightly smaller than the other (Figure 5–5). The subunits of ribosomes are assembled in the cell nucleus, within the nucleolus, and transported to the cytoplasm. Other cell parts are involved in protein synthesis, usually in modifying the finished proteins, but the primary sites are the nucleus, with its chromatin and nucleoli, and the cytoplasmic ribosomes.

Since the nucleic acids are so important in protein synthesis, let's examine them in greater detail and see how they fit into the picture of protein formation in the cell.

The Nucleic Acids

Possibly the most important discovery in the past 50 years was that of the structure of DNA. The chemical details of this work were discussed in Chapter 2. Here we will

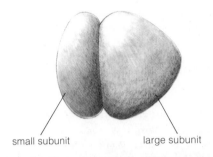

small subunit   large subunit

a

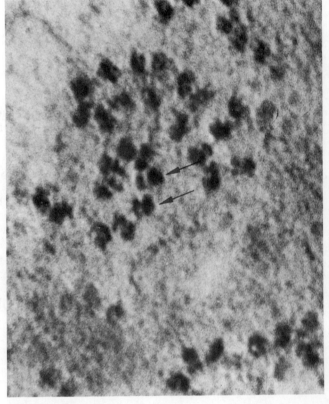

b

Figure 5–5 (**a**) Ribosome structure in bacteria; (**b**) ribosomes (arrows) in thin-sectioned rat liver. The large and small subunits and the cleft marking the division between the subunits are visible in many of the ribosomes. ×200,000.

N. T. Florendo

consider the results of the research. The critical discovery was that, in living cells, DNA occurs in the form of double molecules twisted together in the form of a helix (Essay 5–1). This double helix can be visualized as a twisted rope ladder in which the sugar-phosphate groups make up the side ropes and the nitrogen bases are paired to

Essay 5–1

The Double Helix

One of the truly great discoveries of the last 30 years was the determination of the structure of DNA. James D. Watson, a young postdoctoral fellow working in the laboratory of Francis Crick at Cambridge University, first conceived the basic structure of the molecule in 1953. Watson and Crick knew the basic facts necessary to construct a model of the molecule, but the essential key to the problem escaped them. This is one of the most difficult of all times for a scientist, and one that Watson describes well in his book The Double Helix.

I went ahead spending most evenings at the films, vaguely dreaming that any moment the answer would suddenly hit me. Occasionally my wild pursuit of the celluloid backfired, the worst occasion being an evening set aside for *Ecstasy*. Peter and I had both been too young to observe the original showings of Hedy Lamarr's romps in the nude, and so on the long-awaited night we collected Elizabeth and went up to the Rex. However, the only swimming scene left intact by the English censor was an inverted reflection from a pool of water. Before the film was half over we joined the violent booing of the disgusted undergraduates as the dubbed voices uttered words of uncontrolled passion.

Even during good films I found it almost impossible to forget the bases. . . .

Generally it was late in the evening after I got back to my rooms that I tried to puzzle out the mystery of the bases. Their formulas were written out in J. N. Davidson's little book *The Biochemistry of Nucleic Acids*, a copy of which I kept in Clare. So I could be sure that I had the correct structures when I drew tiny pictures of the bases on sheets of Cavendish notepaper. My aim was somehow to arrange the centrally located bases in such a way that the backbones on the outside were completely regular—that is, giving the sugar-phosphate groups of each nucleotide identical three-dimensional configurations. But each time I tried to come up with a solution I ran into the obstacle that the four bases each had a quite different shape. Moreover, there were many reasons to believe that the sequences of the bases of a given polynucleotide chain were very irregular. Thus, unless some very special trick existed, randomly twisting two polynucleotide chains around one another should result in a mess. In some places the bigger bases must touch each other, while in other regions, where the smaller bases would lie opposite each other, there would exist a gap or else their backbone regions must buckle in. . . .

My doodling of the bases on paper at first got nowhere, regardless of whether or not I had been to a film. Even the necessity to expunge *Ecstasy* from my mind did not lead to passable hydrogen bonds, and I fell asleep hoping that an undergraduate party the next afternoon at Downing would be full of pretty girls. . . .

Not until the middle of the next week, however, did a nontrivial idea emerge. It came while I was drawing

---

form the rungs. The two nucleotide chains of the DNA molecule are held together in a fairly rigid configuration within the double helix by the hydrogen bonds formed between the base pairs. The great numbers of such bonds formed in DNA give the double helix great stability at the temperatures occurring in living cells.

Also discovered was the critical fact that the bases which form hydrogen bonds are always specific pairs. Thus, if adenine is found on one strand, thymine is always its pair on the other, and if guanine is found on one, cytosine is always on the other. The pairs then are A with T and G with C. Along the chain, any sequence of bases is possible; but whatever the arrangement, the opposite strand is always complementary:

```
A T T G C G G T T A C C G T
T A A C G C C A A T G G C A
```

Knowing this immediately suggests a way in which the double molecules could reproduce. Assume that the two strands separate, and free nucleotides of all four kinds are available nearby. Then all that need happen is that the free nucleotides match up with their complementary bases in the old (or parent) strand and two new double-the fused rings of adenine on paper. Suddenly I realized the potentially profound implications of a DNA structure in which the adenine residue formed hydrogen bonds similar to those found in crystals of pure adenine. If DNA was like this, each adenine residue would form two hydrogen bonds to an adenine residue related to it by a 180-degree rotation. Most important, two symmetrical hydrogen bonds could also hold together pairs of guanine, cytosine, or thymine. I thus started wondering whether each DNA molecule consisted of two chains with identical base sequences held together by hydrogen bonds between pairs of identical bases. There was the complication, however, that such a structure could not have a regular backbone, since the purines (adenine and guanine) and the pyrimidines (thymine and cytosine) have different shapes. The resulting backbone would have to show minor in-and-out buckles depending on whether pairs of purines or pyrimidines were in the center.

Despite the messy backbone, my pulse began to race. If this was DNA, I should create a bombshell by announcing its discovery. The existence of two intertwined chains with identical base sequences could not be a chance matter. Instead it would strongly suggest that one chain in each molecule had at some earlier stage served as the template for the synthesis of the other chain. Under this scheme, gene replication starts with the separation of its two identical chains. Then two new daughter strands are made on the two parental templates, thereby forming two DNA molecules identical to the original molecule. Thus, the essential trick of gene replication could come

from the requirement that each base in the newly synthesized chain always hydrogen-bonds to an identical base. That night, however, I could not see why the common tautomeric form of guanine would not hydrogen-bond to adenine. Likewise, several other pairing mistakes should also occur. But since there was no reason to rule out the participation of specific enzymes, I saw no need to be unduly disturbed. For example, there might exist an enzyme specific for adenine that caused adenine always to be inserted opposite an adenine residue on the template strands.

As the clock went past midnight I was becoming more and more pleased. There had been far too many days when Francis and I worried that the DNA structure might turn out to be superficially very dull, suggesting nothing about either its replication or its function in controlling cell biochemistry. But now, to my delight and amazement, the answer was turning out to be profoundly interesting. For over two hours I happily lay awake with pairs of adenine residues whirling in front of my closed eyes. Only for brief moments did the fear shoot through me that an idea this good could be wrong.

And, of course, it was wrong—at least as far as the bases that actually pair are concerned. But the concept was right, and this provided the key for the correct model of DNA that Watson worked out a few days later.

---

helix DNA molecules are formed:

A T T G C G G T T A C C G T	old strand	new helix
T A A C G C C A A T G G C A	new strand	
A T T G C G G T T A C C G T	new strand	new helix
T A A C G C C A A T G G C A	old strand	

How is the detailed structure of DNA related to the

two key questions we raised earlier: Which amino acid goes where in a protein, and what ensures that the right one will go in the right place?

Let's return to the nucleus and the chromatin it contains. Decades of research have shown that protein synthesis is under the control of the genes on the chromosomes. One of the great unifying concepts that emerged from research done in the late 1940s and the 1950s was the realization of a strict relationship between genes and enzymes. For each enzyme there is a gene. This one gene–one enzyme concept has been modified over the years, but it is still generally accepted. At the same time, other research was showing that genes are made of DNA.

Now, if the genetic material is the DNA of the chromosomes, how can a series of nucleotides in a chain contain information? The answer, as we hinted at the beginning, is this: The amino acids are linearly arranged in the protein, just as the bases are linearly arranged in the DNA. If a certain arrangement of bases in the DNA indicates a specific amino acid, then that arrangement itself could determine—be a "code" for—the way the amino acids are aligned in the protein. We will see what this code actually is when we discuss messenger RNA.

RNA Synthesis in the Nucleus

Transcription The information determining the arrangement of amino acids in a protein is located in the DNA of the chromatin in the cell nucleus. But protein synthesis occurs in the cytoplasm, on the ribosomes. How does the message get from the DNA in the nucleus to the ribosomes in the cytoplasm? It is carried by messenger RNA, a special RNA made in conjunction with DNA.

Transcription is the process in which the sequence of nucleotides in DNA is used to specify a complementary sequence of nucleotides in an RNA molecule. This is an enzymatic process and requires specific enzymes. It is based on the fact that the two nucleotide chains of DNA are complementary, and that the sequence in one chain determines the sequence in the opposite chain. Assume that one of the two nucleotide chains carries the coded directions for synthesizing a protein. This nucleotide chain is called the "sense" chain of the two; the opposite chain, for reasons not yet understood, is not transcribed and is called the "missense" chain.

Information is transcribed from DNA to RNA as follows. The two nucleotide chains of DNA unwind and separate, and the sense chain serves as a template for RNA synthesis (Figure 5–6). As the RNA chain is assembled on the DNA template, bases link to it according to the rule for complementary pairing. Wherever a cytosine (C) occurs in the DNA template chain, a guanine (G) attaches opposite it in the growing RNA chain; wherever thymine (T) occurs, an adenine (A) enters the RNA chain; and wherever guanine (G) occurs, a cytosine (C) enters the RNA. But wherever adenine (A) occurs in the DNA template, the base uracil (U) is linked to the RNA chain:

A T T G C G G T T A C C G T   sense chain DNA

T A A C G C C A A U G G C A   RNA chain

T A A C G C C A A T G G C A   missense chain DNA

As the bases are paired, those of the RNA chain are joined together through the action of the enzyme RNA polymerase.

The segment of DNA that is transcribed is a gene, and it is precisely the segment necessary to make a single specific protein. But the DNA of a chromosome is continuous; every chromosome contains thousands of genes, each capable of transcribing RNA; and there are no physical breaks marking the boundaries of a gene. How, then, does RNA polymerase, which ties the bases of RNA together, know when to stop transcribing? We are not sure, but of the many hypotheses advanced, two seem to enjoy the widest favor. The first proposes that special initiator and terminator DNA code sequences mark the beginning and end of the gene and are recognized as such by the enzyme. The second major hypothesis states that transcription is controlled by localized unwinding of the DNA—that only a single gene segment unwinds and becomes available to be copied.

When chromatin fibers are isolated from eukaryotic cells, they always contain two classes of protein molecules besides DNA: the *histones* and the *nonhistones*. These structural proteins are believed to have important functions in regulating the activity of DNA in RNA transcription.

Chromatin contains as much histone protein, by weight, as it does DNA. At the pH ranges occurring in living nuclei, histones carry strong positive charges. This makes them easy to distinguish from other cellular proteins, almost all of which carry a negative or neutral charge at these pHs. Electrostatic attraction binds the positively charged histones to the negatively charged DNA. Histones can be separated into five major fractions, differing in the strength of their attraction and binding to DNA. Four of these five fractions are remarkable in that their amino acid sequences are almost the same in all eukaryotic cells. This indicates that the function of histones in the DNA-protein complex is critical to life and that even single changes in the histones' amino acid sequence are lethal.

The nonhistone, or basic, proteins are scarcely well enough known to be called a class. Estimates of the total number of such proteins in chromosomes range from 3 or 4 to more than 100, with the best current evidence favoring fewer than 20.

Ribosomal RNA Transcription Before we leave the question of transcription and follow messenger RNA out of the nucleus and into the cytoplasm, we must note that more than this one kind of RNA is produced by DNA in the nucleus. Also made are **ribosomal RNA** (rRNA) and **transfer RNA** (tRNA), which plays an important part in protein synthesis. Of the RNA reaching the cytoplasm, about 70 or 80 percent is ribosomal, 10 to

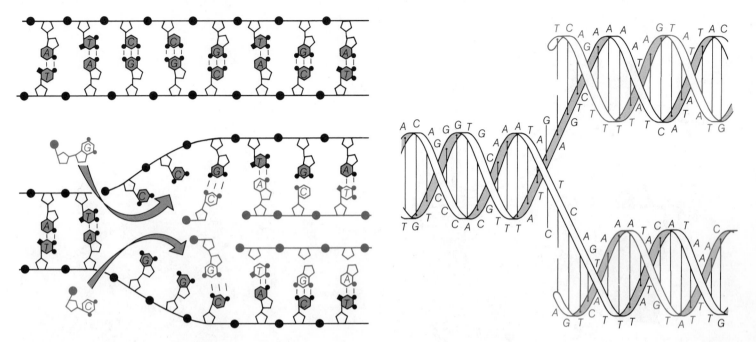

Figure 5–6 The nucleotide chains of the DNA double helix "straightened out" and flattened to show the complementary bonding. The bottom half of the figure shows the mechanism of reproduction by splitting of the hydrogen bonds and addition of the free nucleotides from the medium to produce the two pairs of chains, each pair identical with the first. The figure to the right shows the process in a schematic, three-dimensional way. In either case, the process is controlled by special enzymes. A = adenine, T = thymine, C = cytosine, G = guanine.

15 percent is transfer, and only about 5 percent is messenger RNA.

The RNA in ribosomes is transcribed in several places within the nucleus. Some is made within the nucleolus and transcribed from chromatin that extends into the nucleolar structure. This region of the chromatin, known as the *nucleolar organizer,* forms only a small part of the total volume of the nucleolus. Most of the structure of the nucleolus represents successive stages in the construction of ribosomes. Two of the three types of ribosomal RNA present in mature ribosomes are large molecules, and these are most likely transcribed on the nucleolar organizer chromatin. The third type of ribosomal RNA, a much smaller molecule, is evidently transcribed on parts of the chromatin outside the nucleolus.

Protein Synthesis in the Cytoplasm: Translation

The site of action now moves from the nucleus to the cytoplasm. The messenger RNA becomes associated with a ribosome, while the small transfer RNAs link to amino acids before joining the messenger RNA at the ribosome (Figure 5–7). The position of the transfer RNA–amino acid units is determined by what is coded on the messenger RNA. Once in place, the amino acids are separated from the transfer RNA and joined together to form the chain of the protein molecule, which we call a polypeptide, and the process is essentially complete.

This is the overall picture. Let's now look at the various stages in greater detail, starting with a consideration of transfer RNA.

Transfer RNA Transfer RNAs are relatively small nucleic acid molecules. They are short molecules, compared to other types of nucleic acids, averaging between 75 and 80 nucleotides in length. Most of the transfer RNAs have been isolated and purified; a number have been completely analyzed and "sequenced" so that the precise number and order of nucleotides in the molecules are known. One such tRNA molecule that specifically links to the amino acid alanine is shown in Figure 5–8.

At least 60 different transfer RNAs can be identified in the cytoplasm of eukaryotes. This is of more than passing interest, because only approximately 20 different

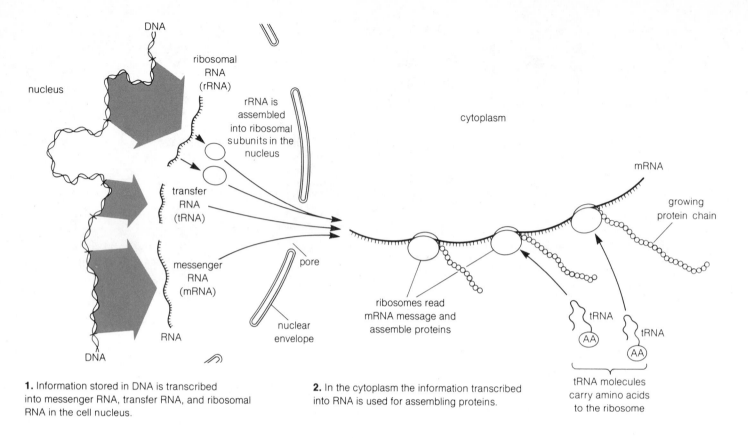

DNA

nucleus

ribosomal
RNA
(rRNA)

rRNA is
assembled
into ribosomal
subunits in the
nucleus

cytoplasm

mRNA

transfer
RNA
(tRNA)

growing
protein chain

messenger
RNA
(mRNA)

pore

nuclear
envelope

RNA

DNA

ribosomes read
mRNA message and
assemble proteins

tRNA

tRNA

AA

AA

tRNA molecules
carry amino acids
to the ribosome

1. Information stored in DNA is transcribed into messenger RNA, transfer RNA, and ribosomal RNA in the cell nucleus.

2. In the cytoplasm the information transcribed into RNA is used for assembling proteins.

Figure 5–7 The major steps of cellular synthesis.

amino acids are coded for and enter protein synthesis at the ribosome. This means that each amino acid can be carried to the ribosome by more than one type of tRNA. A given transfer RNA can, however, bond to only one type of amino acid. Because transfer RNAs are the "translator" molecules in protein synthesis, matching amino acids with "words" in the code of messenger RNA, each amino acid must therefore be coded for by more than one word in the message. This is actually the case. The amino acid serine, for example, is found linked to any one of five different tRNA molecules in the cytoplasm and is coded for by six different "words" in the nucleic acid code.

Amino Acid Activation   In the reaction between a transfer RNA and its amino acid, the final product, the tRNA–amino acid complex, is a high-energy compound. Because the overall energy level has increased as the reaction proceeds, at the expense of ATP, the entire process is usually called *amino acid activation*. But the reaction sequence is really more extensive than this name suggests, because

in addition to increasing in energy level, the amino acids are also matched with their "correct" tRNA molecules. The process works in two steps.

In the first step, an amino acid interacts directly with ATP, in an enzyme-catalyzed reaction that removes two phosphate groups from ATP and attaches the resultant AMP (adenosine monophosphate) molecule to the amino acid:

$$AA + ATP \rightarrow AA\text{-}AMP + \text{inorganic phosphate}$$

amino
acid

aminoacyl
adenylic acid

At this point in the sequence of amino acid activation, the amino acid–AMP complex remains attached to the enzyme.

In the next step, the amino acid is transferred from AMP to a "correct" molecule of transfer RNA:

$$AA\text{-}AMP + tRNA \rightarrow \quad AA\text{-}tRNA \quad + AMP$$

aminoacyl
adenylic acid

aminoacyl-tRNA

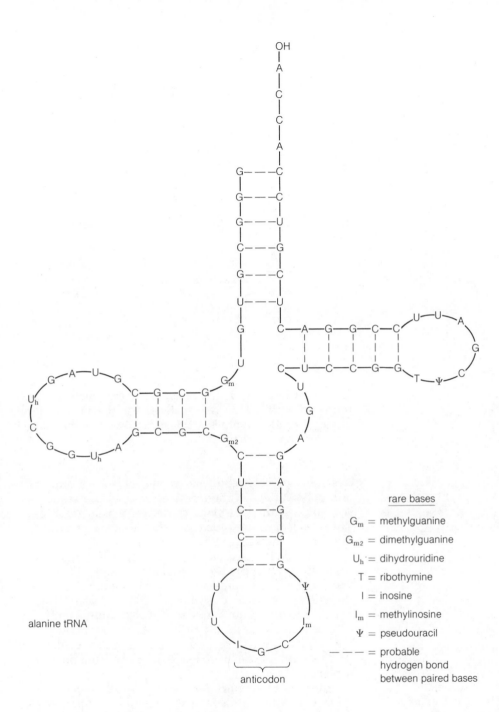

rare bases

G_m = methylguanine

G_{m2} = dimethylguanine

U_h = dihydrouridine

T = ribothymine

I = inosine

I_m = methylinosine

Ψ = pseudouracil

— — — = probable hydrogen bond between paired bases

alanine tRNA

anticodon

Figure 5–8 Alanine tRNA, a transfer RNA molecule containing 77 nucleotides that links specifically to the amino acid alanine. Alanine tRNA, like all tRNA molecules, has a high percentage of rare bases—chemical modifications of the regular G, C, A, and U bases after the tRNA is transcribed in the nucleus. The folding arrangement shown is hypothetical. Redrawn courtesy of J. T. Madison.

The AA-tRNA product is released from the enzyme when this reaction is completed. The products on the right side of this equation contain less order and energy than the reactants, and the overall reaction releases free energy as it proceeds to completion. Much of the energy obtained by the conversion of ATP to AMP is transferred to the aminoacyl-tRNA complex. As a result, this complex may be regarded as a high-energy compound. This energy is eventually used to drive formation of a peptide bond in protein synthesis at the ribosome, as we shall see shortly.

A single enzyme, aminoacyl-tRNA synthetase, catalyzes both of the reactions just shown. There are 20 different forms of the synthetase enzyme, each specific in its activity for one of the 20 amino acids. There is little doubt that a part of the "recognition" between tRNA second base of codon

<table>
first base of codon	U		C		A		G		third base of codon
U	UUU UUC	phe	UCU UCC UCA UCG	ser	UAU UAC	tyr	UGU UGC	cys	U C
	UUA UUG	leu			UAA UAG	* *	UGA UGG	* try	A G
C	CUU CUC CUA CUG	leu	CCU CCC CCA CCG	pro	CAU CAC	his	CGU CGC CGA CGG	arg	U C A G
					CAA CAG	gln			
A	AUU AUC AUA	ileu	ACU ACC ACA ACG	thr	AAU AAC	asn	AGU AGC	ser	U C A G
	AUG	met			AAA AAG	lys	AGA AGG	arg	
G	GUU GUC GUA GUG	val	GCU GCC GCA GCG	ala	GAU GAC	asp	GGU GGC GGA GGG	gly	U C A G
					GAA GAG	glu			
</table>

ala = alanine
arg = arginine
asn = asparagine
asp = aspartic acid
cys = cysteine
gln = glutamine
glu = glutamic acid
gly = glycine
his = histidine
ileu = isoleucine
leu = leucine
lys = lysine
met = methionine
phe = phenylalanine
pro = proline
ser = serine
thr = threonine
try = tryptophan
tyr = tyrosine
val = valine

Figure 5–9 The genetic code in RNA codewords. To find the DNA codeword equivalents, let A = T, U = A, G = C, and C = G. For example, the RNA codeword UCA is equivalent and complementary to the DNA codeword AGT. The triplets marked with an asterisk are terminator codons that signal the end of a message and cause the ribosome to release a finished protein.

anticodons

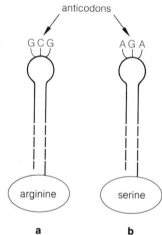

G C G   A G A

arginine   serine

a   b

Figure 5–10 Hypothetical tRNA molecules for arginine (**a**) and serine (**b**).

molecules and their specific amino acids depends on the activity of the 20 synthetase enzymes. Thus, just as a nucleic acid (tRNA) serves as a link between the nucleic acid code and the amino acid sequence of proteins, a protein (the synthetase enzyme) is intimately involved in matching the different transfer RNAs and their amino acids. This close interdependence between the activity of enzymatic proteins and nucleic acids is characteristic of all parts of the mechanism, from the initial transcription of messenger RNA to the final translation of the message during protein synthesis at the ribosome.

The aminoacyl-tRNA complexes formed as the final product of amino acid activation are the immediate energy source as well as the raw material for the next major phase of protein synthesis, the assembly of amino acids into polypeptide chains.

Assembly of Polypeptides Amino acids are assembled into polypeptides at the ribosome in a precise sequence according to the directions carried in messenger RNA. These directions are coded into the sequence of nucleotides in the messenger RNA molecules.

The code is based on the arrangement of three bases per amino acid. The entire code and the amino acids specified by the code words are shown in Figure 5–9. Each three-letter word of the code is called a *codon*. There are 64 possible codons and roughly 20 amino acids to be coded for. This means that there may be more than one codon for the same amino acid. The amino acid proline, for example, can be coded for by any of four different triplets.

In order to simplify our description of protein synthesis, we will consider the assembly of a hypothetical polypeptide containing only two different amino acids, arginine and serine, alternating in the sequence arg-ser-arg-ser-arg-ser. . . . The mRNA carrying the directions for synthesizing this protein would carry only the codons for these two amino acids, alternating in the same sequence. One of the mRNA codons for arginine is known to be the triplet CGC, and one of the codons for serine is the triplet UCU. A messenger RNA molecule for this hypothetical arginine-serine polypeptide would therefore

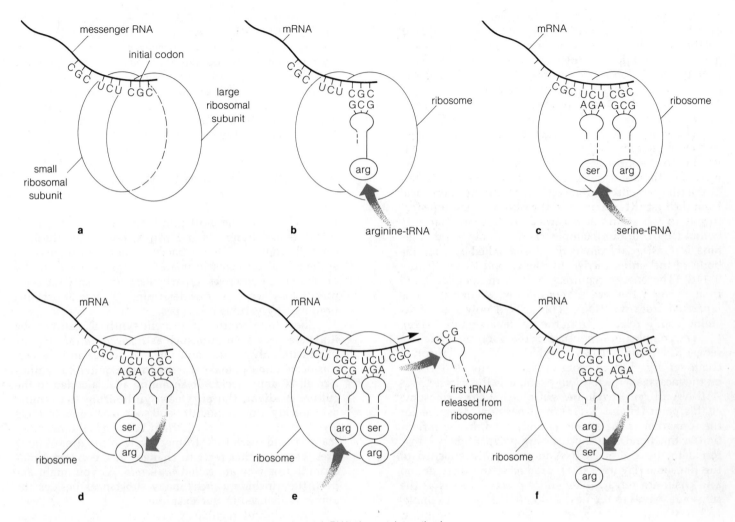

Figure 5–11 The interaction of mRNA, ribosomes, and aminoacyl-tRNAs in protein synthesis.

have as one possible code the sequence . . . CGC/UCU/ CGC/UCU/. . . .

After transcription in the nucleus, the hypothetical mRNA enters the cytoplasm and becomes attached to a ribosome. In the medium surrounding this mRNA-ribosome complex are all of the various molecules of transfer RNA, each attached to its specific amino acid. These elements—mRNA, the ribosome, and the various tRNA–amino acid complexes—form the basic parts of the protein-synthesis mechanism.

Among the tRNAs are several types carrying the amino acids arginine and serine. The ones of interest are those able to recognize and bind to the CGC and UCU codons in the hypothetical messenger RNA. This recognition capacity depends on part of the tRNA molecule called the *anticodon* (Figure 5–10), a region able to form complementary base pairs with the codons in the message. Sup-

posedly, the tRNA chain is folded in such a way that the anticodon site is exposed at one end of the molecule (see Figure 5–10). This anticodon pairs with the codons on mRNA by forming the appropriate base pairs. We will consider that the tRNA carrying arginine has the anticodon GCG, able to recognize and bind to the CGC codon for arginine in the messenger RNA. The tRNA carrying serine has the anticodon AGA, able to recognize and bind to the UCU codon for serine in the message.

These aminoacyl-tRNAs interact with the messenger at the ribosome, as shown in Figure 5–11. At the ribosome, only a short region of the mRNA is exposed at any instant for the formation of base pairs with tRNA–amino acid complexes from the surrounding medium. In our example, if the mRNA sequence CGC is the first codon exposed at the ribosome (Figure 5–11a), a tRNA carrying arginine will attach to the ribosome in such a position that complementary base pairs are formed between the CGC codon on mRNA and the tRNA GCG anticodon (Figure 5–11b). As the ribosome moves to the next codon on the messenger RNA, the mRNA triplet UCU is exposed. A tRNA with the anticodon AGA, carrying the amino acid serine, pairs with the messenger at this point (Figure 5–11c). The total complex now consists of the mRNA, the ribosome, and the two tRNAs carrying arginine and serine.

The two amino acids now at the ribosome are brought into close proximity by this interaction, in a position favorable for the formation of the first peptide bond. Formation of the bond, catalyzed by an enzyme that is part of the ribosome itself, involves separating the amino acid from the first tRNA bound to the ribosome and transferring it to the amino acid carried by the most recently bound tRNA. In our example, the first amino acid, arginine, is transferred from its tRNA and linked by a peptide bond to the serine carried by the second tRNA (Figure 5–11d). The energy required to form the peptide bond is supplied by the free energy released when arginine is separated from its tRNA. This tRNA, now free of its amino acid, is released from the ribosome (Figure 5–11e).

The complex now consists of the ribosome, the messenger RNA, and a tRNA carrying the short polypeptide chain ser-arg. The ribosome moves to the next codon on the message, which in our example is the triplet CGC. A tRNA carrying arginine will attach to the ribosome at this point (Figure 5–11e), in a position favorable for the formation of the second peptide bond. In the formation of this bond, the short ser-arg polypeptide is transferred to the arginine tRNA most recently attached to the ribosome (Figure 5–11f), producing the three amino acid chain arg-ser-arg. The entire process repeats as the ribosome moves to the next codon (UCU in our example) and continues until the end of the mRNA molecule is reached.

Protein synthesis in the cytoplasm of living cells takes place by the same mechanism as outlined in Figure 5–11, except that coded messages spelling out sequences including any or all of the 20 amino acids are encountered by ribosomes. In addition, a large number of "factors," many of them enzymes, are required in addition to the basic elements (tRNA, mRNA, and the ribosome) just outlined. Some of these factors must be present for protein synthesis to begin, others are required for the process to continue, and still others are required for successful termination of protein synthesis, with release of the finished polypeptide from the ribosome.

Special codons are thought to be involved in the initiation and termination of protein synthesis. In bacteria, one of the two codons AUG or GUG always occurs as the first word in the coded message for protein synthesis. These codons have been called *initiator codons.* Unless one of them is present as the first word in the message, protein synthesis cannot proceed. Likewise, one of the

three *terminator codons* (UAG, UAA, or UGA) must be the last word in the message, or protein synthesis cannot be successfully completed. These terminator codons specify no amino acids and evidently have no corresponding tRNA molecules with complementary anticodon sites. Supposedly, as the ribosome reaches one of these terminator codons, no aminoacyl-tRNA binds to the mRNA. The empty codon then in some manner induces release of the completed polypeptide from the ribosome.

Regulation of Protein Synthesis

In all cells, from the most primitive bacteria to the most highly evolved animals and plants, protein synthesis is carefully controlled. In the bacteria, such control involves systems that turn closely related groups of genes rapidly on and off. In other cases, particularly in eukaryotic cells, precise long-term controls determine the type of proteins produced by various cell types.

Short-term control of protein synthesis can best be illustrated in a bacterium such as *Escherichia coli.* If supplied with only an inorganic nitrogen source and glucose, this bacterium can make the enzymes required to synthesize all 20 amino acids. If an amino acid is added to the culture medium, the enzymes synthesizing that amino acid quickly drop in quantity and soon reach undetectable levels. If the amino acid is then removed, the enzymes reappear and reach their former levels in a matter of minutes. Enzymes that react to changes in the external medium in this way are called *inducible.* As you might expect, they include a great many substances besides the amino acids used in our example.

This kind of regulatory mechanism in bacteria was the subject of intensive investigation that led, in the 1950s, to the development of the *operon hypothesis* by François Jacob and Jacques Monod at the Pasteur Institute in Paris. They found that the genes for the induced enzymes in a particular group of related reactions are located close together on the bacterial chromosome and adjacent to a block of DNA that appeared to be the "on-off switch" for the entire collection of genes needed for these enzymes. The group of genes that act together is called the *operon,* and the adjacent control region is called the *operator* (Figure 5–12). The activity of the operon is controlled by yet another gene, the *regulator,* or R gene, that may be some distance away on the bacterial chromosome. The regulator codes a protein, known as the *repressor,* which controls the operator gene by binding with it. In the case of inducible enzymes, the repressor is controlled by the compound that is the *inducer.* The inducer binds with the repressor so that it is no longer capable of inhibiting the operator, and the operon is turned on. Say the inducer is a substrate that the bacteria

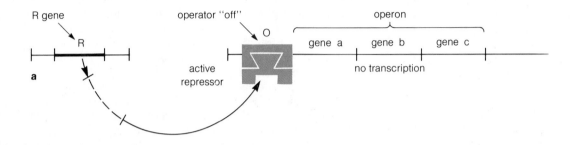

a

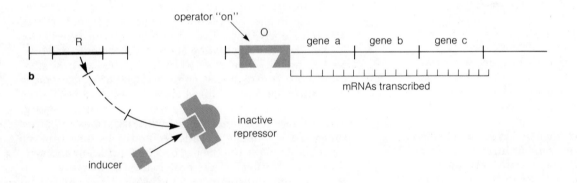

b

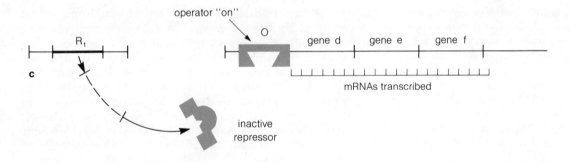

c

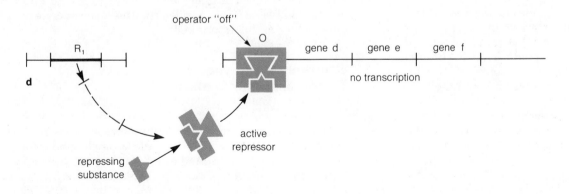

d

Figure 5–12 The operon mechanism of Jacob and Monod.

can break down for energy. The presence of the substrate functions as the inducer, and the cell begins to synthesize the enzymes needed to make use of the substrate. When the inducing substrate is used up, the repressor shuts off the operator, and enzyme synthesis stops. In the case of the amino acid enzymes discussed earlier, the added amino acid acts as a corepressor, the repressor becoming active only in the presence of the added compound. These alternative forms of the mechanism can be summarized as follows:

Induction of enzyme synthesis:

active repressor + inducing substance →
inactive repressor

The inactive repressor does not combine with the operator, and the genes are continuously transcribed (Figure 5–12a and b).

Repression of enzyme synthesis:

inactive repressor + repressing substance →
active repressor

The active repressor combines with the operator, and the genes are turned off (Figure 5–12c and d).

There is ample evidence that this type of scheme is present and important in bacteria, but little evidence exists to substantiate its existence in higher organisms.

Multicellular organisms are faced with a different regulation problem. There is a great deal of evidence that each cell in a higher plant or animal contains a complete complement of genetic information. Yet the way in which this information is used varies greatly from cell type to cell type. The proteins made by the brain are much different from those made by the liver. The maturing red blood cells in bone marrow synthesize little more than hemoglobin and a few enzymatic proteins, but they are inherently capable of making the same proteins as either brain or liver cells. In other words, the DNA in most cells appears to be the same. What regulates the differences in protein synthesis in different tissues?

This has been one of biology's most difficult problems, and only recently has significant progress toward the answer been made. At present this regulation is thought to occur at the level of DNA transcription and to involve the proteins associated with DNA in chromatin. As we noted earlier, two major groups of proteins are associated with the DNA, the histones and the nonhistones. The histones are a well-defined group of proteins remarkably similar in structure from one end of the evolutionary scale to the other and remarkably constant in proportion to DNA from one cell type to another.

In general, histones appear to repress transcription. In other words, they turn genes off. But they are too similar in different tissues, in both chemical character

and amount, to account for the different proteins formed by different tissues. What, then, differentially turns genes on in various parts of our bodies? The answer appears to lie in the nonhistone proteins. These are a much more diverse group of proteins with a wide range of functions. Moreover, they have the characteristic, important for a regulator candidate, of being different in different cell types. Not all the data are in, but the role of the non-histone proteins seems to be to interact with the histones to allow specific parts of the DNA to transcribe messenger RNA. Obviously, even if this hypothesis is correct, many questions remain, the most important of which are the source of the nonhistone proteins and the origin of the differences among various tissues. The intense research now going on in this area should soon bring answers to these questions.

Translational controls acting at the level of protein synthesis in the cytoplasm are important in eukaryotes and have been followed in a number of cell types. The best evidence for regulation at this level comes from studies tracing the amounts of RNA transcribed and the proteins synthesized during various stages of development in fertilized and unfertilized animal eggs. As animal eggs develop, RNA and protein are both synthesized rapidly. The cytoplasm becomes packed with ribosomes and other types of RNA, including mRNA, and large quantities of proteins. At maturity, the egg becomes quiescent and synthesizes no further RNA or protein until fertilization, even though both ribosomes and mRNA are present in the cytoplasm.

At fertilization, protein synthesis begins again almost at once. By exposing egg cells to radioactively labeled uridine it can be shown that in many fertilized eggs this protein synthesis proceeds without transcription of any new RNA in the nucleus—in some cases until the embryo consists of hundreds of cells. Therefore, the mature, unfertilized egg probably has mRNA stored in an inactive form in its cytoplasm. Ribosomes are apparently also inactive before fertilization. Experiments with cytoplasmic RNA and ribosomes isolated from mature but unfertilized eggs have supported the existence of translational control, indicating that a masking protein may inactivate both mRNA and ribosomes. Similar controls may operate in other types of eukaryotic cells.

Ribosomes and the Endoplasmic Reticulum

In our discussion of mRNA and ribosomes, we have dealt with a single strand of mRNA and a single ribosome; but in living cells one strand of mRNA is usually associated with a number of ribosomes. Such a collection of ribosomes held together by mRNA is called a *polysome* and is the functional unit of protein synthesis. The ribosomes appear to move along the strand of mRNA, and a polypeptide chain is formed at each ribosome.

Ribosomes are also found attached to the membrane system known as the endoplasmic reticulum, or ER. When ribosomes are attached to the ER, it is called *rough,* and when they are missing, it is called *smooth.* The membranes of the ER form greatly flattened, saclike interconnected vesicles. The ribosomes are joined to the outside of the ER, often in curved rows or spirals (Figure 5–13).

In cells that synthesize and secrete large quantities of protein, the rough ER grows into masses almost completely filling the cytoplasm. One cell of this type, a pancreatic cell that produces insulin (see Chapter 14), is shown in Figure 5–14.

Experimental evidence indicates that proteins synthesized on the ribosomes of the rough ER are transported across the ER membranes and concentrated inside. Such proteins probably then move into the smooth ER, which pinches off vesicles packed with protein. Such vesicles may either remain in the cytoplasm or fuse with the cell membrane, extruding the enclosed protein from the cell. This is how digestive enzymes reach the gut of animals, including humans. In contrast, free ribosomes produce proteins to be used within the cell.

The Golgi Complex: Dictyosomes

Found in the cytoplasm of both plant and animal cells are stacks of flattened saclike vesicles surrounded by numerous other vesicles that appear to have budded off from the edges of the complex. These units are the *Golgi complex* or *dictyosomes* (Figure 5–15). They have a wide range of shapes and sizes and play an important role in cellular secretion.

The activities of the Golgi complex and the ER in synthesis and secretion are summarized in Figure 5–16. Protein synthesized within the ER is transferred to the Golgi complex—apparently in vesicles that bud off from the edges of the ER and fuse with the outermost membranes of the Golgi complex. Once inside the complex, the proteins are modified and then concentrated to form secretion vesicles that pinch off from the edges of the complex. After budding off, the secretion vesicles usually fuse together into larger structures known as *secretion granules.* The membranes of the secretion granules fuse with the plasma membrane of the cell and the contents of the granules are discharged outside the cell. This is what happens in the mucus-secreting cells in humans. Mucus is a mixture of glycoproteins (a protein with an attached carbohydrate group): the protein is synthesized in the ER, and the carbohydrate is made and added in

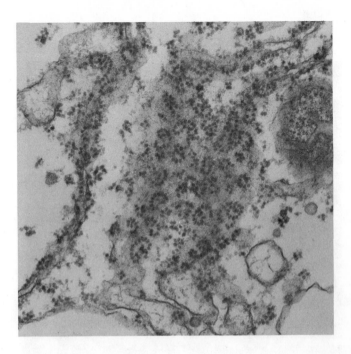

Figure 5–13 A surface view of the ER that shows the ribosomes arranged in curved rows. ×45,000.

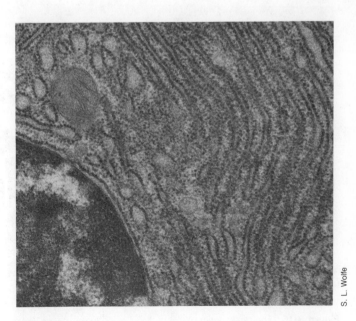

S. L. Wolfe

Figure 5–14 Closely packed rough ER in a pancreatic cell from a rat. ×29,000.

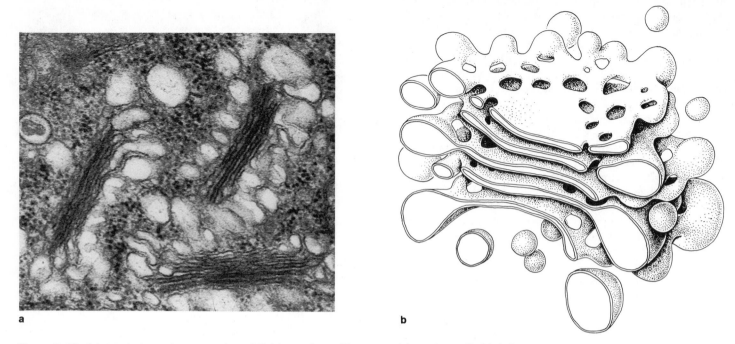

a

b

Figure 5–15 (**a**) An electron microscope view of Golgi complexes (dictyosomes) in a plant cell. (**b**) A three-dimensional drawing of a generalized plant-cell Golgi complex.

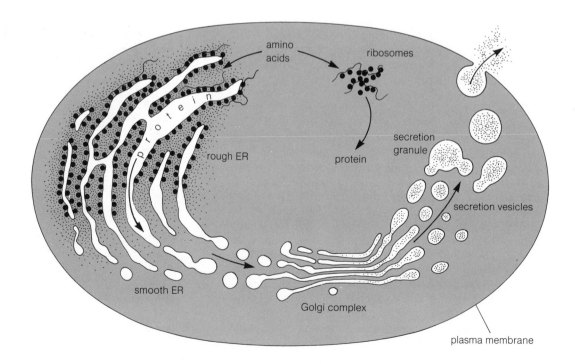

amino acids

ribosomes

protein

secretion granule

rough ER

secretion vesicles

smooth ER

Golgi complex

plasma membrane

Figure 5–16 The relationship of the ER and the Golgi complex in the secretion of materials from cell.

the Golgi complex. The Golgi complex is also concerned with secreting some types of yolk in developing eggs.

In plants, the Golgi complex is extremely important in assembling and secreting cell-wall materials. In some algae, tiny scales of wall material are assembled in the Golgi complex and secreted uniformly over the surface of the cell as the entire cytoplasm rotates within the cell wall. In higher plants, the parts of the cell wall that contain no cellulose are assembled and secreted by the Golgi complex.

Summary

In this chapter we have followed the synthesis of protein in the cell. We have seen how the information for the sequencing of amino acids in proteins is represented in the DNA of the chromatin by the arrangement of the nucleotides. This information is transmitted to the cytoplasm, where protein synthesis takes place, through the formation of messenger RNA. The sequence of nucleotides in mRNA is determined by the sequence of nucleotides in the segment of DNA on which it was formed. In the cytoplasm, the mRNA becomes associated with the ribosomes; transfer RNAs, with activated amino acids attached to them, become aligned according to the code of nucleotide triplets present in the mRNA. The amino acids are then enzymatically connected to one another and are released from the mRNA, forming a polypeptide chain or protein. Various mechanisms for the control of these processes were also discussed. The mechanisms treated in this chapter are important because of the major roles proteins play in cellular membranes and, as enzymes, in the regulation of chemical reactions.

6

Cell Reproduction: Mitosis and Meiosis

When a baby girl is born, the two ovaries in her tiny body contain all the eggs she will ever produce. The eggs are in a state of arrested development, and only after the girl reaches puberty will they begin to mature at the rate of one a month for 30 to 40 years. During this period, the average American woman will have two of these eggs fertilized. Once this has happened, the cells of the new embryo begin a series of rapid divisions that continue through the growth of the fetus, after birth, and, in the case of some cells, until death. In this type of cell division, called **mitosis,** each cell gives rise to two cells that are identical in chromosome number and genetic material. A second type of division, associated with the formation of the eggs and sperm, and called **meiosis,** reduces the chromosome number precisely in half and rearranges the genetic material.

Because it is involved in all normal cell growth and repair, mitosis occurs in all living eukaryotic organisms. It is the mechanism by which the number of chromosomes in each cell is kept constant over the millions of cell divisions necessary to form a multicellular organism. The question of how this is accomplished is one of the major concerns in our discussion of mitosis.

Meiosis takes place only in sexually reproducing organisms. In animals, meiosis is closely associated with the formation of sex cells, but in many plant groups it is not. Meiosis is the answer to the question of why the chromosome number does not double in sexually reproducing organisms each generation, even though a cell of one individual combines with the cell of another individual with an identical chromosome number. The reduction in chromosome number in meiosis is the key to understanding the stability of chromosome number over generations. How this reduction is accomplished and its genetic effects are central to our discussion of meiosis.

Strictly speaking, both mitosis and meiosis refer only to the division of the nucleus and can occur without the division of the cytoplasm of the cell. The actual division of the cytoplasm is called **cytokinesis;** in the vast majority

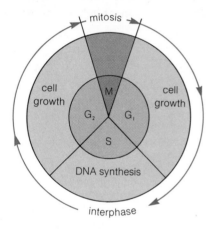

Figure 6–1   The life cycle of the cell.

of cells, it is closely coordinated with mitosis. Because of this close coordination, many biologists make mitosis synonymous with the complete process of cell division that involves both mitosis and cytokinesis.

Cell division is easier to understand if we think of the life cycle of the eukaryotic cell as having two parts (Figure 6–1): **interphase,** during which the cell grows and gets ready to divide; and mitosis (followed by cytokinesis), during which it divides—nucleus first and then cytoplasm. The same sequence of events occurs in meiosis; first interphase and then division. In the discussion of interphase below, we will concentrate on the events associated with mitosis, covering meiosis later.

Interphase

This part of cell division is easily understood if we consider for a moment what has to take place before a cell can divide. A human cell, for example, has 23 pairs of chromosomes (a total of 46) (Figure 6–2), and at the end of cell division there will be *two cells,* each complete with 23 pairs of chromosomes. The two daughter cells will be a little smaller than the parent cell was when division began, but they will have all the structure necessary for life.

Thus, before a cell can divide, it must at the very least duplicate each of its pairs of chromosomes and achieve some growth. Duplicating the chromosomes, of course, involves synthesizing DNA and proteins—both histone and nonhistone. Interphase, then, is a period of *growth* and *synthesis.* It is usually divided into "S" (for synthesis) and "G" (for gap or growth) subphases, according to whether DNA is or is not being synthesized. The usual sequence is shown in Figure 6–1: G_1, an initial period of growth only; S, a period of DNA synthesis; and G_2, a further period of growth. The G_1 stage may last hours, months, or even years. But once a cell has entered the S phase—once it has begun to make DNA—it usually enters mitosis within 9 to 12 hours. An exception is certain animal cells and the cells of higher plants, which may go through one or more S phases without dividing.

Since so many cells are locked into the division sequence at the onset of S, biologists have long been eager to learn what triggers the S phase. One of the earliest theories was that growth itself, the cell reaching a certain size, was the critical factor. This appears to be true for some systems; but cell division takes place in the early development of many animals without any net growth or increase in total mass of the embryo. Thus, other factors besides growth must be involved in initiating cell division. Experiments show that a substance diffusing from a cell already in S can induce a second nucleus to enter the S stage, and that such substances are probably involved in converting cells from G_1 to S under normal circumstances.

Before we leave the subject of interphase, it is important to emphasize that this part of cell division is more than simply the preparation for the next division. During interphase the chromatin—that is, DNA and its associated protein—is literally stretched out in the nucleus. In this condition, the DNA can be transcribed and duplicated. It is during interphase that mRNA is made and that the cell synthesizes protein, not merely for division, but for all the other activities for which that cell is genetically programmed.

The events of interphase can be summarized by emphasizing that this is the period during which the chromatin of the cell is doubled. The critical replication of the genetic material takes place during interphase. All the events of mitosis are simply means by which the replicated chromatin is sorted out into two exactly equal packages—the two new nuclei with equal numbers of chromosomes.

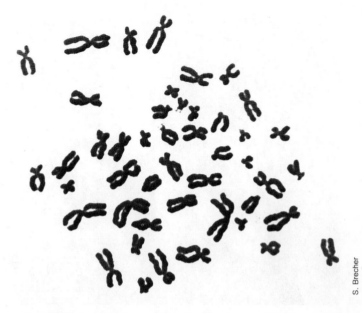

Figure 6–2 Human chromosomes. ×1,500.

Mitosis

During mitosis, the chromosomes of the cell, each with double the amount of DNA and protein it contained at the beginning of interphase, become shorter, thicker, and visible under the light microscope. The nuclear envelope disappears. The chromosomes line up on a plane that runs through the center of the nucleus. Fibers running from the chromosomes to the opposite ends of the nuclear regions appear. Each chromosome then separates longitudinally, and the halves move in opposite directions along the path of the fibers. When the chromosomes reach the end of this path, the fibers disappear and new nuclear envelopes are formed. The result is two new nuclei, each containing the same number of chromosomes and DNA as the original nucleus at the beginning of interphase.

Although cell division takes place with no significant pauses or interruptions, for convenience mitosis is usually described in four stages: **prophase** *(pro* = before), **metaphase** *(meta* = between), **anaphase** *(ana* = back), and **telophase** *(telo* = end). These stages are shown in light-micrographs from a time-lapse movie in Figure 6–3 and in diagrammatic form in Figure 6–4.

Prophase During interphase, the light microscope reveals no structures in the nucleus except a faint chromatin network, scattered blocks of aggregated chromatin, and the nucleolus. As G_2 ends and prophase begins, the nu-

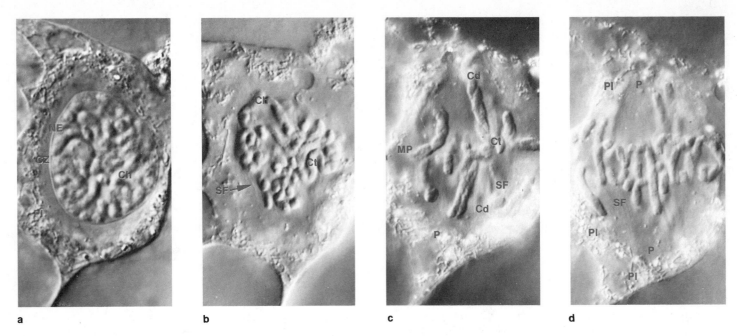

a   b   c   d

Figure 6–3 Mitosis as seen in the light microscope. These cells are from the endosperm of developing seeds and were photographed by a special optical system that gives a three-dimensional effect to the objects. (**a**) Prophase. Chromosomes (Ch) are visible, the nuclear envelope (NE) is still present, and the clear zone (CZ) can be seen with plastids (Pl) and other cell organelles excluded from it. (**b**) Transition to metaphase. Nuclear envelope no longer present, spindle fibers (SF) can be seen, and the chromosomes (Ch) are becoming aligned. (**c**) Metaphase. The centromeres (Ct) are aligned on the metaphase plate (MP), the poles (P) are well defined, and the spindle fibers (SF) are numerous. The chromatids (Cd) can be clearly seen and are unwinding. (**d**) Early anaphase. The

Essay 6–1

Schleiden and Mitosis

One of the most famous and unusual people involved in the early history of cell biology was the German Matthias Schleiden (1804–1881). Schleiden started his career as a lawyer with botany as a hobby. Before long the hobby turned into an obsession, and he devoted all his time to the study of plants. Systematics was then the major field in botany, but Schleiden wanted little to do with it. Instead he became intrigued with the microscope and began a series of studies on the microscopic structure of plants. He managed to secure the professorship of botany at the University of Jena and there wrote in 1838 "Contributions to Phytogenesis." In this work, he stated that the different parts of the plant are composed of cells or their derivatives, stressing that cells are the functional and structural unit of life. He became good friends with

Theodore Schwann, who extended the idea to animals and made the general statement:

Each cell leads a double life: an independent one, pertaining to its own development alone; and another incidental, insofar as it has become an integral part of the plant. It is, however, easy to perceive that the vital process of the individual cells must form the first, absolutely indispensable fundamental basis, both as regards vegetable physiology and comparative physiology and comparative physiology in general; and therefore, in the very first instance, this question presents itself: how does this peculiar little organism, the cell, originate?

This indeed was the question, and Schleiden thought he had the answer: The nucleus multiplies by budding, and a new cell forms around it. Considering the microscopes in use at the time, this was not an unreasonable sugges-

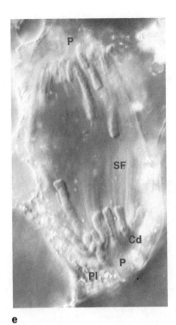

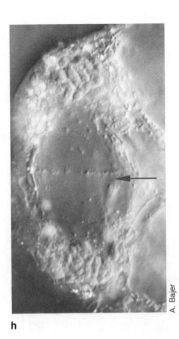

e f g h

chromosomes are moving toward the poles (P), with the centromeres (Ct) first. Spindle fibers (SF) are evident, and the plastids (Pl) and other cell organelles are clustered outside the spindle. (**e**) Late anaphase. The chromosomes, each consisting of two chromatids (Cd), have almost reached the poles (P), around which the plastids (Pl) are clustered. Spindle fibers (SF) are still evident, running from pole to pole. (**f**) Early telophase. The chromosomes have reached the poles, and movement of the chromosomes has stopped. (**g**) Mid-telophase. The chromosomes are more relaxed in appearance, and the nuclei are forming. The cell plate is also forming (arrow). (**h**) Late telophase. The forming cell plate is clearly visible (arrow), and the nuclei (N) are rapidly returning to interphase.

tion. But it was wrong. Slowly, data began to accumulate that would not fit this idea, and the outline of mitosis began to emerge. Schleiden, however, continued to believe that his observations were right and began a vicious campaign against investigators who thought otherwise. He ridiculed their observations and their conclusions, he attacked their intelligence, and when called upon to produce supporting data he would not show up at meetings. In general, he acted in a highly belligerent manner. Given the tenor of the times, it is hard to see why he was not challenged to a duel. Instead he prodded his opponents to greater efforts and more careful observations. Finally, it became clear the cells divided through the process of mitosis and that Schleiden was just plain wrong. Schleiden, as could be predicted, did not take this well, but by now he was engaged in another controversy concerning the role of the pollen tube in plant reproduction—where he was again proved wrong.

By this time, things were getting too much for Schleiden, and he abandoned the study of cells and plants. For a brief period he studied anthropology, where he voiced doubts as to the organic origin of fossils. He eventually ended up in theology and is said to have died a bitter and disillusioned man.

Schleiden is often credited with having provoked more good work by antagonizing other investigators than anyone else in biology. Schleiden himself, however, did do good work. But he did two things that no researcher should: He went beyond his observations in drawing his conclusions, and he would not recognize when contrary views were right. Finally, Schleiden did the inexcusable in science: He attacked his opponents not on scientific grounds but on personal ones. Fortunately, we seem to be past this stage in biology, but it is something that all scientists must guard against, remembering the example of Matthias Jakob Schleiden.

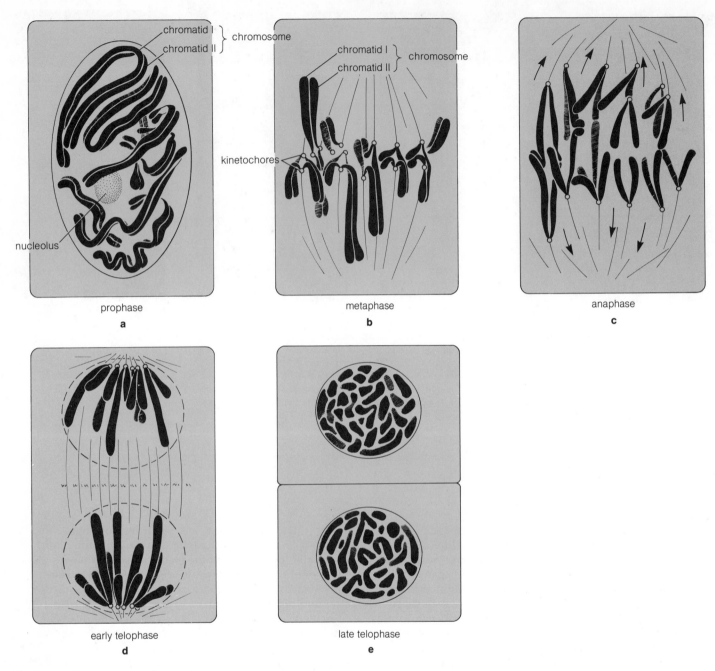

Figure 6–4 Diagram of mitosis in a plant cell.

clear chromatin condenses into threads that are at first greatly elongated and only faintly outlined. The term *mitosis* is derived from the threadlike appearance of the chromosomes as they appear at the onset of prophase *(mitos* = thread). As condensation into distinct threads becomes complete, the scattered blocks of aggregated

chromatin, if present, merge into and become part of the condensing chromosomes (Figure 6–3a).

The reason for the shortening of the chromosomes at the beginning of prophase becomes clear if we consider the differences between the transcription function of the DNA during interphase and the sorting out that takes

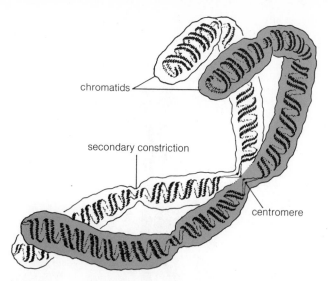

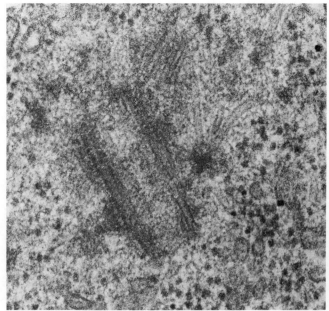

Figure 6–5 The parts of a chromosome. There are two chromatids joined at the centromere or primary constriction. An additional secondary constriction may also be present.

place in mitosis. For transcription to occur, DNA must apparently be in an extended state. But this is clearly not the best physical stage in which to separate out chromosomes and ensure that they are equally distributed to the new cells. It is rather like the difference between trying to separate a mass of long pieces of yarn and sorting a handful of short metal springs. In many nuclei, the chromosomes appear as if by magic under the microscope when mitosis begins, but they have been there all along, in an extended state in which they were below the resolving power of the microscope.

The number of chromosomes appearing in the nucleus at prophase is constant and characteristic for each species. Depending on the life history of an organism, this number is usually either the **haploid** or the **diploid** number for the species: In a haploid nucleus, only one representative of each chromosome is present; diploid nuclei contain two copies of each chromosome. The two copies of each chromosome in diploids make up a pair; these are called **homologous chromosomes,** or **homologs,** because they contain the similar genes in the same sequence along their lengths. Because higher plants and animals are diploid throughout most of their life cycles, and because mitosis follows the same pattern in both haploid and diploid organisms, mitosis is described here as it would occur in a diploid organism.

As prophase progresses, each chromosome is seen to consist of two parts lying side by side and joined at only one point (Figure 6–5). Each part is called a **chromatid,** each is a complete copy of the original chromosome, and each will become a new chromosome in one of the two daughter nuclei. The place where they are joined is called the **centromere** or *primary constriction.* The length and

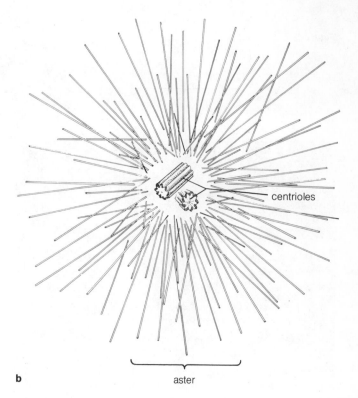

Figure 6–6 The centrioles and aster as seen in animal cells.

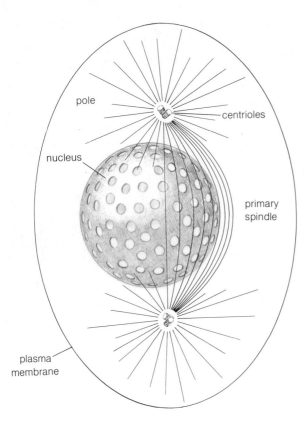

Figure 6–7 The primary spindle.

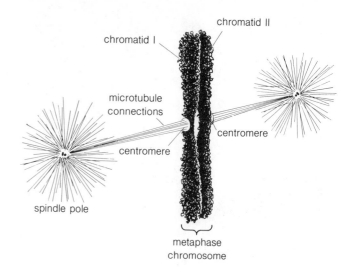

Figure 6–9 Relation of chromosome spindle fibers to the centromeres and the poles. There are also spindle fibers that run directly from pole to pole and are not attached to the chromosomes, but these continuous fibers are not shown here.

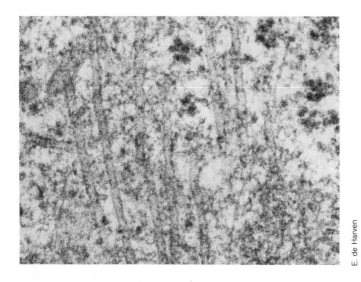

E. de Harven

Figure 6–8 Spindle microtubules. ×110,000.

shape of the chromosomes are often distinctive enough to allow the various chromosomes to be distinguished from one another. Besides differences in length and thickness, the locations of narrow regions called *constrictions* help distinguish chromosomes from one another. Each chromosome has at least one prominent narrowing that is the primary constriction or centromere, and some chromosomes also show secondary constrictions. The collective morphology of the chromosomes of an organism is known as the **karyotype** and is often distinctive enough to allow identification of a species from this information alone.

Almost all animal cells and some plant cells contain a pair of small, barrel-shaped structures in the cytoplasm just outside the nuclear envelope called **centrioles** (Figure 6–6). As the S stage of interphase begins, the centrioles are duplicated, producing two pairs of centrioles. Short lengths of microtubules surround the centrioles, and during centriole reproduction these increase in length and number, forming a starlike array called an *aster*.

During prophase, the two pairs of centrioles separate until they have reached the opposite ends of the nucleus, which are now called the *poles*. As this happens, the microtubules lengthen and fill in the path of the separating centrioles, stretching in the direction of movement. By late prophase, these microtubules form a mass that extends completely around one side of the nucleus and is known as the *primary spindle* (Figure 6–7).

Most plant cells and some animal cells lack centrioles. In these cells, a clear space becomes visible in a narrow zone surrounding the nucleus during prophase, as shown in Figure 6–3a. The clear zone is packed with microtubules extending around the nucleus from pole to pole. Absence of centrioles or asters apparently does not affect spindle formation.

Near the end of prophase the nucleolus grows smaller, and by the onset of metaphase it has usually disappeared. Also during prophase, the nuclear envelope breaks down, so there is no longer a barrier between the chromosomes and other nuclear material and the cytoplasm.

Metaphase As the nuclear envelope disappears, the primary spindle moves into the central position formerly occupied by the nucleus, while the chromosomes, left scattered in this region by the breakdown of the nucleus, move to the midpoint of the spindle, known as the metaphase or equatorial plate (Figures 6–3b and c and 6–4). As they move, the chromosomes become attached at the centromere to bundles of spindle microtubules. The individual spindle microtubules are long, unbranched structures 18 to 20 nanometers in diameter that run from the centromeres to the poles (Figures 6–8 and 6–9).

Anaphase At the beginning of anaphase (Figure 6–3d), the centromere divides and the two chromatids become completely separate structures. The chromatids, now called chromosomes, move toward the poles. This movement is clearly associated with the spindle microtubules, and tension can be seen to be exerted on the centromeres even before the chromatids separate.

The basis for the movement of chromosomes, which takes only 5 to 10 minutes to complete and which happens in close synchrony, is still not understood (see Essay 6–2).

Anaphase movement continues until the separated chromosomes have collected into two masses at the opposite spindle poles, as in Figure 6–3e. Distribution is precise because of the pattern of centromere attachments during metaphase. Because of this precise and equal division, the collections of chromosomes at the two ends of the spindle are exactly equivalent in hereditary information. Once the chromatids have completed their movement to the spindle poles, the cell begins transition from anaphase to telophase.

Telophase During telophase, the chromosomes at the poles uncoil and become indistinct under the light microscope. Segments of new nuclear envelopes (derived from the endoplasmic reticulum) appear at the borders of the uncoiling chromosomes. These segments gradually extend around the mass until each set of chromosomes, now more closely resembling interphase chromatin, is com-

pletely separated from the surrounding cytoplasm by a continuous new nuclear envelope (Figures 6–3f, g, h, 6–4, and 6–10).

The spindle starts to disappear as soon as the transition from anaphase to telophase begins. Asters, if present, become smaller until the pair of centrioles is surrounded by a limited number of relatively short microtubules. These centrioles then take up their characteristic interphase position at one side of the nucleus, just outside the nuclear envelope. All that remains of the spindle when telophase is complete is a layer of short lengths of microtubules that persists at the former midpoint of the spindle.

The entire mitotic sequence, from the onset of prophase to the close of telophase, may occupy as little as 5 to 10 minutes, as in some rapidly developing animal embryos, or may require as much as 3 hours or more in various plant and animal tissues. Of the different stages, prophase is usually the longest, telophase and metaphase the next longest, and anaphase the shortest.

Final Considerations The processes of cell division usually occur in sequence as we have outlined. However, replication, mitosis, and cytokinesis are potentially separable and in some organisms proceed independently. In the salivary glands of *Drosophila* (fruit fly) larvae, for example, chromatid replication takes place without mitosis or cytokinesis, producing large nuclei containing many copies of the basic DNA complement. Or, as happens in some insect embryos, replication and mitosis proceed for a time without cytokinesis, producing a temporarily multinucleate tissue. Eventually, rapid cytokinesis occurs in these embryos, dividing the many daughter nuclei into separate cells.

From the standpoint of results, two events of premitotic interphase and mitosis have the greatest significance for the outcome of cell division. These are the duplication of chromosomes in the S phase of premitotic interphase, and the separation of chromatids by the spindle during mitosis.

Cytokinesis

Up to this point we have been discussing the dividing cell as if it consisted of nothing but a nucleus. However, the cytoplasm of the cell must also divide if two complete daughter cells are to result from mitosis. This process is called *cytokinesis*. When it is complete, the daughter cells are totally enclosed in separate, continuous plasma membranes. Although cytokinesis differs in detail in plants and animals, spindle remnants function in both in a similar way.

Essay 6–2

Microtubules, Movement, and Mitosis

One of the characteristics of life is movement. Most animals move quite obviously, and plants show a variety of subtle movements when carefully observed. Several different underlying mechanisms account for this movement of living organisms, notably muscle contraction in animals and cell growth in plants. Mechanisms found operating on the cellular level in both plants and animals include the "streaming" of cytoplasm, the movement of chromosomes during mitosis and meiosis, and the beating of flagella, as in sperm cells. All these types of cellular movement have a common feature: All the cells involved contain microtubules and microfilaments.

Microtubules are slender tubes some 0.02 micrometers in diameter that may be several micrometers long. They are composed of proteins organized in spherical subunits capable of self-assembly to form microtubules. Microtubules were not discovered until after the invention of the electron microscope. Even then, we did not know they were found in all cells until the 1960s, when the chemicals needed to preserve them were found. Microtubules often seem to function as subcellular structural units, helping to maintain the form of the cell or giving direction to a motion.

Microfilaments are less well known than microtubules and were discovered more recently. They are much less elaborate in structure than microtubules. Microfilaments are present in cells in a variety of sizes, but most attention has focused on arrays of 5 to 8 nanometer microfilaments similar to the filaments of muscle. Their importance lies in the fact that they contain **actin**, a contractural protein that is one of the major components of muscle. Thus, microfilaments must be considered in any theory of cellular movement.

Because microtubules are important in so many cellular processes and have been so actively studied, we will examine their role in certain of these processes. Unfortunately we will say much less about microfilaments, because at present much less is known about them.

As we see in Chapter 6, the precision of mitosis depends on the efficiency of the spindle in separating the chromatids of each chromosome and moving them to the poles at anaphase. We now understand that the force and direction for anaphase movement depend on spindle microtubules, but we do not know the nature of the molecular interactions producing these forces.

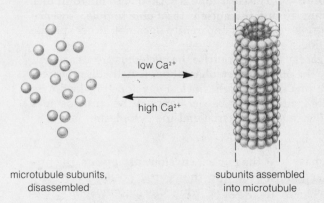

microtubule subunits, disassembled   subunits assembled into microtubule

Figure E6–1 The assembly and disassembly of microtubule subunits in high and low concentrations of calcium ions.

Spindle microtubules apparently form from a pool of cytoplasmic precursors, which turn out to be two closely related molecules of a protein called tubulin. Low temperature, pressure, ultraviolet radiation, and certain drugs can disassemble these chains of molecules and cause the spindle to disappear within seconds. But if the dispersing agent is removed, the spindle quickly reappears, usually within two or three minutes. Since inhibitors of protein synthesis do not keep it from reappearing, no new proteins are apparently required. These findings indicate that the spindle moves readily from its dispersed to its assembled form and back again. R. Weisenberg of Temple University has been able to adjust conditions so that the conversion will occur in a test tube. Weisenberg found that at high calcium concentrations the microtubules came apart, and at low ones they re-formed (Figure E6–1).

Two classes of spindle microtubules are established at metaphase. One class, called the chromosome microtubules, run from the centromeres of the chromosomes to the poles. A second class, called the continuous microtubules, are not connected to the chromosomes in any direct way, but extend from pole to pole.

The two classes of microtubules apparently act differently as mitosis proceeds. The microtubules attached to chromosomes shorten to a fraction of their metaphase length as the chromosomes are pulled to the poles. At the same time, the entire spindle made up of the continuous microtubules stretches or elongates during anaphase, increasing the distance between the poles. Thus, any model of chromosome movement must take into account the fact that one group of spindle microtubules becomes shorter and another becomes longer.

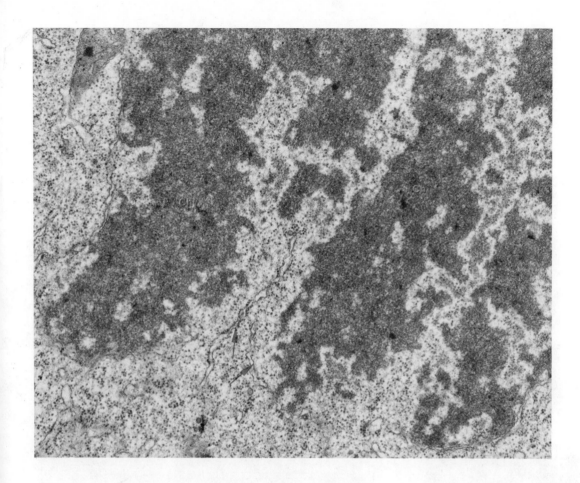

Figure 6–10 Nuclear membrane (arrows) forming around chromosomes (Ch) at poles in telophase. ×24,000.

Cytokinesis in Animals: Furrowing In animals, after anaphase is complete, most of the spindle breaks down and disappears, but short lengths of microtubules persist at the former midpoint of the spindle (and around the centrioles). The microtubules at the former spindle midpoint become surrounded with patches of dense, structureless material. Growth of the dense material continues until the layer, now called the *midbody*, extends completely across the cell (Figure 6–11).

Soon after the midbody develops, a depression or furrow appears in the plasma membrane around the outside of the cell (Figure 6–11b). This furrow sinks into the underlying cytoplasm, following the plane of the midbody, until the two daughter cells are completely pinched off and separated by continuous plasma membranes. As the deepening furrow penetrates into the cell, the midbody is compressed, becomes smaller, and usually disappears. Mitochondria, endoplasmic reticulum, Golgi complex, vesicles, and other cytoplasmic organelles are roughly divided between the daughter cells.

The way the furrow develops makes it obvious that the original position of the spindle determines the plane along which the daughter cells will become separated in cytokinesis. Ordinarily, the spindle midpoint lies at the cell equator, so that the daughter cells are of equal size. However, during egg development in certain species, the spindle is positioned at one side of the dividing cell. The furrow forms opposite the spindle midpoint as usual, cutting the cytoplasm into unequal parts. The factors governing the alignment and position of the animal-cell spindle, which determines the later plane of furrowing, remain unknown.

In cells that are unattached to their neighbors, such as some developing eggs or cells in tissue culture, the appearance of the cell surface during furrowing gives the impression that a drawstring is being tightened around the cell. This has prompted the idea that the furrow contains a contractile element that progressively squeezes off the dividing cytoplasm into two parts. This hypothesis is directly supported by electron micrographs, which re-

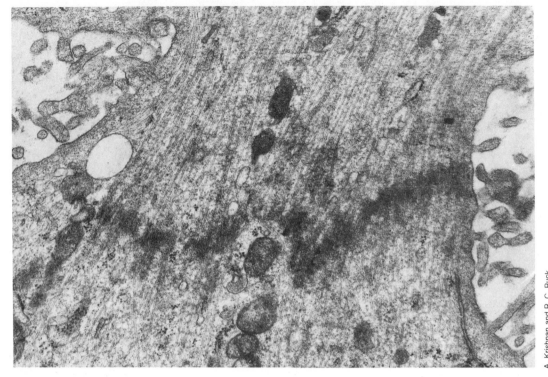

Figure 6–11 Cytokinesis in animals. (**a**) Midbody of a human cell at late telophase. ×35,000. (**b**) Formation of the furrow in a human cell. ×30,000. (**c**) Microfilaments at the edge of the advancing furrow in a dividing egg cell of a rat. ×35,000.

a

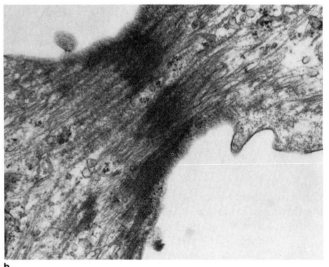

b

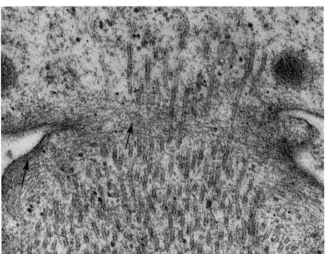

c

veal large numbers of microfilaments at the furrow edge, as shown in Figure 6–11c. These microfilaments, which are also found in many motile systems in cells, are believed to produce contractile force (see Essay 6–2).

Cytokinesis in Plants: Phragmoplast and Cell-Plate Formation The initial stages in plant cytokinesis resemble midbody formation in animals. During telophase, portions of the spindle microtubules persist at the spindle

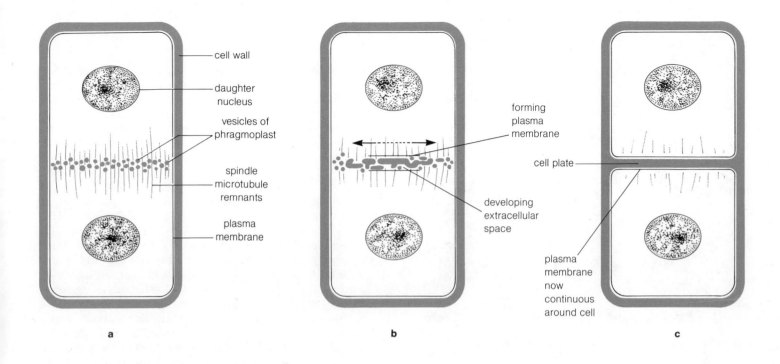

cell wall

daughter nucleus

vesicles of phragmoplast

spindle microtubule remnants

plasma membrane

forming plasma membrane

developing extracellular space

cell plate

plasma membrane now continuous around cell

a

b

c

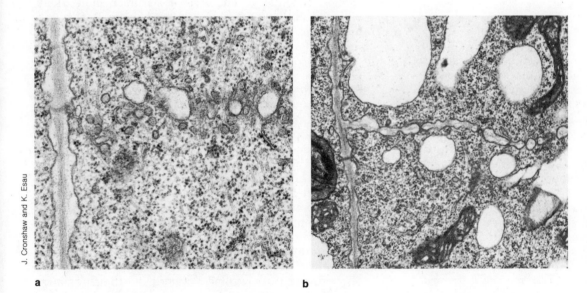

J. Cronshaw and K. Esau

a

b

Figure 6–12 Cytokinesis in plants. The diagrams illustrate major stages in phragmoplast and cell-plate formation. The electron micrographs correspond to (**a**) and (**b**) of diagram.

midpoint. These microtubules become surrounded by a layer of dense material (Figure 6–12), as in midbody formation in animals. However, much of this dense material in plants is enclosed in membrane-bound vesicles that originate from the Golgi complex. These vesicles gradually increase in number until a continuous layer extends across the equator of the cell, following the former spindle midpoint. This layer of microtubules and vesicles, the

phragmoplast, determines the position of the new wall that will separate the daughter cells.

Cell-wall formation begins in the cell interior and gradually extends outward toward the plasma membrane, in a direction opposite to furrowing in animals. This takes place as the vesicles in the central area of the phragmoplast fuse together, forming two layers of plasma membrane. As the vesicles fuse (Figure 6–12a), their electron-

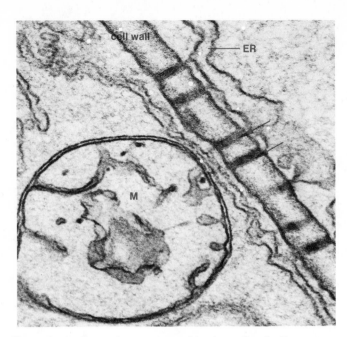

Figure 6–13 Plasmodesmata (arrows) in plant-cell walls. These structures connect the cytoplasm of adjacent cells.

dense contents extrude into the developing extracellular space between the daughter cells. Progressive fusion of the phragmoplast vesicles continues outward toward the parent cell walls (Figure 6–12b) until the daughter cells are completely separated by two continuous plasma membranes and an extracellular space. This space is then filled with a layer of dense material originating from the fused vesicles. When fully formed, the layer of dense material separating the daughter cells is called the *cell plate* (Figure 6–12c).

Once formed, the cell plate is progressively thickened and strengthened by the deposition of new cell-wall material between the membranes separating the daughter cells. At points, this new wall is perforated by cytoplasmic connections that remain intact between the daughter cells (see Figure 6–13). These narrow connections, termed **plasmodesmata,** persist and evidently serve as channels of communication between the cells of the plant tissue.

The plane of division in plant cells, as in animal cells, is determined by the position taken up by the spindle at metaphase. Usually, the spindle occupies the cell midpoint at metaphase, so that the subsequent cytokinesis divides the mitotic cell into two equal parts. In some plant tissues, however, the spindle takes up an eccentric position (as in some animal eggs), and the cytoplasmic division that follows is unequal. The forces that determine the plane of cytoplasmic division are unknown.

Meiosis

Although mitotic divisions are by far the more common in both plants and animals, meiotic divisions play an important role in the life history of most organisms. *Meiosis* comes from the Greek *meioun,* "to diminish," because during meiosis, which occurs only in diploid cells, the chromosome number of each cell is reduced by half. In humans and most animals, meiosis is directly associated with the formation of eggs and sperm. In most plants, however, meiosis is associated with the formation of spores that germinate to form a haploid plant that in turn forms eggs and sperms through mitotic divisions.

The critical point about meiosis is the reduction in chromosome number, which allows the original (diploid) number to be restored when the haploid egg and sperm unite at fertilization. But why should such an elaborate mechanism evolve in the first place? What advantage does sexual reproduction have for an organism? The answers to these questions will occupy us later in this book, when we consider genetics and evolution. For now we need only accept the notions that genetic diversity in a population may be extremely important in the survival of the species, and that genetic diversity is much more likely with sexual than with asexual reproduction. But we can understand this better when we understand meiosis.

Overview As in mitosis, meiosis is preceded by an interphase during which DNA and other necessary cell components are synthesized. Meiosis is actually two coordinated divisions, called meiosis I and meiosis II, one always following the other. During the first division the homologous chromosomes pair, and there is an exchange of parts in the paired chromatids. At anaphase of meiosis I, the centromeres of the chromosomes do not divide, so when the paired chromosomes separate, it is the homologous chromosomes that move to the poles. In consequence, each new nucleus formed during telophase I has half the number of chromosomes of the original nucleus. Meiosis I, therefore, is the reduction division of meiosis. Meiosis II is much like a regular mitotic division. At anaphase of meiosis II, the centromere divides and the chromatids separate to form two new sets of chromosomes. These chromosomes are not identical, as in the case of mitosis, because of the exchange of chromatid segments between the homologous chromosomes during meiosis I. At the end of meiosis we have, then, four nuclei with half the number of chromosomes of the nucleus entering the division, and chromosomes with altered arrangements of genes.

To appreciate what is happening during meiosis and the significance of the elaborate mechanism involved, let's look at this important division in greater detail.

Meiosis I Meiosis I begins with prophase I, which includes highly significant chromosome rearrangements. Meiotic prophase I lasts much longer than mitotic prophase in the same organism, and may in fact extend over weeks, months, or even years. Although actually more or less continuous, it is divided for convenience into five stages, which we shall name according to the activities of the chromosomes: (1) condensation, (2) pairing, (3) recombination, (4) synthesis, and (5) recondensation (the traditional names for these stages are included in the descriptions that follow). These stages are shown in the series of photographs in Figure 6–14 and in the diagrams in Figure 6–15.

Condensation is traditionally called *leptotene,* from the Greek *leptos,* "thin." This initial stage of meiotic prophase I begins as the chromosomes first condense into visible threads (Figures 6–14a and 6–15a). In the condensation mechanism, as in mitotic prophase, the 10 nanometer chromosome fibers fold down into compact structures that become thicker as the folding becomes more extensive.

Pairing is also called *zygotene,* from the Greek *zygon,* "yoke," or "joining," taken from the appearance of the chromosomes (Figures 6–14b and 6–15b). To understand this stage, it is first necessary to appreciate the fact that the homologous chromosomes are identical in size, shape, and arrangement of their genes. During this stage of meiosis I, the homologous chromosomes come together, arranging themselves in a precise manner that results in the exact pairing of the two chromosomes. The pairing mechanism, called *synapsis,* is so precise that chromosomes become aligned in exact register, gene by gene. The nature of the forces bringing homologous chromosomes together in synapsis has not been determined, and no intermolecular forces are known to operate over the relatively long distances that separate the homologs when synapsis begins.

Recombination (or *pachytene,* from the Greek *pachus,* "thick") begins as pairing of homologous chromosomes is completed (Figures 6–14c and 6–15c). During the recombination stage, folding of the individual chromatin fibers of the chromosomes continues until the chromosomes have been reduced to a fraction of their length at the beginning of prophase I. Each of the paired chromosomes consists of two chromatids, making a total of four closely associated chromatids, called a **tetrad,** in each paired structure during the recombination stage.

The critical point about recombination is that breaks occur in the tightly packed chromatids, and that these breaks are followed by fusions of the broken ends. But the ends that fuse may be from different chromatids, resulting in an exchange of segments among chromatids. Since the chromatids are linear arrangements of genes, such exchanges mean that there is now a different arrangement of genes than there was at the beginning of

meiosis. We know little of how such breaks and recombinations take place except that they are extremely precise and that some DNA synthesis is involved.

Synthesis (or *diplotene,* from the Greek *diplos,* "double") occurs next. As the recombination stage comes to a close, the separation between homologs widens. Also, the division between chromatids becomes visible in the light microscope for the first time, and each of the four chromatids can be seen individually. This loosening of the pairs signals the beginning of the synthesis stage (Figures 6–14d and 6–15d).

The homologous chromosomes become so widely separated at this stage that they almost seem to repel each other except at scattered attachment points (arrows in Figure 6–15d). On close inspection, these points prove to be regions in which two of the four chromatids cross over between homologous chromatids (see magnified circle in Figure 6–15d). These crossing places, called **crossovers** or **chiasmata** (from the Greek *chiasma,* "cross piece"), are remnants of the molecular exchanges of chromatid segments that took place during genetic recombination in the previous stage.

Recondensation (or *diakinesis,* from the Greek *dia,* "across," and *kinesis,* "motion") occurs in a gradual transition from synthesis that is difficult to identify in many species (Figures 6–14e and 6–15e). At this final and usually relatively brief stage of meiotic prophase I, the chromosomes, if uncoiled during the previous stage, condense again into tightly coiled, dense structures. Any chiasmata present are held together only by the remnants at the tips of the chromosomes (arrows in Figure 6–15d). This movement of chiasmata to the tips of the chromosomes is probably a mechanical consequence of the increased coiling at this time. In some organisms, the chromosomes coil so tightly that they become almost spherical.

During this final stage of prophase, the spindle for the first meiotic division forms much as it does in mitosis. The centrioles, if present, have replicated during the previous interphase, and the two pairs, within the asters, migrate to opposite sides of the nucleus to form the spindle poles just as in mitosis. This pattern is most typical of the meiotic divisions leading to sperm formation in male animals. In animal eggs (those of insects and many vertebrates), the centrioles disappear and cannot be detected in developing eggs at any stage of meiosis. In these eggs, the spindle forms as it does in plants without centrioles.

Just as in mitosis, breakdown of the nuclear envelope in meiosis I provides a convenient reference point for the transition from prophase I to metaphase I (Figures 6–14f and 6–15f). The spindle moves to the position formerly occupied by the nucleus, and the tetrads, left scattered at positions around the spindle by the breakdown of the nuclear envelope, make their way to the spindle midpoint. Except for the pairing of homologous chromo-

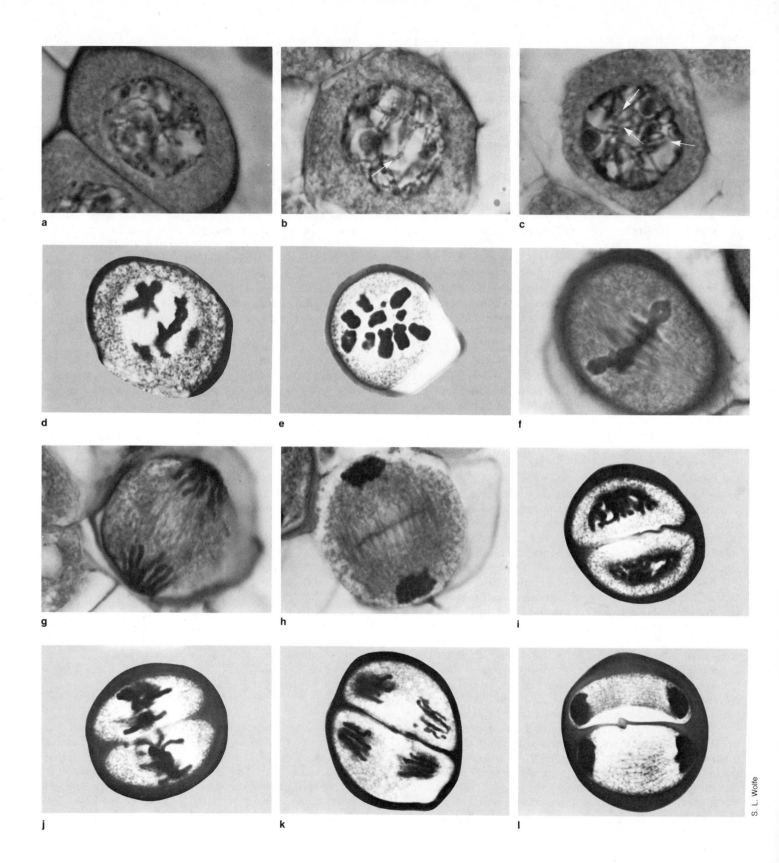

a

b

c

d

e

f

g

h

i

j

k

l

S. L. Wolfe

114 Two The Cell

Figure 6–14 Meiosis in the lily. (**a**) Condensation stage. (**b**) Pairing stage. Pairing is in progress at the points marked by arrows. (**c**) Recombination. The narrow separation between the paired homologs is just visible at some points (arrows). (**d**) Synthesis. In the lily, uncoiling during the synthesis stage is not pronounced. (**e**) Recondensation. (**f**) Metaphase I. (**g**) Anaphase I. (**h**) Telophase I. (**i**) Prophase II. (**j**) Metaphase II. (**k**) Anaphase II. (**l**) Telophase II. ×1,000.

somes and maintenance of the tetrads, meiosis at this point closely matches the transition from prophase to metaphase in mitosis.

The major change from the pattern observed in mitosis, and in fact the distinctive event of the meiotic divisions following prophase I, occurs as the tetrads attach to the spindle microtubules. Each of the four has a centromere. Recent electron micrographs have shown that in meiosis the two centromeres of a homologous chromosome connect only to one pole of the spindle; the centromeres of the other chromosome of the pair make microtubule connections leading to the opposite pole (Figure 6–16). Thus, the centromeres of each homologous chromosome function as a single unit at metaphase I of meiosis.

Anaphase I of meiosis (Figures 6–14g and 6–15g) begins as the chromosomes are pulled apart by the spindle microtubules and start their movement to opposite ends of the cell. The connections made by the centromeres during metaphase ensure that homologous chromosomes are separated at this point. Any remaining chiasmata are pulled loose as movement to the poles begins. As a result of the centromere connections, the poles at the completion of anaphase I contain the *haploid* number of chromosomes. Because each of these chromosomes still contains two chromatids (as a result of interphase replications), the same amount of DNA is present at each pole as would normally be there at the end of mitosis. *Haploid,* then, refers to the number of chromosomes rather than the mass of chromatin present.

A well-defined telophase I (Figures 6–14h and 6–15h) does not always intercede between anaphase of meiosis I and the second meiotic division. All possible gradations are found in nature, from organisms with no detectable uncoiling of the chromosomes at telophase (some insects), to species with almost complete reversion to a state like interphase. In the latter organisms (corn and some other plants), the chromosomes partly uncoil and a nuclear envelope temporarily surrounds the polar masses of chromatin. However, in most species telophase I is a transitory stage between the first and second meiotic divisions in which the meiotic cells pause only briefly before entering prophase of meiosis II. No DNA replication occurs at this stage in any known organism, since the duplicated chromatids are still present in each of the polar masses.

During telophase I, the single metaphase spindle is reorganized into two spindles that form in the regions of the telophase I spindle poles. Centrioles and asters, if present, also divide at this time, placing a single centriole at each pole of the two spindles. These events complete the cellular rearrangements for the second meiotic division.

Meiosis II The second meiotic division follows essentially the same pattern as an ordinary mitotic division. After a brief or even nonexistent interphase II and prophase II, the chromosomes left at the two poles by the first meiotic division move to the midpoints of the two newly formed spindles (Figures 6–14i and 6–15i). If uncoiling has occurred during telophase I or interphase II, the chromosomes condense back into tightly coiled rodlets as they move. The coiled chromosomes attach themselves to microtubules of the metaphase II spindle (Figures 6–14j and 6–15j) exactly as in mitotic metaphase. Each chromosome at this stage contains two chromatids, where centromeres attach to opposite poles of the spindle, as in mitosis.

At anaphase II (Figures 6–14k and 6–15k), the two chromatids of each chromosome separate and move to opposite poles of the spindle. This separation and movement delivers the haploid number of chromatids to each pole of the spindle; as a result, each pole contains one-fourth of the DNA present at G_2 in the original cell entering the two sequential meiotic divisions.

At telophase II (Figures 6–14l and 6–15l), the polar masses of chromosomes uncoil and are gradually surrounded by continuous nuclear envelopes. The four nuclei formed have widely divergent fates in various species of plants and animals. In the males of animal species, all four nuclei are enclosed in separate cells by cytoplasmic divisions and differentiate into functional *gametes,* or sex cells—the sperm. In females, only one of the nuclei resulting from meiosis becomes functional as the egg nucleus. The other three are compartmented by unequal division of the egg cytoplasm into small, nonfunctional cells called *polar bodies.* This unequal division concentrates most of the cytoplasm into the single large egg cell and saves the majority of the nutrient-rich cytoplasm for the functional egg.

The fate of the products of meiosis in plants is not as direct as in most animals. Frequently, the haploid cells do not function as gametes, dividing mitotically instead to give rise to haploid plants. Eventually, gametes are

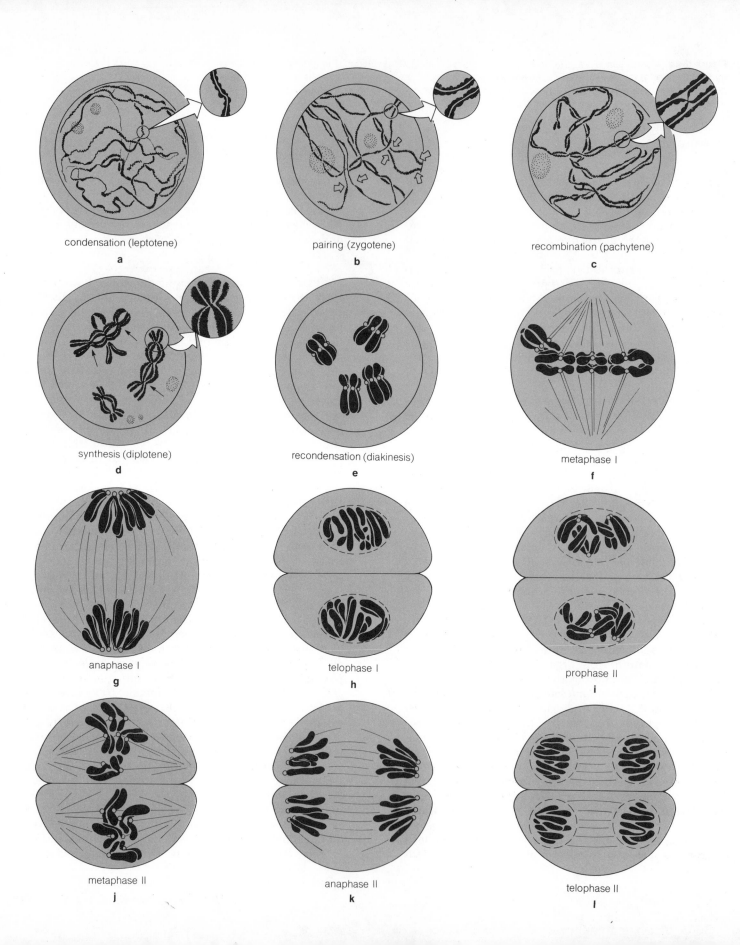

condensation (leptotene)

a

pairing (zygotene)

b

recombination (pachytene)

c

synthesis (diplotene)

d

recondensation (diakinesis)

e

metaphase I

f

anaphase I

g

telophase I

h

prophase II

i

metaphase II

j

anaphase II

k

telophase II

l

Figure 6–15 Meiosis in diagrammatic form. Although both chromatids are shown for diagrammatic purposes in the magnified circles of diagrams (**a**), (**b**), and (**c**), the split between the two chromatids of a chromosome is not actually visible during these stages. (**a**) Condensation stage. (**b**) Pairing stage. Pairing is in progress at several points (arrows). (**c**) Recombination. (**d**) Synthesis. The chromosomes are held in pairs only by crossover points (arrows) remaining from the condensation stage. In most organisms, all four chromatids usually become visible in the tetrads at this time. (**e**) Recondensation. (**f**) Metaphase I. (**g**) Anaphase I. (**h**) Telophase I. (**i**) Prophase II. (**j**) Metaphase II. (**k**) Anaphase II. (**l**) Telophase II.

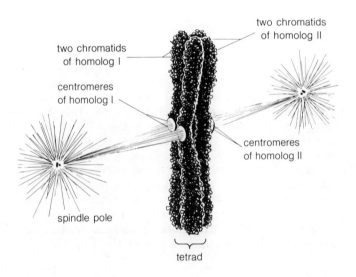

Figure 6–16 Attachment of spindle fibers to centromeres of the homologous chromosomes (tetrads) at metaphase I. (Compare with Figure 6–10.)

formed by mitotic divisions of specialized cells of these haploid plants. Thus, in plants there may be a complete haploid generation between the diploid generations, instead of a few cells as in animals.

Genetic Importance of Meiosis

The baby girl born at the beginning of this chapter will resemble her mother and father, as well as her brothers and sisters, but she will be different from all of them. She and her siblings both received one set of chromosomes from their father and one set from their mother; yet they are not identical because of the meiotic divisions that gave rise to the eggs produced by her mother and the sperm produced by her father. The girl may differ from her brothers and sisters in eye color, hair color, blood type, and a host of other more or less obvious characteristics.

Each of these characteristics is determined by a gene carried on a chromosome; and, since she has two sets of chromosomes, her cell nuclei each contain two copies of each gene for each character. But there are more than two kinds of hair and eye color, and there are four distinct blood types in the human population. This means that there are different versions of any given gene, and these are called **alleles.** Only one or two alleles of a gene are present on the two homologous chromosomes of a single individual. However, many individuals of a species, if taken together, may possess several different versions of the same gene, just as in the case of our baby girl and her brothers and sisters.

The fusion of haploid gamete nuclei to produce a *zygote* nucleus at fertilization, as in the case of our baby girl, brings the pairs of homologous chromosomes together: one set from the egg nucleus that is maternal in origin and one set from the sperm that is paternal in origin.

Independent Assortment of Chromosomes In meiosis, these tetrads come together and synapse during prophase I. At metaphase I, the pairs line up on the spindle and make attachments to microtubules of the spindle as shown in Figure 6–17. These attachments are made in such a way that the homologous pairs are separated at anaphase I and move to opposite poles of the spindle. As a result, for any *one* pair of homologs, one pole receives a chromosome of paternal origin, and the opposite pole receives a chromosome of maternal origin. The different pairs separate independently from each other at anaphase I, and not all paternal chromosomes go to the same pole. In fact, although the pairs are equally divided, any combination of chromosomes of maternal and paternal origin may be delivered to a given pole. After the second meiotic division, then, the gamete nuclei may receive any possible combination of chromatids of maternal and paternal origin.

This feature of meiosis is one of the mechanisms producing variability in the gametes. Although both members of any homologous pair have the same genes at corresponding points, different alleles of these genes may be present. The homologs, therefore, may differ in

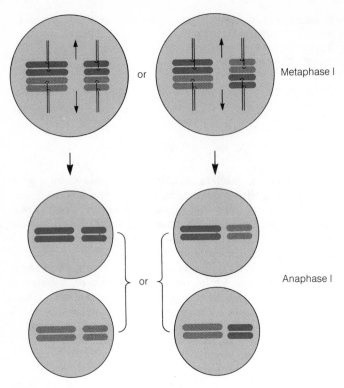

or — Metaphase I

or — Anaphase I

Figure 6–17 Independent assortment of chromosomes at meiotic anaphase I.

genetic information, and the different combinations of paternal and maternal chromosomes at the spindle poles at anaphase II result in genetic variability between the gametes.

The probability that any two gametes will receive the same combinations of maternal and paternal chromosomes is low and decreases with increasing numbers of chromosome pairs. For example, in humans, with 23 pairs of chromosomes, the probability that any two sperm or egg cells will receive the same combination of maternal and paternal chromosomes is one in $1/(2^{23})^2 = 1/2^{46}$— or about 1 in 70 trillion!

Recombination The independent assortment of chromosomes at meiosis I shuffles the maternal and paternal chromosomes, but it does not change the relationship of the genes on a particular chromosome. This occurs during the recombination stage of meiosis I. Consider a pair of homologous chromosomes containing the two genes A and B (Figure 6–18a). Gene A has two alleles, A and a; gene B has the alleles B and b. After replication and pairing (Figure 6–18 b and c), the chromosomes are as shown in Figure 6–18d when recombination (e) takes place. During the two meiotic divisions, the four chromatids are separated; at the close of meiosis each of the four resulting nuclei contains one of the four chromo-

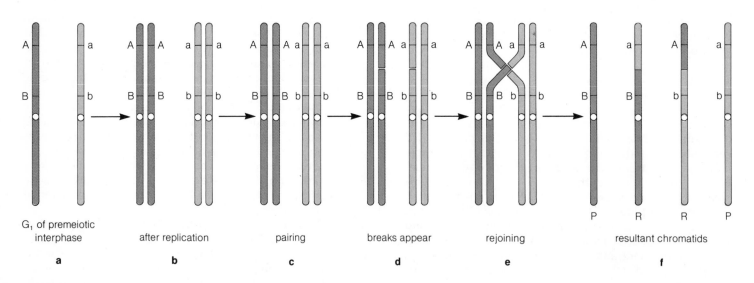

G₁ of premeiotic interphase	after replication	pairing	breaks appear	rejoining	resultant chromatids
a	b	c	d	e	f

Figure 6–18 Results of crossing over on the distribution of genes on homologous chromosomes.

somes shown in Figure 6–18f. Two are identical to the original parent chromosomes: *AB,* and *ab.* Two have the new combinations *aB* and *Ab.*

Recombination and independent assortment create a high degree of genetic variability, but within a framework of basic genetic stability. Thus, although we are not identical to our sisters, brothers, or parents, neither are we wildly different from them. This type of variability is of great importance in evolution.

Summary

Each organism has a specific number of chromosomes in its nuclei. As the organism grows and develops, each cell divides many times, yet the number of chromosomes per nucleus remains the same. The reason for this constancy of chromosome number is the remarkably precise method of chromosome replication and separation known as mitosis. The DNA and protein of the chromosomes are doubled before mitosis actually begins. Mitosis itself is marked by the shortening of the chromosomes, their alignment in the center of the nuclear region shortly after the disappearance of the nuclear membrane, the formation of a spindle, the division of the centromeres, the movement of the chromosomes to the poles, and the reforming of the nuclear membrane.

During sexual reproduction, two nuclei with the same number of chromosomes fuse. The chromosome number does not double each generation because of the process of meiosis. During meiosis, the number of chromosomes per nucleus is reduced by one half. Meiosis is actually two divisions: The first results in the halving of the chromosome number, while the second is much like mitosis. In meiosis there is independent assortment of the chromosomes and recombination of the genes.

Suggested Readings

Books

Baker, J. W., and G. E. Allen. 1974. Matter, Energy and Life, 3rd ed. Addison-Wesley, Reading, Mass.

Branton, D., and D. W. Deamer. 1972. Membrane Structure. Springer-Verlag, New York.

Calvin, M., and J. A. Bassham. 1962. The Photosynthesis of Carbon Compounds. W. A. Benjamin, Menlo Park, Calif.

Dickerson, R. E., and I. Geis. 1969. The Structure and Action of Proteins. Harper & Row, New York.

Dupraw, E. J. 1968. Cell and Molecular Biology. Academic Press, New York.

Ingram, V. M. 1972. The Biosynthesis of Macromolecules, 2nd ed. W. A. Benjamin, Menlo Park, Calif.

Jensen, W. A., and R. B. Park. 1967. Cell Ultrastructure. Wadsworth, Belmont, Calif.

Jensen, W. A. 1970. The Plant Cell, 2nd ed. Wadsworth, Belmont, Calif.

Kirk, J. T. O., and R. A. E. Tilney-Bassett. 1967. The Plastids. W. H. Freeman, San Francisco.

Ledbetter, M. C., and K. R. Porter. 1970. Introduction to the Fine Structure of Plant Cells. Springer-Verlag, New York.

Lehninger, A. L. 1975. Biochemistry, 2nd ed. Worth, New York.

Lehninger, A. L. 1971. Bioenergetics: The Molecular Basis of Biological Energy Transformation, 2nd ed. W. A. Benjamin, Menlo Park, Calif.

Lehninger, A. L. 1964. The Mitochondrion: Molecular Basis of Structure and Function. W. A. Benjamin, Menlo Park, Calif.

Loewy, A. G., and P. Siekevitz. 1970. Cell Structure and Function, 2nd ed. Holt, Rinehart and Winston, New York.

Mazia, D. 1961. Mitosis and the physiology of cell division. In J. Brachet and A. E. Mirsky, eds. The Cell, vol. 3. Academic Press, New York. Pp. 77–412.

Miller, G. T., Jr. 1971. Energetics, Kinetics, and Life: An Ecological Approach. Wadsworth, Belmont, Calif.

Novikoff, A. B., and E. Holtzman. 1970. Cells and Organelles. Holt, Rinehart and Winston, New York.

Porter, Keith R., and Mary A. Bonneville. 1968. An Introduction to the Fine Structure of Cells and Tissues, 3rd ed. Lea and Febiger, Philadelphia,

Rabinowitch, E., and Govindjee. 1969. Photosynthesis. Wiley, New York.

Rhoades, M. M. 1961. Meiosis. In J. Brachet and A. E. Mirsky, eds. The Cell, vol 3. Academic Press, New York. Pp 1–75.

Schrader, F. 1953. Mitosis. Columbia University Press, New York.

Swanson, C. P. 1969. The Cell, 3rd. ed. Prentice-Hall, Englewood Cliffs, N.J.

Troughton, J., and L. A. Donaldson. 1972. Probing Plant Structure. A. H. and A. W. Reed Ltd, Auckland, New Zealand.

Watson, J. D. 1968. The Double Helix. Atheneum, New York.

Watson, J. D. 1976. Molecular Biology of the Gene, 3rd ed. W. A. Benjamin, Menlo Park, Calif.

White, E. H. 1970. Chemical Background for the Biological Sciences, 2nd ed. Prentice-Hall, Englewood Cliffs, N.J.

Wolfe, S. L. 1972. Biology of the Cell. Wadsworth, Belmont, Calif.

Zelitch, I. 1971. Photosynthesis, Photorespiration and Plant Production. Academic Press, New York.

Articles

Allison, A. 1967. Lysosomes and disease. Scientific American 217(5):62–72.

Arnon, D. I. 1960. The role of light in photosynthesis. Scientific American 203(5):104–118.

Bassham, J. A. 1962. The path of carbon in photosynthesis. Scientific American 206(6):88–100.

Bjorkman, O., and J. Berry. 1973. High-efficiency photosynthesis. Scientific American 229(4):80–92.

Dickerson, R. E. 1972. The structure and history of the ancient protein. Scientific American 226(4):58–75.

Eisenberg, M., and S. McLaughlin. 1976. Lipid bilayers as models of biological membranes. BioScience 26:436–445.

Engelman, D. M., and P. B. Moore. 1976. Neutron-scattering studies of the ribosome. Scientific American 235(4):44–54.

Everhart, T. E., and T. L. Hayes. 1972. The scanning electron microscope. Scientific American 226(4):54–57.

Frieden, E. 1972. The chemical elements of life. Scientific American 227(1):52–60.

Fox, C. F. 1972. The structure of cell membranes. Scientific American 226(4):30–36.

Goodenough, U. W., and R. P. Levine. 1970. The genetic activity of mitochondria and chloroplasts. Scientific American 223(6):22–30.

Govindjee, and R. Govindjee. 1974. The absorption of light in photosynthesis. Scientific American, 231(6):68–82.

Green, D. E. 1964. The mitochondrion. Scientific American 210(1):63–78.

Kornberg, A. 1968. The synthesis of DNA. Scientific American 219(4):64–79.

Koshland, D. E., Jr. 1973. Protein shape and biological control. Scientific American 229(4):52–64.

Kurland, C. G. 1970. Ribosome structure and function emergent. Science 169(10):1171–1177.

Levine, R. P. 1969. The mechanism of photosynthesis. Scientific American 221(6):58–71.

Mirsky, A. E. 1968. The discovery of DNA. Scientific American 218(6):78–88.

Nomura, M. 1969. Ribosomes. Scientific American 221(4):28–35.

Pickett-Heaps, J. 1976. Cell division in eucaryotic algae. BioScience 26(7):445–450.

Satir, B. 1975. The final steps in secretion. Scientific American 233(4):28–37.

Satir, P. 1974. How cilia move. Scientific American 231(4):44–52.

Stein, G. S., J. S. Stein, and L. J. Kleinsmith. 1975. Chromosomal proteins and gene regulation. Scientific American 232(2):46–57.

THREE

Genetics

7

Inheritance and Genes

Until the late 1400s, the House of Hapsburg was a relatively small, unimportant dynasty in Austria. Then, through a series of brilliantly planned marriages, the Hapsburgs acquired more and more land and power until, at their zenith under Charles V, they controlled Austria, Bohemia, Hungary, the Netherlands, Luxembourg, Burgundy, Spain, Naples, Sicily, and Sardinia—as well as New World possessions. The success of these marriages was the subject of a famous hexameter: "Let others wage wars; you, fortunate Austria, marry." The Hapsburgs were well aware of the source of their power and strove to consolidate it through intermarriages within the family. The result of the increasingly close intermarriages was the appearance of physical and mental defects that brought the male line of Charles V to extinction. Before this happened, however, Spain was subjected to a series of mad and half-mad kings that materially aided in the decline of that once great power. On a less consequential level, the intermarriages tended to preserve the famous "Hapsburg lip," in which the lower jaw and lip were prominent.

The Hapsburg lip was an inherited feature that marked the family. It was a pronounced example of what we all have in common, namely, features and coloring obviously related to those of our parents. We say we have our mother's hair or our father's eyes, and everyone knows what we mean.

But how are such characteristics passed on from parent to child? Our common experiences make it seem only logical to explain heredity as a blending of traits in the offspring through the mixing of parental blood. This is the concept expressed when we hear someone talk of "mixed blood," or "half-breeds."

Until about 1900, such ideas were also common among scientists working on the question of inheritance, but then the work of a remarkable man named Gregor Mendel was rediscovered and gave impetus to a new and vigorous examination of the question. Mendel's research established that characteristics are passed on in the form of hereditary units, the genes. Mendel published his work in 1865, but his conclusions were so advanced for his time that 30 years were to pass before his paper was rediscovered by researchers who were already coming to similar conclusions. Because work had now progressed far enough to allow full appreciation of Mendel's findings, his work inspired a burst of interest and work in genetics, as the new field came to be called, that continues to the present day.

In Chapter 6 we defined the gene as that segment of DNA necessary to code a polypeptide chain and discussed the relationship between genes and enzymes. In this chapter we will describe the basic work that established our present understanding of heredity, concentrating on the transmission and hereditary activity of genes as whole units. This is known as *Mendelian, classical,* or *transmission* genetics.

Mendel's Garden Peas: The Beginnings of Genetics

It is rare in the history of science that development of a major area of study can be traced largely to the work of one man. However, this is the case with the science of genetics and Gregor Mendel. As a result, most of our discussion in this chapter of genes as units of heredity will center on Mendel's research and the modifications of his theories that came about through later work.

Gregor Mendel was an Austrian priest who lived and worked in a monastery in Brunn (now a part of Czechoslovakia). Mendel had had an unusual education for a priest. He had studied mathematics, physics, chemistry, zoology, and botany at the University of Vienna, numbering among his teachers some of the foremost scientists of his day. Mendel got started in his work through curiosity about the manner in which hereditary characters are transmitted from parents to offspring. He chose the ordinary garden pea for his research because pea plants could be grown easily at the monastery without elaborate equipment, and because many pea varieties were known that passed their characteristics on without change from one generation to the next. Mendel was able to order more than 30 of these true-breeding varieties from commercial suppliers, eventually using seven of them in his experiments (Figure 7–1).

His success in analyzing the transmission of hereditary traits was partly based on a good choice of experimental organism. But more than this, Mendel, because of his natural inclination and his training, analyzed his results in statistical terms, a radical departure for his time. By using statistics, he was able to describe the characteristics of the unseen hereditary determiners passed on in the peas and, incidentally, to establish the basic methods to be used in all future work in genetics.

Mendel's Experiments Among the inherited traits Mendel selected for study was a characteristic affecting the shape of seeds. One variety of pea produced pods with round, smooth seeds; another, pods with wrinkled seeds. If these lines were self-pollinated, they bred true. That is, if plants of the variety that bore wrinkled seeds were pollinated only with pollen from their own variety, all the offspring would bear wrinkled seeds; the same was true of the smooth-seeded variety. Mendel wondered what would happen if the two varieties of pea plants were crossed. Would the two traits blend evenly, producing an intermediate plant with lightly wrinkled seeds, or not?

Mendel decided to try the cross. Taking pollen from a plant of a variety that produced round, smooth seeds, he placed it on the flowers of a variety that produced wrinkled seeds. He also made the reverse cross. The first generation of offspring (called the F_1, for "first *filial* generation," from the Latin *filius*, "son") all had round, smooth seeds. Mendel then crossed some of these F_1 plants among themselves. Among the offspring of this cross, called the F_2 generation, Mendel counted 5,474 plants with smooth seeds and 1,850 with wrinkled seeds, in an approximate ratio of three to one, or about 75 percent round-seeded and 25 percent wrinkled-seeded plants.

He tested six other pairs of traits in the same way (Table 7–1). In all cases, all the F_1 plants looked as if they had inherited just one of the two traits; but in the F_2 generation both traits were always evident, in the same approximate ratio of 3:1, with the character evident in the F_1 generation always about three times as prevalent as the other. The statistical picture was totally unlike what would be seen with an even blending of traits.

Mendel's interpretation of these results was nothing short of brilliant. He saw that the results could be ex-

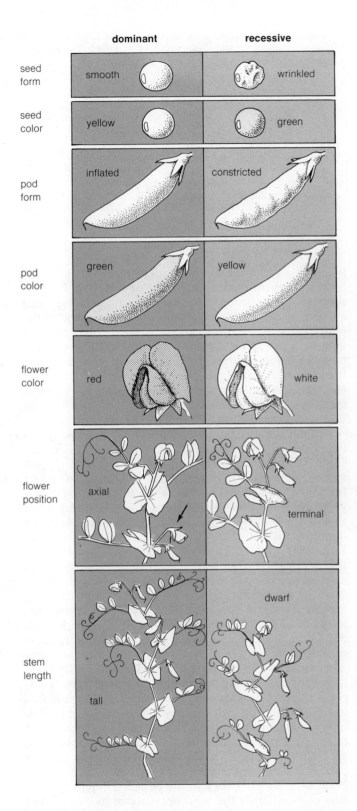

Figure 7–1 A diagrammatic summary of the characteristics Mendel selected for his experiments.

Table 7–1 Results of Mendel's Crosses Using Seven Different Characters in Peas		
Character	F₁	F₂
Round × wrinkled seeds	All round	5,474 round; 1,850 wrinkled
Yellow × green seeds	All yellow	6,022 yellow; 2,001 green
Red × white flowers	All red	705 red; 224 white
Green × yellow pods	All green	428 green; 152 yellow
Large × small pods	All large	882 large; 299 small
Tall × short plants	All tall	787 tall; 277 short
Axial × terminal flowers	All axial	651 axial; 207 terminal

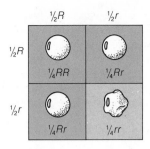

Figure 7–2 The results of Mendel's cross between plants with round and wrinkled seeds in the F₂ generation.

plained by assuming that each adult plant carries a *pair of factors* governing the inheritance of each trait. He proposed that the pair separates as gametes (eggs and sperms) are formed, so that each gamete receives only one of the two factors. At fertilization, the maternal and paternal gametes fuse, and the resulting zygote nucleus receives one factor for the trait from the sperm and one from the egg, once again a pair. Mendel's conclusion that gametes must contain only one member of each pair of factors governing a trait is called *Mendel's first law,* or the *principle of segregation.*

When Mendel found wrinkled seeds reappearing in the F₁ generation, he reasoned that the factor for wrinkled seeds, although undetected in the F₁, was still present and merely masked or dominated by the "stronger" factor for round, smooth seeds. He called this stronger factor *dominant* and its weaker counterpart *recessive,* and stated that when both were present in an organism only the dominant factor had effect. The recessive factor for wrinkled seeds could have effect (be expressed) only when the dominant factor was absent—that is, when both members of the pair were wrinkled-seed factors.

Today, we call the factors Mendel discovered *genes,* and we designate dominant genes by capital letters and recessive genes by small letters, just as Mendel himself did. Often the letters are logical abbreviations of the traits controlled by the genes, as *R* for the round-seeded gene and *r* for the wrinkled-seeded gene. This gives us a shorthand way of referring to an organism's genes. For example, a pea plant possessing a gene pair with identical factors for round, smooth seeds would be designated *RR,* whereas a plant that bore wrinkled seeds and thus must have two genes for wrinkled seeds would be designated *rr.* Both these plants, having the same genes in a given pair, are called **homozygotes.** If they had different genes (such as an *Rr* pair), they would be called **heterozygotes.** It is also clear that the way an organism looks may not reflect its genetic makeup. For example, a pea plant with round, smooth seeds may be either *RR* or *Rr,* where *R*

is dominant. We now speak of the way an organism appears as its **phenotype,** and of its genetic constitution as its **genotype.**

Mendel's experiment with smooth- and wrinkled-seeded peas is illustrated in Figure 7–2. The genes are separated at the time of gamete formation and then can be recombined as shown. Granted Mendel's assumptions—that the hereditary factors occur in pairs, that one is dominant over the other, and that the pairs are separated during gamete formation—his data are easily understood.

Mendel's assumptions also explained the proportions of different types of offspring observed in the F₂ generation. To produce an *RR* zygote, for example, two *R* gametes must be selected from among the available gametes and combined. The likelihood of this is predicted by a simple law of probability: The chance of two independent events occurring together is determined by multiplying their individual probabilities. When two F₁ parents are crossed (each of *Rr* genotype), the chance is ½ that the offspring will receive the *R* gene from one parent and ½ that the offspring will receive the *R* gene from the second parent. Therefore, according to the law of probability, the chance that the offspring will receive *R* genes from both parents is ½ × ½, or ¼. Thus, one-quarter of the F₂ offspring of the F₁ cross *Rr × Rr* would be expected to be *RR,* with round, smooth seeds. By the same line of reasoning, a quarter of these same offspring will be *rr,* with wrinkled seeds. Figure 7–2 shows that there are two ways in which a ¼ probability of *Rr* zygotes can occur. When we add these two probabilities, we see that ½ of the F₂ offspring will have the genotype *Rr.*

The three types of gene combinations, or genotypes, produce only two phenotypes, because one of the two genes, as we have seen, is dominant over the other. Thus, both the *Rr* half and the *RR* quarter of the generation—a total of three-quarters—will have round, smooth seeds, and only one-quarter will have wrinkled seeds, a ratio

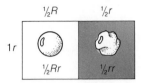

Figure 7-3 The results of Mendel's backcross, proving his hypothesis.

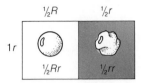

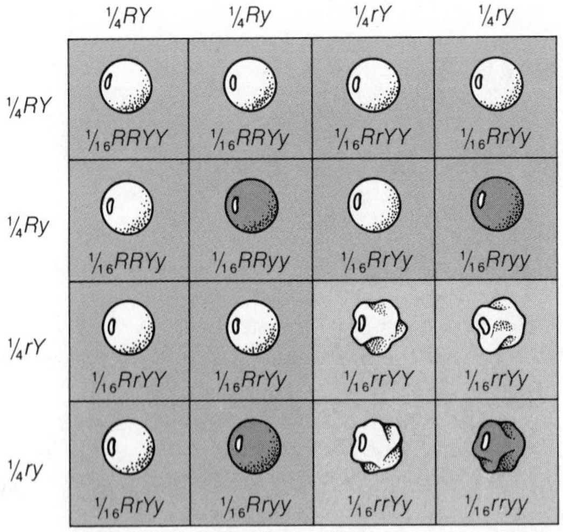

Figure 7-4 The results of crossing F_1 *RrYy* plants.

of 3:1. As Table 7-1 shows, this is the approximate ratio Mendel obtained in this cross (and for all the other factor pairs used in his experiments).

Mendel's Experimental Proofs Mendel saw clearly that the best way to test his assumptions was to make further crosses and see how well their outcomes matched his predictions. Thus, Mendel crossed an F_1 plant with round, smooth seeds, assumed to be *Rr* (a hybrid), with a wrinkled-seeded plant of the original parental type, assumed to be *rr* (a homozygous recessive). We can visualize the predicted outcome of this *Rr* × *rr* cross by putting it on a grid devised for the purpose by the geneticist R. C. Punnett, called a Punnett square. According to the hypothesis, all the gametes of the *rr* plant will contain a single *r* factor. Therefore, the probability of selecting an *r* gamete from this parent should be 1 (that is, certain). The gamete and its probability are entered in the left heading in the Punnett square in Figure 7-3. The *Rr* parent will produce two types of gametes, ½ containing the *R* factor and ½ the *r* factor. These are entered along the top heading of the square. When the possible combinations are completed, the two classes *Rr* and *rr* are obtained, both with the expected frequency of ½.

Therefore, in the predicted outcome of this type of cross (hybrid × homozygous recessive), called a *testcross* or *backcross,* half of the offspring should have round seeds and half, wrinkled seeds—a 1:1 ratio. Mendel performed this testcross, and his results were 49 plants with round seeds and 47 plants with wrinkled seeds.

Dihybrid Crosses All of Mendel's initial crosses, outlined in Table 7-1, involved single pairs of factors, each pair determining a single trait such as seed form or seed color. Crosses of this type are called *monohybrid* crosses. Mendel was also curious about the effects that might be obtained by crossing parental stocks exhibiting differences in *two* sets of hereditary traits controlled by separate and unrelated factors. For a cross of this type, called a *dihybrid* cross, Mendel chose the factors controlling seed shape and seed color.

In Mendel's experiment, plants having round, yellow seeds were crossed with plants having wrinkled, green seeds. All of the F_1 generation had round, yellow seeds. Crossing individuals of this F_1 generation among themselves, Mendel got 315 plants with round, yellow seeds; 101 plants with wrinkled, yellow seeds; 108 plants with round, green seeds; and 32 plants with wrinkled, green seeds, approximating a 9:3:3:1 ratio.

This ratio was consistent with his basic hypothesis if it was assumed that the pairs of factors for seed form and seed color were completely independent of each other during their separation in the formation of gametes; that is, separation of the pair of factors for seed shape in gamete production had no effect on the separation of the pair of factors for seed color. Mendel called this assumption his *principle of independent assortment.*

To understand the effect of independent assortment in the cross, assume that the parental types were *RRYY* (round, yellow seeds) and *rryy* (wrinkled, green seeds). The round-, yellow-seeded parent, as these factors separate, is expected to produce only *RY* gametes. The wrinkled-, green-seeded parent will produce only gametes containing *ry* factors. Combinations of these gametes to produce the F_1 generation would yield only one class of offspring, with the combination *RrYy.* Assuming that the factor for round seeds is dominant over that for wrinkled seeds, and that the factor for yellow seeds is dominant over that for green, all of the F_1 would have, as observed, round, yellow seeds.

If these factor pairs for shape and color are segregated independently in gamete formation, each of the heterozy-

Essay 7–1

The Priest and the Professor

In Nature and Man's Fate, *Garrett Hardin reviews the story of Gregor Mendel with a lively style and in some detail. At one point he speculates about who might have received reprints of Mendel's "Versuch uber Pflanzenhybriden" ("Experiments on Plant Hybrids"), the now famous 1866 paper. Did Darwin receive a copy? Perhaps, since Mendel greatly admired his work; but a search of Darwin's papers at Down failed to unearth a copy. Hardin then continues:*

Of only one distribution of this paper are we sure, and that is the copy Mendel sent, with an accompanying letter to Karl Wilhelm von Nageli, Professor of Botany in the University of Munich. The choice of referee was, on the surface a wise one. Nageli, only a few years older than Mendel, was a most eminent botanist, and he was particularly interested in heredity. Surely he would appreciate the new theory! Let us read part of Nageli's first letter and see. (This reply was sent by return post—two months later.)

It seems to me that the experiments with *Pisum* [peas], far from being finished, are only beginning. The mistake made by all the more recent experimenters is that they have shown so much less perseverance than Kolreuter and Gartner. I note with pleasure that you are not making this mistake, and that you are treading in the footsteps of your famous predecessors. You should, however, try to excel them, and in my view this will only be possible (and thus alone can any advance be made in the theory of hybridization) if experiments of an exhaustive character are made upon one single object in every conceivable direction. No such complete series of experiments, providing irrefutable proofs for the most momentous conclusions, has ever yet been made. . . . Your design to experiment on plants of other kinds is excellent, and I am convinced that with these different forms you will get notably different results [in respect to the inherited characters]. It would seem to me especially valuable if you were to effect hybrid fertilizations in *Hieracium* [hawkweed], for this will soon be the species about whose intermediate forms we shall have the most precise knowledge. What I should especially recommend for experiments are *H. pilosella, H. auricula, H. praealtum, H. pratence, H. murorum, H. tiacum, H. cymosum;* and, on the other hand, *H. vulgatum, H. glaucum, H. alpinum, H. amplexicaule, H. prenanthoides, H. tridentatum.*

A modern biologist, reading Nageli's condescending letter, is filled with something very like rage. Could the Herr Doktor Professor not see that this little pamphlet of forty-four pages made obsolete several thousand pages of Kolreuter, Gartner, Herber, Lecoq, Wichura, Kolliker, and Lucas? Could he not read? Apparently not—not in the sense that such a Euclid-bare work as Mendel's demands reading. Nageli's letter made only the slightest

gous F_1 pea plants would be expected to produce *four* types of gametes. The R factor, when separated from r and delivered to a gamete, could combine with either the Y or the y factor of the pair for color. Similarly, the r factor for seed shape could be delivered to a gamete with either the Y or y of the color pair. Therefore, four types of gametes, $RY, Ry, rY,$ and $ry,$ would be produced by $RrYy$ individuals if the factors separate independently; these should have equal chances of appearing, amounting to $\frac{1}{4} RY, \frac{1}{4} Ry, \frac{1}{4} rY,$ and $\frac{1}{4} ry.$ These gametes, and their expected frequencies, are entered for both parents in the headings to Figure 7–4.

We can see all the possible combinations by filling in the squares of Figure 7–4. We find 16 possible combinations, each with a frequency of 1 in 16 offspring. However, since the R and Y factors are dominant, the classes $RRYY, RRYy, RrYY,$ and $RrYy$ will all have round, yellow seeds. These combinations occupy 9 of the 16 squares in the diagram, giving round, yellow seeds a total frequency of $\frac{9}{16}.$ The combinations producing wrinkled, yellow seeds *(rrYY and rrYy),* like those producing round, green seeds *(RRyy and Rryy),* occupy 3 squares, for a total expected frequency for each of these phenotypes of $\frac{3}{16}.$ Finally, the combination *rryy,* producing plants with wrinkled, green seeds, occupies only 1 square and thus will be expected in only $\frac{1}{16}$ of the offspring of the cross.

These expected frequencies and the $9:3:3:1$ ratio for

reference to the technical contents of Mendel's paper, a reference as devoid of understanding as a professional book reviewer's.

If Nageli did not in the least understand Mendel's work, why did he bother to reply at all? From kindness, perhaps; and noblesse oblige? The closing part of the passage quoted suggests a less distinterested motive as well. The great professor had for many years been struggling with the riddle of inheritance in the hawkweeds, without success. It took a lot of work. Here was a country priest who plainly loved hard work: why not get him to help out with the hawkweed research, thus advancing the professor's program, and giving the little priest a chance to make a modest name for himself in science, perhaps? It looked like a good bargain and, over a period of eight years, the professor contributed many hours of his invaluable time to writing encouraging letters of misunderstanding to Father Gregor. We must grant that Nageli meant well. Unfortunately, he could hardly have done Mendel a greater disservice than getting him to study the hawkweeds, for these plants often reproduce in a very irregular way, by what has come to be known as apomixis. In this process a plant may go through the motions of sexual reproduction—pollination and all that—but the embryo that develops in the seed is derived from maternal tissue only. It is obvious that if an investigator doesn't know what the real parentage of a plant is (but thinks he does), he is going to have the devil's own time unscrambling the heredity. This was the puzzle Nageli quite innocently handed Mendel, and Mendel, of course, failed to solve it. It was many years after his death before the secret of apomictic reproduction was uncovered.

Why was Mendel's work unappreciated in his own lifetime? There are many reasons. For one thing, it was based on simple algebra, an unheard-of approach in that golden age of the descriptive naturalist. For another, the author was unknown, and he said his piece only once. It is exceedingly improbable that a radical new truth will be heard unless it is repeated many times. But events seemed to conspire against the repetition of this particular truth. Two years after the publication of this paper, Mendel was made abbot of his monastery. From that time on he was the busy executive, with only snatches of time available for research. Such research as he did carry on may have made him doubt the generality of the laws he had discovered with the peas, for hawkweeds did not appear to follow them. It is also known that he worked with honeybees, and if he tried to work out their heredity, he must have been similarly discouraged, for the males (drones) have no fathers—a fact not known in Mendel's time. He must have often wondered if his laws of heredity were really true. Darwin said that no belief is vivid until shared by others—and Mendel found no others to share his belief, to make it so vivid that he could muster courage to shout it. He died unknown.

In the year of Mendel's death, 1884, Nageli published his magnum opus on heredity, a weighty tome of 822 dispensable pages. In it he did not once refer to the work of the priest at Brunn.

phenotypes derived from them closely approximate Mendel's actual results of 315:101:108:32 for these combinations. Thus Mendel's hypothesis, with the added assumption that pairs of factors separate independently of each other, explains the observed results. To be absolutely certain, Mendel performed testcrosses—$Rryy \times rrYy$, for example. This cross gave the expected 1:1:1:1 ratio in the offspring. The other testcrosses also completely confirmed the hypothesis.

Mendel's Conclusions Mendel's work can be summarized in terms of the four major concepts introduced by his work with peas:

Heredity can only be explained in terms of units (genes) and these units occur as pairs—the concept of unit inheritance.

During sexual reproduction, the gene pairs separate so that each gamete receives only one gene of each pair—the concept of segregation.

One gene can be dominant over another, masking its presence—the concept of dominance.

Genes are transmitted independently during reproduction—the concept of independent assortment.

Mendel's hypothesis and his methods for deriving and testing it are all the more remarkable if you remember that his results amounted to a completely abstract view of the hereditary material, based on mathematical reasoning rather than simple description, in direct contrast to the techniques and ideas of biology in his day.

The Impact of Mendel's Research   Mendel's work was published in 1866 and sent to more than 120 scientific societies and organizations in Europe. Undoubtedly, it was read by many, though not all. Was Mendel's work immediately acclaimed? On the contrary, it was completely ignored. The value of Mendel's work was not appreciated until 1900, and he was not given the credit he deserved until long after he died in 1884. Some of the reasons for this neglect of Mendel's work are explored in Essay 7–1, written by Garrett Hardin.

The Chromosome Theory of Inheritance

By the time Mendel's work was rediscovered in 1900, most of the details of chromosome behavior in meiosis had been worked out, and it was not long before the similarities between chromosomes and Mendel's factors were pointed out. In a historic paper published in 1903, Walter Sutton, then a graduate student at Columbia University, drew attention to the fact that chromosomes occur in pairs in sexually reproducing organisms, as do Mendel's factors. Further, Sutton noted, one member of each chromosome pair is derived at fertilization from the male parent and one from the female parent—again, an exact parallel to Mendel's factors. The two chromosomes of each pair are separated and delivered singly to gametes, as are Mendel's factors. Finally, the separation of any one pair of chromosomes in gamete formation is independent of the separation of any other pair, as in the behavior of different pairs of factors in Mendel's dihybrid crosses. On the basis of this total coincidence in behavior, Sutton concluded that Mendel's factors, the genes, are carried on the chromosomes.

Linkage   Mendel studied seven pairs of factors that separated independently of each other in the formation of gametes. According to Sutton's chromosome theory of inheritance, each of those pairs, since they separated independently to form gametes, had to be carried on different chromosomes. And indeed, by chance, or perhaps by good management, the total number of pairs Mendel chose for his experiments exactly matches the number of chromosome pairs in his experimental organism: Peas have seven pairs of chromosomes. If Mendel had extended his study to eight pairs, his hypothesis that pairs always separate independently would have required some modification.

This was first pointed out shortly after Sutton published his hypothesis of the equivalence of behavior between genes and chromosomes. Several people at this time, including Sutton himself, drew attention to the fact that, unless the number of gene pairs was limited to the total number of chromosome pairs in an organism, not all of the pairs could possibly separate independently of each other. Two or more gene pairs carried on the same chromosome pair would be carried as a unit into gametes, and would thus be linked together. Because it seemed obvious that there were probably thousands of genes in most organisms, many sets of genes should be linked together in their transmission to offspring. This was more than an educated guess, because patterns of inheritance were already known that suggested such **linkage** of several sets of genes on the same chromosome.

The first of these patterns was discovered in peas in 1906 by the English scientists W. Bateson and R. C. Punnett. These investigators found that two pairs of genes—one controlling flower color (purple versus red) and designated *C,* and one controlling length of pollen grains (long versus short) and designated *S,* were transmitted to offspring in the same combinations and rarely separated independently. In other words, if one parent had purple flowers and long pollen grains, its offspring also tended to have *both* purple flowers and long pollen, as if flower color and pollen length were controlled by a single gene. But obviously this could not be true, because individual pea flowers could be found with purple flowers and short pollen grains. And in this case, the purple color and short pollen also tended to be transmitted together. But most disturbing of all to anyone trying to form a valid hypothesis to cover all these events was the observation that the linked characters sometimes separated. For example, a parent with purple flowers and long pollen grains occasionally gave rise to offspring with purple flowers but short pollen grains.

No satisfactory explanation either for linked inheritance or for its exceptions was advanced until 1910. In that year, Thomas H. Morgan of Columbia University, studying linked genes in the fruit fly, *Drosophila melanogaster,* noted that linkage can be explained if the genes usually inherited together are carried on the same pair of chromosomes.

Recombination   As for the exceptions, Morgan suggested that the unexpected combinations probably arose in some way from the chiasmata, or crossovers, visible between chromatids of the tetrad in prophase of meiosis (see p. 113), possibly by the mechanism shown in Figure 7–5. In most of the cells of the purple-flowered, long-pollen *(CcSs)* parent entering meiosis, no chiasmata were formed between the points on the tetrad bearing the alleles for pollen size and flower color (Figure 7–5a). These

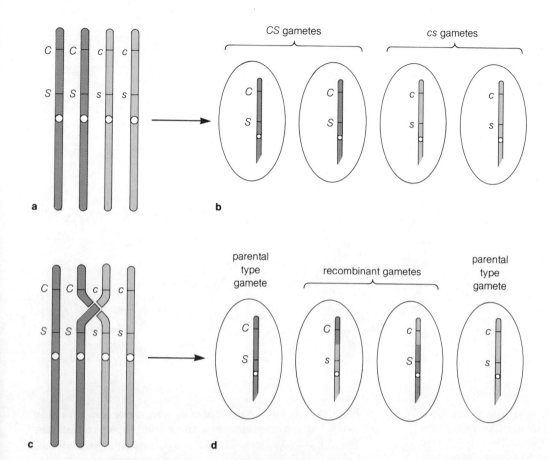

Figure 7–5 Morgan's explanation of linkage and recombination.

four chromatids separated and entered the gametes singly, producing the two gamete types noted in Figure 7–5b. In a small percentage of the cells entering meiosis in the *CcSs* parent, a crossover or chiasma formed between the two genes (Figure 7–5c). This crossover, by breakage and exchange of parts of two of the four chromatids, linked together the new combinations *C* and *s* in one chromatid, and *c* and *S* in the other. In short, the genes were *recombined*, and the affected chromatids were called *recombinants*. Note, however, that two of the four chromatids of the tetrad remain unchanged, and these are called *parentals*. In other words, when the four chromatids separate during the two meiotic divisions and enter separate gametes, they give rise to *four* different types: two parentals, *CS* and *cs*, and two recombinants, *cS* and *Cs* (Figure 7–5d). The recombinants, when fused with the *cs* gametes from the *ccss* parent fertilization, gave rise to the two additional classes, *ccSs* and *Ccss*, which together comprised 3 percent of the offspring. The mechanism underlying this phenomenon came to be called *recombination*.

Morgan's hypothesis of linkage and recombination thus allowed the newer observations to be reconciled with Mendel's hypothesis that pairs of genes separate independently in the formation of gametes. Independent assortment applied only to genes carried on separate chromosomes; linked genes, carried on the same chromosome, would be inherited according to Mendel's pattern for single genes, except for recombination of the linkage due to crossovers.

Mapping The observation that 3 percent of the offspring were recombinants means that 3 percent of the gametes originating from the *CcSs* parent contained recombined chromosomes, or that for every 100 gametes formed, 3 contain recombinants. To produce an average of 3 percent recombinants, *C* and *S* genes must recombine in 6 percent of the cells because, as we have seen, recombination always yields two unrecombined or parental chromosomes as well as two recombinants. In general, therefore, the percentage of cells entering recombination is always twice the percentage of recombinant gametes.

Morgan noted that recombination was much more frequent in some pairs of genes than in others, varying from less than 1 percent of all gametes to rates so high

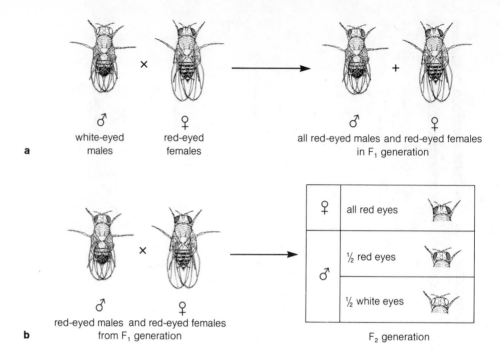

Figure 7–6 Sex linkage in a cross with *Drosophila*. (**a**) Cross between white-eyed males and red-eyed females. (**b**) F₁ cross between the red-eyed males and red-eyed females resulting from (**a**).

a — white-eyed males × red-eyed females → all red-eyed males and red-eyed females in F₁ generation

b — red-eyed males and red-eyed females from F₁ generation ×

♀ all red eyes
♂ ½ red eyes
½ white eyes

F₂ generation

that it was hard to be sure the genes were linked together. Thinking about the physical mechanism of breakage and crossover, Morgan reasoned that the farther apart two genes were on a chromosome, the more easily a crossover could form between them, and the more likely they would be to recombine. Thus, Morgan hypothesized, the recombination frequency of any two genes is proportional to the distance between them on the chromosome.

It was quickly realized that, if Morgan was right, recombination frequencies could be used to locate linked genes relative to each other on the chromosome—that is, to *map* the chromosome. For example, assume that three genes, A, B, and C, are known to lie on the same chromosome. Recombination is detected between genes A and B in 10 percent of the cells, between A and C in 8 percent, and between B and C in 2 percent. These frequencies are compatible with only one possible arrangement of these genes on the chromosome:

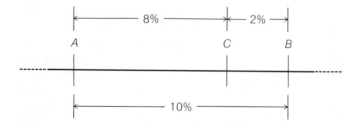

Note that not only the order in which the genes appear along the chromosome, but their relative separation can theoretically be determined by the crossover frequencies. Using this method, most of the known genes of the best-studied organisms, the fruit fly *Drosophila*, corn, the mold *Neurospora*, the bacterium *Escherichia coli*, and various other species have been mapped and assigned positions on the chromosomes.

These positions are only relative, however, and the genes are fixed in respect to each other only by recombination percentage distances (called *crossover units*), not actual physical distances in micrometers or nanometers. In addition, because some regions of chromosomes undergo recombination more frequently than others, the relationship between the distances separating genes and crossover frequencies is not exactly proportional. Therefore, the "maps" provided by recombination studies are abstract charts of the linear order of genes rather than maps of the actual physical positions of the genes on the chromosomes.

Sex Linkage In 1910 Morgan, working with *Drosophila*, noted a curious pattern of inheritance in eye color that seemed to depend on the sex of the fly. All the genetic traits studied up to this time had been transmitted indiscriminately to male and female offspring, and sex could be ignored in analyzing the frequency of various classes. Morgan first detected sex linkage in a line of flies

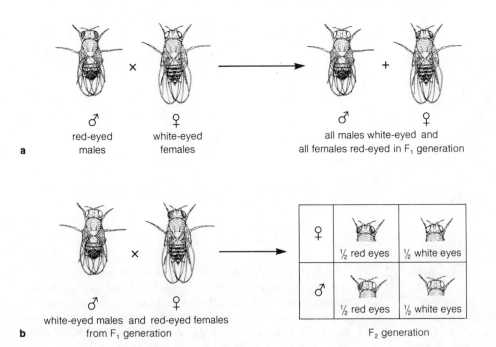

a
♂ red-eyed males × ♀ white-eyed females → ♂ + ♀ all males white-eyed and all females red-eyed in F₁ generation

b
♂ white-eyed males and ♀ red-eyed females from F₁ generation × → | ♀ | ½ red eyes | ½ white eyes |
| ♂ | ½ red eyes | ½ white eyes |

F₂ generation

Figure 7–7 Sex linkage in the reciprocal cross of Figure 7–6. (**a**) Cross between red-eyed males and white-eyed females. (**b**) F₁ cross between the red-eyed females and white-eyed males resulting from (**a**).

that had white eyes instead of the usual red. If white-eyed males were crossed with red-eyed females, the entire F₁ generation, both males and females, had red eyes. This suggested that eye color in the fruit fly is controlled by two alleles of a single gene, with red dominant over white. In the offspring of a cross between individuals of this F₁ generation, these alleles would therefore be expected to segregate without respect to sex, according to a ratio of three red-eyed flies to every white-eyed fly. Surprisingly, Morgan found that all the F₂ females were red-eyed, whereas half the males were red-eyed and half white-eyed (Figure 7–6). All the white-eyed F₂ flies were male.

The reciprocal cross—white-eyed females bred to red-eyed males—gave different results: All the F₁ females were red-eyed and all the males white-eyed. But in the F₂ generation, both sexes were evenly divided between red-eyed and white-eyed types (Figure 7–7).

Sex Determination Morgan realized that this initially baffling pattern of inheritance followed the transmission of sex chromosomes in *Drosophila*. In many organisms, sex is determined by a special pair of chromosomes, appropriately called the **sex chromosomes.** The chromosomes that are not sex chromosomes are called **autosomes.** In diploid organisms, the autosomes are present in homologous pairs. In the female *Drosophila,* there is also a pair of homologous sex chromosomes, called

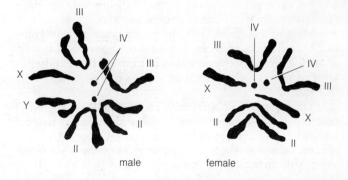

male   female

Figure 7–8 The chromosomes of *Drosophila melanogaster* in the diploid condition, showing the difference between female (XX) and male (XY) sets of chromosomes. The other chromosomes are present in pairs: two number II, two number III, two number IV.

the XX chromosomes. In the male, however, the sex chromosomes are nonhomologous, for one is an X chromosome and the second is different in shape and known as the Y chromosome (Figure 7–8). The presence of a pair of X chromosomes in a zygote determines that the organism will be a female, while an XY combination results in a male. The Y chromosome of the male contains only a fraction of the genes of the X chromosome, and for most genes the Y chromosome can be considered inert.

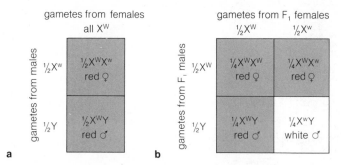

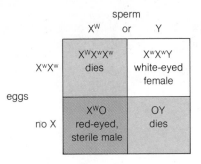

Figure 7–9 Morgan's explanation of sex linkage. In this figure W and w indicate the condition of the gene and X and Y the sex chromosomes.

Figure 7–10 A cross between white-eyed females and red-eyed males in which nondisjunction occurs.

This same pattern of XX and XY sex determination is found in humans. In both humans and *Drosophila,* the type of sex chromosome carried by the sperm determines the sex of the offspring; in the male offspring, the X chromosome is always donated by the mother. (One wonders how different history might have been if this fact had been discovered in the fifteenth century instead of the twentieth—England might still be a Catholic country, for example, had King Henry VIII known that only sperm contain the Y chromosome for making sons.)

Morgan reasoned that the inheritance of eye color in *Drosophila* could be explained if the gene for this trait is carried on the X chromosome but not on the Y. Inheritance of eye color would then follow the pattern shown in Figure 7–9.

This system of XX females and XY males is the most common arrangement for sex determination in nature, holding true for most animals and some plants. In some species, however, there is no equivalent of the Y chromosome; this chromosome is simply absent in males. These species are therefore XX (female), and XO (male; 0 = no chromosome). This occurs in bees, where the males are produced from eggs that the queen fails or refuses to fertilize. In a number of groups, such as butterflies and moths, the sex chromosomes are distributed in the opposite way, with XY females and XX males. There are also organisms, particularly plants, in which no difference can be detected between the chromosome set of males and females; that is, there are no identifiable sex chromosomes.

The sex chromosomes, predictably, have a direct developmental role in determining the sex of an individual. In species with the XY or XO system, unusual genetic effects may occur because the genes on the single X chromosomes of males so often have no counterpart, as they would if they were on an autosome or in a female. Thus, in effect, a gene that would be recessive elsewhere is expressed.

Nondisjunction In another part of Morgan's work, he crossed white-eyed females with red-eyed males. He expected that all F$_1$ females would have red eyes and all F$_1$ males white eyes. He gained the predicted distribution with the following rare exceptions: One red-eyed male or white-eyed female appeared in every 2,000 to 3,000 flies. These exceptional individuals, rather than disproving Morgan's hypothesis, actually supported it when correctly interpreted.

C. B. Bridges, one of Morgan's co-workers at Columbia University, reasoned that the exceptional flies could be explained by assuming that, very rarely, the two X chromosomes of a female failed to separate—or *nondisjoined,* as Bridges termed it—during meiosis, and were delivered to the same egg. Because of this **nondisjunction,** for every egg produced with two X chromosomes, another egg would be formed with no X at all. If this happened in Morgan's white-eyed females, half of these exceptional eggs would contain two X chromosomes, both carrying the recessive allele for white eyes. The other half, deprived of an X chromosome as a result of nondisjunction, would carry no gene for eye color at all (Figure 7–10). All the other chromosomes of the egg, the autosomes, were assumed to be normally distributed. This is precisely what was found when testcrosses were made and the chromosomes counted.

Sex-Linked Genes in Humans Several important human traits and conditions are transmitted through genes located on the sex chromosomes. Two of the best-studied examples of sex-linked inheritance through the XX-XY system in people are certain types of color blindness and hemophilia.

Red-green color blindness in humans follows a pattern of inheritance very similar to the eye-color pattern Morgan studied in *Drosophila.* Although testcrosses cannot be carried out with humans, inheritance over several

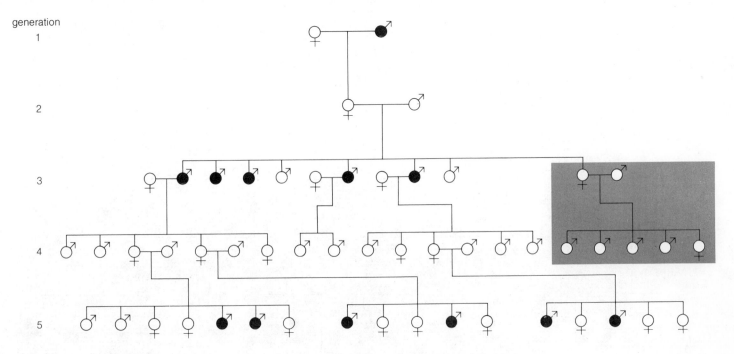

Figure 7-11 Pedigree of a family with a history of color blindness. Females are designated by the symbol ♀ and males by ♂; solid black circles indicate the presence of color blindness. Marriages are indicated by a horizontal bar between two circles.

generations can often be analyzed by looking into family records and constructing a *pedigree,* a chart showing all marriages and offspring for as many generations as possible, the sex of individuals in the different generations, and presence or absence of the trait being studied.

A pedigree of a family with a history of color blindness is shown in Figure 7-11. Females are designated by the symbol ♀ and males by ♂; solid black circles or squares in the pedigree indicate presence of the trait. At the very top of the pedigree (generation 1), the earliest recorded marriage was between a color-blind male and a normal female. The single child of this marriage (generation 2), a daughter, married a normal male. This female, who received an X chromosome bearing the recessive allele for color blindness from her father, passed the trait to five of her seven sons (generation 3). The two remaining sons received the X originating from their grandmother, carrying the dominant allele for normal color vision. Three of the color-blind males in this generation married normal females. The genotypes of the wives cannot be deduced with certainty from the pedigree, but it seems probable that none of these females carried a recessive allele for color blindness, because none of their children was color-blind. In fact, not one of the sons or daughters in generation 4 shows the trait.

All of the daughters of color-blind fathers are *carriers*

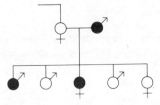

Figure 7-12 Possible outcome of the marriage enclosed in broken lines in Figure 7-11 if the daughter carrying the trait had married a color-blind male. Both color-blind and normal sons and daughters could arise from this marriage.

of the trait, however, because each received an X chromosome carrying the recessive allele from her father. Thus, the trait shows up again in some of the sons in generation 5, who receive the X with the recessive allele from their mothers. Note that the presence and absence of the trait alternate from generation to generation in the males. This is because a father cannot pass a trait carried on his X chromosome directly to his sons. Such alternation, or "skipping," of a trait in a human pedigree is a useful clue; it suggests that the allele under study is recessive and is carried on the X chromosome.

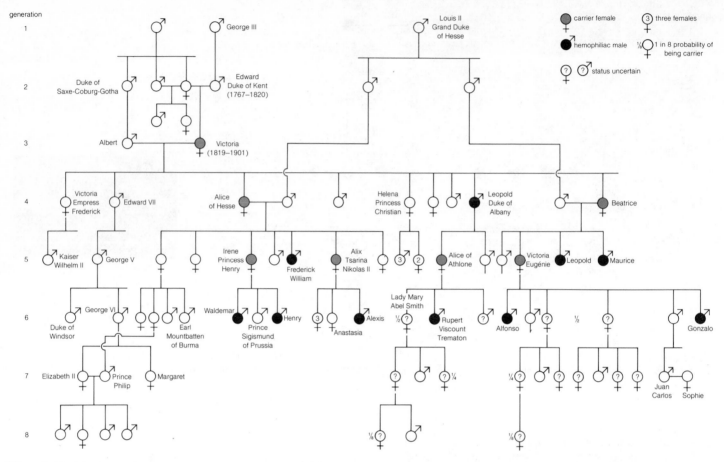

Figure 7–13 Inheritance of hemophilia in descendants of Queen Victoria of England. From V. A. McKusick, *Human Genetics*, 2nd ed., © 1969. Reprinted by permission of of V. A. McKusick and Prentice-Hall, Inc.

None of the females in the pedigree of Figure 7–11 is color-blind. But a color-blind female could have been born if, somewhere in the series of generations, a female with normal vision who carried a recessive allele for color blindness on one X chromosome married a color-blind male. This possibility is shown in Figure 7–12, in which the marriage shown enclosed in broken lines in Figure 7–11 is considered as if the daughter of the line had married a color-blind male. Some of the sons would be likely to receive from the mother the X carrying the dominant allele for normal vision, and some the X with the recessive allele for color blindness. Thus, both color-blind and normal sons could arise from this marriage. The same possibilities hold true for the daughters. The father's X chromosome brings one recessive allele to each daughter. This allele will be combined with one of the two kinds of X chromosomes from the mother— the one bearing the recessive allele to produce a color-blind daughter, or the

one bearing the dominant allele to produce a daughter with normal color vision.

Hemophilia is another human disability that is recessive and carried on the X chromosome. Hemophiliac individuals are called "bleeders," because they lack the normal mechanism for clotting of the blood. Although afflicted males, with luck and good care, can reach maturity, their lives are tightly circumscribed by the necessity to avoid injury of any kind; even slight bruises can cause internal bleeding and prove fatal. In recent years, the plight of hemophiliacs has been eased by daily injections of the factor missing in their blood. One of the most famous cases of hemophilia occurred in the royal families of Europe descended from Queen Victoria of England (Figure 7–13). Because hemophilia had never been recorded among Victoria's ancestors, the recessive trait probably arose in the queen (or in her father's sperm) as a mutation. Because so many sons of the various families of European royalty were affected, the presence of the trait has had an incalculable influence on the course of history. In Russia, for example, the preoccupation of the royal family with their hemophiliac son was a major factor in the events leading to the Bolshevik Revolution and the eventual installation of a Communist government.

The condition affected only sons in the royal lines, but it presumably could have affected daughters if a hemophiliac son had married a female carrying the recessive allele on one of her two X chromosomes. Because the disease is very rare, the chance of a hemophiliac male's marrying a female unaffected but carrying the trait is exceedingly low, and only a very few marriages of this type have ever been recorded. For a discussion of other inherited diseases and traits, see Supplement 7–1.

Nondisjunction and Human Heredity Various abnormal combinations of sex chromosomes have been observed among humans, the most common being XO females and XXY males. These combinations probably arise through nondisjunction of sex chromosomes in meiosis by mechanisms similar to those in *Drosophila* described earlier. In XO females (about 1 in 5,000 births) the ovaries are underdeveloped, and the affected individuals, although female in appearance, with normal external genitalia, are sterile. XO females are also typically short in stature, with undeveloped breasts. This group of traits is known as *Turner's syndrome.* XXY males *(Klinefelter's syndrome,* occurring in about 1 in 400 to 600 births) have male external genitalia but very small and undeveloped testes. Body hair is sparse, and some development of the breasts is usually noted. These XXY males are also sterile. Males with an extra Y chromosome (XYY males) are apparently normal physically.

Mongoloid idiocy *(Down's syndrome)* may arise through nondisjunction of the twenty-first pair of autosomal chromosomes; that is, mongoloids have three copies of one chromosome of the set. (Eggs deprived of the chromosome by nondisjunction evidently fail to develop if fertilized.) These individuals are short in stature and mentally retarded, with eyes once thought to resemble those of the Mongoloid races of China and Japan. (Among the Chinese, victims of Down's syndrome are thought to resemble Caucasians.) More than half are dead by the age of 10. For unknown reasons, the autosomal nondisjunction responsible for Down's syndrome is far more common among older mothers. The syndrome is relatively common in general, occurring once in every 500 to 600 births among women under 30, but it increases fivefold among mothers aged 40 to 45 (the age of the father does not seem to affect the frequency of Down's syndrome births). At these frequencies, it is the major cause of severe mental retardation in humans. Unfortunately, the disrup-

tions caused by additions or deletions of whole chromosomes through nondisjunction, such as Turner's, Klinefelter's, and Down's syndromes, are incurable and are likely to remain so. There are now techniques available for the *in utero* testing of a fetus early in development if such syndromes are suspected.

Sex-Influenced Traits Not all characteristics that differ in expression between the sexes are controlled by genes located on sex chromosomes. Many traits, such as balding in men, growth and distribution of body hair, enlargement of the breasts in women, voice pitch, and even the form of the external sex organs are controlled by genes and alleles present in both males and females. These genes, carried on autosomes rather than on the sex chromosomes, produce different external effects in men and women because of the influence of other factors, primarily the sex hormones, and not because a gene is present in one sex and absent in the other. Imbalance in the hormones affecting these genes can cause the appearance of characteristics of the opposite sex in males and females, such as the enlargement of the breasts in men or growth of facial hair in women.

Summary

Mendel, in a series of brilliant experiments, showed that heredity can only be explained in terms of units that we now call genes. He found that one gene can be dominant over another, masking its presence. Moreover, a pair of genes is present for each characteristic in an individual, but genes are transmitted independently during reproduction. Mendel's work, ignored during his lifetime, was rediscovered in 1900. Not long after, it was found that the units or genes discovered by Mendel were actually part of the chromosomes. Genes on the same chromosome were not transmitted independently during reproduction; instead, they acted as if they were linked together—which, of course, they are. It was also found that sometimes this linkage is not absolute and that genes appeared to have changed places. This discovery, called recombination, was found to depend on the exchange of parts of the chromatids during meiosis. Some characters were also found to be associated with the sex of the organism. Genes controlling these characters are located on the sex chromosomes, which are also responsible for the sex of the individual. Occasionally, a pair of homologous chromosomes will fail to separate at meiosis—the phenomenon of nondisjunction. In humans, nondisjunction is the basis of several genetic diseases.

Problems

1. Some people can roll their tongues, and others cannot. A biology student polled his family to discover how the trait is inherited and found the following pedigree:

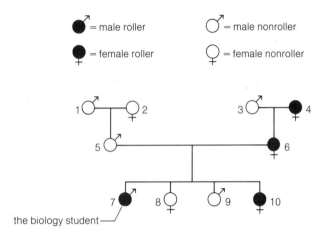

● = male roller   ♂ = male nonroller

● = female roller   ♀ = female nonroller

the biology student

a. Is the marriage between #5 and #6 a monohybrid or dihybrid cross?

b. What is the genotype of #7? #4?

c. Is individual #6 homozygous or heterozygous?

d. Is rolling dominant or recessive? Is the trait autosomal or sex-linked?

e. Which of Mendel's laws does this pedigree illustrate?

f. If #7, the biology student, marries a woman who cannot roll her tongue, what is the probability that they will have a child who cannot roll its tongue?

2. **a.** For each of the parental genotypes *Aa, AaBb, AaBbCc,* and *AaBbCcDd,* what are the possible gamete types?

b. Do you see an arithmetic trend to the number of possible gamete types in **a?** If you do, try to derive a formula you can use to calculate the maximum number of gamete types for any given parental genotype.

c. What are all of the possible gamete types for the genotypes *AABb, AaBBCc, AABBCc, AABBCC?*

d. On the basis of all the above examples, when is it appropriate to use the formula you derived in part **b?**

3. The roundworm *Ascaris megalocephala* has only one pair of chromosomes. If Mendel had worked with this worm rather than with peas, is it likely that he would have made all his discoveries? Explain your answer.

4. Fur color in rats is a completely dominant autosomal trait, with brown dominant over white. A rat breeder wants to find out if a brown female rat he has is homozygous. Propose a cross he could perform to find out whether this rat is homozygous, and describe what the phenotypic ratio of the F_1 generation would be if the mother were homozygous. What is this type of cross called?

5. What is the probability that each of the following will occur:

a. a color-blind girl born from a carrier mother and a color-blind father?

b. a girl in a family that already has six boys?

c. in flipping two coins, obtaining one head and one tail?

d. having five boys and one girl in a family?

6. A man you know is color-blind. Knowing that you are a biology student studying genetics, he comes to you to find out whether he inherited the trait from his mother or his father. How do you respond to him?

7. A mouse parent is homozygous for fur color *(AA)* and heterozygous for bristle length *(Bb)* and eye color *(Cc).*

a. How many different gamete types are possible, assuming independent assortment?

b. How many different gamete types are possible if genes A and B are linked together on one chromosome, and C is on a separate chromosome?

8. Given the following distances between four linked genes arranged in a line with A and D on the ends, calculate the correct order on the chromosome: A to B = 31 map units, A to D = 36 map units, A to C = 14 map units.

9. Diagram crossing over between the B and C loci here, showing the stages at synapsis, at crossing over, at the end of meiosis I, and at the end of meiosis II:

A	b	C
A	b	C
a	B	c
a	B	c

10. A man with normal color vision whose father was color-blind marries a woman with normal color vision whose father was also color-blind.

a. What are the genotypes of the man and woman?

b. What are the chances that the first child from this marriage will be a color-blind boy?

c. Of all the children (sex unspecified) from these parents, what proportion is expected to be normal?

11. Striatonigral disease, or Joseph's disease, is an autosomal, dominant nervous-system disorder that begins to show symptoms at about 25 years of age and usually causes death within 13 years after that.

a. A seemingly healthy couple marry and have a son when the husband is 21. At the age of 26, the husband is diagnosed as having Joseph's disease. What is the probability that their son has the disease, assuming the wife is normal?

b. Would it be possible to be a carrier of Joseph's disease? Why or why not?

c. Would it be possible to have generations in family pedigrees skip the disease, as in sex-linked traits? Why or why not?

12. A Turner's syndrome female, with only one X chromosome, occurs in about one out of every 25,000 births. Diagram how a normal couple could have a Turner's syndrome child.

13. Females from a true-breeding strain of mice having long ears and green eyes were crossed with males from a true-breeding strain having short ears and brown eyes. In the F_1 generation, all the female mice had brown eyes and long ears, while all the males had long ears and green eyes. The F_1 generation was allowed to interbreed. The characteristics of the F_2 generation were as follows:

Females	
long ears, brown eyes	98
long ears, green eyes	102

Males	
long ears, brown eyes	23
short ears, green eyes	27
long ears, green eyes	75
short ears, brown eyes	74

a. How is each trait inherited (autosomal or sex-linked, dominant or recessive)?

b. If you think the two traits are linked, calculate the map distance between the two genes.

14. In a cross between two parents heterozygous for two traits, what will be the ratio of offspring if the genes are not linked?

15. You are presented with a vial of *Drosophila*. Some of the flies have vermilion eyes, and others have red eyes. Some flies have bristles on their bodies which look singed, others have straight bristles. You mate a singed-bristled, vermilion-eyed male with a straight-bristled, red-eyed female and obtain an F_1 generation with all straight bristles and red eyes. You mate the F_1 males with singed-bristled, vermilion-eyed females in order to determine how the traits are inherited, and obtain the following phenotypes:

Females	
red eyes, straight bristles	240

Males	
vermilion eyes, straight bristles	119
red eyes, straight bristles	120

a. How is the trait for bristle shape inherited?

b. How is the trait for eye color inherited?

c. Do the two traits assort independently or are they linked?

d. If the two traits are linked, what is the map distance between them?

Supplement 7–1
Genetically Determined Human Traits and Diseases and Their Control

Many inherited traits in humans cause disease or are otherwise interesting because of their effects on human life. Several of these traits and their patterns of inheritance were discussed in this chapter, including color blindness and hemophilia (sex-linked traits), and Down's, Turner's, and Klinefelter's syndromes (caused by additions or deletions of whole chromosomes). In this supplement, we will describe some other human traits and diseases that have been intensively studied and are likely to be of interest.

Genetically Determined Diseases

Sickle-cell anemia, which affects the red blood cells and impairs their ability to carry oxygen, is found almost entirely in people descended from African blacks. The trait, caused by a recessive allele of a gene coding for part of the hemoglobin molecule, results in the change of a single amino acid in the oxygen-carrying molecule. Homozygotes (people with recessive alleles on both chromosomes) are very seriously affected and rarely survive past late adolescence. The red blood cells in these individuals are quite fragile and take on a characteristic "sickle" shape when blood samples are deprived of oxygen. Carriers of the trait, about 10 percent of the black population in the United States, have one normal and one recessive allele of the gene. Although carriers are essentially unaffected and produce enough normal hemoglobin through the activity of the dominant allele, sickling hemoglobin produced by the recessive allele can still be detected in their bloodstreams. Matings between carriers have about one chance in four of producing a child with sickle-cell anemia. In the U.S. black population, matings of this type result in the fatal anemia in about 1 of every 500 children born. If only one parent is a carrier, none of the children would be expected to have the disease. However, each child of a marriage between a carrier and a noncarrier of the sickling allele would have a 50 percent chance of being a carrier, without symptoms of the disease.

The recessive allele has the unusual side effect of protecting carriers from malaria. Even though these carriers possess some normal hemoblobin in their red blood cells (the exact ratio between normal and sickle hemoglobin in the blood cells of carriers is controlled by modifying genes), blood parasites causing malaria cannot survive in the bloodstream because the life-span of sickle cells is shorter than that of normal red blood cells, and the parasite doesn't have time to mature. The protection conferred by the gene provides a definite survival advantage in highly malarial areas. As a result, even though the homozygous recessive combination is usually fatal, the gene persists in high frequency in malarial districts such as the Congo River basin in Africa because heterozygous individuals are favored.

Phenylketonuria, caused by a recessive gene that affects the metabolism of the amino acid phenylalinine, frequently leads to severe mental retardation. Fortunately, the recessive allele is rare, and infants carrying both alleles for the trait, and thus afflicted by the disease, number only about 4 in every 100,000 births. Affected individuals lack an enzyme essential for converting phenylalanine to the amino acid tyrosine, so that phenylalanine accumulates in the bloodstream. Mental retardation evidently results through damage caused by the high concentrations of this amino acid in the blood. Phenylketonuria can be detected by analysis of urine samples within a few weeks of birth. Careful regulation of the diet to eliminate foods containing phenylalanine can then reduce the amount of the amino acid in the bloodstream and prevent the mental retardation that would otherwise follow.

Galactosemia is similar in effect, also resulting in severe mental retardation. The disease is caused by a recessive allele that, when present in both chromosomes, leads to deficiency of an enzyme for the metabolism of galactose, a sugar found in quantity in milk. Accumulation of galactose in the bloodstream causes damage to the nervous system that eventually leads to mental retardation and death. If identified soon after birth, infants with the disease can be effectively treated, with no significant ill effects, by substituting foods that contain no galactose for milk in the diet.

Control of Genetically Based Human Diseases: Eugenics

Many genetically determined human diseases, such as diabetes, hemophilia, and phenylketonuria, were once fatal or so seriously handicapped affected individuals that they were unlikely to reach breeding age. Thus, the frequency of the alleles causing these diseases remained constant at low levels in the human population as a whole. Recently, however, medical science has provided treatments for most of these diseases, with the happy result that affected people can lead a normal life. Unfortunately, since a normal life usually includes marrying and having children, the harmful alleles are passed on to offspring and thus are increasing, although slowly, in the population. Consequently, in contrast to the pattern for infective agents such as bacteria and viruses, developing treatments for hereditary diseases is likely to result in an *increase* of these maladies in the human population.

What can be done about the situation? Probably very little. Selective breeding, called *eugenics,* is possible among domestic animals but difficult to accomplish in human society. Involuntary sterilization, although proposed now and then as a solution, is a serious violation of human rights that few people can accept even as a punitive measure. Such controls can be practiced if carriers of a harmful trait are identified and agree voluntarily not to have children. Controls of this type, either imposed or voluntary, are sometimes referred to as *negative eugenics,* because they decrease the frequency of unfavorable genes in the population. With the gradually increasing frequency of hereditary diseases, tests to identify carriers and genetic counseling on the advisability of having children will probably become a standard part of marriage practices.

On the other side of the coin, *positive eugenics* has sometimes been proposed as a method to increase the frequency of favorable traits such as high intelligence, health, and physical beauty in the population. Under this plan, individuals favored with these qualities would be deliberately selected as breeders or as suppliers of stock in sperm banks. Farfetched as it might sound, this plan is regularly proposed and in recent history was actually adopted as official policy by at least one political system—that of Nazi Germany. In practice, it is doubtful that much improvement in the overall population could result from positive eugenics as long as negative eugenics remained voluntary. And, as we have pointed out, negative eugenics through involuntary control of carriers by sterilization is not likely to be accepted by the human race. As a result, human difficulties caused by genetically determined diseases are likely to increase gradually unless some method can be found to alter the genotype of affected individuals in a controlled way.

8

Mutation, Alleles, and Gene Expression

In 1933 on a fox farm in Norway there appeared a single male pup with platinum-colored fur. This unique color was discovered to result from a change in a single gene. A new allele had appeared. And because this allele was dominant rather than recessive, the owners of the fox got rich. Soon everyone wanted a platinum fox coat, and a single pelt sold for $11,000 in 1939.

Mutations

Such changes in genes, producing new alleles, are called **mutations.** The term is complex and has a distinctly odd history. It was first used by the Dutch botanist Hugo De Vries, one of Mendel's rediscoverers. Working with the evening primrose *(Oenothera),* De Vries found characteristics appearing that were not present in either parental line. He believed that these new characteristics represented heritable changes in genes. De Vries was right in thinking genes could change and that the changes could be inherited, but his explanation did not happen to fit the events he observed in the evening primrose. Unfortunately for him, that flower has a highly anomalous type of chromosome behavior. Virtually none of the changes he saw were true gene mutations. Not long thereafter, though, Morgan found true mutations in his fruit flies.

The most common meaning associated with the term *mutation* is a specific change in a gene. However, the term can also refer to structural changes in chromosomes.

Mutations occur through changes in the nucleotide sequence of the DNA of the gene. Agents or conditions that make such changes easier or more likely can produce mutations. The first such agent—called a **mutagen**—was discovered by Hermann J. Müller, who found in 1927 that treating *Drosophila* with X rays greatly increased the rate of appearance of mutations. (He received the 1946 Nobel Prize for his discovery.) L. J. Stadler achieved

the same result with barley the following year. Exposure to radioactive materials has the same effect. Both X rays and radioactive materials produce ionizing radiation, and its effect on the rate of mutation is proportional to the dose of radiation the organism receives. The manner in which ionizing radiations produce mutations is not fully understood. The action may be direct, altering or destroying a part of the DNA chain, or indirect, with the ionizing radiation producing chemical changes within the cell. Ultraviolet light, which is not an ionizing radiation, can also induce mutations. The most effective wavelength is the one DNA absorbs most strongly.

In one-celled organisms, any mutation that does not kill the cell will be "inherited." In multicellular organisms, however, the mutagen must act on germ cells (cells that will form eggs and sperms) if the mutation is to be transmitted to future generations. Mutations in other cells (somatic, or body-cell, mutations) may cause recognizable phenotypic changes, but these will not be transmitted to offspring.

Some of these somatic effects may be serious. Certain forms of cancer, for example, may arise from somatic mutations. This is one of the reasons for the current ecological concern that continued use of aerosol spray cans may eventually destroy the ozone layer. Without this layer, significantly higher amounts of ultraviolet radiation would reach the surface of our planet, increasing the rate of somatic mutations in human beings, some of which would show up as skin cancers.

Many chemicals induce mutations. Some chemical mutagens are as simple as hydrogen peroxide (H_2O_2), whereas others are more complicated, such as the nitrogen mustard gases used as weapons during World War I. One mutagen effective in microorganisms is 5-bromouracil (BU), which is an analogue to thymine. (An *analogue* is a compound with a molecular structure similar to that of another compound.) When bacteria are deprived of thymine and grown in the presence of BU, the BU is incorporated into the DNA being synthesized, in every

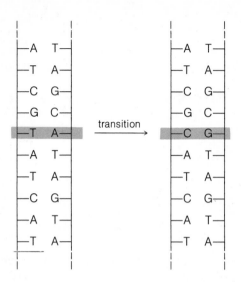

Figure 8–1 A transition, a form of DNA mutation in which a different base pair is substituted for the correct one.

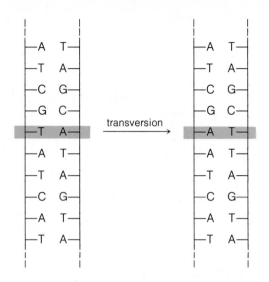

Figure 8–3 A change in the DNA sequence of a gene through a transversion. In this case, the correct base pair is retained, but the purine and pyrimidine bases of the pair exchange places in the DNA double helix.

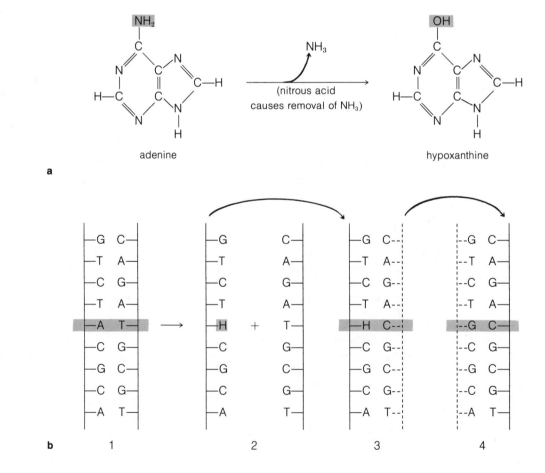

Figure 8–2 A possible mechanism for generating a transition.

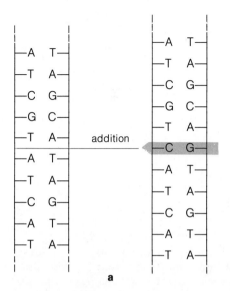

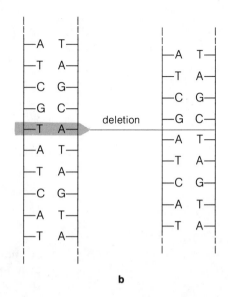

Figure 8–4 Mutations through addition (**a**) and deletion (**b**) of single base pairs.

position where thymine would normally go. At reproduction, then, errors will occur in the synthesis of new DNA—errors we call mutations.

Point Mutations If the change in the DNA of a gene involves a difference in only one nucleotide pair, we call it a *point mutation.* Such mutations occur at a regular but very low frequency during DNA replication. All result in "mistakes" in the copy made from template DNA during replication. In one form, called a *transition* (Figure 8–1), one purine is substituted for another, or one pyrimidine for another. One of the several known mechanisms by which transitions may occur is shown in Figure 8–2. Nitrous acid (a chemical that can induce mutations) may convert an adenine in the template to another purine, hypoxanthine (Figure 8–2a). At a subsequent replication, instead of pairing with thymine, as adenine would have, hypoxanthine pairs with cytosine (Figure 8–2b). Thus C is substituted for T in the newly synthesized chain. Subsequently when the new chain is replicated, C is paired with G. Several other kinds of transitions occur spontaneously in natural systems.

In a second type of change, called a *transversion,* a purine is substituted for a pyrimidine, or vice versa, in either the template or the newly synthesized chain (Figure 8–3). Although transversions are known to take place spontaneously in living cells, no biochemical mechanism is known by which this type of change might occur. This is the type of mutation that resulted in sickle-cell anemia.

Depending on the position of a substituted base in a coding triplet, both transitions and transversions may induce a change in the amino acid normally coded for, or may convert a coding triplet into a terminator codon (see p. 94). These changes, in turn, may affect the polypeptide coded for by the transcription unit. If the sequence of amino acids changes at the active site of an enzyme, the protein may be so altered that its catalytic activity is completely destroyed. Such changes elsewhere in the enzyme, however, may have no significant effects.

The remaining kinds of point mutations that occur during DNA replication are deletions or insertions of single bases in the template or in newly synthesized nucleotide chains (Figure 8–4). Both of these changes are known to occur in nature.

Insertions of single bases can be induced by adding the chemical acridine to replicating cells. This chemical is linked into the newly synthesized DNA chain between nucleotides, creating a wider separation between bases than normally occurs. At the next replication, when this altered chain serves as a template, a new base may be inserted in the "empty" space opposite the acridine.

Deletions and insertions cause a change in the *reading frame* of the DNA code. In other words, from the position of the insertion or deletion to the end of the transcription unit, the second or third base of each triplet will now be "read" as the *first* base of each triplet (see Figure 8–5). Single deletions or insertions, especially if located near the beginning of a gene, cause extreme changes in the amino acid sequence of the encoded protein, usually with complete loss of activity.

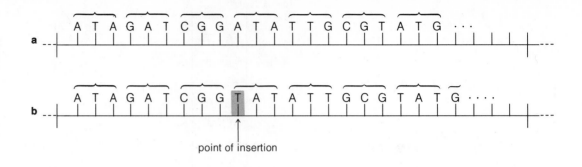

Figure 8–5 A change in reading frame of a DNA sequence because of an addition. (**a**) The unmutated gene; triplets are read in sets of three as shown. (**b**) An insertion of a base pair at the point shown causes a change in reading frame. All the triplets after the point of insertion are read incorrectly.

point of insertion

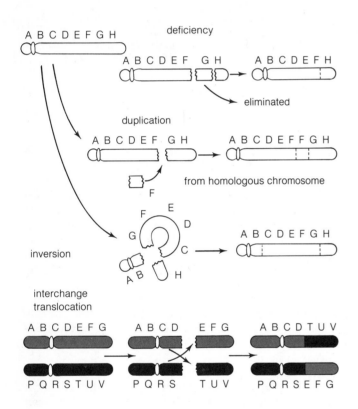

Figure 8–6 Diagrams showing how chromosome breakage and reunion can give rise to the four principal structural changes that chromosomes undergo: deficiency, duplication, inversion, and translocation.

Gene and Chromosome Mutations Alterations in larger segments of the DNA, the gene and chromosome mutations, usually arise through deficiency, duplication, inversion, and translocation of whole segments of chromosome arms following breaks that are introduced in some way (Figure 8–6). Breaks in chromosomes may occur spontaneously at any stage of the life cycle, or they can be induced by exposure to various kinds of radiation, including ultraviolet light and X rays, that contain sufficient energy to open breaks in the DNA chains of chromosomes. (Some chemicals may also cause chromosome damage; see Essay 8–1.) The broken segments may rejoin the same chromosome, either in correct order or reversed, may become linked to other chromosomes, or may be completely lost from the nucleus during cell division. Complete loss of segments is usually lethal to the cell. Transfer of a segment from one chromosome to another, or reversal of a segment on the same chromosome, may have a range of effects. These changes are not necessarily lethal if the break occurs precisely between transcriptional units and entire genes are moved away to the new location. However, if a break occurs within the boundaries of a transcription unit, so that only part of a gene is transferred to a new chromosome, the mutation is likely to be lethal.

Mutations and Heredity In single-celled organisms, mutations may alter the biochemical activities with effects ranging from the almost undetectable to the lethal. As we might expect, mutations causing extensive alterations are the most likely to upset the cell's delicate biochemical balance and kill it. Rarely, a mutation is favorable, inducing changes in cellular activity that improve an organism's interaction with the environment and increase the chances for survival of the species.

In multicellular organisms, mutations in germ cells have more serious effects than mutations in somatic or body cells because the genetic constitution of offspring may be altered. Mutations here have essentially the same range of effects on the survival of the species as do mutations in single-celled organisms. Most of these mutations are likely to be disruptive and to reduce the capacity of offspring to survive in the environment. Rarely, however, mutations may occur that produce more favorably adapted individuals.

Essay 8–1

Of Grass and Chromosomes

Marijuana, or grass, is casually smoked by thousands of individuals every day. Considered harmless by many, it is the first step on the road to hell in the eyes of others.

In 1969, the U.S. government began making marijuana of controlled quality available to research scientists, and evidence has since accumulated indicating that grass is not so innocuous as first believed. It is also clear, however, that the data are far from convincing and that controversy surrounds many of the proposed detrimental effects of marijuana. One of the most interesting of these proposed effects is chromosome damage. Radiation and many chemicals cause chromosome breakage. Such breaks may leave pieces of chromosomes outside the nucleus at cell division, cause chromosomes to adhere to one another, or result in the outright loss of chromosomes from the genome. Ionizing radiation from X rays and from radioactive sources causes serious chromosome damage and is therefore considered extremely hazardous to humans. The question has arisen whether marijuana also causes such chromosome damage and may indeed be as dangerous as fairly high levels of ionizing radiation.

Evidence in favor of such an effect has been gathered by Morton A. Stenchever and his associates at the University of Utah School of Medicine. He examined white blood cells (leucocytes) from 49 individuals who had used grass for an average of three years. Looking at cell division in these white blood cells, Stenchever found an average of 3.4 chromosome breaks per 100 cells per subject. This figure is more than twice as high as the average (1.2 per 100) in control individuals. It is comparable to the damage associated with doses of ionizing radiation and presents the same potential hazard. Similar, although not identical, results were found by a number of other groups, including Douglas G. Gilmour and his associates at the New York University School of Medicine.

Working at the Columbia University College of Physicians and Surgeons in New York City, Akira Morishima found, again looking at white blood cells from heavy marijuana users, that 30 percent of the cells contained only 5 to 30 chromosomes, compared to the normal human complement of 46. Morishima found similar abnormal chromosomes in only 7 percent of the leucocytes from control individuals. There is also evidence from Gabriel Nahas of the Columbia University College of Physicians and Surgeons that, in heavy users of marijuana, there was a decrease in DNA synthesis. This would cause a large number of abnormalities and would surely be hazardous.

As convincing as this evidence is, there is equally convincing evidence that grass has no effect. For example, Warren Nichols of the Institute of Medical Research, Camden, New Jersey, examined the leucocytes of 24 individuals who had been given marijuana under controlled conditions for as long as 12 days (compared, however, to three years in Stenchever's work). He found no evidence of an increased incidence of abnormalities. Likewise, Henry B. Pace and his associates at the University of Mississippi found no evidence of abnormality in cultured rat cells exposed to marijuana smoke. Similar negative results have been found by other researchers.

At the present time, the data give no clear-cut answer to the question of whether marijuana does in fact cause chromosome damage. It seems, however, that any smoke, whether from tobacco or grass, is injurious to one's health and a good thing to avoid.

Multiple Alleles

One of the Mendel's basic assumptions was that each hereditary trait is controlled by pairs of factors, the alternative alleles of each gene. This concept has withstood the test of time with respect to single diploid individuals, but it soon became apparent that if whole populations are considered, more than two alleles of a gene may occur. Within a few years after the rediscovery of Mendel's work in 1900, several genes with more than two allelic forms were discovered. By the 1920s, many examples of this phenomenon, called *multiple allelism,* had been recorded.

We now understand that most genes probably have many allelic forms. However, in diploids there is a maximum of two alleles in any one individual because only two copies of each chromosome are present. (In homozygotes, these two alleles are the same, and in heterozygotes, they are different.) In the cells of haploid organisms, only one copy of each chromosome, and thus only one allele of each gene, can be present in any one individual, although many alleles may be present in the population as a whole. In contemporary terms, we know that these alleles all differ from one another because of small variations at one or more points in the sequence of nucleotides making up the gene.

One of the best examples of the inheritance of multiple alleles is coat color in rabbits. Common wild rabbits have a brownish coat of fur that is uniformly colored over the whole animal. At the other end of the spectrum is the all-white fur of the albino domestic rabbit. Crosses between wild-type and albino rabbits reveal that these colors are controlled by a single gene, designated C, the allele causing brownish fur color being dominant and the albino form being recessive. Thus, CC or Cc rabbits have brown fur, and cc rabbits are albino. (Albinism in higher animals, in general, is due to the presence of a recessive gene in the homozygous cc combination.)

Several other fur-color types were discovered in rabbits that also gave a 3:1 ratio if crossed with the wild type. Chinchilla rabbits have silver fur instead of brown. If crossed with wild-type rabbits, the offspring (F_1) are brown; crosses produce an F_2 generation that is ¾ brown and ¼ chinchilla, indicating that chinchilla is recessive to wild type. However, crosses between chinchilla and albino rabbits reveal that the chinchilla allele is dominant to albino. Another interesting allele, called Himalayan, produces a white rabbit with pigmented nose, ears, tail, and legs. This coat color is also a recessive form of the gene C for wild-type fur: Himalayan rabbits crossed with the wild type give 3 wild type to 1 Himalayan in the F_2 of standard crosses. Crosses between different combinations of the Himalayan, chinchilla, and albino alleles reveal that Himalayan is recessive to chinchilla but dominant to albino. These alleles are identified in the genetic shorthand as C = wild type, c^{ch} = chinchilla, c^w = albino, and c^h = Himalayan. Their dominance relationships are summarized as follows:

1. C with C, c^{ch}, c^h, or c^w yields wild type with brownish fur

2. c^{ch} with c^{ch}, c^h, or c^w yields chinchilla

3. c^h with c^h or c^w yields Himalayan

4. c^w with c^w yields albino

In other words, brownish fur is dominant over chinchilla, which in turn is dominant over Himalayan, which is dominant over albino.

Most genes exist in a series of alleles with dominance rankings similar to this system for coat color in rabbits. This additional complexity presents no real difficulty in genetic analysis, however, because for any given individual, as already mentioned, only two of the alleles will be present.

Incomplete Dominance among the Alleles of a Gene Work with multiple alleles has also revealed that the effects of recessive alleles are not always completely masked by dominant forms of the gene as they were in the traits Mendel studied. This *incomplete dominance* means simply that in a diploid, heterozygous individual, the effects of both alleles can be detected. We will first describe incomplete dominance of this type in a simple two-allele system to illustrate the basic principles involved. An excellent example of such a system has been noted in the inheritance of feather color in chickens. A color called blue Andalusian is much admired. Breeders trying to increase their stock noted that this color would never breed true. Blue Andalusians, if crossed among themselves, always gave two new colors, black and speckled white, in addition to blue, in the ratio 1 black:2 blue:1 speckled white. Both the black and the speckled white always bred true. Black and speckled-white matings, however, produced all blue Andalusian offspring.

These results suggest that the three colors are the effects of different combinations of two alleles of the gene for feather color—C = black and c = speckled white. Since black is not completely dominant to white, Cc individuals have a mixture of effects. Thus, a cross between black *(CC)* and white *(cc)* individuals yields all Cc, or blue Andalusian, chickens in the F_1, as actually observed. Crossing these F_1 individuals would produce the combinations CC, Cc, and cc in the F_2 offspring, in the ratio 1:2:1. The hypothesis can be tested by predicting and observing the outcome of backcrosses between the individuals of the F_2 generation and the original black and speckled-white parents. The predicted results:

black parent × blue F_2 *(CC × Cc)*
    = ½ black, ½ blue offspring
white parent × blue F_2 *(cc × Cc)*
    = ½ white, ½ blue offspring

are actually obtained.

Incomplete versus complete dominance, in molecular terms, probably reflects differences in the activity of proteins synthesized under the direction of the various alleles of a gene. In *complete* dominance, the dominant allele directs, through DNA transcription, the synthesis of an active form of the protein; for example, the protein involved may be an enzyme. The recessive allele directs the synthesis of an inactive protein; presumably, the difference in nucleotide sequence of the recessive allele DNA coding for the protein causes an amino acid substitution in the polypeptide chain that destroys the activity of the active site of the enzyme. Assume that the enzyme is responsible for catalyzing a step in a biochemical series producing a pigment molecule. This pigment is localized in the surface hair of an organism and is recognized as color. In *CC* individuals, color is present; in *Cc* individuals, if dominance is complete, the single dominant allele directs the synthesis of enough of the required enzyme to produce normal color. Individuals possessing recessive forms of the allele on both chromosomes of the pair synthesize no active enzyme, and are unable to convert precursor substances into the required pigment. As a result, these individuals are colorless or albino.

In *incomplete* dominance, both forms of the allele produce active enzymes. In this case, the differences in DNA sequence between the two alleles cause amino acid changes in the polypeptides synthesized that alter, but do not destroy, the activity of the proteins. In the enzyme for color used in the last example, the altered forms of the enzyme may cause changes in the biochemical pathway, yielding pigments with slightly different colors as a result. In *CC* individuals, only one pigment is present. *Cc* individuals synthesize both enzymes and have colors representing a mixture of the two pigments; *cc* individuals have only the second pigment. All three classes would be separately recognizable in the offspring.

Multiple Alleles: The ABO Blood Groups in Humans The human ABO blood groups were discovered in 1900 by Karl Landsteiner as a result of his work on transfusions. Wondering why transfusions sometimes killed people, Landsteiner found that only certain combinations of four blood types, designated by A, B, AB, and O, can be successfully mixed.

Within human blood, red blood cells are suspended in a colorless fluid called *plasma.* They must remain suspended and freely circulating in order to carry oxygen effectively. Landsteiner determined that, in the wrong transfusion combinations, an unknown agent in the plasma of one blood group caused the red cells of the other blood group to clump together and fall out of suspension, clogging up small vessels and starving body tissues for oxygen. Obviously, if this happened on any large scale or to any vital organ, the person receiving the transfusion would die.

We now know that the erratic killer in transfusions is an immune reaction analogous to the one that protects us from certain diseases. The unknown agent in the plasma is a protein called an **antibody,** programmed to recognize and inactivate certain foreign substances called **antigens** that enter the blood. (Like antibodies, antigens are usually proteins.) The difference is that we usually do not become immune to a disease without having had it once, whereas we seem to have antibodies to transfusions of certain types of blood simply by virtue of having another type in our veins. Type A blood, for instance,

Table 8–1 Human ABO Blood Types and Their Donor-Recipient Relationships

Blood Type	Has This Antigen on Red Blood Cells	Has Antibodies against	Can Receive Blood Cells from
A	A	B, AB	A or O but not AB or B
B	B	A, AB	B or O but not AB or A
AB	AB	None	A, B, AB, or O
O	O	A, B, AB	O only

has natural antibodies against the red blood cells of type B blood, and vice versa.

The four blood groups, however, are named not for their antibodies but for their antigens, or lack of antigens. There are two antigens, called A and B, whose presence, absence, or combination is controlled by three alleles of a single gene, I (for immune), as follows:

$I^A I^A$ or $I^A i$ (A antigens on red cells) = type A blood
$I^B I^B$ or $I^B i$ (B antigens on red cells) = type B blood
$I^A I^B$ (A and B antigens on red cells) = type AB blood
      ii (no antigens on red cells) = type O blood

Note that the I^A and I^B alleles are both dominant over the *i* allele and that there is a lack of dominance between the I^A and I^B alleles. People with both of these alleles, one on each of the two chromosomes carrying the I gene, produce blood cells with both the A and B surface antigens. Thus, there are no natural antibodies against A or B cells in type AB individuals, and these people can receive red blood cells (but not plasma) from A or B individuals. Red blood cells from *ii* (type O) individuals carry no antigens and can be introduced into the blood of A, B, or AB people without danger of coagulation. Other permissible combinations are outlined in Table 8–1.

Besides helping to prevent transfusion deaths, our knowledge of blood groups has a legal application. Inheritance patterns of the ABO groups, and of other blood antigens, can rule out some men as possible fathers of a child in paternity cases. For example, a man with type O blood could not possibly be the father of an AB child. Obviously, however, the proof is only negative: Coincidence of blood types cannot be used to prove paternity.

Polygenic Inheritance: Traits Controlled by Multiple Genes

Some traits follow a pattern of inheritance that seems to mimic the blending of characteristics disproven by Mendel. This is especially true of such human characteris-

Figure 8–7 with Punnett square:

	$\frac{1}{4}P_1P_2$	$\frac{1}{4}P_1p_2$	$\frac{1}{4}p_1P_2$	$\frac{1}{4}p_1p_2$
$\frac{1}{4}P_1P_2$	$\frac{1}{16}P_1P_1P_2P_2$	$\frac{1}{16}P_1P_1P_2p_2$	$\frac{1}{16}P_1p_1P_2P_2$	$\frac{1}{16}P_1p_1P_2p_2$
$\frac{1}{4}P_1p_2$	$\frac{1}{16}P_1P_1P_2p_2$	$\frac{1}{16}P_1P_1p_2p_2$	$\frac{1}{16}P_1p_1P_2p_2$	$\frac{1}{16}P_1p_1p_2p_2$
$\frac{1}{4}p_1P_2$	$\frac{1}{16}P_1p_1P_2P_2$	$\frac{1}{16}P_1p_1P_2p_2$	$\frac{1}{16}p_1p_1P_2P_2$	$\frac{1}{16}p_1p_1P_2p_2$
$\frac{1}{4}p_1p_2$	$\frac{1}{16}P_1p_1P_2p_2$	$\frac{1}{16}P_1p_1p_2p_2$	$\frac{1}{16}p_1p_1P_2p_2$	$\frac{1}{16}p_1p_1p_2p_2$

Legend:

$\frac{1}{16}$	$P_1P_1P_2P_2$	black
$\frac{4}{16}$	$P_1p_1P_2P_2$ $P_1P_1P_2p_2$	dark brown
$\frac{6}{16}$	$P_1P_1p_2p_2$ $P_1p_1P_2p_2$ $p_1p_1P_2P_2$	mulatto
$\frac{4}{16}$	$P_1p_1p_2p_2$ $p_1p_1P_2p_2$	tan
$\frac{1}{16}$	$p_1p_1p_2p_2$	white

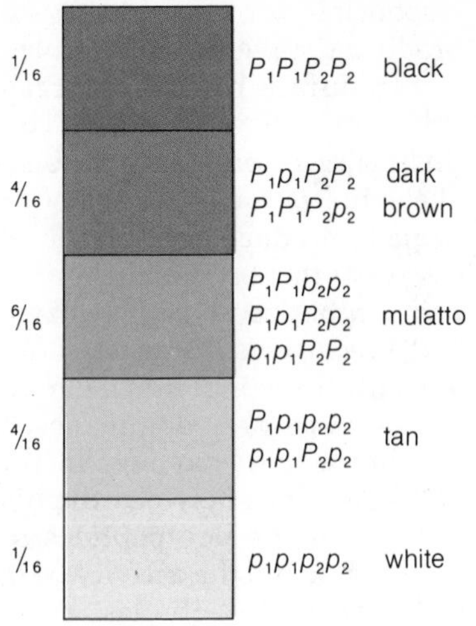

Figure 8–7 Inheritance of skin color in the children of mulattoes.

tics as skin color, body size, and intelligence, which at first glance seem intermediate in the offspring of parents who differ markedly in these traits. Careful analysis, however, indicates that the apparent blending results from interactions between the alleles of two or more genes controlling the same inherited characteristics. One of the best studies of these patterns is the inheritance of skin color in humans, first investigated by C. B. Davenport among mulattoes in Bermuda and Jamaica.

Davenport's careful analysis of family pedigrees showed that offspring of intermediate color were the usual, but not the only, result of matings between mulattoes. Such matings also occasionally produced children who were either darker or lighter than the parents. In fact, matings between mulattoes produced a spectrum of skin pigmentation ranging from a color apparently as dark as that of pure blacks to colors indistinguishable from the various white skin tones.

Davenport saw that this range could be explained by assuming that more than one gene is involved. He proposed that two different genes, which he designated by P for *pigmentation (P_1 and P_2),* control human skin color. Pure African blacks, with no white ancestors, should carry the four alleles $P_1P_1P_2P_2$ (note that the capital *P,* used to indicate dark pigmentation, does *not* indicate dominance). Pure whites on the other hand, should carry the four alleles $p_1p_1p_2p_2$, for light pigmentation. Matings between such homozygotes would produce the genotype $P_1p_1P_2p_2$, which is assumed to be a mulatto with an intermediate brown skin. Since neither allele of either of the two genes is dominant, the heterozygotes can be distinguished in outward appearance, or phenotype, from either type of homozygote.

A mating between two $P_1p_1P_2p_2$ mulattoes with the same degree of pigmentation, by contrast, would produce different results—some of them unexpected if blending were assumed. The possible combinations of gametes from the two individuals are shown in the Punnett square in Figure 8–7. Both parents will produce the four gametes with equal chances of ¼. These, when combined in the squares of the figure, produce a range of pigment combination varying more or less evenly from pure black, $P_1P_1P_2P_2$, through a range of intermediate values (indicated by the shading in Figure 8–7), to pure white, $p_1p_1p_2p_2$. A range of pigments of this sort would not be expected if parental pigments blended evenly in offspring. If that happened, two mulatto parents of intermediate color would always produce mulatto children of intermediate color.

More recent studies have confirmed Davenport's theory that skin color is controlled by more than one gene. The children of mulatto parents show the expected range of color from black to white (Figure 8–7), with intermediate values, also as expected, arising more frequently than either extreme. But these later studies have also pinpointed a difficulty with Davenport's work. If pigmentation were controlled by only two genes, as Davenport suggested, the white or black extreme would be expected in 1 out of every 16 births. Actually, the extremes are much rarer than this—only about 1 in every 1,000 births. This low frequency suggests that skin pigmentation may instead be controlled by as many as five genes, each with at least two alleles, *P* and *p*.

Graphing the distribution of skin pigmentation in offspring of mulattoes would produce a bell-shaped curve, which is typical of traits controlled by more than one gene. A graph of measured frequencies of body size in men (Figure 8–8) produces the same kind of bell-shaped curve, indicating that this human trait also is controlled by more than one gene. In such cases we say inheritance

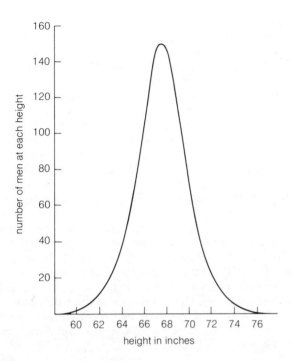

Figure 8–8 Graph of numbers of men of different heights, ranging from 58 to 77 inches. The distribution of frequencies produces a typical bell-shaped curve, indicating that height is controlled by more than one pair of genes.

is *polygenic.* Polygenic control of characteristics is common in both plants and animals.

Hybrid Vigor

From the standpoint of profits to mankind, perhaps the most valuable discovery of genetics was that of hybrid vigor. Joseph Gottlieb Kölreuter, one of the most important of the pre-Mendelian investigators of inheritance, discovered in 1763 that hybrids were often more vigorous than either parent. Darwin, in his 1876 book on plant reproduction, concluded that "cross-fertilization is generally beneficial and self-fertilization injurious." When pure lines were bred together for generations, they became progressively less vigorous, yielding less and less. When two of these purebred lines were crossed, the progeny showed a greatly increased vigor, and yields went up 25 to 40 percent or more. In 1910, George Shull coined the term *heterosis* to describe this situation.

Corn agriculture in the United States is based completely on hybrid corn. Primarily for this reason, U.S. yields per acre are the highest in the world. Seed compa-

nies produce hybrid seed corn under carefully controlled conditions with carefully selected parents. Farmers buy such hybrid seed and plant it. They will not save any of the crop for next year's seed, because the second year's crop would not be so good as the first. It would include genotypic variants ranging between the extremes of the parents of the hybrid corn, as in the skin color example just discussed. This means that farmers are forced to buy new seed each year, but the increase in production more than offsets the cost. It also means that farmers can take advantage of the new strains being developed by plant breeders. One of the major factors in the continued high productivity on U.S. farms has been the extensive and successful plant-breeding programs undertaken by the U.S. Department of Agriculture.

Geneticists still do not understand the basis for hybrid vigor, but two theories are usually cited. One is the overdominance theory, which suggests that both dominant and recessive alleles contribute to the growth and vigor of the individual. This theory says that two dominant alleles are too much of a good thing, two recessive alleles are too little, and the combination of a dominant and a recessive allele is just right. Recessive alleles, in other words, are not totally passive but contribute something positive.

The second theory, called the dominance theory, is best explained by an example. Several known varieties of dwarf corn are unable to synthesize the growth hormone gibberellin. Each variety lacks this ability for a different reason—that is, each is unable to carry out a different step in the synthesis (but able to carry out all the other steps)—and each is homozygous recessive for the gene involved. If any two of the dwarfs are crossed, each supplies the missing dominant effective allele (and hence the missing enzyme) for the other; gibberellin is synthesized; and the offspring are much taller and more vigorous than either parent. According to this theory, in other words, different pure lines contain different homozygous recessives; chances are, then, that crossbreeding will pair each such recessive allele with a corrective dominant and thus give the offspring greater vigor.

The Expression of the Genome: Interaction of Heredity and Environment

The presence of a specific gene sets the limits of development but does not necessarily determine the extent to which this limit is reached. In the case of the dwarf corn, the right genetic combination is necessary for gibberellin production, but if the plant lacks an essential nutrient, such as nitrogen, it may remain stunted even if the genes are present. Environment and the genetic composition of the organism always interact in a given individual.

One of the recurring questions in human genetics is the relation between heredity and environment in intelligence. There is no question that some types of mental deficiencies are genetic, and it is also true that malnutrition will result in arrested mental development. But what of intelligence within the normal range? Is it determined by genes, nutrition, or both? And are these factors related to race? Such questions have been raised time and again, but the answers are often so clouded by rhetoric and emotion that they are hardly science.

In any consideration of gene expression, it is important to remember that no gene acts in isolation. Its action is always influenced by the overall genetic makeup, or genotype, of the individual in which it occurs. The expression of some genes depends upon, or is modified by, other genes known as *modifiers* or *epistatic* genes. Two terms have been devised to describe the complex interaction between such modifying genes and their targets. *Penetrance* is the likelihood that a given gene, carried in a combination where it could be expressed, will be expressed. Penetrance is usually given as the percentage of individuals carrying the gene who actually show the expected phenotype. *Expressivity* denotes the manner in which the phenotype is expressed, for example, the extent to which the expression of the gene is influenced by the external environment. For instance, the genes Mendel was studying showed 100 percent penetrance and clear-cut expressivity.

Not only the rest of the genotype, but the external environment as well, may influence both penetrance and expressivity. Himalayan rabbits bear a dominant gene for black fur color, but only the tips of their feet, ears, nose, and tail are black, because the expression of this gene is highly sensitive to temperature. If we shave a patch on a Himalayan rabbit's back and keep the patch covered with an ice pack, the new fur there will be black. For the gene to be expressed, the temperature must be below about 39°C, as it normally is at the tips of the extremities. Siamese cats show a similar pattern.

Genes always act within some environment—the environment created by all the other genes as well as the surrounding physical environment. The total complex of genes in an organism is called the **genome.** Understanding the functioning genome—why genes act only at certain places and at certain times—is one of the most complex, important, and challenging problems of biology.

Summary

Genes are subject to changes that result in new forms, or alleles. Such changes, called mutations, are of great importance both to the individual and to the species. Mutations may be caused by chemical or physical agents. Mutations are changes in the sequence or type of nitrogen bases found in the DNA of the cell. A second type of mutation, called a chromosome mutation, results from rearrangements of whole segments of chromosomes.

In a diploid individual, there can be no more than two alleles present for a given gene, but in the population as a whole there may be many alleles of the same gene. Often, the effects of both alleles in a diploid heterozygous individual can be detected, as a result of the phenomenon of incomplete dominance. Multiple alleles are important in the understanding of genetic traits in many organisms, including humans.

Hybrid vigor, or heterosis, occurs when the offspring of two purebred lines are crossed and the resulting hybrid is more vigorous than either of the parents. Hybrid vigor is particularly important in agricultural crops.

Finally, genes set the limit on the development of the organism, but the environment may determine the extent to which the gene can be expressed. Environment and the genetic composition of the organism always interact in a given individual.

Problems

1. A man is suing his wife for divorce on the grounds of infidelity. He claims that their third child is not his own. Their first child and second child have blood types O and AB respectively. The third child, whom the man disclaims, has type B. Can this information be used to support the man's case?

2. In mice, there is a set of multiple alleles of the gene for coat color. Three of these alleles are: $A^B =$ black, $A^C =$ chinchilla, and $A^Y =$ yellow. The following crosses were performed:

black male × chinchilla female
offspring: 45 male, 42 female black; 41 male, 46 female chinchilla

chinchilla male × yellow female
offspring: 52 male, 47 female chinchilla; 49 male, 52 female yellow

black male × yellow female
offspring: 100 male, 105 female black

What is the relationship among the alleles with respect to dominance and recessiveness?

3. In a breed of cats, genes for black or white fur color do not show dominance or recessiveness. If a cat carrying only white-fur genes is bred to a cat carrying only black-fur genes, all of the offspring have gray fur. If two of these gray cats reproduce, what color do you predict their offspring will be?

4. There is a gene for colored feathers *(C)* in chickens that is completely dominant over a gene *(c)* for lack of color. Chickens may also carry a gene that prevents the colored-feather gene from exerting its effect *(I)*. This gene is completely dominant over another gene that allows for normal color development *(i)*.

  a. Feather color in chickens is an example of what type of inheritance?

  b. How many colored-feathered offspring will result from a cross between an *IiCc* male and an *IICc* female?

  c. How many colored-feathered offspring will result from a cross between an *IiCc* male and an *IiCc* female?

5. In *Drosophila* flies, wings may be long (wild type) or stunted (vestigial). You perform a cross between wild-type heterozygous males and vestigial females to study the inheritance of this trait. To ensure your sample size is large enough, you have six test vials, each containing five male and five female parents. You place them on a row on top of your desk, which is next to a radiator.

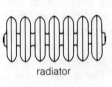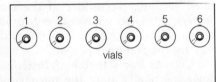

The results of the F$_1$ generation are as follows:

Vial	Percentage with Vestigial Wings	Percentage Wild Type
1	2	98
2	10	90
3	25	75
4	40	60
5	46	54
6	48	52

  a. What would you have expected the phenotypic ratios to have been in the F$_1$ generation?

  b. On the basis of the observed data, how can you explain the deviation from the expected?

6. A soybean farmer finds that some of the soybeans produced on his field are larger than others. Since larger soybeans mean higher yield and thus more money, he selects some large beans and saves them for seed for the following growing season. He keeps track of which plants grow from these selected large seeds and measures the size of the soybeans they yield. He discovers that only half of these new beans are the large size, and comes to you for an explanation. How would you explain the results of his experiment to him?

7. You are a genetics counselor, and a woman who works in an X-ray laboratory comes to you for advice. She and her husband are planning to start a family soon, but she is concerned because her hand was exposed to a high dosage of X rays a few months earlier. She would like to know whether this exposure will result in her passing on mutations to her offspring. How do you advise her?

8. A dwarf bull calf is born to apparently normal parents in a herd of cattle. How might you decide whether the dwarf is the immediate result of a mutation, the result from a chance mating of carriers of a recessive dwarfism, or the result of environmental modification?

9. The fungus *Neurospora* synthesizes the essential metabolite A via a series of precursor substances B, C, and D. The transformations of precursors are dependent upon enzymes which, in turn, are under genetic control:

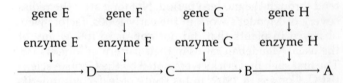

Compare the consequences of a mutation in gene H with those of a mutation in gene F in terms of survival of the *Neurospora* mutants.

10. Identify the type of mutation illustrated in each example below as a transition, a transversion, a deletion, or a chromosome mutation.

  a. A chemical mutagen, hydroxylamine, that causes the change G-C to A-T, but not A-T to G-C

  b. The mutation type that causes sickle-cell anemia

  c. A transfer of a segment of a transcriptional unit

Supplement 8–1
Cytoplasmic Inheritance

Discoveries of an unusual pattern of inheritance, based on genes carried in the cytoplasm rather than in the nucleus, date from the work of classical geneticists and ex-

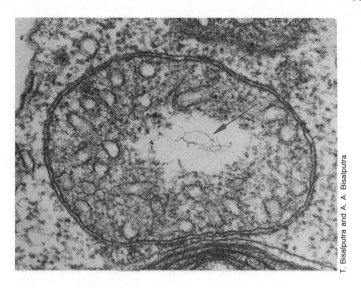

Figure 8–9 A DNA deposit (arrow) in the matrix of a mitochondrion of the alga *Egregia.* ×66,000.

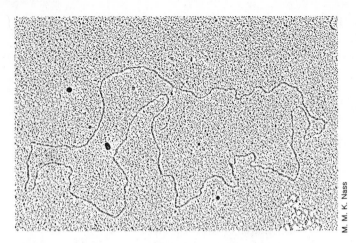

Figure 8–10 A DNA circle isolated from a mouse mitochondrion. ×34,100.

tend through the modern period. Not long after the rediscovery of Mendel's work in the early 1900s, factors were observed in plants that did not appear to follow any of the usual Mendelian ratios in their pattern of inheritance. The first such mavericks were noted by the German scientist C. Correns in 1909, in his study of leaf-color inheritance in the four-o'clock, *Mirabilis.* In *Mirabilis,* individuals sometimes show patterns of white and green segments (called variegation) on their leaves and stems. Chloroplasts within the cells of the bleached segments are colorless and do not turn green. If enough normal, green tissue is also present, plants containing the white segments are viable and may produce flowers on both the white and green segments.

By choosing flowers from different regions of these variegated plants, Correns was able to obtain pollen and eggs originating from entirely white or green portions of the plants, or from regions in which the white and green segments were intimately mixed. Crosses between pollen and eggs from these segments revealed that the offspring, instead of following predictable Mendelian ratios, always reflected the color of the tissue from which the egg was derived. The color of the parts of the plants contributing the pollen had no effect on the outcome of the cross. For example, eggs in white-segment flowers fertilized by pollen from green-segment flowers produced completely white seedlings. (This condition was lethal and the seedlings died.) The reverse cross (green-segment eggs fertilized by white-segment pollen) produced normal green seedlings with no variegation. Eggs in flowers on segments of the plant with closely mixed white and green

(variegated) segments produced white, green, or variegated seedlings according to the exact location of the flowers, no matter what segments produced the fertilizing pollen. Thus, color inheritance was determined in some way by the maternal parent.

Correns knew that in *Mirabilis,* as in virtually all flowering plants, only the nucleus of the pollen enters the egg at fertilization; all of the pollen-tube cytoplasm is left outside. Since all the genetic material from the pollen nucleus was present, but had no effect at all on leaf color, and since only the egg's cytoplasm was present, Correns reasoned that the gene for the greening of the chloroplasts was carried in the cytoplasm and not in the nucleus of *Mirabilis* gametes.

This hereditary pattern, called *cytoplasmic inheritance,* has since been observed in hundreds of species of plants, animals, and microorganisms. As linkage was discovered and chromosome maps were developed, geneticists noted that the cytoplasmically inherited factors, as expected, showed no linkage to any genes carried on the chromosomes of the cell nucleus.

Many of the traits carried by cytoplasmic genes affect primarily the biochemistry and function of mitochondria and chloroplasts. Inheritance often follows the pattern noted in *Mirabilis,* in which most or all of the cytoplasm of the offspring is derived from one of the two parents. Once DNA was established as the molecular basis of heredity in the 1950s, scientists began to search for DNA in the affected mitochondria and chloroplasts. If cytoplasmic genes are actually carried in these structures, and if DNA is the genetic material, they reasoned, then close

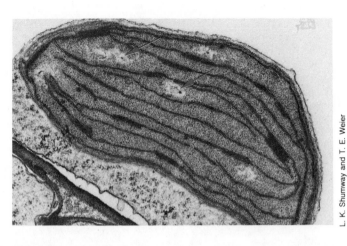

L. K. Shumway and T. E. Weier

Figure 8–11 DNA deposits (arrows) in the stroma of thin-sectioned corn tissue. ×20,000.

inspection of these organelles ought to reveal the presence of DNA. By the 1960s, techniques for detecting nucleic acids had been refined enough to reveal very small quantities of DNA in both mitochondria and chloroplasts.

Mitochondrial DNA is the best-studied of the cytoplasmic genes. Electron microscopy shows the DNA as finely fibrillar deposits in the matrix of the mitochondrion (Figure 8–9). When isolated, the DNA proves to be in the form of a closed, irregular circle (Figure 8–10), much like viral and bacterial DNA. The quantity of DNA in the mitochondrial circle is small (it contains only about 4.5 to 5.5 micrometers of DNA), only a fraction of the DNA in a bacterial nucleus.

Chloroplasts also show DNA deposits in the stroma (Figure 8–11) and appear to contain molecules of greater length than mitochondria. Isolating purified DNA from chloroplasts is difficult, but most studies indicate that chloroplast DNA exists as linear or possibly circular molecules perhaps 35 to 40 micrometers in length. Even at this length, the chloroplast DNA molecule contains only 2 to 3 percent as much DNA as a bacterial nucleus.

Biologists are uncertain about the possible functions coded for by mitochondrial and chloroplast DNA. They have learned about certain functions by studying mutations that seem to have been inherited through the cytoplasm. Although many such mutations involve enzymes present in mitochondria and chloroplasts, all the affected enzymes identified so far are assembled under the control of nuclear rather than cytoplasmic genes. Thus, the mutations must somehow alter the capacity of mitochondria or chloroplasts to incorporate and use proteins coded for by nuclear genes.

Both mitochondria and chloroplasts have also been shown to contain ribosomes and all of the biochemical machinery necessary for protein synthesis. The ribosomes inside these organelles differ from the ribosomes in the cytoplasm outside these organelles, resembling bacterial, rather than eukaryotic ribosomes, both in structure and in properties. Tests for similarities in nucleotide sequence show that the RNA of these ribosomes is coded for by mitochondrial and chloroplast DNA rather than nuclear DNA. The DNA deposits of these organelles probably also contain genes coding for all the transfer RNAs required for protein synthesis.

9

The Molecular Biology of the Gene

One of the most important decisions a research biologist faces is what experimental organism to use. Mendel picked the right organism for his work, but De Vries did not. Research in classical genetics was greatly speeded by Morgan's selection of *Drosophila,* and biochemical genetics was largely founded on Beadle and Tatum's work with the mold *Neurospora.* People outside research often question the use of such esoteric organisms, but there are usually excellent reasons for selecting them. The important point is that what is found to be true for *Drosophila* may also be true for humans.

In the 1950s, when genetics began to concentrate on the molecular aspects of the gene, it became clear that new types of experimental organisms would be needed, because new kinds of questions were being asked. The most helpful organism that was found—the bacterium *Escherichia coli*—was a humble one indeed, and close to home. *E. coli* is one of the many bacteria that normally inhabit our intestines.

One significant advantage of *E. coli* as an experimental organism is that the composition of the culture medium on which it grows can be precisely controlled. Thus, researchers know exactly what goes into the bacterial cells. By comparing this with what comes out, they have a good chance of learning what changes occurred inside. As a result, the effects of genetic changes can often be traced in precise biochemical terms. Such studies have led to a more complete and accurate definition of the gene and its activities at the biochemical and molecular level.

The same methods were soon extended to eukaryotes, particularly yeasts and other fungi such as *Neurospora* and *Aspergillus,* in which large numbers of offspring can be grown and analyzed biochemically. Molecular studies of this type have even been carried out with *Drosophila* and corn. As a result, the findings from prokaryotes concerning the molecular nature of the gene are known to apply equally well to eukaryotes.

Sexual Reproduction and Recombination in Bacteria

Until the 1940s, sexual reproduction and recombination were not known to occur in bacteria. The possibility that these processes might take place in microorganisms was of great interest, however, because of the obvious advantages in studying recombination in organisms that could be bred by the billions, with complete generations passing in only 20 minutes. In comparison, even *Drosophila* takes two weeks to grow from zygote to sexual maturity, and geneticists working with corn are lucky to get two generations a year.

In 1946, J. Lederberg and E. L. Tatum, working with *E. coli* at Yale University, investigated the possibility that recombination might occur in bacteria. Wild-type *E. coli* bacteria can grow on a minimal medium containing only inorganic salts and an energy source such as glucose. Using X rays and ultraviolet light as mutagens, Lederberg and Tatum developed several mutant strains that could not grow on this minimal medium. One strain could grow only if the vitamin biotin and the amino acid methionine were added to the minimal medium, meaning that the mutations had obviously affected at least one of the enzymes required to synthesize each of these substances. A second strain could make biotin and methionine, but could not make threonine and leucine and could grow only if these amino acids were added to the medium. These wild-type and mutant genotypes can be represented in genetic shorthand as:

Wild type:	bio$^+$	met$^+$	thr$^+$	leu$^+$
Mutant strain 1:	bio$^-$	met$^-$	thr$^+$	leu$^+$
Mutant strain 2:	bio$^+$	met$^+$	thr$^-$	leu$^-$

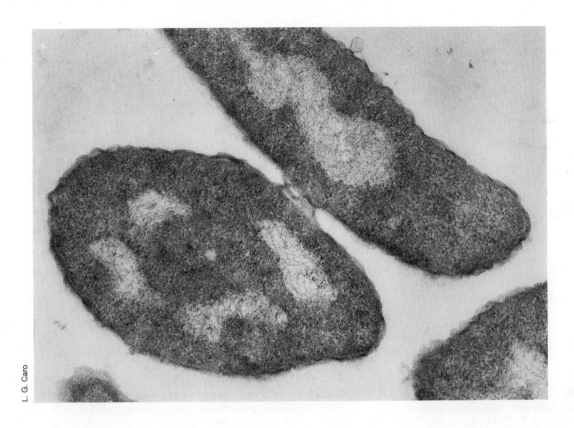

Figure 9–1 Conjugating cells of *E. coli*. A bridge of cytoplasm has formed between the cells in the region of contact. ×68,000.

A plus sign indicates the ability to synthesize the substance in question; a minus sign, the lack of that ability.

Lederberg and Tatum mixed about 100 million cells of the two mutant strains together and placed them on the minimal medium. Most of the cells died, but a few hundred bacteria survived and divided to form colonies containing cells descended from individuals that were able to grow on the minimal medium. These new cells must have had all the wild-type alleles, or they would not have survived. Lederberg and Tatum established that these cells were recombinants. Thus, the bacterial cells had undergone a form of sexual reproduction—or at least had achieved the kind of exchange of genetic material that usually occurs only in sexual reproduction.

The process by which this exchange took place was not immediately clear. At first, two bacterial cells were thought to have fused together, producing a diploid zygote (bacterial cells are normally haploid and contain only one copy of a single chromosome). However, it was later established that, instead of actually fusing, the cells had formed a temporary cytoplasmic bridge between them (Figure 9–1). During this process, called **conjugation,** part or all of the chromosome of one cell moved into the other through the bridge.

Only a few of the cells in a bacterial culture are able to donate DNA. Those few were found to contain, in addition to the single normal chromosome, a small supplementary piece of DNA. This extra piece was called the F factor (F = fertility). Cells possessing the factor, designated F⁺ cells, conjugate with cells not possessing the factor (F⁻). In this process, the F factor is usually transferred to the F⁻ cell, transforming it into F⁺. A part of the chromosome is transferred from the F⁺ to the F⁻ cell along with the F factor. Once this occurs, pairing and recombination may take place between the chromosome fragment introduced from F⁺ and the recipient cell chromosome.

Some conjugating strains of *E. coli* were found to undergo recombination at a much higher rate than the F⁺, F⁻ cells. In these strains, designated Hfr (for "high-frequency recombination"), the F factor DNA was found to be directly linked into the chromosome by covalent bonds rather than separately suspended in the cytoplasm. These Hfr strains, because of their greatly increased recombination frequency, were to prove extremely useful in later experiments.

In recombination experiments with *E. coli,* it appeared that the entire chromosome was rarely transferred from the Hfr to F⁻ cells, because some of the alleles being studied seemed never to be transferred to the F⁻ cell at

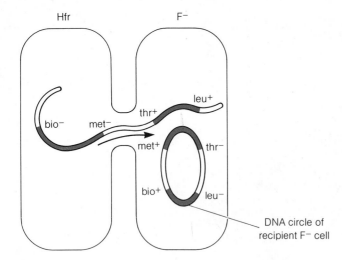

Figure 9–2 Injection of genes from an Hfr to an F⁻ bacterium. Part or all of the Hfr chromosome passes through the bridge between the conjugating cells to enter the F⁻ bacterium.

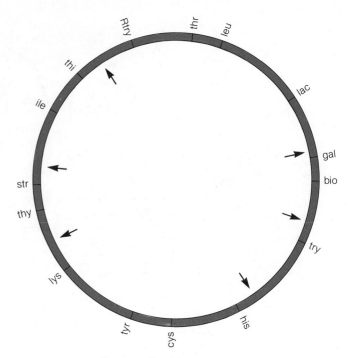

thr = requires threonine
leu = requires leucine
lac = cannot metabolize lactose
gal = cannot metabolize galactose
bio = requires biotin
try = requires tryptophan
his = requires histidine
cys = requires cysteine
tyr = requires tyrosine
lys = requires lysine
thy = requires thymine
str = resistant to streptomycin
ile = requires isoleucine
thi = requires thiamine
Rtry = uncontrolled synthesis of tryptophan

Figure 9–3 Partial map of the *E. coli* chromosome. The arrows mark points at which the F⁺ factor may be inserted into the chromosome.

all. The basis for this characteristic of bacterial recombination and the mechanism underlying conjugation itself were worked out by F. Jacob and E. L. Wollman of the Pasteur Institute in Paris, using the Hfr strains of *E. coli.*

These investigators mixed Hfr and F⁻ bacteria together and, at regular intervals after mixing, removed some of the cells and separated the conjugants by agitating them in an ordinary household blender. Jacob and Wollman found that the amount or length of the chromosome transferred was directly proportional to the length of time the cells were allowed to remain in conjugation, indicating that the Hfr cell slowly injects its chromosome, as a linear molecule of DNA, into the F⁻ cell (Figure 9–2). Full transfer of the entire chromosome from Hfr to F⁻ cells, which evidently occurs rarely in nature, required about 90 minutes.

Experiments of this type with *E. coli,* using a large number of mutant alleles, allowed the location of the known genes of this bacterium to be mapped on the chromosome. Surprisingly, the genes were discovered to be arranged in a closed circle. (Figure 9–3 shows a partial map of the *E. coli* chromosome.) Obviously, the circle must break if segments of various lengths are transferred from Hfr to F⁻ cells. Fortunately for the mapping experiments, the point of breakage is uniform within many strains of *E. coli.* The breaks occur regularly at the point of insertion of the F⁺ factor into the chromosome in Hfr strains; the F⁺ factor may be integrated at any of several points around the circle (shown by the arrows in Figure 9–3).

Transformation and Transduction We have just seen that transfers of genetic material can take place by conjugation, but other mechanisms are also known. In one of these, called **transformation,** pieces of DNA released from fragmenting cells are absorbed by healthy cells from the medium; recombination may occur between segments of the absorbed DNA and the chromosome of the absorbing cell.

Transformation was first thought to be an unusual phenomenon, but more recent evidence indicates that it

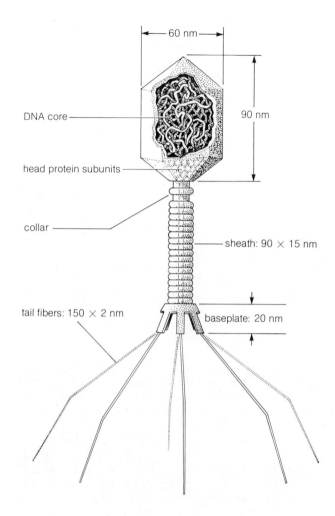

DNA core

head protein subunits

collar

tail fibers: 150 × 2 nm

60 nm

90 nm

sheath: 90 × 15 nm

baseplate: 20 nm

Figure 9–4 The structure of a T-even bacteriophage particle.

is the general property of bacterial cells. Any part of the chromosome may be transferred from donor to recipient cells by this mechanism. Transformation is best known through laboratory studies, but it probably also occurs in nature.

In the third known pathway for the transfer of chromosome fragments from one cell to another, called **transduction,** the transferring agent is a virus that infects bacteria. The virus grows and multiplies in an infected cell, and during growth may include a fragment of the host chromosome within a virus particle. These virus particles, released from the host cell, may infect a second cell and transfer the included chromosome fragment to the recipient, where recombination takes place between the DNA of the recipient cell and that of the introduced fragment.

Viruses, Bacterial Viruses, and Viral Recombination

Free virus particles consist of a core of nucleic acid, either DNA or RNA, surrounded by a shell, or coat, of protein. All degrees of complexity exist in viruses, from simple particles consisting of a nucleic acid core and a shell of protein molecules of a single type, to more complex particles with shells made up of several different kinds of protein, lipid, and carbohydrate molecules. In most viruses, the core contains a single, continuous nucleic acid molecule. The nucleic acid core may be either a single nucleotide chain or two chains twisted into a double helix.

Virus particles are probably best classified as nonliving material *when outside the host cell.* In this form, suspended in solution, viruses carry out none of the activities of life and are inert except for their capacity to attach to host cells. Once attached, the viral nucleic acid enters the host and converts the cell's machinery to the production of viral protein and nucleic acid. Thus, the viral DNA acts as a chromosome, able to survive either inside or outside the host cell and, when coated with a protective layer of protein, transmissible through the surrounding medium. Viruses of different kinds infect bacterial, plant, and animal cells, and have no activity except when inside the host.

The Bacterial Viruses Most of the bacterial viruses, called *bacteriophages,* are relatively complex (Figure 9–4). The best known bacteriophages are those that attack the bacterium *E. coli* (Figure 9–4 shows one such virus). These include the T-even types (bacteriophages T2, T4, and T6), and the lambda type. The many-sided "head" of these viruses encloses the single molecule of nucleic acid or DNA. Attached to the head is a long tail containing several different proteins, each distinct from the head protein. In the T-even bacteriophages, the tail is complex and consists of a collar at the point of attachment to the head, a long cylindrical sheath, and a baseplate. The baseplate carries a number of long, hairlike extensions of protein, the tail fibers. The lambda bacteriophage resembles the T-even types but has a less complex tail.

Life Cycle of the T-Even Bacteriophages Free bacteriophage particles collide randomly with bacterial cells in the medium. As soon as a T-even virus particle touches a host cell, the tail fibers "recognize" and specifically bind to sites on the bacterial cell wall (Figure 9–5a). The head and tail sheath then inject the DNA core into the bacterial cell. The proteins of the shell remain outside (Figure 9–5b).

Once this core is inside the bacterial cell, a part of

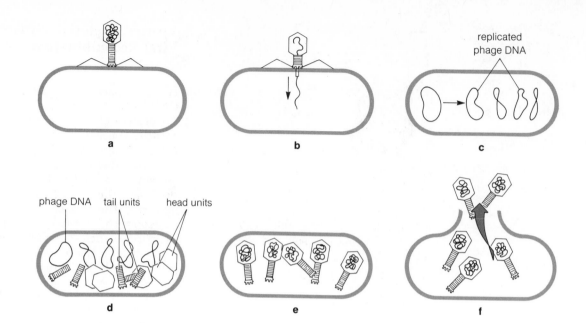

replicated
phage DNA

a

b

c

phage DNA   tail units   head units

Figure 9–5 Life cycle of the T-even bacteriophages.

d

e

f

the bacteriophage DNA is immediately transcribed by bacterial RNA polymerase. These first messenger RNAs direct the ribosomes of the infected bacterium in synthesizing viral proteins. The first proteins made include all the enzymes necessary for replicating the viral DNA. The bacterial cell, now converted into a mechanism for producing viral DNA, enters into a period of bacteriophage DNA replication until from 100 to 1,000 new bacteriophage DNA molecules are synthesized (Figure 9–5c).

Soon after DNA replication begins, additional segments of the viral DNA are transcribed into messenger RNAs. These "late" viral messengers direct the bacterial cell machinery to synthesize proteins of the bacteriophage and assemble them into heads and tails. As the head and tail segments accumulate in the bacterial cytoplasm (Figure 9–5d), the newly synthesized bacteriophage DNA condenses into tightly packed masses inside the heads, and finally the finished tails are attached to the heads (Figure 9–5e). As synthesis of the virus particles is completed, another "late" mRNA directs the synthesis of an enzyme that breaks down the bacterial cell wall. The bacterial cell ruptures, releasing the newly synthesized bacteriophage particles to the surrounding medium (Figure 9–5f).

Lysogenic Bacteriophage Viral infection does not always proceed to completion, with immediate production of new virus particles and rupture of the host bacterium. Some bacterial viruses, such as the lambda bacteriophage, enter the host cell and exist for generations in a form that does not hurt the host. Viral proteins are not synthe-

sized during this stage of infection, and host cells do not rupture. This type of infection is called **lysogeny.**

In lysogenic infection, the lambda particle attaches to the surface of an *E. coli* cell and enters it by a mechanism similar to that used by the T-even phages (Figure 9–6a,b). Instead of immediately converting the bacterial cell machinery to production of new lambda particles, however, the viral DNA is incorporated into the bacterial DNA circle of the *E. coli* chromosome by covalent bonds. This linkage occurs as follows: The lambda DNA, which forms a closed circle, pairs closely with a region of the *E. coli* DNA circle (Figure 9–6c). Breaks are introduced in both the bacteriophage and bacterial DNA, (Figure 9–6d), and the broken ends rejoin in such a way that the ends of the viral DNA are now joined with the ends of the bacterial DNA. The rejoined ends are then covalently linked together; when the process is complete, the *lambda* DNA is integrated as a segment of the bacterial DNA circle (Figure 9–6e). In this form of infection, in which the lambda DNA is called a *prophage,* the viral DNA is replicated along with the bacterial DNA and is transmitted to daughter cells at each division (Figure 9–6f). Replication and transmission, without activation of the genes for production of new virus particles, may continue through many bacterial generations. Eventually, in some of the descendants containing the integrated lambda chromosome, the viral chromosome is released from the bacterial DNA and becomes completely active. (One agent known to cause release and complete activation of the lambda DNA is ultraviolet radiation.) After the phage DNA has been released from the bacterial chromosome, by reversal of the insertion process, the lambda

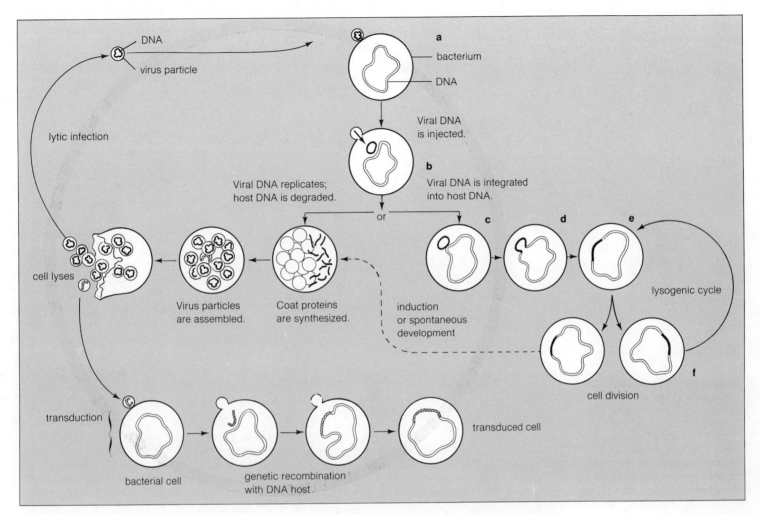

Figure 9–6 In the case of lysogenic phages, two developmental cycles are possible. The one shown on the left is the same as that of a normal phage and results in the death of the bacterium and the formation of new viruses. The one on the right shows how the viral DNA is incorporated into the bacterial DNA and is replicated every time the bacteria divide. But under some conditions—even after many generations of bacteria—the virus may become active and destroy the cell. When the new virus particles are being assembled in the bacteria, a mistake may occur, and a piece of bacterial DNA may be coated with viral protein. This is capable of infecting a new bacterium, and the old bacterial DNA fragment becomes a part of the new host bacterium's DNA. This is transduction, shown at the bottom of the figure.

DNA directs replication of new viral particles, rupturing the bacterial cell and releasing the replicated bacteriophages to the surrounding medium as in ordinary bacteriophage infection.

Integration of "foreign" DNA into host chromosomes by the lysogenic process is not limited to the lambda bacteriophage infecting *E. coli*. Other DNA particles in bacteria, such as the F factor we just studied, can also be integrated into the cell chromosome in this way. As already noted, integration of the F factor, normally carried

in the cytoplasm as a separate piece of DNA, into the bacterial chromosome converts the cell from F⁺ to Hfr. Similar mechanisms also occur among the viruses infecting eukaryotes.

This type of viral infection is of particular interest in the case of eukaryotic cells because of the possibility that it may be involved in some forms of cancer. Lysogenic viruses form the basis of the **oncogene** (from the Greek *oncos*, "tumor") theory of cancer. Among the viruses known to be oncogenic in birds and mammals, some

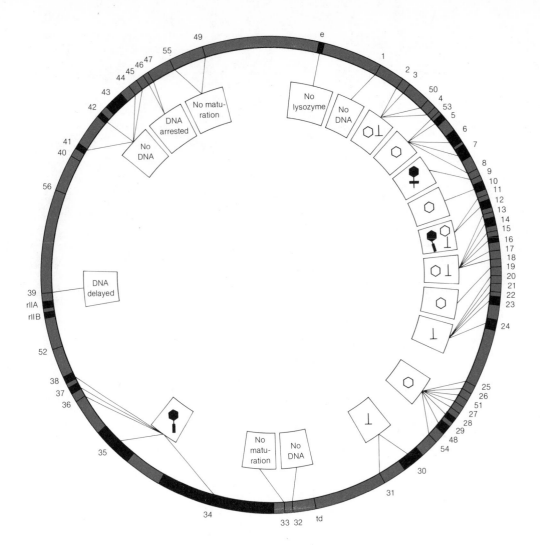

Figure 9–7 A partial map of the DNA circle of the T4 bacterio- phage. The numbers identify the locations of 56 genes. Redrawn from "The Genetics of a Bacterial Virus," by R. S. Edgar and R. H. Epstein. Copyright © 1965 by Sci- entific American, Inc. All rights reserved.

contain DNA as their genetic material and others contain RNA. RNA-containing tumor viruses are thought to in- duce certain human leukemias and breast cancers. They act in the following way: The RNA of the virus acts in the host cell as a template for the formation of DNA. This so-called **reverse transcription** is possible because of an RNA-dependent DNA polymerase that the RNA virus brings into the host at the time of infection. The newly made DNA is then incorporated into the DNA of the host's chromosomes where, when the time comes, it can act as the template for more viral RNA. That some cancers appear to be caused by viruses does not imply that all cancers are caused by them.

Recombination in Bacteriophages Alleles of genes in the bacteriophages infecting *E. coli* can be recognized by the activity they show against *E. coli* when the bacteria are grown into a solid, thick layer on culture medium

solidified in a flat dish. Infective viral particles are added to the dish. Where infection occurs, with repeated rupture of bacterial cells, the bacterial layer is clarified into a circular spot or *plaque,* which is easily recognized. Differ- ent alleles of the bacteriophage genes are identified ac- cording to which strain of *E. coli* they can kill, and at what rate.

Recombination can occur between phage particles if two or more of them infect the same bacterial cell at the same time. At some point during infection, replication of the phage DNA, or incorporation of the replicated DNA into viral particles, viral DNA molecules from the different parental types may pair and cross over by break- age and exchange. Any recombinant viral chromosomes produced from this breakage and exchange are then packed into protein coats, and, upon rupture of the host cell, are released to the medium to infect further hosts.

The frequency of recombinant types can be deter- mined by counting plaques; this technique reveals that

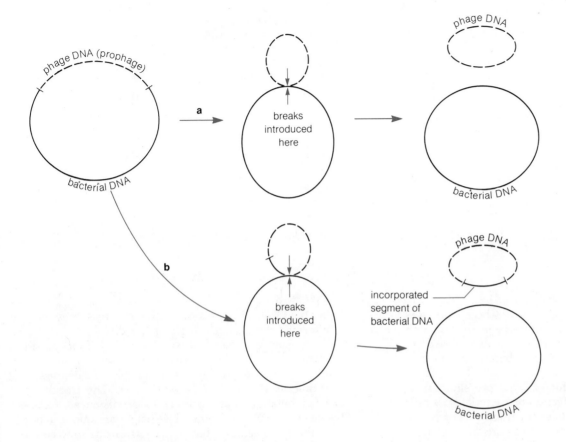

Figure 9–8 Transduction of parts of bacterial chromosomes by bacteriophages.

different genes have different recombination frequencies. These frequencies follow the same pattern as in eukaryotes: Widely separated genes recombine more often than closely spaced genes. Comparison of these frequencies has allowed us to "map" the bacteriophage DNA. These maps reveal that bacteriophage DNA molecules are circular, as is the chromosome of *E. coli.* A partial map of the best-known *E. coli* bacteriophage, T4, is shown in Figure 9–7. More than 50 genes have been assigned locations on the map.

Transduction of Parts of Bacterial Chromosomes by Bacteriophages During replication inside the host bacterium, as noted earlier, bacterial viruses may incorporate a part of the host chromosome into the finished viral particles. As a result, some of the viruses released when the host cells rupture may contain fragments of bacterial DNA. The DNA of the particle may include an incomplete bacteriophage chromosome, covalently linked to the bacterial DNA fragment, or it may contain only bacterial DNA.

Particles of the first type, containing both phage and bacterial DNA, arise from lysogenic viruses, like lambda, that for a time become incorporated into the host chromosome as a prophage. Here bacteriophage chromosomes are linked into the bacterial chromosome by a breakage and exchange mechanism. On release of the viral chromosome by a reversal of this mechanism, usually after many generations of existence as a prophage, the breakage ordinarily takes place so that no additions or deletions are made in either the viral or the host chromosome (Figure 9–8a). Rarely (about one in every 100,000 releases of viral prophage chromosomes), the position of the breaks is imperfect, so that some of the bacterial DNA is split off and incorporated within the viral circle (Figure 9–8b). This composite DNA circle, containing both bacteriophage and bacterial genes, is coated with viral protein and released along with normally constituted phage particles when the host cell ruptures.

Phage particles of the second type, containing bacterial DNA but no viral DNA, are formed during *lytic* cycles of growth. In these lytic cycles, replication of the viral DNA and formation of new protein coats begins immediately. No lysogenic phase, with insertion of the viral DNA into the host chromosome, takes place. As the viral parti-

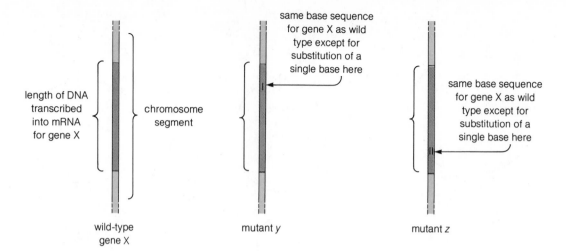

Figure 9–9 DNA containing wild-type and mutant sequences within gene X.

length of DNA transcribed into mRNA for gene X

chromosome segment

wild-type gene X

same base sequence for gene X as wild type except for substitution of a single base here

mutant *y*

same base sequence for gene X as wild type except for substitution of a single base here

mutant *z*

cles form in the bacterial cytoplasm, a few become packed with fragments of the host chromosome instead of with bacteriophage DNA. These particles are also released to the medium along with the normal bacteriophage particles.

In both types of transduction, the protein coat of the particles is intact and infection of bacterial host cells takes place normally. Injection of the fragments of bacterial DNA, either "pure" or linked to a portion of a bacteriophage genome, supplies new bacterial alleles to the recipient bacterial cell. These fragments may pair with homologous regions of the bacterial circle and enter into recombination by breakage and exchange.

The Gene

The results described in Chapters 7 and 8, including monohybrid and dihybrid crosses, linkage and recombination, sex linkage, incomplete dominance, and polygenic inheritance, are usually grouped together as the findings of "classical" genetics. They were essentially complete by 1920.

From these classical studies, certain characteristics of the physical nature of the gene could be predicted. Bridges's study of nondisjunction had proved the chromosome theory of heredity, firmly locating the genes on the chromosomes and thus inside the nucleus of the cell. Recombination studies, and calculation of the frequency and manner of crossing over, had determined that the genes are aligned in unbranched, linear order on the chromosomes. Although the chemical nature of the gene was unknown in the 1920s and 1930s, scientists believed that

either the DNA or the histone of the chromosomes actually formed the chemical basis of heredity.

The gene was believed to be a unit of function, in some manner controlling more or less clearly defined characteristics of an organism. The gene was also defined at this time as the *unit of recombination.* The frequency of recombination is proportional to the distance separating genes on a chromosome; close neighbors among genes recombined so seldom that thousands of offspring could be produced with no detectable recombinant types. Immediately adjacent genes showed the lowest frequencies of all. These findings, obtained with genetic crosses in *Drosophila* and corn, seemed to indicate that recombination occurred only between genes, and not within their boundaries. Thus, classical genetics had defined the gene as the smallest unit of the chromosome entering into recombination or crossing over. As a part of this hypothesis, it was proposed that the physical breaks and exchanges between chromatids in crossing over would take place only in the boundary zones between genes on the chromosome, and that the smallest length of the chromosome that could be studied and mapped by recombination frequencies is a single gene.

With the sensitive techniques of molecular biology, however, using bacteria and viruses, recombination has been detected within the boundaries of the gene. One of the best examples is in the *r* region of the T4 bacteriophage. This region has been mapped and has been shown to contain two adjacent genes, designated A and B. More than 1,000 mutant alleles have been detected within the A and B genes. The closest of these are estimated to lie only one or two nucleotides apart. Thus, it is obvious that the unit of recombination is much smaller than the gene and approaches the dimensions of single nucleotide pairs along the DNA molecule. Crossing over of this type,

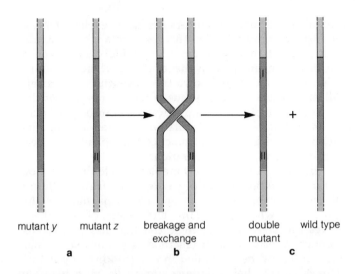

mutant *y*   mutant *z*   breakage and   double   wild type
exchange   mutant

         a   **b**   **c**

Figure 9–10 A possible result of recombination between the mutant forms of gene X shown in Figure 9–9.

between sites located within the same gene, is called *intragenic recombination.*

Figures 9–9 and 9–10 illustrate the pattern of intragenic recombination and its possible effects on the protein coded for by a gene (*y* and *z* in Figure 9–9) in which differences in the nucleotide sequences appear at different points. Mutant *y* has a base change in position I, and mutant *z* has a base change in position II. Assume that the two mutant alleles *y* and *z* cause substitutions in the sequence of amino acids making up the protein coded for by the gene. Although the two mutations cause amino acid changes at different points in the polypeptide chain, both change the active site of the protein significantly and destroy its biochemical function.

Figure 9–10 shows a possible result of recombination between the two mutant forms of the gene. In Figure 9–10a, DNA molecules containing the two alleles have been brought into close proximity by pairing. A physical breakage and exchange, if located between the positions of the two mutants (Figure 9–10b), will produce two recombinants, one containing both mutations in the same chromosome, and one converted to the active, wild-type gene (Figure 9–10c).

Intragenic recombination of this type probably occurs in all organisms, both prokaryotic and eukaryotic. Because it occurs at a very low frequency, it has been detected primarily in prokaryotes, and in a few eukaryotes such as the yeasts and other fungi that can be grown in large quantities. However, laborious experiments with *Drosophila* and corn, in which many thousands of offspring were grown and counted, have revealed that intragenic recombination also occurs in these organisms.

Genes as Units of Function The molecular and biochemical work of the 1950s and 1960s established the DNA code and the mechanism of protein synthesis. At this time all genes were considered to be coding units that spelled out the directions for making enzymes. As a result, a "one gene–one enzyme" concept was proposed by Beadle and Tatum. Realization that not all proteins are enzymes, and that many proteins contain more than one polypeptide chain, forced scientists to modify this statement to assert that each gene codes for a single polypeptide ("one gene–one polypeptide").

In recent years, extensive work with the DNA of the cell nucleolus established that these genes, instead of coding for polypeptides, code for ribosomal RNA. These RNAs are not messengers and are never translated into polypeptides, but form a part of ribosomes instead. Other genes code for transfer RNA instead of proteins or polypeptides. Therefore, even the "one gene–one polypeptide" concept had to be abandoned.

The gene is now best described as *a length of DNA in the chromosome that contains the information required for a unit of function in the cell.* This unit of function may be either one polypeptide of a complex protein containing more than one protein chain, or the entire polypeptide chain of a simpler protein. These proteins may be either enzymatic or structural. Alternatively, the unit of function may represent ribosomal or transfer RNA. Finally the units of function may be concerned with regulation of areas of the chromosome that are never transcribed into RNA of any type, such as the operator region of prokaryotic chromosomes. Recombination may take place within any region of these genes.

Summary

Using bacteria and viruses, molecular biologists have added to our knowledge of the gene. Geneticists had established that genetic recombination occurs in bacteria. It is now known that bacterial cells form small bridges through which a portion of the DNA of one cell enters another. This DNA fragment may then recombine with the DNA of the host cell, changing its genetic makeup. DNA may also be transferred when pieces of DNA released from fragmenting cells are absorbed by healthy ones. This is called transformation.

Viruses may be involved in the transfer of bacterial DNA from one cell to another. Viruses consist of a core of either DNA or RNA surrounded by a coat of protein. Only the nucleic acid of the virus enters the cell, but this is able to direct the machinery of the cell to make many new, complete virus particles. Viruses infect both prokaryotic and eukaryotic cells. Some viruses cause immediate disruption of the host cell, while others may have their DNA incorporated into the chromosome of the host and not become active for long periods. Some RNA-containing viruses can cause the production of DNA based on their RNA, which may then become incorporated into the chromosomes of the host.

Finally, in this chapter, we reviewed the evidence on the nature of the gene and concluded that the best definition at present is that the gene is a length of DNA in a chromosome that contains the information required for a unit of function in the cell.

Supplement 9–1
Of Chromosomes, Clones, and Plasmids, or Science Fiction Come True

Any devotee of science fiction knows of races on distant planets generated from pots full of identical single cells, genetically engineered to develop into individuals ideally suited to their tasks in life. Until recently, such ideas seemed as bizarre and farfetched as the mythical monsters our ancestors believed in, with bodies of lions, wings of eagles, and heads of humans. Yet recent evidence in genetics and molecular biology have brought some of these possibilities close to reality.

Almost as soon as Watson and Crick proposed the helical model of DNA in 1953, people began predicting that some day scientists would be able to modify DNA into any form they desired. Moreover, this specially tailored DNA would be introduced into cells, and new organisms would develop according to the molecular biologists' specifications. Though discussed at length in the popular press, and at times even in learned journals, it seemed far from achievement. Many natural barriers stood in the way. Substances do not move in and out of cells readily, and the cell has mechanisms for dealing with the intrusion of foreign particles. Moreover, the DNA itself is protected and not easily manipulated. Finally, there seem to be insurmountable problems in reproducing organisms with altered genetics, as it is difficult to see how sexual reproduction would tolerate such changes in DNA. In recent years, however, dramatic new developments have made these doubts no longer serious obstacles and have brought closer the day that genetic manipulations will be commonplace.

One of these developments is a modification of an old procedure, *cloning*. Many plants reproduce sexually only under special conditions. Other plants have special desirable features that would be lost through sexual re-

production. Thus, horticulturalists have developed means of vegetative reproduction that result in a whole population of plants derived from a single individual. One of the most dramatic cases is that of the navel orange, which originated on a single known tree in southern California. All the hundreds of thousands of navel-orange trees that are now producing have come from that one original plant or its cuttings. Such a population is called a clone, and cloning as an agricultural practice with plants has been used for many, many generations.

However, there is a new method for cloning that does not involve whole pieces of an organism. Instead, clumps of cells may be broken up and each cell, individually, grown to a whole new plant. The clumps of cells themselves can start from a single cell that has proliferated through mitosis. By such a method, any change in a single cell can easily be magnified into a working population. This is already being done commercially with plants in new varieties of, for example, chrysanthemums. Instead of taking 8 to 10 years before they are marketable, these new varieties can be gotten to the growers in 3 or 4. The procedure is less successful with animal cells than with plants, but similar manipulations can be made in certain cases with animal cells.

This has led to a resurrection of another science-fiction idea, that of people cloning themselves. In this scenario, a small piece of tissue from a person, or perhaps even a single cell, is used to generate a new person. Since the genetic information is the same as in the original individual because no sexual reproduction has taken place, the cloned offspring should be identical to the original. This kind of thinking has led to bizarre suggestions and fears in the popular press. Writers can imagine the development of whole legions of identical beings and the mass production of geniuses and supercreative people. The flaw in this fantasy lies not in the potential of cloning, but in the obvious fact that we are all the result of much more than genetic programming in our development. Genetics is only a part of the picture of the development of any organism. Our personalities—and even our physical appearances—are dependent, to a large extent, on influences from our environment. Over the last 50 years, American men and women have steadily grown taller. This is not a genetic phenomenon, but one based on nutrition and the achievement of genetic potential. Even individuals with identical genetic backgrounds, as in the case of identical twins, do not have identical personalities, particularly when raised in different environments. We are who we are only partially because of the genes we carry and at least as much because of external influences.

However, another aspect of cloning is enormously important and has pushed forward the state of genetic engineering. The key point here was mentioned earlier— that a change arising in a single cell can be duplicated many times by cloning.

How, then, are changes in genetic composition to be achieved in the laboratory without sexual reproduction and mutation? The answer is through the *plasmid.* Plasmids are circular bits of DNA resembling small editions of bacterial chromosomes, which exist apart from the chromosomes in some bacteria. In their natural circumstances, plasmids are interesting in themselves, but they become fascinating when we consider certain characteristics they have. For example, the plasmid can sometimes pick up a short segment of DNA from the chromosome of a cell and transfer it to another cell. This transplanted segment of chromosomal DNA can become integrated into the recipient cell and thus allow the transfer of genes between species.

The laboratory procedures involved in the transfer of plasmids and their incorporation into the recipient cells' DNA are fairly complicated and have resulted from a series of independent discoveries made in rapid succession in the late 1960s and early 1970s. One of the most important of these was made in 1967, when a series of enzymes called DNA ligases was discovered. These enzymes were found to repair breaks in DNA molecules and, under certain circumstances, to join together loose ends of DNA strands. Other enzymes were found that would break DNA molecules into small units. Thus, means were available to produce segments of DNA, but such segments are not self-replicating.

The discovery of plasmids meant that the molecular biologists had a self-replicating segment of DNA to which they could selectively attach other pieces of DNA. Using the newly discovered DNA ligases, molecular biologists were able to open the plasmid, insert a new segment of DNA, and close the ring again. The bacterium was allowed to reproduce, and masses of new plasmids were formed. The daughter bacteria can be broken open and the plasmids isolated through differential centrifugation.

Means were then discovered by which the new plasmids could be introduced into a different cell. This is done by damaging the cell membrane enough that the plasmid can penetrate but not enough that the cell is killed. This is obviously a tricky procedure, but it has been worked out successfully in bacteria.

Finally, the plasmid and its extra DNA replicate in the new closed cell, conveying to that cell the properties contained in the DNA that has been transmitted.

At present, plasmid transfer of DNA has been restricted to bacterial cells. Nevertheless, the possibilities of work with bacteria are almost staggering. One group of researchers was able to selectively transfer specific antibiotic resistance from one strain of bacteria to another. In another case, DNA from a eukaryotic cell, the toad *Xenopus laevis,* was introduced into a bacterium. The genes that were transposed coded for nucleolar RNA. Since bacteria have no nucleoli, being prokaryotic cells, it was easy to find the resulting RNA that was originally part of the toad's nucleolar component. Because of such phenomena, we can use plasmids to explore specific properties of chromosome genes in eukaryotes like *Drosophila,* the common fruit fly that is so widely used in genetic studies. We can also tailor bacteria to do certain things. Bacteria are extremely important in many industrial processes and are the source of many compounds used in industry. Being able to make such bacteria is of great importance.

While it is clear that we are not in a position to transport specific genes into human cells and have individuals of specific characteristics result, we are coming closer to such a goal. That we should reach it, if such a goal is desirable, seems highly problematic. It is also clear that society would probably put restrictions on scientific research before it reached such a point. The scientists involved are themselves extremely sensitive to the moral considerations involved. Well aware of the dangers involved in plasmid research, they called a moratorium on it until certain basic safety considerations could be worked out. It is possible, for example, that while a known gene was being transferred through a plasmid, an unknown or cryptic gene might also be transported that would produce a toxin lethal to higher animals. If such a gene escaped into the natural environment, it could cause havoc. Considerations such as these led to a dramatic meeting at the beginning of 1975 of 86 U.S. biologists and 53 foreign scientists to develop guidelines for plasmid research. They established four categories of experiments with different degrees of risk. Certain safety standards were set, including the rule that bacteria should be used that were so specific in their nutritional requirements that they could not live in natural surroundings.

Where all this will lead us is still not clear. We are entering an area of biological research that is every bit as tricky and dangerous as nuclear physics was in the 1940s and 1950s. We hope that basic discoveries will result in more productive agricultural plants and an improved quality of life on earth. At the same time, we run a risk of releasing new diseases far more potent than any we have ever seen. Once again, as long as these developments are used for peaceful purposes, humanity as a whole will benefit in the long run. If, however, a government were to decide to use such biological materials for its own ends, as in biological warfare, the results could be devastating. It seems unrealistic to think that scientists or society will allow the research to stop. The conference of biologists has already been criticized by Senator Edward M. Kennedy of Massachusetts for not taking the opinions of the nonscientific community into account when setting aims and goals. Clearly, the social impact of this kind of science is so great that scientists will not be allowed to decide alone where the research will carry them. At the same time, only scientists have the background and the data to understand and predict the course of research underlying the development the nonscientists fear. It is an interesting situation and one that will be with us from now on.

Suggested Readings

Books

Goodenough, U., and R. P. Levine. 1974. Genetics. Holt, Rinehart and Winston, New York.

Levine, R. P. 1968. Genetics, 2nd ed. Holt, Rinehart and Winston, New York.

Sinnott, E. W., L. C. Dunn, and T. Dobzhansky. 1958. Principles of Genetics, 5th ed. McGraw-Hill, New York.

Stanier, R. Y., M. Doudoroff, and E. A. Adelburg. 1970. The Microbial World. Prentice Hall, Englewood Cliffs, N.J.

Sturtevant, A. H. 1965. A History of Genetics. Harper and Row, New York.

Voeller, B. R. 1968. The Chromosome Theory of Inheritance. Appleton-Century-Crofts, New York.

Watson, J. D. 1976. Molecular Biology of the Gene, 3rd ed. W. A. Benjamin, New York.

Articles

Campbell, A. M. 1976. How viruses insert their DNA into the DNA of the host cell. Scientific American 235(6):102–113.

Cohen, S. N. 1975. The manipulation of genes. Scientific American 233(1):24–33.

Day, P. R., 1977. Somatic cell genetic manipulation in plants. BioScience 27:116.

Grobstein, C. 1977. The recombinant-DNA debate. Scientific American 237(1):22–33.

Miller, O. L., Jr. 1973. The visualization of genes in action. Scientific American 288(9):34–42.

Ruddle, F. H., and R. S. Kucherlapati. 1974. Hybrid cells and human genes. Scientific American 231(1):36–44.

FOUR

Development

10

Reproduction in Plants

Plants fill the world around us. Flowers grow in meadows and outside our windows, scent the air we breathe, and fill our eyes with their beauty. Plants also fill our stomachs, clothe our bodies, and help cure our ills. Ever since human beings domesticated plants and established agriculture, we have been directly dependent on plants and plant reproduction to survive. The surplus of foodstuffs that resulted from agriculture has enabled our population to increase—to a point where it threatens to outrun the maximum food supply the earth can produce. Thus, in a very real sense, the question of how plants reproduce is a matter of survival for all of us.

Plants, like animals, reproduce both sexually and asexually. Sexual reproduction involves meiosis, which recombines the genes so that considerable variation exists among the offspring of any set of parents. This variation, which is central to the process of evolution (as we will see in Chapter 28), is necessary if a species is to adapt to changes in the environment.

But many plants reproduce entirely or primarily by asexual means. In all the many types of asexual reproduction in plants, a somatic cell or portion of the parent plant produces a whole new plant. All offspring are genetically identical to the parent, and genetic recombination has not occurred. If combination is so necessary to survival, why would any organism reproduce asexually? The answer is that some environments are not changing, or at least are not changing significantly. In a stable environment, the lack of variation is an advantage: If the parent has survived, then those offspring that most closely resemble it will be most likely to survive in turn.

Moreover, many plants reproduce *both* sexually and asexually, and in so doing have the double advantage of the variability of sexual reproduction and the stability of asexual reproduction. To see how these two modes of reproduction are fitted to the life of different types of plants, we will first consider the modes of asexual reproduction found in plants. We will then examine sexual reproduction and, finally, the entire life history of

representative groups of plants. In discussing the life history of each type of plant, we will consider its environment and its adaptations to that environment.

Forms of Asexual Reproduction

The simplest form of asexual reproduction is by cell division (mitosis), which can be illustrated by algae growing in a pond. In the spring, the pond is an environment ready to be filled and exploited. The one-celled forms of algae multiply by dividing at astonishing rates and can fill the pond in a matter of days. Filamentous algae (long, threadlike, many-celled forms anchored to one place by modified basal cells or *holdfasts*) also reproduce by mitotic division. Fragments of such algae break loose and float to the surface, where they drift about, increasing in length through cell divisions. Eventually they form entire new filaments, each with its own holdfast.

Most filamentous algae have evolved a second means of asexual reproduction that both increases the numbers of their offspring and scatters them more widely: They form motile cells called *zoospores* (Figure 10–1). These swim to new sites, become attached to a suitable substrate, and produce new filaments by mitotic divisions. A single filament may produce thousands of zoospores at a time. Various environmental factors cause algae to form zoospores. In decreased light, as when a part of the pond is shaded, algae in the shaded portion will form zoospores. The zoospores then establish new plants in the parts of the pond with more light. Increasing the available nitrogen also stimulates zoospore formation. Since nitrogen is an essential element of plant growth, it makes survival sense to reproduce heavily when nitrogen is plentiful. Thus, when late spring rains carry fresh supplies of nitrogen compounds from fields to the pond, zoospores are formed and new filaments are established.

Many fungi living in the pond also produce zoo-

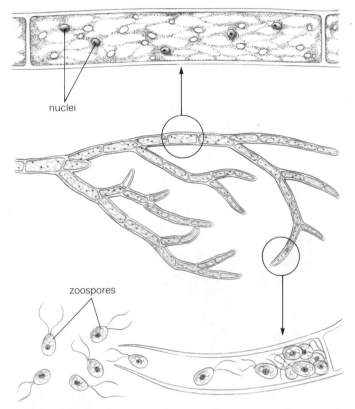

nuclei

zoospores

Figure 10–1 Zoospore formation in a filamentous alga with multinucleate cells. Each nucleus is the basis for a single zoospore.

spores. One common aquatic fungus is *Saprolegnia*, growing on dead organic matter such as flies or frogs that may be found in the pond (Figure 10–2). Here the zoospore released by the attached plant swims until it becomes attached to a suitable substrate—again, some form of dead organic matter. The organism has evolved sensitive chemical sensing mechanisms to determine what is a suitable substrate and what is not. Once attached, the zoospore develops filaments of multinucleate cells, called **hyphae** (singular **hypha**), that penetrate the substrate. These hyphae secrete enzymes that break down the substrate into compounds the fungus can absorb and use for growth. When the substrate is exhausted, cytoplasm and nuclei accumulate in the ends of some of the hyphae, and more zoospores are formed.

Figure 10–2 (**a**) A dead frog covered with lush growth of fungal hyphae, in this case, *Saprolegnia*. (**b**) Such fungi can produce large numbers of spores that are rapidly discharged when mature. This series of photographs of the fungus *Dictyuchus* was taken at 3-minute intervals and gives a good impression of the rate of discharge.

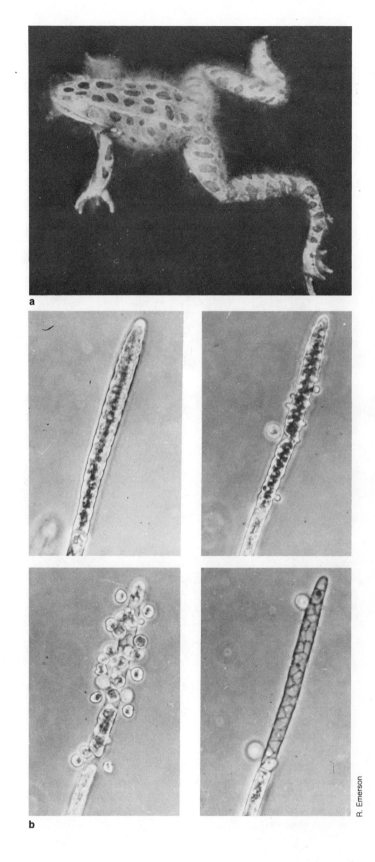

a

b

R. Emerson

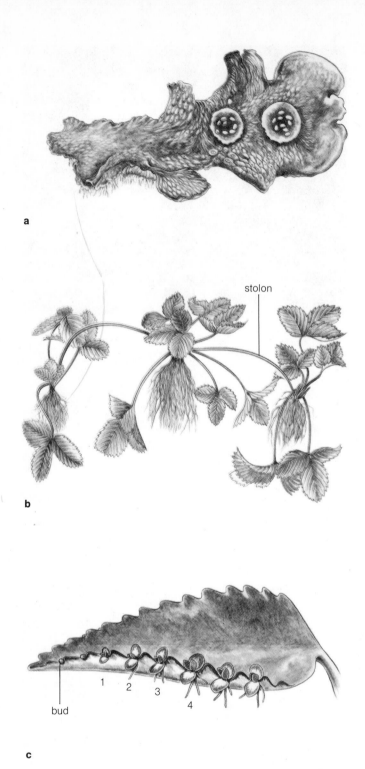

a

stolon

b

bud
1 2 3
4

c

Figure 10–3 (**a**) Two gemmae cups in liverworts, with the gemmae visible in the bottoms. (**b**) Asexual reproduction in the strawberry by the formation of stolons or runners. (**c**) A *Bryophyllum* leaf forming a new individual by budding. The stages in bud development into a new plant are numbered.

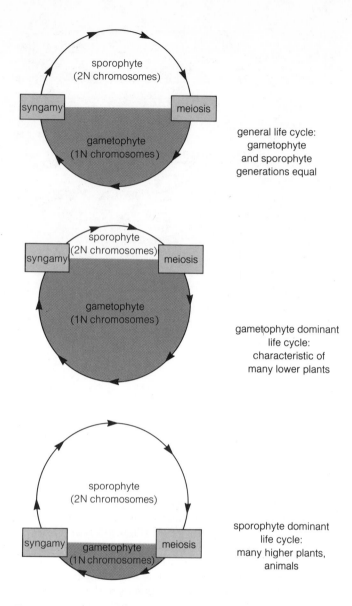

sporophyte
(2N chromosomes)

syngamy

meiosis

gametophyte
(1N chromosomes)

general life cycle:
gametophyte
and sporophyte
generations equal

sporophyte
(2N chromosomes)

syngamy

meiosis

gametophyte
(1N chromosomes)

gametophyte dominant
life cycle:
characteristic of
many lower plants

sporophyte
(2N chromosomes)

syngamy

meiosis

gametophyte
(1N chromosomes)

sporophyte dominant
life cycle:
many higher plants,
animals

Figure 10–4 Generalized plant life cycles.

Terrestrial fungi have evolved asexual spores that are resistant to drying. Millions of such spores can be formed by an individual during a growing period and can be carried by the wind for many miles. Many plant diseases are communicated in this way.

Spores, whether zoospores or resistant spores, are single cells. There are, however, other forms of asexual reproduction, particularly in the higher plants, that involve masses of cells. For example, among the liverworts, a group of plants related to the mosses, small masses of tissue called *gemmae* are formed in a cup-shaped receptacle (Figure 10–3a). When the cups are flooded with water, the gemmae are released and float away from the parent plant. Raindrops splashing into the cups may throw gemmae as far as a meter from the parent plant. If the gemmae land on favorable substrates, they will rapidly grow into new plants.

In some ferns, the leaves elongate and bend so that their tips touch the ground. These tips can then form a new shoot and roots, thus producing a new plant. Strawberries have stems (called stolons or runners) that are capable of developing a new plant (Figure 10–3b). Some species of succulents develop tiny plantlets on their fleshy leaves (Figure 10–3c). The plantlets drop loose and fall to the ground, where they grow into mature plants. The bulbs of such plants as daffodils, tulips, lilies, and onions give rise to smaller bulbs that reproduce the plant. The horizontal roots of many seed plants (such as poplar, sweet gum, beech, and poison ivy) produce new plants from buds. And everybody has read about cuttings taken from prize rosebushes, fuchsia hybrids, or valuable fruit trees. These are but a few of the many ways plants can reproduce asexually.

Sexual Reproduction in Plants

The life of an organism from the formation of the zygote to the subsequent development of gametes and their fusion to form another zygote is called a *life cycle* or *life history*. The pattern found in most animals is that the zygote gives rise to a diploid individual in which meiosis occurs only in the cells that will form the eggs or sperm. In plants, the patterns of sexual reproduction are much more complex and varied than in animals, usually involving complete haploid as well as diploid generations, one following the other. This change back and forth between haploid (N) and diploid (2N) generations is called the *alternation of generations,* and an understanding of the phenomenon is fundamental to an understanding of plant reproduction (Figure 10–4).

Alternation of Generations   In the most generalized case of alternation of generations, the zygote gives rise to a diploid individual whose cells undergo meiosis. These meiotic divisions result in haploid spores that germinate to form haploid plants. Certain cells of these haploid plants then produce eggs and sperm through mitotic divisions. An egg and a sperm then unite in a process known as *syngamy* to form a diploid zygote. The zygote divides, giving rise to a new diploid generation, and the cycle is back at its starting point. Since the haploid generation produces gametes, it is called the **gametophyte** or **gametophytic generation;** and since the diploid genera-

tion produces spores, it is called the **sporophyte** or **sporophytic generation.** The spores produced by the sporophyte are sexual spores, because they are produced by meiosis. They should not be confused with the asexual spores, such as zoospores, produced as a result of mitotic divisions of cells in either sporophyte or gametophyte.

There is great variation in the haploid generations among different plant groups. In some plants, the diploid generation may be nothing more than the zygote. At the opposite extreme, the haploid generation may consist of only a few cells completely dependent on the diploid generation. The general trend in the green plants is from life cycles (in the algae) in which the haploid is the photosynthetic, independent stage with the diploid small and frequently nutritionally dependent, to life cycles (in the flowering plants) in which the diploid is the independent form and the haploid is reduced to a few cells dependent on the sporophyte for food.

Life Histories and the Environment

Organisms evolve in response to changes in the environment and become adapted to a particular set of environmental conditions. This is as true for their reproductive systems as it is for the other aspects of their lives. A method of reproduction that works in the water may be completely unsuccessful on land. A type of life history that means continuation of the species in one habitat may mean extinction in another. Therefore, when we look at the reproductive patterns in plants and at the various life histories that have evolved, we should also look at the environmental conditions.

We will present examples of various reproductive patterns of plants and relate these differences to the places in which these plants are found. Starting with the fresh-water environment, then moving to the marine environment, and finally to the land, we will examine how plants have adapted to their diverse environments in the ways they reproduce.

Fresh-Water Environments

In an aquatic environment, a plant is surrounded by water. This water contains all the dissolved minerals the plant needs, as well as carbon dioxide and oxygen. The surrounding water also supports the plant, since it is a relatively dense medium in which the plant is buoyant. Most fresh-water environments are suitable for plant growth and reproduction. Light is usually abundant and other conditions are generally good. At the same time, there

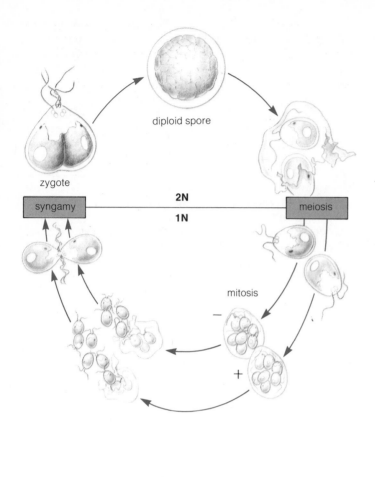

Figure 10–5 Life cycle of *Chlamydomonas*.

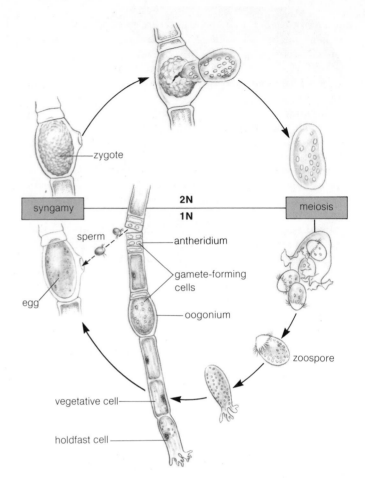

Figure 10–6 Sexual reproduction in *Oedogonium*.

are limitations. One is temperature: In winter, many ponds, lakes, and streams freeze over and what water remains liquid is too cold for active growth. In many locations, ponds and streams dry up for portions of the year, and the organisms growing there must have ways to survive these dry periods.

As a result, we find that most fresh-water environments have many algae and fungi living in them. The species are usually small in size and simple in structure. They have short life-spans, and their life histories include a stage in which the organism is resistant to adverse environmental conditions.

Fresh-Water Algae Among the most common algae in fresh water are the green algae. The vast majority of their life is haploid, with the only diploid cell being the zygote. All fresh-water green algae follow this general pattern, but within the pattern there is considerable variation,

as we can see by examining two common fresh-water green algae, *Chlamydomonas* and *Oedogonium*.

All the *Chlamydomonas* species are one-celled and have a similar structure, consisting of a nucleus, a single chloroplast, an eye spot, and two flagella. The nucleus of the mature cell is haploid, and asexual reproduction by mitotic cell division is common. Gametes are produced by mitotic division of the parent cell (Figure 10–5). The flagellate gametes are similar to the original cell, but smaller. Two morphologically similar gametes fuse into a diploid zygote. The zygote forms a thick, often strikingly sculptured wall. This thick-walled zygote is resistant to extremes of temperature and moisture and can survive dry or cold periods. When water is again available and the temperature is high enough, the zygote germinates. The first division of the zygote upon germination is meiotic, resulting in four small, haploid motile cells. These rapidly grow to the normal size of vegetative cells, and the life history is complete.

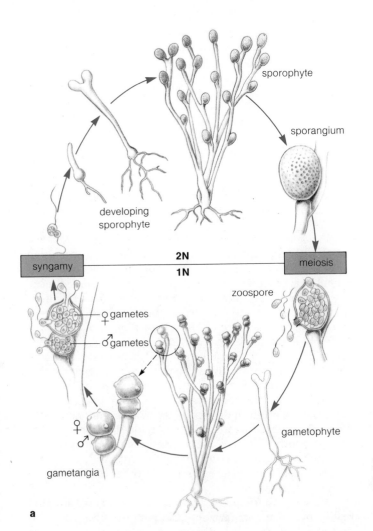

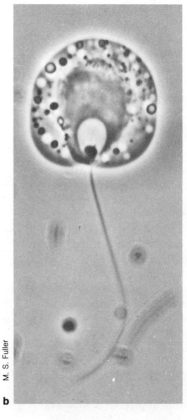

Figure 10–7 (a) The life history of *Allomyces*. (b) This beautiful photograph of a living zoospore of *Allomyces* shows the single flagellum, the single nucleus (the larger, round, light area containing a dark nucleolus near the point of attachment of the flagellum), and the surrounding nuclear cap.

One important feature of this relatively simple history is that, although the gametes are morphologically similar, two produced by the same cell will not fuse. In fact, experiments have shown that there are two mating strains in some species of *Chlamydomonas* (usually indicated by + and −), and only when gametes from these two strains are present does fusion take place. This assures the organism maximum variation upon recombination.

Oedogonium is a common, filamentous green alga composed of filaments of cylinder-shaped cells, each containing a single nucleus (Figure 10–6). In this organism, one kind of gamete, called the egg, is large and immobile; the other gamete, called the sperm, is small and motile. Both eggs and sperm arise from cells produced by mitotic divisions. One filament produces eggs similar to the vegetative cells except that the cell contents are pulled away from the wall, the cell itself is swollen and barrel-shaped, and a pore develops in the wall surrounding the egg. The sperm are produced on separate filaments through

mitotic division of specialized cells. Each sperm has a ring of short flagella at one end and, when released by the breakdown of the surrounding wall, swims to the egg-containing filament.

The sperm are attracted by a substance the egg produces. When a sperm reaches the pore, a fingerlike projection of the egg extends through the opening, attaches to the sperm, and pulls it through the opening, where fusion occurs. The zygote, which is now diploid, forms a thick wall and enters a resistant resting stage. Upon germination, the zygote produces, through meiotic division, four multiflagellate haploid cells, each capable of forming a new filament.

Fresh-Water Fungi Also living in fresh-water ponds are numerous fungi that consume dead plants and animals as food. These have a variety of life cycles. One pattern is that of *Allomyces* (Figure 10–7a). Here, the haploid and diploid generations are equal in size and length of life.

Allomyces is a genus of filamentous water molds that grows on decaying plant matter. Some hyphae penetrate the substrate and absorb nutrients, while others extend away from the substrate and produce reproductive structures.

The cytoplasm in the hyphae of *Allomyces* contains many nuclei and is not divided into cells by cross-walls until reproduction is about to occur. Then both cross-walls and gamete-producing structures called **gametangia** are formed. Male and female gametangia are borne on the same filament. The gametes are formed through cleavage of the multinucleate cytoplasm into uninucleate cells. The small, orange male gamete or sperm swims rapidly through the water; the larger, colorless female gamete swims also, but less actively. The female gamete attracts the male by secreting a hormone; through syngamy, they fuse and form a flagellate zygote. Since each gamete is haploid, the zygote is diploid. For awhile, the zygote swims by means of its flagella, but eventually it settles down on a suitable substrate and germinates into a new hypha. The hypha looks like the organism that produced the gametes, but its nuclei are diploid.

The diploid generation also has reproductive structures, but these produce spores and are called **sporangia.** Actually, two types of spores are produced: one with diploid nuclei following mitotic division, and one with haploid nuclei following meiosis. Both kinds of spores are motile. When the diploid zoospores germinate, they give rise to diploid organisms similar in all respects to their parents. The haploid zoospores produced by meiosis develop into a plant similar in appearance to the parent but containing haploid nuclei and capable of forming differentiated male and female gametangia that will produce male and female gametes.

In *Allomyces* it is the sporangia, not the spores or zygotes, that are resistant to drying out. Each resistant sporangium usually contains 12 diploid nuclei. When conditions are favorable, each of the nuclei undergoes meiosis to form four haploid zoospores (Figure 10–7b). They emerge from the ruptured sporangium and swim around until they make contact with a suitable substrate, eventually producing haploid organisms capable of forming gametes.

Plants of the Intertidal Region

The region between high and low tide is a complex environment for life, yet life abounds. Water is present at varying depths—or not at all—depending on the tide. Along most of the North American coast there are two tides daily. The tides are not uniform but vary throughout the month and the year, governed by the relative positions of earth, moon, and sun. This means that the extent and depth of the water covering the beach and the organisms living there change constantly, although within more or less fixed limits. Thus, for algae growing fairly high on the beach, there is a twice-daily period of drying when the tide goes out. Plants not so high on the beach may be exposed to the air only once or twice a month, and the water over them may at times be 10 to 12 feet deep or more. Since more than 25 percent of the sunlight is lost at the surface of the water, some special adaptation is needed for photosynthetic organisms to live below the surface.

What types of plants are found in the intertidal region? Although many one-celled and some filamentous algae are found in this environment, intertidal algae are often large, complex structures with well-developed root-like *holdfasts,* stemlike *stipes,* and leaflike *blades* (Figure 10–8). Moreover, although some of the algae are green, many are brown or red. The green algae are generally found higher up in the intertidal zone, where they are subject to more drying and covered to a lesser depth at high tide. The brown algae usually grow in deeper water and often have elaborate floats to carry their blades near the surface, where photosynthesis is more efficient. The red algae are often in deeper water, where their accessory pigments enable them to use the feeble light penetrating to that depth. We will use a brown algal genus, *Fucus,* to illustrate how structure, development, and reproduction fit together to ensure survival in the intertidal region (Figure 10–9).

Fucus, commonly known as rockweed, is well adapted to life in the intertidal zone. Growing on the intertidal rocks in large numbers, it is a somewhat flattened, rather coarse-bodied, equally branched alga with a distinct, dark olive-drab color. It has special substances to protect it against drying out and an accessory pigment that helps it to absorb light when submerged.

Since the haploid generation is greatly reduced in size (down to single cells), the plant we normally see is the diploid generation. The thick, leathery blades contain small cavities from which eggs and sperm resulting from meiotic division are released into the surrounding water.

The flagellate sperm swim to the large, nonmotile egg. One sperm fuses with the egg, syngamy occurs, the zygote forms and produces a wall. The zygote sinks to the bottom, where its adhesive outer wall allows it to stick to rocks and other suitable substrates. If it fails to adhere, the zygote may be carried far up onto the beach or far out to sea.

The spherical zygote now becomes differentiated into two sections: a basal section that will form the holdfast and an apical section that will form the stemlike blade. This differentiation is a response to a number of environmental factors, the most important apparently being light,

Figure 10–8 The marine brown alga *Postelsia palmaeformis*, showing the various parts found in many larger marine algae (one-quarter natural size).

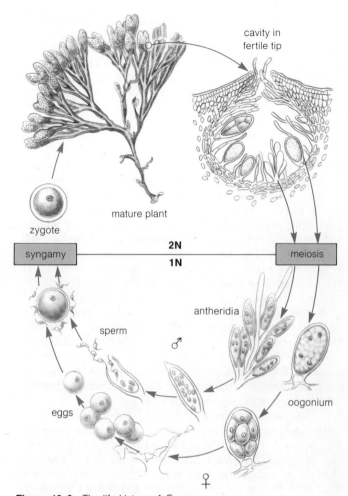

Figure 10–9 The life history of *Fucus*.

since the basal end always forms away from light. Divisions of the zygote result in a small, pear-shaped plant that will develop into the mature plant.

Land Plants

The algae and fungi examined thus far all live in an aquatic environment and therefore, despite the extreme contrasts between placid pond and roaring surf, share common features. They all essentially lack rigid internal structural elements; in the mature stage none can survive prolonged exposure to drying; and all have reproductive cells that require water for transport. The striking difference is the relative importance of the gametophyte versus the sporophyte generations in the fresh-water versus the marine algae.

For several obvious reasons, land is a more difficult environment for plants than water. The problem of drying out is constant, and air offers no support. Water is present in the soil, but to carry on photosynthesis the green parts of the plant must be in the light. Thus, all land plants have evolved a protective outer covering to prevent drying, structures to allow them to absorb water from the soil and transport it to the above-ground parts, and means to hold the photosynthetic parts of the plant in the light.

Modifications in both reproductive and vegetative structures have occurred in response to the limitations of the land environment. Clearly, motile, swimming zoospores would have limited value for the land plant. In contrast, spores with walls that prevent drying of the cell contents are of great survival value. All terrestrial fungi and green plants have developed some type of reproductive structure resistant to drying out, whether simple spores or the more elaborate seeds. Finally, in terms of sexual reproduction, a universal feature of green plants is the enclosure of the cells that will form the gametes and the spores in layers of cells that will not form such reproductive cells. The function of these sterile cells is to protect the reproductive cells from drying out.

To see how plants have met the challenge of the land environment, let us examine the reproductive methods of some of the major terrestrial groups.

Bryophytes: The Mosses The mosses belong to a group of primitive land plants known as the bryophytes. This group also includes the liverworts and hornworts.

In the mosses, shoots develop from filaments of haploid cells produced by the germination of the spores. When the shoot is mature, **antheridia** and **archegonia** are produced. The motile sperm are produced by mitosis within the antheridium, and the eggs are produced in the archegonium (Figure 10–10). In the archegonium, the nonreproductive cells are formed in the shape of a flask or bottle. A column of cells differentiates in the center of the bottle. The bottom cell of the column is the egg; the cells occupying the neck are called the neck-canal cells. When the egg is mature, the neck-canal cells degenerate, and the nonreproductive cells at the top of the neck separate so that a clear canal is formed through which the sperm can swim to the egg.

The mature sperm are released from the antheridium only when water is available. The sperm are attracted by chemicals released from the archegonium to the opening in its neck, down which they swim to the egg. Although several sperm may reach the egg, only one will fuse with it. The fusion of the egg and the sperm results in the formation of the zygote and the establishment of the diploid or sporophyte generation (Figure 10–10).

The moss sporophyte develops on top of the leafy shoot of the gametophyte (Figure 10–10) and depends on the gametophyte for at least some of its food. The sporophyte consists of a foot that functions in the transfer of material from the gametophyte to the sporophyte and a long, slender stalk that ends in a cylindrical sporangium. Here spores are produced through meiotic division. The haploid spores are surrounded by a wall that protects the cell contents from drying out. The tip of the capsule is covered by a lid. When the spores mature, the lid falls away, revealing a ring of teethlike units. The tips of these teeth move back and forth in response to changes in the moisture content of the air and to wetting and drying. The teeth close in wet conditions, preventing the release of spores at a time when they could not be widely dispersed. During dry, windy periods the teeth open, releasing the spores to be carried long distances by the wind. These spores will germinate if they land in a suitable habitat, completing the cycle (Figure 10–10).

The mosses have successfully adapted to land, but they do not fully exploit the environment. Lacking tissue to transport water, and having motile sperm, they are limited to relatively moist regions where free water is available at certain times. The gametophyte is the conspicuous generation, as in the green algae, with which the mosses share many other characteristics.

Ferns Another important variety of land plant is the ferns. When you look at a fern, you usually see only large, compound leaves, for the stem is underground or near ground level. The roots are relatively simple and inconspicuous, arising from the stem near the leaves.

The sporophyte is the conspicuous generation. The spores are produced by meiosis in sporangia located on the leaves (Figure 10–11). The sporangia, usually grouped together in clusters called **sori,** are biological catapults that launch the spores into the air.

The spores are light and can be carried for miles by the wind. Germinating readily on damp soil, they form small, usually green, gametophytes. The type commonly seen in laboratories has two flattened lateral wings of tissue, giving the plant a heart shape. Antheridia and archegonia similar to those found in the mosses are formed on the lower surface of the gametophytes. The multiflagellate sperm need water in which to swim to the archegonia. They swim down the canal of the archegonium, where one will fuse with the egg, forming the zygote (Figure 10–11).

The zygote first divides within 24 to 48 hours, and further divisions produce an embryo (an immature organism) that has a foot as well as a root, shoot, and leaves (Figure 10–11). The foot absorbs materials from the gametophyte during the early stages of development, when the sporophyte is essentially a parasite on the gametophyte. Soon, however, the leaves of the sporophyte develop to the point where it is no longer dependent on the gametophyte.

Ferns have adapted more successfully to life on land than the bryophytes, but they too require water for fertilization. They have shifted to a condition where the sporophyte is the conspicuous, independent generation; but the gametophyte is still an independent generation—although greatly reduced in size.

Gymnosperms: Pines In the gymnosperms, a large and ancient group of plants, the sporophyte is even more dominant and water is not necessary for fertilization. The pines are but one example of the gymnosperms.

Like the ferns and other vascular plants, the gymnosperms produce spores through meiosis, but in this case *two* kinds of spores are formed in male and female cones that are actually modified leaves.

The life history of a pine is illustrated in Figure 10–

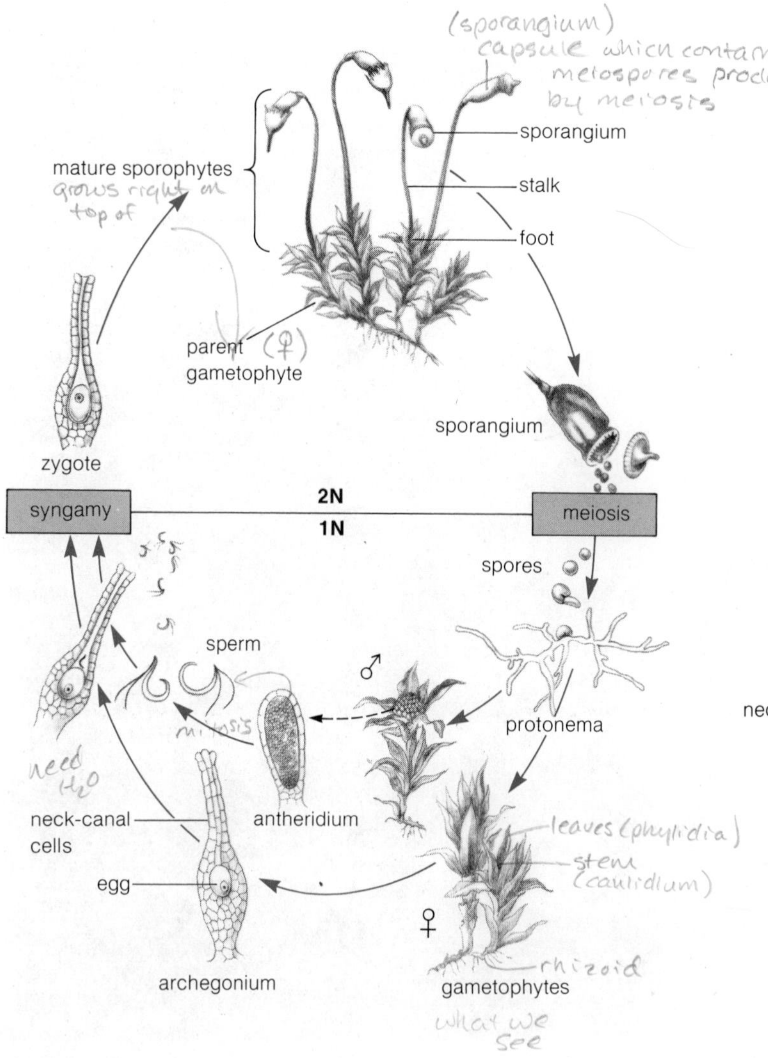

sporangium
stalk
foot
mature sporophytes
parent
gametophyte
zygote
sporangium
syngamy
2N
1N
meiosis
spores
sperm
protonema
neck-canal
cells
antheridium
egg
archegonium
gametophytes

Figure 10–10 The life history of a moss.

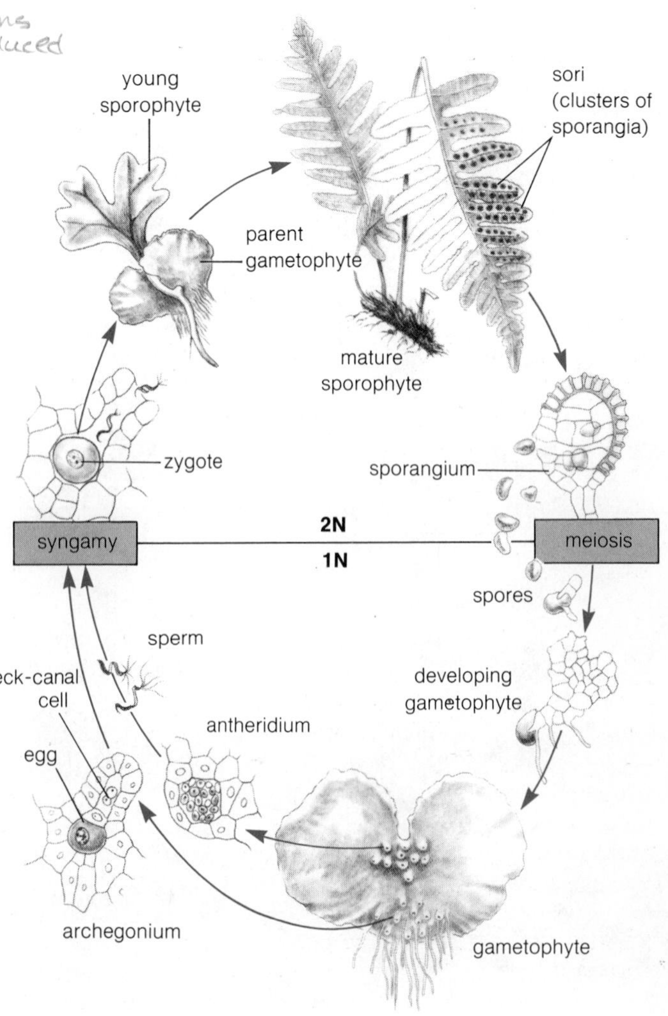

young
sporophyte
sori
(clusters of
sporangia)
parent
gametophyte
mature
sporophyte
zygote
sporangium
syngamy
2N
1N
meiosis
spores
sperm
neck-canal
cell
developing
gametophyte
antheridium
egg
archegonium
gametophyte

Figure 10–11 The life history of a fern.

12. The male cone, which is smaller than the female and much shorter-lived, contains a collection of microsporangia. Within these, cells differentiate and undergo meiosis to form four haploid **microspores.** The nucleus of the microspore divides so that a small gametophyte of only two cells is produced. This microgametophyte is the pollen.

The female cone is a group of modified leaves on which the megasporangia develop (Figure 10–12). A cell in the megasporangium undergoes meiosis to form four haploid **megaspores;** three of these degenerate, leaving a functional megaspore. The nucleus of the megaspore divides mitotically to produce a multinucleate megagametophyte. Cell walls eventually form, and several archegonia differentiate within the megagametophyte, each consisting of an enlarged egg cell and two or four neck cells.

The pollen is carried with the aid of tiny wings by the wind to the female cone, where it germinates and produces a tube that grows slowly through the sporophyte tissue that surrounds the megagametophyte. During this development, the cells within the pollen divide, and eventually two sperm are formed. The pollen tube grows directly to the egg and discharges the sperm into it (Figure 10–13). Since several eggs can differentiate in each megagametophyte, more than one embryo is often formed. However, the developing embryos compete for nutrients, and usually only one survives in the mature seed.

The seed is composed of a resistant outer coat formed by the sporophyte. This surrounds the remains of the

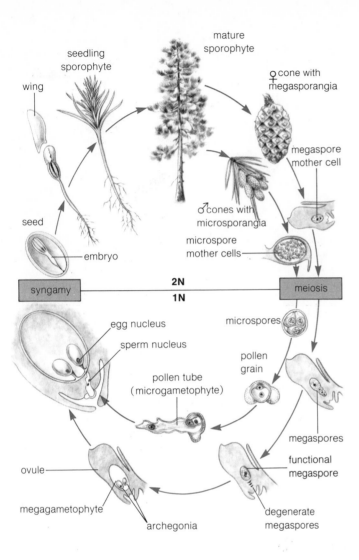

Figure 10–12 The life history of a pine.

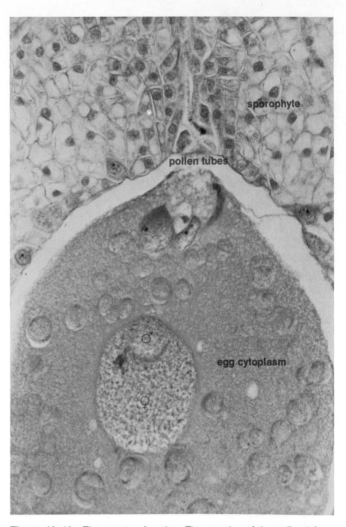

Figure 10–13 The zygote of a pine. The remains of the pollen tubes can be seen in the sporophyte at the upper end of the megagametophyte; within the zygote nucleus, the outline of the male nucleus is still visible.

female gametophyte, which in turn encompasses the new sporophyte, the embryo.

The adaptive significance of this life cycle for survival on land is tremendous. Consider just three major points: (1) The gametophytes now draw on the resources of the mature sporophytes to live. Wherever the sporophyte can live, so can the gametophyte. (2) Water is not necessary for fertilization. (3) The major reproductive agent is the seed—a most remarkable structure. The seed contains not only an embryo of the sporophyte ready to grow, but a built-in food supply and a protective outer covering to shield the embryo from adverse environmental conditions. These features have made the pines and other gymnosperms highly successful land plants. They have sur-

vived two periods of massive plant and animal extinctions during their 345 million years of existence and are still a major component of the vegetation around us.

Angiosperms The flowering plants, or angiosperms, evolved later than the gymnosperms and are in some ways even better adapted to land. Angiosperm reproduction continues the trend we have seen in gymnosperm reproduction: There are two types of gametophytes, both microscopic and both wholly dependent on the sporophyte; a pollen tube carries the sperm directly to the egg; and an elaborate seed is produced (Figure 10–14). Animals are important in the distribution of the pollen and seed

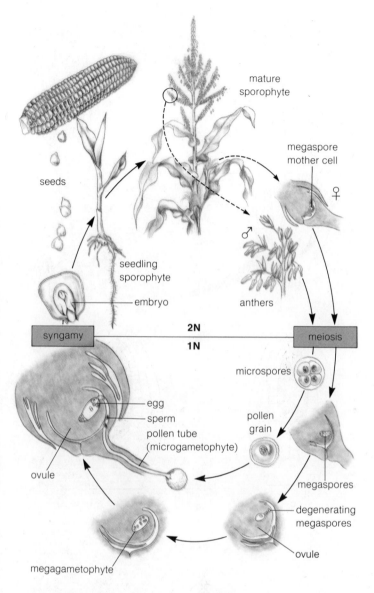

seeds

mature
sporophyte

megaspore
mother cell

♀

seedling
sporophyte

embryo

anthers

♂

syngamy

2N

1N

meiosis

egg
sperm
pollen tube
(microgametophyte)

microspores

pollen
grain

ovule

megaspores

megaspores

degenerating
megaspores

ovule

megagametophyte

Figure 10–14 The life cycle of an angiosperm.

sists of a usually slender stalk, the filament, bearing at its end a boxlike structure, the anther. Meiosis occurs within the anther, producing the microspores. These divide to form the two-celled microgametophyte, or pollen.

At the center of the flower are the **carpels.** Each carpel consists of a **stigma** at the top of which the pollen lands, a **style** in the center through which the pollen tube grows, and an **ovary** at the base containing the **ovules,** with their megaspores and megagametophyte. The carpels and parts of carpels may be free or joined together in a wide range of ways.

Within the ovary at the base of the carpel are the ovules. The organization and development of the ovule are shown in Figure 10–16. There is a small hole (the micropyle) at the top of the ovules through which the pollen tube will grow.

Megagametophyte development begins with the differentiation of a cell that enlarges and undergoes meiosis. Of the four megaspores produced, three degenerate. The fourth enlarges and undergoes a series of mitotic divisions to produce the megagametophyte. These divisions result in eight haploid nuclei that are organized into seven cells. Of these cells, three are located near the micropyle: the egg and two synergids. Two nuclei, called the polar nuclei, are present in the large central cell. The three cells at the opposite end of the megagametophyte are the antipodals. This is the most general pattern of megagametophyte organization.

After germinating on the stigma, the pollen tube grows down the style. During this period, the generative cell in the pollen tube divides to form two sperm. Eventually, the pollen tube reaches the megagametophyte. Here the tip of the pollen tube grows into a synergid where the tip breaks open and the contents of the pollen tube are discharged. One of the sperm nuclei enters the egg, and a zygote is formed. The second sperm nucleus enters the central cell, where the polar nuclei and the sperm nucleus fuse to form the *primary endosperm nucleus.* This is termed **double fertilization,** a phenomenon that occurs only in the angiosperms.

The diploid zygote divides to form a small globular embryo. Meanwhile, the primary endosperm nucleus, which is triploid (two polar nuclei plus sperm nucleus), divides rapidly to form the **endosperm,** a multinucleate mass that may eventually contain hundreds of nuclei. Cell-wall synthesis then takes place more or less simultaneously throughout the endosperm. The endosperm appears to be involved in the nutrition of the embryo, for if it fails to develop the embryo will not mature. The endosperm is generally used up during the development of the embryo, but in some cases, as in wheat and barley, the endosperm remains a significant portion of the seed. Most angiosperm embryos grow from a globular shape to a heart shape as the cotyledons, which are modified storage leaves, form (Figure 10–17). Continued division

of the angiosperms. In fact, the evolution of the fruit is directly related to seed distribution by animals.

A complete flower consists of four rings of parts set on the end of a stem (Figure 10–15). The outermost ring of parts is composed of **sepals;** the next, of **petals.** The sepals and petals are not directly involved in reproduction, but they may be vital in cases where the brightly colored petals attract animal pollinators. In general, petals are more brightly colored and prominent than sepals, which tend to be inconspicuous, green, and leaflike.

Inside the petals are the **stamens.** Each stamen con-

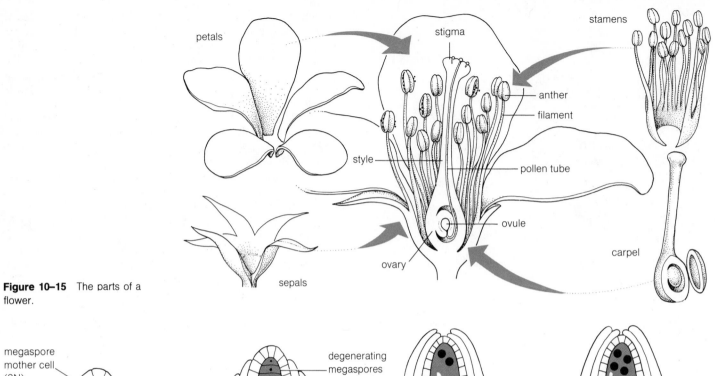

Figure 10–15 The parts of a flower.

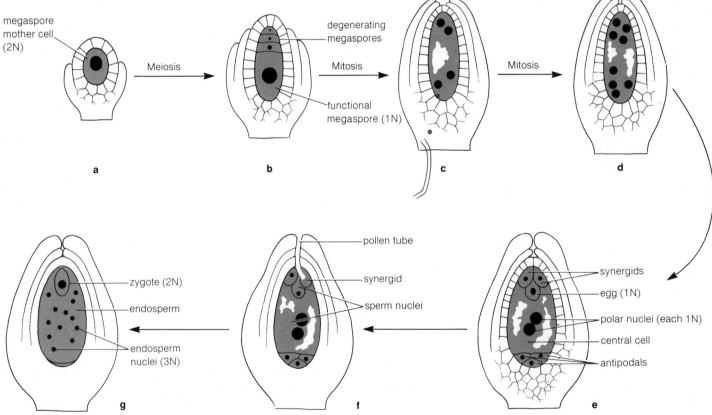

Figure 10–16 The development of the megagametophyte and fertilization in a flowering plant. (**a**) The megaspore mother cell, (**b**) the four megaspores (three of which degenerate), (**c**) the four-nucleate megagametophyte that forms from the mitotic division of the functional megaspore, (**d**) the eight-nucleate megagametophyte, (**e**) the mature megagametophyte, (**f**) double fertilization, and (**g**) the zygote and the young endosperm.

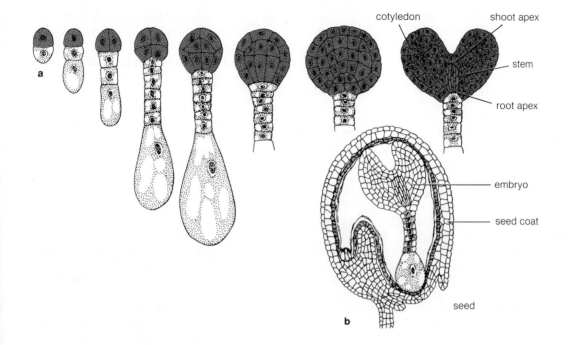

cotyledon   shoot apex

stem

root apex

embryo

seed coat

seed

b

a

Figure 10–17 (**a**) Embryonic development in the plant shepherd's purse. Embryo-forming cells are shaded. (**b**) The embryo plant within the seed.

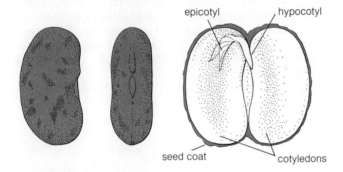

epicotyl   hypocotyl

seed coat   cotyledons

Figure 10–18 Structure of a bean seed showing the two cotyledons, the hypocotyl or root, and the plumule or first leaves.

and some elongation of cells result in an elongate embryo with clearly defined apical regions of both shoot and root.

The mature embryo may have either one or two well-developed cotyledons; a shoot tip with several tiny leaves is formed; and a root tip is present. In most seeds, the cotyledons are the main food-storage organs and make up the bulk of the seed, as in beans, peas, and peanuts.

The angiosperm seed consists of the embryo (the new sporophyte generation), surrounded by the remains of the endosperm (a special tissue that is neither gametophyte nor sporophyte), and a seed coat, produced by the parent sporophyte (Figure 10–18). Seeds may range in size from the dustlike seeds of orchids (close to 4 million may occur in a single fruit) to the double coconut coco-de-ner that weighs 20 kilograms.

While the seed is developing from the ovule, the fruit is developing from the ovary. Pollination followed by fertilization is usually necessary for fruit formation, but in some cases, fruit will form without fertilization—as with bananas, seedless grapes, pineapples, and some other plants.

The fruit protects the seed and helps to disperse it. The fact that the fruits of many flowering plants are edible means that animals will ingest them and the seeds they contain. The seeds pass through the animal and germinate after they are excreted, sometimes many miles from where they are eaten. Many fruits also develop spines that become entangled in the hair of mammals and are thus widely distributed.

Some Conclusions

What general conclusions can be drawn from this brief tour of reproduction in the plant kingdom? The most obvious one is that the structure and reproduction of a given group of plants are molded by its interactions with the environment. We also see that, as plants evolve a particular reproductive strategy, they can survive only where their strategy works. The differences between the fresh-water and the marine algae illustrate this point. The fresh-water forms are all relatively small haploid plants with simple structure, motile zoospores, and resistant zygotes. The marine forms are relatively large with elaborate plant bodies that are often diploid; although motile zoospores are present, there are few resistant cells. These differences fit the environmental conditions of the two aquatic environments: Ponds, streams, and even lakes dry up periodically, but the ocean has not dried up within the evolutionary memory of these plants.

Some observations about general trends in evolution and development can also be drawn from our tour, especially from the green plants—green algae, bryophytes, ferns, gymnosperms, and angiosperms. These plants vary markedly in structure and in the environments they exploit, but they all have the same basic photosynthetic pigments, the same form of food storage (starch), and the same basic cell-wall composition. If we compare their reproductive behavior, beginning with the earliest to evolve and moving to the latest, we find that in the green algae the gametophyte generation is dominant and independent; in the mosses, the gametophyte is still dominant and independent, but the sporophyte is large and well developed; in the ferns, the sporophyte is dominant, but the small gametophyte is still independent; in the gymnosperms, the gametophyte is small, wholly dependent, and encompassed by the large, dominant sporophyte; and in the angiosperms, the gametophyte is reduced to a few cells.

The most obvious explanation of the adaptiveness of such a shift from a predominant gametophyte to a predominant sporophyte is based on the actual amount of genetic material present. The doubled chromosome number means twice as many genes per cell and a stability of the genome that is presumably greater than in haploid organisms.

Another aspect of plants' reproductive behavior in which differences can be seen is the nature and behavior of the gametes. In the algae there is a progression from motile gametes of equal size, to motile gametes of unequal size, to motile sperm and nonmotile eggs. The larger egg has a better food supply for the development of the embryo, and the smaller, more energetic sperm can get around more easily—all in all, a more efficient system of fertilization.

This same pattern is found in animals; but there, huge eggs may be formed, far larger than those of any plant. The reason for this difference is fairly obvious: A chick in its egg is dependent on the food laid down before it was conceived and enclosed within the shell, but the embryo in a flowering plant receives food from the leaves of the mature plant throughout its development.

In plants that have evolved to take advantage of the terrestrial environment, a definite trend away from swimming gametes can be observed. As an adaptation to life on land, any mechanism that frees an organism from needing water to complete its reproductive cycle is obviously a tremendous advantage, and such mechanisms are found in both gymnosperms and angiosperms. The development of pollen, the pollen tube, and seed are major reproductive features that have helped make these groups predominant today.

11

Plant Development

In 1951, a Neolithic canoe was found buried 18 feet deep in a peat bog near Tokyo. Carbon-14 analysis of the wood in the canoe showed that it was at least 2,000 years old. Beneath the canoe were found three Oriental lotus seeds, surrounded by the hard coat characteristic of the species. When this coat was cut through and the seeds were placed in water, they germinated and grew into healthy plants that flowered when mature. These three ancient seeds dramatically illustrate one of the primary functions of seeds: to preserve the plant when the environment is unsuitable for growth and allow growth to continue when conditions are right.

The seed contains the embryo, which is essentially a miniature plant complete with a food supply and surrounded by a seed coat. In nonseed plants, the embryo usually grows immediately into the sporophyte. If conditions are not right for further growth, the plant could disappear from that location. Seeds provide much more flexibility. Some seeds are quiescent only until moisture is ample and temperatures are right; others remain dormant for varying lengths of time, although the extreme longevity of the three lotus seeds is so far unique. Seeds of some plants growing in temperate zones must be cooled severely before they will germinate, thus ensuring that in nature they will germinate only after the snow and ice of winter have given way to spring. Seeds of certain desert plants will germinate only after an inch of rain has leached an inhibitor from the seed coat—wetting the soil to a considerable depth and increasing the possibility that the plant will survive to flower. Some seeds germinate only after forest fires, and still others, only after they have passed through the digestive system of a bird. These dependencies have evolved because they ensure that the developing seedling will have optimum survival conditions.

Once the seed has sprouted, the embryo or seedling responds to gravity by extending its roots into the earth and its shoot system into the atmosphere. Cells at the root and shoot tips actively divide, enlarge, and specialize.

The roots branch, and the shoots produce leaves. On the stem above each leaf stalk (petiole) is a bud that can grow into a branch. Often, the stems become thicker in diameter through cell division, enlargement, and specialization, producing wood and bark. The result of all this is a complex *vegetative plant* with an intricate system of branching roots and leaf-covered stems (Figure 11–1).

At some point in its life—during the first year in all annuals (plants that live for a single year) and many other plants, but only after 20 to 30 years in many trees—the plant switches from the vegetative to the reproductive state, producing sporangia, cones, or flowers. For the plant to do so, the cells at the stem or branch tips must begin dividing, enlarging, and specializing in different ways according to individual sets of regulations contained in their genome. These changes occur in many species in response to such environmental changes as rainfall, the low temperatures of winter, the lengthening days of spring, or the shorter days of late summer. After flowering or reproducing by some other method, plants usually either die (annuals), become dormant until conditions are right for beginning another vegetative cycle (temperate-zone perennials), or immediately begin another such cycle (tropical perennials). The preparations for dormancy in certain hardwood trees of northeastern North America are so spectacular that tourists flock to see them.

The remarkable thing about plant growth is that it can produce such elaborate structures by three simple means: cell division, elongation, and differentiation. Cell division increases the number of cells in the plant. Most of these divisions take place in special regions of the plant known as **meristems** (Figure 11–2). But cell division accounts for only a small amount of the total growth of the plant: Most of the growth in length or breadth is the result of cell elongation and enlargement. The entire early growth of a seedling, for example, is a result of elongation of existing cells, with cell division in the emerging root starting only after many hours. Finally,

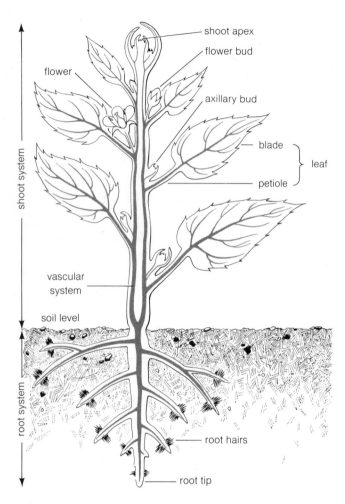

Figure 11-1 The general body plan and parts of a vascular flowering plant.

Labels on figure: shoot apex, flower bud, flower, axillary bud, blade, leaf, petiole, vascular system, soil level, root hairs, root tip, shoot system, root system

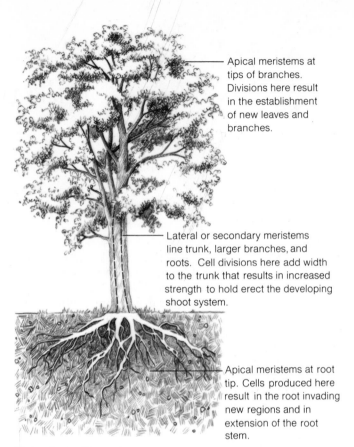

Labels on figure: Apical meristems at tips of branches. Divisions here result in the establishment of new leaves and branches. Lateral or secondary meristems line trunk, larger branches, and roots. Cell divisions here add width to the trunk that results in increased strength to hold erect the developing shoot system. Apical meristems at root tip. Cells produced here result in the root invading new regions and in extension of the root stem.

Figure 11-2 The position of the apical and lateral meristems in a tree.

differentiation encompasses changes that occur in the cell as it assumes its final function. The changes taking place during differentiation are often drastic and elaborate.

The cells of all land plants are surrounded by walls. During the growth of the cell, the composition of the wall changes. The basic chemical constitutents of the wall are the pectic substances, the hemicelluloses, cellulose, and lignin. The first three are all composed of carbohydrates—hexose and pentose sugars or compounds derived from them. Lignin, in contrast, is not a carbohydrate but a much more complex compound.

Cells of the apical meristem are surrounded by *primary cell walls* (Figure 11–3). These are usually thin and composed of pectic substances, hemicellulose, and cellulose. The region of the wall that is common to two cells is called the *middle lamella* and is composed of pectic materials. As elongation ceases and the cells mature, the walls usually become thicker and a *secondary wall* may form between the cytoplasm and the primary wall. Frequently, both the primary wall and the secondary wall will become impregnated with lignin. The presence of lignin makes the wall stronger and less flexible.

With this brief summary of the life cycles of seed plants in mind, let us look at growth in more detail and see what forces control the cycle.

Cell Division: Plant Meristems

All growth originates in the meristematic tissue of the plant. Meristems consist of cells that are capable of repeated cell divisions; other cells usually lose the capacity for division during differentiation. There are two distinct types (Figure 11–2): the **primary** or **apical meristems**

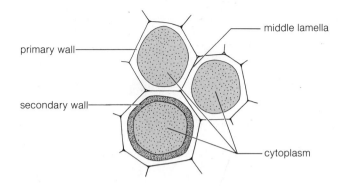

Figure 11–3 Primary and secondary walls surrounding plant cells. The middle lamella holds the cells together.

located at the tip (apex) of stems and roots, and the **secondary** or **lateral meristems** located all along the stems and roots. The apical meristems give rise to the *primary tissues* of the plant and account for the growth in length (elongation) of the stems and roots, as well as the formation of the leaves and flowers. In contrast, the lateral meristems give rise to the *secondary tissues* that account for growth in diameter. The two different kinds of growth can be seen in the life cycle of the oak. All the growth associated with the germination of the acorn and the growth of the young oak seedling results from the development of cells produced by the apical meristems of the root and shoot. Later, as the tree continues to grow, the lateral meristems are formed and the cells they produce add to the thickness of the tree, resulting in the formation of wood and bark. In many short-lived plants (such as most spring wild flowers), no secondary growth ever occurs; the entire plant is the result of cell divisions associated with the apical meristems.

Apical meristems contain cells that continually divide, enlarge, elongate, and differentiate (Figure 11–4). As new cells are produced and grow longer, they carry the meristematic region downward in roots and upward in shoots. The cells that result from divisions of the meristematic cells are initially more or less similar in form, but through differentiation they are converted to the mature tissues of the root and shoot. Secondary or branch roots are continually formed as new meristems arise relatively deep within the root. Leaves are formed from cells near the surface and on the flanks of the shoot meristem (leaf primordia), and branch buds (axillary buds) form in association with them (Figure 11–4).

Two kinds of lateral meristems occur in both roots and shoots: the **vascular cambium,** which forms the wood, and the **cork cambium,** which forms the bark or

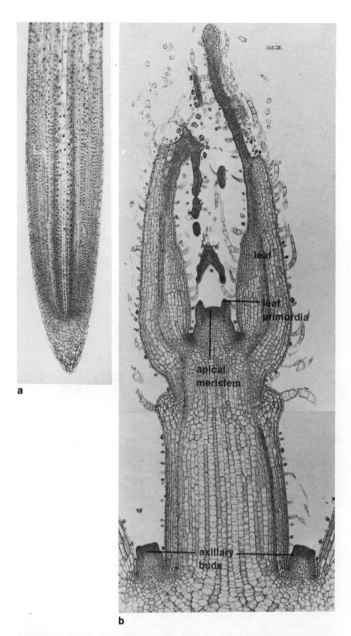

Figure 11–4 The apical meristem of the root (**a**) and the shoot (**b**). The root apical meristem is surrounded by a protective layer of cells (the root cap). The shoot tip produces cells that form the stem and leaves. As the leaves develop they form a protective cover for the meristem itself.

outer covering of the plant. A *cambium* is a sheath of dividing cells that forms an inside layer the full length of the plant and, by cell division, produces cells toward both the inside and the outside. In the vascular cambium, the inside cells differentiate into the **xylem,** or water-

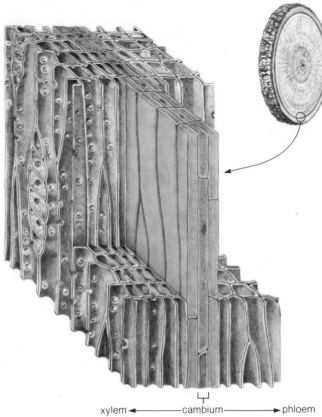

Figure 11–5 The vascular cambium differentiates into xylem and phloem.

xylem ← cambium → phloem

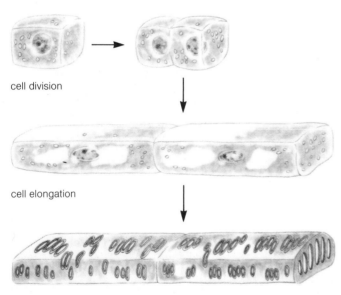

cell division

cell elongation

cell differentiation and maturation

Figure 11–6 Stages in the development of plant cells.

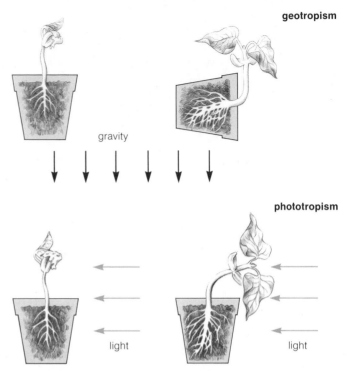

geotropism

gravity

phototropism

light   light

Figure 11–7 Effect of gravity and light on the growth of plants.

carrying tissue system of the plant, whereas the outside cells differentiate into the **phloem,** the tissue system that carries the products of photosynthesis through the plant (Figure 11–5). As the cells of the vascular cambium continue to divide, the diameter of the plant increases, and the outermost cells eventually are crushed and die. Thus, xylem accumulates as wood, but phloem, newly produced each year, lives for only a few years before it is crushed and dies. The cork cambium operates in much the same manner as the vascular cambium, except that the outside cells differentiate into tissue known as cork that protects the cells beneath it from the surrounding environment. For more complete descriptions of these and other types of plant cells, see Supplement 11–1.

Cytokinins Meristematic activity, then, increases the number of cells in the plant, but what controls cell division within the meristem? At least one group of compounds, the **cytokinins,** is known to regulate plant cell division. Since the group was discovered in 1955, about a dozen naturally occurring cytokinins have been found. These compounds, which are related to adenine, are known to cause cell division in a variety of plant systems. But the discovery of cytokinins has not by any means resolved the problem of the control of cell division. Other

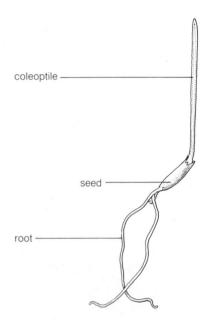

Figure 11–8 An oat seedling. The coleoptile is a sheath enclosing the first leaves.

compounds can cause cell divisions, and there are strong interactions between other growth substances and cytokinins. Still more substances will probably be found that influence cell division.

Cytokinins have other effects on plant cells besides causing cells to divide. For example, they retard aging in leaves and affect the translocation of amino acids.

Elongation

Cells produced by the apical meristem usually are relatively small and roughly the same in all dimensions. In most cells, the stage of growth following division involves elongation, with some cells growing to 10 or more times their original length (Figure 11–6). The results of cell elongation in the overall growth of the plant are often spectacular. Many a small child has been taken into a corn field in the Midwest on a night without a trace of wind to hear the corn grow. And this is exactly what they do hear. The elongation of the cells of the stem is so rapid at night that the growing stem causes the leaves to rustle softly in the cool night air!

Before plant cells can grow longer, the cell wall must become softer and less rigid so that it will stretch. The stretching force is water pressure within the cell, maintained by osmosis. The physical properties of the wall determine the direction of elongation. The potential size of the cell is a function of its capacity to synthesize the necessary wall and cytoplasmic materials, but the time when elongation will occur and the size the cell will actually reach are controlled by the organism and by environmental conditions.

However, to learn what specifically controls elongation, we must look to the growth substances, or plant growth hormones. The two best-known groups are the auxins and the gibberellins.

Auxins Auxins are produced by the apical meristem. They diffuse away from the apex and stimulate cells at some distance from it to elongate. The most important natural auxin is a compound called **indoleacetic acid,** or IAA. Other natural auxins are known, and many artificial auxins have been made.

The distribution of auxin in stems is influenced by both light and gravity. IAA accumulates on the dark side and the lower side of stems. This explains the growth reactions, or **tropisms,** of stems to gravity **(geotropism)** and light **(phototropism)** (Figure 11–7). Stems are positively phototropic (they grow toward light) because IAA actually moves to the dark side of the shoot tip and the cells on that side elongate more. Stems are negatively geotropic (they grow upward) because IAA accumulating on the lower side results in increased elongation of these cells, curving the stem upward.

Roots react to these environmental stimuli—and to auxin—in precisely the opposite way: They grow downward (are positively geotropic) and away from light (are negatively phototropic). Auxin accumulates in the same way in roots as in stems, but in roots the cells with greater auxin concentration elongate less, not more. We do not understand precisely why this happens.

If light is to affect the distribution of auxin, it must be absorbed by a pigment. One way to understand phototropism better would be to find out which pigment is involved. And one way to do that is to determine which wavelengths of light are most effective in causing the biological reaction.

A favorite experimental subject for such work is the oat *coleoptile,* a modified cylindrical leaf that covers and protects the emerging primary leaves of grass seedlings (Figure 11–8). Coleoptiles are widely used in the study of auxin physiology because they grow only through cell elongation, the cells having been formed by cell divisions earlier in the development of the seed. Tests using coleoptiles show that blue light is most important in their positive phototropic response. If we compare the possible absorbers of blue light with pigments known to occur in the coleoptile, we find a yellow pigment called riboflavin

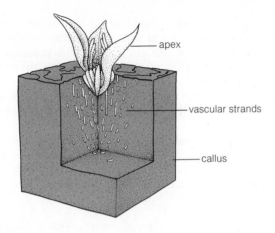

Figure 11-9 The effect of grafting a shoot apex onto a piece of callus. Note the vascular strands (developing xylem and phloem) in the callus and their relationship to the apex.

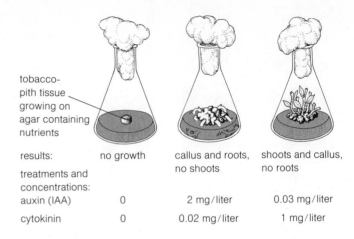

Figure 11-10 A demonstration of the effects of auxin (IAA) and cytokinin on the formation of roots and shoots in tobacco-pith tissue cultures.

the likeliest candidate for involvement in phototropism, but we still do not know for certain. Neither do we know how the pigment is linked to a system that changes the amount of auxin in the stem.

In trying to understand geotropism, plant physiologists have looked for particles in the cell that would be redistributed by the force of gravity. These hypothetical particles are called **statoliths.** Since auxin molecules themselves are too small to be affected by gravity in the plant, researchers looked for larger bodies. Starch grains were soon nominated as the logical candidate, and for many cells they may indeed be the answer. However, we are not sure that starch grains act as statoliths in all cells or that only starch grains are statoliths, and the search goes on.

Ethylene Tied to the mechanism of IAA in plants is the activity of ethylene ($H_2C{=}CH_2$). This gas—which you can smell when fruit is ripening in a closed space—is known to cause a number of changes in plant cells, including changes in the response to auxin. Ethylene seems to cause cells to expand (but not divide), and often to expand in all directions rather than to elongate. Some investigators have suggested that auxins merely induce the formation of ethylene, which then causes change in the cell. We do not know whether this is true, but plant physiologists consider ethylene an important naturally occurring growth substance. Ethylene certainly hastens the ripening of lemons, apples, and many other fruits. Moreover, the fruits themselves give off the gas so that, for example, ripening apples in the same room with green

bananas greatly hastens the ripening of the bananas. And, as the old saying goes, one rotten apple can spoil a whole barrel—by causing overripening.

Gibberellins As important as the auxins in plant growth are the **gibberellins.** Three dozen or more gibberellins are known. Plants treated with a gibberellin may elongate at six times the normal rate. Genetically dwarfed corn and peas, whose cells normally elongate only a little, will grow as tall as the normal varieties after gibberellin treatments, which indicates that they lack gibberellin. Short, squat "rosette" plants, which usually do not elongate at all, often elongate or "bolt" after gibberellin treatment.

Gibberellins are produced largely in young leaves, in roots, and in developing seeds. They affect primarily cell elongation, but they influence many other processes in the plant as well. Fruit development, germination, and flowering in certain plants are all affected by gibberellins. In many cases, the effects of gibberellins and auxins are much the same.

Differentiation

Earlier we said that plant cell growth can be divided into three stages: cell division, elongation, and differentiation-maturation. Actually, however, these stages overlap, with cells in active elongation often still dividing or already beginning to differentiate. Moreover, the processes by which the relatively undifferentiated cells of the meristems are transformed into the many different and complex cells of the mature tissues of the plant are still not fully understood.

One way of studying the complex phenomenon of cell differentiation is to use tissue cultures. Pieces of plant tissues can be grown, or "cultured," on agar (a gelatinous substance) containing sugars, nitrogen compounds, vitamins, and growth substances. Such tissue will grow into an unorganized mass of cells called a **callus.** Implanting a bud on this callus will cause xylem and phloem to differentiate within it (Figure 11–9). This effect can be duplicated by adding sugar and IAA directly to the callus. Low levels of sugar result in xylem formation; high levels cause phloem formation; and intermediate levels produce both xylem and phloem with cambium between. However, these results are obtained only if IAA is present. Additional evidence for the role of IAA in xylem and phloem differentiation has been obtained from studies on roots and stems. For example, both qualitative and quantitative relationships exist between the formation of xylem in *Coleus* stems and the IAA produced in the leaves.

Callus cultures can also be used to study the differentiation of roots and shoots. In cultures derived from tobacco plants, it is possible to control the formation of roots and shoots by adjusting the ratio of auxin to cytokinin (Figure 11–10). When auxin levels are high and cytokinin levels are low, roots but no shoots are produced, and vice versa. In general, experiments have tended to show that IAA induces root formation, while cytokinins produce shoots.

Another shoot phenomenon related to cytokinins and auxins is apical dominance. The terminal bud at the apex is dominant over the buds below it, producing a regular growth pattern found, for example, in the pine. In many plants, the removal of the actively growing tip will cause the other buds to grow. Horticulturalists use this phenomenon to regulate the growth patterns of their plants. They snip back the tips of *Coleus,* fuchsias, and other plants to encourage the growth of side shoots and produce a bushy rather than "leggy" growth. Both auxins and cytokinins are apparently involved in this reaction—auxin inhibiting the secondary buds while the terminal bud is present, the cytokinins stimulating them to grow when the terminal bud is removed as a producer of auxin.

Dormancy

Related to the problems of differentiation is **dormancy.** We are all aware that in autumn trees shed their leaves and the shoot meristems become dormant until the following spring, when new leaves are produced. We are also aware that most seeds do not germinate immediately upon leaving the parent plant but remain dormant for

weeks, months, or even years. Phenomena such as these are controlled by still another growth-regulating substance, abscissic acid, also known as ABA or dormin. Many of the actions of abscissic acid appear to be antagonistic to those of auxins and cytokinins. For example, in the case of leaf fall in autumn, a layer of cells differentiates at the base of the petiole. These cells eventually die, weakening the petiole so that it breaks and the leaf falls from the plant. Both auxins and cytokinins inhibit the formation of this layer (known as the abscission layer), whereas abscissic acid promotes its formation.

Recent research by Leon Dure at the University of Georgia has revealed some of the action of abscissic acid at a molecular level in the case of seed dormancy. Dure has found that mRNA is synthesized late in embryogenesis, shortly before the seed is mature. The translation of this mRNA will inaugurate germination, and abscissic acid controls this translation process. As long as the concentration of abscissic acid is high, translation is inhibited and the seed will not germinate. For germination to occur, the abscissic acid concentration must be lowered and translation must begin. Dure has also shown that, during the early stages of seed germination, protein synthesis is controlled by the mRNA produced before the embryo became dormant. Thus, the critical stage of seed germination can occur without the complete involvement of the genetic store itself.

Much of plant growth is regulated by a balance of growth-regulating substances, both promoters and inhibitors. Gibberellins, IAA, cytokinins, abscissic acid, and probably other growth substances are all involved, and shifts in the balances among them are reflected in developmental responses on the part of the organism.

Flowering

One of the most spectacular aspects of plant development is flowering. Most of us are struck by the beauty of the flowers that bloom in the spring or welcome those that open in the fall. Flowers are produced by apical meristems that change from forming leaves to making the flower parts. This transformation is apparently controlled by a hormone that is produced in the leaves and moves to the apical meristem of shoots, where it induces flower formation. Although this hormone has not yet been isolated, so its chemical nature is unknown, it has been given a name, *florigen.*

Most plants do not bloom all the time, but only at certain seasons. Clearly, then, the production of the flowering hormone must be related to some change in the environment. This turns out to be day length, or, more precisely, the number of hours of light versus dark the plant receives. Plants that respond in this way (whose

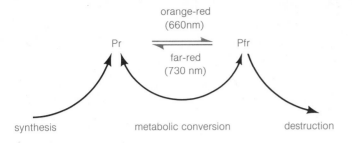

orange-red
(660nm)

Pr ⇄ Pfr

far-red
(730 nm)

synthesis   metabolic conversion   destruction

Figure 11–11 The relationship of Pr and Pfr. The double arrow shows that the reaction goes to the right in orange-red light and to the left in far-red light.

flowering is controlled by light) are called **photoperiodic;** those that do not are called **day-neutral.** Photoperiodic plants can be roughly grouped into two categories: short-day plants that bloom when the days are shorter than a fixed length, and long-day plants that bloom only when the days are longer than a fixed length.

Photoperiodism is a widespread phenomenon: Many responses in both plants and animals are controlled by day length, or by the amount of light received. Leaf drop in trees and mating and migration responses in some animals are photoperiodic.

Photoperiodic flowering is a good example of a successful adaptation to an environment. It ensures flowering at the proper time for the species—in the case of spring annuals, for example, when there is enough light and heat for seed and fruit formation but before the forest canopy becomes too dense for photosynthesis on the forest floor. Photoperiodism is also used by commercial growers, who alter day length artificially to produce flowers when people want them in a given region rather than when they would naturally flower there.

The term *photoperiodism*—and our discussion to this point—might lead you to think that light (or day length) alone is involved in controlling flowering, but both the light and the dark period are important. In a short-day plant, such as the late-summer-blooming cocklebur, as the nights become longer and longer, a critical point is reached and florigen is produced. If this critical dark period is interrupted even briefly by light, the flowering hormone will not be formed.

If flowering is influenced by the number of hours of light (or dark), and if florigen is formed in the leaves in response, then leaves must be able in some sense to measure time. But how? This question still has not been definitely answered, and we shall return to it in Chapter 25, when we consider the nature of biological clocks. We do know that the measurement is somehow related to an important plant pigment known as *phytochrome.*

Several plant-growth responses are related to phytochrome: germination of certain seeds that require light for germination, the formation of the anthocyanins and carotenoids that color autumn leaves and many flowers and fruits, etiolation (depressed leaf development and elongated stems) in plants developing in the dark, and flowering, among others. In all cases, red light (620–670 nanometers) causes the effect, whereas far-red light (710–750 nanometers) nullifies or reverses it. The pigment exists in one form, Pr; when illuminated with red light, it shifts to a second form, Pfr; far-red light converts the Pfr back to the Pr form (Figure 11–11). It is known that Pfr is converted metabolically within the plant, even in darkness, to the Pr form.

We still do not know how phytochrome can control growth responses or what its precise role in photoperiodism is, but knowing that it is involved is the first giant step toward understanding how it operates.

The Mechanism of Regulation

Throughout this chapter we have been skirting the question of how growth substances work, for the simple reason that the answer or answers are not yet clear. Growth substances may act through changes in cell membranes. This is a particularly likely mechanism in the case of auxins, since the cell may react to it within seconds after application.

Growth regulators may also be closely associated with gene function; that is, the regulating substances may activate or deactivate genes. Gibberellins apparently work this way during seed germination in barley. A gibberellin released by the embryo causes the DNA of a special layer of endosperm cells to produce mRNA that results in the formation of the enzyme amylase. The amylase in turn breaks down starch stored in other endosperm cells. The sugars released from the breakdown of the starch are then used by the embryo for growth and development. Thus, in this case, a growth substance seems to be activating a gene. There is also some evidence that abscissic acid may work by deactivating genes. Whether all growth regulators function by turning genes on or off remains for future research to tell us.

Totipotency

Before leaving plant development, let's return to the zygote from which all plants grow. This cell clearly contains within it all the genetic information needed for the development of an entire plant—that is, it is **totipotent.** As the zygote divides and the plant forms, a process of differ-

Essay 11–1

From a Single Cell: The Opportunities of Cell and Tissue Culture

Perhaps the most exciting development in recent years in the general field of plant research is the ability to grow plants from single cells. Considering that all plants and animals grow from zygotes, which are single cells, this may not seem to be the most important of discoveries, but consider the following situations. After 10 or 20 years of work, a plant breeder has finally developed a new color in a carnation that will be widely sought by growers; yet another 5 to 10 years will pass before adequate seed supplies can be obtained. Or consider a chemical manufacturer who hopes to develop a new herbicide without the undesirable side effects of the compounds now in use, or for use on different kinds of plants. Dozens of compounds can be readily synthesized, but it will take many years of field testing even to narrow the search for a commercially useful one. Another case is the long-felt need to use **nitrogen-fixing bacteria** (bacteria capable of converting atmospheric nitrogen to organic nitrogen compounds) in commercial agriculture; yet the most basic crops, such as wheat, corn, and rice, do not have nitrogen-fixing bacteria associated with them. Conventional breeding programs have failed to get close to solving the problem. In all these cases, research with the culture of single cells is promising.

The carnation grower can use single-cell culture procedures in two ways. First, after the variety has been genetically established through conventional breeding techniques, a small segment of the plant is placed in a culture medium that stimulates the formation of a mass of cells known as a callus. Next, the callus is treated so that the cells separate from one another. These individual cells are now cultured, and they in turn form calluses. The culture conditions are changed, and the calluses form roots and shoots, in effect becoming seedlings that can be planted. Such a procedure produces thousands of seedlings, each identical to the parent, in a matter of months; conventional seed production might take years to reach the same point. The culture procedures also allow the propagation of species that do not normally reproduce sexually, such as pineapples and bananas.

The second way in which culture procedures can help a breeder is through the establishment of haploid plants. All higher plants are diploid except for the few cells of the gametophyte, yet haploid cells are useful in breeding programs: Since the genes in haploid cells are not present in pairs, the effect of a change in a gene is expressed directly and cannot be masked by the other member of the pair. A haploid culture can be established by allowing pollen to germinate and causing it to form a small callus. Pollen is the male gametophyte, and its nuclei are haploid. These haploid calluses can be separated into single cells and grown into plants, as we saw in the first method. They can be treated with mutagenic agents or selected by any number of methods.

If a useful strain is developed, a new problem arises: Haploid plants are not so robust as diploids. This means that the haploids should be converted back to diploids. This can be done by treating a haploid callus with a chemical that inhibits chromosome movement at metaphase. Then the chromosomes stay together, and the resulting nucleus is diploid. It also means that the chromosomes are completely homozygous in the genes present and that the plants will breed true.

In developing a new herbicide, the chemical company can use the cell-culture technique to screen the chemicals and determine their effect on the species under consideration. In this case the higher plant cells are treated essentially as bacteria and the chemical is tested on them. This may not give the final answer, and field tests will be necessary, of course; but it is a way of reducing the time and money needed.

The problem of breeding nitrogen-fixing into species that cannot do it themselves, or of coopting bacteria for the job, involves still a different set of techniques. For example, the goal may be to produce wheat plants that have nodules of nitrogen-fixing bacteria, as does soybean. If one could combine the genes of wheat and soybean, it is possible this goal could be achieved. Researchers are now using procedures known to fuse the protoplasts of different species.

Cell suspensions, taken either directly from mature plants or from callus cultures, are treated with a series of enzymes that remove the cell walls, leaving the cells as naked protoplasts. Such protoplasts will shortly synthesize new walls. In the time that they are without walls, however, they can be treated with chemicals that alter their cell membranes so that the membranes fuse. This means that two protoplasts and their nuclei come together to form a single cell. Thus, the genes of two different species can be mixed and expressed in a new plant developed from that cell. A wheat species that would have nitrogen-fixing nodules of bacteria in its roots has not yet been produced, but the work goes on.

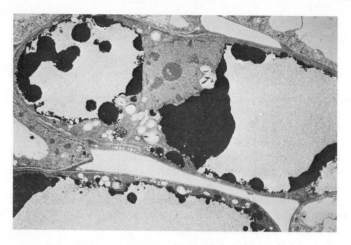

Figure 11–12   Electron micrograph of a parenchyma cell.

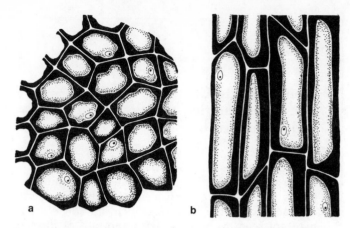

Figure 11–13   Collenchyma cells seen in cross section (**a**) and longitudinal section (**b**).

entiation provides the plant with myriads of specialized cells.

In the course of this development, are genes lost or turned off permanently? The answer is apparently no, for under certain conditions a single cell of a mature plant can form a complete, new individual. The cell must, of course, be living and have a nucleus. It must also dedifferentiate; that is, it must essentially return to the zygote stage before the developmental pattern can be expressed. This expression is influenced by the environment surrounding the developing embryo.

The field of plant development is important and interesting. It still contains a great many important unanswered questions (see Essay 11–1).

Summary

Most plants continue to grow throughout their life-span. Growth in length occurs by division of cells of the apical meristems and the subsequent elongation of these cells, followed by differentiation. Growth in diameter is the result of division and differentiation of cells in the lateral meristems of the plant.

Cell division and elongation are controlled by plant growth hormones. Cytokinins and auxins are important in the regulation of cell division, and the auxins and gibberellins are central to the regulation of elongation. Interactions of these and other plant growth hormones are involved in cell differentiation.

Plant growth is also controlled by various environ-

mental factors such as gravity and light. Light is of particular importance. Plants respond to both the amount and direction of light, as in phototropism, and to the length of time they are exposed to light or dark, as in photoperiodism and flowering. Responses to these various physical factors are translated into growth responses in the plant through the action of receptors and the growth hormones. We are only now beginning to understand the nature of the action of plant growth hormones on the cellular and biochemical levels.

Finally, plant cells show remarkable totipotency and, under proper conditions, single cells from mature plants can be made to produce complete, new plants.

Supplement 11–1
Tissues and Tissue Systems: The End Results of Differentiation

All plant growth is based on cell division, elongation, and differentiation controlled by chemicals. The results of this controlled growth are the plants we see around us. They are beautiful structures composed of millions of cells organized into tissues and organs. Let's now look at the types of cells that are found in vascular plants and how they are integrated into the familiar parts of the plant.

Parenchyma cells are totipotent. Not only can they develop into other cell types even when mature, but they retain the capacity, present in the zygote, to develop into new plants. Single parenchyma cells have been induced to grow into entire mature plants.

Collenchyma   **Collenchyma** cells are essentially long-celled parenchyma with thickened walls (Figure 11–13). Strong yet plastic, collenchyma cells function as support elements in the plant. They are found particularly in young, rapidly growing organs.

Sclerenchyma   **Sclerenchyma** is a supportive tissue more specialized than collenchyma. It can be divided into two types of cells: *sclereids* and *fibers* (Figure 11–14). Sclereids are highly variable in size and shape, ranging from isodiametric cells to H-shaped cells to cells with many long, oddly shaped arms. Fibers, in contrast, are long, narrow, uniformly thick-walled cells. The two kinds of cells are quite unlike in shape but similar in their other features. Both sclereids and fibers are dead when they are functionally mature. They have thick, lignified secondary walls and are usually quite long.

Some of the fibers we use for cloth are sclerenchyma fibers. Linen is woven from fibers obtained from the flax plant *(Linum usitatissimum)*. In this case, the fibers are from 9 to 70 millimeters long and provide a tough yet pleasant-feeling material.

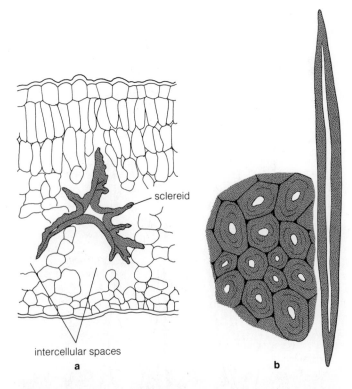

labels: sclereid, intercellular spaces, a, b

Figure 11–14   Two common forms of sclerenchyma cells: (**a**) an example of the multiarmed sclereids found in leaves; (**b**) fiber-type sclerenchyma cells in cross section and longitudinal section.

Types of Plant Cells

Parenchyma   The most basic of all cell types or tissues found in vascular plants is the **parenchyma** (Figure 11–12). This is the "typical plant cell" so often drawn and discussed in biology books. Parenchyma cells have a relatively simple cylindrical shape and thin walls. The center of the cell is often occupied by a large vacuole crossed by strands of cytoplasm and pressing the bulk of the cytoplasm and nucleus against the wall.

Many parenchyma cells contain numerous chloroplasts. In leaves, these cells may be arranged in regular, upright rows, the *palisade parenchyma,* or in loose, irregular arrangements known as *spongy parenchyma.* Both types are usually present in the same leaf.

Parenchyma cells may also be filled with starch-containing plastids. When you eat a potato, you are eating starch-filled parenchyma cells. Starch-storing parenchyma cells frequently are roughly the same in all dimensions and may be so filled with starch that the vacuoles can hardly be seen.

Xylem   The xylem is the water-conducting system of the plant. Within the xylem the *tracheids* (Figure 11–15a) are the cells through which the water moves, except in the most advanced land plants, where tracheids are supplemented or replaced by *vessels* (Figure 11–15b).

Tracheids are relatively long, narrow cells with lignified secondary walls. The oblique end walls are covered only by a thin membrane, and the cells are aligned end to end, facilitating the flow of water. The side walls contain numerous pits. Tracheids found in primary tissue have walls with characteristic bands of wall thickening. The nucleus and cytoplasm of a tracheid degenerate when the cell reaches functional maturity.

Vessel elements (formed only in angiosperms) are shorter and larger in diameter than tracheids. They, too, have lignified secondary walls and lose their cytoplasm and nuclei at maturity. They also lose major portions of their end walls, so that they form a continuous tube for conducting a column of water.

Pits (Figure 11–16) are important structures in the walls of xylem cells, for it is through the pits that water and dissolved substances pass from cell to cell. In pits, a thin portion of the walls of two cells is adjacent, but

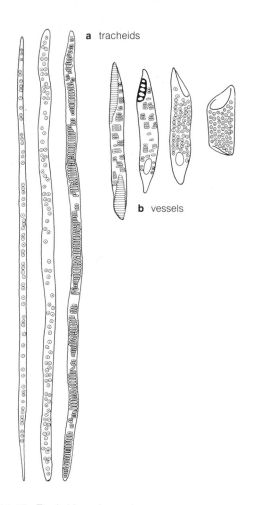

Figure 11-15 Tracheids and vessels.

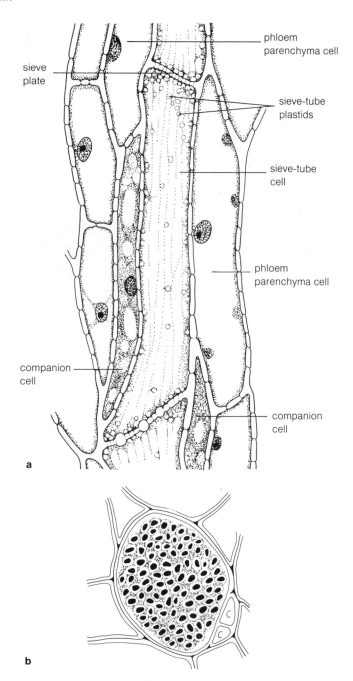

Figure 11-17 (**a**) A longitudinal view of a mature sieve-tube cell and companion cell. (**b**) A face view of a sieve plate; the black areas are actually holes in the wall.

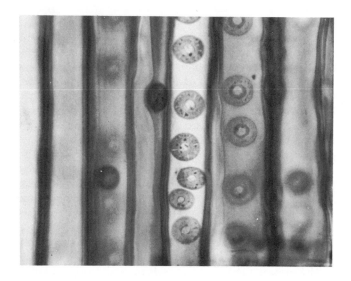

Figure 11-16 Bordered pits in the tracheids of pine.

there is no actual hole in the walls. Instead, a pit membrane, comprising the primary walls and the middle lamella of the two cells, remains.

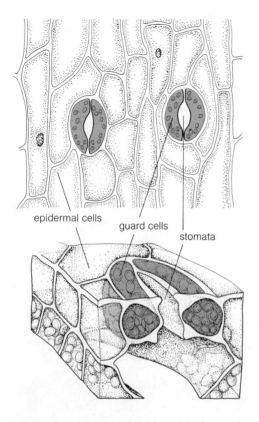

Figure 11-18   Guard cells surrounding leaf stomata.

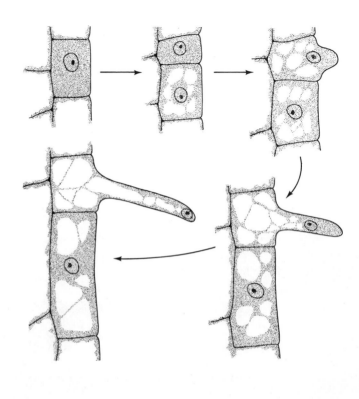

Figure 11-19   Stages in the development of a root hair.

Phloem   The phloem is the functional counterpart of the xylem, transporting solutions of sugars and other organic materials. The sieve cells and sieve tubes of the phloem (Figure 11–17), so called because the end walls (sieve plates) of the cells actually look like sieves, are analogous to the tracheids and vessels of the xylem. Associated with sieve tubes are cells with dense cytoplasm not directly involved in transport but necessary for the functioning of the sieve tubes. These are called companion cells.

Epidermis   The epidermis is the layer of cells covering the outside of the vascular plant. In general, epidermal cells have relatively thin walls, with the external wall being thicker than the others. The epidermis of pine needles is an example of extremely thick-walled epidermal cells. A characteristic feature of epidermal cells found on the above-ground parts of the plant is the **cuticle.** This layer of the fatty substance *cutin* is secreted by the epidermal cells and covers the outer surface of the cell as a separate layer. In many plants, the cuticle is covered with a deposit of wax.

The epidermal cells are alive when mature and contain nuclei capable of division. In flax and begonias, a single epidermal cell has been shown to give rise to a bud that can form a complete plant. Plastids are present, including chloroplasts in some plants. The vacuoles are usually large and may contain pigments or tannins. Tannins are bitter-tasting compounds that appear to repel insects that might feed on the plant.

Various specialized cells are found in the epidermis. These include guard cells, epidermal cell hairs, and secretory cells.

The **guard cells** (Figure 11–18) are a pair of specialized cells that surround the openings, or **stomata,** in the leaf. Changes in the shape of the guard cells regulate the size of the opening, which in turn regulates the amount of water lost from the leaf. The wall of the guard cells is typically uneven in thickness. These uneven, thickened walls play a role in the opening and closing mechanism of the stomata.

Root hairs are simple extensions of the root epidermal cells (Figure 11–19), being initiated some 100 to 300 micrometers behind the root tip. Sometimes they reach 2 millimeters in length. The presence of root hairs greatly

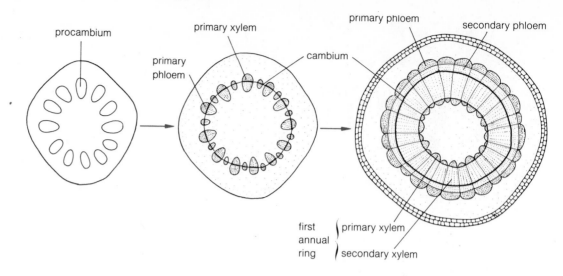

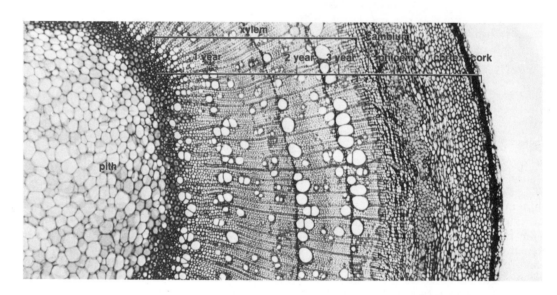

Figure 11–20 The development of the vascular cambium and the secondary vascular tissues.

increases the area of the absorbing surface of the plant in its contact with the soil. A single rye plant was found to have over 13 million roots with a surface area of 2,500 square feet. On these were 14 billion root hairs having a total surface area of 4,300 square feet. The combined surface area of roots plus root hairs was 6,800 square feet—all packed into less than 2 cubic feet of soil!

Similar hairs may be found on stems, leaves, flowers, and fruit. A familiar example is the cotton fiber, which is actually an epidermal hair formed on the surface of the cotton seed. The walls of these cells are much thicker than those of root hairs and are extremely high in cellulose. They may grow from 6 to 60 millimeters long, depending on the variety of cotton.

Some above-ground hairs are glandular, secreting compounds such as nectar and various essential oils. In *Cannabis sativa,* secretory hair cells produce a resin containing a series of complex alcohols, including tetrahydrocannabinols, the active ingredient in the drug marijuana. Such hairs are found only on the upper parts of the female plants.

One of the most unusual plant hairs is the stinging hair found in nettle *(Urtica).* The hair is like a fine capillary tube with calcium at the lower end and silica at the upper end of the wall. A bulblike base is embedded in epidermal cells. A spherical tip breaks off along a predetermined line when the hair comes in contact with an object such as human skin. The sharp edge that is left

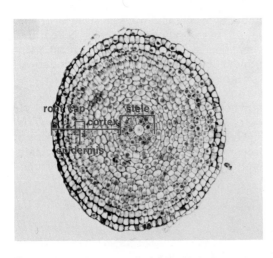

a

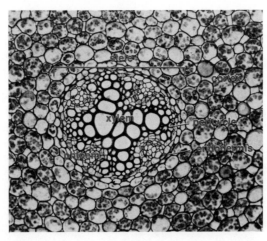

b

Figure 11–21 Cross section of a root. (**a**) Entire young root. (**b**) The center of an older root (at higher magnification).

readily penetrates the skin, and the highly irritating cell contents are forced into the wound.

Nectaries are collections of cells that secrete the carbohydrate-rich solution known as nectar. These structures are usually part of a flower, but they may also occur on stems and parts of leaves. The nectar attracts insects, which are necessary for pollination.

Not all secretory cells in the plant are on the surface or secrete substances external to the plant. Some are within the plant, and their secretions collect in internal cavities. This is true of the resin ducts found in pines and related conifers. Myrrh, frankincense, and resin are products of such internal secretory systems.

The Vascular Cambium and Wood   As the tissues differentiate behind the apical meristems, the cells that will form the primary xylem and primary phloem are first seen as the procambium. Frequently, the xylem and phloem are not present as a complete ring, but as vascular bundles. Each bundle has phloem on the side toward the center and xylem on the outside. In many plants the bundles are arranged in a ring separated by parenchyma cells. As xylem and phloem develop from the procambium in plants with secondary growth, a group of cells that does not differentiate into either xylem or phloem is left in the center of the vascular bundle. These cells retain the ability to divide and eventually develop into the vascular cambium. Cambium cells also differentiate in the parenchyma, between the vascular bundles. Ultimately, the cambium in and between the bundles will form a continuous cylinder of cells around the stem (Figure 11–20).

As we saw earlier, the cells produced toward the center of the stem differentiate into xylem, and those toward the outside develop into phloem. The secondary xylem

is the wood; the secondary phloem forms a part of the bark.

In temperate-zone plants, the vascular cambium is active only part of the year. Growth begins in the spring, when water is available, and continues into the summer. Relatively large cells are produced as early or spring wood, and small cells are produced as late or summer wood. The boundary between the spring wood of one year and the summer wood of the preceding year is quite sharp. The result is a series of clearly defined annual rings that, when counted, tell us the age of the tree.

This procedure is not uniformly reliable. Trees growing by a river, for example, or in a tropical rain forest, may scarcely show annual rings because they are always amply supplied with moisture. Temperature differences alone, moreover, can produce faint annual rings, even when other growth conditions are uniform. And two rings may be produced during a single year, if an unusually wet and cool period occurs in midsummer in a temperate forest.

One of the many remarkable features of the vascular cambium is its ability to reactivate division for many years—3,000 to 4,000 years in the case of the redwoods.

Now let's consider how these cells and tissues are assembled into the main parts of the vascular plant—the root, the stem, and the leaf.

The Root

The root (Figure 11–21) is covered by an epidermis, usually one cell thick. Just inside the epidermis is a relatively thick layer of parenchyma cells, constituting the **cortex.** These cells usually contain starch and sometimes particles

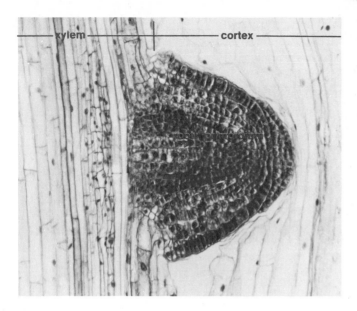

Figure 11–22 Secondary roots originate with cell divisions in the pericycle that form small masses of cells called root primordia, which grow through the cortex to the outside.

of protein or globules of fat. The innermost layer of the cortex is made up of a cylinder of cells called the **endodermis.** This layer of cells is in direct contact with the vascular cylinder inside—the *stele.* The endodermal cells are recognizable because their radial walls are thickened.

The stele is where materials move along the axis of the plant. Its outer layer, just inside the endodermis, is called the **pericycle.** This parenchymous tissue retains a meristematic potential, giving rise in virtually all plants to the lateral secondary roots. As the root meristem forms

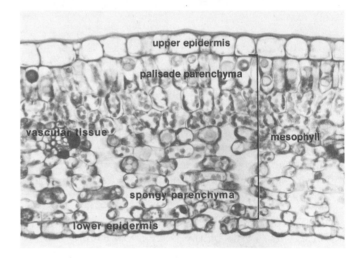

a

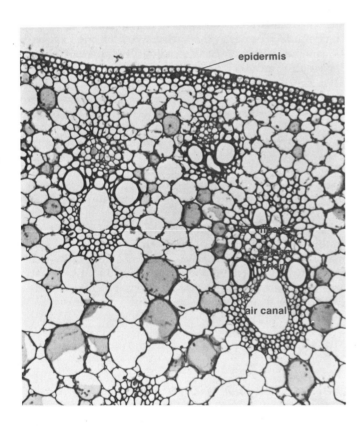

Figure 11–23 Vascular bundles in a corn stem.

b

Figure 11–24 Microscopic structure of the leaf. (**a**) Note the tightly packed cells below the upper epidermis (palisade parenchyma) and the loose cells below (spongy parenchyma). (**b**) A section cut parallel to a leaf. No cell is far from a vein.

in the pericycle, it grows through the peripheral tissues of the mature root (cortex and epidermis), finally reaching the soil (Figure 11–22).

Inside the pericycle are the xylem and phloem. The xylem often forms a solid core in the center of the stele with ridgelike projections extending toward the pericycle, and the phloem alternates with the ridges of the xylem.

The Stem

The apical meristem of the stem is enclosed in a bud made up of *leaf primordia* and young leaves. Various types of hairs are usually associated with the bud. New meristems do not develop internally to form branches in the way that secondary roots are formed. Rather, the stem has an intricate system of leaves and buds. Leaves are attached at **nodes;** the part of the stem between nodes is called the *internode;* the angle between the leaf stem (the petiole) and the main stem is called an *axil,* and the bud that occurs in the axil is called an *axillary bud.*

Different species and groups of plants have developed characteristic patterns of leaf attachment. Sometimes only one leaf is attached at a node (called *alternate* arrangement because the next leaf will be on the other side of the stem); in other cases, two *(opposite* arrangement) or more than two *(whorled* arrangement) may be attached at one node. In some species, the axillary buds remain dormant, so that the plant consists only of a main stem with its leaves; in others the buds become active, producing highly branched plants. Unlike roots, stem apical meristems can form reproductive structures—flowers and fruits.

The vascular tissue of the stem may be present in continuous cylinders, as a ring of vascular bundles, or have the vascular bundles scattered throughout the stem

(Figure 11–23). In many stems, the cells of the outer cortical layers are rich in chloroplasts and are important to the photosynthetic capacity of the plant. Collenchyma, sclerenchyma, and ducts of various kinds are common in stems. Unlike roots, the stems of most seed plants lack a well-defined endodermis.

The Leaf

The basic structure of the leaves of most higher vascular plants is a blade and a petiole (the leaf stem). Sometimes there are several leaflets on a single petiole, producing a *compound leaf.* In grasses, the leaf is long and narrow, lacking a petiole but having a modified basal portion called the *leaf sheath* that wraps around the stem.

The leaf has an upper and lower epidermis containing guard cells and stomata. Below the upper epidermis, the palisade parenchyma is usually found. Cross sections such as that in Figure 11–24a make the cells look tightly packed together, but there is actually considerable space between them, as shown in a section cut parallel to the leaf surface (Figure 11–24b). The spongy parenchyma below the palisade consists of cells more irregular in shape than those in the palisade and includes intercellular spaces. The palisade and spongy parenchyma form the *mesophyll.* The leaf *veins* are relatively elaborate vascular bundles, including phloem, xylem, and sclerenchyma. The veins subdivide into *veinlets,* and each veinlet ends in a single tracheid.

Leaves are usually determinate organs; that is, their lifetimes are limited and fixed in advance. Even in the tropics, leaves eventually grow old and die (sometimes only after several years). This is also true of "evergreen" leaves in conifers.

12

Reproduction in Animals

A clump of fertilized frog eggs floats in the shallow water of a pond. The female frog who released the eggs and the male frog who released the sperm have returned to their usual home on land. The unattended embryos soon hatch as free-swimming tadpoles. The tadpoles live in the pond as aquatic animals until they mature and develop into their adult form. They then leave the pond and return only to mate. This familiar story gives only the barest hint of the complex nature of animal reproduction and development. It does give an idea of the interaction between the organism and its environment, an interaction we must constantly keep in mind.

We will consider animal reproduction first in aquatic and then in terrestrial environments. Important points to consider include the fact that water is the transport medium for the sperm of animals in either aquatic or terrestrial environments, and that water is also the medium in which the embryos of all species develop. As we shall see, terrestrial animals have a host of adaptations to provide the small but necessary quantities of water for these functions while moving about on land, independent of bodies of water. The theme of the first part of the chapter is the evolution of life cycles from the aquatic to the terrestrial.

A wide variety of animals inhabits the two major kinds of aquatic environments. Many simple one-celled organisms, some sponges, the many-armed hydra, many kinds of worms, rotifers, some clams, many kinds of crustaceans such as water fleas and copepods, fish, and some amphibians live their entire lives in fresh water. Another vast array of organisms live their entire lives in salt-water environments. Some one-celled organisms, many kinds of sponges, a host of coelenterates (sea anemones, jellyfish, corals), and several kinds of worms are marine. Various arthropods (such as crabs, shrimp, and barnacles), many mollusks (clams, snails, limpets, octopus, squid), and all of the echinoderms (starfish, sea lilies, sea urchins) live their entire lives in salt water. Marine vertebrates are well represented by fish and a few other forms such as sea snakes, sea turtles, dolphins and porpoises, whales, manatees, and dugongs. Figures 27–14 and 27–15 show the diversity of marine animals.

Reproduction in the Aquatic Environment

Many species reproduce by *fission,* or division of one organism to form two, with replication of all structures (Figure 12–1a, c). This method, used primarily by one-celled organisms, is *asexual,* not involving the contribution of different genetic material from two parents, as you saw in several algae in Chapter 10. However, some one-celled organisms practice both sexual and asexual reproduction. The importance of sexual reproduction in providing a mechanism for genetic recombination and its consequent significance in the evolution of these organisms are discussed in Chapter 30.

Reproduction in some forms, such as some sea anemones, occurs by another form of asexual reproduction. *Budding* is the process whereby a new organism develops from cells of the old and breaks off to become a new, independent organism (Figure 12–1b).

Some multicellular animals also reproduce both sexually and asexually. Sponges and coelenterates, such as some kinds of jellyfish, can reproduce either sexually or asexually, by *vegetative reproduction.* A mass of cells broken from the organism develops into a new sponge by means of cell division and differentiation. When sponges and jellyfish are about to reproduce sexually, specialized cells in their bodies produce haploid gametes through meiosis and release them into the water. These meet, fuse (to restore the diploid number), and grow into a new organism.

Many other multicellular forms reproduce by fission or fragmentation just as sponges do (Figure 12–1c). Flatworms, roundworms, leeches, and some kinds of aquatic segmented worms can regenerate into whole animals if they are cut crosswise into chunks. This is basically a repair mechanism, but it does result in new individuals. These animals are also capable of sexual reproduction. In fact, individuals may have both male and female reproductive organs in their bodies (a condition called *hermaphrodism),* but they do pair up and exchange gametes for fertilization. The earthworms shown in Figure 12–2 are hermaphrodites.

Many other aquatic animals reproduce sexually only, despite problems of getting the gametes together. The arthropods (copepods, water fleas, insects, and others), the clams and snails, and the vertebrates require male and female gametes contributed by separate organisms, and we see in the pond many different ways of facilitating gamete transfer to ensure fertilization. Some male animals, such as insects and some fish, have special body structures for copulation and for transferring sperm directly into the body of the female (facilitating *internal fertilization* of the eggs). Others have behavioral mechanisms that cause males and females to pair up, so that eggs and sperm, though shed into the water, are shed very close to each other. Many species practice such *external fertilization.* Courtship in fish and in frogs helps males and females to pair up before fertilization. (Courtship behavior is discussed in more detail in Chapter 21.)

Some Life Cycles in the Aquatic Environment

Water supports many organisms throughout their lives, providing food, oxygen, temperature regulation, and other components of the environmental medium. Protozoans can continue dividing, and their populations continue growing, as long as the water provides enough nutrients to support growth. In fact, abundance of nutrients in a stable predator-prey or competition situation, or in both (see Chapter 28), allows the continuation of a great diversity of species.

Many forms have more complex life cycles than those just described. The water flea, *Daphnia,* produces many eggs. Minute organisms with markedly different body form from that of the adults hatch from the fertilized eggs. These are **larvae.** They feed and grow, and the larval body form changes until the adult form is attained. Many animals undergo a pre-adult period characterized by several distinct larval stages.

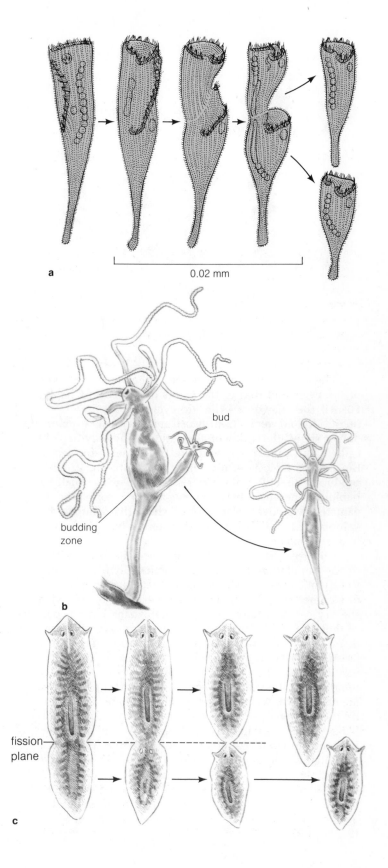

a   0.02 mm

bud

budding zone

b

fission plane

c

Figure 12–1 Asexual reproduction in animals: (**a**) fission in *Stentor,* a ciliated protozoan; (**b**) budding in *Hydra;* (**c**) fission in the flatworm *Planaria.*

Figure 12-2 Earthworms copulating.

Figure 12-3 Mated pair of frogs.

There are important differences between this type of life cycle and that of the plants described in Chapter 10. All the changes in the life cycle of *Daphnia* occur in the diploid generation. Most animals, unlike plants, do not show alternation of generations. The only haploid cells in the animal life cycle are the gametes formed at the end of meiosis. Only rarely in plants are gametes produced directly from meiosis, and there is usually a haploid generation between meiosis and formation of gametes. Moreover, plants rarely show the complex development within one generation, either haploid or diploid, that is common in animals.

Climate and season also influence the composition of the population and its reproduction. If, for example, the aquatic environment is a pond that dries up in winter or summer, its inhabitants must have a way of surviving the dry period. Many algae and protozoans have as part of their life cycle a stage called a *spore,* or *cyst.* The spore is a hard-walled structure containing the organisms in a "resting," or noneating, stage characterized by a very low metabolic rate. It is highly resistant to dryness and high temperature. The fertilized eggs of some arthropod species are also able to withstand desiccation and are "activated" by the next wet period. However, most arthropods, mollusks, and vertebrates lay their eggs so that hatchlings will emerge while the pond is full of water. The young will then be old enough to leave the pond or to burrow deep under it in a permanently moist area when the pond dries. In addition, many organisms have such short lives that one "season" of adulthood is their full span. The eggs or spores they leave as the pond dries perpetuate the species.

Water and Gamete Transfer The great majority of animal species require water as a mechanism for gamete transfer even though some live most of their lives on land. The return of many kinds of amphibians (from the Greek *amphibios,* "both lives," referring to the ability to live both on land and in water) such as frogs, toads, and salamanders to ponds to breed each spring is an example. Some frogs, for instance, emerge from winter hiding places with the first warm rains. Males arrive at the pond before females, usually choosing the same pond year after year. Females are attracted to the pond by the males' call. Each species has its own call, and the din at the pond on a warm spring night when several species are calling is almost deafening! Females find males of their species by recognizing the call and seeking out the nearest calling male; the male then courts the answering female, eventually clasping her body with his forelegs, as in Figure 12-3. The female sheds hundreds of eggs into the water as the male emits sperm over them. This is external fertilization, but with a relatively high probability of success. The eggs are then left to develop as best they can.

Salt water also is a medium for gamete transfer. Many invertebrates shed huge numbers of eggs and sperm into the water, where fertilization takes place as in the freshwater pond. Many fish use this external fertilization mechanism as well, though some, such as sharks and surf-

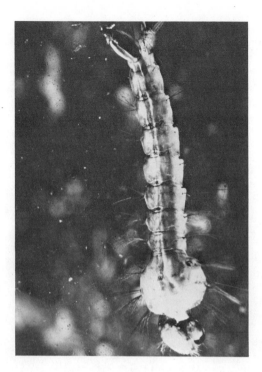

Figure 12–4 A mosquito larva or "wriggler."

perch, practice internal fertilization in which modified fin structures of the male are used to transfer sperm directly into the body of the female. Octopuses, too, use this mechanism, one arm being modified as a sperm-transfer organ. The marine mammals (such as whales and dolphins) practice internal fertilization by copulation. Although several kinds of animals return from land to fresh water to breed, no such seasonal migration from land to salt water occurs. However, some species of fish, such as eels, go from fresh to salt water to breed; salmon and lampreys go from salt to fresh water. Many marine organisms migrate within the oceans to find places where salinity, temperature, and other factors are optimal for reproduction.

Larval Development in Water Many animals, among them some frogs, salamanders, and insects, spend a larval period in the water in which they hatch. Frog tadpoles are usually **herbivores** (plant eaters) in the pond, and only after **metamorphosis**, the change from larva to adult, become **carnivores** (animal eaters). Insect larvae may live freely or in cases or compartments secreted by the female or developed as the egg matures. They often have many segments and a wormlike appearance, such as that of mosquito wrigglers (Figure 12–4). After a larval period,

they may metamorphose directly into adults or may have an additional pre-adult period of organization and growth, the pupal period. Their development is discussed in Chapter 13.

Reproduction in Terrestrial Environments

Reproduction on land has the disadvantage that water is rarely available for gamete transfer and larval development. For a species to succeed on land, it must have evolved a way to overcome this disadvantage. Some invertebrates and amphibians have found a partial solution, laying their eggs in very moist places. If near a body of water, hatchlings may wriggle overland to spend an aquatic larval period. More often, however, the hatchlings will be adapted to life on land. The terrestrially adapted insect nymphs are an example of this way of life. Several species of tropical frogs have simply eliminated the larval stage. They breed in the axils of leaves that have trapped water at their bases. Fertilization is external, and the eggs are laid in the water. In a very short time, relative to that needed by pond-breeding frogs, fully metamorphosed froglets emerge from the egg membranes. This kind of life cycle, without a larval period, is called *direct development.*

The *amniote* egg is a highly significant development in vertebrate evolution, for it frees the higher vertebrates—reptiles, birds, and mammals—from the necessity of returning to water to breed. In the amniote egg, membranes develop that enclose the embryo in a fluid-filled sac. The human fetus, for example, develops in the "bag of waters"—the amniotic sac. Amniote eggs contain all the water and nutrients necessary to the embryo before hatching and can be laid in a very dry environment. Many reptiles, all birds, and some insects lay eggs that have watertight outer membranes (although the eggs of insects are not amniote), unlike fish and amphibians.

In order for watertight membranes to enclose the developing egg, of course, internal fertilization must have taken place. All reptiles, birds, and mammals, as well as all insects that lay "land" eggs, practice internal fertilization through male-female copulation. In several of these groups, the females retain the eggs and the developing embryos in their bodies after fertilization. When eggs and developing embryos are retained, fewer offspring are conceived, but their chances for survival seem to be increased.

In some species, females nourish the embryo inside their bodies. The organ that allows them to do this is called a **placenta.** Derived from both maternal and fetal tissues, the placenta provides the developing embryo with dissolved nutrients and oxygen and carries away its meta-

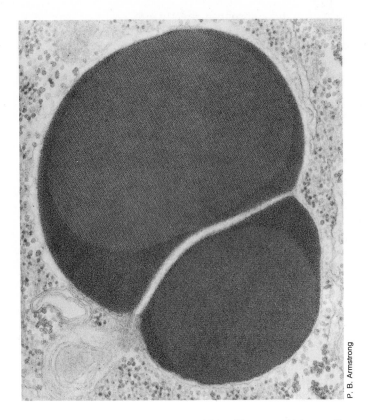

Figure 12–5 A yolk body from the amphibian *Xenopus* at high magnification. ×49,600.

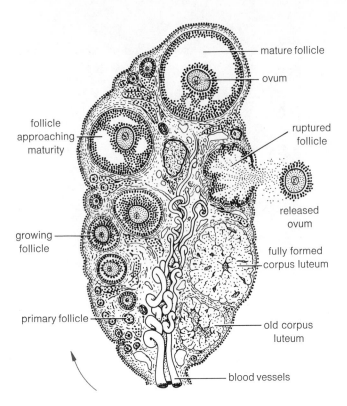

Figure 12–7 Life cycle of an egg and its follicle shown in a diagram of the mammalian ovary. Beginning at the arrow, the stages follow clockwise around the ovary.

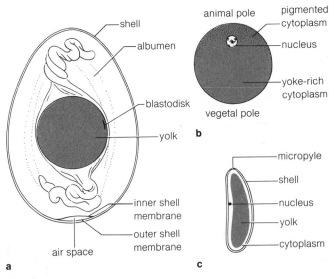

Figure. 12–6 Types of eggs produced by different animals. (**a**) Hen; (**b**) frog (×15); and (**c**) fly (×15).

bolic wastes. (The structure and function of the placenta are discussed in detail in Chapter 13.) This organ is found in virtually all mammals, some reptiles, a few amphibians and fish, the arthropod *Peripatus,* and some insects. It is found primarily among terrestrial animals and a few highly evolved aquatic forms.

In summary, several evolutionary trends can be found among terrestrial animals: from external to internal fertilization; toward acquiring a copulatory mechanism, usually involving a male structure for introducing sperm into the body of the female; from permeable to impermeable outer egg membranes; from egg-laying to egg retention; from the absence of maternal nutrition to placentation.

Time of reproduction—indeed the whole life cycle— is influenced in the pond, in the ocean, or on land by many environmental factors. All animal species are affected, to greater or lesser degrees, by aspects of the physical environment (temperature, minerals, moisture, length of sunlight) and the biotic environment (such as plant or animal species, food, intra- and interspecies competition, and beneficial associations).

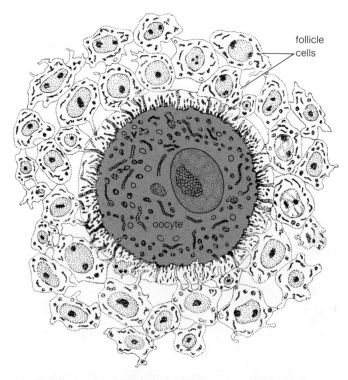

follicle cells

oocyte

Figure 12–8 Relationships between oocyte and follicle cells in the mammalian ovary. Protoplasmic projections from the follicle cells penetrate the outer zone of the oocyte and make contact with its surface.

Gametogenesis

The Development of the Egg The whole point of the sexual reproductive process is to ensure the meeting of egg and sperm to form new young and thereby perpetuate the species. Let us therefore consider the way the egg and sperm develop and mature, and some details of the process of **fertilization,** or union of egg and sperm.

Egg growth and development, **oogenesis,** begins early in the life of a female animal. To summarize the events in egg development that are discussed in detail in Chapter 6, the **oogonia,** potential egg cells, undergo mitotic divisions that increase their numbers. They then undergo a growth period during which they are called **oocytes.** This period may be a matter of days (as in hens) or of years (as in humans).

The size of the egg depends largely on the amount of food reserve stored in it. The most common reserve is what we call the *yolk* of the egg, comprising proteins, phospholipids, and neutral fats (Figure 12–5). The yolk

occurs as granules, and its distribution is a feature used to characterize eggs. There are three main types of yolk distribution in eggs. One type is represented by eggs with rather few fine yolk granules that are evenly distributed and compose less than 30 percent of the total volume of the egg, as in echinoderms (starfish, sea urchins), protochordates (acorn worms, sea squirts), and mammals.

Some eggs, such as those of sharks, lungfish, and amphibians, contain much more yolk than the first type, occurring in large flattened granules called *yolk platelets.* The platelets are densest in the lower part of the egg. This gives rise to the designation of that part of the egg as the **vegetal pole** (containing the nutrient yolk); the cells of the **animal pole** opposite the vegetal pole divide faster, establishing primordial cells for the embryo. This unequal distribution of yolk is the second type. Yolk may be distributed through much of the cytoplasm, as in amphibians, or its mass may restrict the cytoplasm to a cap containing the nucleus, as in birds, reptiles, and bony fish. Cell division restricted to the cap results in the formation of a **blastodisc.** Among invertebrates, many mollusks (clams, snails) have such eggs.

Arthropods, especially insects, have a third type of yolk distribution in which the cytoplasm is restricted to the center of the egg. The central cytoplasm contains the nucleus and is surrounded by yolk. Figure 12–6 illustrates the three egg types resulting from these three patterns of yolk distribution.

In many groups of animals, the oocytes are produced in special organs, the *ovaries,* and are surrounded by special ovarian cells during growth and maturation. The lining or epithelium of the ovary gives rise to *follicle* cells. In mammals, the follicle is one cell-layer deep before the ovum begins to mature; then it thickens to become several layers deep (Figure 12–7). A cavity, the *antrum,* appears in the mass of follicle cells; it is filled with fluid, presumably secreted by the follicle cells. Initially, the follicle cells are close together, but as the oocyte matures, the space between the cells increases. The cytoplasm of follicle cells and of the oocyte has fingerlike extensions that intermesh (Figure 12–8). This greatly increases surface area and facilitates metabolic turnover in the oocyte. Figure 12–9 summarizes the relationships between follicle cells and the ovum. When the egg is nearly mature, a denser material appears between egg and follicle cells. It consolidates and forms the **vitelline membrane.**

The egg nucleus, whether located centrally or peripherally in the cell, enters the prophase of meiosis as the whole cell begins to grow. Homologous pairing of chromosomes occurs, but meiosis proceeds no further until the end of the growth period. The nucleus enlarges, and the chromosomes may throw out transverse extensions of replicated DNA. Active synthesis of RNA and proteins is occurring inside the oocyte, so that new materials are being created as well as brought in from outside the cell.

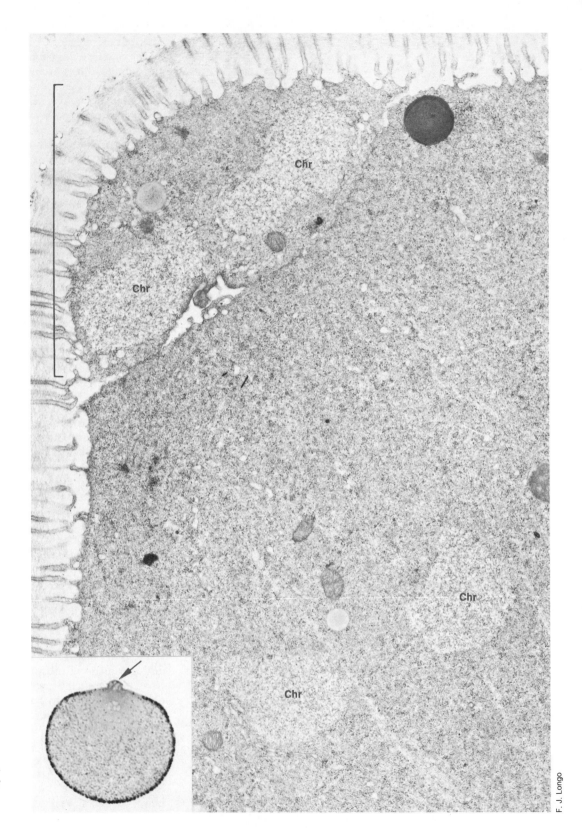

Figure 12–9 An egg of the surf clam *Spisula*, with its first polar body (indicated by the bracket in the electron micrograph and the arrow in the light micrograph shown in the inset). The masses of chromatin (Chr) have been separated into the egg and polar body by the first meiotic division. ×19,500.

Chr

Chr

Chr

Chr

F. J. Longo

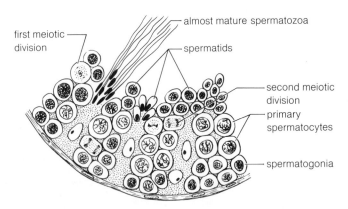

Figure 12–11 Part of a seminiferous tubule in a mammal, showing the various stages of sperm development.

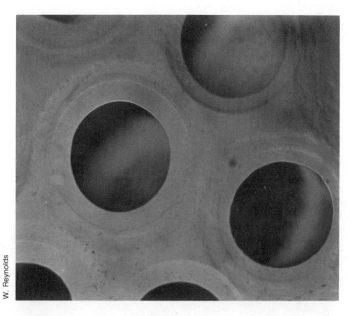

Figure 12–10 Amphibian eggs, from *Xenopus*. The clear jelly coat around the eggs is secreted as an additional surface coat as the eggs pass through the oviducts on their way to the exterior. ×15.

One to many nucleoli also are involved in protein synthesis in the oocyte. The cytoplasm of the oocyte has distinct regions and chemical characteristics. Besides containing yolk, it may also have peripheral pigment granules, ions, and enzymes in specific regions and concentrations.

When growth is finished, the egg is ready to continue meiosis. However, it does not divide into equal-sized cells. Much of the growth of the oocyte has involved accumulating food reserves, and cytoplasm and yolk are partitioned unequally at each division. The nuclear membrane breaks up; the chromosomes concentrate and move to the edge of the egg. A bulge forms on the surface of the egg; half of the chromosomes of the first division enter it. The bulge, containing chromosomes and a little cytoplasm, pinches off from the oocyte but remains associated with it (Figure 12–9). This bulge is the *first polar body*. During the second meiotic division, which may be delayed until fertilization, the process is repeated, and the *second polar body* is formed. The first polar body also divides, and the three polar bodies that result from meiosis degenerate. The egg is then mature and contains half of the chromatids (is haploid), most of the cytoplasm, and all of the yolk. Thus only one cell, the egg, is produced by these divisions, not the four cells we will see produced in male gametogenesis.

The egg is surrounded by a series of membranes. There are two classes of egg membranes. The first class includes the primary membrane, the plasma membrane,

and secondary membranes such as the vitelline membrane of insects, mollusks, and many vertebrates, the **chorion** of insects, and the *corona radiata* (follicular cells that surround the egg in the follicle and after ovulation) of mammals. These membranes are all formed in the ovary by the egg itself or by follicle cells.

The second class of membrane includes those made in the oviducts or other accessory structures. The "jelly coat" of amphibian eggs is an example. Secreted by the oviducts (Figure 12–10), it protects the egg when it is laid; it may also help the eggs adhere to one another and to the substrate. The jelly swells with water when the egg is laid. Five membranes are distinguishable in birds: a vitelline membrane (in part laid down by the oviduct), the "white" of the egg (which is more nutrient material for the embryo), two thin shell membranes, and finally the shell (see Figure 12–6c). These membranes are laid down serially as the egg passes by regions of specialized glands in the oviduct during its 24-hour descent from the ovary.

The Development of the Sperm **Spermatogenesis,** or sperm formation, resembles oogenesis but differs in some important details. Sperm development takes place in the male **gonads** (reproductive glands), the *testes*. It is a continuous process throughout the adult lives of many animals, and many phases of sperm development can sometimes be seen simultaneously in a single testis. The testes of insects and higher vertebrates consist of a series of tubes, the *seminiferous tubules*. In them, one often sees the earliest stage of development, the **spermatogonia,** at the periphery of the tube; the next stages, the **spermatocytes,** more medially; and later stages, **spermatids** and occasionally mature **spermatozoa,** at the innermost layer of cells of the tubule (see Figure 12–11).

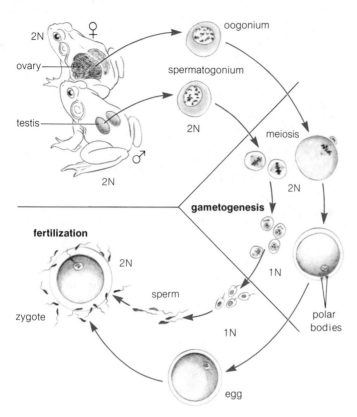

Figure 12–12 Formation of the egg and sperm in a typical vertebrate.

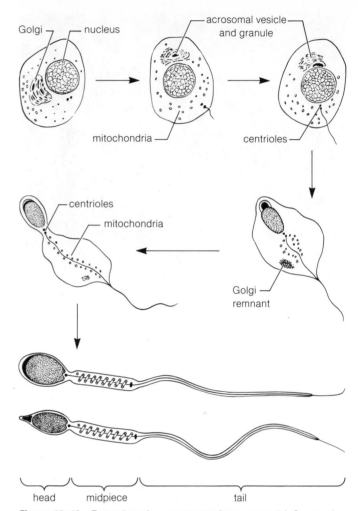

Figure 12–13 Formation of spermatozoon from spermatid. Successive stages show the formation of the acrosome by the Golgi complex, elimination of the cytoplasm, and formation of the midpiece and tail.

The three major differences between spermatogenesis and oogenesis are that there is little or no growth in spermatocytes before or during meiosis; meiotic divisions result in equal-sized cells; and maturation (differentiation) of sperm takes place after meiosis. The chromosome movements in spermatogenesis are similar to those of oogenesis but are central in the cell, and cytoplasm is partitioned equally as the chromosome number is reduced from diploid to haploid. Meiosis in one spermatogonium produces four sperm instead of the one egg produced in oogenesis. Figure 12–12 summarizes oogenesis and spermatogenesis in the frog.

The spermatozoon is initially a haploid nucleus in cytoplasm with no accessory structures. However, the spermatozoon must find the egg and fuse with it, and as it matures, it gains a locomotor system (Figures 12–11 and 12–13). A mature sperm has three major parts: the head, the midpiece, and the tail or flagellum. Most of the head is composed of the nucleus and its haploid set of chromosomes. The front end of the head is modified as the *acrosome,* derived from Golgi bodies. This structure contains granules and enzymes that "dissolve" the egg membranes so that the sperm can penetrate and fertiliza-

tion can take place. The back part of the head contains the centrioles, which will take part in division in the fertilized egg. The midpiece contains the base of the flagellum and mitochondria that provide the spermatozoon with energy for movement. The flagellum is the longest part of the cell and may bear a membrane, a knob, or other structures depending on the species. All the spermatozoon's "dead weight," primarily cytoplasm, is lost during maturation, lightening the sperm and making it more easily motile. The nucleus loses water and RNA and becomes elongate. The shape of the sperm head varies a great deal according to species. These various shapes are usually considered modifications to increase the efficiency of movement in water.

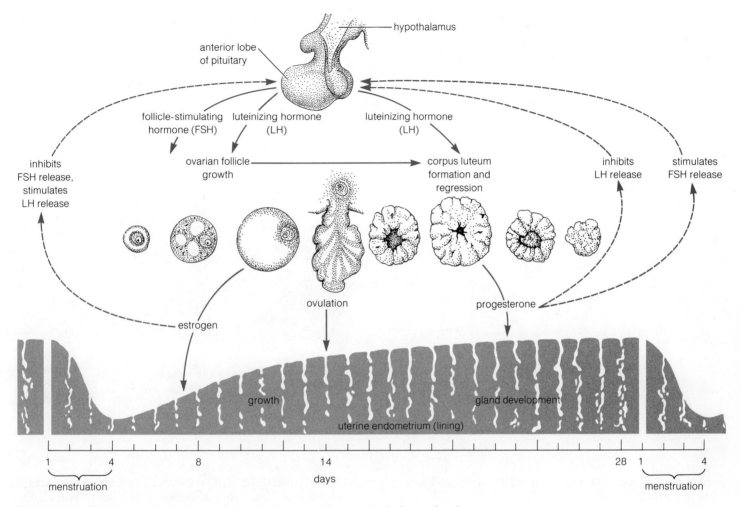

Figure 12–14 Pituitary and ovarian hormone control of the menstrual cycle in the human female.

Regulation of Gametogenesis We have mentioned that environmental conditions trigger reproduction in many animal species. To be more precise, the environment moderates hormonal responses that more directly control production of eggs and sperm. For instance, let us consider the example at the beginning of this chapter—frogs in a fresh-water pond. During the summer, the females' ovaries are empty of eggs and the males' testes contain no sperm. During fall and winter, oocytes develop, culminating with storage of food reserves, or "yolking up." Shortly before egg-laying times, the oviducts that will carry the eggs from the ovaries to the environment enlarge and become convoluted. The testes have their own parallel cycle. The hormones that regulate these cycles, includ-

ing prolactin, progesterone, and testosterone, are common to many vertebrates. The cyclic increase of sex-hormone secretion, gamete "ripening," and associated morphological (structural) change are collectively called **estrus,** or *heat,* particularly in mammals. This time of peak levels of sex hormones is the only time most mammalian females are willing for, or capable of, copulation. Humans, however, do not have a "heat" period and are usually receptive to copulation throughout the cycle.

Such cycles seem to ensure that young will be produced under the most favorable conditions for their development. For example, fawns are born in spring, so that much of their development occurs during the summer, when food is plentiful. The well-nourished juvenile

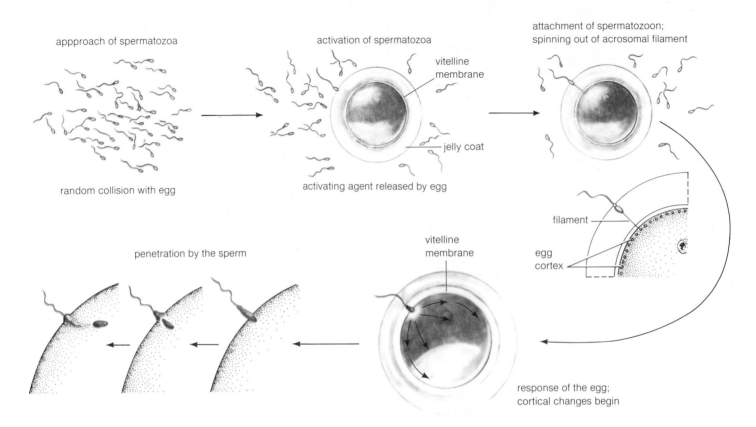

appproach of spermatozoa

random collision with egg

activation of spermatozoa

vitelline membrane

jelly coat

activating agent released by egg

attachment of spermatozoon; spinning out of acrosomal filament

filament

egg cortex

penetration by the sperm

vitelline membrane

response of the egg; cortical changes begin

Figure 12–15   Major events in fertilization of an animal egg.

deer is then better able to withstand the cold and lack of food during winter.

Better studied is the hormonal cycle in humans and the way it affects gamete production. In men, sperm is produced at a nearly constant rate from puberty onward, rather than on a yearly cycle. The pituitary gland produces LH (luteinizing hormone) and FSH (follicle-stimulating hormone), both identical in chemical structure to the female pituitary hormones, at nearly fixed levels. LH acts upon the **interstitial** tissue of the testis (tissue between sperm-producing cell masses) to produce **testosterone,** which regulates the maturation of sperm and also produces the male secondary sex characteristics, such as hairiness and deeper voice; FSH also regulates spermatogenesis.

Women, however, have a finely orchestrated balance of hormones that produces a cycle averaging 28 days (Figure 12–14). Egg production begins even before birth. At sexual maturity, one ripe egg is ovulated each month, rather than hundreds once a year.

The human cycle begins with menstruation (the shedding of the uterine lining, or endometrium, that had been prepared for implantation of a fertilized egg). Following

menstruation, the pituitary gland steps up its production of follicle-stimulating hormone (FSH) as hypothalamic releasing factors are produced. FSH influences the ovary to initiate the growth of a follicle containing an ovum. The follicle secretes the hormone **estrogen** during the 14 days it is growing (the follicular phase). At this point the pituitary stops producing FSH because of the presence of high amounts of estrogen, and begins to produce luteinizing hormone (LH) instead. The ovum bursts from the mature follicle *(ovulation)* and is swept into the oviduct. The follicle cells fold into the ovum cavity (under the influence of LH) to form the *corpus luteum.* The corpus luteum secretes the hormone **progesterone,** which causes the uterine lining to thicken and have an increased capillary supply. This part of the cycle is the luteal phase.

If the ovum produced in this cycle is fertilized, progesterone keeps the uterine lining in this thick and blood-rich condition as the implanted zygote (fertilized egg) begins development and secretes a hormone (chorionic gonadotropin) which maintains the corpus luteum. Chorionic gonadotropin level is the basis for many pregnancy tests—if the hormone is present, pregnancy is too. If the egg is not fertilized in the oviduct (Fallopian tube) within

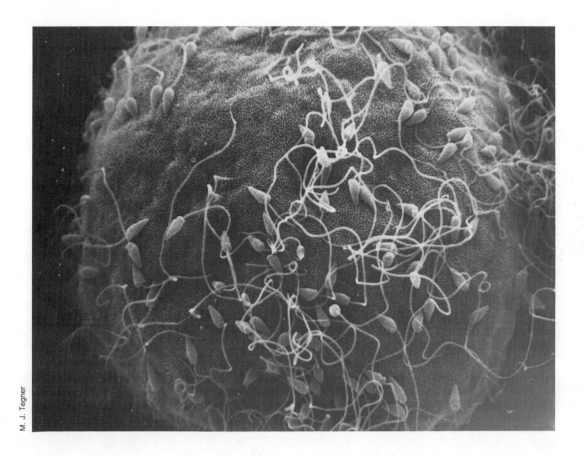

M. J. Tegner

Figure 12–16 Attachment of sperm cells to the egg of the sea urchin *Strongylocentrotus*, viewed with the scanning electron microscope. ×2,300.

24 to 48 hours, the corpus luteum function begins to degenerate after about seven days. It stops producing progesterone, the pituitary stops producing LH, and the uterine lining, no longer maintained by these hormones, sloughs off—another menstrual period begins. The interplay of hormones and tissue responses is summarized in Figure 12–14.

This regulation of oogenesis depends upon negative feedback among several hormones—when one hormone is at high levels of production, others are suppressed. FSH is inhibited by high estrogen levels (Figure 12–14) and LH by progesterone, for example. The optimum time for fertilization, then, is about midway between menstrual periods, during the 48 hours after the ovum is shed from the follicle.

Fertilization: The Fusion of Egg and Sperm

Fertilization begins when the head of the sperm penetrates the egg membranes so that the sperm nucleus becomes included in the egg. Two distinct events are part of fertilization: activation of the egg to start development, and joining of paternal and maternal chromosomes to restore the diploid number of the species and determine the genetic direction of the organism. In many species, fertilization also triggers completion of meiosis in the egg. Development into an embryo does not always follow directly upon fertilization. For example, some protozoans go into a dormant state after fertilization, and development is triggered later by a change in the environment.

Let's return to frogs as an example to describe the complex events of fertilization (Figure 12–15). Frog sperm begin swimming as soon as they enter the water. Their movements are random, but they are shed in huge numbers close to the eggs, and the eggs are large targets. The jelly coat of each egg begins to swell when it hits the water and releases a glycoprotein (see Chapter 2) sometimes called *fertilizin* that causes sperm to stick to the egg's surface (Figure 12–16 shows this phenomenon in the sea urchin). This mechanism occurs in many animal species, but the specific amino acids and sugar groups that make up the glycoprotein vary from species to species. Sperm produce a substance sometimes called

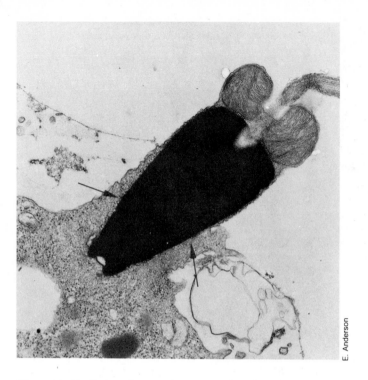

Figure 12–17 Fusion of egg and sperm membranes in fertilization of the sea urchin *Arbacia*. The egg cytoplasm (arrows) has begun to flow around the sperm nucleus. ×20,000.

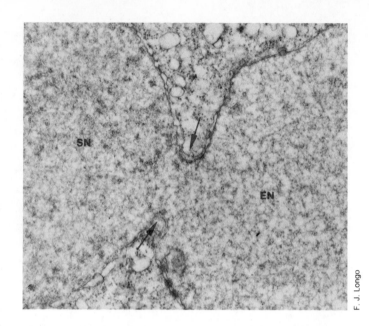

Figure 12–18 Fusion of the egg nucleus (EN) and sperm nucleus (SN) during fertilization in the sea urchin *Arbacia*. The nuclear envelopes of the two nuclei have become continuous at the points marked by the arrows. ×15,000.

antifertilizin. These two chemicals may effect a bond, as in an antigen-antibody reaction. This chemistry is species-specific, so sperm from one species are rarely able to fertilize eggs from a different species.

In many species of both vertebrates and invertebrates, the presence of the egg—specifically, the presence of certain substances in the outer membrane or egg jelly, or secreted by granules in the protoplasm at the surface of the egg—stimulates the sperm to begin swimming, and probably attracts them to the egg.

Penetration of the egg is effected chemically in virtually all species of animals that have been investigated. The sperm releases lytic enzymes from the acrosome that dissolve the jelly coat, or the material binding cells together (as of the mammalian follicle cells forming the corona radiata), or both. In some forms, such as starfish, a filament makes the initial contact with the plasma membrane of the egg. Immediately upon contact, the cytoplasm of the egg bulges out, forming the *fertilization cone* (Figure 12–17 shows this occurring in the sea urchin). The cone engulfs the sperm and then retracts, pulling the sperm into the egg. Usually only one sperm penetrates, but if the sperm are unusually concentrated or the egg is abnormal, more than one may get in. The resultant development is almost always abnormal. However, in some mollusks, sharks, reptiles, and birds, several sperm may enter the egg without harmful effect. These animals have a mechanism that eliminates the extra sperm nuclei, and only one sperm is involved in constituting the embryo.

Changes in the Egg as Fertilization Takes Place When an egg has been fertilized, we can see many changes in it almost at once. The first is a series of *cortical reactions,* cytoplasmic changes near the cortex (surface) of the egg (Figure 12–15). In the frog, for example, certain granules in the cytoplasm swell and burst, releasing fluid to the egg's surface. The *fertilization membrane* begins to form and is pushed up from the surface of the egg as the fluid released by the granules absorbs water. The granules also are thought to produce a substance that helps cement together the cells produced during division. Sperm penetration, from surface contact to incorporation of the head of the sperm, may be complete in 9 seconds. It was once thought that the cortical reaction and the elevation of the fertilization membrane kept out additional sperm, but it is now known that those first reactions take some 60 seconds, and sperm are bombarding the egg all that time. The mechanism that prevents other sperm from penetrating after the first has entered is not yet known.

Several other reactions take place in the egg upon fertilization. The meiotic divisions are completed, oxygen consumption increases drastically, a number of ionic changes take place, and protein-digesting enzymes become more active. Perhaps most significantly, amino acids are activated and a new protein is synthesized, apparently on the mRNA of the egg. In the frog, the pigmented surface cytoplasm contracts toward both the dorsal and the ventral poles of the egg, and the cortical layer rotates on the inner mass of cytoplasm. Subsequent events in development indicate that the point of penetration of the sperm may influence the symmetry of the fertilized egg. The pigment pattern of the frog egg allows these events to be followed.

The Role of the Sperm inside the Egg   In some animals (such as mammals), the entire sperm enters the egg; in others, only the head or the head and midpiece. In many animals, the egg nucleus must complete meiosis before fusion can occur. During this time, the sperm nucleus and centriole reorient themselves 180° so that the centriole precedes the sperm nucleus in its path toward the egg nucleus. The sperm nucleus, now called the male **pronucleus,** starts swelling, and its chromatin becomes concentrated and granular. Figure 12–15 summarizes the events of fertilization to this stage.

Pronuclear Fusion   As the male pronucleus approaches the female pronucleus (as the egg nucleus is called at this stage), which has completed meiosis, its centriole becomes surrounded by an aster, as in mitosis. The pronuclei fuse in the center of an egg with little yolk, and in the area of cytoplasm in yolky eggs. In some animals, the pronuclei may literally fuse; in others, they are placed in close proximity. Nuclear membranes fuse at the contact point, and the contents of the two pronuclei merge and become surrounded by a common nuclear membrane (Figure 12–18). In other forms, nuclear membranes dissolve and chromosomes are freed in the cytoplasm. The centriole and aster of the male divide in two, so that chromosomes from male and female pronuclei align on the spindle. Then, after the first mitotic or cleavage division, the chromosomes become enclosed in a common nuclear membrane in each of the two daughter cells formed by division.

Some Unanswered Questions about the Mechanism of Fertilization   As we discussed earlier, several animals—rotifers, aphids, and bees—can develop from unfertilized eggs in a process called **parthenogenesis.** Ripe eggs that normally require fertilization to activate them can be made to develop without fertilization—a process called *artificial parthenogenesis.* Weak acids, various salts, fat solvents, temperature changes, electric shocks, shaking, and pricking all can activate an egg. Since the agent in each case is known, its specific effects can be pursued. All these agents cause sublethal damage to the egg. It seems likely, then, that activation in fertilization results partly from mechanical damage done to the cortex of the egg by the sperm. The fact that the activating sperm brings male chromosomes along in the same package makes natural fertilization an efficient system for triggering development. However, many questions remain about the activation of eggs as well as about precisely how sperm penetrate and how nuclei fuse.

Summary

Animal reproduction is accomplished both sexually and asexually, the latter by fission or vegetative reproduction. Many organisms go through their entire life cycles in an aquatic environment. Even those that are terrestrial require water for sperm transport, and some also require it for egg and larval development. Major evolutionary trends in reproduction among terrestrial animals include acquisition of internal fertilization, an impermeable outer egg membrane, and egg retention and nutrition by the mother.

Egg development occurs over a considerable span of time. Eggs can be characterized by the amount and distribution of yolk. Egg (oocyte) maturation includes much growth and the characteristic divisions of meiosis so that *one* large haploid cell is produced from each immature oocyte.

Sperm development results in four equal-sized haploid cells from each meiotic event. The mature sperm cell is composed of a head containing the nucleus, a midbody containing mitochondria, and a tail or flagellum.

The reproductive cycle is mediated by hormones. The menstrual cycle in humans is an example of hormonal control by negative feedback.

Fertilization—union of egg and sperm—includes two events: activation of the egg to start development, and joining of male and female chromosomes to restore the diploid number. Penetration of the egg by sperm is usually effected chemically. The egg surface undergoes a series of reactions, and the fertilization membrane lifts from the surface. The sperm nucleus (male pronucleus) enters the egg and approaches the egg nucleus (female pronucleus). Their nuclear membranes fuse and the contents merge. The centriole and aster from the sperm divide in two, so male and female chromosomes align on the spindle. The first cell division ensues.

Supplement 12–1
Population Regulation by
Birth Control

Letters similar to the following frequently appear in newspaper advice columns. "Dear Mrs. Smith, I am thirteen years old. I am afraid I am going to have a baby. What can I do?"—an old and frequent story. A person who isn't fully enough acquainted with her physiology and her emotions is in the frightening position of possibly bearing a baby. She didn't really know how not to, and she doesn't want it for all kinds of reasons. At the same time, in an early 1976 issue of the journal *Science,* these papers were published: "The decline of unplanned births in the United States," "Population control: Pakistan tries a major new experiment," and "Phytoestrogens: adverse effects on reproduction in California quail." Both these situations relate to our societal concern for regulating the sizes of populations, our personal concerns about limiting family size, and the quest for effective ways of accomplishing these goals. There remains an appalling gap between information about the technology and sociology of birth control and practical use of that information by those who need it. Let's consider some aspects of the practice of birth control, historically, currently, and in the future.

Human populations have tried throughout recorded history, and before that, to regulate their numbers. Although there have been a few historical exceptions, this has generally meant attempting to limit population growth. If we analyze population dynamics, we see that there are only two ways of limiting numbers: by increasing the number of deaths in a population, or by reducing the number of births. War, famine, and disease increase deaths, but few groups have consciously used them as devices for population regulation. Many have turned to infanticide at various times in their history, however, and infanticide is still more common than is generally known. But reducing the number of births has been the alternative preferred by most cultures. Birth control is considered undesirable by some populations, but it is less undesirable than control by death.

Historically, abstinence from sexual intercourse has been a favored means of birth control. Celibacy and chastity (not having intercourse throughout life or for long periods of time) have been practiced by segments of many populations. Abstinence from intercourse for short periods, such as the time of presumed ovulation, has also been practiced and is the basis for the present-day "rhythm method" of contraception. Late marriage for women, as is common in Ireland and China today, is

also a means of reducing the numbers of births in a population.

Today, however, social attitudes among most human populations are such that these methods of birth control are rejected. People do not wish to give up the pleasure of intercourse, but they do not want a baby each year either. Several methods of birth control are now available to meet this need.

Two ancient methods place primary responsibility on the male partner for prevention of fertilization. These are the use of *withdrawal* and of a *condom*. Withdrawal is the removal of the penis from the vagina just before ejaculation, or sperm excretion. As soon as the association between intercourse and the male ejaculate, and then birth, was figured out, various societies used withdrawal to prevent conception. Use of a sheath around the penis, a condom, to contain the ejaculate and to prevent its deposition in the woman's vagina is nearly as ancient. For centuries thin lambskin, the tissue of pig's bladder, and other sorts of thin but tough membranes were used as sheaths during intercourse. Recently, condoms have been made of thin rubber or of plastic. Of the many birth-control methods, the condom is the only one that also prevents the spread of venereal disease. However, both of these methods interrupt intercourse and may reduce pleasurable sensations for one or both partners. Especially in withdrawal, there is a large chance of error— that is, pregnancy—because some sperm escape before ejaculation.

Methods that place the primary responsibility on the female partner and prevent fertilization include use of a *diaphragm, spermicidal jelly,* or *spermicidal foam*. These, too, are barriers to sperm penetration and have been used in some form throughout recorded time, and doubtless before. A papyrus records in 1850 B.C. that a mass of crocodile dung and honey was prescribed for insertion into the vagina. Greek women used oiled cloth, and English and Italian women of the Renaissance are known to have put sponges in the vagina to absorb the ejaculate and, they hoped, prevent fertilization. Obviously, all sorts of materials may have been used for this purpose. The modern diaphragm is a thin rubber or plastic disk attached at its rim to a spring. It is inserted into the vagina to cover the cervix, or neck of the uterus. It may be used alone or with a spermicidal (sperm-killing) jelly and may be inserted an hour or more before intercourse. It cannot be removed for several hours after intercourse, for sperm in the vagina must be dead or pregnancy is risked. The spermicidal foam contains both a sperm killer and a substance that liberates carbon dioxide bubbles in the vagina, blocking the passage of sperm. The jelly may also be used alone but is less effective than when used in conjunction with a diaphragm or condom. Some kinds of foams and jellies must be inserted just before intercourse, others a few hours before.

The rhythm method requires abstention from intercourse at the time of ovulation. As seen earlier in this chapter, ovulation usually occurs at about the fourteenth day of a 28-day menstrual cycle. The egg can be shed on day 13, 14, or 15 in such a cycle and is viable for up to 48 hours. Since sperm are also viable for about 48 hours, intercourse with this method should not occur on days 11 through 17. Further, few women have an exact 28-day cycle with regular ovulation. Most women's cycles vary. For example, if a woman varies between 26- and 30-day cycles, she should not have intercourse from day 8 through day 20 of her menstrual cycle. In fact, she can safely have intercourse for about one week each month. This irregularity, the uncertainty about the exact time of ovulation, the number of celibate days demanded, and the constant worry and fear of pregnancy make the rhythm method not only unreliable but emotionally unsatisfactory for most people.

Another method commonly used today is "the pill." The compounds lumped together under this popular name are usually a combination of estrogens and progesterones that suppresses ovulation. The levels of these hormones prevent the normal feedback mechanism described in Chapter 12. The pituitary produces no FSH, so no egg follicle can mature. LH production is also suppressed. The female partner must take the pill on days 5 through 25 of her menstrual cycle. After day 25, in the absence of pill hormones, the estrogen and progesterone levels fall and the endometrium of the uterus is shed as in normal menstruation.

Some women experience unpleasant side effects when using the pill. Weight gain, tender breasts, headaches, lethargy, and decreased interest in sex have all been cited (but so has increased interest in sex). These frequently subside after a few weeks of taking the pill. More serious are incidences of increased blood-clotting and of liver disturbances associated with the pill. Women with histories of blood or vascular disorders, migraine headaches, or liver disease are usually told to consider the pill with caution. Indeed, anyone taking the pill should be monitored by a physician.

Much research continues on the use of hormones in birth control. "Mini-pills," containing very small quantities of the hormones, seem to have nearly the same high rate of success as "the pill," and they cause fewer side effects. "Sequentials," pills that administer estrogens the first several days of the cycle and progesterones the last few, are thought to more nearly simulate a normal hormonal cycle. But they have recently been implicated in higher rates of vascular disease and, possibly, uterine cancer, and some have been withdrawn from the market. The "morning-after" pill is essentially a big dose of estrogen injected within a few days after intercourse. It induces menstruation, and therefore prevents implantation if fertilization has occurred. It cannot be used routinely, for it places considerable stress on one's system. Further, diethylstilbestrol (DES), a synthetic estrogen used for this purpose, has recently been linked to vaginal cancer in teenaged women and testicular cancer in young men whose mothers had taken the hormone to prevent miscarriage. Use of DES as a "morning-after" contraceptive is no longer allowed.

Being tested, but not available at this writing, are several other means of using hormones to prevent conception. One is a hormone injection to be given every three months, whose action would also be cyclical. Another is the implantation under the skin, or in the uterine wall, of a capsule that would slowly dispense small quantities of hormone. It is thought that these might function for months, years, or perhaps the entire childbearing period if desired. They could be removed if the bearer wished to have a child.

Attempts are being made to find a male equivalent of the pill that will suppress spermatogenesis. Some progress is being made using immunological techniques, so that the male would suppress his sperm production by an internal antibody reaction. It is anticipated that a male contraceptive of this sort will soon be available.

A birth-control method that prevents implantation of the fertilized egg in the uterus is the use of an *intrauterine device* (IUD). An IUD is essentially a foreign body in the uterus, but we do not know precisely how it acts. It may change the uterine lining by irritating it, or it may speed up the contractions of the Fallopian tubes so that the egg arrives in the uterus before it is ready to implant. IUDs come in many shapes—loops, coils, shields, and bows. Some are plated with copper, which seems for some unknown reason to enhance their effect. Some 20 percent of women who have tried IUDs cannot use them because they expel them or have excessive menstrual bleeding. Occasionally the IUD will pierce the uterine wall with drastic results. An excessively high rate of pregnancy with the IUD in place has occurred with such IUDs as the Dalkon Shield. This has too often resulted in miscarriage and in uterine infection and death of the woman. Pregnancy with an IUD in place has recently been implicated in the birth of a child with abnormal limb development.

More permanent methods are also available. These are surgical procedures that prevent fertilization by removing or closing parts of the ducts that transport eggs or sperm. Since these procedures are rarely reversible (though progress is being made in surgical duct restoration), the person who undergoes one of them is rendered sterile. For a woman, the surgery is called a *tubal ligation*, or *laparoscopy*, and involves severing the Fallopian tubes and tying or cauterizing them. In the past, this always had to be done through an abdominal incision under a general anesthetic. It was classified as major surgery. Now some operations of this type can be done through an

Table 12-1 A Comparison of Birth-Control Methods

Birth-Control Method	Pregnancies per 100 Woman-Years	Advantages	Disadvantages
None	40–50	—	—
Rhythm	24	No cost; acceptable to various religions	Requires much motivation, regular cycle
Withdrawal	18	No preparation or cost	Frustration
Spermicidal foam, jellies, tablets	16–18	Easy to use, no prescription necessary	Continual cost; unattractive or irritating for some people
Condom	10–15	Easy to use; helps prevent venereal disease	Expense; interrupts sexual activity
Diaphragm	8–12*	No cost except for initial fitting and purchase; need not interrupt sexual activity	Not usable by all women; insertion difficulties
IUD	2–5	No expense after initial insertion; needs little attention	Side effects; increased bleeding; problems if pregnancy occurs; expulsion
Contraceptive pill	1	Easy to use, aesthetically acceptable	Continual cost; prescription required; daily attention needed; side effects
Tubal ligation or vasectomy	0	Permanent; no further attention needed	Permanent; relatively high cost for one-time procedure

*Note: Pregnancy rate less than indicated when diaphragm is used with jelly.

incision in the vagina or in the umbilicus—called "band-aid surgery" since the only external sign of surgery is a band-aid covering the half-inch incision. Although the procedure is far from trivial, it is less traumatic than the older way, and the procedure takes only a few hours in a physician's office or an outpatient clinic. The comparable surgery for a male is a *vasectomy,* in which the vasa deferentia, or sperm ducts, are cut and their ends tied off. These ducts are just under the skin near the base of the penis, and the surgery is usually performed under a local anesthetic. The male ducts are more easily accessible than those of the female, and the surgery is therefore less complicated.

Table 12-1 compares the effectiveness, advantages, and disadvantages of the various birth-control methods.

A method of birth control once pregnancy has begun is removing the embryo, or *abortion.* This is an ancient method. A nearly endless variety of physical and chemical agents thought to induce abortion could be listed, ranging from drinking the juices of certain plants to having other women jump up and down on the pregnant woman's abdomen until the blood gushes out her vagina. The only 100 percent reliable methods, though, involve direct interference with the implanted embryo. Legal abortions done in the first 12 weeks of pregnancy involve dilation of the cervix and scraping the uterine lining to remove the

embryo *(curettage)* or, recently, use of a thin tube attached to a vacuum line whose air pressure *aspirates* (sucks out) the embryo and uterine lining. Legal abortions from 12 to 20 weeks of pregnancy often involve "salting" the amniotic fluid. A hypertonic saline solution is injected into the amniotic cavity, stimulating the uterus to contract and expel the embryo. At any time after about 8 weeks the embryo or fetus can be removed by abdominal surgery in which the uterus is opened and the fetus removed. Progress is also being made in the use of prostaglandins, which are secreted by both males and females and are involved in a number of internal regulatory processes, to induce abortion. Putting ethical and moral questions aside, since we cannot settle them here, let us look at abortion for its value as a method of birth control. It removes a particular pregnancy. Short of infanticide, it is the only method available after a pregnancy is verified. Thus, it is probably necessary as a back-up in our set of weapons against runaway population. But it is expensive, it is not approved by all members of society, and it is physiologically stressful and potentially dangerous enough so that it is not considered a useful long-term method of birth control for an individual.

Why are we interested in population regulation and birth control? We have before us in the world several examples of less affluent populations who cannot feed, clothe, or educate their offspring. They do not limit births and, as both cause and effect, have very high infant mortality rates and incidence of disease. From our point of view, it seems obvious that regulating these populations would ensure more food and shelter for all, since fewer people would share the same resources. Such stressed populations, however, are not the only ones for which population regulation is desirable. The peoples of more

affluent, industrially developed countries consume the earth's resources at a much faster rate than less affluent people. These resources must be conserved, and we must all help conserve them. A major way to do so is to limit the number of consumers. Further, limiting births ensures that each child will be planned for and wanted, and that parents are both willing and able to provide the tangible and intangible factors needed to nurture their offspring.

13

Animal Development

The process of development—one cell giving rise to an intricately organized body of different kinds of cells, tissues, and organs, functioning harmoniously in the individual—proceeds according to a definite timetable. As William Harvey observed in 1651, "no part of the future offspring exists *de facto,* but all parts inhere *in potentia* . . . ; all parts are not fashioned simultaneously, but emerge in their due succession and order." The course of development is a dynamic continuum, for cell numbers, sizes, shapes, and functions progressively change, making use of the genetic information in the cells and of the products of the cells themselves. To get a clearer idea of the events of the developmental process, we will outline the patterns found in some representative animals, especially the frog and the chick, on which much of the definitive research was done. We will emphasize certain

landmark phases in the process, but never lose sight of the fact that we are really dealing with a continuum.

Following fertilization, the process includes **cleavage,** which begins with the first division of the zygote and continues through a period of rapid cell divisions, then **blastulation,** in which a fluid-filled cavity develops in the solid ball of cells. Blastulation is followed by **gastrulation,** a complex reorganization of the cells that establishes the basic cell layers from which all subsequent cells are derived. **Neurulation** establishes the unit that becomes the brain and the spinal cord; differentiation, or modification of structure and function of cells and tissues, proceeds until the pattern characteristics of adults of the species is developed. This chapter discusses these events and their consequences. We will see that a frog tadpole and a newborn baby share more than just a

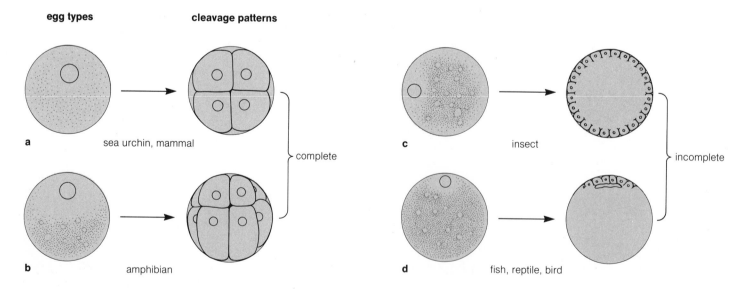

egg types   **cleavage patterns**

a   sea urchin, mammal

b   amphibian

c   insect

d   fish, reptile, bird

complete

incomplete

Figure 13–1   Egg types and corresponding cleavage patterns. Shading indicates distribution and amount of yolk.

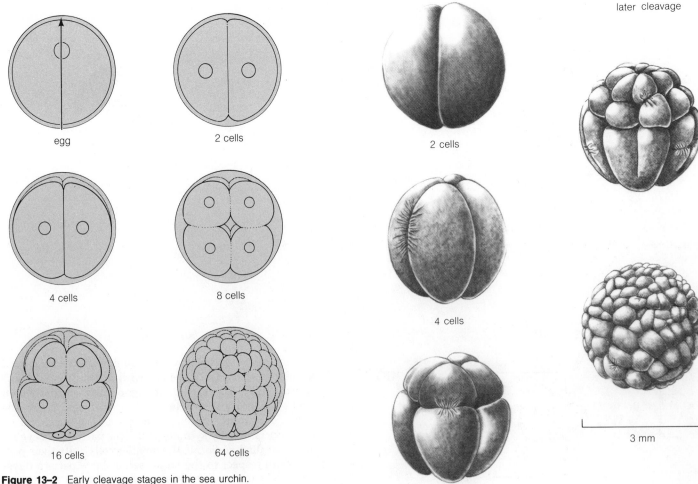

Figure 13–2 Early cleavage stages in the sea urchin.

Figure 13–3 Surface views of cleavage in the frog egg.

youthful state; they both reflect patterns of development characteristic of virtually all animals. Supplement 13–1 illustrates the development of the human fetus.

Cleavage

Cleavage is an extremely active period of growth by cell division, yet in most animals the volume of egg and embryo remains relatively constant. The usual pattern in cell division is for each of the daughter cells to enlarge to the size of the original cell before division. In early embryonic development, however, the cells become progressively smaller until a minimum cell volume characteristic of the cell type and species is reached. For example, in the mature egg of the sea urchin before cleavage, the volume of the cytoplasm is 550 times that of the nucleus. At the end of cleavage, the cytoplasmic volume is only

6 times that of the nucleus in each of the several cells. Cleavage, then, is a progressive reduction of cell size. Early divisions are synchronous (they occur together), but synchrony is lost as cell number and complexity of form increase.

The structure of the animal egg—and thus the pattern of cleavage—varies greatly, depending to a large extent on the amount of yolk present. Cleavage patterns in turn influence patterns of later development.

In organisms like the sea urchin and mammals, the cells produced at each division are equal in size, and each is completely surrounded by a plasma membrane (Figures 13–1a and 13–2). This is called *complete cleavage*. Many amphibians have a similar pattern but, because their eggs contain more yolk, and it is unequally distributed, the

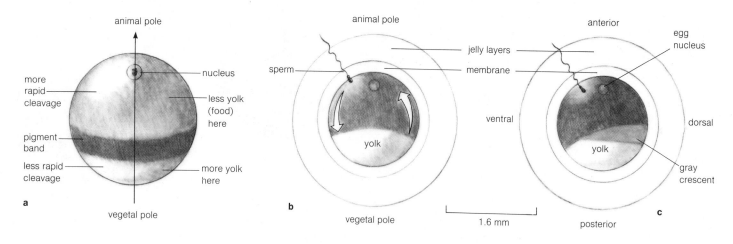

animal pole

more
rapid
cleavage

nucleus

less yolk
(food)
here

pigment
band

less rapid
cleavage

more yolk
here

a

vegetal pole

animal pole

jelly layers

sperm

membrane

yolk

b

vegetal pole

1.6 mm

anterior

egg
nucleus

ventral

dorsal

yolk

gray
crescent

c

posterior

Figure 13–4 (**a**) Polarity in the animal zygote; (**b, c**) establishment of bilateral symmetry following fertilization.

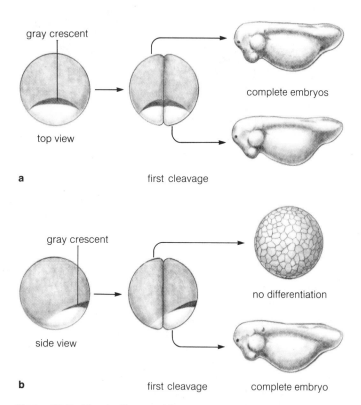

gray crescent

top view

a

first cleavage

complete embryos

gray crescent

side view

b

first cleavage

no differentiation

complete embryo

Figure 13–5 The significance of the gray crescent as revealed by separating the cells at the two-cell stage. (**a**) When the gray crescent is divided between the two cells, each cell will form a complete embryo. (**b**) When one cell contains all of the crescent and the cells are separated, the one lacking the crescent fails to develop.

division of the cytoplasm is unequal, with the larger cells forming in that part of the egg with the most yolk (Figure 13–1b and 13–3). Cleavage in these cases is usually complete. When division is unequal, as in frogs, the smaller cells are arranged directly over the larger ones, and the structure is radially symmetrical. This pattern of development is called *radial*. If the smaller cells shift their position with respect to the larger set, a *spiral* pattern develops. Such unequal division occurs because the mitotic spindles lie obliquely to the central axis of the embryo. Spiral cleavage, with some exceptions, is characteristic of mollusks, annelids, and certain other groups of worms.

In eggs with great quantities of yolk, nuclear division is normal. Following the end of mitosis, however, plasma membranes form but do not completely surround the daughter cells. Thus, *incomplete cleavage* takes place, with the cells maintaining a common cytoplasm and all having equal access to the yolk.

There are two patterns of incomplete cleavage. The first pattern is found in insects (Figure 13–1c). Here, the nucleus divides several times and the resulting daughter nuclei move to the periphery of the cell, where incomplete cleavage takes place. The second pattern is found in birds, reptiles, and some fishes (Figure 13–1d). Here cleavage is incomplete and localized in one region of the egg. This localization is the result of the separation of the yolk from the active cytoplasm of the egg. Division occurs on top of the yolk and early cleavages are vertical to the surface of the egg, so the cells all lie in one plane. The yolk is not partitioned, but each cell is in contact with it.

Cleavage is important in animal development because it subdivides the egg in such a way that the resultant cells contain different samples of the egg cytoplasm. This is a type of primary cellular differentiation.

Cleavage may also be *determinate* or *indeterminate*. If components of the cytoplasm are divided nearly equally among cells formed during early cleavage, the pattern is indeterminate. If, however, some components are segregated into specific cells by early divisions, cleavage is said to be determinate. Such segregation was observed in frogs' eggs in the 1930s. In some frog species, upon fertilization, the pigmented and unpigmented regions of the egg cytoplasm rotate, and a region called the *gray crescent* forms (Figure 13–4). The first cleavage may result in division through the gray crescent so that both cells contain part of the cytoplasm of the crescent (Figure 13–5a). If the cells resulting from this indeterminate cleavage are separated and grown, *both* cells develop into normal embryos. If, however, cleavage is determinate—that is, results in gray-crescent cytoplasm being restricted to only one of the two products of cleavage (Figure 13–5b)—only the cell having gray-crescent cytoplasm will develop normally when grown alone. The other cell divides irregularly for a time, and then its daughter cells die. Therefore, gray-crescent cells play a role in normal development. We will consider that role shortly.

The geometrical regularity of the cleavage pattern in many embryos is impressive, especially in the early stages. This feature has made it possible to trace cells formed early in cleavage to tissues and organs that are developed later.

Blastulation

In organisms with complete cleavage, the 64-cell stage (called a *morula,* from the Latin for "mulberry") is a spherical cluster of cells (Figure 13–6a). Some yolk has been used for energy and as raw material for protein synthesis taking place in the cells. Synchrony of division is lost, especially in yolky eggs where cells divide faster at one pole than at the other. As the cells undergo further cleavage, some begin to lose internal contact with each other and a cavity forms (Figure 13–6b). In many organisms, the cells adhere to form a cell layer surrounding the cavity (Figure 13–6c). An embryo at this stage is called a **blastula** and the cavity, the **blastocoel.**

The shape of the blastula varies from organism to organism. In starfish and sea urchins, the blastula is a single-layered sphere of equal-sized cells (Figure 13–6d). In frogs, the presence of yolk and the unequal-sized cells produced by the early cleavages result in a hemispherical blastocoel (Figures 13–6e, 13–7).

Blastulae also form in eggs with incomplete cleavage.

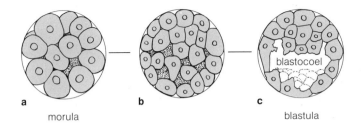

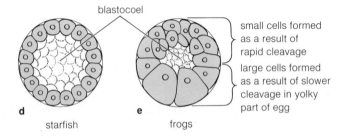

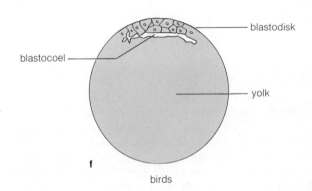

Figure 13–6 Formation of the blastula and some of the various types of animal blastulae. Each figure shows the developing organisms sliced in half from top to bottom, illustrating the middle of the structure.

In birds, cleavage occurs on top of the yolk; the blastula is a disk *(blastodisk)* over the yolk (Figure 13–6f). Fishes form a blastodisk that is a single cell layer thick, but most reptiles form a blastodisk two cell layers thick.. Below the two layers is a cavity analogous to the blastocoel.

Development in the Blastula The next stage of development is characterized by cell movements. To understand this stage better, researchers have constructed so-called fate maps of blastulae, showing the initial positions of the cells and what eventually happens to them. Since some frog embryo cells are highly pigmented, early workers tried to map their movements relative to less highly

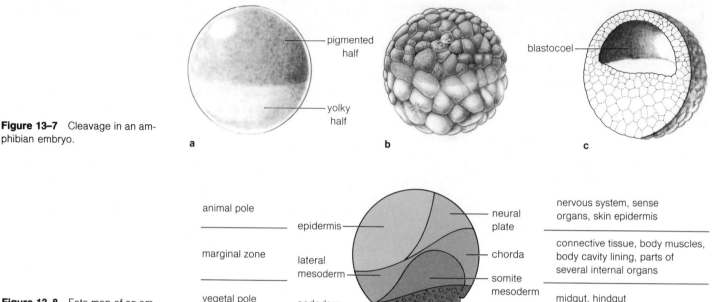

Figure 13–7 Cleavage in an amphibian embryo.

a   b   c

pigmented half

yolky half

blastocoel

animal pole

epidermis

marginal zone

lateral mesoderm

vegetal pole

endoderm

neural plate

chorda

somite mesoderm

nervous system, sense organs, skin epidermis

connective tissue, body muscles, body cavity lining, parts of several internal organs

midgut, hindgut

Figure 13–8 Fate map of an amphibian embryo.

pigmented cells. When this failed, they developed techniques for marking cells with vital dyes (which do not interfere with normal cell activities) or carbon particles (which stick to cell surfaces). These were first used on amphibian embryos, so let us consider the fate map of a frog (Figure 13–8). Specific regions on the surface of the blastula eventually develop into specific organs. There are three main regions: (1) a region of small cells with little yolk called the animal pole; (2) a region of intermediate-sized cells called the marginal zone; and (3) a region of large, yolk-containing cells, called the vegetal pole. These names arose from observations of development in frog eggs. The dorsal cells divide rapidly and become the primordial tissue of the embryo, hence they are "animal." The ventral cells supply nutrient yolk to the "animal" cells and therefore are called "vegetal" cells. (This is a good example of the tradition behind some biological terminology: Terms are kept for convenience even when the concepts on which they are based are archaic.)

The small cells of the animal pole give rise to the nervous system, the sense organs, and the skin epidermis. The marginal zone gives rise to connective tissue, body muscles, body cavity lining, and parts of several internal organs. The large cells of the vegetal pole give rise to the midgut and hindgut. In this development, as the fate maps show, the polarity originally established in the egg is expressed in the embryo. These fates are further discussed on p. 224.

Gastrulation

In starfish and frogs, for example, all the cells that will become the **gastrula** (involuted, two-cell-layered stage) are on the surface of the blastula, yet some of the organs and tissues they form will eventually be located deep within the animal. Thus, a drastic relocation of the cells of the blastula must take place so that organs are made internal, eyes are positioned, skin covers the whole animal, and the organ systems are located as in the adult. This complex reshuffling process is called gastrulation. The word *gastrulation* means "formation of a little stomach," and the process was so named because in some animals the cell movements initially form a stomachlike pouch. In modern embryology, it has come to mean the formation of the primordial tissues of the embryo. As one might expect from what we have seen of the influence of egg type on cleavage patterns, and the influence of cleavage and yolk distribution on blastula patterns, patterns of gastrulation vary among animals, reflecting these earlier developmental stages.

One of the simplest types of gastrulation is found in *Branchiostoma* (sometimes called amphioxus). These small marine animals are members of the phylum Chordata. They do not develop vertebrae, and the **notochord,** an embryonic support structure in vertebrates, is present in the adult animal. In *Branchiostoma,* as in many chor-

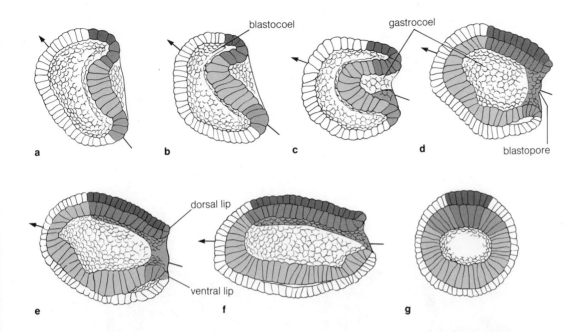

blastocoel

gastrocoel

a

b

c

d

blastopore

dorsal lip

ventral lip

e

f

g

Figure 13–9 Gastrulation in *Branchiostoma* (amphioxus). (**a**) through (**f**) are longitudinal sections; (**g**) is a transverse section through the midbody of (**f**). Arrows indicate anteroposterior (top to bottom) axes. White = ectoderm of epidermis; light color = mesoderm; gray = endoderm; dark color = neuroectoderm; medium color = notochord.

dates and in echinoderms, the blastula is one cell layer thick. Gastrulation begins when the vegetal cells flatten and begin to bend into the blastocoel (Figure 13–9a). This inward bending continues until the vegetal cells touch the animal cells, forming a double-walled cup (Figure 13–9c-f). The vegetal cells form the inner lining of the cup, and the rim of the cup outlines a newly formed opening, the **blastopore,** encompassing the **gastrocoel,** the internal cavity of the gastrula. Its dorsal lip is important in developmental sequencing, as we shall see.

At this stage, the embryo is a gastrula. The *primordial germ layers,* the tissues from which all adult structures are derived, are formed in gastrulation. The outer layer of the double-walled cup of the gastrula is the **ectoderm.** It gives rise to skin and its derivatives, such as glands, feathers, horns, and hair. A tissue derived from ectoderm, the *neurectoderm,* forms the neural plate (Figure 13–8) and hence the brain, spinal cord, associated nerves, and the sense organs, such as eyes and ears. The inner layer of the gastrula is the **endoderm,** which gives rise to the midgut and hindgut of the embryo, the linings of the digestive tract, of the lungs, and of some of the internal organs, such as the ducts of the liver. Endoderm also gives rise to the primary sex cells.

A third primordial tissue layer, the **mesoderm,** soon forms between the ectoderm and the endoderm. In *Branchiostoma,* the mesoderm originates as pouches formed from endoderm. The mesoderm forms the chorda, lateral and somite (muscle block) mesoderm (Figure 13–8), which give rise to connective tissue, including cartilage and bone, blood cells, muscle, the lining of the internal

Table 13–1 Primary Tissues and Their Derivatives

I. Ectoderm
A. Skin
   1. Glands
   2. Feathers
   3. Hair
   4. Horns, etc.
B. Neurectoderm
   1. Brain
   2. Spinal cord
   3. Nerves
   4. Special sensory tissues
      a. Retina of eye
      b. Olfactory receptors
      c. Hearing receptors
      d. Touch, taste, position, pain, and other receptors
II. Endoderm
A. Midgut and hindgut of embryo
B. Lining of esophagus, stomach, intestine
C. Ducts of liver, pancreas
D. Primary sex cells (oogonia, spermatogonia)
III. Mesoderm
A. Connective tissue
   1. Adipose
   2. Fibrous, etc.
B. Muscle
C. Bone, cartilage
D. Blood cells
E. Peritoneum
F. Mass of lungs, liver, pancreas, kidneys, gonads, spleen, stomach, intestines, and so on, in distinction to their other components, formed of endoderm (see IIB, C, D)

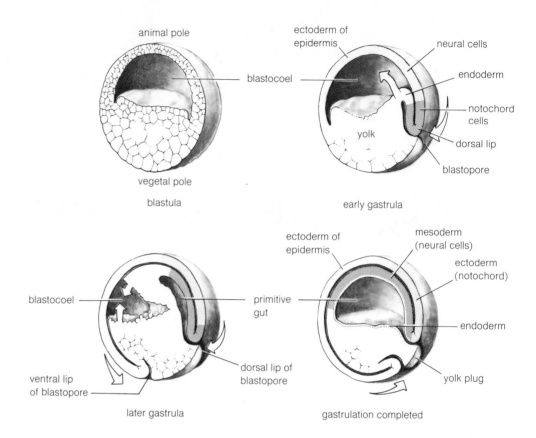

animal pole

blastocoel

vegetal pole

blastula

ectoderm of
epidermis

neural cells

endoderm

notochord
cells

yolk

dorsal lip

blastopore

early gastrula

blastocoel

ventral lip
of blastopore

dorsal lip of
blastopore

later gastrula

ectoderm of
epidermis

mesoderm
(neural cells)

ectoderm
(notochord)

primitive
gut

endoderm

yolk plug

gastrulation completed

Figure 13–10 Gastrulation in the frog egg, seen in cross section. The arrows indicate the direction of cell movements.

body cavity, and the mass of such organs as lungs, heart, liver, pancreas, kidneys, spleen, stomach, and intestines. The primary tissues and their derivatives are summarized in Table 13–1.

Gastrulation is less straightforward in the frog, largely because of the presence of yolk. Simple invagination at the vegetal pole is not mechanically feasible because of the large mass of inert yolk within the vegetal cells. Instead, portions of the cell layer of the animal pole move down around the yolk mass and then fold in at its edge. This involution (inward curling) begins at what will be the dorsal side of the yolk mass, to form what is at first a crescent-shaped blastopore. This infolding slowly spreads to all sides of the yolk, so that the crescent is converted into a circle. Movement of the other cells around the yolk eventually encloses this material almost completely within the cavity of the gastrula. The yolk plug emerges at the blastopore. These changes during frog gastrulation are shown in Figure 13–10.

The most extreme case of gastrulation influenced by the amount and distribution of yolk is found in birds' eggs, of which the chick is a classic example. The blastodisk is dwarfed by the yolk, and the pattern of gastrulation is drastically modified from that of *Branchiostoma*

or frogs. In the chicken, fertilization takes place in the upper part of the oviduct near the ovary and before the surrounding layers (including the white and the shell) are laid down. Embryonic development begins at once and progresses as the embryo moves down the oviduct. By the time the "egg" is laid, the embryo has reached a late blastula stage containing some 20,000 to 60,000 cells. The blastodisk is roughly two layers thick; the top layer is the ectoderm and the bottom is the endoderm (Figure 13–11a). The endoderm splits away from the ectoderm, and cells from the upper layer involute by changing their shapes along a longitudinal midline of the embryo to form the mesoderm. This involution gives rise to a clearly visible line, called the *primitive streak*.

Cell Migration Gastrulation is accomplished by the movement of cells to new areas of the developing organism. These movements change the shape of the embryo, and new structural elements are formed. As important as cell movement is in gastrulation, surprisingly little is known about the mechanisms involved.

Microfilaments appear to be important in several cases of changes in cell shape related to embryo develop-

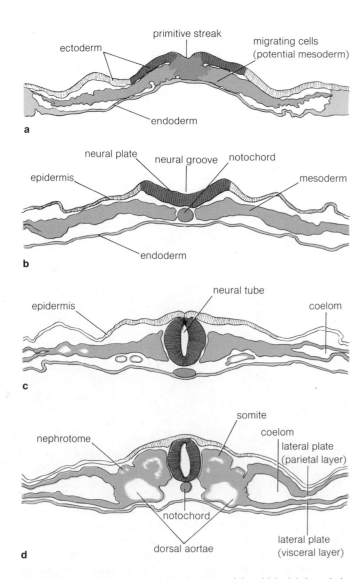

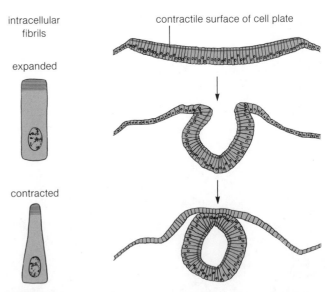

Figure 13–12 The probable role of microfibrils in changes in cell shape and consequent rolling-up of a flat plate of cells to form a tube.

Figure 13–11 Early stages of development of the chick: (**a**) the primitive-streak stage; (**b**) the neural-plate stage; (**c**) the neural-tube stage; and (**d**) primary organ rudiments.

ment. In the tree frog, for example, the end of the cell near the outer surface of the embryo contains microfilaments oriented parallel to the surface (Figure 13–12a). When these microfilaments contract, the cells take on a pyramidal shape and the sheet of cells curves, changing the relative position of the cells.

Movement of the endoderm and mesoderm into the embryo's interior is often achieved by the migration of individual cells. These movements also appear to be related to changes in cell shape and to involve microfilaments. In the sea urchin, the cells of the **mesenchyme,** which form from the mesoderm, migrate singly within the embryo. In the newly formed, nearly spherical primary mesenchymal cells, the microfilaments are randomly oriented. Later these cells become asymmetrical; correlated with this change in shape is a reorientation of the microfilaments so that they are parallel to the long axis of the cell. In this form, the cells migrate, and it has been suggested that the movement and microfilaments position are related.

The movements undertaken by the primordial tissues are characteristic of each type: Ectoderm spreads as a sheet; endoderm forms a tube. It can also be shown experimentally, using tissue culture techniques, that these movements are independent of other cells in the organism.

In these procedures, tissue and cells are removed from the organism and grown (cultured) outside the body. When a mass of dissociated cells from several primordial tissues (ectoderm, mesoderm, and endoderm, for example) is cultured, the cells first form a ball and then sort themselves out, ectoderm cells adhering to other ectoderm cells, mesoderm to mesoderm, and endoderm to endoderm. Moreover, the aggregated cells show the same spatial relationship to one another as in the embryo—ectodermal cells surrounding endodermal cells, with mesoderm between them.

Other Changes during Gastrulation Since cell movement requires energy, it is hardly surprising that respiration in the embryo increases during gastrulation. This

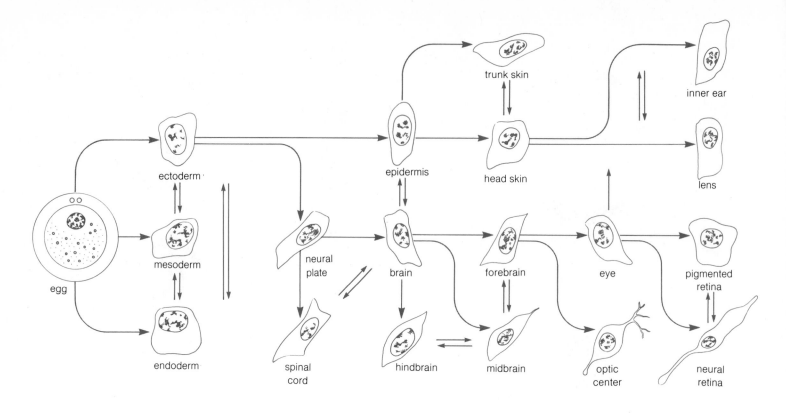

Figure 13–13 The progressive determination (differentiation) of some cell types derived from the ectodermal germ layer. The symbol ↕ indicates that the cells in question have interchangeable fates, if they are experimentally manipulated in the right way at the right time.

means that nutrient reserves are broken down and restructured, often with a reduction in cell size. In amphibians, glycogen is the reserve carbohydrate oxidized to obtain energy for gastrulation. Chemical analyses of various cell areas of the amphibian gastrula have shown that glycogen diminishes in nearly all cells, but the cells of the dorsal lip of the blastopore lose 31 percent of their glycogen, contrasted with between 1 and 9 percent for other cells. These data indicate that the dorsal lip cells are respiring most actively, a conclusion that correlates with their rapid and fairly extensive movement.

Besides increased respiration during gastrulation, there is a marked increase in protein synthesis and breakdown. The protein of yolk cells is the major source of material. Experiments have shown that protein synthesis in the embryo increases fivefold from beginning to end of gastrulation in a species of salamander and threefold in a sea urchin.

Transplantation and Tissue "Fate" At the beginning of gastrulation, the embryo is little more than a hollow ball of cells. By the late gastrula stage it has an outer covering of ectoderm, a central layer of mesoderm, and an inner layer of endoderm. We have already discussed the adult fates of these tissues; we must now consider how early in development this fate is decided. For example, can we alter the developmental direction of a group of cells or a tissue? If so, how late in development can we do so? These and similar questions can be approached through transplant experiments.

Transplant experiments done early in gastrulation show that the cells' fates are not yet fixed. If potential ectoderm is transplanted into a part of the gastrula that will involute—into mesoderm—it will develop into mesoderm. Specifically, it will develop into the body part associated with its surrounding tissue; for example, ectoderm transplanted into a patch of potential kidney mesoderm will develop into kidney.

If the same transplants are performed on a late gastrula, the results are quite different. A piece of ectoderm transplanted to another part of the gastrula at this later stage will develop as epidermis no matter what kind of tissue surrounds it. This fixing of the fate of embryonic animal cells—which is unlike what happens in plant cells—is called *determination* (see Figure 13–13). Cell determination appears to be inherited, for certain genetically programmed metabolic pathways are a part of the specialization process of developing cells.

Transplantation experiments have also shown that the contact of one kind of cell with another causes determination. A piece of the dorsal lip of the blastopore of one species of salamander was transplanted to the lateral lip area of a gastrula from a different species (Figure 13–14). The graft invaginated normally. Later in development an almost complete secondary embryo was found on the site of the graft. Some cells in the secondary embryo were those of the graft; however, many others, forming several organs, were cells of the host embryo. (The origin of the tissue could be recognized by size and color differences between the two species.) Since the host cells would not have developed as they did without the presence of the graft, it was concluded that the graft induced development in the host. This process is called **induction.**

Other experiments showed that only the dorsal lip of the blastopore can induce a complete organism when transplanted. For this reason, the dorsal lip is called the *primary organizer.*

We have already seen that, in some frog species, the region of the egg distinguishable as the *gray crescent* before cleavage begins has special properties that are involved in normal cleavage. Experimental evidence shows that even at this early stage the gray-crescent cells possess these special organizing properties. Furthermore, it has been demonstrated that the properties of the organizer are inherent in the cortex of the egg. We know, then, that the "determinate" properties of cell division have to do with the partitioning of the cortex of the egg during cleavage.

Properties of the Primary Organizer   The dorsal lip tissue does not have to be "alive" to induce orderly development. It can be dried, frozen, boiled, even treated with certain chemicals, and still function when transplanted. This suggests that a chemical inducer is present in these cells. Scientists have learned several of its characteristics but have not yet succeeded in isolating it.

We know, for example, through an elegant series of experiments done by Clifford Grobstein in the late 1950s and early 1960s, that the chemical can pass through a nonliving substance and still induce development. Grobstein placed salamander dorsal lip tissue on one side of a millipore filter and potential ectoderm tissue from an early gastrula of a salamander on the other. (A millipore filter is an artificial material with a uniform pore size that allows only substances of a specific maximum size to pass; in Grobstein's case the pore size was 0.8 micrometers.) The pore size prohibited direct cell contact between

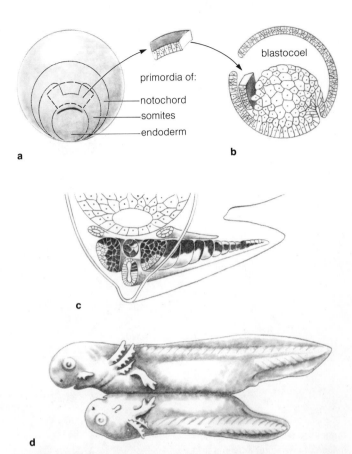

Figure 13–14   (**a**) and (**b**) Transplantation of a piece of the dorsal blastopore lip into the blastocoel of another gastrula. (**c**) and (**d**) Self-differentiation and induction by the graft. In (**c**), the tissues derived from the graft are shown in color, the induced tissues of the host in white.

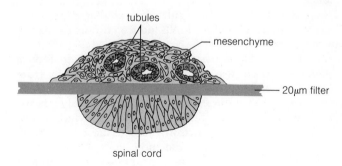

Figure 13–15   The transfilter induction of tubules in mouse embryo mesenchyme by mouse spinal cord cells.

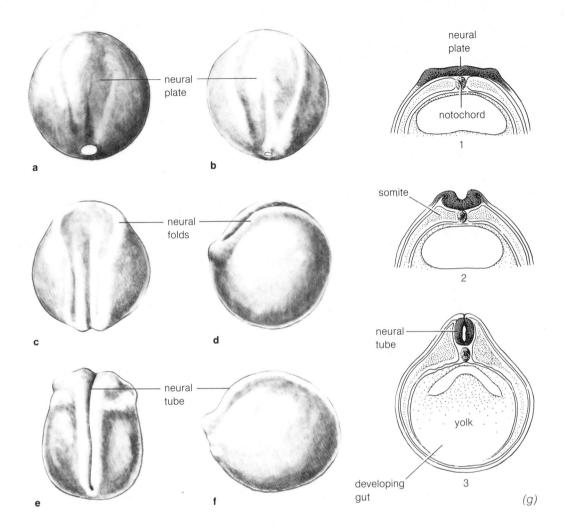

Figure 13–16 **(a–f)** Stages in formation of the neural tube in a frog embryo. **(g)** Diagrams of cross sections showing folding of the neural plate to form the neural tube.

the two pieces of tissue, but large molecules could diffuse. The ectoderm rounded up, formed a neural tube and, in one case, an eye with a lens, even though it was separated from the dorsal lip tissue by the filter. Obviously, the induction observed must have depended on a substance diffusing through the filter from the dorsal tissue. Figure 13–15 shows the results of a similar experiment in which mouse spinal cord cells induced tubule formation in mouse mesenchyme.

Many kinds of tissue have proved able to induce development, but responses vary. In fact, especially from frogs, salamanders, chicks, and mice, almost any differentiated tissue can induce if transplanted into early gastrulae; liver, kidney, and muscle are all good inducers. The normal inducing substance in these tissues is probably a protein, but its exact composition has not been determined. Moreover, there are probably several different kinds of inducers, perhaps including other substances—such as alcohols—in addition to proteins.

Organ Formation

Let's look at the early embryology of the frog and follow the development of a particular structure, the neural tube, which with further development will give rise to the brain and spinal cord. Ectoderm cells differentiate to form a *neural plate* at the end of gastrulation. Then a *neural groove* forms as the edges of the neural plate thicken and the plate depresses. The thickenings fold upward and finally fuse. Two tissue layers result. The upper is a layer of ectoderm continuous with that surrounding the entire embryo; the lower is the hollow, thin-walled neural tube, formed by the fusion of the neural plate folds. These movements, called *neurulation,* are summarized in Figures 13–11, 13–16, and 13–17.

Further development occurs by similar kinds of movements. The brain forms as a series of outpocketings and constrictions. The brain, sense organs, and other fea-

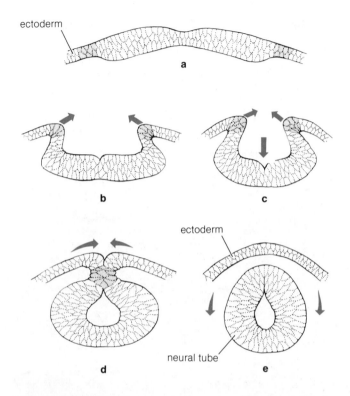

ectoderm

a

b

c

ectoderm

neural tube

d

e

Figure 13-17 Cell movements in conversion of the neural plate to the neural tube.

tures of the head develop earlier and faster than more posterior regions of the body. This phenomenon is called *cephalization* (from the Greek *cephalos,* "head").

The sequence of morphological change during development has been observed for many species. "Normal tables" that classify developmental changes into arbitrary stages have been prepared for some species, including several species of frogs. Figure 13–18 is the normal table for *Rana pipiens.* We can see the pattern of external change through the early stages of development, including neurulation. We will refer to it again later in this chapter to see more of the course of development.

Various kinds of cell movements are involved in organ formation, or **organogenesis.** These *morphogenetic movements* occur in different combinations to form the organs of the embryo. They include: (1) thickening of a layer, by cell division and migration of cells to that region: (2) separation of cell layers, as in the separation of the mesodermal mass to form discrete muscle blocks or *somites;* (3) folding, as occurs at the edges of the neural plate to form the neural tube; (4) thickening followed by excavation; for example, the neural tube of bony fish arises as a solid structure whose cells then part centrally to form a fluid-filled cavity; (5) the breakup of layers,

followed by cellular migration, as in blood-cell formation; (6) aggregation, as when a mesodermal structure forms around an endodermal organ such as the fibrous capsule of the liver, or the condensation of kidney tissue; and (8) degeneration, which takes place when cellular components become unnecessary to further development, as in the loss of webbing between fingers.

Let us look at a complex developmental sequence, that of the formation of a limb of a frog or a chick, to examine development and tissue interactions, summarized in Figure 13–19. The first appearance of tissue destined to become the forelimb of the chick arises as a localized thickening called a *limb bud.* The limb bud is first seen as a proliferation of mesenchyme cells (from mesoderm) under the epidermis (from ectoderm). The bud grows until it is longer than it is wide, and tissues begin to differentiate. The far end of the bud flattens and becomes circular, the "hand plate." Five projections extend outward and thicken—these will form the "fingers" of the frog. In the chick, regional cell death occurs, resulting in the formation of three major elements of the end of the wing. The mesenchyme segregates inside the bud—some areas aggregate firmly, others loosely. The packed areas will become *cartilage,* a special type of dense connective tissue in which the intercellular material, or matrix, has a rubbery consistency.

The loosely packed mesenchyme contributes to the formation of blood vessels. Most of the muscle tissue is formed from cells that migrate into the limb bud and form a single large mass. The forelimb muscles differentiate from this mass. Nerve cells also migrate to the bud, resulting in the *innervation* of the limb. The remaining mesenchyme contributes to fibrous tissue that will become tendons and other structures.

At this stage of development, the forelimb is a flattened, three- (chick) or five-pointed structure (frog) with cartilage blocks, nerves, blood vessels, and incipient muscles. The cartilage is converted to bone by the replacement of the original matrix with one composed of calcium carbonate and calcium phosphate. Blood vessels become tubular and are connected to the main vessel channels. Complex associations of bone, muscle, and connective tissue form joints. Finally, the adult structure of the frog limb, involving 25 bones, 34 muscles, and hundreds of nerves, appears as a result of the complex interaction, differentiation, migrations, and loss of various cell types. The chick wing has lost most of the digital elements but has nearly as many muscles and nerves as the frog limb.

Differentiation

We saw earlier that the fates of cells become fixed or determined at a certain point in an organism's embryonic

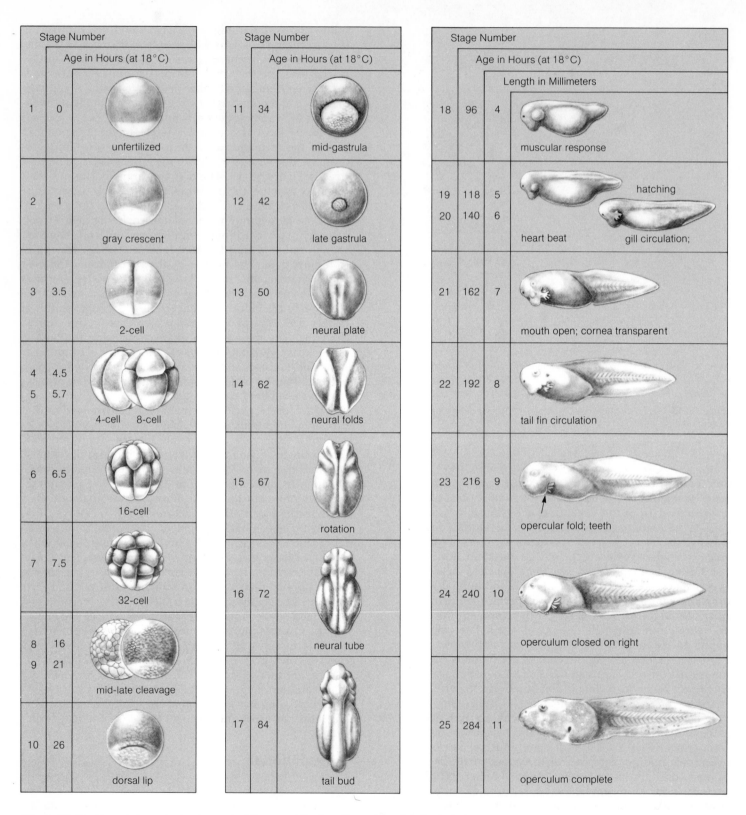

Figure 13–18   Normal stages in development of the egg of the frog *Rana pipiens* into the tadpole.

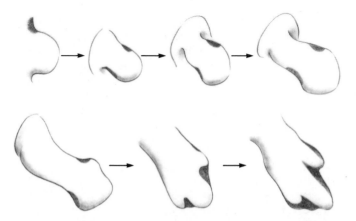

Figure 13-19 Stages in development of the chick embryo wing bud showing regions of massive cell death and breakdown in the superficial mesoderm.

development. This happens in part because cells gradually become different from one another and acquire new properties of structure and function, thus losing their ability to substitute for one another. This process is called *differentiation*. The patterns of neurulation and of limb formation just described are the results of differentiation. Supplement 13–2 describes the differentiation of human reproductive organs, and Supplement 13–3 outlines the general development of animal tissues. Cartilage, bone, tendons, blood vessels, and blood cells are all derived from the primary embryonic mesoderm. Mesodermal cells give rise to cartilage *or* blood cells *or* other components. At this stage, the fate of cells is irreversibly fixed, and the cells that normally give rise to cartilage cannot be induced experimentally to give rise to blood cells.

The fertilized egg is considered totipotent because that single cell can form any of the kinds of specialized cells found in the body of an adult organism. Embryonic development can thus be seen as a progressive loss of potentiality, but a gain of efficiency in a specialized function.

We know that all the cells descended from a single fertilized egg have the same genetic complement. How, then, do different kinds of cells arise? The theory of gene control was discussed in Chapters 7 and 8. Development, too, is genetically controlled. You have seen that certain genes direct different kinds of biochemical pathways. Further, it is known that certain genes are especially active at certain times. Analysis of the giant chromosomes found in the salivary glands of several species of flies has shown that there are sharp bands on the chromosomes, and these bands have been correlated with genetic maps in the fruit fly *Drosophila*. Looking in detail at these chromosomes,

several researchers noticed regions in which the usual banding was obscured. These regions of the chromosome were diffuse and "puffed." The puffs are the result of the separation of DNA strands in that part of the chromosome to form many loops. In the early 1950s it was concluded that the puffs indicate regions of gene activity. It was later shown that this activity was mRNA synthesis. Studies of the puffs show that their locations vary in different kinds of tissues and at different stages of development of any one kind of tissue. Finally, all the cells of one stage in any one tissue have the same puff pattern.

The concept of variable gene activity is of great importance in considering the regulation of development. It is thought that some genes are activated and function for only limited periods during early development—that is, from cleavage through neurulation. Other genes may be activated at the blastula or gastrula stage and remain active as long as the animal lives. Finally, after neurulation, some gene regions are active for limited periods. They may be "turned on and off," or may function for only a short time. The suppression of genes that occurs when cell families (the descendants of a differentiated cell) are established seems to be irreversible.

The cell families are in part the result of metabolism and thus genetically directed. The partitioning of the cytoplasm in early cleavage may be a key to differentiation. Such components of the cytoplasm as yolk and cortical granules are partitioned unequally during the early cell divisions. These different quantities of metabolic substrates in the cells of early cleavages could induce different enzymes by favoring the expression of different genes. It is known that the quantity of substrate can influence enzyme induction, as explained by the Jacob-Monod model of operator and repressor genes (Chapter 5). The same mechanism is involved in developing cells. Thus, daughter cells with different constitutions could arise from the two kinds of early cleavage cells. As certain genes cease to function, the overall potentialities of the cells are lost. Cell specialization (differentiation) is now irreversible.

Embryonic Maintenance during Development

A survey of the animal kingdom reveals wide variation in the relation of the parent to the embryo. Within this diversity, several major evolutionary trends in embryonic maintenance can be seen: a trend from the laying of eggs that contain the embryo (**oviparity**) to retention of the embryo with nutrition provided by the parent (**viviparity**); a trend from abandonment of eggs to parental care of eggs and young; and a trend from large numbers of eggs to few young nurtured by parents. These trends appear in both vertebrate and invertebrate groups.

Many species lay great numbers of small, yolky eggs that are fertilized by sperm discharged into the external environment near them. Fertilization in these cases depends to a large extent on chance. Moreover, the yolk is the only nutrient material supplied to the growing embryo, and development must be carefully timed so that when the yolk is exhausted, the young animal will be able to obtain food for itself. Such species often produce large numbers of eggs, because so many will never be fertilized, and so many of those that are fertilized will be killed by predators or the rigors of the environment. Many species lay fewer eggs but provide them with some parental care, though not with additional nutrition. Among the arthropods, some females such as *Daphnia* carry developing eggs in sacs on their bodies; many other arthropods, fish, amphibians, reptiles, and birds prepare nests and guard the eggs after laying them; and some even guard the young until they can fend for themselves. Some species, especially among the lower vertebrates, retain eggs and young in the body of the parent, though, again, they do not nourish them. For example, some fish carry developing eggs in their mouths until hatching. One species of frog broods its eggs in its stomach and finally coughs up fully metamorphosed froglets. Several frogs carry eggs on their backs until the tadpoles emerge. Certain snakes and lizards retain their eggs in the females' oviducts, releasing them at different stages of development according to the species. In some forms, gilled larvae are born; in others, development is essentially completed without a larval stage, and juvenile individuals are born.

Parental care can be extended much further. Birds feed their nestlings, and many mammals train their offspring for survival. The amount of care that parents give seems to be correlated with the number of young: The fewer the young, the more the care. This makes sense, because the more protection the young receive, the better their survival is ensured. The culmination of these trends is represented by the situation in which the few eggs produced are fertilized within the body of the female. They stay in the oviduct or uterus and develop there. Nutrients produced by the parent are transferred to the developing embryos by a special organ called the placenta.

This advanced mechanism for nourishing embryos inside their mothers' bodies is found in some members of all vertebrate groups except the birds. The placenta acts as an exchange organ between parent and embryo by bringing capillaries of the two into close contact, thus permitting gases and organic molecules to diffuse from one to the other. There is no actual fusion of the vessels between parent and embryo, and the parent's blood is never mixed with that of the embryo. Nutrients and oxygen pass into the embryo, and various waste products of metabolism that accumulate in the embryo's blood-

stream diffuse into the parent's circulation. The formation of the placenta is discussed in Supplement 13–1.

Although the placenta is the most common mechanism for nourishing the young inside the mother's body, parents do provide nutrition in other ways. Some sharks have special glands in the oviducts that produce a nutrient substance, the "uterine milk," that is ingested by the embryos. Females of one order of amphibians (the caecilians) produce a similar substance that their young scrape from the oviduct wall, at the same time ingesting epithelium and some smooth muscle fibers from the wall!

The Larval Period

As we saw in Chapter 12, many animal species, both vertebrates and invertebrates, undergo a free-living aquatic larval period after hatching. During this period the young are able to feed and make their own way. The larvae are not simply smaller versions of the adult. In fact, both their body form and way of life are different from those of adults. For example, the limbless, tailed frog tadpole is often herbivorous, feeding on algae and other plant material in the pond by means of its horny mouthparts; as an adult frog, it will be a long-limbed jumper that catches insects by flicking its tongue. The adaptive significance of the larval stage may be to allow an animal to exploit varied resources available during different parts of the year, as in the frogs, or to facilitate dispersal through the habitat, as in the crabs and other marine crustaceans. Water currents scatter crab larvae far from their hatching sites. Then they settle to the substrate and complete their development, moving little thereafter.

Arthropods, such as insects and lobsters, have a hard outer layer, the **exoskeleton,** that is the main structural support of the body. As larvae, these animals may pass through several different stages, each with a distinct body form or morphology. This is a more gradual metamorphosis than that of the frog. At each stage in the life of an arthropod, up to 15 stages in some crustaceans, the exoskeleton is shed to permit growth. Some insects have only one or two larval stages. The caterpillar is the familiar larval stage of moths and butterflies. Echinoderms, such as the sea urchin, have motile, translucent, soft-bodied blastulae, gastrulae, and larvae (Figure 13–20).

Let us look at the frog to see how the tadpole stage is involved in its life cycle. The frog egg hatches in water, and the larva, commonly called the tadpole, must maintain itself in this medium. The tadpole has no limbs at first but has a tail for swimming and a streamlined body. It has gills for respiration, specialized toughened epidermal mouthparts for feeding, and a long, coiled intestine. Also present are adhesive organs consisting of long column-shaped cells that produce a large amount of muco-

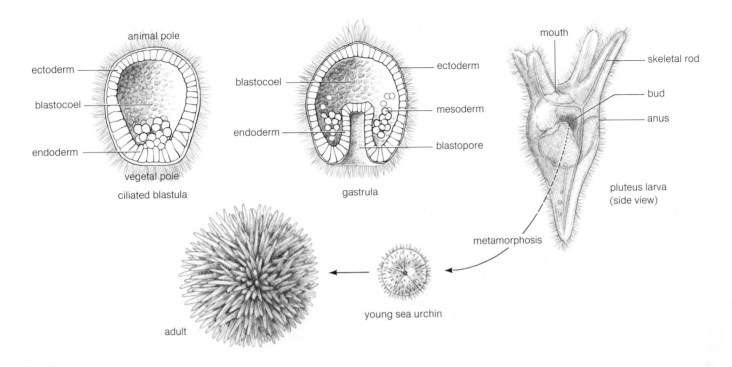

Figure 13–20 Later developmental stages in the sea urchin.

protein. This sticky secretion helps newly hatched tadpoles, who have limited swimming powers, to attach themselves to submerged objects. This saves them from lying for long periods on the bottom, where the oxygen supply may be considerably depleted. Physiologically, the larva has several distinctive biochemical pathways adapted for an aquatic existence. The tadpole produces ammonia as its nitrogenous waste product. It is formed in active cells and is transported to the blood. In freshwater animals, this highly toxic material is "flushed out" of the animal with large amounts of water, whereas saltwater and land animals convert it to the less harmful **urea** by another biochemical pathway. The tadpole has a different visual pigment and a different type of hemoglobin than the adult frog has. These are particularly adapted to the tadpole's aquatic life. The sequence of development to this stage is shown in the normal table in Figure 13–18.

Metamorphosis

Metamorphosis is a drastic morphological and physiological reorganization of the organism that takes place in many functional units in a short time. In frogs, limbs form and the feeding apparatus reorganizes so that jaws and the tongue develop. The middle ear and tympanum form, enabling the adult to detect airborne sound. The intestine shortens, reflecting the change to a carnivorous diet. (A smaller volume of food is required, and its processing is quite different from that of plant material.) The skin becomes keratinized (its outer cell layers acquire a proteinaceous hardening and water-retardant material) and its color changes. Several biochemical changes also take place. Among the most striking is that the adult begins to transform its nitrogenous wastes into urea, which can be excreted with far less water than ammonia—an important consideration in a land animal. Finally comes the maturation of the reproductive system. All these elaborate changes are under strict hormonal control, particularly thyroxine and prolactin (see Chapter 14).

For the frog, as for most animals in which it occurs, metamorphosis is a particularly perilous time. The changeling cannot eat. It is under severe metabolic stress induced by its overall bodily reorganization. Moreover, it is ill equipped to defend itself, for it is losing its ability to behave like a larva but has not yet developed its adult behavior. As a result, many transitional individuals die. Since biological fitness depends upon the number of offspring that survive to adulthood, the advantage to the lineage of having larval and adult stages must be great enough to compensate for the disadvantages of metamorphosis.

The important thing to recognize from the frog example is that the theme of metamorphosis is reorganization of structure from larva to adult. This includes new ways of moving and eating and, most important, it includes sexual maturity. The alternate larval-adult life history and the elaborate mechanisms that make it possible are a successful adaptation by many animals to a variety of environmental conditions.

Growth in Animals

Growth is basically an increase in total mass. It can occur by enlarging existing units (cells or organs) or by multiplying them to produce more units. The ovum grows as it accumulates yolk, whereas a gland grows by increasing the numbers of its cells. Growth can be measured in many ways: Increase in cell numbers, change in weight, volume, or chemical content can all be measured and plotted against time. The curve derived from such measurements is called the growth curve and is S-shaped, or sigmoid. This kind of growth curve is typical of cells, organisms, and populations (see Figure 28–6). Such a curve tells us that growth starts slowly, increases rapidly to a steady rate indicated by the slope of the middle portion of the curve, and eventually slows again. In the developing animal, all organs show such curves, even though not all of them grow at the same rate. Growth is regulated to ensure balance and to limit size.

The embryo must compensate for changes in the surface-to-volume ratio that occur during growth. As a sphere increases in size, its volume increases faster than its surface area. In a developing embryo, the surface area determines the rate at which gases are exchanged, food is taken up, and waste products are released. Many of the changes in the embryo result from a physiological necessity to increase surface area. The development of gills, lungs, and intestines, for example, is a means of increasing surface area.

Some organisms stop growing at maturity; others continue to grow throughout their lives. The apical meristems in plants produce cells as long as the plants live. Animals more characteristically stop growing at maturity, although cells may be replaced throughout life. Fish and amphibians grow rapidly during the larval period and just after metamorphosis. With maturity, growth slows down but continues.

Many organs have characteristic growth patterns. Skin and blood cells can multiply throughout life. However, nerve, heart, and muscle cells cannot increase their numbers after maturity. They can grow only through an increase in size. Further, certain organs cannot increase the number of functional subunits. Consider the human kidney. At maturity, it has a fixed number of **nephrons,** the multicellular functional units where exchange of ions and waste products takes place. Humans have two kidneys. If one of these is removed or, even more drastically, one and half of the other are removed, the remaining kidney or part of one begins to grow. The cell number increases greatly, but the nephrons merely increase in size, not in number. Such differences in growth patterns are apparently under genetic control.

What stops growth? This is as important a question as what starts growth, for too much growth is as bad for an organism as too little. Several phenomena, physical and chemical, inhibit growth. Space within an organism is usually finite, as determined by surface contacts. When cells can no longer move freely, they often stop increasing in number. They will also stop growing when they become crowded, or reach a saturation density. This is called *contact inhibition* of growth. Contact inhibition can readily be seen in cell cultures.

The availability of nutrients is also important in cell growth. Again, in cell cultures, the amount of medium in a culture plate or tube is finite; when it is gone, growth will stop. Similarly, development and growth within an organism are limited by the amount of available food.

Aging

The ability of cells to divide also determines the possibility of growth. One of the causes of aging is the progressive reduction in the numbers of cells that can divide or even continue their synthetic function. Not only do organs in the body eventually have fewer cells able to divide, but many organs lose weight with age as well. Figure 13–21 shows changes with age in proportions of the human body. It is thought that the rate of cell division is genetically controlled and diminishes with age.

Recent work suggests that the genetic control of cell division may be chemically mediated. Scientists at the University of California, Berkeley, have recently demonstrated that the number of generations through which cultured human lung cells will survive can be increased from 50 to 120 by adding vitamin E to the culture. This line of research has promise for an eventual understanding of the aging process. In pursuing this understanding, we should keep in mind that diminished function is not confined to the aging of the organism but is an important factor in growth and development in selected tissues and organs. For example, in the metamorphosis of the frog, the formation of the hand involves cell death and degeneration, as does the loss of larval gills and tail.

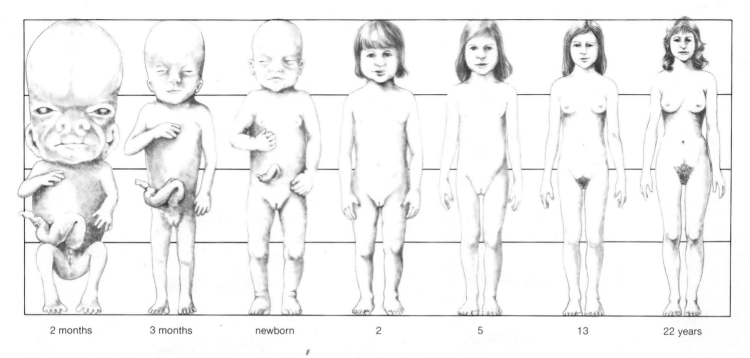

| 2 months | 3 months | newborn | 2 | 5 | 13 | 22 years |

Figure 13–21 Changing proportions of the human body during prenatal and postnatal growth. All stages are drawn to the same total height.

Regeneration

Many animals show remarkable powers to heal damage to their bodies. There are two major kinds of repair that an animal can perform. One is the healing of a cut or lesion of its surface, a broken skeletal element, or an internal bruise or hemorrhage, provided it is not so severe as to cause death. The second is the **regeneration** of a body structure, such as a limb, that has been lost—a much more drastic kind of repair. Almost all animals are capable of healing wounds; fewer are capable of regenerating body parts. This ability is confined to certain groups and, in fact, often to certain ages within those groups.

Regeneration is more likely to occur in simple organisms than in complex ones. A sponge can regenerate from only a few cells; a flatworm or an earthworm can grow back half its body when chopped in two; a starfish can not only grow a new arm, but—if one arm is attached to the central part of the body—regenerate the rest of its body plus all its other arms!

Regeneration is also found among vertebrates, but to a lesser degree. Amphibians can regenerate whole limbs, tails, jaws, and even brains, but this ability is reduced as the animal ages. For instance, a young frog can regenerate an entire forelimb if its original limb is ampu-

tated at the shoulder, but an old frog regenerates only a nonfunctional bud following the same surgery. However, an old frog can regenerate a functional toe. There is, then, an age restriction on the degree of regeneration. Some lizards can regenerate their tails, but the new tail will not be the same as the old one. The original tail had vertebrae, muscles, nerves, and blood vessels; the regenerated tail is a stout rod that is not differentiated into vertebrae and lacks muscles, nerves, and vascular supply. Birds and mammals are not capable of even that degree of regeneration. However, certain avian and mammalian tissues *can* regenerate, some readily (such as red blood cells), and some very slowly (such as peripheral nerves). Some people have had the experience of having feeling suddenly return to a fingertip whose nerve supply was accidentally severed long before. Nerve tissue can take years to regenerate to the point that connections are reestablished.

Regeneration of the amputated salamander limb has been well described by several researchers. During the first phase, the wound heals and is closed over by epidermal cells. The damaged tissue breaks down. Then, differentiation of the remaining tissues begins when the material holding together the cells of bone, cartilage, and connective tissue breaks down. These cells aggregate at the wound site, where they form a mound of tissue called

Figure 13–22 (**a**) Regeneration in a salamander of amputated part of forelimb; (**b**) stages in that regeneration.

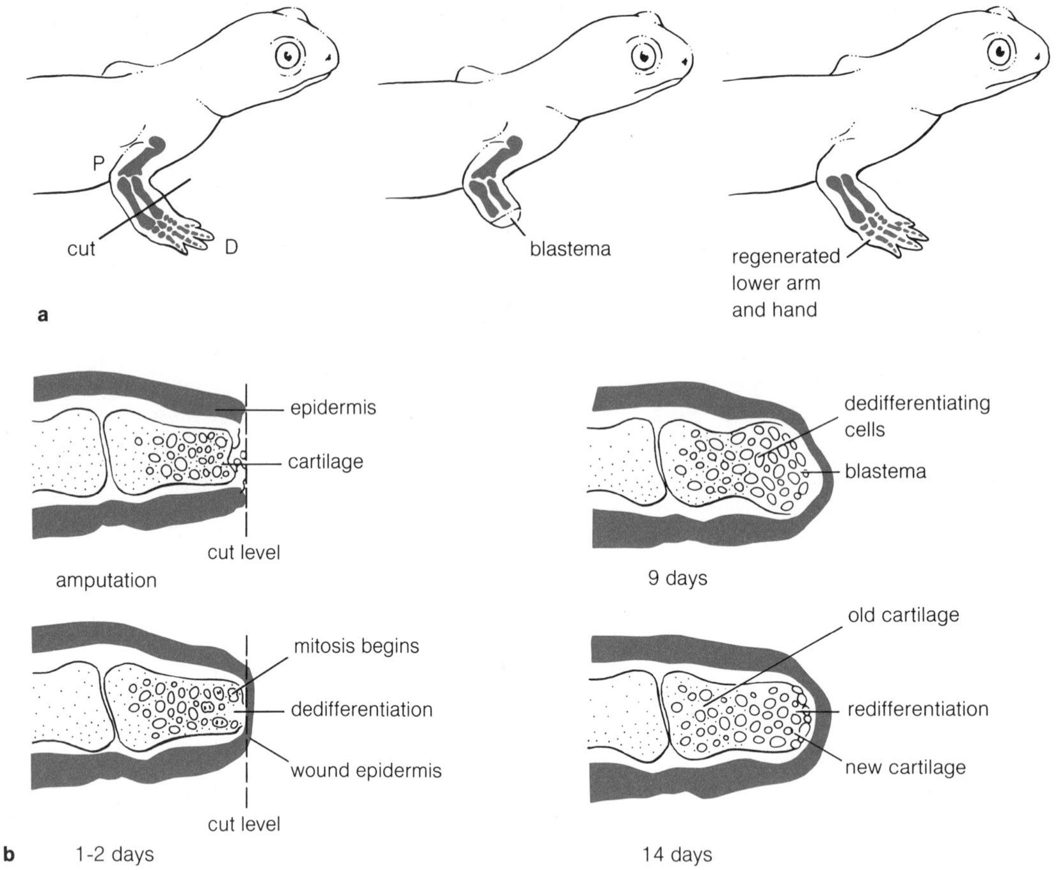

the *blastema* and where they will redifferentiate into the various tissues (Figure 13–22). The blastema enlarges, forming a bud, and morphogenetic movements take place as in the development of the original limb, described earlier. There is one major difference in this growth pattern—nerve fibers must be present or invade the amputation site for regeneration to proceed. Nerves experimentally transplanted to an amputation site increase the rate of regeneration. There is, in fact, a critical volume of nerve tissue that must be exceeded before growth will take place. It is thought that a chemical nerve-growth factor is involved. Growth, morphogenesis, and degeneration together show a unique pattern in the process of regeneration.

Summary

Following fertilization, the process of development involves an increase in the number of cells with little change in volume for some time, and the reorganization of these cells. Cleavage, or division of the egg and subsequent cells, proceeds rapidly, forming a solid ball of cells. A cavity forms in the ball of cells, so that a hollow ball (the blastula) forms in many species. The blastula reorganizes through complex cell movements (gastrulation) so that the basic cell layers are in a particular position. Cells of the dorsal region of the gastrula proliferate and reorganize to form the precursors of the brain and spinal cord (neurulation). Differentiation, or modification of cell and tissue structure and function, proceeds until adult form is attained. Egg type and the amount of yolk influence the pattern of development of blastula and gastrula.

Tissue culture techniques are providing information on cell movements, cell associations, and induction, or the production of one part under the influence of another. The potency of cells is also being analyzed, especially through transplant techniques. In many species, the products of early cell divisions can each produce an intact embryo when they are separated. Later cells cannot do so; they can only produce certain kinds of tissues—their fates are determined. Organ formation proceeds as morphogenetic movements, differentiation, and cell death proceed.

Many animal species undergo a larval period in which

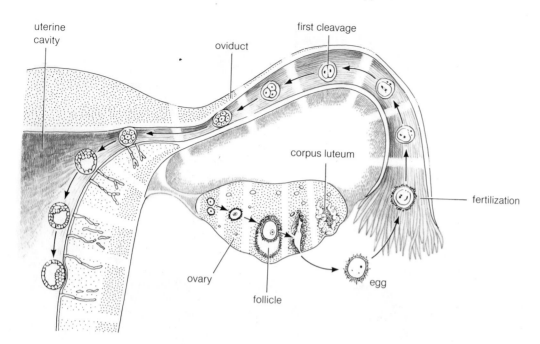

uterine cavity

oviduct

first cleavage

corpus luteum

fertilization

ovary

follicle

egg

Figure 13–23 Human egg formation: ovulation, fertilization, cleavage, blastocyst formation, and implantation.

they are able to obtain their own nutrition, are sexually immature, and usually have a different form from the adult. Often major components of physiology and biochemistry distinguish larvae from adults. Larvae undergo metamorphosis, a drastic morphological and physiological reorganization in a short time, to achieve adult form. Changes in the frog from tadpole to adult are an example of this process. Metamorphosis is hormonally controlled. Thyroxine is the hormone primarily responsible for frog metamorphosis.

Many organisms are capable of regeneration, or the restoration of lost parts. This capacity diminishes as organisms become more complex and as individuals age.

Supplement 13–1
Human Development

Human development before birth is a magnificent example of the complex interplay of growth and differentiation, and their controlling factors. Further, it results in *us*—our own genes via our gametes give rise to those blue-or brown-eyed, bald or curly-haired means of perpetuating our own species.

The events of human development begin high in the oviduct with fertilization of the egg by one of several

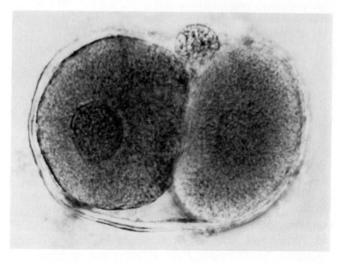

A. T. Hertig

Figure 13–24 A human embryo at the two-cell stage, × 1,500. At this time, the embryo is in transit through the oviduct leading from an ovary to the uterus.

million sperm swimming up the tract. Fertilization triggers the completion of meiosis, which had been quiescent since before the birth of the maternal parent (see Chapter 6). Cleavage takes place as the ovum travels down the oviduct to the uterus. Rather little is known of cleavage stages in human embryos (Figures 13–23 and 13–24), but

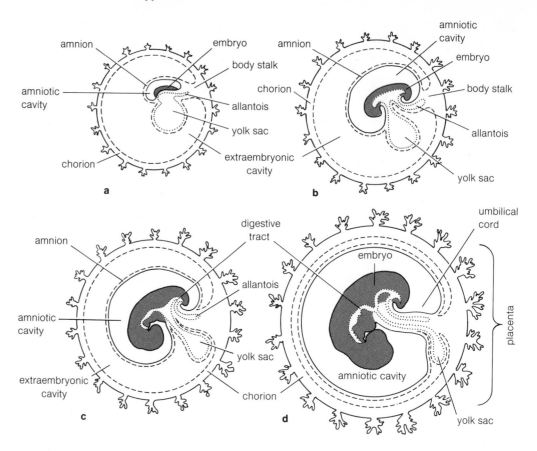

Figure 13–25 Stages in the development of the umbilical cord and body form in the human embryo. The brace indicates the region of the chorion where the placenta forms.

96 hours after ovulation the 16-cell stage is reached, and this dividing **blastocyst** arrives in the uterus.

Early divisions are slow and irregular. The rate of division then increases rapidly, so that by 120 hours, 100 cells compose the developing organism. Fluid from the uterine cavity infiltrates between the cells of the mass, as a central cavity forms. The cells are arranged in a hollow outer layer, the **trophoblast,** with an inner cluster, the *inner cell mass,* pressed against one side of the outer layer. When the region of the trophoblast in contact with the inner cell mass contacts the lining of the uterus, *implantation,* the association with the maternal structure, takes place. Trophoblast cells multiply and insinuate themselves between the cells that line the uterus. The trophoblast differentiates into two layers. The outer layer absorbs fluids and blood released as the uterine lining disintegrates and is digested following implantation.

The inner cell mass differentiates to form the embryo. At 7.5 days, a space that will become the amniotic cavity forms between the disk of the inner cell mass and a layer of cells derived from the trophoblast. By 9 days the blastocyst is completely embedded in the uterine lining. The *amnion* (the extraembryonic membrane surrounding the embryo and the fluid in which it lies) is organized, the

amniotic cavity enlarges, and the embryonic region becomes a two-layered disk. The lower layer is associated with the developing *primary yolk sac,* which lines the cavity of the blastocyst.

At 11 to 12 days of development, the two layers of the trophoblast have a more complex organization. Extraembryonic (outside the embryo) membranes have developed, establishing the environment for the embryo and a means for its nutrition and gaseous exchange. These layers of the trophoblast comprise the *primary villi,* which increase the surface area for chemical exchange. *Secondary villi* with extensive blood-vessel systems form from the mesodermal cores. The trophoblast has become the *chorion,* or outer extraembryonic membrane. Part of the extraembryonic mesoderm remains as the *body stalk,* connecting the amnion and the chorion. A projection of endoderm, the **allantois,** extends into the body stalk (see Figure 13–25). Also during this period, the embryonic disk forms three layers as mesoderm and notochord move from the upper layer to assume a "central" position.

After the twentieth day of development, the embryo grows rapidly. The blastocyst on that day is completely embedded in the uterine lining. With growth and enlargement, the embryo and its membranes, covered by layers

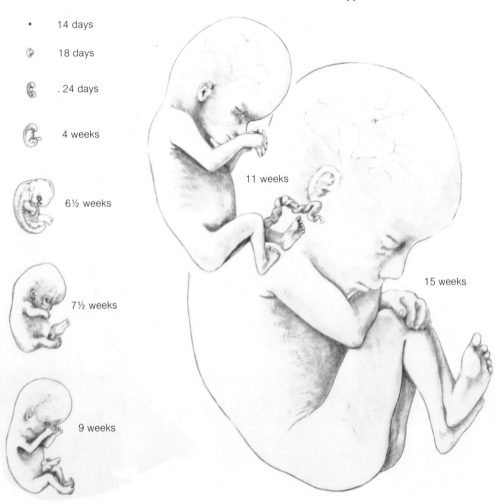

14 days

18 days

. 24 days

4 weeks

6½ weeks

7½ weeks

9 weeks

11 weeks

15 weeks

Figure 13–26 A series of human embryos of approximately natural size.

of uterine cells, bulge out into the cavity of the uterus. As development progresses, the chorionic villi, once covering the entire blastocyst, become confined to the area immediately beneath the blastocyst on the uterine wall. The part of the chorion that has villi and the part of the uterine lining with which the villi are in contact form the placenta (Figure 13–26d). The placenta has both maternal and fetal components and is the structure for gaseous and nutrient exchange for the fetus. Recall that early in development, villi protrude into blood-filled spaces in the uterine lining. As these spaces become confined, the capillaries of the uterine lining and those of the villi come into close contact so that exchange of materials can occur. However, maternal and fetal circulations never fuse, although they are separated by only two cell-layers—the maternal capillary wall and the fetal capillary wall in the villi. The amnion continues to expand and presses the body stalk, containing the rudimentary allantois and yolk sac, into a cordlike structure. This cord

elongates with continued embryonic growth, and its central region contains arteries and veins. It is the *umbilical cord.*

As growth continues, the chorion and its covering layer of uterine cells expands into the cavity of the uterus. It finally becomes so enlarged that it fills the uterus and meets the lining of the rest of the uterus. The layer of uterine cells covering the chorion then fuses with the uterine lining. From the fourth month of growth on, the amnion membrane containing the fetus and the fluid in which it rests dominates the uterus.

Development of the embryo can best be seen in Figures 13–26 and 13–27. Notice particularly the early development of the brain and eye, which are organized by the twenty-fourth day. The muscle blocks, or somites, also are organized by then (Figure 13–27d); the bulges of the heart chambers are visible at 3.5 weeks. The arm buds appear at 28 days (Figure 13–27h), hindlimbs at 32. The heart begins to beat at 30 days; the sex of the

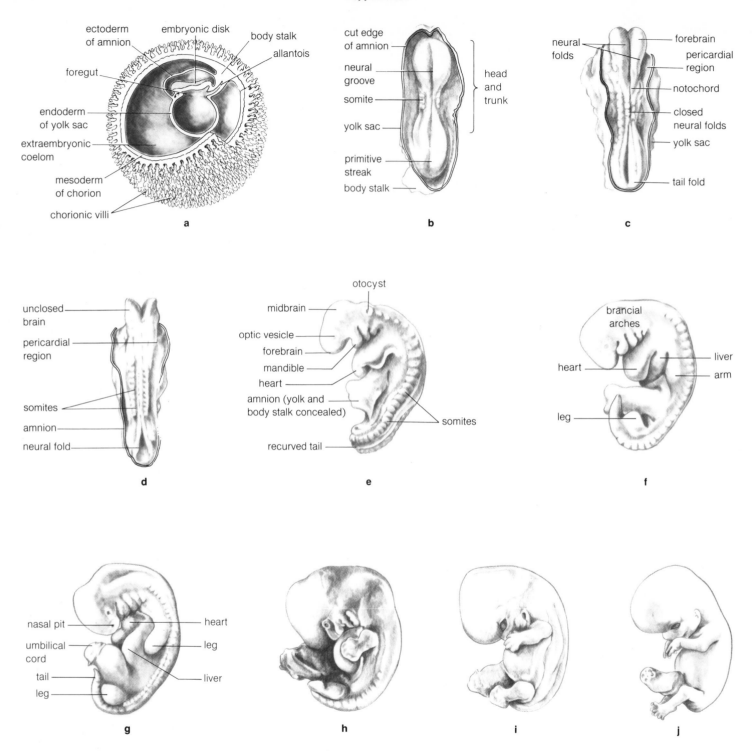

Figure 13–27 Diagrams of human embryos: (**a**) 10 days old; (**b**) 20 days old; (**c**) at 18 days with amnion cut away, viewed from above (×45); (**d**) at 21 days (×42); (**e**) at about 3.5 weeks (×27) (note that the embryo body develops in front of the primitive streak); (**f**) at 22 days, showing 10 pairs of somites (×34); (**g**) at 26 days, with 25 somites viewed from the left side (×16); (**h**) at 4 weeks (×12); (**i**) at 5 weeks (×7.5) (note the arm and leg buds and umbilical cord); (**j**) at 37 days; (**k**) at 41 days; (**l**) at 47 days.

Table 13–2 Events of Early Human Development

Age in Weeks	Size in mm	Description
2.5	1.5	Neural groove indicated, blood islands formed, embryonic disk flat
4	5.0	All somites present, limb buds indicated, heart prominent, eye, jaws, thyroid present; gut tube differentiated, liver, pancreas, lungs, kidney, nerves forming
5	8.0	Tail prominent, umbilical cord organized; mouth and pharyngeal glands form, intestine elongates, genital ridge forms
6	12.0	Head becomes dominant in size, limbs recognizable, gonads form; heart has general definitive form; cartilage formation begins
8	23.0	Nose forms, digits recognizable; tongue muscles well formed, gut structure further differentiates, liver large; testis or ovary recognizable as such; main blood vessels have typical plan, first indication of bone formation, definitive muscles of head, trunk, limbs formed
12	56.0	Sex readily determined by external inspection, tooth primordia formed, fusion of palate complete; blood formation in bone marrow begins, blood vessels well formed; ossification spreading with many bones outlined; brain and spinal cord attain general structural form
16	112.0	Face looks human, head hair appears, muscles become spontaneously active; gastric glands formed, kidney has typical form; skin glands form; eye, ear, nose approach typical appearance
20–40 (5–9 mo.)	160.0– 350.0	Body lean but has "baby" proportions (month 6), eyelids open (7), testes descend into scrotum (8), fat collects (8–9); tooth formation continues; tonsils, appendix, spleen acquire typical structure; fingernails form (8); spinal cord myelinization begins (5), brain myelinization begins (9.5); retina of eye complete and light-sensitive (7), taste sense present (8)

embryo can be recognized by examining the external genitalia at 10 weeks. By 8 weeks muscles are organized; fetal movement may be felt by the mother as early as 10 to 12 weeks. Follow the sequence of development of the arm in Figure 13–27h–l. The bud differentiates into a paddlelike structure, then into a strut with a webbed, five-pronged terminal pad. As differentiation progresses, upper and lower arm regions form, digits extend, and webbing is lost. After 2 months of development, the fetus is recognizably formed. Its large head has a nose, mouth, eyes, and ears; limbs are discrete. Early development is summarized in Table 13–2.

As growth continues, differentiation progresses. Bones mineralize and harden, blood vessels proliferate, visceral organs become discrete and functional. By the end of the seventh month, only the refinement of function remains to be completed, and life outside the mother is possible with clinical help. The fetus grows longer and adds considerable weight during the eighth and ninth months. It acquires an energy-storing, insulating fatty layer. At birth, the baby is a vocal, motile, hungry little being ready to make its demands of the world.

Supplement 13–2
Human Reproductive Organs

The development of male and female reproductive organs is determined by the genetic sex of the developing fetus. For the first 6 to 8 weeks of development in humans, the organs are undifferentiated, and sex cannot be recognized except by examination of the chromatin in the nucleus. Embryos of both sexes have genital ridges that will differentate into gonads and ducts, and the beginnings of other glands (Figure 13–28). At about the eighth week, differentiation occurs, and testes can be seen to be structurally different from ovaries. In the female, the ducts that will become the oviducts enlarge; they begin to disappear in the male. Ducts that drained the embryonic kidney shift to drain the testes in the male; these ducts disappear in the female. New ducts develop in both sexes to drain the kidneys. By the twelfth week, the sex of the developing human can be recognized externally. The genital tubercle, or phallus, has become the male penis

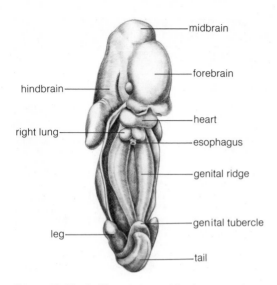

Figure 13–28 Indifferent stage of the human embryo.

Table 13–3 Human Reproductive Organ Homologies

Male	Indifferent Stage	Female
Testis	Gonads	Ovary
Vestigial	Genital ligaments	Ligaments of ovaries and uterus
Sperm ducts, ureter and pelvis of kidney	Wolffian ducts	Ureter and pelvis of kidney
Vestigial	Mullerian ducts	Oviducts, uterus, upper part of vagina
Penis	Phallus	Clitoris
Scrotum	Genital swellings	Labia majora

and the female clitoris; the swellings in that region become the male scrotum and the female labia majora. The ovaries remain pelvic in the female; the male testes usually descend into the scrotum in the eighth month of development. The scrotal position of the testes provides a lower temperature than in the pelvis. This is thought necessary to keep the sperm from dying before they are ejaculated. It has been demonstrated that temperatures equivalent to that of the midbody substantially reduce the numbers of viable sperm.

Both male and female organs, internal and external, develop from the same tissue site and have similar structure. They therefore are said to be *homologous.* Table 13–3 lists the primary reproductive organs that are homologous in males and females. Figure 13–28 illustrates the undifferentiated state; Figures 13–29 and 13–30 show the male and female consequences of development. Compare these illustrations with the table to evaluate the structures that developed from the same origin.

The adult testes and ovaries produce sperm and ova, respectively, and various hormones. The development of sperm and ova is discussed in Chapter 12; the functions of the hormones, in Chapter 14. The ducts that bear sperm (Figure 13–29) go from the testes into the pelvis to join the urethra, the duct in the penis that bears both semen (sperm plus various glandular secretions) and urine. During intercourse, valves at the entrance of the urinary ducts close, so that sperm and urine are not mixed. The prostate gland and the seminal vesicles open into the sperm passage and secrete a substance rich in fructose and other products. It is thought that this may be a source of energy for sperm.

The body of the penis (Figure 13–29) and that of the clitoris comprise spongy tissue that fills with blood upon sexual stimulation. This swelling effects erection of the penis (and clitoris) so that the penis can be inserted into the vagina during intercourse. Further sexual stimulation during intercourse causes a sympathetic nervous system response so that the male ejaculates semen into the vagina. The combination of nervous, vascular, and psychological responses produces the pleasurable effect of sex, including climax in both male and female.

The female reproductive system is characterized in part by the close association of the openings of the oviducts with the ovaries (Figure 13–30). At ovulation, the ovum emerges from the ovary into the coelom and is swept into the oviduct by currents created by the tips of the ducts. The ovum then passes down the duct, where fertilization may or may not take place. The functions of the uterus, formed by the fused posterior parts of the paired oviducts, are described in Chapter 12. In contrast to that of the male, the female reproductive system is completely separate from the urinary system, and the vagina and the urethra have separate openings between the labia.

Supplement 13–3
Animal Tissues and Their Origins

Several different kinds of tissues were mentioned in the outline of the development of the frog limb. Each kind arises from one of the three primary tissues, but differen-

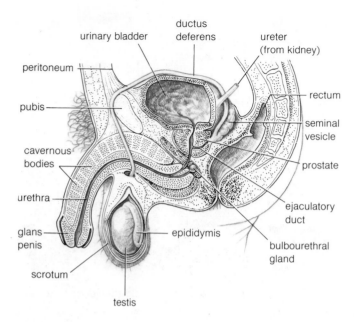

Figure 13–29 The human male genital system shown as if divided into right and left halves.

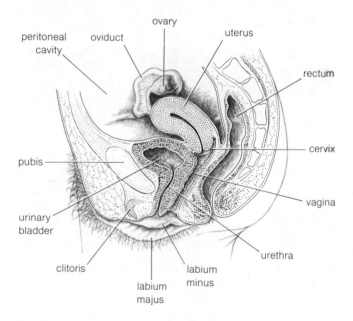

Figure 13–30 The human female genital system shown as if divided into right and left halves.

tiates and acquires a more specialized morphology and function. The adult tissues of vertebrates can be classified into four groups, each with subtypes (see Table 13–1).

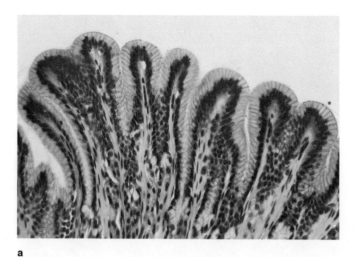

a

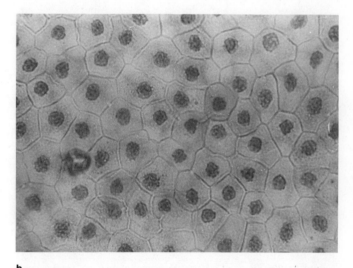

b

Figure 13–31 Types of epithelial (covering or lining) cells; (**a**) columnar; (**b**) squamous (flattened). Epithelia are *simple* if they are one layer thick, *stratified* if they are two or more layers thick.

The first group includes tissues of the **epithelium.** These tissues cover the body's surface, line its inner cavities, or proliferate to form glands. The skin is a covering epithelium; it is derived from embryonic ectoderm, as are its glands. The lining of the intestine and its secretory cells are derived from endoderm. The linings of the urinary and reproductive systems are derived from mesoderm. Therefore, it is not embryonic origin that causes a tissue to be called epithelium, but its structure, position, and function.

Epithelial cells that cover or line structures have one of three shapes—*cuboidal,* or cube-shaped in section, as in the kidney; *columnar,* elongate cells including those

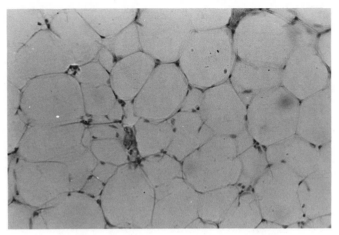

a

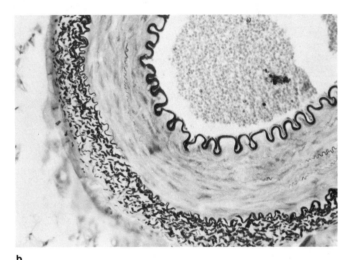

b

Figure 13–32 Types of connective tissue: (**a**) adipose (fatty) tissue; (**b**) fibrous tissue in an artery wall.

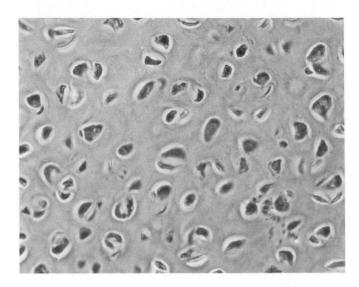

Figure 13–33 Cartilage, a support tissue. Cells are embedded in inter-cellular material.

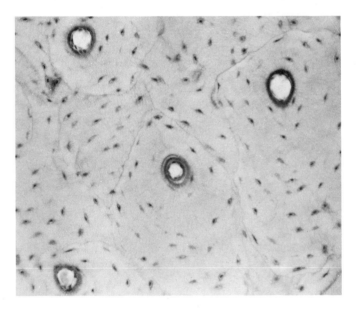

Figure 13–34 A cross section of bone, another support tissue. Note that the cells communicate by tiny canals.

lining the intestine (Figure 13–31a); or *squamous,* flattened cells such as those lining the inside of the cheek (Figure 13–31b). If an epithelial layer is only one cell-layer thick, it is called *simple;* if it is two or more layers thick, it is *stratified.* Epithelial cells may bear cilia or have smooth surfaces. Epithelia provide protection to organs and the outer part of the body; in the intestine they absorb some digestive products; and some secrete fluid to keep inner surfaces moist and lubricated.

Glands are aggregations of epithelial cells that are highly specialized for secretion. *Exocrine* glands, such as the sweat glands of the skin, secrete their products to the outside of the body and have ducts to carry the secretions away. *Endocrine* glands (also called ductless

glands), like the pituitary, thyroid, and adrenal glands, secrete directly into capillaries. Their functions are discussed in Chapter 14. Glands can be simple tubes *(tubular)* or pockets *(alveolar),* or compound aggregations of the two structures.

Connective tissues compose the second major tissue

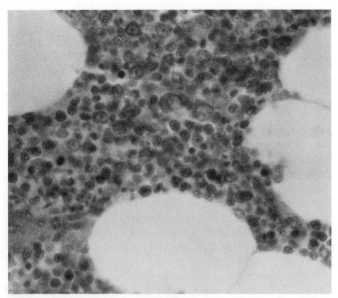

Figure 13-35   A section of bone marrow.

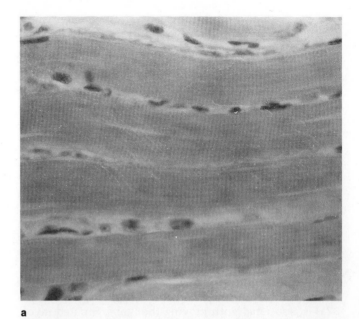

a

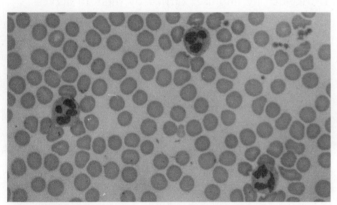

Figure 13-36   Blood cells.

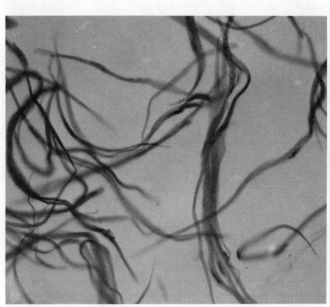

b

Figure 13-37   Examples of the kinds of muscle cells: (a) striated (voluntary); (b) smooth (involuntary).

group. Most kinds of connective tissue cells produce *intercellular* substances, so called because they occur between connective tissue cells. The intercellular material (matrix) is nonliving and can be very strong. The main functions of ordinary connective tissues are holding, binding, connection, and support. Fatty (adipose) and fibrous tissues (Figure 13–32) are binding tissues; cartilage and bone are support tissues. Cartilage cells are embedded in intercellular material (Figure 13–33) that can bear weight, some kinds without bending. Nutritive substances from tissue fluids diffuse through the matrix of the living cartilage cells. If the matrix becomes impreg-

nated with calcium salts, diffusion cannot take place, and the cells die. This is a normal part of the developmental sequence of bones, in which cartilage forms, then dies and is replaced by bone. Bone cells also lie in an intercellular matrix, but the cells are connected to each other and to the tissues at the surface of the bone by a system of tiny canals (Figure 13–34). Thus they can be nourished. The bone matrix is impregnated with calcium salts, primarily calcium phosphate, in a way that makes it strong and rigid.

Another group of connective tissues is made up of blood-forming cells and the blood cells themselves. They are "connecting" tissues only in a loose sense, but like cartilage, bone, fat, and fibrous tissue, they arise from mesenchyme, a derivative of mesoderm. In this case, the classification is based on embryonic origin. The cells of the lymph nodes, spleen, thymus, and bone marrow (Figure 13–35) are blood-forming structures. The red and white blood cells (Figure 13–36) are specialized cells whose functions are discussed in Chapter 17.

The third class of tissue is muscle. Muscle is derived from mesoderm, and its most noticeable property is its ability to contract. This ability is what makes possible all sorts of movement—locomotion, digestion, respiration, and other kinds of organ movements. There are three kinds of muscle. *Striated* voluntary muscle (Figure 13–37a) is associated with moving the limbs. Striated involuntary, or *cardiac,* muscle is heart muscle. *Smooth* involuntary muscle is found in the walls of the digestive tract, blood vessels, lungs, and other organs. Smooth muscle cells usually look spindle-shaped with central nuclei under the light microscope (Figure 13–37b). Striated muscle cells (both voluntary and cardiac) are long and multinucleated. The nuclei are positioned at the boundary of the fiber, and the body of the cell shows the characteristic stripes or striations.

The major class of animal tissue is nerve tissue. All nerve cells are derived from neurectoderm. Nerve cells compose the gray and white matter of the brain and spinal cord (central nervous system), and the nerves, ganglia (aggregations of nerve-cell bodies), and special nerve endings for nerve–muscle junctions and for touch, temperature, and other sensory reception. The varied structure of nerves follows one basic plan: The nerve-cell nucleus is located in the *cell body;* impulses are brought to the body by branched structures called **dendrites** and are conducted from the cell body by a single structure called an **axon.**

All the organs of the body consist of various arrangements of these four kinds of tissue. The stomach, for example, has inner and outer epithelial linings. These linings are bound to layers of smooth muscle by connective tissues, and the contraction of the smooth muscle cells is triggered by impulses carried by them to nerves.

14

Hormones and Chemical Coordination

In the fall, as the days grow colder and the birds start south, the woolly bears go on the march in the eastern United States. These bright black-and-brown caterpillars, common but generally retiring, crawl out from hiding and appear everywhere, traveling with all the urgency so slow-moving a creature can muster. Parking lots, streets, sidewalks, and playgrounds are often dotted with woolly bears, many of them crushed by passing people or vehicles. The survivors stay on the move until they find suitable wintering places. After emerging from their wintering places in the spring, they feed for awhile, and then attach themselves to a fixed object, weave a protective coat of silk about their bodies, and undergo a complete physical change. When summer comes, the cocoon splits, and what emerges is not a woolly bear but a tiger moth. The new moth spreads and dries its wings and then flies off to feed and reproduce, producing or fertilizing the eggs that will become more woolly bears and, in another winter's time, more tiger moths.

The annual drama of the woolly bear points up the pervasiveness of chemical communication in living things. Practically every aspect of the moth's life cycle—hatching of the eggs into caterpillars, metamorphosis, emergence of the adult moth from the cocoon, mate-finding, reproduction—is orchestrated by messages transmitted by organic molecules. These chemicals link the various actions of organs and tissues with changes in the internal and external environments, and they underlie many of the phenomena we observe in nature. The spectacular reproductive display of a cock sage grouse, the growth of a human child into a midget, a giant, or an adult of normal stature, the development of a frog from a tadpole, milk production in the breasts of a woman who has given birth—all these phenomena, and literally millions more, depend on the timely production and release of special chemicals.

In the last chapter we saw how animals developed from fertilized eggs to adults. We mentioned briefly that this development was under the control of various hor-mones. In this chapter we will concentrate on these hormones and others found in animals, stressing the vertebrate endocrine system. But first let us examine the whole question of chemical messengers, particularly in animals.

Kinds of Chemical Communication

The various chemical messengers in animals can be grouped according to the sites where they are produced and the sites where they act, as follows: (1) intracellular messengers, (2) neurotransmitters, (3) neurohormones, (4) hormones, and (5) pheromones (see Chapter 23). The phytohormones, or plant hormones, were discussed in Chapter 11.

No true intracellular messenger has been identified, yet we assume that such messengers exist, because the workings of the cell's organelles are regulated somehow, most likely by the nucleus. We do know of a group of so-called secondary messengers, primarily cyclic adenosine monophosphate (cAMP) and cyclic guanine monophosphate (cGMP). When a hormone impinges on the cell membrane, these messengers appear. They are produced by reactions in the membrane itself, and they initiate specific responses within the cell. We also know that various metabolites within the cell control enzyme synthesis by affecting transcription from the genome (Chapter 5).

The chemical messengers that are synthesized within one cell and affect other cells make up a diverse group. Some work close to the site of their synthesis, while others act at a distance. The **neurotransmitters** (Chapter 21) are locally active, typically traveling only 20 to 30 nanometers across a synapse from the nerve that released them to the nerve or muscle they affect. The neurotransmitters are all relatively small molecules derived from amino acids.

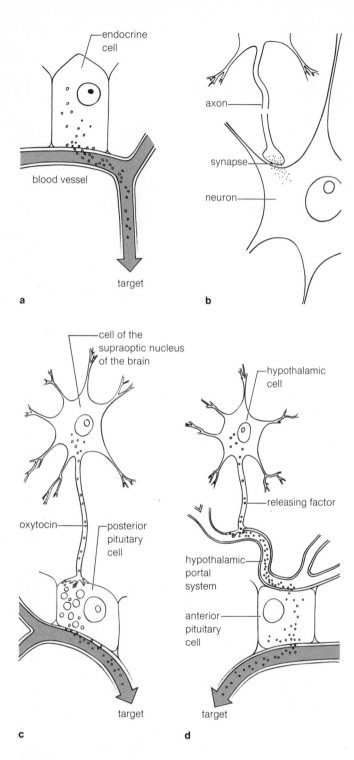

endocrine cell

blood vessel

target

a

axon

synapse

neuron

b

cell of the supraoptic nucleus of the brain

oxytocin

posterior pituitary cell

target

c

hypothalamic cell

releasing factor

hypothalamic portal system

anterior pituitary cell

target

d

Figure 14–1 Neurosecretory cells (**c, d**) transport their product by nerve processes to the blood, in contrast to the endocrine cells (**a**) that dump it into the blood directly, and nerve cells (**b**) that release their product into the synapse, where it acts on an adjacent neuron.

Certain nerve cells are specialized for secreting chemical messengers into the bloodstream. These cells are called **neurosecretory cells,** and their secretions are known as **neurohormones.** The neurosecretory cells differ from other nerve cells in that they typically terminate on or near blood vessels (Figure 14–1). Some such nerve terminals (axons) also contain storage and release centers called *neurohemal organs.* Chemically, the neurohormones are amino acid derivatives, oligopeptides (chains of 20 or fewer amino acids), or proteins.

Norepinephrine, synthesized in the adrenal glands, is derived from an amino acid and, curiously, it is chemically identical to the neurotransmitter of the sympathetic nervous system (see Chapter 20). Traveling through the body in the blood, norepinephrine affects organs and tissues, called *targets,* well distant from its point of origin in the adrenal glands.

The **hormones** are generally defined as organic molecules produced in one part of the body but affecting some other, separate part. Some hormones are produced by organs that have other functions as well. The stomach wall, intestine, and liver, for example, produce hormones that regulate digestion. However, many of the hormones are produced in specialized organs, called **endocrine glands.** The endocrine glands are ductless, liberating their products directly into the blood, and they have lent their name to the study of hormones, *endocrinology.* (By contrast, the **exocrine glands** release their secretions to the exterior or into a body cavity through a duct or tube. Examples are the mucus, sweat, digestive, and salivary glands.)

Hormones are chemically of two main types: Some are amino acid derivatives, oligopeptides, or proteins; others are fatty acid derivatives or members of the class of lipids called steroids. The physiological effects of each of these chemical messengers are closely related to its molecular structure. By comparing the effects of organic molecules synthesized in the laboratory with the effects of natural hormones, endocrinologists can determine the location of the active site on the hormone. Typically, these sites are remarkably small. Since many hormones from different vertebrates are chemically similar if not identical, human medicine uses hormones derived from pigs, sheep, and cattle. Synthetic hormones are also used, and they are actually more powerful than their natural analogues. Insulin is derived from either pigs or cattle. The two are slightly different structurally, but both are effective in humans. (Insulin is too expensive to synthesize as yet.) Thyroid hormone used in human therapy is derived from cattle.

The nervous system and the endocrine system often work together, with neurohormones connecting the two. This arrangement provides a link between the external environment and the hormonal states that control a wide range of behavioral, physiological, and even developmental changes. The nervous system perceives certain stimuli

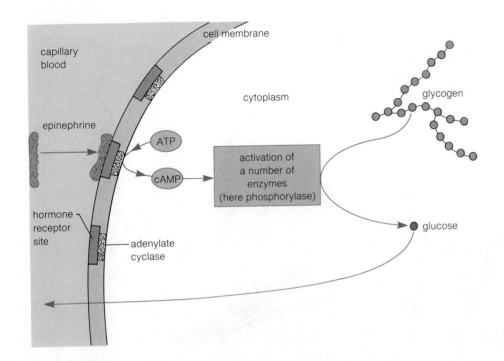

cell membrane

capillary blood

cytoplasm

glycogen

epinephrine

ATP

cAMP

activation of a number of enzymes (here phosphorylase)

hormone receptor site

adenylate cyclase

glucose

Figure 14–2 Model of the mechanism of action of a peptide and protein hormone on a target cell (in this case, the control of the release of glucose from a liver cell by the action of epinephrine).

and directs the release of specific neurohormones that in turn affect the action of certain endocrine glands. For example, when the human nervous system perceives a threat—say, a possible mugger lurking in a doorway—it releases neurohormones that order the adrenal glands on the kidneys to release epinephrine. This hormone then readies the body either to meet the threat head-on or to flee from it. Epinephrine increases heartbeat and blood pressure, speeds the respiratory rate, raises the blood sugar level, and prompts the liver to transform its stored glycogen into ready glucose. As research uncovers many more such linkages between the nervous and endocrine systems, many investigators are coming to consider the two together as the *neuroendocrine system.*

Phytohormones and hormones all have their effects within a single individual. **Pheromones,** by contrast, are chemical communicators between individuals of the same species. Found in many vertebrates, they are most developed among the insects, where specialized chemicals attract mates, mark trails, tell friend from foe, and, among termites and other social insects, prevent the development of a new queen as long as the old one is productive.

Hormone Activity and Control

Of all the chemical messengers, the hormones are the most fascinating. They control a wide range of physiologi-

cal processes, and their function is intrinsic to the organization of complex living systems.

How Hormones Work Identical or similar hormones can have vastly different physical effects on different targets. Testosterone is chemically similar in men, deer, and birds, yet it produces a beard in the first, antlers in the second, and nuptial garb feathers in the third. Likewise, thyroid hormone affects metamorphosis in amphibians, maze-learning in rats, metabolic rate of birds and mammals, schooling behavior in fishes, tooth eruption in mammals, and pigment deposition in the feathers of developing birds. Obviously, the same hormone affects various cells and tissues differently.

Another important point about endocrine function is that most bodily processes are regulated not by one hormone, but by several. For example, systematic studies by numerous investigators have shown that the proper development and maintenance of the rat's mammary glands require at least five hormones from the pituitary gland plus several steroid hormones from the ovary and the placenta. These facts indicate that a hormone's effects depend not only on the action of the hormone itself but on the specific nature of the target tissue.

Our current concept of hormone activity is that the target cells "recognize" specific hormones, which then activate them. If radioactively labeled estradiol, a female hormone, is injected into a female rat, the hormone con-

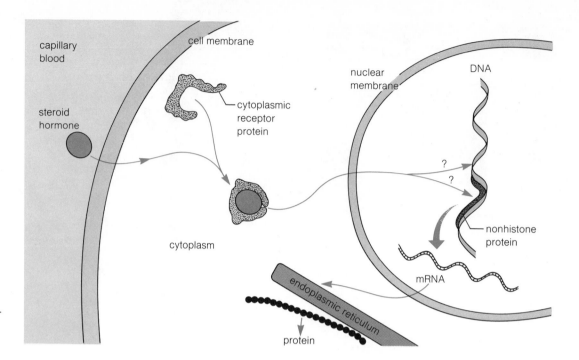

Figure 14–3 Model of the mechanism of action of a steroid hormone on a target cell.

centrates in those tissues most affected by the hormone, namely the uterus, ovaries, oviducts, and vagina. Similar experiments with other hormones and targets yield much the same result.

It is thought that the target tissue's "recognition" of the hormone resides in receptor sites, probably specific proteins, that bind the hormone to the surface of the tissue. Thus, thyroid cells have receptors specific for thyroid-stimulating hormone, and kidney cells for antidiuretic hormone (which reduces transport of water). Despite intensive experimental work, no receptor protein has been purified and chemically defined, but research continues.

This research has revealed two general patterns of hormone action. A pattern typical of the hormones that are proteins or derivatives of amino acids is thought to involve a receptor site on the cell's outer membrane and a secondary messenger—the cAMP and cGMP mentioned earlier. The binding of the hormone to its specific receptor site on the membrane activates the enzyme adenylate cyclase. This enzyme catalyzes the breakdown of an appropriate triphosphate, such as ATP, to the corresponding cyclic nucleotide (AMP). The secondary messenger (cAMP) then diffuses into the cytoplasm and either activates or deactivates the enzyme systems for which the cell is specialized (Figure 14–2). Earl W. Sutherland of Washington University won the 1971 Nobel prize in medicine and physiology for working out the general outlines of this model. Many of the specifics are still unknown.

The second mechanism of hormone action, the one typical of fatty acid and steroid hormones, bypasses the cell membrane. Radioactively labeled steroid hormones have been observed to enter the cell, where the hormone is recognized by cytoplasmic receptor proteins that bond with the steroid and move it into the nucleus. The hormone-protein complex interacts with the genome to affect RNA and then enzyme synthesis, by bonding with a specific acceptor site either on the DNA or on the proteins surrounding it. This model is outlined in Figure 14–3.

How Hormone Release Is Triggered The timing and quantity of hormone release are controlled by what can be called *neuroendocrine reflex arcs.* That is, a precise stimulus triggers an involuntary response in the nervous system, which in turn initiates endocrine activity by releasing a neurohormone.

In rabbits, copulation triggers the neuroendocrine reflex arc that leads to ovulation. The sensory stimuli of mating travel through the spinal cord to a region of the brain called the hypothalamus, which responds by releasing neurohormones called *releasing factors.* These neurohormones travel to the pituitary gland through a portal vein (a vein that travels from one capillary bed to another) and stimulate the secretion of pituitary hormones into the blood. Changes in the ovaries brought on by the hormones trigger ovulation within 10 hours of copulation. Since rabbit sperm survive for 24 hours, pregnancy is

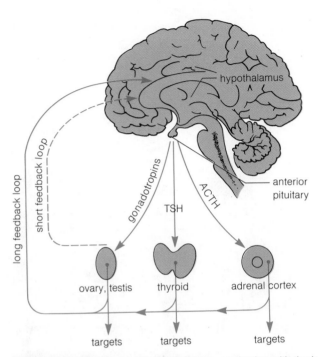

Figure 14–4 Feedback control in hormone production with the brain. The gonads, the thyroid, and the adrenal cortex are stimulated to secrete their respective hormones by chemical messengers (tropic hormones) from the anterior pituitary at the base of the brain. The hormones they produce act on specific target tissues but also circulate to the brain and repress the output of the tropic hormones. In this way (the "long" feedback loop), the output of the endocrine glands is regulated. In the "short" feedback loop (broken lines), the tropic hormones affect the output of the hypothalamus that produces them.

almost sure-fire, giving use to our stereotype "breeding like rabbits."

Certain external breeding stimuli are known as well. The African weaverbird and several Australian waterfowl breed only after heavy rains, an adaptation that ensures the young an adequate supply of water in generally arid habitats. Such seabirds as the elegant tern and Heerman's gull, which breed in colonies, do not reproduce until many birds are crowded together. Apparently, close social contact triggers the breeding response in these birds. The probable reason behind this colonial synchronization is that eggs laid much before or after the rest can be picked off more easily by predators. Natural selection thus favors a limited colonial breeding period.

In humans, cows, and other mammals, similar sensory stimuli trigger milk "let-down." The nervous stimulus of the young sucking on the nipple or teat is relayed through the brain to the pituitary gland, where neurosecretory cells release the hormone oxytocin. This hormone

causes cells around the milk-producing mammary alveoli to contract, squeezing the milk from them. This response, like many others, can be conditioned. In dairy cattle, the sights and sounds of milking time replace the sucking stimulus of the calf and release the milk. In addition, emotional states can affect the endocrine response. It is, in fact, true that "contented" cows give more milk. And women who are anxious about nursing their newborns find that their anxiety inhibits the very response they want to encourage.

Internal stimuli also trigger hormonal responses. For example, certain nerve cells in the hypothalamus monitor the osmotic pressure of the blood. If the pressure increases beyond a certain point, indicating a loss of water, the pituitary releases vasopressin, or antidiuretic hormone (ADH). The hormone acts directly on the kidney tubules, causing them to take up more water, and thus reduces losses to excretion (see Chapter 18).

How Hormone Levels Are Controlled There is much more to the working of a hormone than the timing of its release. Obviously it is also important to shut the hormone off when it is no longer needed. For example, the rabbit that has ovulated because of copulation will not benefit from the continuing release of ovulatory hormones. Similarly, continued production of ADH could actually be dangerous: Continued water uptake would thin the blood and so increase the amount of lymph leaking out into the tissues that severe and even fatal edema (tissue-swelling) might result. To control just such happenings, the body has evolved *feedback loops.*

To see how a feedback loop works, we will look at the interaction between thyroxine and body temperature in mammals. At low temperatures, nervous stimuli trigger a response in a temperature-regulation center in the hypothalamus. The hypothalamus secretes a neurohormone, thyrotropin-releasing factor, into a portal vein that connects it to the anterior pituitary gland. Stimulated by the neurohormone, the pituitary releases thyroid-stimulating hormone (TSH) into the bloodstream. When the TSH reaches its target organ, the thyroid, it triggers the release of thyroxine, a hormone that acts on the cells of the body to accelerate their rate of heat production. This increase in body heat decreases the stimulation relayed to the heat center in the hypothalamus and thus dampens the secretion of TSH and of thyroxine. This mechanism is called a long feedback loop because the response of the target organ is involved. There may also be a short feedback loop, in which some pituitary hormones (like TSH) act directly on the hypothalamus, without the target organ response being involved in the feedback loop, as shown in Figure 14–4.

In order to stay effective, a hormone must maintain a certain concentration in the blood. Therefore, the rate at which a hormone reacts and is deactivated and the rate at which it is excreted have much to do with the sum of its effects. It is generally to the body's benefit to have hormones that do not last too long, so they will not continue to build up after they have outlived their usefulness. Hormones are cleared out of the blood in the kidneys and excreted. Also, enzymes in the blood, liver, and kidneys degrade hormones chemically and render them inactive. As a result, hormones have short half-lives. Insulin, as an example, is destroyed by the enzyme insulinase in the liver and has a half-life of only 15 to 35 minutes in the systemic circulation (that is, half of the hormone is destroyed or excreted in 15 to 35 minutes). Half-lives generally vary from a few minutes for the protein hormones like insulin to a few hours for the steroids, which are carried in the blood bonded to carrier proteins that protect them against chemical degradation.

The Vertebrate Endocrine System

With our model of hormonal action in mind, we can survey the endocrine glands of vertebrate animals (Figure 14–5) and examine in more detail the hormones they produce. Table 14–1 lists the endocrine glands and their hormones.

Testes and Ovaries It has long been known that castrating a male animal—that is, removing its testes—changes it profoundly. (Removing a female's ovaries has drastic effects, too, but since the ovaries are hidden, they remained mysterious much longer.) Aristotle first recorded the effects of castration in the fourth century B.C., but that knowledge had long been part of the practical wisdom of herdsmen, who castrated pack animals to calm them and food animals to fatten them. Humans have even castrated their own kind. Middle Eastern potentates used eunuchs (human castrates) as harem guards, and in post-Renaissance Europe, where the church forbade women to perform in public, castrates supplied the soprano voices of choirs and theater companies. Castration has also been used as a punishment for rape, among other crimes.

Modern endocrinology began with castration experiments. In 1849, A. A. Berthold, a German physician and professor, removed the testes of immature roosters and observed, as Aristotle had, that the birds' combs atrophied. He then transplanted testicular tissue into the birds, and they grew normal combs. Obviously, the transplants caused the growth of the combs. The significant question was how. Knowing that the transplants had no nervous connections with any other tissues in the bird,

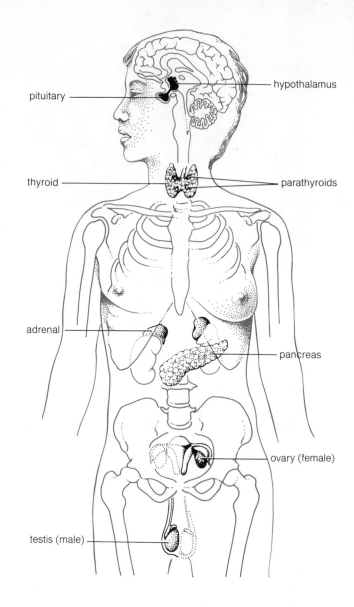

Figure 14–5 The major endocrine glands of a human being.

Berthold concluded correctly that they must have released chemical substances that stimulated the combs to grow normally.

The substances Berthold proposed are the sex hormones. Testes produce male sex hormones (androgens), and ovaries produce female sex hormones (estrogens). Developmentally, these hormones play a crucial role in the appearance and growth of sexual characteristics. In the human male, the androgens, particularly testosterone, promote the growth of a beard, the deepening of the voice, the growth and development of the genitals, and a muscular body build. Estrogens in the female lead to the growth of the breasts, build-up of fat in the hips

Table 14–1 Summary of Mammalian Hormones

Source	Hormone	Principal Effects
Testes	Androgen	Stimulates development and maintenance of male secondary sexual characteristics
Ovaries	Estrogen	Stimulates development and maintenance of female secondary sexual characteristics
	Progestin	Maintains pregnancy; stimulates development and maintenance of female secondary sexual characteristics
	Relaxin	Stimulates relaxation of maternal symphysis ligaments at birth
Thyroid	Thyroxine, triiodothyronine	Necessary for normal development and maintenance of nervous system; stimulates oxidative metabolism
	Calcitonin	Prevents excessive rise in blood calcium
Parathyroids	Parathormone	Regulates calcium-phosphate metabolism
Thymus	Thymosin	Stimulates immunologic competence in lymphoid tissues
Adrenal medulla	Epinephrine (Adrenalin)	Stimulates cardiovascular system; stimulates conversion of glycogen to glucose and increase in blood glucose
	Norepinephrine (Noradrenalin)	Similar in action to adrenalin; causes more vasoconstriction and less conversion of glycogen to glucose
Adrenal cortex	Glucocorticoids (corticosterone, cortisone, hydrocortisone, etc.)	Stimulate formation of glucose and storage of glycogen; help maintain normal blood-sugar level; inhibit incorporation of amino acids into proteins of muscles
	Mineralocorticoids (aldosterone, deoxycorticosterone, etc.)	Regulate Na-K metabolism, osmolality of tissue fluid
	Cortical sex hormones (adrenosterone, etc.)	Stimulate secondary sexual characteristics
Hypothalamus	Releasing factors	Regulate hormone secretion by anterior pituitary
	Oxytocin, vasopressin	See posterior pituitary
Anterior pituitary	Growth hormone	Influences incorporation of amino acids into protein; stimulates growth
	Thyrotropic hormone	Stimulates thyroid
	Adrenocorticotropic hormone (ACTH)	Stimulates adrenal cortex
	Follicle-stimulating hormone (FSH)	Stimulates growth of ovarian follicles and seminiferous tubules of testes.
	Luteinizing hormone	Stimulates conversion of follicles into corpora lutea; stimulates secretion of sex hormones by ovaries and testes
	Prolactin	Stimulates milk secretion by mammary glands; has variable effect on pituitary, gonads, sex accessories, secondary sex characteristics; also stimulates osmoregulation
Posterior pituitary	Oxytocin	Stimulates contraction of uterine muscles; stimulates release of milk by mammary glands
	Vasopressin	Stimulates increased water reabsorption by kidneys; at high doses stimulates constriction of blood vessels
Pineal	Melatonin	May help regulate pituitary
Pyloric mucosa of stomach	Gastrin	Stimulates secretion of gastric juice
Mucosa of duodenum	Secretin	Stimulates secretion of pancreatic juice
	Cholecystokinin	Stimulates release of bile by gallbladder and enzymes by pancreas
	Gastric inhibitory peptide (GIA)	Inhibits secretion of gastric juice
Pancreas	Insulin	Stimulates glycogen formation and storage; stimulates glucose transport into cells from blood
	Glucagon	Stimulates conversion of glycogen into glucose; stimulates increase in blood glucose

and breasts, and the maturation of the vagina and uterus. They are also involved in the menstrual cycle (see Chapter 12). Many of the characteristics of one sex can be readily induced in the other by injecting the appropriate hormone. Estrogens and androgens are chemically similar. They are classified as steroids, characterized by a backbone of carbon atoms in four rings. An androgen, testosterone, and an estrogen, estradiol, are shown in Figure 14–6.

Sex hormones affect behavior as well as physical characteristics. In most animals they appear to be crucial to the sex drive and to other sex-related behavior. Female canaries do not normally sing, but if given testosterone, they will begin to sing like males. If a genetically female rat, guinea pig, or monkey is given a small amount of testosterone during a critical period just before or after birth, the animal will behave like a male as an adult. Similarly, depriving a genetically male animal of androgens at this same period produces an adult with characteristically female behavior. We noted earlier that sex hormones are known to concentrate in the reproductive organs. The same sort of concentration occurs in regions of the hypothalamus known to play a role in sexual behavior, and this concentration probably alters the nervous pathways permanently and influences behavior.

As to how these phenomena apply to humans, little is known. It is obvious that "male" behaviors occur commonly among females, and vice versa, but the biological reason for our apparent difference from other mammals is not known. Human sexuality is, however, as much cultural as it is physiological.

Sometimes sex hormones have the unfortunate effect of promoting the growth of abnormal or tumorous tissue. Certain types of breast cancer depend on estrogens, and prostate cancer may be advanced by a steady supply of testosterone. Treating such cancers may entail removing the ovaries or testes. This therapy often brings about a remission in the cancer, but the tumor may also mutate into a hormone-independent form and continue growing.

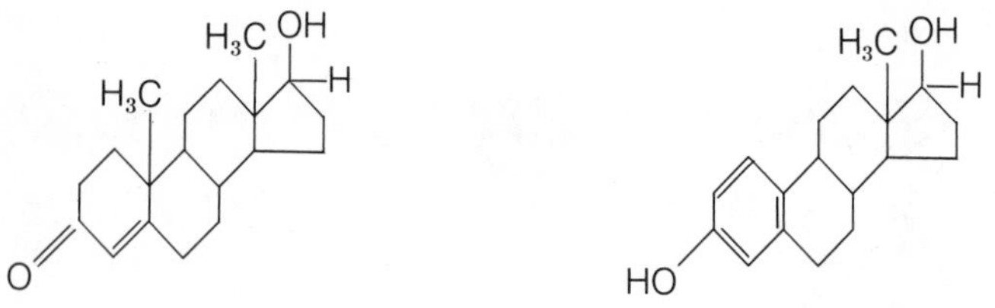

Thyroid   The paired **thyroid glands** found in many vertebrates have fused in humans to become a single organ located in the front of the throat under the larynx. The thyroid produces **thyroxine,** a simple hormone composed of an amino acid with four atoms of iodine (Figure 14–7). The thyroid also produces *triiodothyronine,* which is thyroxine minus one of the iodines. Its effects are the same as thyroxine's, only more pronounced, and it is produced in smaller amounts.

Thyroxine and triiodothyronine affect the body cells as a whole, speeding their overall metabolism. Overproduction of the thyroid hormones (hyperthyroidism) increases body temperature and blood pressure, with resulting weight loss and muscular weakness. Underproduction

Figure 14–6   Chemical structures of the natural male sex steroid testosterone and the natural female hormone estradiol.

of the hormones (hypothyroidism) causes obesity and lethargy. The thyroid hormones are particularly important during development, and children are therefore most susceptible to the effects of hypothyroidism. Left untreated, such children develop cretinism, a condition in which growth, sexual development, and intelligence are retarded. Fortunately, synthetic thyroxine has been available since 1927 and can be administered to children with thyroid problems.

Thyroxine has particularly dramatic developmental effects in frogs. The German biologist Friedrich Gudersnatch in 1912 discovered that when he fed parts of several kinds of frog body organs to tadpoles, those eating the thyroid gland seemed to start metamorphosis first. A series of experiments followed in which thyroids were removed from tadpoles; those later supplied with thyroxine metamorphosed; the others did not.

These experiments were continued by William Etkin at the City College of New York. Etkin found that if he administered a slight amount of thyroxine, legs developed, but later events, such as resorption of the tail, were slowed. A large dose caused the later events to take place on schedule, but leg development was so fast that the limb developed abnormally. A normal rate for both kinds of development could be achieved by starting with a low dose, then increasing it until it was 20 times as concentrated at the end of metamorphosis as at the beginning.

Etkin correlated this finding with the structure of the thyroid gland itself. Before metamorphosis, it is small and has few cells; these few are inactive in producing thyroxine. Throughout metamorphosis the thyroid grows, becoming largest and most active at the end of metamorphosis. The growth of the thyroid is itself under the control of a hormone produced by the pituitary. This is **thyroid-stimulating hormone** (TSH), which at the beginning of metamorphosis causes the thyroid gland to produce thyroxine. This hormone in low doses starts metamorphosis. Its release is mediated by thyroid releasing factor (TRF) secreted by the hypothalamus. The pituitary becomes sensitive to thyroxine at about this time,

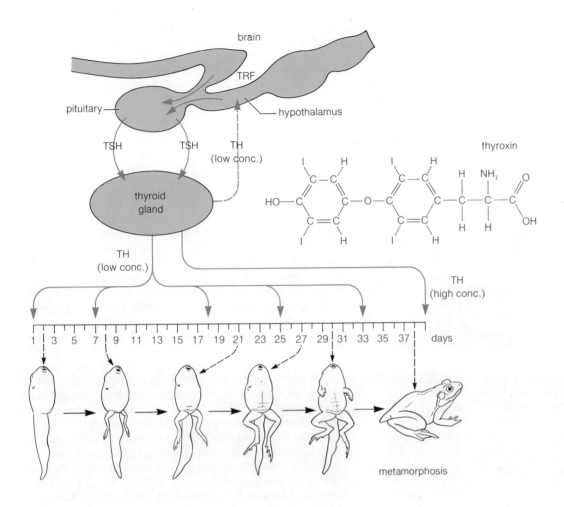

Figure 14-7 Hormonal interrelations in amphibian metamorphosis.

so that growth of the thyroid and its secretion are regulated. These interrelations are shown in Figure 14–7.

The word *goiter* describes a thyroid so enlarged that it disfigures the neck. Goiter used to be common in animals and humans in certain areas of the world, particularly in inland regions. In 1905, David Marine of Western Reserve University discovered that goiter was caused by a lack of iodine in the diet. When iodine is deficient, thyroxine production falls off, and the anterior pituitary releases large amounts of thyroid-stimulating hormone into the blood. This hormone stimulates the thyroid cells to grow and divide as a way of raising the level of thyroxine, and the result is goiter. Today few Americans suffer from goiter because table salt contains added iodine and makes up any dietary deficiency.

Parathyroids The human **parathyroids** are four identical, pea-sized organs located on the rear surface of the thyroid. These glands produce a protein hormone called *parathormone,* or *parathyroid hormone* (PTH), which regu-

lates the levels of calcium and phosphate in the blood. This control is necessary for the proper working of nerves and muscles. Even minor variations of Ca^{2+} concentrations profoundly affect neuromuscular activity and the function of various enzyme systems.

PTH maintains a high level of calcium in the blood by promoting the release of calcium from the bones, the principal reservoir of the mineral in the body, and by inhibiting the excretion of calcium from the kidneys and intestines. Vitamin D adds to the effect of PTH by promoting the absorption of calcium from the intestines.

Overproduction of PTH weakens and erodes the bones by withdrawing excess calcium. Insufficient PTH changes ion concentrations in fluids surrounding muscles and nerves, producing high irritability, convulsions, and eventually death.

Pancreas The **pancreas** releases digestive enzymes into the small intestine through a small duct. In the late 1880s, Johann von Mering and Oskar Minkowski studied the

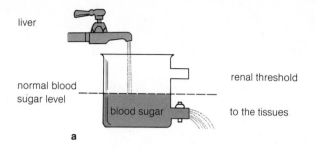

liver

normal blood
sugar level

blood sugar

renal threshold

to the tissues

a

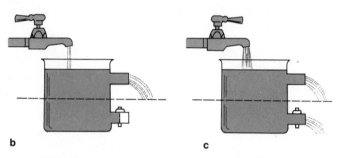

b

c

Figure 14–8 Mechanical analogy illustrating normal glucose regulation (**a**) and two ways the renal threshold for glucose excretion may be exceeded. (**b**) The utilization of glucose is impaired. (**c**) The liver is unable to store glucose or glycogen. Diabetes probably involves a combination of defective metabolism in both the tissues and the liver.

digestive functions of the pancreas by removing the organ from dogs and observing the effects. One thing they noticed was that ants were attracted to the urine of the experimental dogs. Such behavior was abnormal, and von Mering and Minkowski sought a reason for it. The urine, they found, contained sugar, as does the urine of humans suffering from the disease diabetes mellitus. Investigating further, the two scientists tied off the pancreatic ducts of other dogs. No sugar appeared in the urine of these animals. Apparently, the pancreas was releasing directly into the blood a substance that controlled sugar metabolism.

It was not until 1922 that this substance was identified as **insulin** by F. Banting and C. H. Best of the University of Toronto. Insulin is produced by clusters of pancreatic cells called the islets of Langerhans, after the anatomist who discovered them. These cells represent only 1 or 2 percent of the total weight of the pancreas.

Insulin is one of the simplest proteins, consisting of two polypeptide chains. In fact, insulin is the first protein whose primary structure was elucidated. Insulin affects

almost all cells by increasing their use of glucose. It also stimulates liver and muscle cells to increase the synthesis of glycogen and lipid and to inhibit the breakdown of these same molecules. The net effect of insulin's hormonal action is to lower blood glucose, helping to keep it in the normal human range of 60 to 100 mg per 100 ml of blood.

The immediate cause of diabetes mellitus is too little insulin. As a result of low levels of insulin in the blood, glucose levels in the blood rise. If sugar in the blood exceeds the renal threshold (the level at which the kidney is unable to reabsorb all of the glucose in the filtrate) of 160 to 180 mg per 100 ml of blood, the excess glucose is excreted from the kidney in the urine (Figure 14–8). Unable to use sugar, the diabetic's body consumes its stores of protein and fats, turning them into the very glucose it cannot use. As a result, diabetics literally waste away. They are also continuously thirsty, since they need an immense quantity of water to excrete the nitrogen from protein metabolism, the ketones from incompletely oxidized fats, and the huge amounts of glucose.

Diabetes can be treated with daily injections of insulin. However, the hormone must be used judiciously, since too much can lower blood glucose dangerously, producing tremors, lapses of consciousness, and coma. Insulin overdose can be treated by giving sugar orally or intravenously or by administering epinephrine to spur the conversion of stored glycogen to glucose.

The islets of Langerhans contain alpha cells and beta cells. The beta cells produce insulin. The less numerous alpha cells secrete a second hormone, called *glucagon*. Glucagon, also a protein, increases blood sugar and stimulates the breakdown of glycogen in the liver. Glucagon is therefore an antagonist to insulin, as are several hormones from the adrenal glands, such as epinephrine. Blood-sugar level is a good example of an internal condition regulated by the interaction of several antagonistic hormones.

Adrenal Glands The two human **adrenal glands** sit atop the kidneys like pyramidal caps. Although each adrenal appears to be but one organ, it is in fact two: an inner region, or *medulla,* and an outer covering, or *cortex.* The medulla and cortex arise from different tissues in the embryo which migrate together but perform completely separate endocrine functions in the adult. In some fishes, they are fully separate organs. In birds and reptiles, the tissues are combined, but their functions remain distinct.

The adrenal medulla secretes the hormones *epinephrine* (trade name, Adrenalin) and *norepinephrine,* or Noradrenalin. These hormones are both derivatives of the amino acid tyrosine (see Chapter 2). The two hormones have similar effects: increasing heartbeat and blood pressure, constricting skin capillaries, erecting the hair, and decreasing blood flow to the intestine. Epinephrine has the added effect of elevating blood sugar, thus acting as an antagonist to insulin, and of dilating the arteries to the muscles, liver, and brain, thus increasing the blood flow to these organs. The overall effect of the two medullary hormones is to ready the body to meet stress in the fight-or-flight response mentioned earlier. However, experimental animals subjected to predators can survive even when the whole adrenal medulla has been removed.

The cortex of the adrenals produces an array of hormones entirely different from those of the medulla, and far more numerous. At least 50 different cortical hormones or hormonelike substances have been isolated to date. Only about 10 of them are known to be active in the body, but some of the remaining 40 may be precursors of the others. All 50 are steroids.

The cortical hormones are divided into three groups according to the nature of their activity. The first group, the mineral corticoids, regulates salt and water balance by acting on the kidney tubules. This group includes the hormone aldosterone, which triggers increased uptake of water and sodium in the kidney.

The second group, the glucocorticoids, includes some of the best-known cortical hormones, such as cortisone, hydrocortisone, and corticosterone. These hormones regulate carbohydrate and protein metabolism, inhibiting the synthesis of proteins from amino acids and diverting the acids instead into the production of carbohydrates, including glycogen in the liver, a process known as *gluconeogenesis.*

The glucocorticoids are well known because they are so widely used in medicine. They prepare the body to handle stress or trauma, and they sometimes produce a state of euphoria or well-being, in addition to being anti-inflammatory agents. Veterinarians regularly give glucocorticoids intravenously to dogs bitten by poisonous snakes; the hormones reduce the inflammation, restrict damage, and make the suffering animal feel much better. Recently two physicians in San Francisco reported that a synthetic hydrocortisone called betamethasone can save some premature babies. Such infants often die because their underdeveloped lungs cannot yet handle the strain of breathing air, but betamethasone administered to the mother before birth crosses the placenta into the infant and prepares its body for the stress ahead. Cortisone is used to treat a great number of maladies, including arthritis, muscle and joint injuries in athletes, and severe allergic reactions, all of which are variations of the inflammatory response.

The hormone's effects on these conditions remain mysterious. No one knows precisely why injecting cortisone into the pulled thigh muscle of a football player should reduce the pain. Moreover, cortisone should not be seen as a magic potion, since it may have serious side effects. Taken over a long time, cortisone makes bones brittle and easily broken, greatly slows the healing of wounds, and reduces natural resistance to infection.

The third group of cortical hormones functions as sex hormones. Both androgens and estrogens are produced, but androgens predominate. In both men and women, these hormones add to the effects of the hormones produced in the gonads. Like the testicular androgens, the adrenal androgens are also responsible for some of the physical changes, such as a deepened voice and increased facial hair, seen in women who have malfunctioning adrenals.

Pituitary Gland The **pituitary,** or **hypophysis,** is a small gland suspended on a stalk from the part of the brain called the **hypothalamus** (Figure 14–9). Like the adrenals, the pituitary is really two organs distinct in origin and in function.

The *posterior lobe* of the pituitary, or *neurohypophysis,* actually an outgrowth of the hypothalamus, arises in the embryo as an outpocketing of the brain. The posterior lobe stores and releases two hormones, oxytocin and vasopressin. In females, oxytocin stimulates the muscles of the uterus to cause labor at childbirth. Sucking stimuli also trigger the release of oxytocin, which in turn initiates the milk "let-down" response mentioned earlier. Vasopressin, also known as antidiuretic hormone (ADH), raises blood pressure by constricting certain kidney arterioles (see Chapter 18) and reducing water excretion in the distal portion of the kidney tubule. In the frog, ADH also promotes the reabsorption of water from the bladder and skin.

Both these hormones are merely released by the posterior pituitary, not produced there. They are synthesized in the hypothalamus, transported on nerve axons to the posterior pituitary, and stored there until needed. Thus, vasopressin and oxytocin are fittingly called neurohormones.

The *anterior lobe* of the pituitary, or *adenohypophysis,* although intimately associated with the posterior lobe, arises embryologically not from the brain but from the roof of the pharynx. It is an extremely important endocrine gland. Since it produces several hormones that regulate the functions of other endocrine glands, it was once commonly called the "master gland." We know now, however, that the adenohypophysis is itself controlled by the brain.

The anterior pituitary's several hormones, all proteins or glycoproteins, have a variety of side effects. Thyroid-stimulating hormone (TSH) stimulates the thyroid gland to produce thyroxine, and adrenocorticotropic hormone (ACTH) stimulates the adrenal cortex to produce its hormones. The two gonadotropic hormones, as their name implies, affect the gonads. Follicle-stimulating hormone

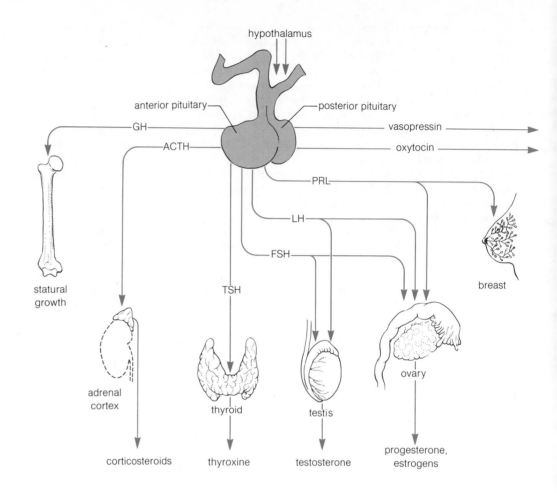

Figure 14–9 The pituitary gland, its hormones, their target glands, and the effects or products of these target tissues.

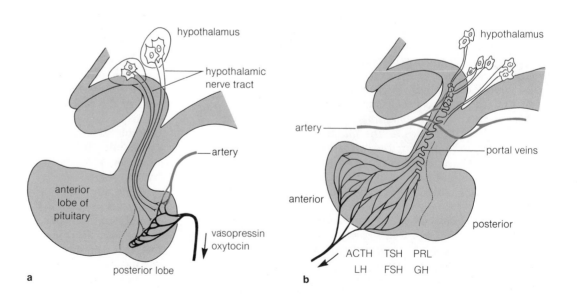

Figure 14–10 The release of hormones from the posterior pituitary (**a**) and the anterior pituitary (**b**).

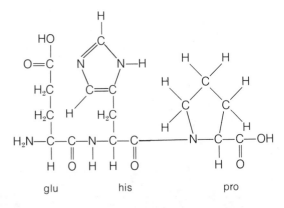

Figure 14-11 The amino acid content of TRF, the releasing factor for thyroid-stimulating hormone (TSH), is glutamic acid, histidine, and proline in equal proportions.

(FSH) controls the development and maturation of germ cells, and luteinizing hormone (LH) controls the gonads' production of sex hormones and triggers ovulation in the female (for details, see Chapter 12). Three additional hormones produced by the anterior pituitary are prolactin, or lactogenic hormone (PRL), which stimulates the production of milk in the mammary gland; growth hormone, or somatotropic hormone (STH), which promotes body growth; and melanophore-stimulating hormone (MSH), or intermedin, whose function in mammals, which lack the skin-darkening melanophores of lower vertebrates, is unknown.

The functions just given for these hormones actually constitute only a small fraction of their effects. In many cases, these pituitary hormones act together with each other, producing a wide variety of effects. For example, prolactin, best known for its role in stimulating the milk glands of mammals and the crop glands of pigeons and doves, has also been identified as part of at least 40 other functions, including metabolism, growth, energy metabolism, osmoregulation in the skin of fishes, and such behaviors as reproduction and parental care. Prolactin overlaps considerably with STH in function.

The anterior pituitary hormones affect the body so basically that any imbalance in production can have disastrous effects. Take STH as an example. Too little of this in a child, and the child becomes a midget; too much, and the child becomes a giant. Adults need only a limited amount of STH for the proper activity of many tissues. Too much causes certain bones to grow after others have stopped, producing a condition known as *acromegaly.* Acromegalics have disporportionately large lower jaws, noses, eyebrow ridges, hands, and feet.

Unlike the posterior pituitary hormones, the anterior pituitary hormones are not neurohormones produced in the brain. They are synthesized within the anterior lobe. However, these hormones are produced and released in response to neurohormones from the brain. No nerve cells connect the brain and the anterior pituitary, but extracts of tissue from the hypothalamus can stimulate the secretion of the anterior pituitary hormones. Obviously some chemical or chemicals in the extract trigger the secretion. In 1945, G. W. Harris proposed that these chemicals travel from the brain to the anterior pituitary through the bloodstream. Endocrinologists now agree. The neurohormones are believed to travel throuch the portal vein, which connects the capillary beds of the hypothalamus and the anterior pituitary (Figure 14–10).

Recently two of these brain-to-pituitary hormones have been isolated and synthesized. The first was identified by two separate research teams: A. V. Schally and his colleagues at the Tulane University School of Medicine, and Roger Guillemin, Roger Burgus, and their research group. Their task was staggering. Schally and his group made extracts from 2 million pig brains, and Guillemin and Rogers processed 5 million sheep brains—500 tons—to get 1 milligram of relatively pure thyroid-stimulating hormone releasing factor (TRF). In 1969, Guillemin and Burgus learned that its structure consists of only three amino acids: glutamic acid, histidine, and proline (Figure 14–11). Sheep and pig TRF were found to be identical and to be just as active in humans as in pigs and sheep. As soon as its structure was known, TRF could be synthesized in unlimited amounts. Shortly thereafter, in 1971, luteinizing hormone releasing factor (LRF) was isolated, purified, and synthesized. Guillemin and Schally received the Nobel prize in 1977 for their research.

This research—which still goes on—is already proving medically significant. Pathological conditions long thought to originate in the pituitary may actually be caused by abnormalities in the hypothalamus. In the past, these conditions were treated by administering the deficient pituitary hormones, but they can now be corrected by providing the appropriate hypothalamic releasing factor. Also, the newly understood importance of the hypothalamic releasing factors may lead to the development of a safer and even more effective birth-control pill.

Control of the anterior pituitary does not end with the hypothalamus. The hypothalamus itself is controlled by another region of the brain, the **thalamus,** which coordinates information from both internal and sense organs. Neurosecretions from the thalamus affect the hypothalamus, which in turn affects the pituitary, whose secretions stimulate the other endocrine glands, which in turn affect the hypothalamus through the feedback loops described earlier.

Pineal Gland In some primitive vertebrates, like lizards, the **pineal gland** is an eyelike structure that monitors

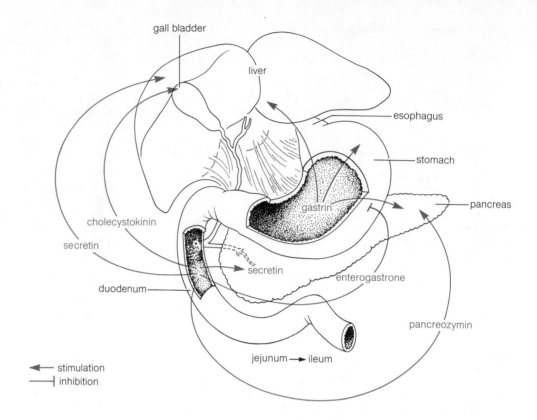

gall bladder

liver

esophagus

stomach

pancreas

gastrin

cholecystokinin

secretin

secretin

enterogastrone

duodenum

pancreozymin

jejunum → ileum

Figure 14-12 The source and action of the gastrointestinal hormones. The hormones are produced entirely within the digestive tract, and they act there; but they are transported by the systemic blood circulation.

← stimulation
⊣ inhibition

light. When enough light falls on the pineal to stimulate it, the gland secretes the hormone melatonin, which concentrates the pigment granules in the skin pigment cells, or melanophores, increasing the distance between pigmented areas and lightening the animal's overall color.

In mammals, the pineal is something of an enigma. The mammalian pineal does not respond directly to light but apparently receives information about light conditions, namely day length, relayed from the brain along nervous pathways. The pineal then translates this information into melatonin production or release. Mammals, however, as mentioned earlier, lack the melanophores affected by melatonin in lower vertebrates, and the hormone apparently serves other purposes. Melatonin is thought to inhibit the anterior pituitary's output of gonadotropic hormones, and it may also play a role in regulating other patterns of behavior besides reproduction. Beyond this limited bit of knowledge, the pineal remains mysterious.

Gastrointestinal Hormones Digestion is regulated not only by the autonomic nervous system, but also by a number of hormones produced in localized areas of the digestive tract. Essentially, these hormones act as a signal sent from one organ to another to trigger some needed digestive function in the target organ (Figure 14-12).

When food enters the stomach, the internal mucosa of the stomach releases **gastrin** into the blood. When the gastrin reaches the cells of the upper stomach through the circulatory system, it stimulates the secretion of hydrochloric acid. Later, the passage of the partially digested material from the stomach into the duodenum triggers the release of four duodenal hormones. **Secretin** stimulates the flow of digestive juices from the pancreas and the liver. **Pancreozymin** also stimulates the pancreas. **Cholecystokinin** stimulates the gallbladder to release bile. **Enterogastrone,** whose secretion is triggered by the presence of fats in the digested material, inhibits the production of hydrochloric acid in the stomach. All these hormones are short-chain proteins, and two of them, gastrin and secretin, have been sythesized.

Secretin deserves a special place in endocrinology's hall of fame because it was the first chemical regulator demonstrated to work without the nervous system. Secretin was first identified by Bayliss and Starling in 1902. Looking for a proper term to describe the physical activity of their discovery, they invented the term *hormone* from the Greek *hormon,* "to excite."

Other Endocrine Glands Like the gastrointestinal system, both the placenta and the kidney produce hormones, besides performing their primary functions. The placenta produces a variety of hormones, including estrogens and progesterone, that affect the metabolism of the pregnant female. Another placental hormone, human chorionic gonadotropin (HCG), is produced in large amounts from roughly the fifteenth day of pregnancy until a little after the second month. Its physiological effect is similar to that of LH, and it maintains the corpus luteum, which continues to produce progesterone. HCG is excreted in the urine, so one way to tell whether a woman is pregnant, particularly before more obvious symptoms appear, is to analyze the urine for HCG. HCG has also earned a certain measure of notoriety for its use by quick-weight-loss clinics. Supposedly, daily injections of the hormone combined with a low-calorie, fat-free diet cause the body to consume its own fat reserves and shed weight rapidly. However, little or no medical evidence supports this claim.

The kidney produces a pair of hormones that affect the workings of the circulatory system. Usually, new red blood cells are produced at about the same rate that aged red blood cells are processed out of the blood in the liver and spleen. Sometimes, however, production falls behind. The resulting oxygen deficiency stimulates the kidneys to release an enzyme which, in combination with a plasma protein, forms erythrogenin, a hormone that stimulates the bone marrow to synthesize red blood cells. The second hormone, renin, is released whenever specialized cells in the renal (kidney) arterioles detect that blood pressure has dropped below a certain threshold. The renin converts a protein released by the liver into angiotensin, a hormone that constricts arteries, raises blood pressure, and signals the adrenal cortex to release aldosterone, which constricts the blood vessels still more and raises blood pressure even further.

The **thymus gland,** located in the center of the chest near the lower end of the **trachea** (the tube that leads down toward the lungs), is another of the body's mystery organs. Large in childhood, the thymus shrinks with age, apparently growing less crucial to the body's normal functioning. If the thymus is removed surgically from a newborn mammal, the animal will die of infections within a few weeks. But a juvenile or adult animal subjected to the same surgery survives with little ill effect. The thymus apparently plays a role in developing the immune responses of the growing animal, and there is some evidence that it produces a hormone essential for the development of lymphocytes (see Chapter 17). Little more than this meager bit is known about the workings of the thymus, and it is the object of continuing research.

The Frontiers of Endocrinology

The thymus is hardly the last unknown tile in the mosaic of the body's endocrine functions. Recent research has uncovered several new groups of hormones or hormonelike substances that are still poorly understood.

One group of little-known chemical messengers is the *chalones.* These proteins, which inhibit cell mitosis, have been studied most in the epidermal tissues, where they apparently serve as growth regulators and as important agents in healing and regeneration. A lack of chalones stimulates cell division and growth. For example, when a portion of the liver of a rat is removed, the level of serum proteins—including chalones—drops, stimulating cell division. In time, the liver actually replaces all the removed tissue. Since a lack of chalones causes cells to proliferate, an abundance of these proteins should stop cell division. If this supposition proves true, the chalones may prove an effective way of treating cancer. Cancer cells proliferate at rates hundreds and even thousands of times greater than those of normal cells. Giving a cancer patient chalones might arrest the growth of the tumor before it has a chance to do fatal damage.

Damaged tissues produce the hormonelike chemical *histamine.* Histamine dilates blood vessels in and around the damaged area, making the vessels leaky and allowing fluid into the tissues to swell them. The precise benefits of this defense mechanism are not certain, but some researchers feel that it enables white blood cells and antibodies to escape from the vessels more easily and travel into the tissues where they are most needed.

Histamines are, however, best known not for their role in healing, but for their role in allergies. In an allergy victim, the localized immune response to the allergen (the sensitizing agent) destroys not only the allergen but also some of the cells in the tissue where the reaction occurs. As a result, histamine is released in large amounts. In the case of hay fever, the swelling and fluid loss in the nose, throat, and eyes produce symptoms much like those of a cold. Continued release of histamine gradually constricts the bronchioles of the lungs and may cause asthma. Therefore, allergic reactions are often treated by administering antihistamines, compounds that counteract the histamine response.

Since the 1930s, semen has been known to contain substances that cause uterine tissues to contract and that, when injected, lower blood pressure markedly. On the assumption that these substances arise in the prostate gland, they were called *prostaglandins.* As it turned out, these semen-borne prostaglandins actually come from the seminal vesicles. Further research has shown that prostaglandins are produced by most, if not all, cells in the body.

The prostaglandins are a unique group of hormonelike chemicals. To begin with, they are fatty acids, the only fatty acids known to have hormonal effects. Prostaglandins are produced in minute quantities and probably only as needed, since they degrade quickly and are hard to store. The chemical precursors are found in the cell membrane, but the prostaglandins are probably synthesized inside the cell.

Research on the effects of these chemicals has centered on prostaglandins synthesized in the laboratory or extracted from sea corals, an abundant source. Tiny amounts of prostaglandins have been shown to have dramatic and varied clinical effects. They cause smooth muscle to contract. Most likely, the prostaglandins in semen stimulate contractions of the uterine wall that hurry the sperm cells through the uterus and into the Fallopian tubes in search of an ovum to fertilize. They also lower blood pressure and affect blood flow. In dogs and cats, they inhibit gastric secretions and prevent peptic ulcers. They have been used medically to promote labor contractions in women during childbirth, as well as to induce abortion. The prostaglandins are also thought to play a role in controlling transmissions in the sympathetic nervous system by slowing the release of norepinephrine and to affect the action of all other hormones, possibly by activating adenylate cyclase, the enzyme that catalyzes the synthesis of cAMP.

It is more or less obvious that substances with effects as powerful as those of the prostaglandins play an important physiological role. However, little is known about that role, and research on this group of intriguing chemicals continues.

Chemical Messengers among Individuals

Vertebrate hormones serve largely to bind the body into a working whole, to make sure that the right hand knows what the left hand is doing and that they do it together.

In a similar fashion, chemical messengers among individual organisms that live in colonies or aggregates guarantee the precise coordination of the group.

cAMP and the Cellular Slime Molds   The cellular slime molds are fungi that inhabit moist soil and feed on bacteria. Typically, these molds occur as single, free-living, amoeboid organisms; but once an area has been cleared of bacteria and food becomes scarce, the individual slime mold cells begin to aggregate in a sluglike mass of 100,000 or more individuals. The "slug," which responds to light and heat as if it were a single organism, migrates to a new feeding area, leaving a characteristic trail of slime. After the slug reaches its destination, it stands on end and forms a stalk. The cells on the very pinnacle of the stalk become spores, which diffuse out from the stalk and fall onto the soil. If conditions are right, the spores germinate and develop into the amoebae that are the first stage of the cycle (Figure 14–13).

The prime scientific question about this remarkable life cycle is: How do thousands of free-living cells coordinate themselves into a single mass? For many years, John Tyler Bonner and his associates at Princeton University have studied this question. They observed that free slime mold amoebae approached the aggregating mass only if they were downstream from it. Thus, whatever signaled aggregation was apparently water-borne.

This chemical signal has now been isolated and identified as cAMP. Curiously, the aggregating cellular slime mold cells also produce an extracellular enzyme that destroys cAMP. Since cAMP is the stimulus to aggregation, one wonders what function could be served by an enzyme that destroys the stimulus itself. It is now thought that this seemingly contradictory way of doing things actually

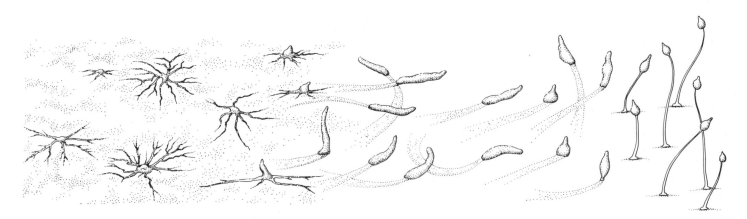

Figure 14–13   Life cycle of the cellular slime molds.

aids recognition. The small cAMP molecules diffuse rapidly away from the cell that secretes them, while the larger enzyme molecules diffuse slowly and thus remain in the vicinity of the cell. As a result, most of the cAMP released by a cell diffuses away, and any that stays behind is destroyed. This increases the concentration gradient for cAMP into the cell. When another cAMP-secreting cell is nearby, its cAMP diffuses inside rapidly and draws the second cell upstream toward it. The effect is to gather all the free-living cells together faster than would be the case if they secreted only cAMP.

Pheromones Pheromones are chemical messengers released by one individual of a species that elicit a specific behavior in another individual of the same species. The term covers many kinds of chemicals that elicit many kinds of behavior in many kinds of animals.

Humans are pheromonal lightweights compared to social insects. Just as hormones bind the vertebrate body into a functional whole, pheromones make a termite mound, an ant colony, or a beehive a single entity. Among their many functions, they serve as trail markers, alarm signals, sex attractants, and cues in caring for the brood, regulating the temperature of the nest, and telling living nestmates from dead. Insect pheromones may be passed orally from one individual to another, or they may be volatile chemicals carried by the wind to target organisms kilometers away. Like the water-borne cAMP of the cellular slime molds, wind-carried pheromones can work only on insects that are downwind.

Summary

This chapter is primarily about hormones—the chemical messengers that regulate the body's responses to a variety of internal and external stimuli. The role of hormones in development and morphogenesis in plants and animals was discussed in Chapters 11 and 12.

Functionally, the nervous and endocrine systems are closely integrated. Some neurons produce hormones, and the hormones, in turn, stimulate other specialized tissues to produce their own products, including other hormones. Most hormone levels are controlled within narrow concentrations and are produced only in response to specific conditions. In this chapter we discussed the interactions between the nervous and the endocrine systems, the regulation of hormone levels, and the probable modes of hormone action. We then surveyed the various endocrine organs and their products in vertebrate animals, using man as a model. The role of chemical messengers among individuals, rather than among cells and tissues, was briefly discussed.

Supplement 14–1
Metamorphosis

Metamorphosis occurs in both insects and amphibians. In insects it is divided into two categories—complete and incomplete (Figure 14–14). Moths are said to undergo complete metamorphosis—the immature form is strikingly different from the adult, and usually four stages

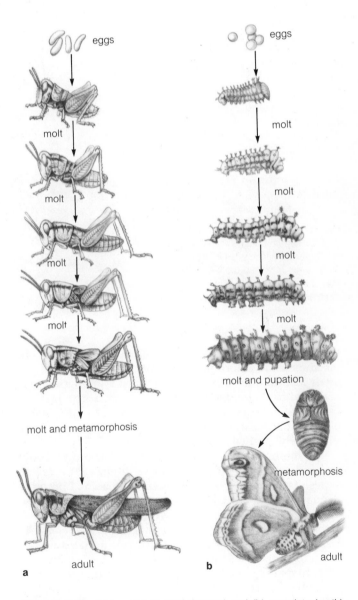

Figure 14–14 (**a**) Incomplete (grasshopper) and (**b**) complete (moth) metamorphosis.

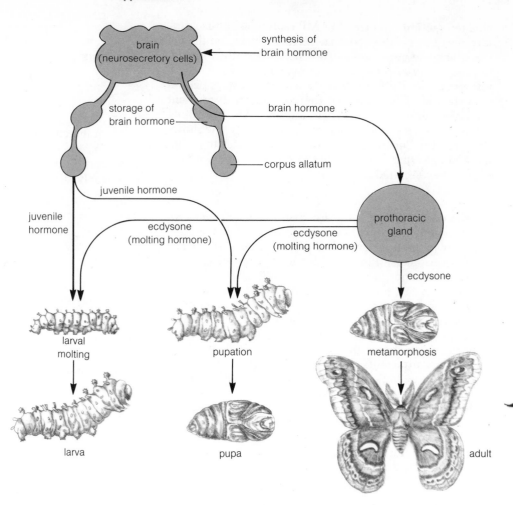

Figure 14–15 Complete metamorphosis of the moth *Platysamia cecropia* showing the relationship of the three major hormones involved in metamorphosis of the moth and other insects.

(egg, larva, pupa, and adult) occur in the life cycle. In incomplete metamorphosis, by contrast, the immature animals look like the adults but are smaller and lack wings and mature reproductive systems. Relatively little is known about the mechanisms involved in incomplete metamorphosis, but complete metamorphosis has been intensively studied.

In moths, the larva is a caterpillar that moves about and actively feeds on plant material. It grows through a series of *molts*—a process by which the animal becomes detached from the old hard **cuticle** that surrounds it. While the new cuticle is still soft, the animal can increase in size. But the cuticle soon hardens, and to increase in size any further, the larva must molt again. After a series of molts, the larva spins a fibrous cocoon around its body and enters the stationary pupal stage. Inside the cocoon, the pupa undergoes a drastic biochemical and morphological reorganization. Larval structures are destroyed, and new adult structures develop from "reserve" cells. Finally

the adult emerges, complete with wings, three pairs of legs, antennae, and remodeled eyes, head, and body. The adult moth has an entirely different way of life from that of the caterpillar. It flies about and usually feeds on nectar by extending its long tongue into the bases of flowers. Rather than cutting off pieces of leaf to grind up for food as the larva did, the adult moth drinks concentrated sugar solutions (or it may not eat at all). These several stages and their hormonal interactions are shown in Figure 14–15.

The model for the control of complete metamorphosis that has emerged from many studies is as follows: Molting is triggered by the release of brain hormone that activates the prothoracic glands to produce ecdysone (molting hormone). If the amount of juvenile hormone, produced by the corpora allata (a pair of bulblike glands connected to the brain) in the blood is high, a larval molt will occur. If the amount is low, an adult will be formed. In the adult insect, the prothoracic glands degenerate so that no more molts occur. However, the corpora allata are reactivated in the adult, and the juvenile hormone produced takes on an entirely different function. In the adult, it functions in reproduction, notably egg maturation.

During amphibian metamorphosis, thyroxine affects different tissues as the amount of hormone in the bloodstream changes and as particular tissues respond to thyroxine in different ways. For instance, hindlimbs begin to develop well before forelimbs, and the tail is resorbed as the lysosomes in tail cells increase their activity in response to thyroxine. These changes can be seen in Figure 14–7, which shows frog development. The conversion of the excreta from ammonia to urea and the changes in the hemoglobin and in the visual pigments are all responses to thyroxine. This one hormone affects almost every body cell, but each kind of cell has a different response at a different threshold.

Suggested Readings

Books

Arey, L. B. 1965. Developmental Anatomy, 7th ed. Saunders, Philadelphia.

Austin, C. R. 1968. Ultrastructure of Fertilization. Holt, Rinehart and Winston, New York.

Balinsky, B. I. 1975. An Introduction to Embryology, 4th ed. Saunders. Philadelphia.

Ehert, J. D., and S. M. Sussex. 1970. Interacting Systems in Development, 2nd ed. Holt, Rinehart and Winston, New York.

Galston, Arthur W., and Peter J. Davies. 1970. Control Mechanisms in Plant Development. Prentice-Hall, Englewood Cliffs, N.J.

Jensen, William A., and Frank B. Salisbury. 1972. Botany: An Ecological Approach. Wadsworth, Belmont, Calif.

O'Brien, T. P., and Margaret E. McCully. 1969. Plant Structure and Development. Macmillan, London.

Salisbury, Frank B. 1963. The Flowering Process. Pergamon Press, New York.

Spratt, Nelson T., Jr. 1971. Developmental Biology. Wadsworth, Belmont, Calif.

Torrey, John G. 1967. Development in Flowering Plants. Macmillan, New York.

Wareing, P. F., and I. D. J. Phillips. 1970. The Control of Growth and Differentiation in Plants. Pergamon Press, Oxford.

Articles

Albersheim, Peter. 1975. The walls of growing plant cells. Scientific American 232(4):80–95.

Beer, Alan E., and Rupert E. Billingham. 1974. The embryo as a transplant. Scientific American 230(4):36–48.

Bryant, Peter J., Susan V. Bryant, and Vernon French. 1977. Biological regeneration and pattern formation. Scientific American 237(1):66–81.

Edwards, R. G., and Ruth E. Fowler. 1970. Human embryos in the laboratory. Scientific American 223(6):44–57.

Edwards, R. G. 1966. Mammalian eggs in the laboratory. Scientific American 215(2):72–81.

Gierer, Alfred. 1974. Hydra as a model for the development of biological form. Scientific American 231(6):44–67.

Gray, G. W. 1957. The organizer. Scientific American 195(5):58–79.

Gurdon, J. B. 1968. Transplanted nuclei and cell differentiation. Scientific American 219(6):24–35.

Gustafson, T., and M. I. Toneby. 1971. How genes control morphogenesis. American Scientist 59:452–462.

Hendricks, Sterling B. 1968. How light interacts with living matter. Scientific American 219(3):174–189.

Jensen, W. A. 1973. Fertilization in flowering plants. BioScience 23:21–27.

Konigsberg, I. R. 1964. The embryological origin of muscle. Scientific American 211(2):61–67.

Segal, Sheldon J. 1974. The physiology of human reproduction. Scientific American 231(3):52–79.

Taylorson, R. B., and S. B. Hendricks. 1976. Aspects of dormancy in vascular plants. BioScience 26:95–101.

Wessells, Norman K. 1971. How living cells change shape. Scientific American 225(4):76–85.

FIVE

Organism Form and Function

15

Nutrition

In 1930 an estimated 400,000 Americans, primarily in southern states, suffered from a disease known as pellagra, from the Latin word meaning "rough skin." The disease, however, was much more than rough skin; it also involved extreme diarrhea and a variety of mental disorders, and it led to death. Uncounted cases could be found in Spain, Italy, Egypt, parts of India, and South Africa as well. The feature all the victims had in common was that they were poor—not moderately poor, but "dirt" poor. Because of economic pressures, their diets were uniformly bad and highly restricted in terms of the foods they consumed. They ate only the cheapest food, which consisted almost completely of corn.

Because of the hundreds of thousands of cases of pellagra, the cause of the disease had been sought since the late 1700s. That corn was an important part of the cause was recognized early, but it was not fully accepted by physicians. Many believed pellagra was a contagious disease, which some thought was transmitted by flies. Others noted that families had histories of the affliction and made elaborate genealogies to trace the genes responsible for the disease. This interpretation was particularly pleasing to some politicians of the countries in which pellagra was found because it meant that there was nothing they could do to eradicate the plague of corn.

The major research that established the cause of the disease was carried out between 1913 and 1925 by an American physician, Joseph Goldberger. In a series of brilliant investigations done while he was in the United States Public Health Service, he showed beyond a shadow of a doubt that the cause of pellagra was the corn diet consumed by the poor. Corn lacked some important nutrient necessary to support human life. Shortly after Goldberger's death, other researchers found the missing nutrient to be nicotinic acid, or niacin. Niacin, a B vitamin, is present in low amounts in corn. Equally important is the fact that corn is also extremely low in tryptophan, an amino acid the body can convert to niacin. Thus, a diet consisting only or mainly of corn results in pellagra and death.

The control of pellagra in the United States came in the 1940s, when fortified bread was introduced. Adding various vitamins and minerals to commercial processed flour (from which the germ and bran with niacin had been removed) and bread ensured an adequate minimum level of niacin in the diet of all. The disease has been brought under control in most countries where it earlier posed such a problem, but it has persisted in India and is appearing in various parts of Africa where processed grain has been introduced. It will continue to be a problem in any nation where food is limited in quality and the bulk of the diet consists of corn or other cereals low in niacin and tryptophan.

This brief discussion of pellagra makes several points concerning the whole topic of nutrition. *Nutrition,* in the broadest sense, is the process or processes whereby organisms secure and incorporate materials they need for energy, for building and repairing their structures, and for all the activities of life. These materials may be organic or inorganic in origin and may take the form of compounds, elements, or ions. With regard to nutritional requirements, organisms can be grouped into two large classes. On the one hand are the green plants, called *autotrophs* ("self-feeders"). These organisms require only minerals, CO_2, H_2O, O_2, and light, which they use to synthesize all the other compounds they need to live. Nongreen plants and animals, on the other hand, are *heterotrophs* ("other-feeders"). Animals depend ultimately on green plants for their nutrition. They cannot (with a few minor exceptions) carry on photosynthesis, fix nitrogen from the atmosphere, or absorb mineral elements from the soil. Moreover, animals usually require more complex nutrients than do plants. In many cases, they need to absorb amino acids, not simply nitrogen, carbon, and oxygen from which to assemble their own. To avoid pellagra, for example, humans must have access to niacin or tryptophan.

This also means that the requirements of green plants and animals are quite different. In plants the problem is usually one of getting minerals, including nitrogen, from the soil and carbon dioxide and oxygen from the atmosphere. Therefore, our discussion of plant nutrition will center on soil, the nitrogen cycle, and other aspects of the mineral nutrition of plants. For animals, we will have to consider how animals catch, digest, and use their food. We will also have to consider their nutritional requirements and the diseases—such as pellagra—that result when these requirements are not met.

Plant Nutrition

Mineral Nutrition in Plants Until relatively recently, no one knew how plants obtain food or the other materials they need for life. The history of our knowledge of plant nutrition is an excellent example of how, as we accumulate data, our ideas change and mature. The early Greek philosophers, who thought the world was made of four elements—fire, water, earth, and air—debated in vain whether plants were made of earth or of water. Early in the seventeenth century, the Belgian scientist Jan van Helmont sought to test the question experimentally. He planted a 5-pound willow tree in 200 pounds of dry earth and let the plant grow for five years, adding only water. At the end of this period, he weighed the tree and the soil and found that the tree had gained 164 pounds, but the soil had lost only 2 ounces! From what van Helmont knew at the time (he believed in the four Greek elements and did not know about photosynthesis), he logically concluded that plants were made of water and not earth.

This conclusion stood until the late seventeenth century, when John Woodward in England found that mint plants grew best in water that had been in contact with soil, second best in Thames River water, and worst in pure rainwater. These observations suggested to Woodward that plants needed something besides water in order to grow.

In the late nineteenth century in Germany, Julius Sachs and W. Knop experimented with growing plants in solutions of different salts and salt mixtures. They were able to carry out such experiments because the developing chemical industry in Germany made such compounds available in reasonable quantities. They found that plants could be grown in a solution containing three salts: calcium nitrate ($CaNO_3$), potassium phosphate (KH_2PO_4), and magnesium sulfate ($MgSO_4$). Plants grown in this solution seemed to require nothing more than water and carbon dioxide. The elements present in this solution—calcium, nitrogen, potassium, phosphorus, magnesium, and sulfur—came to be known as the *major nutrient elements.*

As the synthesis of chemical compounds became more precise and as the chemicals used in the experiments became more and more pure, it became apparent that other elements besides these six were needed for plant growth. (In Sachs' and Knop's experiments, these other elements had been present, hidden as impurities in their chemicals.) These *minor nutrient elements* include iron, manganese, boron, zinc, copper, molybdenum, and chlorine. Table 15–1 summarizes the major and minor elements in plant nutrition.

But what do the major and minor nutrient elements mean in practical agriculture and gardening? Every home owner knows (or soon finds out) that a healthy lawn requires fertilizing several times a year. The fertilizers most widely used contain nitrogen, phosphorus, and potassium. Nitrogen is particularly important to most lawns, for without adequate amounts (10–12 lb of commercial fertilizer per 1,000 ft^2 per month) they will soon turn brown and die. The reason lawns require so much fertilizer is that the leaves of their grasses are continually being cut and removed. Consequently, the nitrogen incorporated into those leaves is never returned to the soil. In addition, watering may leach chemicals out of the soil. The same is true of crop plants, and for fields to remain fertile over years of cultivation, nitrogen and phosphorus must be added. How this is accomplished depends on the soil and the money available for farming. In all highly developed countries practicing intensive agriculture, chemical fertilizers are used. Nitrogen is added to the soil in the form of nitrates or ammonia that has been obtained through industrial chemical processes. In less developed countries, these compounds are obtained from the manure produced by domestic animals and humans. We will see in a moment how bacteria are involved in the whole problem of nitrogen content of the soil.

The minor nutrient elements are needed in very small amounts, but when they are lacking the results can be serious indeed. One example is the "little-leaf" disease of peach trees. This disease baffled California orchardists because they found no sign of bacteria or viruses. Finally it was noted that trees treated with commercial iron sulfate (a common treatment for cases of poor leaf development) showed improvement, not because of the iron or the sulfur, but because zinc was present as an impurity. When zinc was applied directly to the trees, the disease disappeared. Several other diseases of fruit trees, including an important citrus disease, also turned out to be caused by a lack of zinc. Surprisingly enough, in most cases the orchard soils contained enough zinc to last the trees for hundreds of years. The problem was that the zinc was in a form they could not use.

Too much of a good thing can also be a problem. For example, boron is necessary for plant growth and development. In some areas of the world farmers must add it to the soil to raise crops. But too much boron can be toxic and damage the plants. In many arid regions where irrigation is practiced, the water used is high in

Table 15–1 Essential Mineral Elements of Higher Plants

Mineral	Form Available to Plants	Lbs. Needed to Grow 100 Bushels Corn	Functions of Mineral Elements	Symptoms of Mineral Deficiencies
Nitrogen (N)	NO_3^-, NH_4^+	160	Part of amino acids, proteins, coenzymes, nucleic acids, chlorophyll, ATP	Lack of growth or limited growth with chlorosis (yellowing) or loss of leaves in severe cases. Purplish coloration due to accumulation of anthocyanin pigments. Entire plant affected, older leaves most
Phosphorus (P)	$H_2PO_4^-$, HPO_4^{2-}	40	Part of sugar phosphates, nucleotides, nucleic acids, phospholipids, coenzymes	Dark green, stunted plants. Accumulation of anthocyanin pigments. Delayed maturity. Entire plant affected, older leaves most
Potassium (K)	K^+	125	Coenzyme. Necessary for protein synthesis, essential for stomatal function	Mottled chlorosis, necrosis (spots of dead tissue, especially at tips and margins, between veins). Older leaves most affected. Weak stalks, roots more susceptible to disease
Sulfur (S)	SO_4^{2-}	75	Part of proteins, coenzyme A, thiamine, and biotin	Chlorosis of young leaves, usually no necrosis. Veins remain green, tissue between light green
Magnesium (Mg)	Mg^{2+}	50	Part of chlorophyll. Activates numerous enzymes. Maintains structure of ribosomes	Mottled or chlorotic leaves, may redden. Leaf tips turned upward. Older leaves most affected
Calcium (Ca)	Ca^{2+}	50	May be essential in synthesis and stability of middle lamella. Maintains structure and permeability of membranes. Part of α-amylase	Inhibition of root development and death of shoot and root tips. Young leaves and shoots most affected
Iron (Fe)	Fe^{3+}, Fe^{2+}	2	Essential part of cytochromes, ferredoxin. May activate certain enzymes	Interveinal chlorosis of young leaves, stems short and slender. Buds remain alive
Chlorine (Cl)	Cl^-	0.06	Activates photosynthetic enzymes	Wilted leaves, chlorosis, necrosis. Stunted, thickened roots or club-shaped roots near tips
Manganese (Mn)	Mn^{2+}	0.3	Formation of amino acids. Activates many enzymes. Electron carrier, catalyst	Chlorosis of young leaves, necrosis between veins; smallest veins remain green. Disorganization of lamellar membrane
Boron (B)	BO_3^{2-}, $B_4O_7^{2-}$	0.06	Carbohydrate translocation	Death of stem and root apical meristems. Leaves twisted, pale at bases. Swollen, discolored root tips. Young tissues most affected
Zinc (Zn)	Zn^{2+}	Trace	Chlorophyll formation, indoleacetic acid production, part of certain enzymes	Reduction in leaf size and length of internodes. Distorted leaf margins. Chlorosis. Older leaves most affected
Copper (Cu)	Cu^+, Cu^{2+}	Trace	Electron carrier, part of certain enzymes. Nitrate reduction	Young leaves dark green, twisted, wilted, misshapen; tip remains alive
Molybdenum (Mo)	MoO_4^{2-}	Trace	Electron carrier. Essential in nitrogen fixation	Chlorosis or twisting and death of young leaves

dissolved boron. As the boron accumulates in the soil, it stunts plant growth, and if it cannot be leached out, the land may be lost to cultivation.

Essential Elements and Soil   Except for carbon and oxygen, which they get from carbon dioxide in the air, land plants get all the elements essential for life from the soil. Vascular plants normally take these elements in through the roots. Thus, in order to understand the role of essential elements in plant nutrition, we must consider first soil and then its relation to roots.

Soil is a mixture of inorganic materials derived from a parent rock layer, plant and animal remains, and organisms growing in these remains. Soils vary greatly from one region to another (Figure 15–1). The rich black soils of the American high prairie and the Russian steppe differ markedly from the thin, hard soils of the tropics. This

difference is due less to the parent rocks from which the soils arose than to the moisture, temperature, and vegetation typical of the region. If the rainfall is high, as in the tropics, many of the dissolved minerals are quickly leached out of the soil into rivers and streams. Those remaining are trapped in the clay particles of the soil, where they interact with the carbonic acid, which is carbon dioxide dissolved in rainwater. This produces an acidic soil low in nutrients. In contrast, in regions where rainfall is low and evaporation high, minerals are brought to the surface by capillary action. Such high-nutrient soils characterize deserts and semiarid grasslands.

Temperature and moisture also profoundly affect the organic material and organisms in the soil. At relatively low temperatures, bacteria and fungi are slow at breaking down litter from trees and other plants, and a deep layer of partially decayed organic matter called *humus* builds up on the surface. At high temperatures and high humidi-

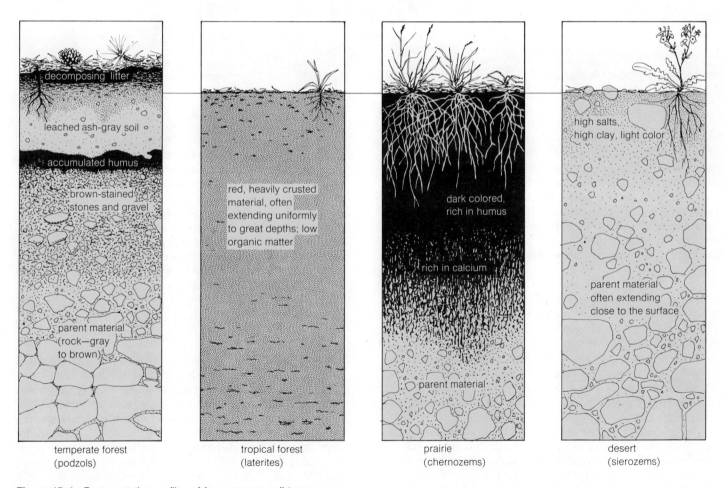

Figure 15–1   Representative profiles of four common soil types.

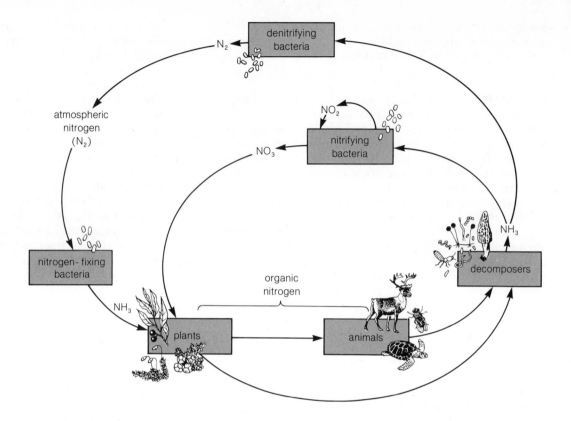

Figure 15–2 The nitrogen cycle.

ties, soil microorganisms grow rapidly and break down organic material. Little organic material accumulates in tropical soils. Where water is abundant in cool, anaerobic conditions, the growth of microorganisms is prevented and little decay occurs. Such conditions are found in certain bogs. As a result, large amounts of organic material accumulate and form peat. In Ireland, Scotland, and some northern European countries, the peat accumulated over hundreds of centuries is exploited as fuel.

Microorganisms, Soil, and Plant Nutrition Every square centimeter of soil is permeated with bacteria and fungi. These are critical agents in the nutrition of higher plants, and indirectly in that of higher animals, for they release elements incorporated into the organic molecules of plants and animals. Moreover, some of the bacteria and blue-green algae can use atmospheric nitrogen directly and thus provide this essential element to other plants and animals.

The nutrients are never tied up in any one body. They cycle through bodies and through the soil. Let's examine this cycling by beginning with some microorganisms found in the soil interacting with a leaf that has fallen to the forest floor. Once fallen, the leaf is quickly covered with fungal spores, some of which germinate within minutes. They form filaments that secrete enzymes to break down the polysaccharides and proteins of the nearby leaf cells to sugars and amino acids. The fungi absorb some of the sugars and amino acids and use them for energy and for synthesizing their own protoplasm. Such *extracellular digestion* is relatively inefficient, because part of the digested material diffuses away and is lost to the fungi. However, this diffused material is important to the various nutrient cycles because it is taken up by the other plants in the vicinity. Eventually the fungi and plants die, and their remains are, in turn, decomposed by bacteria secreting their own enzymes.

The Nitrogen Cycle Some bacteria break down the protein of fallen leaves, as well as that of dead fungi and other organisms, into amino acids and other organic compounds, releasing energy for their own growth. Other bacteria obtain energy by converting amino acids to ammonia, which can be absorbed by plants and used to synthesize amino acids and proteins. In most soils, oxygen is present and different bacteria change the ammonia into nitrites, NO_2^-. Plants cannot use nitrites, but other soil bacteria can, converting nitrites into nitrates, NO_3^-, and

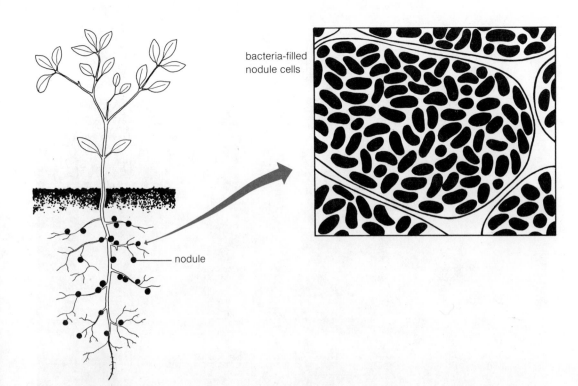

bacteria-filled
nodule cells

nodule

Figure 15–3 The root nodules in clover, and the relationship of the bacteria to the root cells.

gaining energy in the process. The bacteria involved in the various conversions of ammonia to nitrate are known as **nitrifying bacteria.** To the higher plants, the important fact about these conversions is that nitrogen becomes available to them in the form of nitrates and ammonia—the only forms in which they can absorb and use it (Figure 15–2).

Animals contribute to the nitrogen cycle by eating plants and breaking their protein down into amino acids or polypeptide groups. They use some of these to make their own proteins, but others are broken down further to ammonia, urea, or uric acid and excreted. This excreted material is also used by bacteria and converted to nitrates.

We have been treating the recycling of nitrogen compounds without asking where nitrogen comes from in the first place. The answer is that we live in a sea of nitrogen: The earth's atmosphere is about 80 percent gaseous nitrogen. This atmospheric nitrogen cannot be used either by higher plants or by animals, but only by certain bacteria and blue-green algae. These organisms are **nitrogen fixers,** meaning that they can take nitrogen from the air and use it to make amino acids and other nitrogen compounds. To do this they must expend energy. Some of these organisms live freely in the soil, and others live within special clumps of cells in the roots of higher plants,

primarily those of the legume family (Figure 15–3). This is what makes such crops as clover and beans so important to farmers in many parts of the world. By rotating crops so that a field is planted one year in legumes—clover, beans, peas, or alfalfa—and then several years in wheat, oats, or barley, farmers keep the fields fertilized with little or no chemical fertilizer. In intensive agriculture, where it is not profitable to rotate crops, or where the amount of nitrogen added to the soil by rotation is not great enough for maximum yield, farmers add nitrogen fixed from the atmosphere electrochemically. Yet, of the 100 million tons of nitrogen that are calculated to be added to the soil each year worldwide, less than 10 percent is in the form of chemical fertilizer.

Before we leave the question of nitrogen and plants, we must mention the *denitrifying bacteria.* Found in soil in which oxygen is lacking, these bacteria convert nitrate to gaseous nitrogen, using the oxygen thus obtained for their own metabolism.

The reactions we have just been discussing are known collectively as the *nitrogen cycle.* This cycle is immensely important. Just as life could not exist on earth without photosynthesis, so it could not exist without nitrogen-fixing bacteria and the other organisms involved in making nitrogen available to plants. The nitrogen cycle,

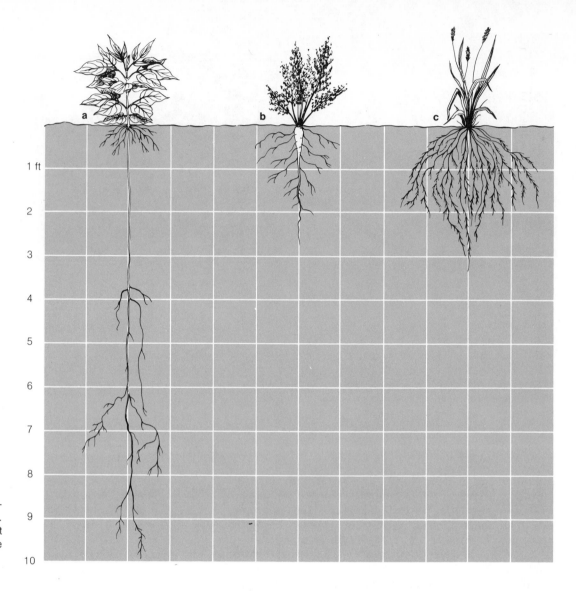

Figure 15–4 Types of root systems. (**a**) Poison ivy. Taproots adapted to reach moisture deep in the ground. (**b**) Dandelion. Taproot system adapted to store food. (**c**) Blue grama grass. Fibrous root system adapted to take up surface water.

however, is not the only cycle present in the soil. Also important are cycles in which phosphorus, calcium, and sulfur are taken from and returned to the soil. All of these cycles also involve microorganisms. Ecological aspects of the cycles will be discussed in Chapter 26.

Soil and Roots The major pathway of nutrients into the plant is through the root system. Roots penetrate every part of the soil, forming a living network that anchors the plant and absorbs water and minerals. Root structure is highly variable. Some plants have a single large, vertical root known as the taproot (Figure 15–4a, b). In carrots and beets, the taproot stores sugars and starches as well as supporting the plant and taking in

water and minerals. Another common root type is the fibrous system, typical of grasses (Figure 15–4c). Fibrous roots lack a single dominant root; instead, numerous fine roots form a dense network in the soil.

Within each type, the form of root system a plant develops depends to a large extent on the conditions in which it grows. The root systems of apple trees and corn plants both go deep into the soil and spread out shallowly in it. Such a system appears well adapted to conditions of moderate rainfall. Desert plants appear to have one of two different types of root system. Mesquite taproots, for example, may go almost straight down for 25 meters and reach deep water tables. Many cacti, on the other hand, have extremely shallow but widespread root systems that appear adapted to occasional rains.

Figure 15–5 Root hairs extend into the soil from near the tip of the root.

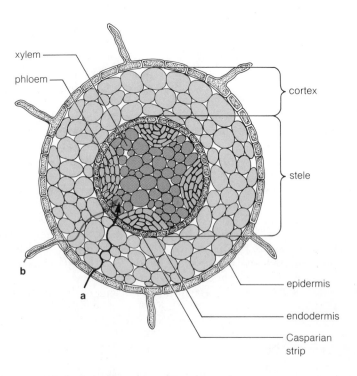

Figure 15–6 Movement of materials through the root. (**a**) Water moves freely through the cell walls until it reaches the endodermis, where it is forced through the cytoplasm. (**b**) Minerals can be absorbed into the root hairs and moved from cell to cell through the cytoplasm.

The possible extent of a plant's root system was demonstrated by a rye plant that was shown to have a root system some 7,000 square feet in surface area. This huge area is made possible by the presence of root hairs near the tip of the root (Figure 15–5). These hairs are tiny extensions of the epidermal cells that cover the root for a centimeter or two immediately behind the meristematic region of the root tip. Each root system has millions of them. It is the root hairs that actually absorb mineral elements from the soil.

Mineral Absorption from the Soil   Mineral elements in the soil are available to the plant only when they are in solution. Soil water surrounding both the soil particles and the root hairs could have dissolved in it all the minerals available in that soil. What is actually in the soil water will depend not only on the soil itself but on the pH of the water. As we saw in the case of the zinc deficiency in fruit trees, a necessary element may be available in the soil in a form inaccessible to the plant. This applies particularly to iron; a soil may be bright red from the iron present, yet the plants in it show marked iron-deficiency symptoms. The extent to which most elements

will enter the soil water from the soil particles is largely determined by the amount of H⁺ generated during the normal respiration of the cells of the root.

The elements dissolved in the soil water pass readily into the root through the cell walls. In the epidermis and cortex, the walls form a continuous system that is freely permeable to water and the dissolved minerals (Figure 15–6, pathway a). Between the cortex and the vascular tissue lies the endodermis. The cells of the endodermis have a special water-resistant thickening in their radial walls, called the *casparian strip.* Thus, when the water in the walls reaches the casparian strip, movement toward the center is no longer possible. The only way to the xylem at the center of the stele is through the cytoplasm of the endodermal cell. Since the cytoplasms of all the cells are connected through plasmodesmata (see Chapter 3), once water and minerals have entered the cytoplasm of any cell of the epidermis or cortex they can pass through the cytoplasm of the endodermis (Figure 15–6, pathway b). The endodermis appears to control the movement of materials into the stele by controlling the supply of energy necessary to transport the materials across the cell membrane.

The next question is how the minerals enter the cytoplasm. In some cases, they are merely responding to concentration gradients and diffusing from regions of higher concentration to lower ones. However, mineral elements are usually more concentrated within the cytoplasm than in the soil water, so substances coming into the cell are moving against the concentration gradient. This means that there must be active uptake of the minerals, and there is evidence that this is the case. Active transport (see Chapter 3) across cell membranes requires energy, and it has been shown that mineral uptake will occur only if respiration is taking place. However, the mechanism of this active transport is not yet clearly understood.

Once beyond the endodermis, the minerals are transferred to the xylem, which carries them to the rest of the plant. The minerals may leave the cytoplasm and enter the xylem by the reverse of the mechanism by which they entered—they are currently thought to be actively secreted. The mechanism by which water and dissolved minerals move up the xylem will be discussed in Chapter 17.

Animal Nutrition

Animals are ultimately dependent on plants for their nutrition. Animals are grouped into broad categories on the basis of what they eat. Animals that feed directly on plants are called *herbivores*. Those that catch and eat other animals are known as *carnivores*. **Omnivores** eat both plants and animals.

Before we consider some of the diverse ways animals have of procuring food, we will look at the chemical substances required in animal nutrition and examine the reasons they are needed. Special problems of human nutrition will be discussed in Supplement 15–1.

Carbohydrates and Fats Carbohydrates and fats (Chapter 2) are the most common sources of food energy and carbon for animals. Metabolizing 1 milligram of fat yields about 9 calories, whereas an equal amount of most carbohydrates yields only about half that much—as any dieter will tell you. Many animals, including humans, can use either carbohydrates or fats for energy and readily convert one to the other. Since the body is in constant need of energy for maintenance, temperature regulation, activity, and the like, carbohydrates and fats are the foods needed in the greatest bulk. At least in the short run, it matters not whether the calories come from carbohydrates or from fats. Arctic Eskimos, for example, derive most of their food energy from animal fats, whereas many Asians get theirs from the starch of rice.

Different animals have evolved to use different sources of food energy, and some animals are much more restricted than others in the types of food energy they can use. Some insects metabolize only carbohydrates, and others use only fats. Bees metabolize only sucrose, fructose, glucose, and several other sugars found in the nectar of flowers. The sphinx moth ingests the same nectar sugars, but it converts them to fats for use during flight.

Because of its high energy yield per unit of weight, fat is an excellent fuel for flying animals. Migratory insects and birds put on great stores of fat before traveling. In fact, among such small migratory birds as the warblers, stored fat may represent as much as 50 percent of total body weight. Migratory mammals likewise store up fat to fuel their journeys.

Proteins The bulk of our bodies, from enzyme to muscle, is composed of protein. The large protein molecules we eat cannot be absorbed whole but must be disassembled into their component amino acids, which can in turn be reassembled in a potentially infinite variety of protein combinations. Most animals can build some amino acids from simpler organic molecules or convert one amino acid to another. The remaining amino acids necessary for protein synthesis must be provided in the diet. Adult humans can build all but 8, and they require these 8, the so-called essential amino acids, in their diets—isoleucine, leucine, lysine, methionine, phenylalanine, threonine, tryptophan, and valine. Children require 2 more—arginine and histidine. Rats need the same 10 amino acids as immature humans, but chickens and turkeys need yet an additional 2.

Proteins (Chapter 2) are not all alike. They contain many different amino acids in various proportions, and plant proteins particularly may lack one or more of the essential acids altogether. This means that just eating your 70 grams or so of protein a day is not good enough; you have to eat the *right* proteins with the *right* amino acids. In some underdeveloped countries, cereal grains, which are low in lysine, provide more than two-thirds of the dietary protein. Most of the remaining one-third comes from legumes, which are low in methionine. We already mentioned that corn is low in tryptophan. Plant breeders have tried to increase the lysine content of grains, since lysine is the amino acid least abundant in plant proteins and the one most likely to be missing for people on a low-protein diet. Among their successes was a high-lysine corn. Lately, they have succeeded in crossing wheat *Triticum* and rye *Secale* to produce a new genus, *Triticale*. This new grain combines the desirable characteristics of both its parents: the high protein of wheat and the high lysine of rye.

Despite this and other hopeful developments, protein deficiency, already common among the world's poor, is increasing. A protein-deficiency syndrome so widespread that it has some 32 different names is commonly known by its Ghanaian name, *kwashiorkor,* or "red baby." Another African name, translating as "what happens to the first baby when the second baby comes," reveals the basic problem. The disease is particularly prevalent where babies are weaned from protein-rich mother's milk to a diet that is almost entirely starch. Usually a baby is not weaned until another baby is born, which means that population growth has added to an existing dietary problem. Children with kwashiorkor are anemic, listless, and mentally retarded. Their livers degenerate and their abdomens enlarge, producing the grotesquely swollen belly of protein-deficiency starvation. Sores cover their skin, and they fall easy victim to infection and disease. Unless their diet is improved, they die—usually from disease rather than protein starvation. The amino acid content of the world's food plants is no mere biochemical detail, but a matter of life or death to millions of humans, as well as to other animals with specialized diets.

Vitamins   Vitamins are organic compounds that are essential in small amounts for a variety of biological processes. For example, in some cases they form structural parts of molecules that the organism cannot synthesize. Examples are the respiratory coenzymes (B vitamins) and specific functional molecules participating in vision (vitamin A), calcium metabolism (vitamin D), and blood-clotting (vitamin K). What unifies vitamins, different as they are in structure and function, into a class of compounds readily discussed in television commercials? The fact that the animal in question cannot synthesize them itself. Animals do in fact sometimes synthesize the final vitamin, but they can do so only if supplied with a complex molecule that they modify. This can be illustrated with vitamin A.

Vitamin A is synthesized from the pigment carotene that gives carrots, winter squash, and cantaloupes their characteristic orange to yellow colors. Vitamin A is a major source of retinene, the visual pigment of the eyes. A deficiency of vitamin A leads to night-blindness. Plants are the only source of carotene. Eskimos, whose diet lacks vegetables, get their vitamin A or its derivatives mostly from the livers of fish, which in turn obtain the carotene from the single-celled algae they consume. The polar bear's fish diet concentrates such large amounts of vitamin A in its liver that a meal of this liver can give a person a toxic overdose of the vitamin.

The B vitamins include a complex of about eight compounds that play roles in cellular respiration. The best known are thiamine (B_1), which is part of the coenzyme catalyzing the oxidation of pyruvic acid. Another is riboflavin (B_2), a component of the cytochrome system. Others include biotin, niacin (a component of NAD and NADP), pyridoxine (vitamin B_6), cobalamin (B_{12}), folic acid, and pantothenic acid. Many animals, including in-

sects and cattle, get some B vitamins from the synthetic activities of microorganisms living in their gut. Diseases caused by vitamin B deficiencies include pernicious anemia (insufficient B_{12}), beriberi (insufficient B_1), and pellagra (insufficient niacin). In humans, niacin is synthesized by the gut bacteria and the liver, and no deficiency disease appears unless the diet is deficient in both the niacin and tryptophan, a precursor in the synthesis of the vitamin. B vitamin intake is not always simple. Egg white contains abundant biotin, for example, but it also contains a protein, avidin, that binds biotin into an inactive complex. Consequently, biotin will not be absorbed from the gut unless the egg is cooked, inactivating the avidin and making the biotin available.

Most mammals can synthesize vitamin C in their livers; reptiles and amphibians produce it in their kidneys; but invertebrates (including insects), fishes, and such highly evolved mammals as the primates must get all their vitamin C from their diet. In humans, a diet lacking in vitamin C (ascorbic acid) causes scurvy, a disease whose symptoms include bleeding gums, loss of teeth, slow healing, and anemia. In the days of sailing ships, scurvy was common among sailors because their rations for voyages lasting a year or more consisted of salt pork, beans, and flour, with no fruit. When, in the eighteenth century, physicians of the British navy observed that eating fruit prevented scurvy, limes were added to the usual naval diet. Scurvy ended and the nickname "Limey" began for the British sailor.

Vitamin D and its derivatives are involved in calcium transport within the body, including absorption of the mineral from the gut. A deficiency of vitamin D produces rickets, characterized by stunted bones and bowed legs. Vitamin D is produced in the skin from sterols (7-dehydrocholesterol) under ultraviolet radiation (sunlight), but it can also be taken in the diet from egg yolk, milk, and fish oils. Vitamin D deficiency is almost unknown in the tropics because of the constant sunshine. However, it was common in English cities at the height of nineteenth-century industrialization. Children were brought up in narrow streets where the sun could scarcely have penetrated even if it had not been dimmed by air pollution. Moreover, many children worked indoors all day.

Recently, vitamin D has been implicated in theories explaining the evolution of skin pigments. The human species probably evolved in the tropics, where dark pigmentation protected the body from too much ultraviolet light. But as humans moved north into less sunny climates, dark skin color became a liability, because it prevented the penetration of what sunlight there was and resulted in a deficiency of vitamin D. Some investigators speculate that this effect produced a selection pressure favoring light-skinned humans.

Vitamin requirements are generally similar among animals. However, there are differences. For example, both

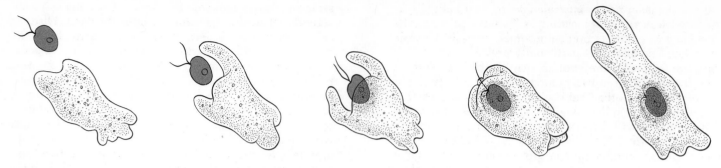

Figure 15–7 *Amoeba* engulfs food particles by flowing around them with its pseudopodia.

humans and rats need ascorbic acid, but the rat can synthesize all it needs. Thus, to the rat, ascorbic acid is not a vitamin. Both humans and insects need cholesterol (insects need it or some other sterol to make molting hormone), but humans synthesize their own. Thus, to humans, cholesterol is not a vitamin.

Since vitamins are so necessary to the working of the body, we might think that the more we take, the better. But there is no evidence that amounts in excess of the requirements do any good. In fact, for some of the vitamins that are soluble in fats, such as vitamin A, large amounts can be toxic. Large amounts of the other vitamins, such as vitamin C, are simply excreted. It is extremely difficult to establish minimum vitamin requirements, since they vary with age, size, physical activity, the rest of the diet, and many other factors. Also, not all deficiencies result in outwardly visible disease symptoms.

Minerals Plants and animals require almost exactly the same minerals. Some of the minerals animals require include sodium, potassium, magnesium, phosphorus, calcium, iodine, magnesium, and chlorine. The calcium and phosphorus are primary components of bones and teeth. Phosphorus is used in nucleotides and high-energy phosphate bonds, just as in plants. Sodium and chloride ions are essential in nerve and muscle excitation, and as osmotic components in the blood. Iodine becomes a part of the hormone thyroxine. Iron is used in hemoglobin, myoglobin, and cytochrome molecules, and cobalt is a structural component of vitamin B_{12}.

Some of these minerals, such as sodium and calcium, are needed in relatively large amounts. Those required in minute amounts include cobalt, zinc, copper, molybdenum, chromium, bromine, vanadium, and selenium. The functions of these trace minerals are unknown. Lithium may also be required in trace amounts, but if so its exact function remains a mystery. We do know that lithium

administered in minute doses reduces the symptoms of people suffering from mania, a mental disturbance characterized by sudden bursts of physical and mental activity followed by severe depression.

Recently there has been renewed interest in the relation of minerals to the growth of invading bacteria in host organisms. This is based on an observation made in 1944 by A. Schade and L. Caroline that strong iron-binding proteins in egg white and plasma withhold the metal from invading bacteria. It is now known that, following bacterial invasion in humans, the blood rapidly becomes depleted of iron. The intestines stop absorbing it, and the liver begins to store it. Individuals who have great amounts of iron in their bodies because of liver destruction, blood diseases, or diet, are often highly susceptible to even small numbers of invading bacteria. This attempt to withhold iron—and perhaps other minerals such as zinc and phosphorus—from invading bacteria is called *nutritional immunity*. The role of iron in bacterial invasion makes one wonder about the benefits to health of taking the iron supplements so heavily advertised in the media.

Nutritional Symbiosis Many heterotrophs have evolved associations with autotrophs that compensate for their biosynthetic deficiencies. Many sponges, hydra, and sea anemones look green because of the unicellular algae within them. Some of the sugars the algae make during photosynthesis leak out, providing the host animals with carbon and energy. Most reef-building corals are restricted to waters in which the algae they contain can flourish. The familiar lichens that grow around us on rocks, fence posts, buildings, and trees are a symbiotic association of filamentous fungi and unicellular algae. This association is so highly evolved that the form of the two organisms growing together is unlike either growing alone.

Nutritional associations of bacteria, fungi, or proto-

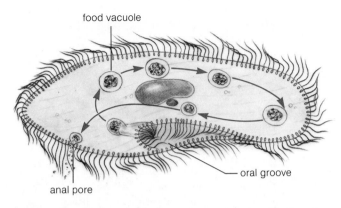

food vacuole

oral groove

anal pore

Figure 15–8 In *Paramecium* food is ingested via the ciliated oral groove. It is digested by enzymes secreted into a food vacuole, and undigested food is expelled from an anal pore.

zoans with larger animals are common. The fat-body cells of cockroaches and the guts of many insects contain bacteria that produce essential vitamins and amino acids. Termites, the householder's nemesis, are not themselves able to digest the wood they eat, but their digestive tracts are densely packed with protozoans whose enzymes digest cellulose, the principal constituent of wood. If these symbionts are killed, the termite will starve to death, no matter how much wood is available.

Feeding Mechanisms   Clearly the capture and intake of food is critical to animals' survival. It is therefore no surprise to find that each animal is a specialist, to one degree or another, in its feeding mechanism. Indeed, many aspects of the structure and behavior of animals can best be explained in terms of food-catching and food-processing techniques. The elaborate behaviors required to locate and capture food rely heavily on sharpened vision, on the sense of smell, and on hearing. Bats and porpoises have sonar, rattlesnakes have infrared temperature sensors to detect small warm-blooded mammals, and some fishes have an electrical sense used to detect prey. Some praying mantises and spiders deceive their insect prey by looking like flowers. Many spiders build nets to capture insects, while others stalk and pounce. At first glance there seems no end to the variety of mechanisms used to get food. Looking more closely, however, we can see patterns in the feeding behavior of diverse animal groups, from the single-celled to those composed of billions of cells.

A few protozoans can absorb nutrients directly through the body wall, but most must surround and engulf their prey. Amoebae do this by extending *pseudopodia* ("false feet") around bacteria and other small objects that

form their food source (Figure 15–7). The organism takes up the food by flowing around a particle and enclosing it in a temporary, membrane-bound cavity, the *food vacuole*. The ciliate protozoan *Paramecium* also engulfs particles by forming food vacuoles (Figure 15–8). However, here the particle is guided down a specialized *oral groove* by cilia.

In both *Amoeba* and *Paramecium,* the foodstuffs in the food vacuole enter the cytoplasm only after they are digested. *Digestion* is the breakdown of large, complex molecules into smaller ones. This process is speeded by digestive enzymes, each of which is specialized to break a specific chemical bond or class of chemical bonds. In the protozoans, the digestive enzymes are contained in small membrane-bound bags, the lysosomes, which empty their enzymes into the food vacuole and digest its contents (see Essay 4–1). Once digestion is complete, the small nutrient molecules can be absorbed into the cytoplasm and the wastes eliminated when the membrane of the vacuole fuses with the plasma membrane of the cell and the contents of the vacuole are discharged.

Digestion within the cell is called *intracellular.* We noted earlier that the extracellular digestion found in the fungi entails secretion of enzymes to the outside of the cell. Extracellular digestion in single-celled organisms has the advantage of simplicity, in that it requires no specialized digestive organs, but it is inefficient, since only a small percentage of the digested food molecules can be absorbed.

In multicellular animals, filter-feeding mechanisms, even for similar sorts of food, are incredibly diverse. The tiny larvae of certain mosquitoes use a filter near the mouth to strain bacteria from the few spoonfuls of water in the tree cavity in which they live. The clam and its relatives use their gills to filter particles from the water. The gills are composed of vertical filaments that hang down into the mantle cavity. Cilia on the gill surface propel streams of water over the gills. The food particles strained out by the gills are transported onward to the mouth by subsequent ciliary action in the marginal groove (Figure 15–9). The world's largest animals, the baleen whales, are also filter feeders. They feed primarily on tiny shrimp, a major component of the ocean's passively drifting animal life, the zooplankton. The whale takes in great volumes of water and passes it through the fringes of the whalebone plates in the roof of its mouth. Then it scrapes the trapped shrimp off the baleen with its tongue and swallows them.

Sucking is another common means of feeding. The suctorian protozoans, as their name implies, suck in entire bacterial cells after skewering them with a lancelike appendage. Sucking is particularly common among insects. In the mosquitoes, paired stylets that form the proboscis or "beak" are grooved so that they form a tube when brought together. The sharp stylets are inserted into the

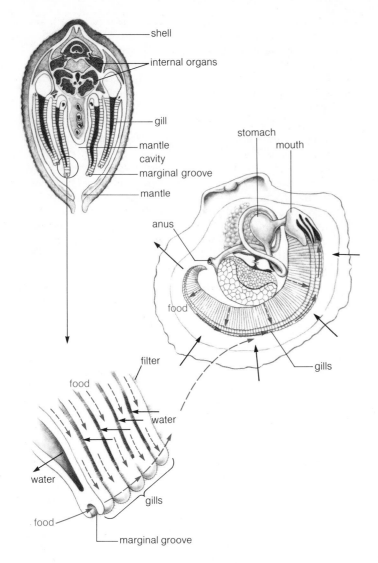

Figure 15–9 The gills and direction of water flow (black arrows) and food movement by way of the cilia (colored arrows) during filter feeding in a clam.

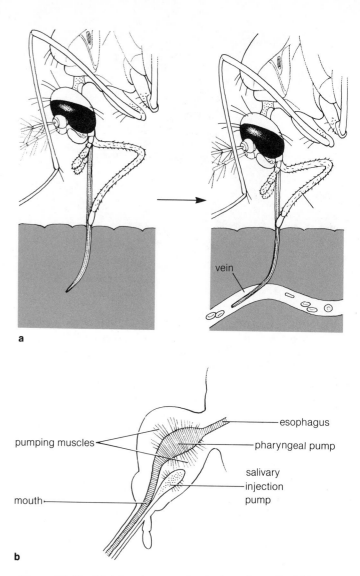

Figure 15–10 (**a**) A female mosquito piercing a host. The labium (sheath for the mouthparts), which is not inserted, is folded as the other mouthparts penetrate the skin. (**b**) Location of the chambers with attached muscles used to create negative pressure and suction in the head.

victim, and a muscular pump in the mosquito's head creates a negative pressure that draws up the blood (Figure 15–10). The butterfly's sucking mechanism is similar to the mosquito's, a proboscis composed of paired and grooved mouthparts held tightly together to make a tube. It is used to suck in nectar from flowers. One pump in the butterfly's head pulls the nectar in, while a second extends and retracts the proboscis by regulating its blood pressure. The beak of the hummingbird also serves as a device for sucking in flower nectar, but the mechanism is not well understood. Certainly its workings are not

so clear-cut as the process a child uses in sucking soda up through a straw.

Insects have mouthparts that are adapted to biting (grasshopper), sucking (butterfly), or two sets of mouthparts so that they can do both (bees). Such insects as the grasshopper have a pair of toothed horny jaws, or *mandibles,* that are used to cut leaves into small pieces.

Animals that feed on big pieces of solid food have evolved ways of tearing it into pieces small enough to be ingested. A cheetah can run down and kill an antelope, but if it had no way of dividing this prey into bits small

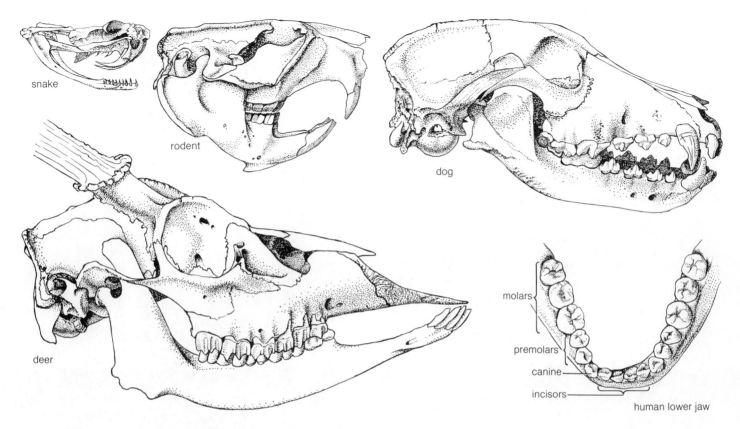

Figure 15–11 Structure and arrangement of teeth in different animals.

enough to be swallowed, it might just as well not have gone to the trouble. Hence the importance of teeth in mammals, both in securing food and in grinding it up before it is swallowed.

Vertebrate teeth have several functions, and their structure varies accordingly. The chisel-like front teeth, or *incisors,* are used for biting. They are prominent in humans but are most highly adapted in such rodents as the beaver, which fells trees with them. Next to the incisors are the *canines.* This pair of teeth is absent in rodents and quite reduced in humans but is prominent in the monkeys and the carnivores, including the dog, from which the name for these teeth derives. Carnivores use the canines largely for biting and holding prey. In the extinct sabor-toothed tiger, they were used for stabbing. The male baboon employs its impressive fangs for defense against leopards and for aggressive display directed toward other males. The prominent tusks of the elephant and the wild hog are modified upper incisors. In the back of the mouth are the *premolars* and *molars.* In humans, these teeth are flat and shallowly ridged, adapted for crushing and grinding. Cats and dogs have peaked, sharp

molars for shearing meat, whereas the herbivores have complex ridged molars suited to grinding tough plant matter. The dentition of several representative vertebrates is shown in Figure 15–11.

Snails have a different approach. They have an organ called the *radula* that works much like a piece of sandpaper, scraping plant tissue into small pieces.

Digestion and Digestive Tracts No matter how an animal eats, it is taking in material in a form that its body cannot use. The food must be digested, and animals have evolved a variety of ways to do this. When we look at the more primitive multicellular animals, such as hydra and jellyfish, we find that food is held in a central cavity with only one opening. Digestive juices are secreted into this cavity, and the food is broken into smaller pieces that are engulfed and digested in much the same manner as in *Amoeba.* In the flatworms, the cavity is more complex in shape, allowing more time for digestion and better absorption of materials by the surrounding cells; but still a single opening serves as both mouth and anus. Since

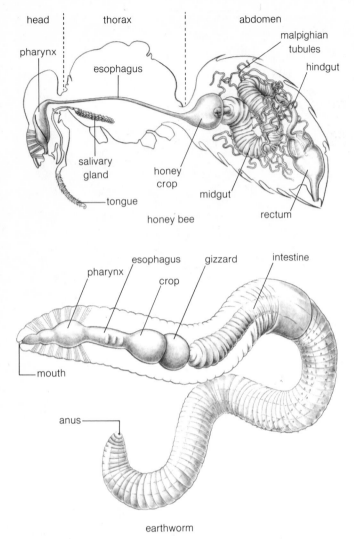

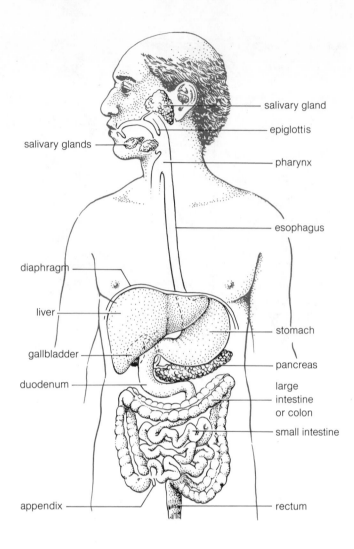

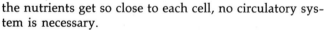

Figure 15–12 The one-way digestive tracts of the honeybee and the earthworm have areas of specialized function much like those of vertebrates.

Figure 15–13 The anatomy of the human digestive system.

the nutrients get so close to each cell, no circulatory system is necessary.

The earthworm and the honeybee have all the major features present in our own digestive tract (Figures 15–12 and 15–13): a mouth where food is received and processing begun, a stomach (the crop and gizzard) where the food is further broken down, a long intestine in which digestion continues and absorption takes place, and, finally, the anus, through which the unused material is excreted. This one-way system is much more efficient for processing food. It allows for continual intake of food, separation of the various activities associated with its

breakdown and absorption, and a convenient way to eliminate indigestible food material. This pattern is found, with many variations, in all higher animals. We will concentrate on the details of the pattern found in humans.

Taste receptors (see Chapter 22) on the tongue test the suitability of all entering food. If it "passes inspection," the teeth grind the food into smaller bits. The food is mixed with saliva as it is chewed. Secreted by several glands in the mouth, the saliva lubricates the food. It also contains the first of the many digestive secretions listed in Table 15–2: amylase, which hydrolyzes starch into maltose.

Table 15–2 Digestive Juices in Humans (Active Components)

Digestive Organ	Source	Action
Mouth Cavity		
Amylase	Salivary glands	Splits starch into the disaccharide maltose
Stomach		
Pepsin (pepsinogen)	Stomach mucosa	Splits peptide bonds of tyrosine and phenylalanine at amino ends
Rennin	Stomach mucosa	Curdles milk (taking the proteins out of solution)
Hydrochloric acid	Stomach mucosa	Softens food; activates pepsinogen
Small intestine		
Endopeptidases		
Trypsin (tripsynogen)	Pancreas	Splits peptide bonds adjacent to lysine and arginine
Chymotrypsin (chymotripsinogen)		Splits peptide bonds of tyrosine, phenylalanine, leucine, and methionine at carboxyl ends
Exopeptidases		
Various carboxypeptidases	Intestinal glands and pancreas	Splits amino acid linkages at free carboxyl ends of amino acid chains
Various aminopeptidases	Intestinal glands and pancreas	Splits amino acid linkages at free amino ends of amino acid chains
Lipase	Pancreas	Breaks down fats to fatty acids and glycerol
Amylase	Pancreas	Splits starch into the disaccharide maltose
Disaccharidases, including		
Sucrase	Intestinal glands	Splits sucrose to glucose and fructose
Maltase	Intestinal glands	Splits maltose to 2 glucose units
Lactase	Intestinal glands	Splits lactose to 2 glucose units
Bile (bile salts, bile pigments, and cholesterol)	Liver	Emulsifies fats

From the mouth, the food moves backward into the pharynx with the aid of the tongue (Figure 15–13). The trachea (windpipe), which leads down into the lungs, also opens into the pharynx. To keep food from clogging the windpipe, the muscular flap of the glottis called the epiglottis closes down over the opening of the windpipe as the food is swallowed. If this mechanism fails, food entering the windpipe stimulates coughing, which expels the potentially lethal food back into the pharynx.

Food moves through the esophagus and into the stomach by **peristalsis,** a wavelike constriction of circular muscles in the walls of the passageway that pushes the food mass along in front of it. Peristalsis continues through the whole length of the digestive tract.

Most of the time, the junction between the esophagus and stomach is closed off by a constricted **sphincter** muscle (a ringlike muscle that acts like a valve in a tube), which prevents materials in the stomach from moving back up into the esophagus (Figure 15–13). This sphincter muscle relaxes only when food is swallowed, allowing the food to enter the stomach.

The stomach is essentially a large muscular sac that breaks down the food mechanically by churning. However, some chemical digestion occurs here as well. The gastric juice secreted by the glandular mucosa lining the stomach wall has two main ingredients: the enzyme pepsinogen, produced by the "chief cells," and hydrochloric acid, produced by the "oxynctic" cells. This acid, one of the most powerful occurring in nature, softens the fibrous food material, inactivates salivary amylase, and transforms the inactive pepsinogen to its active form, *pepsin:*

$$\text{pepsinogen} \xrightarrow{\text{HCl}} \text{pepsin}$$

Since HCl at the concentration of human digestive fluid could eat away the cells of the stomach wall, and since pepsin is a protein-digesting enzyme, one wonders

Essay 15–1

The Evils We Do to Ourselves: Ulcers

The human stomach is truly a remarkable organ. Secreting copious amounts of hydrochloric acid and pepsin, which together are capable of reducing most proteins to soluble components, it does not digest itself. One of the reasons we do not digest our own stomachs is that we also secrete large amounts of mucus, a mucopolysaccharide compound that helps protect the cells beneath it. But the system is not foolproof. We can, frequently it seems by our own initiative, overcome the safeguards inherent in the system and begin to digest the walls of our stomachs and the part of the small intestine immediately attached to it, the duodenum. Holes created by such digestion are called ulcers, and as much as 10 percent of the U.S. population may be afflicted with them.

Many ulcers are relatively minor, with the hole extending only through the mucous membrane. In more serious cases, the hole extends into the muscular coat of the stomach or intestine. Finally, in the most acute form, the hole goes completely through the wall of the organ and the digestive juices enter the peritoneal cavity. All forms are accompanied by pain, usually several hours after eating, when the stomach has emptied its contents and acid begins to accumulate. Bleeding is also a common symptom, and blood may be passed through the bowels or by vomiting.

For the less serious forms of ulcers, controlling the diet is a means of allowing the stomach to heal itself. The idea is to free the stomach of HCl, and a regime of bland foods, such as lean meat, cooked fruits and vegetables, and milk, is prescribed. Coffee, alcohol, cigarettes, and highly seasoned foods, all of which stimulate gastric-juice secretion, must be avoided.

There are other ways to stop or slow the secretion of the gastric juices. One is through the use of drugs that suppress its production. Another is to remove half to three-quarters of the stomach surgically. The area of the stomach producing the acid is thus reduced, and the problem is eliminated. Another procedure is to cut the vagus nerve that goes to the stomach. This is effective because the production of gastric juice is partially under nervous control. Although cutting the nerve does not interfere with the functioning of the stomach, it does reduce production of gastric juices. It also prevents a patient from knowing when his stomach is full, and people with severed vagus nerves have trouble determining when to stop eating, except by visual cues. A new technique is to reduce gastric-juice formation by freezing the lining of the stomach. A balloon is placed in the stomach and joined to a freezing apparatus. The ballon is then filled with 95 percent alcohol at a temperature low enough to freeze the lining of the stomach superficially.

The important feature of ulcers is that in so many cases they are brought on by the individuals themselves. There is a high correlation between personality type and the incidence of ulcers. The hard-driving person, particularly when faced with difficult or frustrating decisions, is the prime example of the ulcer victim, but others include people who turn all their fears and frustrations inward. In the healthy stomach, there is a delicate balance between the action of the acid and pepsin, on the one hand, and the resistance of the mucous membranes, on the other. This is the balance that our fears and frustrations can upset.

---

why the stomach does not digest itself. Sometimes it does, of course, and an ulcer results (Essay 15–1), but several safeguards reduce this danger. For one thing, mucus on the stomach wall inactivates the enzyme there. However, the chief danger to the stomach lining is from the hydrogen ions of the ionized acid. The acid does not ionize, however, until it is in the cavity, or *lumen,* of the stomach, and tight junctions between the gland cells keep the ionized acid out. When these junctions are broken, the hydrogen ions enter the cells and may destroy them, ultimately rupturing the capillaries and causing bleeding.

Ethyl alcohol and aspirin can break down the mucosal barrier, particularly when acting together. In most people, the bleeding is slight, but others are particularly sensitive and may become anemic if they use aspirin heavily. Despite its excellent defenses, the stomach surface normally loses about a half million cells per minute, and is normally renewed with new cells every three days.

The secretion of gastric juice is regulated in several ways. Among them is nervous stimulation of the cells lining the stomach. In addition, food in the stomach stimulates the mucosa to secrete a hormone, gastrin, that cir-

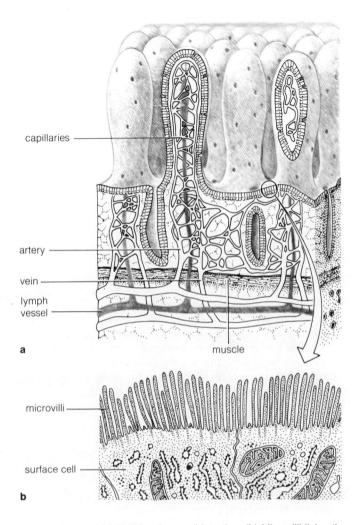

capillaries

artery

vein

lymph vessel

a

muscle

microvilli

surface cell

b

Figure 15–14 (**a**) Villi lining the small intestine. (**b**) Microvilli lining the surface of each villus.

culates in the blood back to the stomach wall and there promotes the production of gastric juice (see Chapter 14 for details).

The soupy food *(chyme)* leaves the stomach a little at a time through an opening controlled by a sphincter muscle and enters the small intestine. Most digestion occurs here, as enzymes from the liver, pancreas, and gut wall mix with the food and break it down. We do not know all the steps involved, but we do know that the process is initiated by contact between the acidic chyme and the intestine wall. The intestine wall releases a hormone, secretin, into the bloodstream (Chapter 14). Secretin promotes the secretion of bile from the liver, via the bile duct, and of pancreatic juice from the pancreas (Figure 15–13). The bile may be stored temporarily in the **gallbladder** to provide the large amounts needed after

a big meal. Bile is composed of *bile salts* that dissolve fat droplets much as soap does, making the fat molecules available to the action of the lipases that digest them. These lipases are enzymes secreted by the pancreas, and they enter the small intestine through the pancreatic duct. The pancreas also secretes proteolytic (protein-digesting) enzymes and an amylase like that found in saliva. The proteolytic enzymes from the pancreas and from the gut wall split only specific kinds of chemical bonds, as summarized in Table 15–2.

The small nutrient molecules released by the enzymes and bile salts are absorbed in the small intestine. Some of this absorption occurs by way of direct diffusion. In vertebrates, however, sugars and amino acids are taken up from the gut by active transport into a system of blood vessels, which carries them to the liver for further chemical processing, whereas fat is emulsified and absorbed into the lymphatic system (Chapter 17) as neutral triglycerides. The small intestine's structure is well adapted to its absorptive tasks. For one thing, it is long— some 23 feet in humans—so that the total surface area available for absorption is great. But length is only one factor in the enormous surface area of the gut. The inner intestinal wall is folded, and these folds are covered with projections, the **villi**, which are themselves covered with microscopic projections, the *microvilli* (Figure 15–14). The effect of these repeated evaginations (protrusions) is to increase the available surface area exponentially.

The walls of the intestine secrete large amounts of fluid into the gut. If all this fluid were lost, we would quickly dehydrate. Recovering this water is a principal function of the large intestine, also called the *colon*. In addition, the colon absorbs food molecules and nutrient elements, including many released by the bacteria and protozoans in the gut itself. These microorganisms are particularly important to such herbivores as termites, rabbits, and cows, in which they secrete cellulase, an enzyme that breaks down the cellulose of the ingested plant material. In humans, the microorganisms produce vitamin K and some of the B vitamins, digest proteins and amino acids, and prevent serious infections by invading disease-causing bacteria. The microorganisms can be killed off by prolonged use of antibiotics, upsetting the normal workings of the intestine, but once the antibiotic assault ceases, the microorganisms reenter the intestines and colonize them quickly.

The product of the large intestine, the feces, is stored briefly in the rectum and then voided through the anus. About 20 percent of the fecal mass is made up of live bacteria and protozoans. The rest is the undigested remains of food.

Digestion in Herbivores Before we leave the topic of digestion and nutrition, let's consider the special case of the herbivores. These animals face a definite digestive problem: They must consume enormous quantities of plant cells that are primarily cellulose to extract even a small amount of nutrients. We have already noted that herbivores have longer intestines than carnivores, creating additional surface area for absorbing the low concentration of nutrients, and that their digestive tracts contain microorganisms capable of digesting cellulose. However, these adaptations are not sufficient. Among large, land-dwelling herbivorous mammals, two digestive patterns are found.

The first is seen in the **ruminants,** such as cattle, sheep, deer, and antelopes. The ruminants have four-chambered stomachs. The first and largest chamber, the *rumen,* is followed by the *reticulum.* These chambers are thought to have evolved as expansions of the esophagus. The rumen and the reticulum contain a "liquor" (brothy fluid) harboring cellulose-digesting bacteria and protozoans. Large particles float on the surface of the liquor and are regurgitated and chewed ("chewing the cud") until they either dissolve or are small enough to stay in solution. The bacteria and protozoans living in these fermentation tanks produce large amounts of formic, acetic, propionic, and butyric acids. These volatile fatty acids, which may supply as much as 90 percent of a cow's energy needs, are absorbed directly through the stomach wall. The food solution is then swallowed into the third chamber, the *psalterium,* where excess water is removed. The resulting pulp passes into the fourth chamber, the *abomasum,* where conventional digestion largely takes place.

The advantage of the ruminant system is that most of the ingested plant material is ultimately broken down so that it yields nearly all the nutrients available both within the plant cell and in the cell wall. Obviously, the woodier the food the animal ingests, the more times it must be recycled as cud, and the slower the overall digestive process. Since most plant material contains little protein, the animal is forced to ingest large amounts of food. As a result, the ruminant rarely lacks for carbohydrates as an energy source, but the slow digestion of woody foods may produce a protein deficiency.

A nonruminant herbivore like the horse processes and uses food much differently. Some cellulose is digested by microbes in the colon and in the pouch of the colon called the *caecum,* but this occurs too far back in the digestive tract for the horse to be able to use the carbohydrates produced. This means that a horse has to get its carbohydrates much as a human does. Food passes through the horse's digestive tract at twice the speed it moves through the cow's (from 30 to 45 hours compared to 70 to 100 hours), allowing twice as much food to be consumed for the smaller amounts that are used. The nonruminant digests plant material much less completely than the ruminant—a fact that can be readily verified by comparing horse and cow manure—and the nonruminant extracts less of the available carbohydrates.

Summary

All organisms must acquire nutrients from the environment and biochemically transform them into usable forms. The major nutrients include a source of raw materials for building blocks of the body (principally carbon and nitrogen), sources of energy, and minerals. Plants derive their carbon from the air, their energy from sunlight, and their minerals from the soil, and we examined some of the relationships between soil and plant nutrition. Directly or indirectly, animals derive their chemical building blocks and their energy, as well as most of their minerals, from plants. Symbiotic microorganisms may also be used as carbon and energy sources. In addition, animals must have various organic molecules (vitamins) that they are unable to synthesize. The role of the various nutrients and the effects of their deficiencies were also discussed in this chapter. Finally, we surveyed some of the diverse mechanisms whereby animals derive food from the environment and how, through digestion, they convert ingested food to molecules that the cells can use.

Supplement 15–1
Human Nutrition

In this chapter we have discussed the question of human nutrition in general terms, but what of the specifics? What do you and I, as residents of a highly industrialized Western nation, need in our diets? What do we get? What effect does food-processing have on our nutrition? These and other questions are the subject of this supplement, written by Katherine Ware of the Department of Nutrition, University of California, Berkeley.

Most of us are concerned about our health. We want to feel well and live long. Such good health is the product of many important factors, not the least of which is good nutrition. Strong physical and mental growth, body maintenance, and general well-being stand upon the foundation of good nutrition. The fact is both obvious and inescapable. Less obvious, however, are the ways in which one can achieve a consistently good diet. Our first nutritional requirement is energy, and without sufficient calories the other nutrients are of little help. On the other hand, calories by themselves cannot keep a person healthy, or even alive, for very long. So the first step toward good nutrition must be to establish the individual's requirements, then find a wholesome diet that will satisfy them.

The most up-to-date information available on the nutritional requirements of the U.S. population is compiled by the Food and Nutrition Board of the National Research Council and published under the title *Recommended Dietary Allowances.* This publication is revised about every five years to keep pace with advances in the field. Its recommendations are applicable to virtually all healthy Americans. For the general public ("average" people of "average" age in "average" situations), the Food and Drug Administration has condensed this volume into a simpler form called the *U. S. Recommended Daily Allowances* or RDA. Legislation now requires all food manufacturers to print on package labels the percentage of this recommended daily allowance their products supply. Consumers can use this information to figure their daily nutrient consumption and, incidentally, to compare similar products and shop for the best nutrient buys.

Normal people will meet all of their requirements in a varied, well-balanced diet without the aid of nutrient supplements. A great deal is written, however, about the quality of food available in this country, the possibility of maintaining a good diet with such food, and the alternative solutions to the problem—if there is a problem. So let us examine briefly some aspects of this situation.

Are Processed Foods Unwholesome? For someone in our society to eat even a single meal without consuming any processed foods is unlikely. Many foods we seldom think of as processed are, in fact, processed. For instance, milk bought at the store is processed by pasteurization to remove possible pathogens. Beef is aged—another form of processing. Most people regard these kinds of processing as good and wholesome. Another sort of processing includes the addition of coloring, taste modifiers, and preservatives whose immediate effects on consumers are neutral at best.

But what about highly processed foods, where substantial changes are made in the product? What about dehydrated mashed potatoes, canned fruits and vegetables, corn chips, and soft drinks? Such highly processed foods sacrifice nutrients for form and storage ability. In the processing of white flour, some nutrients are removed to improve shelf life or to appeal to the eye of the consumer ("white" flour is considered more attractive than brown). That is, some foods are *deliberately* made less nutritious so that other organisms will not consume them and thus cause the product to spoil. In other cases, the destruction of many nutrients is unfortunate, but inciden-

tal to the processing. One thinks of canned fruits and vegetables, whose nutrients often leach into the packing liquid or are destroyed by high temperatures during sterilization.

To judge the value of extensive processing, one should balance these losses against the value of having available many foods that would not be in our diet at all if they were not processed. However, it can safely be said that, in general, fresh foods are superior to highly processed foods, and we should try to eat at least half our food in an unprocessed or lightly processed state.

How Good Is a Vegetarian Diet? The Western diet, with its heavy consumption of animal products, is the exception rather than the rule among the world's cultures. Most populations rely on grains and legumes as their staple foods. People can survive in good health on a vegetarian diet if they go about it in the right fashion. In the United States, where being a vegetarian is usually not a desperate matter of survival, people have the choice of becoming lacto-ovo-vegetarians (including dairy products and eggs in their diet) or strict vegetarians. Lacto-ovo-vegetarians can meet all of their nutritional requirements if they follow a sensibly varied diet that includes plenty of milk, cheese, eggs, and legumes. They should conscientiously select foods that are high in iron and zinc, as these are the two nutrients most likely to be lacking in a vegetarian diet. (Some foods high in iron are soybeans, lima beans, green leafy vegetables, dried apricots, and dried prunes; foods high in zinc include legumes, nuts, and whole grains, especially the germ and bran.) Even a growing child can survive and grow well on a lacto-ovo-vegetarian diet.

Strict vegetarians face a more difficult task in trying to meet all their nutritional needs. A child on this diet is likely to suffer irreparable mental retardation or die of malnutrition unless given a supplementary source of high-quality protein. No one fruit or vegetable product can supply all the essential amino acids in the proper balance. Foods must be thoughtfully combined to compensate for amino acid deficiencies (soy flour mixed with wheat flour, for instance, or soy with rice). The possibility of iron and zinc deficiencies exists here as well. In addition, the strict vegetarian may also get too little calcium. But—most important—the strict vegetarian will have no source of vitamin B_{12}, which is found only in animal products. A B_{12} deficiency manifests itself in such symptoms as a sore tongue, loss of weight, weakness, back pains, apathy, and mental and neural abnormalities. Ultimately, a lack of B_{12} will result in death. Vitamin B_{12} supplementation is necessary to a strict vegetarian diet and need not seriously disrupt the regimen, for B_{12} supplements are derived from special bacteria and fungus cultures grown just for this purpose.

It may be prudent to note here the so-called macrobiotic diet, a strict vegetarian diet. In this diet, one works by steps toward a goal, diet number seven, consisting solely of brown rice. *A person who strictly follows this diet will die of malnutrition* due to the lack of necessary nutrients. The first such death, that of a young woman, occurred in 1965. The macrobiotic diet was labeled as Zen Buddhist by its perpetrator, George Ohsawa, to increase its allure, but the Zen Buddhists deny that the diet is part of their religion.

Synthetic Versus Natural Vitamins   Chemically there is no difference between a pure vitamin isolated from an organic source and one synthesized in the laboratory. Both produce the same results when fed to an animal. Synthetic vitamin pills are, of course, chemically pure, and people who insist upon natural vitamins often believe that other nutrients are present in the crude materials that make up a small portion of the natural vitamin pills. However, as we already said, you can get a full supply of vitamins plus all the other nutrients in sufficient quantities for optimal health, growth, and reproduction by eating the foods themselves, at which point taking either kind of pill becomes redundant and, if done to excess, could become more a problem than a help.

Can Vitamins Be Harmful in Large Doses?   Anything taken in a large enough dose can be harmful. Vitamins are no exception. The water-soluble vitamins C, B_{12}, riboflavin, niacin, thiamine, pantothenic acid, biotin, folacin, and pyridoxine and the fat-soluble vitamin E have little known toxicity even at high dosages. However, the fat-soluble vitamins A, D, and K can be highly toxic.

Doses of 50,000 IU (International Units) or more of vitamin A taken daily (the U.S. RDA is 5,000 IU per day) for six months or longer are known to produce hypervitaminosis A. The symptoms in children are irritability, swellings over the long bones, and dry, itching skin. In adults the symptoms are headache, nausea, diarrhea, and possible bone fragility. However, if the source of vitamin A is carotene, there is slight chance of toxicity even at high doses. All that will be observed is an orange tint to the skin.

Vitamin D is extremely toxic in high doses. The U.S. RDA is 400 IU per day. Intakes of even 3,000 or 4,000 IU per day can be harmful. Vitamin D is intimately related to calcium metabolism. Excess Vitamin D can lead to decalcification of the bones, resulting in increased levels of serum calcium with deposition of calcium in the soft tissues (blood vessels and organs). Toxic symptoms in adults are vomiting, diarrhea, weakness, loss of weight, kidney damage, and bone fragility. In addition, infants suffer retarded growth.

Vitamin K_3 (menadione, a synthetic derivative of the natural form of vitamin K) is used in medical practice as a blood coagulant. However, even low doses (5 mg per day) have caused such drastic symptoms in rats that its sale as a food supplement has been prohibited in the United States. Vitamin K_3 is known to cause jaundice in infants. The natural forms of the vitamin (K_1, K_2) have not exhibited signs of toxicity even at high doses.

These few issues have been of concern to nutritionists for many years now and are likely to remain popular topics for some time to come. A great deal of research will be done before they are laid to rest. But for the present most of the conclusions drawn regarding these everyday issues of diet tend toward a moderate stance, emphasizing the importance of educated judgment as a guide to good nutrition. The familiar well-balanced diet is still the most practical one we know. Special diets like the lacto-ovo-vegetarian diet are adequate when sensibly followed. Knowledge should rule one's use of food supplements and restrain one's inclination to follow food fads.

16

Gas Exchange in Plants and Animals

The concepts of life and fire are closely associated in the human mind. We speak of the spark of life in a new-born baby, a warmhearted person, acting in the heat of passion, and the flame of life going out at death. These ideas are right. We saw in the second law of thermodynamics that energy must be put into living systems to maintain their high degree of organization. This energy is available only through the controlled burning of the products of photosynthesis during cellular respiration. Through cellular respiration, the energy of the molecular bonds of organic molecules is released for use in cellular work. During respiration, as during a fire, oxygen is used up and carbon dioxide and heat are given off.

The need for oxygen presents organisms with a problem, because these energy transformations occur inside all the organism's living cells. Thus, cells in the organism's interior as well as those on its surface need a way of receiving oxygen and removing the carbon dioxide that is a toxic by-product of metabolism. Some organisms with low rates of cellular respiration, such as plants, flatworms, earthworms, and sea anemones, meet their energy needs with gas exchange through their surfaces. But in large, active multicellular animals, with their high rates of cellular respiration and great distances separating interior and surface, direct diffusion falls well short. In this chapter, we will compare and discuss the various ways in which animals have solved the problem of providing their cells with oxygen and removing the carbon dioxide produced by metabolism.

Gas exchange, also known as *external respiration* to distinguish it from cellular respiration, comprises two processes. In the first, called *ventilation,* air or water laden with oxygen is moved from the organism's surface to areas of the body specialized for taking in oxygen and giving up carbon dioxide. The organs of gas exchange, such as lungs and gills, are thin, moist membranes with their supporting structures. It is essential that the membranes be moist, since living organisms can absorb sufficient oxygen only when it is dissolved in water. The membranes must also be thin enough to allow the gases to cross back and forth quickly.

In the second part of external respiration, *oxygen and carbon dioxide transport,* the gases travel between the areas of gas exchange and the cells. Then, in the cells, the oxygen is used in the metabolic reactions of cellular, or internal, respiration. These same reactions give off carbon dioxide, which is then transported back to the organs of gas exchange for release to the outside.

Plants and Gas Exchange

Plants do not require the specialized ventilatory mechanisms of active multicellular animals. Much of a plant's bulk consists of supportive tissues, which are frequently inert metabolically. As a result, the plant's overall rate of cellular respiration is low. Photosynthesis takes in carbon dioxide and gives off oxygen, complementing cellular respiration and providing for the recycling of gases within the plant itself. Under certain conditions, there is no gas exchange between the plant and the environment. However, usually one process or the other does predominate, and some gas exchange occurs. At night, for example, photosynthesis ceases, but cellular respiration continues, with the plant taking in oxygen and giving off carbon dioxide (Figure 16–1).

The leaf surface is pocked with numerous openings known as *stomata* (singular, *stoma*), or *stomates* (Figure 16–2). Each stoma opens into the intercellular spaces of the mesophyll tissues, and gases move into these spaces by diffusion. All the cells within the leaf are covered by a film of water into which gases dissolve and then diffuse into (and out of) the cells and into the mitochondria and chloroplasts within them. Because of the recycling of gases in the plant, the large number of stomata, the high ratio of leaf surface to volume, and the plant's low metabolic rate, no special pumping mechanisms are

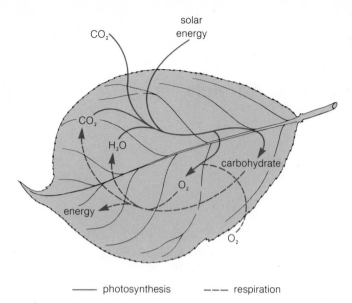

Figure 16–1 Basic schema of gas exchange during energy capture and release in a photosynthesizing leaf. At very low rates of photosynthesis or very high rates of respiration, there is presumably internal cycling of gases between the two processes.

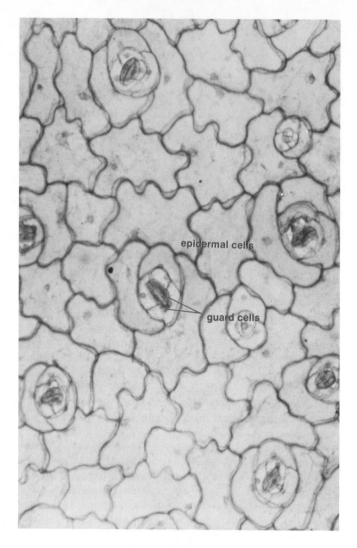

Figure 16–2 Stomata on a leaf.

needed to move gases about inside the leaf. Diffusion alone suffices.

Basically, the stoma is an opening between two highly specialized epidermal cells known as *guard cells.* As Figure 16–3 shows, the action of the guard cells regulates the size of the stoma, opening or closing it as the plant's needs and activities dictate. The guard cells respond to changes in turgor. In many plant species, the guard-cell wall bordering the stoma is more rigid than the far side. To understand how the guard cells work, imagine a long, half-inflated, cigar-shaped balloon. If you tape a piece of rigid cardboard to one side of the balloon and inflate the balloon the rest of the way, the entire balloon stretches under the pressure of the air, but the rigid side obviously stretches less than its flexible counterparts. As a result, the whole balloon is drawn into an arc bending away from the rigid side. Similarly, when the guard cell takes in water osmotically (after taking in sugar actively) and its turgor increases, much the same thing happens: The rigid cell wall bows away from the stoma, opening it. When turgor decreases, the rigid cell wall returns to its original shape and closes down the stoma.

In plants, the problem of water loss complicates gas exchange. The stomata must be open a great deal of the time to allow gas exchange to proceed freely; yet whenever they are open, water escapes. Water loss to evaporation in the aerial parts of a plant is called **transpiration.** Transpiration is no small entry in the plant's biochemical

balance. An adult corn plant contains around 2 liters of water, yet in its few months of life 100 times that amount will be lost to transpiration.

The guard cells hold transpiration to an acceptable level. When water is available and the plant cells are turgid, the stomata will be open. Should the supply of water fall off and the plant lose turgidity, the guard cells become flaccid and close the stomata. This prevents further transpiration and also stops the entry of more carbon dioxide, thus depressing the plant's rate of photosynthesis and conserving even more of its water.

Other adaptations to reduce transpiration have evolved, most noticeably in desert plants, where water loss is a particularly pressing problem. In some species, the stomata appear only on the ventral side of the leaf,

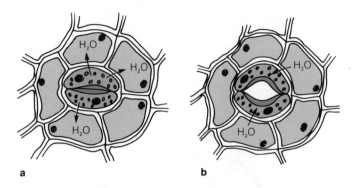

a   b

Figure 16–3 Operation of guard cells. (**a**) Stoma is closed after water has been withdrawn from the guard cells. (**b**) Stoma is open when turgor pressure bends guard cells into curved shape.

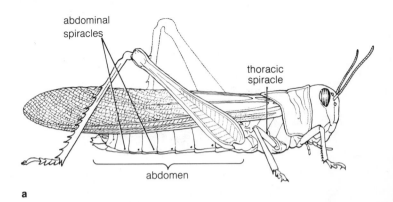

a

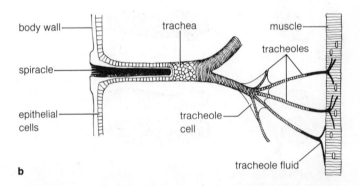

b

Figure 16–4 (**a**) The body parts of an insect, showing the location of the spiracles. (**b**) The diagram shows the relationship of the spiracle and tracheoles to the tracheae.

in the shade. The stomata are also often located in depressions in the leaf surface, reducing air movement across the stomatal opening and lowering the rate of evaporation. The yuccas and the cacti have added a biochemical wrinkle. In most plants, the stomata are open during the day, when sunlight is available to power photosynthesis and CO_2 fixation. The yuccas and the cacti have evolved the ability to fix CO_2 at night, store it until day, and release it then for photosynthesis. These plants can thus photosynthesize with their stomata closed, greatly reducing the amount of water lost to the heat and aridity of the desert.

Animals and Gas Exchange

Most terrestrial animals respire at rates 100 or more times higher than those of the plants, and their gas-exchange needs are correspondingly greater. Direct diffusion through the surface fulfills the ventilatory requirements of a few land dwellers, among them the common earthworm. In most others, however, invaginations, or infoldings, of the animal's exterior lead inside the body to the cells or to specialized organs. In the insects, these channels connect respiring cells directly to the outside atmosphere, and gases are exchanged by diffusion between cells and channels. In higher animals, the circulatory system is the means by which gases are transported from the respiratory surface to the individual cells. (The other functions of the circulatory system will be examined in Chapter 17.)

Terrestrial Insects Flying insects have the highest weight-specific metabolic rates known in the animal king-

dom, sometimes using more than 200 milliliters of oxygen per gram of muscle tissue per hour in flight. However, 99 percent of this metabolic activity occurs in the flight muscles of the thorax, the portion of the insect's body between head and abdomen, where the wings are attached. The abdominal segments behind the thorax contain no large muscles and therefore have minimal respiratory needs. The insect's hard, chitinous outer covering, or cuticle, protects against mechanical damage and greatly reduces water loss to evaporation, but it also prevents gas exchange through the surface. Specialized organs of external respiration are needed.

These organs are the *tracheae,* a system of finely branched air tubes (Figure 16–4). The smallest of these branches, the *tracheoles,* terminate at or inside the individual cells, where oxygen and carbon dioxide are exchanged. The tracheae are connected to the outside by small openings called *spiracles.*There are three pairs of spiracles in the thorax and six to eight pairs in the abdomen, one pair for each segment of the insect's body. A device analo-

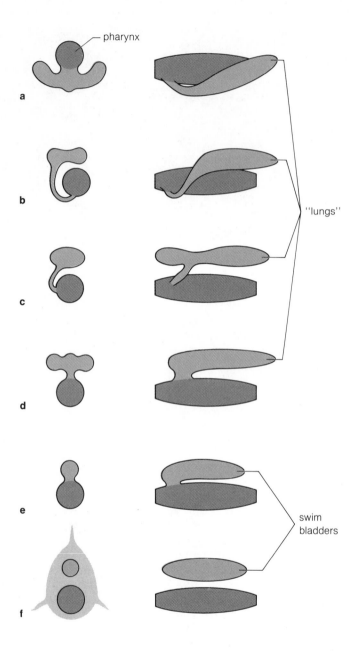

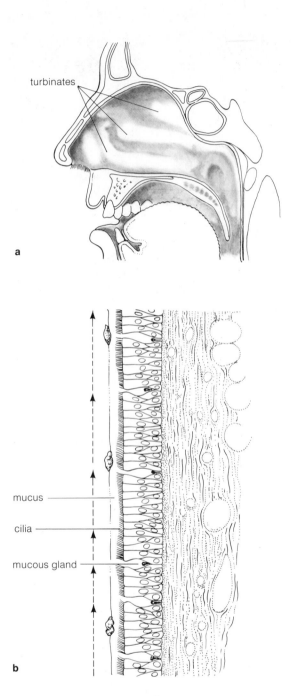

Figure 16–5 The evolution of lungs and of the swim bladder. Shown are successive stages in the possible evolution, in which evaginations of the pharynx gave rise to lungs and then to a swim bladder. The arrangement seen in (**a**) is observed in some present-day lungfish. (**b, c**) Hypothetical stages. The arrangements seen in (**d**), (**e**), and (**f**) are observed in modern fishes. In modern vertebrate animals, the trachea leading to the lungs is a dorsal evagination from the pharynx, suggesting that it may have arisen most directly from (**d**).

Figure 16–6 (**a**) The nasopharynx protects the lungs from foreign matter and large temperature fluctuations. Hairs in the nose and the convolutions of the turbinate bones help entrap particles. (**b**) Smaller particles settle on the walls of the trachea, bronchi, or bronchioles, where mucus-covered cilia beat in waves and carry the particles up to the pharynx for expulsion.

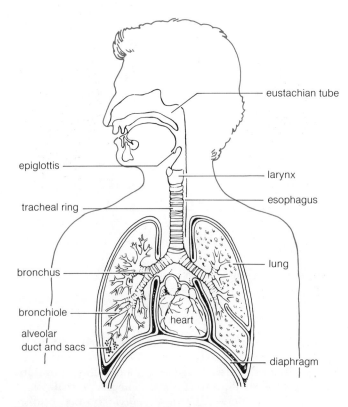

epiglottis

tracheal ring

bronchus

bronchiole

alveolar
duct and sacs

eustachian tube

larynx

esophagus

lung

heart

diaphragm

Figure 16–7 The ventilatory system in humans.

gous to the guard cells of the stomata moderates the size of the spiracle, but it functions differently, opening or closing with muscular action directed by the central nervous system. In addition, active insects breathe by actively pumping air.

Insects, like all terrestrial animals, face the never-ending problem of desiccation. One of the advantages of an internal respiratory system is that it conserves water by enclosing the moist membranes of gas exchange within the body, where they can be sealed from the exterior air by the spiracles when the insect is inactive. Insects also cut water losses by reducing their resting metabolism to 1 percent or less the flight rate and by shutting their spiracles.

Terrestrial Vertebrates Most land-dwelling vertebrates are covered with a relatively impermeable coat that protects them against damage and desiccation and, as a side effect, precludes direct gas exchange at the body surface. Vertebrates with mosit, thin skins, such as the frog, do ventilate through the skin, but even these animals also have an active ventilatory mechanism involving lungs.

The lungs of the vertebrate, like the tracheae of the

insect, evolved as invaginations into the body cavity (Figure 16–5). The invaginations that eventually evolved into lungs probably arose first in primitive fishes adapting to a semiterrestrial way of life. The respiratory sac developed originally as a ventral outpocketing of the pharynx, the part of the digestive tract directly behind the mouth. In time, the sac migrated to the dorsal side, its location in the modern lungfish, which follows the semiterrestrial lifestyle of its ancestors and can breathe air. The pharyngeal respiratory sac of the ancient fishes that left the water for the land evolved into modified and elaborated organs that led ultimately to the lungs of the amphibians, reptiles, birds, and mammals. In many of the totally aquatic modern fishes, the original connection between the air sac and the pharynx has been lost. This isolated air sac acts not as a respiratory mechanism but as a hydrostatic organ—the *swim bladder*—helping the fish maintain its balance and buoyancy.

Internal respiratory surfaces confer numerous advantages on the organism. As we have seen, their position inside the body reduces water loss. Also, the internal membranes can be folded and looped about to yield a respiratory surface far greater in area than the animal's skin. Finally, internal muscular action can move the gases actively over the respiratory membranes, allowing for more rapid gas exchange and a higher metabolic rate than would be possible with unaided diffusion and convection.

The human respiratory system exemplifies the mammalian way of breathing. Air taken in through the nostrils passes into the nasal cavity, where it is cleaned, warmed, and humidified. The *turbinate bones* of the nasal cavity force the air into numerous channels lined with mucous membranes that warm the air and trap dust particles suspended in it (Figure 16–6a). From there it moves into the pharynx, which also acts as an air conditioner and filter. In the lower pharynx, the air enters the trachea, a rigid tube that leads down toward the lungs (Figure 16–7).

Ingested food and inhaled air both pass through the pharynx. This is an obvious anatomical hazard, since food accidentally entering the trachea can block the respiratory system and cause death in minutes. To prevent such a mishap, swallowing is accompanied by an involuntary reflex that tips a muscular flap, the epiglottis, down over the tracheal opening, or glottis, directing the food mass away from the trachea and into its proper path.

The trachea branches into two *bronchi* (singular, *bronchus*), each of which leads to a lung (Figure 16–7). The tracheal and bronchial linings are coated with cells that secrete mucus that traps any remaining dust and foreign matter. The linings also contain ciliated cells that wave in unison, pushing the mucus up into the pharynx, where it is disposed of by swallowing or spitting (Figure 16–6b).

In the lungs, the bronchi divide into approximately

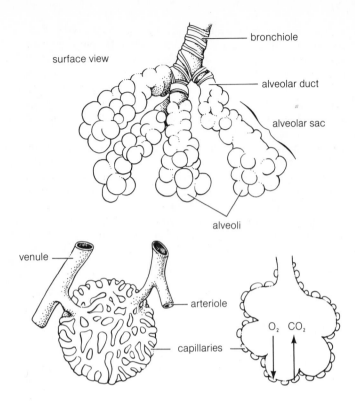

surface view

bronchiole

alveolar duct

alveolar sac

alveoli

venule

arteriole

capillaries

O₂ CO₂

Figure16–8 The alveoli and the capillaries that cover them.

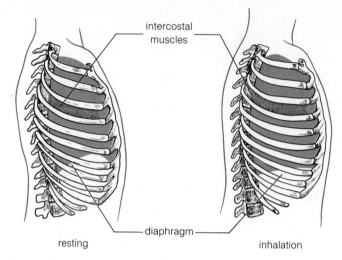

intercostal
muscles

diaphragm

resting

inhalation

Figure 16–9 The breathing mechanism.

a million *bronchioles.* The bronchioles lead ultimately to the tiny bulging air sacs called *alveoli* (Figure 16–8), which in a pair of adult human lungs number about 300 million. The membranous walls of the moist alveoli are surrounded by a network of capillaries, and through these walls the blood in the capillaries takes up oxygen and gives off carbon dioxide. The repeated branchings and pocketings of the bronchioles and alveoli greatly expand the respiratory surface of the lung. In the human, the total area of gas-exchange membrane is some 70 square meters, about 40 times greater than the surface area of the body.

Ventilation of the human lung entails two muscular mechanisms. In the first, the muscles of the rib cage (the *intercostals*) contract and raise the ribs. The second, occurring at the same time, involves the *diaphragm,* a sheet of muscle separating the chest cavity from the abdomen, which contracts and moves downward. Both muscular actions enlarge the volume of the chest (Figure 16–9). Because of this negative pressure, air rushes in from the outside and down into the lungs. The relaxation of the intercostals and the diaphragm, and the rebound of the stretched elastic lung tissue, return the chest to its former, smaller size, squeezing the air out in an exhalation.

The lungs do not fill and empty completely with each breath. The lungs' total maximum capacity, called *vital capacity,* is the difference between maximum inspiration and maximum expiration (for the college-age male, approximately 6.7 liters). The *tidal volume* (normal intake plus exhalation) is 0.5 liter. After exhaling the tidal volume, one can forcibly expire another 1.5 liters or so (the expiratory reserve), but there is still 1 to 1.5 liters *residual volume* in both lungs (total) that cannot be expired. Total lung volume is equal to the vital capacity plus the residual volume.

Birds breathe differently from mammals and much more efficiently. At 3,000 to 4,000 meters above sea level, a human not acclimated to high altitude shows oxygen deficiencies (see Supplement 16–1). At 6,000 meters, a mouse is immobile from lack of oxygen, but at this same altitude, sparrows can still fly because birds extract more oxygen from inhaled air than do mammals. Instead of drawing air in and then squeezing it back out, avian lungs are a one-way system. They are so constructed that air flows through them at both inhalation and exhalation (Figure 16–10), yet air passes only once through the lungs. Large inflatable air sacs extend beyond the lungs into the body cavity and even into the interiors of the wing bones. To allow this one-way air movement and the complete ventilation it affords, tiny loops of membranous tissue in the lungs serve as the gas-exchange surface, instead of the dead-end alveoli of the mammals.

In birds, two breaths are necessary to move a gas molecule, say nitrogen, through the system. On inhalation, the molecule travels through the trachea to the posterior air sacs; on exhalation, the molecule passes from the posterior air sacs into the lung. During subsequent

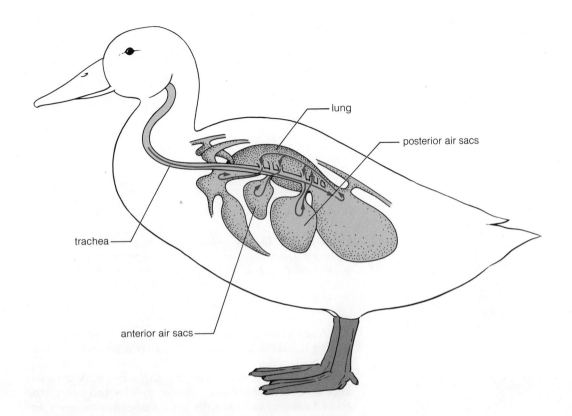

lung

posterior air sacs

trachea

anterior air sacs

Figure 16–10 The avian respiratory system. During inspiration the air sacs are filled, and during exhalation the lungs are filled and the anterior air sacs are evacuated.

inhalation the molecule of nitrogen completes its path through the lung and moves to the anterior air sacs. During the exhalation, the molecule is voided from the anterior air sacs to the outside via the trachea.

Control of Ventilation In all vertebrates, the rate of ventilation is controlled by a respiratory center in the medulla oblongata of the brain stem (Figure 16–11). Acidity resulting from carbon dioxide dissolved in the blood activates sensors in the walls of the arteries. In response to nerve impulses from these sensors, the respiratory center relays nerve signals to the diaphragm and the intercostals. As the muscles contract, the chest expands, the lungs are inflated, and stretch receptors in the lungs are activated. These receptors send nerve impulses to the medulla, which ceases its signals to the chest muscles and diaphragm. The muscles relax, and the breath is exhaled.

Of course, breathing is also under a limited measure of voluntary control—we can consciously stop breathing, or breathe faster. But we can stop only for a short time. Once the concentration of carbon dioxide in the blood builds to a certain point, the respiratory center overrides all commands from higher brain centers and breathing begins again, whether we want it to or not. There is evidence that stretch receptors in the major arteries plus

hormones, wakefulness, and perhaps sensory receptors in the muscles also play a role in breathing, but their precise interaction is not yet understood.

Gas Exchange in Water

Organisms inhabiting aquatic environments face problems different from those faced by the air breathers. Air contains a relatively constant proportion of oxygen, about 21 percent, but the amount of oxygen dissolved in water varies with the degree of atmospheric contact and with temperature. At 5°C, water saturated with air contains 0.92 percent oxygen, but at 20°C, saturated water is only 0.65 percent oxygen. This means that water at 5°C contains 23 times less oxygen than does air, and at 20°C, 32 times less. Oxygen in water is not only scarcer; it is also harder to extract. Oxygen's diffusion rate in water is 300,000 times slower than in air.

In small aquatic animals with low metabolic rates, like the jellyfish and the hydras, diffusion alone provides for external respiration. The same holds for aquatic plants, even such large varieties as the kelps, which are thin and many-branched and therefore have increased surface area available for gas exchange.

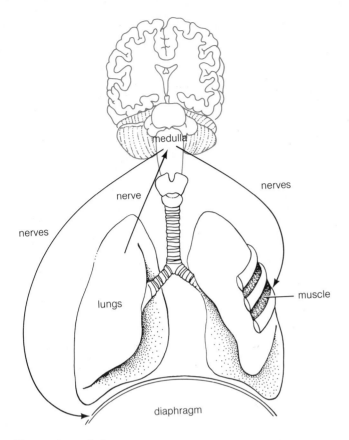

Figure 16–11 Reflex control of respiratory rhythms.

Figure 16–12 A whirligig beetle carrying a bubble of air beneath the surface.

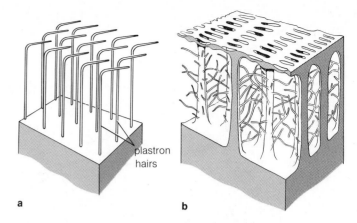

Figure 16–13 Plastrons on some aquatic insects consist of tiny hairs (**a**) or of a framework of microscopic projections (**b**) covering almost any body area, such as legs or abdomen, depending on the species.

Aquatic animals with high metabolic rates face a difficult gas-exchange problem. To get enough oxygen, they must move water actively over their respiratory surface. However, since water has a high viscosity—that is, resistance to flow under force—pumping it in and out of the body interior would require great amounts of energy. Instead, the respiratory surfaces occur on the outside of the body, an adaptation impossible for terrestrial animals because the membranes would dry out. Desiccation is no problem for organisms living in water, and the water itself supports the delicate membranes of gas exchange and protects them against collapse.

The commonest method of ventilation of gas exchange among active aquatic organisms is by the use of *gills.* Gills are exterior evaginations of the body, which, like lungs, are richly supplied with blood. Gills, although all essentially similar in their workings, occur in a wide variety of forms. In the starfish, the gills are little bumps on the skin. In many aquatic insects and annelid worms, the gills are lateral projections that look much like flaps.

Gills are most highly developed in the fishes. Concealed and protected behind the bony flap of the *operculum* (plural, *opercula*), the fringelike gill filaments are subdivided microscopically to yield a large, capillary-rich membrane. Most fresh-water fishes ventilate their gills actively. With its opercula closed and its mouth open, the fish expands the opercula, drawing water in. Then, with its mouth closed, it raises the floor, forcing the water back over the gills and out through the opercula. Certain fishes of the open ocean, however, do not ventilate their gills actively. Instead, the tunas, mackerels, and certain sharks must swim continuously with their mouths open, stopping only briefly, to keep water flowing over their gills. If restrained, these fish quickly suffocate.

Insects have developed respiratory adaptations to aquatic life other than gills. The diving beetle that submerges with a bubble of air in tow is a familiar example (Figure 16–12). The bubble is more than a simple storage tank. As the beetle respires and withdraws the bubble's oxygen, oxygen diffuses from the water into the bubble to replace it. Thus the bubble acts as a *physical gill,* in distinction to the biological gills just examined.

The *plastron,* shown in Figure 16–13, carries the principle of the physical gill even further. The plastron typically consists of a layer of fine hairs or other microscopic projections trapping a thin film of air. These hydrophobic (literally, "water-hating") hairs resist wetting and keep the film of air from collapsing. The air, in contact with the spiracles, works as an air reservoir and a physical gill. The plastron may cover a part or the whole of the insect's body, providing much more gas-exchange surface than the diving beetle's bubble. As long as the water contains enough dissolved oxygen, the plastron can satisfy all the submerged insect's respiratory needs.

One major group of aquatic animals that does not use gills is the marine mammals, such as the seals and sea lions and the whales, porpoises, and dolphins. These animals, like the rest of the mammal class, breathe air. They evolved from land mammals, and they have retained the lungs of their forebears. This is not simple evolutionary stodginess; gills would confer no advantage on the marine mammals. These warm-blooded creatures have high metabolic rates, and ventilating gills rapidly enough to satisfy their needs would require tremendous expenditures of energy, expenditures far in excess of what it costs them to breathe air.

The question arises: Why not ventilate the lungs with water? Actually, there is nothing in the structure of the lung to prevent it. A mouse confined to a laboratory tank with artificially aerated water can remain under the surface for several hours. In most cases, however, the low tidal volume of the lungs and the low oxygen content of water rule out water-breathing for the high-metabolism mammals.

Summary

Plants and animals use oxygen and produce carbon dioxide as a result of cellular respiration. Plants use some of this carbon dioxide in photosynthesis, but animals void it as a waste product. In this chapter we have examined the potential problems and evolutionary solutions of ventilating the cells in the interior of terrestrial and aquatic animals, as well as in plants. One of the primary adaptations for efficient gas exchange, the circulatory system, will be discussed in detail in the next chapter.

Supplement 16–1
Up, Up, and Out:
High-Altitude Hypoxia

When the Spaniards began their conquest of the Incas of Peru, they found that as they sought the cities in the mountains they had increasing problems with their own health and that of their animals. They experienced shortness of breath and found themselves taking huge lungfulls of air with little effect. They became dizzy. Some had headaches and found it difficult to concentrate on even simple tasks, while others felt intoxicated. They showed unexpected irritability and personality changes. It was also clear that their horses were having a variety of similar difficulties. Yet the Incas were having no such problems and seemed well adjusted to the heights at which they lived. Moreover, the Spaniards found that after some time in the mountains their breathing stopped being labored and, for most, the other symptoms disappeared.

What the Spaniards (and many of the athletes in the 1968 Mexico City Olympics) had experienced was the effects of high-altitude hypoxia, or oxygen deficiency. The problem became acute during the early days of aviation for, as improved airplanes could fly higher, the pilots reached air with less oxygen, just as the Spaniards had as they climbed into the Andes. As one goes up in the atmosphere, the concentration of molecules decreases, so that by 10,000 feet (roughly the altitude of Cuzco, Peru), only some 65 percent of the oxygen found at sea level is present.

Yet humans have adapted to life at such altitudes and even higher with few ill effects and are capable of hard physical work. Careful studies of peoples living at such altitudes, either as natives or as newcomers, have revealed some of the many changes that occur in the body to accommodate the decrease in oxygen.

The first and most rapid response to low oxygen concentration is an increase both in the number of breaths and in the depth of breathing. This increases the oxygen available to the lungs and the blood, but it has some negative side effects. At the same time that hyperventilation is increasing the amount of oxygen, it is decreasing the amount of carbon dioxide. If this is not corrected, it may change the pH of the blood from its normal state (pH 7.4) to a more alkaline state, resulting in dizziness, vision difficulties, and nausea.

With time, the rate of breathing decreases as other adaptations occur, but the rate of ventilation of a person not born and raised at high altitudes will always be somewhat higher there than that of a native. One reason for this is that highland natives have greater lung volume, particularly greater alveolar area combined with an increased number of capillaries. This means that they have greater access to the available oxygen than does the lowland person. Moreover, once development has ceased in the child, lung capacity is relatively fixed; marked enlargement does not occur in the adult.

Second, at high altitudes, in response to the lack of oxygen, the bone marrow is stimulated to increase the production of red blood cells. This response begins immediately, but proceeds slowly. As a result, at 10,000 feet, the blood contains some 5 to 8 million blood cells per cubic millimeter, compared to 4.5 million at sea level. Not only are more red blood cells produced, but they are richer in hemoglobin. Through these two mechanisms, the oxygen-carrying capacity of the blood at high altitudes is increased.

Third, for the oxygen in the blood to be of use, it must diffuse from the blood into the cells. At high altitudes this is aided by the dilation of existing capillaries and through the development of additional capillaries. The effect of increasing the capillary network is greater blood flow in the tissues, resulting in more ready diffusion per unit of time. In highland natives, a whole set of changes in the circulatory system aids in effecting more rapid movement of the blood to the lungs and then to the tissues. One of these is the increased size of the right ventricle of the heart (see Chapter 17), that part of the heart immediately involved with blood movement to the lungs.

Finally, there are still other adaptations to high-altitude hypoxia, some of which are not yet clearly understood. There is evidence that different biochemical pathways may be involved in cellular respiration, and other biochemical reactions may also be changed at high altitudes.

Given time, humans can undergo remarkable adaptation to high altitudes and lack of oxygen. The adaptation is most complete when individuals grow up under these conditions. It is never quite as complete or effective if they have matured at a lower altitude and then moved higher. What we should all remember is to allow our bodies to undergo some adaptation to high altitudes before we attempt to be physically active in such areas.

17

Transport Systems

Today's orthodoxy, in science as well as religion, is often yesterday's heresy. Practically every adult American who has had even a minimal education knows that blood circulates through the body. Yet the man who first published this idea met nothing but ridicule from the learned people of his time.

The Roman physician Galen, who died around A.D. 200, maintained that blood was formed from ingested food and consumed in the tissues. In Galen's view, blood moved outward from the heart and liver in tidal movements of ebb and flow. This view, although questioned in its specifics by such thinkers as Leonardo da Vinci and Vesalius, dominated European thought for centuries.

Then William Harvey, a successful royal physician of the early seventeenth century, drew together information gathered from his own experiments on cadavers and animals and from his observations of the movement of blood in wounds and concluded that blood is not consumed. Instead, he was certain that blood circulated from heart to tissues and back again. In his tract on the subject, Harvey wrote, "When, I say, I surveyed all this evidence, I began to think whether there might not be *a motion as it were in a circle.*" The idea of circulation cost Harvey dearly. His medical practice fell off, and people in the street thought him "crack-brained."

Harvey's work, based as it was on careful observations rather than armchair philosophizing, was important in the foundation of modern science and the origin of contemporary physiology. Harvey uncovered the mechanics of a basic physiological fact. How does this fact relate to the needs of the cells?

Whether cells exist as individuals or as parts of a multicellular organism, they must be bathed in a watery medium that provides nutrients and oxygen and removes wastes. This need, as we have seen, is met simply in the one-celled organisms. The larger, more complex organisms, however, whether plant or animal, have more of a problem. Their different cells all have limited (specialized) capabilities, but they all have the same basic needs.

How are these needs met? The answer lies in the circulatory system. Lungs and gills, stomata, the digestive tract, kidneys, and other organs have evolved and specialized to regulate various aspects of the common watery medium that bathes all cells. All the individual cells of the body necessarily lie at some distance from one or the other of these service centers. For this reason, in both animals and plants, the organs and the cells of other tissues are linked by transport systems, carrying the gases, nutrients, and other dissolved molecules back and forth among them. The transport systems have evolved to take on several functions besides transport, some of which we will discuss here.

Since one of the many tasks of the transport system is gas exchange, this chapter extends the discussion of the previous chapter. We will begin our study of transport with plants, then examine circulation in humans and lower animals, and conclude with the steady-state functions of the human circulatory system.

Transport in Plants

Plants face two transport problems. The first is to get water and other nutrients from the soil to the leaves, where most of the plant's photosynthesis takes place. The other is to move the products of photosynthesis from the leaves to the parts of the plant that need and use them. Vascular plants have evolved two tissue systems to handle these problems: the xylem for water movement and the phloem for the translocation of the products of photosynthesis. In Chapter 11 we described the cells present in these tissue systems. Now we will see how they function.

Water Movement in Plants For the leaves of a tree to survive, their cells must be supplied with water and

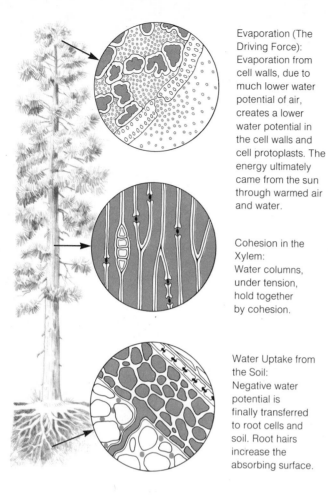

Evaporation (The Driving Force): Evaporation from cell walls, due to much lower water potential of air, creates a lower water potential in the cell walls and cell protoplasts. The energy ultimately came from the sun through warmed air and water.

Cohesion in the Xylem: Water columns, under tension, hold together by cohesion.

Water Uptake from the Soil: Negative water potential is finally transferred to root cells and soil. Root hairs increase the absorbing surface.

Figure 17–1 Water movement in plants according to the transpiration-tension-cohesion theory.

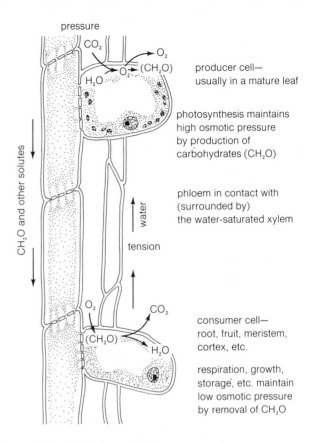

pressure

producer cell— usually in a mature leaf

photosynthesis maintains high osmotic pressure by production of carbohydrates (CH_2O)

phloem in contact with (surrounded by) the water-saturated xylem

consumer cell— root, fruit, meristem, cortex, etc.

respiration, growth, storage, etc. maintain low osmotic pressure by removal of CH_2O

Figure 17–2 Carbohydrates (CH_2O) are produced in leaves and used in other tissues. According to the pressure-flow theory, the difference in osmotic concentration between these two tissues causes water to move through the xylem according to the osmotic gradient that is created. The consuming cells are supplied with carbohydrates and other nutrients by translocation through the phloem.

minerals from the soil. Even in a small tree this is not easy, but when the tree is a redwood the problem would appear to be difficult in the extreme. Yet we know that water does move upward—and at the appreciable rate of some 40 meters (120 feet) per hour, or roughly as fast as the tip of a sweep-second hand on a clock about 25 centimeters (10 inches) in diameter. Despite many years of research we still are not completely sure how plants manage water movement, particularly over long distances, but we have a hypothesis that we think explains most of the known facts.

This is the transpiration-tension-cohesion hypothesis. It is based on known properties of water molecules and facts concerning water in the plant. Simply, the hypothesis states that water in a plant stem is under tension because of transpiration (evaporation of water from leaves), and since water molecules are cohesive (that is, they cling to one another), water will move up the stem

in response to this tension (Figure 17–1). Let's examine various parts of the hypothesis.

Water is conducted through the hollow cells of the xylem (tracheids and vessels). These cells are joined end to end so that continuous columns of water run from the roots to the leaves; in a tall tree, these columns may be 100 meters or more in length. Thus, a plant stem is much like a collection of extremely long, thin drinking straws bundled together. The xylem does not contain simply inert columns of water, but moving streams in which water molecules and substances dissolved in the water are carried from the bottom to the top of the plant, much like a liquid being drawn through a drinking straw. In a straw, the liquid moves up because we reduce the pressure on the top of the straw and atmospheric pressure pushes the liquid up. This mechanism is not found in plants; the height to which water can move by this means is far below that of a good-sized tree. Instead, the motive

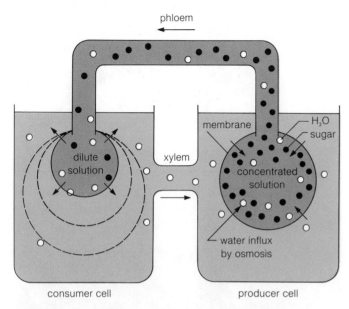

phloem

membrane

H₂O
sugar

dilute
solution

xylem

concentrated
solution

water influx
by osmosis

consumer cell

producer cell

Figure 17–3 A model illustrating the pressure-flow theory of solute translocation. Note that concentration of black particles will control the rate and direction of flow, but white particles (which are much more dilute) will move along in the resulting stream. Broken lines on the left imply that flow may occur due to expansion (growth) of the second osmometer (tissue), as well as movement out through the membrane.

force is generated by water loss from the living cells of the leaves through transpiration. As we saw in Chapter 16, transpiration is the loss of water from these cells by evaporation from their surfaces. This loss of water means that the cells' osmotic potential increases and water will move into them through diffusion from cells with more water. Eventually water moves from the cells of the xylem into the surrounding living leaf cells.

This places the water in the tracheids and vessels under tension (Figure 17–1). At this point, a critical property of water enters the picture, namely its remarkable cohesiveness and adhesiveness to the surface of the xylem lumen. Water molecules exhibit a strong tendency to attract one another, and for this reason a column of water when placed under tension will hold together and not develop bubbles. The small diameter of the tracheids and vessels also strengthens the columns of water. There is evidence that transpiration generates enough of a deficit in pressure to move water to the top of even the tallest trees.

Water removed at the top of the column is ultimately replaced by water that diffuses from root cells, and this water is in turn replaced with that from the soil (Figure 17–1).

Some plants have a second method of water movement, resulting from a phenomenon known as *root*

pressure. In this case, water moves into the root in response to the high osmotic concentration of substances in the root cells, and the entire system acts together to push water up the stem. This is an entirely different method of water movement and is believed to operate only in restricted circumstances and not in trees or the majority of plants.

Translocation Water movement is only one of the functions of the transport system in plants. Carbohydrates and other organic molecules must be moved from one part of the plant to another. These materials—typically sugars (primarily sucrose), amino acids, and hormones—travel in the sieve tubes of the phloem in a movement known as **translocation.** The general direction of translocation is from the point of origin, or *source,* frequently a leaf, to the point of use, or *sink,* usually the cells of the stem, root, or developing fruit. Thus, whereas water commonly moves upward through the plant, translocated substances usually move downward (Figure 17–2). There are, of course, exceptions. In many tree species during early spring, carbohydrates move up from the roots to the developing leaves above.

The mechanism of translocation is not yet fully understood. Of the several current explanations, the most likely—called the *pressure-flow theory*—has dissolved material moving through the sieve-tube elements in response to osmotic pressure gradients that create a mass, or bulk, flow.

Figure 17–3 diagrams an experimental model of the pressure-flow theory. The model comprises two osmometers, which are simply funnels covered with membranes permeable to water but impermeable to solutes. One osmometer contains a highly concentrated solution; the other, a dilute solution. Both osmometers are surrounded by a still more dilute solution, and they are connected by a tube. By osmosis, water moves from the surrounding dilute solution into the osmometer with the highly concentrated solution. The pressure of this added water forces some of the solution in the "producer cell" osmometer through the tube into the "consumer cell" osmometer. In response to this flow, water moves out of the "consumer cell" osmometer into the surrounding solution. Movement from high concentration to low will continue until the solutions in the two osmometers are equally concentrated.

Plants may behave in a similar fashion, but no state of equilibrium is reached. The leaf cells may be analogous to the osmometer with the concentrated solution, for their photosynthesis produces a high concentration of nutrients. The root and other nonphotosynthetic tissues consume nutrients and are thus like the osmometer with the dilute solution. The phloem connects the two systems, and through it moves the bulk flow.

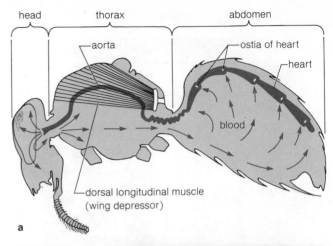

b

b

M. H. Zimmerman

M. H. Zimmerman

Figure 17–4 Study of translocation in the phloem by the use of aphids. (**a**) An aphid hanging upside down on a branch of a tree. Note the droplet of honeydew being exuded from the insect. (**b**) A cross section of the tree showing an aphid stylet that has penetrated to a sieve-tube element.

Figure 17–5 (**a**) Longitudinal section of a honeybee showing the heart extending through the abdomen, continuing in the thorax as the aorta, and emptying in the head. Direction of blood flow is indicated by arrows. (**b**) The heart of a sphinx moth is seen through the transparent body wall as a long dark streak after the scales have been rubbed off.

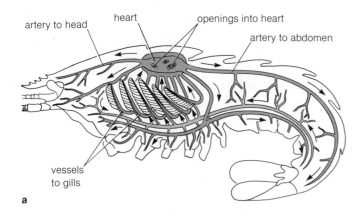

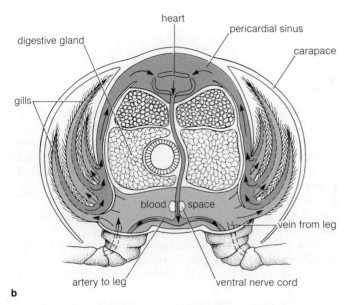

Figure 17–6 (**a**) Longitudinal view and (**b**) cross section of the circulatory system of a crustacean. Arrows indicate direction of blood flow.

Many researchers are still investigating translocation. One difficulty is that translocation is a phenomenon of the whole plant, and almost any manipulation or surgery can alter the very thing one wishes to measure. However, in recent years an unexpected ally has joined the phloem researchers: the aphid.

Aphids feed by piercing plant tissues with their mouthparts (Figure 17–4). Microscopic sections of pierced tissue indicate that the sieve cells themselves are penetrated. Stabbing until they encounter fluid under pressure, the aphids are then force-fed. The aphids extract chiefly amino acids and some sugars from the rapidly passing fluid. The remaining fluid they excrete as liquid fecal droplets (honeydew). Once these droplets were of interest only to ants who "tended" the aphids for this nourish-

ment; now they are collected also by plant physiologists.

One experimental procedure involves supplying the plant with radioactively labeled amino acids or carbohydrates. Then aphids are allowed to feed at various locations, and their fecal droplets are monitored on photographic plates or with radioactivity counters. By noting which aphids (that is, which feeding locations on the plant) contain which radioactive substances, experimenters can determine the rate and direction of translocation without disturbing the plant in more than a minor way. One fact uncovered by this procedure is that material flows both ways in the phloem at the same time.

Despite much excellent research, we remain ignorant about various aspects of the transport system in plants. Transport systems in animals, by contrast, though often apparently more complex, are better understood.

Transport in Animals

Circulation in Invertebrates Animals composed of only a few cell-layers need no special organ system for distributing gases and food molecules to all the cells. Direct diffusion satisfies their needs. In hydra, for example, food molecules and oxygen readily diffuse from the central cavity into the two surrounding layers of cells, and carbon dioxide diffuses from the cells into the cavity or to the outside. No specialized ventilatory or transport systems are required.

More complex animals, however, need greater specialization. In a flying insect, the digestive tract is located in the abdomen, but food molecules are consumed primarily in the flight muscles of the thorax. Nutrients move from the abdomen to the thorax in the blood, which is pumped forward by the heart. The heart is a vessel that extends the length of the abdomen just under the dorsal cuticle (Figure 17–5). The blood enters the heart through holes in its sides called *ostia*. Valves keep the blood from flowing back out through the ostia, and the heart's muscular contractions pump the blood forward into the thorax and head. There the blood leaves the vessel and flows freely over the tissues, following pressure gradients and returning ultimately to the abdomen. This arrangement is called an *open circulatory system*, since the blood is not always confined to a vessel in its course through the body.

In insects, blood transports nutrients and usually plays no role in gas exchange. In a walking insect like the cockroach, the heart can be removed with little harm. In a flying insect like a bee or dragonfly, the heart is much more essential, supplying needed food molecules to the highly active flight muscles. A bee without a heart will not die, but it can only walk, not fly.

Unlike the insect, the lobster and other large crustaceans have specialized local areas for gas exchange, and the blood serves as a transport system for oxygen and carbon dioxide. The lobster's circulatory system thus follows a different design from the insect's (Figure 17–6). From the heart, the blood follows several arteries leading to various parts of the lobster's body. The blood leaves the arteries and bathes the cells, as in the insect's open circulatory system, but it collects in blood spaces and flows through the gills to take up oxygen and give off carbon dioxide before returning to the heart. The lobster's blood contains a special respiratory pigment that increases its capacity to take up oxygen from the water flowing over the gills. We will look more closely at the functions of such pigments later in this chapter.

Unlike the insect and the lobster, the earthworm has a *closed circulatory system,* one in which the blood is confined to vessels for its entire course through the body. The system consists of a dorsal vessel, in which, as in the insect, the blood flows forward (Figure 17–7). The vessel's pumping is aided by the action of five pairs of hearts in the worm's forepart. These hearts pump blood into the ventral vessel. The blood flows backward through the ventral vessel, then filters through a mesh of finer vessels to return to the dorsal vessel.

Circulation in Vertebrates Vertebrates, particularly birds and mammals, require large quantities of oxygen and nutrients to achieve and sustain their high metabolic rates. As a result, these animals have sophisticated plumbing and transport systems. One part of their circulatory systems is specialized to exchange gas at the respiratory surfaces, and the other to distribute gas and transport nutrients to the rest of the body. Here we will examine the functional morphology of vertebrate circulation proper, then move on to the details of gas and nutrient transport.

The heart is a muscular pump that provides the pressure that forces the blood through the body. If you compare the hearts of the various vertebrate orders (Figure 17–8), you can trace the approximate evolution of the organ.

In fishes, the heart consists of two enlarged portions (chambers) of a blood vessel. The first part, the *atrium* (or, sometimes, auricle), is relatively elastic and stretches when it receives blood. From the atrium the blood passes into the second chamber, the *ventricle,* whose muscular walls contract and force the blood into the vessels leading to the gills. At the gills, blood is oxygenated and, still under pressure from the heart, it continues to the rest of the body, ultimately to return again to the atrium.

Most amphibians have lungs supplied with large blood vessels that communicate with the heart. The single atrium of the fishes is now divided in two. One atrium receives oxygenated blood from the lungs and the other

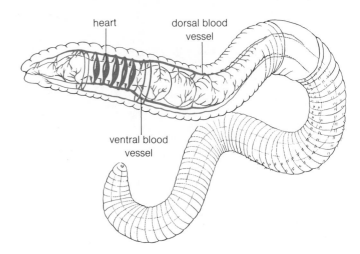

Figure 17–7 The closed circulatory system of an earthworm.

atrium receives unoxygenated blood from the rest of the body. Both atria empty into the same ventricle. Although this mechanism tends to reduce the mixing of oxygenated and unoxygenated blood, some of the already oxygenated blood is returned to the lungs before it has unloaded its oxygen. This system (although adequate for the sluggish metabolism of amphibians and more efficient than that of fishes) is still quite inefficient, since it wastefully recirculates oxygenated blood.

The obvious way to increase the efficiency of the amphibian heart would be to separate the ventricle into two chambers and thus prevent the mixing of oxygenated and unoxygenated blood. The beginnings of this separation can be seen in the ventricular partition characteristic of the reptilian heart. In the mammals and birds, this evolutionary trend culminates in the complete division of the original single ventricle into two. As a result, circulation to and from the lungs (the *pulmonary circulation*) is fully separated from the circulation of the rest of the body (the *systemic circulation*). Mammals and birds contain two essentially distinct circulatory systems, one for unoxygenated blood and one for oxygenated blood, with no mixing between the two.

Circulation in Humans The human circulatory system is characteristic of mammalian circulation, and, like all vertebrate circulatory systems, comprises a heart, arteries, capillaries, and veins (Figure 17–9). **Arteries** are vessels that carry blood away from the heart. The two largest arteries are the *aorta* and the *pulmonary artery*. The aorta receives the blood from the left ventricle of the heart and sends branches to the head, arms, legs, intestine, kidneys, and other organs. The pulmonary artery receives

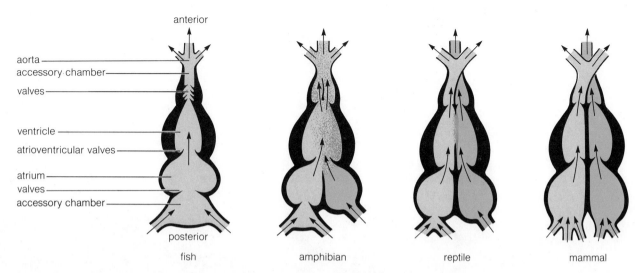

anterior

aorta
accessory chamber
valves

ventricle
atrioventricular valves

atrium
valves
accessory chamber

posterior

fish   amphibian   reptile   mammal

Figure 17–8 There is increasing morphological separation between systemic and pulmonary circulation from the most primitive classes of vertebrates to birds and mammals. The separation between the two sides of the heart presumably decreases mixing of oxygenated and deoxygenated blood. Recent evidence indicates, however, that some reptiles, for example, achieve nearly complete separation of pulmonary from systemic blood flows.

the blood from the right ventricle, divides into two branches, and carries blood only to the two lungs. The heart takes its oxygen and nutrients not from the blood continuously coursing through it but from its own system of arteries, the *coronary arteries* (Figure 17–10).

Veins are vessels that carry blood back to the heart. They have thinner walls than the arteries and are located closer to the body surface. (The vessels visible in your arms and legs are all veins.) Two large veins return blood to the right atrium of the heart. One (the *superior vena cava*) drains the head region, and the other (the *inferior vena cava*) drains the lower body. The left atrium receives blood from the four *pulmonary veins* draining the two lungs.

The tiniest branches of the blood vessels, the ones that connect arteries and veins, are called **capillaries** (Figure 17–11). They are not much wider than the red blood cells that pass through them, and their walls are only one cell thick. They form such a fine and intricate mesh in the tissues that they are in close contact with all the cells. It is through the capillary walls that gases, nutrients, and metabolites are exchanged.

The discovery of capillaries by the Italian anatomist Malpighi proved Harvey's theory of circulation. When Harvey wrote, the capillaries had not been discovered, and he was not sure how blood from the arteries crossed over into the veins. Malpighi's find supplied that missing link. Unfortunately for Harvey, he was already three years in the grave at the time.

We can trace the circulation of the blood by following

one of the oxygen-carrying red blood cells, or **erythrocytes,** on its journey through the body. We take as our starting point a capillary in a leg muscle, where carbon dioxide diffuses from the cells into the liquid portion, or *plasma,* of the blood, and the oxygen released by the erythrocyte moves into the cells. The erythrocyte now travels into the small veins, or venules, then into larger veins, and finally into the inferior vena cava. It enters the heart at the right atrium, and moves through the right atrioventricular valve into the right ventricle. Contraction of the ventricle pumps the erythrocyte and its fellows through the pulmonary artery and into the lungs. There the erythrocyte enters the beds of capillaries surrounding the alveoli. Carbon dioxide picked up by the plasma from the leg muscles diffuses into the alveoli; from which it is expelled to the exterior by the mass flow from breathing. During inhalation the erythrocyte takes on oxygen following a gradient inward. The oxygen binds with the respiratory pigment *hemoglobin* (Hb) in the erythrocyte:

$$O_2 + Hb \rightarrow O_2Hb$$

It is now ready to be carried to the rest of the body.

Leaving the capillaries surrounding the alveoli, the blood bearing the oxygenated erythrocyte travels into veins, finally reaching the large pulmonary veins that empty into the left atrium of the heart. From there it passes through the left atrioventricular valve and into the left ventricle, which contracts and pumps the blood

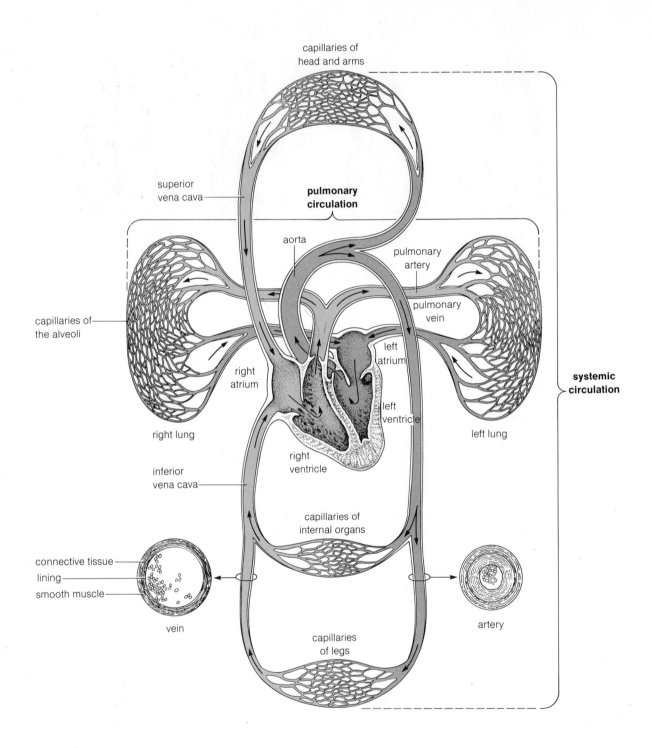

capillaries of
head and arms

superior
vena cava

**pulmonary
circulation**

aorta

pulmonary
artery

pulmonary
vein

capillaries of
the alveoli

left
atrium

left
ventricle

right
atrium

**systemic
circulation**

right lung

left lung

inferior
vena cava

right
ventricle

capillaries of
internal organs

connective tissue

lining

smooth muscle

vein

artery

capillaries
of legs

Figure 17–9 The human heart is a double pump in one housing. The right ventricle provides the power for the pulmonary circulation. The blood from the left ventricle leaves by way of the aorta and supplies the rest of the body.

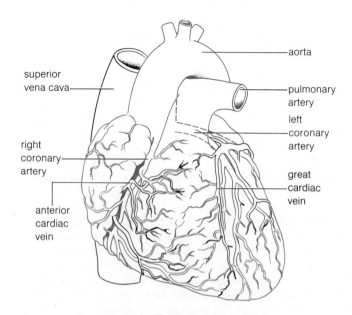

Figure 17-10 The arteries and veins supplying the heart.

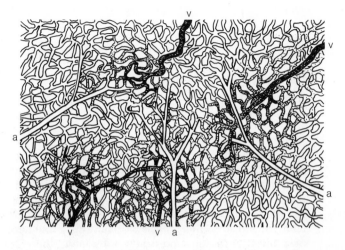

Figure 17-11 Diagram of a capillary bed: *a*-arterioles; *v*-venules.

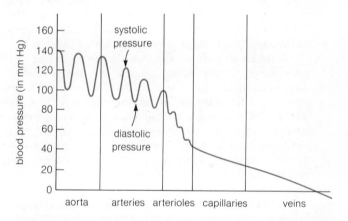

Figure 17-12 Variation of blood pressure in different parts of the human circulatory system.

into the aorta. The aorta branches repeatedly, sending smaller arteries to all regions of the body. Thus our erythrocyte could arrive back at the leg muscle where we began following it. It has passed through the heart twice, once through the pulmonary circulation and once through the systemic circulation.

As Figure 17-9 showed, the heart is essentially two pumps in one housing: The right atrium and ventricle propel the pulmonary circulation, while the left atrium and ventricle propel the systemic circulation. However, the two atria beat together, as do the two ventricles. The atria fill with blood and then contract, forcing the collected blood into the ventricles. The thick, muscular walls of the ventricles then contract in a powerful wave that begins at the tip of the heart. The fluid pressure produced by this contraction closes the atrioventricular valves, preventing any backflow of blood into the atria. The closing of the valves produces a sound that can be picked up by a doctor using a stethoscope and is used to determine the health of the valves. Normal closing causes a sharp, crisp sound, while stiff or leaky valves cause a murmur as blood flows through continuously.

The heartbeat can be divided into two phases: **systole,** when the ventricles contract, and **diastole,** when the ventricles are relaxed and filling with blood pumped from the atria. The fluid pressure of blood leaving the heart at the ventricular contraction is called *systolic pressure.* During systole, healthy arteries stretch slightly, and as they rebound during diastole, they provide an additional pressure called *diastolic pressure.*

Doctors measure blood pressure in the artery of the upper arm with a device that records fluid pressure in millimeters of mercury. In a normal resting person, the systolic pressure is about 120 millimeters, and the diastolic pressure is about 80 millimeters (Figure 17-12). By convention, this value is stated as 120/80. Blood pressure does vary throughout the body. It is highest in the aorta, where the normal systolic and diastolic pressures are 140 and 120 millimeters respectively. In the capillaries, the pressures decline to 40 millimeters and 20 millimeters, and they fall to about 0 millimeters in the venae cavae. This pressure gradient propels the blood on its journey through the body, ensuring a steady flow through the

Essay 17–1

When the Plumbing Goes Bad: Arteriosclerosis

Arteries are essential to the flow of blood in our bodies. They are the vessels that, with one exception, carry oxygenated blood to the tissues. (The pulmonary artery is the exception, as it brings oxygen-depleted blood to the lungs.) Arteries are complex in morphology and, regardless of size, have three layers in their walls. The innermost layer, or tunica intima, consists of a lining, a fine network of connective tissue, and a layer of elastic fibers bound together in a membrane pierced with many openings. The middle layer, or tunica media, is mainly smooth muscle fibers arranged in roughly spiral layers. The muscle fibers in this region are under nerve control. The outermost layer, the tunica adventitia, is a tough layer consisting of collagen, a fibrous protein of great strength. Minute blood vessels supply the artery wall with oxygenated blood, and small veins carry the blood away.

Arteries are, thus, complex structures subject to the same needs as any other organ in the body. Moreover, they must be capable of contraction if they are to function properly. Unlike the water-carrying tissue of the vascular plant, which is composed of dead cells, the cells of the arteries and of the veins are alive. They are thus subject to change, and some of these changes can be detrimental to their function. One such detrimental change results in arteriosclerosis, more commonly known as hardening of the arteries.

Arteriosclerosis starts with small injuries to the inner layer of the artery. A healing process begins with the division of cells of the lining and muscle cells of the middle layer. As these cells proliferate, they may begin to accumulate fat, particularly if the cholesterol level in the blood is high. Cholesterol, a fatty acid normally made by cells of the body, is an important constituent of cell membranes. The amount of cholesterol in the blood can be increased by what we eat, and many foods, such as egg yolks, are exceptionally high in it. At this point the process may stop and the entire mass may be reabsorbed, leaving only a tiny scar. However, the cells may continue

to grow, so that the mass enlarges to form a *plaque.* Such plaques appear to be permanent and protrude into the center of the artery. Plaques restrict the flow of blood in the artery and lessen the elasticity of the wall. The plaque may stabilize at this point, and a tough coating may develop on it, reducing its interference with the function of the artery. Blood vessels also develop into the plaque and supply the cells of the plaque with blood. Calcium may accumulate in the plaque, as well as additional fats. The plaque may become necrotic and begin to break down. A crack may appear in the surrounding cap of the plaque, so that the arterial blood comes in contact with the dead or degenerate parts of the plaque. When this occurs, the arterial blood will clot at the site of the break. This clot may be large enough to block the artery, or it may break free and be carried in the bloodstream to some smaller artery where it becomes lodged, again restricting blood flow. If a blood clot blocks one of the arteries supplying the heart with blood, a heart attack may result. If the clot blocks an artery supplying the brain with blood, a stroke may be the outcome.

This, in general outline, is what happens in the most common form of arteriosclerosis. There are, however, a great many unanswered questions concerning the events described. For example, the importance of the role of cholesterol in plaque formation is far from clear. It is important to remember that this sterol is a normal biochemical constituent of our cells and is found in our cell membranes. It is also a precursor molecule in the formation of many hormones. Yet, it also appears true that in many people, high levels of cholesterol in the blood can be correlated with large numbers of well-developed plaques. A large number of paradoxes are associated with the role of cholesterol in arteriosclerosis. For example, residents of a Swiss Alpine village consumed large amounts of fat, yet the cholesterol level in their blood was comparable to levels observed in populations that consume low-fat diets. The general conclusions appear to be that, although high-cholesterol diets do not mean automatic heart failure and stroke, reduction in the intake of cholesterol is a reasonable precaution to help minimize arteriosclerosis.

capillaries. Numerous valves, particularly in the veins, keep the blood from flowing back the wrong way.

In some people, fat deposits build up on the arterial walls and lessen their elasticity (Essay 17–1). The result

is an abnormally high systolic pressure. High blood pressure (hypertension) over long periods of time weakens the arteries and can lead to heart attack, severe kidney damage and malfunction, and brain hemorrhage (stroke).

The overall rate of circulation—the speed at which the blood moves through the body—depends on a variety of factors. The most important is the pumping capacity of the heart. The human heart pumps about 290 liters per hour during sleep, and five times that much during maximal exertion. Thus, the heart's pumping capacity largely determines the rate at which the tissues can be aerated. The pumping capacity depends in turn on two factors: the pumping rate and the stroke volume (the volume of blood pumped at each contraction of the ventricles).

The stroke volume varies with the demands made on the heart. During vigorous exercise, large volumes of blood return to the heart and stretch the fibers of cardiac muscle in the right ventricle as they flow in from the atrium. This stretching causes the ventricle to contract more powerfully than normal, increasing the amount of blood pumped from the heart. This phenomenon, in which the stretching of the heart increases the stroke volume, is known as Starling's law.

If the heart is stretched repeatedly and regularly, average stroke volume increases. Most people have pulses of 70 or 80 times a minute. In other words, the ventricle has to contract 70 or 80 times to pump the usual 5 liters of blood a minute. By contrast, marathoners, who typically run 100 or more miles a week in training, have pulse rates of 40 to 50. Their hearts are enlarged and more powerful than average, and fewer contractions of the ventricle are required to pump the same quantity of blood. Although there is no conclusive evidence as yet, some medical researchers are convinced that the higher stroke volumes produced by regular exercise increase the health of the heart and lower the risk of heart attack.

Once blood leaves the aorta, its distribution is regulated by involuntary muscles in the walls of the blood vessels that widen or narrow the passageways, allowing more or less blood to enter. Exercise, temperature, and other factors also affect the distribution of blood. For example, in a human at rest, the muscles account for only 20 percent of the oxygen consumed, and the circulation to the muscles is correspondingly low. But if that resting person suddenly jumps up and takes off at a hard run, the metabolic rate of the muscles may increase 50 times. The call for oxygen increases accordingly, and blood is shunted from elsewhere in the body to the muscles.

Temperature regulation may also cause blood flow to be shifted about. If you are outside on a cold day, arterioles (small arteries) in the skin, hands, and feet constrict, keeping blood out and conserving body heat. This is the reason why frostbite usually affects exposed parts of the face and the hands and feet first. By contrast, if you are outside in the bright sun on a hot day, blood is shunted from the core to the periphery to unload excess heat into the environment. This mechanism explains why your face flushes on a hot day. It also explains the problems athletes have when temperatures are high. Since the shifting of blood to the skin makes less blood available to the lungs and muscles, athletes performing in the heat have trouble exerting themselves for very long. Thus, a miler who can break 4 minutes when the temperature is 65°F will find the same feat impossible when the mercury hovers around 100°F.

The one organ that is—and must be—supplied with a relatively uniform supply of blood is the brain. If circulation to the brain is cut off for even 3 to 5 seconds, consciousness is lost. The need for a continous heartbeat is therefore obvious. After 4 minutes without blood circulation, brain cells begin to die. Oxygen starvation for 9 minutes produces irreversible and massive brain damage.

The activity of the heart, like the flow of blood through the vessels, is under involuntary nervous control. However, since the heart in the developing embryo begins to beat even before the nerves appear, and since a heart removed from the body will continue to beat, it is apparent that the heart also has an independent pattern of activity which is modulated, but not initiated, by the cardiac nerves. The heartbeat is stimulated by impulses generated and distributed by a system of specialized muscle cells in the heart. These cells behave in part like muscle, in that they contract, and in part like nerve, in that they conduct electrochemical impulses. These cells occur in two masses, the *sinoatrial node* (SA node) and the *atrioventricular node* (AV node). Figure 17–13 diagrams the workings of the nodes. The SA node, located in the right atrium under the opening of the superior vena cava, initiates each heartbeat and sets the pace of the heartbeat, earning it the name "pacemaker." The impulse from the SA node spreads out over the atria, causing them to contract. The impulse passes on to the AV node, located near the base of the wall separating the atria. From the AV node, the impulse travels along a bundle of specialized fibers, the Purkinje fibers, that branch downward over the walls of both ventricles and stimulate the powerful, wavelike contraction that pumps the blood out into the arteries under such great pressure.

Heart or artery disease can result in an arrhythmic heartbeat. This condition is now commonly corrected by surgically implanting an artificial pacemaker in the chest. The pacemaker is simply an array of long-lived batteries connected to the heart and set to deliver a small electrical charge, usually at a set rate of around 80 times a minute. The pacemaker takes the place of the malfunctioning nodes and regulates the heartbeat.

Superimposed on the activity of the two nodes are nervous impulses. The tempo set by the SA node can be speeded by impulses from the sympathetic nervous system or slowed by the vagus nerve of the parasympathetic nervous system. (We will examine these two systems in Chapter 20). In fact, if the vagus nerve is stimulated strongly enough, the heart can be made to stop. Usually such stoppages are under involuntary control, but practitioners of certain disciplines of Indian yoga can control the heart voluntarily and stop its beating for a few seconds at will. Obviously, this ability has to be used with great care because of the risk of damage to the brain.

Gas Exchange and the Blood   Now that we have examined the physical features of the human circulatory system, we can look more closely at how the blood performs its major role of transporting oxygen from the lungs to the cells and bringing carbon dioxide back.

The circulatory systems and associated tissues of many animals contain special compounds that serve as oxygen carriers. In vertebrates, the main respiratory pigments are **hemoglobin** in the blood and **myoglobin** in the muscle. Hemoglobin greatly increases the oxygen-carrying capacity of the blood, and myoglobin aids in moving oxygen from the blood into the cell. These pigments are crucial to the efficiency of the circulatory system. If human blood contained no hemoglobin and oxygen were simply carried in solution in the blood, the heart would have to pump 83 liters of blood per minute into the lungs to take up the same amount of oxygen carried by 5.5 liters of blood with hemoglobin.

The respiratory pigments have also been important in our contemporary understanding of the chemistry of life. Myoglobin was one of the first proteins for which a three-dimensional model was constructed. The work on myoglobin was done by John C. Kendrew and M. F. Perutz in England. They solved the structure in 1957 and shared the 1962 Nobel prize in chemistry.

The myoglobin molecule consists in part of a single polypeptide chain folded in a specific manner. The chain is called the *globin.* In the center of the globin is a *prosthetic group*—that is, a metal ion or organic group, other than an amino acid, bonded with a protein and serving as the myoglobin's active site. The prosthetic group of myoglobin is *heme,* a red pigment, structurally related closely to chlorophyll, that holds an atom of iron (Figure 17–14). The heme is the protein's active site, and it is here that the transported oxygen binds with the myoglobin molecule.

After succeeding with myoglobin, Perutz moved on to the structure of hemoglobin (Hb). The Hb molecule turned out to be much larger than myoglobin, comprising a number of subgroups. Its backbone consists of four long protein chains, each of them much like the single polypeptide chain of myoglobin. However, the four chains of the Hb molecule, folded together into one complex structure, are of two types. In humans, there are two identical alpha chains and two identical beta chains.

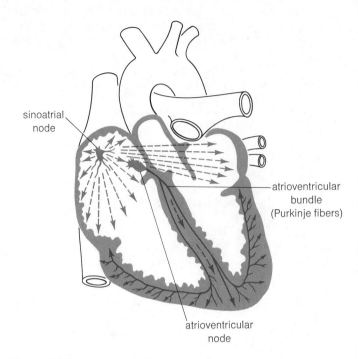

Figure 17–13   A section of the heart, showing the sinoatrial and atrioventricular nodes.

When Hb bonds with oxygen, alpha and beta chains fit together into a spherical molecule of about 6.4 × 5.5 × 5 nanometers. Embedded in separate pockets of the giant molecule are four hemes like those of myoglobin, each of which bonds with a single oxygen molecule. Oxygenated hemoglobin (oxyhemoglobin) has a structure different from that of unoxygenated hemoglobin. Oxyhemoglobin is scarlet, but when the oxygen is released, the molecule turns dark red. Thus, blood leaving the heart through the aorta is bright red, while that returning through the venae cavae is a darker red.

The hemoglobin in vertebrate blood is packaged in erythrocytes. One might wonder whether this arrangement is excessively ornate. Why isn't the hemoglobin simply carried in solution within the blood? The reason is that a high concentration of proteins (in this case, hemoglobin) in the blood would create an osmotic pressure great enough to drain water from the cells into the blood. Thus, animals with free hemoglobin can tolerate only so much of the protein in their blood, and their oxygen-transport abilities soon reach a limit. In the vertebrates, the problem is solved by confining the hemoglobin to red blood cells. Since there are 280 million hemoglobin molecules in each human erythrocyte and some 5 million erythrocytes per cubic millimeter of blood, the amount

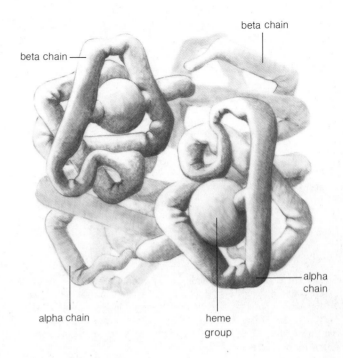

Figure 17–14 The structure of the hemoglobin molecule. The sausagelike globin is a complexly folded protein. Four globin chains, each with an attached heme, make up a hemoglobin molecule. The heme (structure at left) is located in one of the folds of the globin. Each heme binds one oxygen molecule.

of hemoglobin in our blood is far greater than it could be if the protein were in solution.

The sequence and identity of the 140-odd amino acids in each of the two types of chains in the hemoglobin molecule vary from one species to another. This fact has been put to scientific use by biologists interested in evolution. The longer ago two given animals diverged from a common ancestor, the more time their proteins have had to change, and the more their hemoglobins are likely to differ. Thus, the degree of variance in hemoglobins can be used to measure the closeness of the relationship between species or groups of species. For example, hemoglobin taken from apes varies only a little from human hemoglobin but differs greatly from fish hemoglobin.

Even all these various hemoglobins, however, behave in the same fashion. Should the protein sequence vary in a way that affects the basic behavior of the hemoglobin, oxygen transport in the affected animal would be impaired.

This is precisely the case in sickle-cell anemia. A single glutamic acid takes the place of a valine and changes the behavior of the entire hemoglobin molecule. The abnormal hemoglobin of the sickle-cell anemic changes shape after releasing its oxygen, twisting the red blood cell into the sickle-shaped configuration characteristic of

the disease. The sickle cells rupture easily, eventually making victims of the disease anemic. Because of their shape, these cells can clog small blood vessels, shutting off blood flow to internal organs.

Why are the globins necessary at all, since oxygen binds with the heme and not with the protein? The heme must have an affinity for oxygen that is very specific, neither too low nor too high. The globin, depending on its structure, modifies this affinity. With too low an affinity, the blood would pick up too little oxygen during its very brief trip through the capillaries in the lung. Too high an affinity could mean that the oxygen would remain permanently bound and not be released to the cells.

The oxygen-binding capacity of hemoglobin varies considerably with conditions and with species. In humans, hemoglobin is 98 percent saturated at the concentration of oxygen in the lungs, but the bonds are loose enough that 40 percent can be unloaded at the lower concentration of the tissues (Figure 17–15). During exercise, oxygen is consumed rapidly, and the concentration of oxygen in the tissues drops. Under these conditions, the oxyhemoglobin gives up as much as 60 percent of its load.

Hemoglobin's affinity for oxygen also varies with pH and carbon dioxide levels, dropping as acidity and carbon

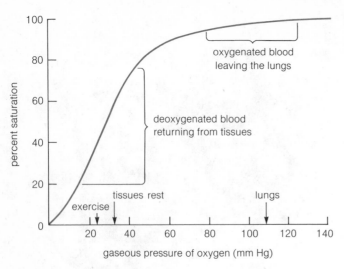

Figure 17–15 Oxygen saturation curve for hemoglobin. When the gaseous pressure of oxygen is about 25 mm Hg, for example (as it is in the tissues during exercise), then the hemoglobin gives up about 60 percent of the oxygen it carries. When the gas pressure is about 110 mm Hg, as in the lungs, oxygen tends to bind to the hemoglobin to nearly full saturation.

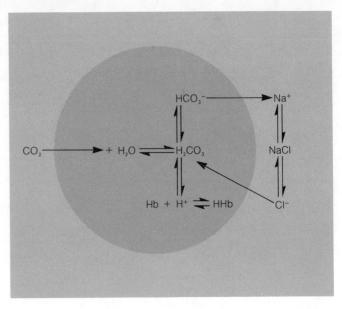

Figure 17–16 The chloride shift in a red corpuscle.

dioxide increase. This *Bohr effect* allows cells to take up oxygen faster during exertion. Ongoing metabolism concentrates carbon dioxide in the cells, and some of the carbon dioxide reacts to form carbonic acid. The rising levels of carbon dioxide and acidity lower the hemoglobin's affinity for oxygen and move the oxygen into the cell that much faster.

The oxygen affinity of hemoglobin varies from one group of animals to another, depending in large part on their characteristic metabolic rates and body temperatures. "Cold-blooded" animals have relatively low oxygen demands. The hemoglobin of such animals bonds more tightly with oxygen than the hemoglobin in the warm-blooded species. Birds are highly active creatures that use large quantities of oxygen in proportion to their weight. As you might expect, then, avian hemoglobin has a low oxygen affinity and gives up its oxygen to the tissues readily.

Hemoglobin may even vary from one developmental stage to another. The hemoglobin of the mammalian fetus has a higher oxygen affinity than does the hemoglobin of the adult. The fetus requires great quantities of oxygen, yet it does not ventilate or share the maternal blood supply. Instead, all the oxygen has to cross a membrane from the mother's blood into the fetal blood. If fetal hemoglobin had the same oxygen affinity as adult hemoglobin, there would be little gradient across the membrane, and the fetus would very likely get too little oxygen. The

higher oxygen affinity in the fetal hemoglobin increases the gradient and assures a proper supply of oxygen.

In effect, an adult who goes from sea level to high altitudes is returning to the fetal condition—that is, to a situation of relatively low oxygen concentration. However, hemoglobin does not change at high altitudes. Instead, the body produces more red blood cells, the oxygen-carrying vehicles (see Supplement 16–1).

The chemistry of hemoglobin is by no means the whole story of oxygen transport from blood to tissues. As we mentioned earlier, myoglobin helps move oxygen into the cells. This protein, which gives meat its characteristic red color, is found in the cytoplasm of muscle cells. Myoglobin has a higher affinity for oxygen than does hemoglobin. Thus, the presence of myoglobin in muscle cells increases the concentration gradient and speeds the diffusion of oxygen from blood to tissue. Myoglobin also serves as a temporary storage site for a small amount of reserve oxygen, enough to power the muscles for a few seconds of vigorous exercise. Since physical training increases the amount of myoglobin in muscle tissues, well-conditioned athletes extract more oxygen from their muscles than do the sedentary sorts who watch them on TV.

Two pollutants adversely affect the blood's ability to transport oxygen. One is carbon monoxide, found in the exhaust of internal combustion engines and in cigarette smoke. Hemoglobin's affinity for carbon monoxide is about 300 times greater than its affinity for oxygen. Carbon monoxide in the atmosphere bonds readily with hemoglobin, reducing the amount of oxygen transported in the blood. Oxygen starvation results. The severity of the starvation depends on the proportion of carbon monoxide in the atmosphere. At concentrations as low as 0.001 percent, the central nervous system is mildly affected, and vision loses some of its sharpness. At 0.01 percent—a level commonly exceeded on busy city streets at rush hour—headache, nausea, and dizziness begin. Exposure to 0.1 percent is quickly fatal.

Another pollutant, nitrite, bonds with hemoglobin to form methemoglobin, also rendering the hemoglobin incapable of carrying oxygen. Nitrite poisoning occurs in areas (such as the San Joaquin Valley of California) where high-intensity agriculture based on nitrate fertilizers is practiced. The nitrates leach into rivers, streams, and ground water and are then ingested in drinking water. Nitrates are not poisonous in themselves, but intestinal bacteria convert them to nitrites, which bond with hemoglobin and cause nitrite poisoning.

Although the respiratory pigments are largely a vertebrate adaptation allowing for higher metabolic rates, they are also found in some invertebrates. The larvae of certain species of midges (small biting flies) have a pink coloration from hemoglobin dissolved in their blood. The hemoglobin helps these animals extract oxygen from the poorly aerated mud they live in. But this hemoglobin releases oxygen to the tissues only when the concentration of oxygen inside the larvae is very low. Thus, hemoglobin in midge larvae acts much as myoglobin does in the muscles of vertebrates.

Many mollusks and aquatic arthropods like the horseshoe crabs *(Limulus)*, the true crabs, and the lobsters have respiratory pigments called *hemocyanins* carried in solution in the blood. The hemocyanins are proteins much like hemoglobin, except that copper replaces the iron of the hemes. When oxygenated, hemocyanins are blue rather than red.

Carbon dioxide also enters red blood cells, but bringing carbon dioxide from the tissues to the lungs involves no special carrier pigments. In a reaction catalyzed by the enzyme carbonic anhydrase in the red blood cells, carbon dioxide combines with the water in blood to form carbonic acid, H_2CO_3. This reaction allows much more carbon dioxide to be carried in the blood than if the gas molecules were left unaltered. Excess H^+ released in the carbon dioxide–water reaction reacts with Hb to form HHb. Some of the HCO_3^- diffuses into the plasma, but the movement of these negative ions is countered by Cl^-, which enters the red blood cells during the so-called chloride shift (Figure 17–16):

$$H_2O + CO_2 \xrightarrow{\text{carbonic anhydrase}} H_2CO_3 \rightleftharpoons H^+ + HCO_3^-$$

The high carbon dioxide concentration in the tissues drives the reaction to the right, putting carbonic acid into solution; the low carbon dioxide concentration in the alveoli of the lungs drives the reaction back to the left. The liberated carbon dioxide diffuses into the alveoli and is removed at exhalation.

Thus far, we have been considering blood primarily as a vehicle for the transport of gases. The circulatory system also plays other roles, some of which we will now examine.

Nutrient Pathways and the Lymphatic System
Nutrients leave the gut by two separate routes. Carbohydrates and amino acids take the *hepatic portal vein* to the liver. A portal vein is a vein that begins and ends in capillaries. The hepatic portal vein begins in the intestinal capillaries and ends in the capillaries of the liver.

The liver is the body's biochemical virtuoso, and the nutrients transported by the hepatic portal system may be processed there in a variety of ways. The sugars may be converted to glycogen, a storage carbohydrate, and held for later use. The liver can store enough glycogen to supply the body's energy needs for about 24 hours. At the proper hormonal signal, the glycogen is converted back to sugar and released into the blood. Amino acids may be converted to sugars or to glycogen, or they may be passed on to be used in other parts of the blood. The liver also detoxifies ingested poisons, among them alcohol, and stores vitamins and minerals.

Fats take a different route from the intestine, entering the *lymphatic system* (Figure 17–17). The lymphatic system is the body's second circulatory system, but it is a one-way street rather than an unending circle. In the intestine, fats are broken down into their component fatty acids. Blood capillaries absorb and transport the short-chain fatty acids, while the long-chain fatty acids enter the lymph capillaries, or *lacteals*. When the lacteals are engorged with fats, they have the white color that gives them their name (from the Latin *laeteus*, "milky white"). From the lacteals, the fats move into progressively larger lymphatic vessels and finally into the thoracic duct (the largest lymphatic vessel of all). The thoracic duct drains most of the body and empties into the left subclavian vein. The right lymph duct drains the heart, lungs, the right upper part of the body, and the right side of the head and neck, emptying into the right subclavian vein.

The fluid in the lymph system is actually a filtrate of the blood. As we mentioned earlier, the vertebrate circulatory system is closed. Nevertheless, fluid does leak from the capillaries into the tissues surrounding them. This fluid is called the *lymph*. Essentially it is the same as the plasma of the blood, except that it contains only a fraction of the proteins found in plasma. The lymph bathes the tissues and collects in lymph capillaries that

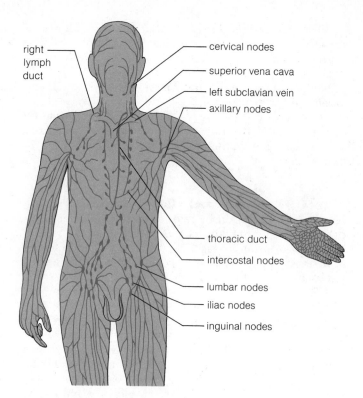

Figure 17-17 The lymphatic system drains the entire body, emptying into the subclavian veins and returning the fluid to the heart and thence to the blood circulatory system. The system is interspersed with nodes that trap bacteria and other foreign matter.

Labels on figure:
right lymph duct
cervical nodes
superior vena cava
left subclavian vein
axillary nodes
thoracic duct
intercostal nodes
lumbar nodes
iliac nodes
inguinal nodes

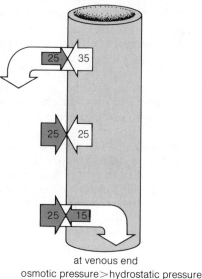

at arterial end
hydrostatic pressure > osmotic pressure

25 35

25 25

25 15

at venous end
osmotic pressure > hydrostatic pressure

Figure 17-18 The fluid flow into and out of capillaries depends on the hydrostatic pressure provided by the heart's pumping and on the osmotic pressure.

empty into larger ducts, and finally the fluid makes its way to the thoracic duct, where it is returned to the bloodstream.

If the lymph were not drained out of the tissues, they would quickly swell. The massive swelling (edema) of the disease elephantiasis is brought on by parasitic worms that block the lymphatic ducts. Lymph collects in the tissues, ballooning them grotesquely. Normally, muscular exercise promotes lymph movement. Thus, exercise is important and helps relieve tired, sluggish feelings because lymph (with wastes) is removed to the blood, from which the wastes are removed by the kidney.

Besides transporting fats, leaked fluids, and wastes, the lymphatic system plays a role in protecting the body against microbial invasion (Supplement 17–1). At specific places in the system, such as the leg and arm axillaries, lymph nodes interrupt the lymphatic vessels (Figure 17–17). The nodes contain white blood cells and antibodies that eliminate bacteria and other foreign particles from the lymph. The work of the nodes against infection is often most noticeable when they are particularly active, during the course of an infectious disease. Then the nodes,

particularly those in the neck and under the arms, swell visibly and become painful to the touch.

Homeostasis and the Circulatory System   Claude Bernard, a great French physiologist of the nineteenth century, once wrote a sentence that has since become an aphorism of biological science: "La fixité du milieu intérieur est la condition de la vie libre" (the price of a free life is a stable internal environment). He meant that, for an organism to act independently of outside conditions, to live and move among a variety of temperatures, pressures, and so on, it first had to be able to maintain the internal conditions necessary for its survival. This ability to regulate internal conditions to produce a constant, steady state is called *homeostasis.*

Virtually every cell of every tissue in every organ of a living thing engages in one or another homeostatic function. Here we will concentrate on certain steady-state roles played by the circulatory system.

As we noted, fluid leaks out of the capillaries into the tissues. Left unchecked, this leakage causes edema and a loss in blood volume and blood pressure. The lymph drains off the filtrate, and, in addition, the blood circulation system itself regulates the volume of the *interstitial fluid,* the lymph that bathes the cells and forms their true environment.

Blood-Clotting

To do its job right, blood must remain free flowing. Any thickening can block a blood vessel and damage the tissues beyond. This is precisely what happens in a stroke, when a blood clot in the brain shuts off the oxygen supply to a part of that organ and kills the starved tissues. However, our need for free-flowing blood makes us vulnerable to another danger: Even a small wound could cause a fatal loss of blood. Normally, of course, this does not happen. The blood forms a plug and seals off the leak. The mechanism underlying this phenomenon is, as you might suspect, very complex, involving checks and balances that no doubt evolved to ensure that clots form only when they are needed.

Among the proteins carried in the blood plasma is one called *fibrinogen,* a large molecule that is the key to clotting. The basic mechanism in the formation of a blood clot is the conversion of fibrinogen to a gel state called *fibrin,* which forms the plug to the exterior and closes off the leak. You might expect the trigger for this reaction to be contact between blood and air. Curiously, this is not the case. For fibrin to form, certain other reactions must first occur. When blood platelets (small disk-shaped bodies arising from cell fragments) come into contact with damaged tissue, they release a group of substances called *thromboplastins.* The thromboplastins catalyze a reaction that converts *prothrombin,* present in the blood, to *thrombin.* Thrombin, in turn, catalyzes the fibrinogen-fibrin reaction, but only when Ca^{2+} is present. In summary, the reactions are:

$$\text{blood platelets} \xrightarrow[\text{tissue}]{\text{injured}} \text{thromboplastins}$$

$$\text{prothrombin} \xrightarrow[\text{Ca}^{2+}]{\text{thromboplastin}} \text{thrombin}$$

$$\text{fibrinogen} \xrightarrow[\text{Ca}^{2+}]{\text{thrombin}} \text{fibrin}$$

Blood-clotting is a sensitive system, and malfunction can be dangerous indeed. Hemophiliacs, or "bleeders," suffer from an inherited disorder in the production of prothrombin. As a result, their blood does not form clots properly, and even a minor blow can produce massive and destructive internal bleeding.

To prevent potentially dangerous internal clots, various parts of the body produce anticlotting substances, among them heparin, a compound synthesized mainly in the liver. Heparin is used medically to treat phlebitis (inflammation and clotting in the veins, usually of the leg) and to prevent clotting after major surgery. Another medical anticoagulant (clot-preventing drug) is dicoumarol. Dicoumarol is actually a toxin produced by certain fungi that infect sweet-clover hay and cause a fatal disease called "bleeds" in cattle that feed on the hay. The toxic effect of dicoumarol has been put to practical use in a number of rat poisons that do their fatal work by causing severe lung hemorrhages in the unwanted rodents.

Blood at the arterial end of the capillaries is still under pressure from the contractions of the left ventricle. This pressure forces water and salts through the thin capillary walls in fairly large amounts, but proteins like albumin and globin leak out only in small quantities. The resulting concentration of proteins in the capillaries raises the osmotic pressure and prevents further leakage.

In the venous capillaries, the net flow of fluid reverses, as shown in Figure 17–18. Since blood pressure here is very low, and since the osmotic forces of the proteins in the blood are dominant, water and salts return from the tissues to the venous capillaries. Thus, the proper "milieu intérieur" is maintained by the chemistry of the blood itself.

Obviously, however, this system is easily upset, as, for example, by abnormally high arterial blood pressure.

In this instance, more fluid leaks from the arterial capillaries, resulting in edema and a loss of blood volume. Problems of this sort are met by dilation or constriction of the arterioles that increases or decreases the blood supply to capillaries and thus regulates the amount of fluid available to the tissues. Blood volume is particularly crucial when the body has been injured. Massive bleeding drops the blood pressure. This results in a decrease of the fluid leaking out into the tissues and keeps the blood volume as high as possible. Another mechanism that protects the body against severe blood loss is blood clotting (Essay 17–2).

Certain parts of the body are more subject to damage from an oversupply of interstitial fluid than others, and capillary function therefore varies with the part of the body in question. The lungs are a good example. Too much fluid leaking out of the pulmonary capillaries could fill the alveoli and cause one literally to drown in one's own juices. As a safeguard against this danger, the right ventricle is only one-fourth the size of the left, with the result that blood pressure—and therefore leakage—is much lower in the pulmonary circulation than in the systemic.

Summary

Whether cells exist as individuals or as part of multicellular organisms, they must be bathed in a watery medium that provides nutrients and oxygen and removes wastes. In large and complex organisms, specific organs have evolved that regulate various specialized aspects of the watery medium bathing the cells. Specialization, in turn, implies interdependence, and the transport systems in both plants and animals effectively link the different service centers of the body and make the specialization possible.

In vascular plants, the transport system consists of two tissue systems: the xylem for water movement and the phloem for the translocation of the products of photosynthesis. Bulk flow is due to osmotic gradients. In most invertebrate animals, flow is due to active pumping through an "open" system. In vertebrates, we find both an open system of lymphatic vessels and a closed system of blood vessels that depends for bulk flow on the pumping of the heart.

In this chapter we surveyed comparative aspects of the transport systems in plants and animals. We also discussed in detail the adaptations of the systems for gas and nutrient transport and for the regulation of tissue fluid volume.

Supplement 17–1
The Immune System

Our world is a microbial broth. The air we breathe, the water we drink, even the surfaces of our skin are alive with billions of viruses and microorganisms. The majority are harmless, but others can parasitize the body and produce serious disease. Despite this constant invasion by foreign organisms, we are more often healthy than sick. The reason lies in the immune system, a remarkable apparatus whereby the body monitors itself and destroys invaders, foreign proteins, and even its own tumors.

The agents of immunity are the white blood cells, or **leucocytes.** Leucocytes are considerably different from erythrocytes. They are far less common (outnumbered at least 700 to 1), larger, without hemoglobin, and capable of moving against the flow of the bloodstream. They can even escape through the capillary walls and patrol the intercellular spaces in the tissues. The leucocytes are actually not so much components of the bloodstream as travelers who use its highways freely.

Leucocytes are grouped into five classes, the two most numerous of which are the *neutrophils* and the *lymphocytes.* The neutrophils arise in the bone marrow. They are the first leucocytes to arrive at the scene of an injury or infection, where they seek out invading bacteria and engulf them much as an amoeba ingests food (see Chapter 15). The lymphocytes, which arise in the lymphatic system, perform more sophisticated immunity functions.

There are two types of lymphocytes: the *T-cells,* which combat microorganisms and reject tumors and foreign tissues directly, and the *B-cells,* which neutralize invaders by synthesizing specific proteins. Both B-cells and T-cells begin in the hemopoietic (blood-forming) stem cells of the developing bird and mammal embryo. As the embryo develops, the stem cells migrate through the bloodstream from the yolk sac (where they first appear) to the spleen, the fetal liver, and the bone marrow. In each of these organs, they multiply and differentiate into various kinds of blood cells. Some of these differentiated cells later reenter the bloodstream and travel to the thymus gland, where they differentiate further to become T-cells. Again returning to the bloodstream, they finally lodge in the lymph nodes, where they are still subject to some hormonal control by the thymus.

In birds, stem cells that pass through a unique intestinal organ called the bursa of Fabricius rather than the thymus develop into B-cells. There must be an equivalent site in mammals where some group of embryonic cells splits off from the rest to become the B-cells, but the identity of that site is not yet known.

The T-cell population is large and heterogeneous. In the thymus, receptor sites on the cells' surfaces have been altered so that there is a wide range of differences among them. The nature of the receptor is unknown, but presumably the site provides a precise fit with a specific molecule that is either alone or part of the coat of a microorganism—one that has invaded the species in question within evolutionary memory. When such an invader, or *antigen,* enters the body, it presumably makes contact with many different T-cells and finally encounters one with a receptor site that fits the antigen and holds it fast. The T-cell is then stimulated to divide repeatedly, producing a rapidly proliferating clone of identical cells. The T-cells force the antigens to clump together (agglutinate) and thus inactivate them.

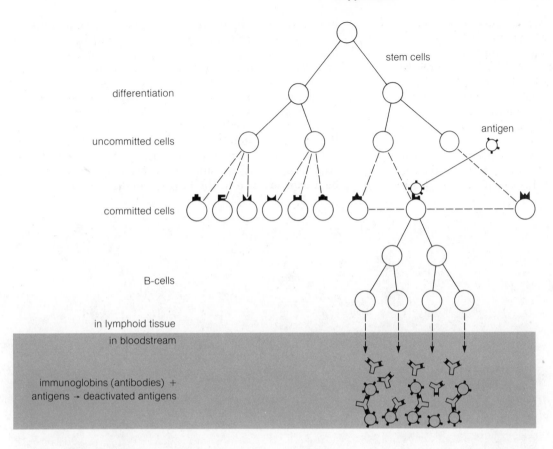

stem cells

differentiation

uncommitted cells

antigen

committed cells

B-cells

in lymphoid tissue

in bloodstream

immunoglobins (antibodies) +
antigens → deactivated antigens

Figure 17–19 The clonal selection theory of immunity holds that the stem cells proliferate and differentiate in embryonic development to produce cells, each of which becomes committed to produce a unique immunoglobin in response to antigen contacting the cell surface. Committed cells may interact with an antigen that fits an active site of the molecule. Having been "recognized," the cell becomes committed and begins to proliferate and produce its specific immunoglobins, which are distributed throughout the body by the bloodstream.

The B-cells are also highly diverse, again because of differences in the binding sites that allow each cell to bind only with a specific antigen. The B-cells' receptors are known to be proteins called *immunoglobins,* or *antibodies.* Each cell makes antibodies specific to its antigen. When a B-cell binds with an antigen, it not only begins to proliferate and produce a clone of identical B-cells, but also synthesizes and secretes identical antibodies at the rate of some 2,000 per second for every second of the several days of its mature life. The free antibodies bind with the invaders and inactivate them. This model, in which an antigen stimulates the rapid production of the very antibodies that stop its advance, is called the *clonal selection theory.* The theory, first proposed by Sir McFarlane Burnet, is portrayed schematically in Figure 17–19.

The past decade has witnessed considerable progress in our understanding of the structure of antibodies. Specific antibodies may well number in the millions, but all have the common structural characteristic of containing two pairs of polypeptide chains, one pair short, or "light" as it is called, and one pair long, or "heavy." The antibodies can be classified by the types of chains

they contain. The amino acid sequences of certain regions of these chains do not vary, whereas the sequences of other regions vary greatly. The specificity of a given antibody for a particular antigen almost certainly depends on these variable regions of the polypeptide chains.

A great many questions remain to be answered by immunological research. For example, the immune system obviously depends on an ability to tell the body's own cells from foreign cells. How this happens is as yet unknown. Also, it is unclear how antigens induce B-cells to divide, nor do we know how the millions of antigen-specific B-cells and T-cells all arise from one batch of similar embryonic stem cells.

Another puzzling aspect of the immune system is "linked" immunity, in which immunity to one antigen also confers immunity to a different antigen. For example, people exposed to *Salmonella typhi,* the bacterium that causes typhoid fever, are immune as well to *Rickettsia prowazekii,* an unrelated bacterium causing epidemic typhus.

Another troubling aspect of immunity is the allergic reaction. Essentially, an allergy is the immune response run amuck. We do not know why some people are allergic to certain antigens—such as pollen, house dust, animal hair, or even carbon paper and money—and others are not. It is thought that the allergic symptoms, such as teary eyes and runny nose, are caused by a rapid release of antibodies to which the affected tissues are sensitive. Allergies are more than mere discomforts. Allergic reactions to bee and wasp stings, to antibiotics like penicillin, and to the horse-serum antivenom used to treat snakebite can be lethal.

Research in immunology has been the bright light of biochemistry during the past decade, and it promises to illuminate still more in the future. If we can learn how the body distinguishes its own cells and proteins from all other cells and proteins, we will have come a long way toward understanding the basic biochemical and cellular organization of complex living systems. In addition, this research promises to be of immense practical significance. Some cancers apparently arise because the immune response fails to recognize and attack them. Immunological research could provide ways of stimulating the immune system to destroy cancers more successfully than our present methods of surgery, X rays, and drugs. And that, given the suffering caused by cancer, would be a true advance in the human condition.

18

Osmoregulation, Excretion, and Water Balance

Without water, life cannot be. The Greek philosopher Thales seemed to know this more than 2,000 years ago, when he said that water was the origin of all living things. Thales had none of our modern proofs for his statement, but his observations and his intuition were right. Actually, the statement should be qualified to "Without water, *life as we know it* cannot be." But exobiologists, who deal with the possibility of life on other worlds, are the only ones who have much need of this quibble. Earthly life is inconceivable without water.

Water is crucial to the workings of the cell, the basic unit of life, in many ways. Photosynthesis, respiration, and enzymatic processes occur only when water is present. Water is the universal solvent of enzymes and other proteins, and it is the medium that transports raw materials and the products and wastes of metabolism. Water also protects the cell against wide and sudden fluctuations of temperature, because it can absorb heat and because heat is dissipated when water evaporates.

Water is abundant in the cell and in the extracellular fluids, and it makes up a large proportion of the weight of a functioning organism. Jellyfish are 95 percent water, humans about 65 percent. These figures, of course, are averages. Water content varies from one part of an organism to another, such as from blood to bone. Since enzymes can function only within narrow limits of osmotic concentration, the right water content, and thus the right osmotic concentration, must be maintained exactly, in every part of the organism's body. Yet water freely penetrates the cell membrane in either direction. The cell must be able to remove water and salts when they are excessive, and take in water and salts when they are lacking. The cell must also be able to rid itself of metabolic wastes. Balance is the important concept, and the subject of this chapter.

From a global point of view, water is abundant on earth, covering three-quarters of the surface of the planet. Organisms, however, do not live in global points of view,

but in limited environments where the abundance, predictability, and osmotic character of the water supply vary greatly. Deserts lack nothing so much as water, yet they are inhabited by many water-loving organisms, including shrimps and toads. Fresh-water fish will die in salt water, but the seas are hospitable environments for innumerable organisms. Likewise, salt-water organisms can survive only a short time in fresh water, because the salts in their bodies leach out into the environment. Each of these habitats makes its own particular demands of the organisms that live there.

The first life-forms on earth almost certainly did not have to contend with problems of salt and water balance. They probably evolved in the uniform and unchanging environment of the seas, a chemical surrounding that has been pretty much the same for the 3 billion years sea life is estimated to have been present. Probably these life-forms handled salt and water exchange passively, as do many marine invertebrates today. This convenient arrangement ended as organisms abandoned the sea for new environments. It is thought that multicellular organisms entered estuaries and made their way upstream into fresh water. Other forms invaded the intertidal zone, and geological events trapped some of them in landlocked seas. From all these habitats, plants and animals ventured onto the dry land itself. As they did so, they evolved mechanisms for maintaining an internal extracellular medium roughly the same as that of their ancestral home in the seas.

In this chapter, we will look at the physiological and anatomical features that plants, single-celled organisms, and multicellular animals have evolved to solve the same problems. We will examine the role of water in salt balance and excretion, and we will see how structurally diverse organisms accomplish the same tasks in different ways and how structurally similar organisms adapt to vastly dissimilar environments.

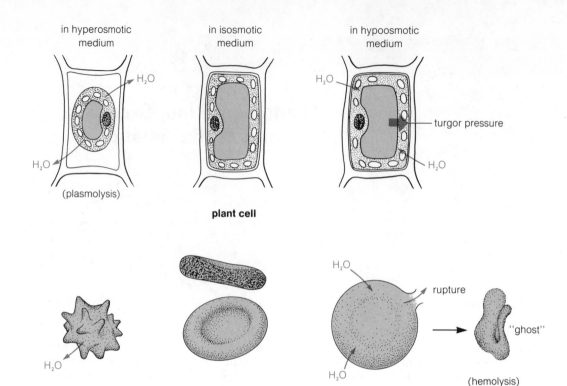

in hyperosmotic medium

in isosmotic medium

in hypoosmotic medium

H_2O

H_2O

(plasmolysis)

H_2O

turgor pressure

H_2O

plant cell

Figure 18–1 Effect of hypo- and hyperosmotic media on a plant cell and on a red blood cell. Water is withdrawn in hyperosmotic medium: The plant cell shows plasmolysis; the red blood cell shrinks. Water enters the cells in hypoosmotic medium: The plant cell increases its turgor pressure; the red blood cell bursts because it is not contained within rigid walls.

H_2O

H_2O

rupture

"ghost"

H_2O

(hemolysis)

red blood cell

Water Balance in Land Plants

Since, as we saw in Chapter 17, water flows through plants from root to leaf, plants would be hard pressed to regulate the osmotic concentrations of their extracellular fluids as precisely as animals must. In a way, plants escape the problem, since their rigid cell walls allow them to withstand much lower osmotic concentrations than animal cells can tolerate.

If a cell from either a plant or an animal is put in a solution with a lower osmotic concentration than its own (see Chapter 3 for discussion), the cell will take up water by osmosis and swell (Figure 18–1). This swelling continues until it is counteracted by an opposing force of equal magnitude. Since the osmotic concentration of animal systems is typically regulated by specific organs, an isolated animal cell can scarcely resist the swelling, sometimes ballooning until it bursts. For example, when human red blood cells are put in water, they break open *(hemolysis)*, leaving behind empty cell membranes called "ghosts."

In the plant cell, however, the rigid cell wall prevents swelling beyond a certain point, mechanical pressure stopping any further osmosis. Generally, a plant cell maintains a high osmotic pressure against the cell wall. This turgor pressure in all the many cells of a plant helps

it stay erect. Without enough turgor pressure, the plant will first wilt and then die. As the cell contents shrink, they pull away from the cell walls, a process called *plasmolysis.*

Survival as a plant means maintaining turgor pressure and resisting plasmolysis—in other words, retaining enough water—and plants have various mechanisms for doing this. The form of the plant is often decisive in water balance. Broad-leaved plants are so subject to evaporation through their large leaf surfaces that they succeed best in tropical rain forests and some temperate zones, where water is abundant. Desert perennials sprout small leaves that stay only part of the year, or they have a few large, fleshy leaves that offer a small evaporative area relative to their volume—or they make do with no leaves at all. The cacti of the New World and the euphorbs of Africa and Asia are good examples of leafless desert plants (Figure 18–2). By separate evolutionary paths,

Figure 18–2 Desert plants have evolved in a variety of ways that conserve water. (**a**) The ocotillo grows small leaves only when water is present. (**b**) Many species of cacti in the New World and (**c**) euphorbs in the Old have pleated stems that can greatly expand to store water. (**d**) The leaves of the creosote bush are waxy and small, features that minimize water loss. (**e**) In contrast, tropical plants can permanently afford to have large leaves since they generally have abundant water.

a

b

c

d

e

these unrelated plants have produced two physically similar solutions to water balance in a desert environment. Leaves are generally reduced in size or converted to non-photosynthetic organs such as spines, and photosynthesis in the cacti and euphorbs occurs only in the green stem. In addition, many of these plants have elastic stems that can expand greatly to store moisture. The towering saguaro of the Arizona desert and the closely related cardon of northern Mexico and southern Arizona have fluted stems that swell with water the way the pleats of an accordion do with air. A drawback to this water-conserving combination of large size and low relative surface area is that heat from solar radiation can build up in the plant until some of the desert succulents (plants with large, thin-walled cells capable of storing water) reach internal temperatures 20°C higher than the unshaded atmosphere.

Although the selective pressures of scarce water have produced plant forms well adapted to conserving moisture against drought, heat poses another problem altogether. High temperatures affect the workings of the cell, principally by deactivating, and ultimately denaturing, the enzymes required for metabolism. In general, temperatures of 50 to 54°C are lethal, but some plants are able, in some unknown way, to tolerate even more severe conditions. The cholla and prickly pear cacti *(Opuntia)* can survive temperatures approaching 63°C. The creosote bush *(Larrea tridentata),* a successful and widespread shrub in North American deserts, keeps its stomata open during the day and retains its leaves all year. It can tolerate both desiccation and high temperatures, abilities that usually go hand in hand.

Not all desert plants survive by resisting desiccation. Some annuals germinate and complete their life cycles in the short time water is available, as after a heavy rain. Others have developed the ability to dry out and become dormant for a time, a strategy pursued successfully by the lichens. A good example is *Ramalina maciformis,* a lichen of Israel's Negev Desert. In the early morning, when relative humidity may reach 80 percent, the plant takes up water and engages in a round of photosynthesis—a botanical breakfast, if you will—before dehydrating in the heat of the day. Photosynthesis has already ceased by about 8 A.M. Lichens can remain dehydrated for years, resuming normal photosynthesis as soon as they are rehydrated. A hydrated lichen cannot survive temperatures higher than 38 or 39°C, but the same plant dehydrated can last through 100°C as well as very low temperatures. This ability permits lichens to occur abundantly not only in hot deserts but also in the Arctic tundra, where water is as scarce as it is in the desert and where deep cold replaces extreme heat. One such plant is the misnamed reindeer moss, a lichen that makes up as much as 60 to 90 percent of the diet of reindeer and other Arctic herbivores.

Osmotic Regulation and Water Balance in Animals

In some marine animals, the body fluids have the same osmotic concentration as the medium in which they live and move about. Such fluids are said to be *isosmotic,* and such animals are called **osmoconformers.** Other animals maintain a steady internal osmotic concentration against a range of external osmotic concentrations. This process is called **osmoregulation.** If the internal osmotic concentration is kept higher than the external, the osmoregulation is *hyperosmotic;* if lower, it is *hypoosmotic.*

Since osmotic concentration depends not on the identity of the solute but on the number of molecules (Chapter 3), osmoregulation can theoretically be achieved by juggling the concentration of any compound or ion available in sufficient quantity. For example, to keep the osmotic concentration of their blood similar to that of their ocean environment, the cartilagenous, or elasmobranch, fishes (sharks, skates, and rays) maintain a high concentration of blood-borne urea. Some other molecule could play the same role. The osmoregulatory importance of urea springs from the peculiarities of the evolution of the elasmobranchs, not from the biochemistry of that particular compound.

In referring to osmoregulation, we mean primarily the active movement of Na^+ and Cl^-. Sodium chloride is one of the most common salts in the sea, and it is usually the salt of highest concentration in living organisms. Osmoregulation entails the active absorption or excretion of Na^+. Cl^- follows passively because of the attraction of the opposite electrical charge. Water moves along the osmotic gradient.

Many other ions move into and out of cells, but this movement entails far fewer molecules than does the osmoregulation of Na^+. However, these other ions also crucially affect the working of the cell, and their concentrations, too, must be kept within narrow limits. Too much K^+, for example, blocks DNA synthesis, but too little stops protein production. Some animals maintain ion concentrations passively, and others use active transport to move the ions in or out against concentration gradients. Both active and passive mechanisms are called *ionic regulation.*

We will now look at specific problems of osmotic and ionic regulation in various animal groups.

Marine Invertebrates Except for a few insect larvae, by far the most numerous and successful aquatic arthropods are the crabs and their allies (class Crustacea). Crustacea inhabit the ocean depths, the estuaries, and fresh water. Some live on dry land. Some species can live only in water at rather precise salinities. One of the reasons some of these animals can inhabit such a wide range of habitats is that they possess formidable abilities to maintain the salt concentration of their body fluids independent of environmental osmotic concentration. That is, they osmoregulate. The deep-sea crustaceans, however, never experience pronounced changes of salinity in their stable habitat, and most of them are osmoconformers. By the same token, they cannot tolerate salinity changes. For example, if the spider crab *Maja* (Figure 18–3) is placed in a medium with 80 percent the osmotic concentration of seawater, the osmotic concentration of the crab's blood changes until it equals the medium. But the crab cannot tolerate this change, and it dies. Animals with such a narrow range of salinity tolerances are said to be *stenohaline* (from the Greek *stenos*, "narrow," and *halos* "salt"). Other osmoconforming stenohaline marine invertebrates include echinoderms (starfish), cephalopods (squid and octopi), and cnidaria (jellyfish).

Generally, only osmoregulators can tolerate large changes in external salinity. Thus, only they can live in both salt and fresh water. The shore crab *Carcinus* keeps its internal fluids at least somewhat independent of external osmotic concentrations (Figure 18–3). When *Carcinus* is placed in salt water, the osmotic concentration of its blood approaches that of the medium. But when the crab is then put in a more dilute solution, the osmotic concentration of the blood remains high, even though osmosis constantly brings water in and diffusion takes salt out. The hard exoskeleton slows this flow, but in addition the crab maintains its high osmotic pressure by actively taking in salts from the environment through its gills and by producing a dilute, copious urine that "bails out" the water.

Fresh-Water Invertebrates Practically all fresh-water invertebrates maintain an osmotic concentration higher than that of the medium, and they follow the pattern of *Carcinus* in osmoregulation. For example, the crayfish's hard exoskeleton limits osmotic flow. The animal's kidney produces a large volume of dilute urine, and salts lost to diffusion are replenished from food and by active absorption from the environment, largely through its gill membranes.

The internal osmotic pressure of many invertebrates is due largely to dissolved salts, but in the aquatic insects, which are also hyperosmotic, the pressure is due almost entirely to dissolved amino acids. This is a characteristic of insects in general. Salts account for only a small part of the blood's osmotic pressure. Since amino acid molecules are larger than salts, they diffuse out less readily. In addition, the impermeable cuticle blocks water and amino acid movement.

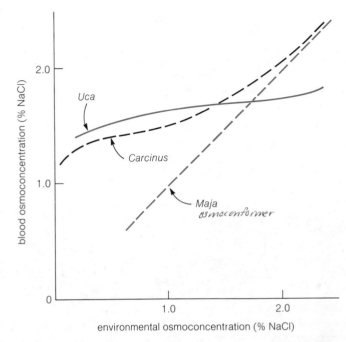

Figure 18–3 The relationship between environmental and blood osmoconcentration in three crabs with different types of osmoregulation. *Maja*, the spider crab, is an osmoconformer. It is a deep-sea animal ordinarily not subjected to different osmotic concentrations. *Uca*, the fiddler crab, inhabits the intertidal zone, where it may be subjected to both fresh and salt water. It is a hyper- as well as a hypoosmoregulator. *Carcinus* is a crab of a shallow water near estuaries where salinities may vary. It is able to hyperosmoregulate, and at higher salinities it becomes an osmoconformer. *Maja* closely follows the isosmotic line.

Fishes The bony fishes appear to have evolved largely in fresh water, where they were hyperosmotic to the medium. Subsequently, certain species reentered salt water. For some unknown reason, these pioneers retained their dilute body fluids and were hypoosmotic to the sea. Both fresh- and salt-water species are protected to some extent by an impermeable body covering, but the respiratory surfaces of the gills cannot be waterproofed if they are to serve as organs of gas exchange. Instead, the gills work like osmotic "windows" for the absorption or excretion of salts.

Since marine bony fishes are hypoosmotic to their surroundings, they tend to lose water and take up salts passively. To keep their osmotic balance, these animals drink seawater and actively excrete excess salts through the *chloride cells* of the gills. The gills also eliminate nitrogenous wastes, primarily in the form of ammonia. The kidneys of most bony fishes are of little use in osmoregulation since they cannot produce a urine more concentrated than the blood.

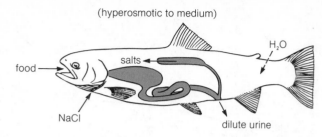

fresh-water fish

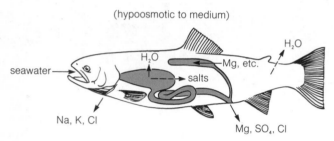

salt-water fish

Figure 18–4 Main paths of water and ion movement during osmo-regulation of bony fish in fresh and salt water. Active transport is indicated by solid arrows, and passive transport (osmosis and diffusion) by broken arrows.

Fresh-water bony fishes face the opposite problem. Since they are hyperosmotic to the medium, water enters and salts leave. To keep themselves hyperosmotic to their surroundings, the fish dump excess water in a dilute and voluminous urine, and they actively take in salts through their gills. As in the marine forms, ammonia is excreted through the gills. Figure 18–4 compares the fresh- and salt-water strategies of the bony fishes.

The marine elasmobranch fishes also evolved first in fresh water, but they moved back to the sea earlier than the bony fishes and developed a different osmoregulatory mechanism. These animals retain some of their nitrogenous wastes and convert them to urea. The urea goes into the blood and makes the blood hyperosmotic to the environment. As a result, the sharks, skates, and rays absorb only small quantities of water.

Reptiles and Birds The reptiles and the birds, which evolved directly from them, are terrestrial animals that face the never-ending problem of losing water to evaporation. They reduce water loss with impermeable body cov-

erings and with internal respiratory organs. In addition, both types of animals have a common chamber, the *cloaca,* in which both feces and urine collect before they are voided. As in the analogous rectum of the insects, water is absorbed from the wastes in the cloaca; and also as in the insects, nitrogenous wastes are in the form of uric acid, which can be excreted with little or no water.

Generally, the kidneys of reptiles and birds cannot produce urine with a salt concentration higher than the blood's. One exception is the savannah sparrow, which often inhabits salt marshes. This bird produces a hyperosmotic urine and can drink salt water.

A different adaptation is found in many sea birds and predatory birds and in marine iguanas, crocodiles, sea snakes, and marine turtles. These animals have a pair of *nasal glands* near their eyes that concentrate excess salt and excrete it. These droplets are the "crocodile tears" the giant reptile sheds even as it consumes its prey. Gulls, too, are often seen with these salty droplets hanging from the tips of their bills.

Mammals Mammals also face the problem of drying out, and their major osmoregulatory task is retaining water. The primary organ of water balance in the mammals is the kidney, which we will examine shortly.

We can best understand how a mammal balances water losses against water gains by looking at an extreme case: desert mammals. These animals have no unique physiological mechanisms that fit them to the desert's regimen. Instead, their adaptations lie in refinements of physiology and behavior.

Knut and Bodil Schmidt-Nielsen studied water balance in Merriam's kangaroo rat of the Arizona desert. They learned that a kangaroo rat feeding on barley seeds could survive indefinitely without drinking water. To find out how, the Schmidt-Nielsens measured all possible avenues of water intake and water loss. These measurements revealed that most of the kangaroo rat's water comes from the oxidation of food carbohydrates and fats, and that the greatest source of loss is exhalation during breathing. If water is available, the rat loses considerable water to urine, but when water supplies fall short, the urine becomes highly concentrated, thus reducing the amount of water used. The kangaroo rat does not sweat, a characteristic that eliminates an entire avenue of water loss. The animal reduces its water losses still further by staying underground during the day and coming out only at night, when the rate of evaporation is lowest.

Curiously, the Schmidt-Nielsens found that a kangaroo rat fed soybeans instead of barley cannot maintain water balance and has to drink to survive. The reason is that the higher protein content of the soybeans produces more urea, which has to be voided in solution with water. Rats fed soybeans lose more water to urine than they gain from oxidation. However, a soybean-fed rat can survive if given salt water, producing a urine two or three times saltier than seawater.

Excretion, Water Balance, and Excretory Mechanisms

As we saw in Chapter 4, the cellular respiration of fats and carbohydrates produces water and carbon dioxide as waste products. The water becomes directly available to the animal; the carbon dioxide dissolves in the extracellular medium and is eliminated as a gas. The central excretory problem revolves around the waste products of protein metabolism. The amino acids of proteins contain nitrogen in the form of amino groups ($-NH_2$). When proteins are oxidized or converted to carbohydrates and lipids, the amino groups are cleaved off in a process called *deamination,* which in humans occurs in the liver. The $-NH_2$ groups are then converted to ammonia, NH_3.

Some animals (principally aquatic inverebrates) may excrete ammonia directly, a measure that saves them from expending energy to convert the ammonia to some other chemical form. However, there are decided drawbacks to ammonia as a nitrogenous waste product when it cannot be leached into the surrounding medium. The compound is extremely toxic, and the body cannot store it without being poisoned. Ammonia is found only in aquatic organisms, such as marine invertebrates and bony fishes, which have ready access to the water needed to flush out the waste.

In many animals, the ammonia is chemically combined with carbon dioxide to produce *urea,* CH_4ON_2 which is much less toxic.

urea

In humans this reaction occurs in the liver. The urea is released into the blood, then filtered out in the kidneys to be voided in solution with water. Since urea is less toxic than ammonia, the body can tolerate higher levels of it and use less water to excrete it. Urea is the nitrogenous waste of mammals and elasmobranch fishes.

Uric acid, $C_5H_4O_3N_4$, is another, more complex nitrogenous waste product:

uric acid

Although complicated to produce, uric acid pays certain physiological dividends. It is relatively inert and nontoxic, thus reducing the drain on sometimes precious stores of water. Also, uric acid is relatively insoluble in water. Since the molecule precipitates out of solution, it can be stored in a dry form indefinitely, further separating it from the body's water reserves.

By examining the excretory mechanisms of various organisms, we can learn how water balance, osmoregulation, and ionic regulation are interwined. How have these various functions been solved by animals varying in size and complexity from the single-celled *Paramecium* to an organism such as man, with billions of cells?

Water-Expulsion Vesicles   Some substances move relatively freely back and forth across the cell membrane of a protozoan. For example, when a *Paramecium* is first placed into a hyperosmotic solution, water flows out into the medium faster than solutes enter. The organism shrinks and begins to dehydrate. In a hypoosmotic solution, on the other hand, water enters faster than solutes leave, and the *Paramecium* swells. However, *Paramecia* and other protozoans can live in both salt and fresh water. In nature, these organisms are roughly isosmotic with salt water, and a salt-water habitat poses little problem. But when the *Paramecium* moves to fresh water, the influx of water must be counteracted or the organism will burst.

The *Paramecium* rids itself of the excess water and nitrogenous wastes with an organelle called a *water-expulsion vesicle,* or *contractile vacuole* (Figure 18–5). The contractile vacuole gathers up water and wastes from inside the cell and empties it to the outside in regular pulsations. The time separating pulsations depends on the environment, varying from about 13 seconds in fresh water to 65 seconds in seawater for *Paramecium woodruffi.* The reason for the difference in interval is the quantity of water that has to be excreted. Measurements on *Paramecium caudatum* show that, per hour, the organism voids 30 times as much in fresh water as in salt water;

this is 5 times the volume of the entire organism. There is evidence that the vesicle's activity is controlled metabolically. Mitochondrialike granules surround the organelle, and chemicals like cyanide that inhibit cellular respiration stop the pulsation of the vesicle, allowing water and nitrogenous wastes to build up inside the organism.

Flame Cells   The flatworms—planaria, flukes, and tapeworms—are relatively large, multicellular animals without circulatory systems. Most of their metabolic wastes leave by diffusion, either from the central body cavity through the mouth or through the body wall. Water balance is handled by a tubular excretory system that branches throughout the body of the flatworm (Figure 18–6). The channels lead to the outside through numerous excretory pores. Excess water and probably some wastes enter the system through the bulbs branching from the tubules. These bulbs are called **flame cells,** because their beating cilia look something like flickering candle flames. The cilia propel the water into the tubules and out into the environment through the excretory pores.

Malpighian Tubules   Insects have an open circulatory system and markedly different excretory organs from most higher animals, including humans. These organs, called **Malpighian tubules,** are actually sacs or pouches branching out from the gut. Numbering from 2 in some species to over 200 in others, the tubules are actually blind sacs that empty into the gut at the junction of the hindgut and midgut (Figure 18–7). The tubules are usually, but not always, loosely suspended in the space surrounding the internal organs, and contractions of muscles in the walls of the tubules move them about. These movements probably increase blood flow over them and speed the movement of wastes into the tubules.

There is no pressure gradient to force materials into the tubules. Various ions and uric acid are taken in by active transport, while water enters along the osmotic gradient. Since the tubules are closed at one end, the solution of wastes can flow only toward the gut. After entering the digestive tract, the urine continues on into the rectum, where most of the water is reabsorbed and the uric acid precipitates as crystals. The mixture of urine and feces the insect finally excretes is often powdery and dry.

Nephridia   Unlike the flatworms and insects, the common earthworm has a closed circulatory system that makes it possible for wastes to be transported to organs set aside for excretion alone. The excretory and circulatory systems are intertwined anatomically, and circulation plays an important role in excretion.

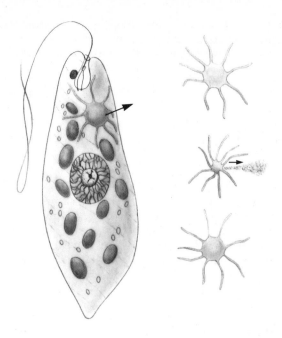

Figure 18–5   Water-expulsion vesicle. Some single-celled organisms, like this *Euglena*, collect excess cellular water by a system of radiating canals that empty into a vacuole. When full, the vacuole expels its fluid to the exterior.

The earthworm's excretory organ is the **nephridium,** and each segment of the worm's body contains two nephridia (Figure 18–8). Body fluids in the coelomic cavity enter the nephridium through an opening called the *nephrostome.* A tubule leads from the nephrostome through the partition between body segments to the *nephridiopore.* As the fluid moves through the coiled tubule, active transport moves chemicals needed by the body out of the fluid and into the blood flowing through capillaries intimately associated with the tubule. Wastes and excess water remain behind and collect in the *bladder.* Accumulated wastes are voided to the outside through the opening of the nephridiopore.

The Mammalian Kidney   Functionally, the mammalian kidney is much like the worm's nephridium. But instead of two nephridia in each of the many body segments, a single pair of kidneys handles the excretory functions of the whole organism. The kidneys are a site of intimate contact between the excretory and circulatory systems (Figure 18–9). Blood follows the renal artery from the aorta into the kidney, which filters the blood under pressure, retaining wastes and returning needed elements to the blood. The blood returns to the circulatory system

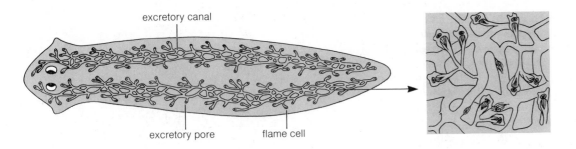

excretory canal

excretory pore   flame cell

Figure 18–6 Flame-cell system in planaria.

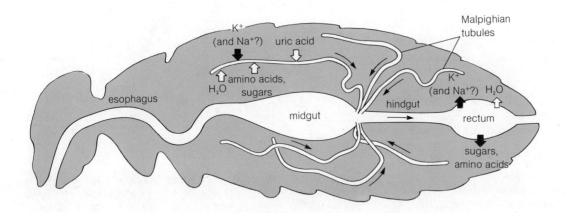

K^+ (and Na^+?)   uric acid

amino acids, sugars
H_2O

esophagus

midgut

hindgut

Malpighian tubules

K^+ (and Na^+?) H_2O

rectum

sugars, amino acids

Figure 18–7 Structure and function of the excretory system of an insect—the Malpighian tubule–rectal system. The heavy black arrows show active transport and the white arrows passive transport or some other unknown, mechanism.

through the renal vein. The wastes collect in the *renal pelvis,* then make their way through the *ureter* to the *bladder,* where the urine is stored until voided to the outside through the *urethra.*

The workings of the kidney are due to the organ's unique morphology. In humans each kidney contains approximately 1 million nephrons, which are its functional units, comparable to the nephridium of the earthworm. The nephron comprises a *Bowman's capsule* and a long tubule (Figure 18–10a). Wedged into the Bowman's capsule is a tuft of capillaries known as the *glomerulus.* When the Italian Marcello Malpighi first discovered these capsules in 1661, he called them "corpuscles." About 200 years later, William Bowman of King's College in London showed that the corpuscles are actually a tuft of capillaries intruding into a tube. In 1854, two years after the publication of Bowman's paper, the German physiologist Carl Ludwig suggested that the glomerulus and Bowman's capsule work together as a mechanical filter. He proposed that the blood's own pressure forces the plasma from the glomerulus into Bowman's capsule, leaving the blood cells and proteins behind. He proposed further that the tubule selectively reabsorbed nutrients and salts and left the wastes behind. Experimental work has established the validity of Ludwig's hypothesis.

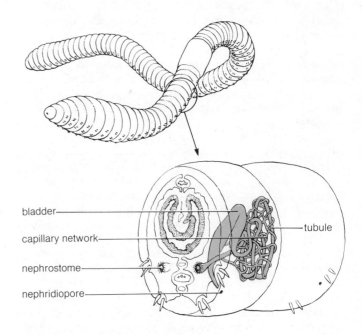

bladder

capillary network

nephrostome

nephridiopore

tubule

Figure 18–8 Nephridium of the earthworm.

The tubule and the associated capillary net beyond Bowman's capsule are specifically fitted for reabsorption. The first portion, called the *proximal convoluted tubule,* is arranged in several loops. The tubule then descends from the outer area, or *cortex,* of the kidney to the organ's interior, or *medulla,* making a long loop called the *loop of Henle.* The tubule ascends to the cortex again to loop repeatedly in the *distal convoluted tubule.* The distal tubule connects with the *collecting tubule,* which empties into the renal pelvis. The entire tubule, from proximal to distal regions, is associated with a network of capillaries joined by an arteriole to the capillaries of the glomerulus. The capillary network leaves the nephron through a venule that leads to the renal vein.

The human kidney filters about 650 milliliters of blood each minute. This large volume of liquid allows the body to rid itself of wastes rapidly. However, voiding the filtrate directly would cause the organism to lose a great amount of water and other molecules. Normally this does not happen. Instead, the tubule reabsorbs some 99 percent of the filtrate, leaving behind a concentrated solution of wastes.

The mechanism for reabsorbing water from the filtrate is a fascinating example of natural engineering (Figure 18–10b). There is no biological method for active transport of water; instead, biological systems for transporting water rely on osmosis, moving salts by active transport to create an osmotic gradient for water to follow passively. This is the essential mechanism in the reabsorption of water in the nephron. Sodium ions are actively transported from the ascending portion of the loop of Henle into the surrounding tissue fluid. The wall of the ascending loop is impermeable to water and thus keeps the water from following the sodium passively. The wall of the descending loop is permeable to sodium, and the sodium extracted from the ascending loop diffuses from the tissue fluid into the filtrate in the descending loop. As the filtrate flows through the loop, more and more sodium enters, and the concentration rises, peaking at the lowest point in the loop. Then, as the filtrate enters the ascending loop, sodium is removed by active transport. By the time the filtrate reaches the distal convoluted tubule, it is once again a dilute solution. Since this mechanism involves a flow in opposite directions—down the descending loop and up the ascending loop—it is called a countercurrent system. And since this system increases the gradient along its length and thus increases the sodium concentration, it is called a *countercurrent multiplier.*

At first glance, this mechanism seems to be an exercise in robbing Peter to pay Paul: Take a dilute fluid, raise its concentration, then dilute it again. But the point of this process is to recover water. The active transport of sodium in the ascending loop creates a high osmotic pressure in the region between the ascending loop and the collecting tubule. The walls of the collecting tubule are

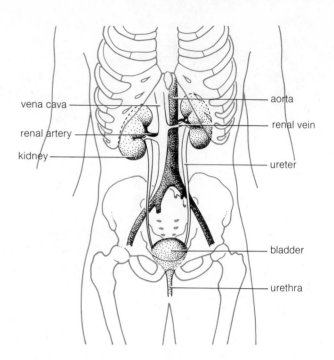

Figure 18–9 The organs of the human excretory system, showing its close association with the blood circulatory system. Blood is forced under pressure into the kidney from the aorta and returns to the posterior vena cava. The urine collects in the pelvis of the kidney and leaves by the ureter.

permeable to water. As the dilute filtrate flows down the collecting tubule, water follows the osmotic gradient out of the tubule and reenters the circulatory system through the capillary bed. The remaining filtrate, now much more highly concentrated because of the loss of water, makes its way to the renal pelvis and then through the ureters to the bladder.

The degree of concentration in the urine depends largely on the length of the loop of Henle. One of the longest loops of Henle is found in the kangaroo rat, which can make urine with concentrations as high as 7 percent salt and 20 percent urea. The human nephron has a shorter loop and can achieve salt concentrations no greater than 2 percent. Thus, drinking seawater, which averages 3.5 percent salt, dehydrates a human, since the kidney produces at least 1.75 liters of urine for every liter of salt water consumed.

The concentration of the urine does vary somewhat, owing to the activity of the posterior pituitary hormone vasopressin, or antidiuretic hormone (ADH). When vasopressin is lacking, reabsorption of water drops to zero, and urine become copious and dilute. Alcohol inhibits the secretion of ADH, and heavy drinking can affect kidney function. However, in the interest of correct physiol-

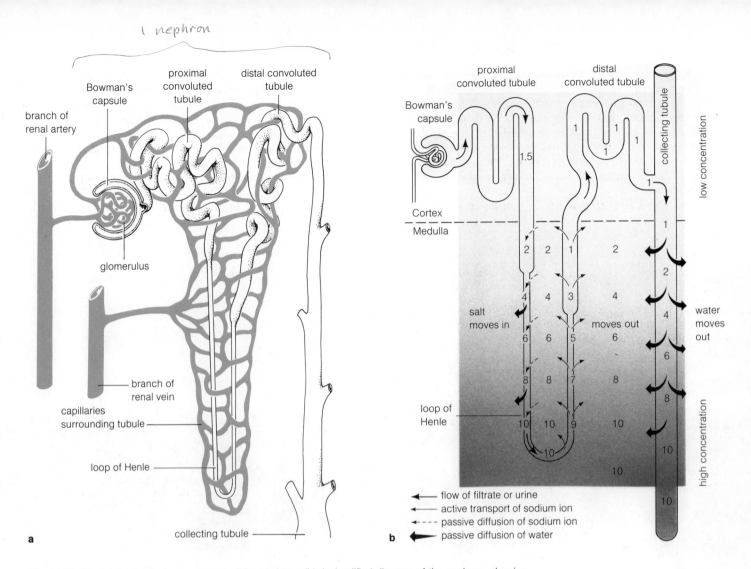

nephron

Figure 18–10 (a) Anatomical arrangement of the nephron. (b) A simplified diagram of the nephron, showing how salt and water move in and out of the tubule. The concentration of salt is represented by the numbers in the figure. It is highest at the bottom of the figure.

ogy, it should be noted that the diuresis (loss of water through urination) that follows drinking a few cans of beer is not usually due to alcohol. Beer contains only about 4 percent alcohol, not enough to affect ADH secretion. The extra urine results from added water intake.

The tubule reabsorbs molecules other than water as well, and in this way the kidney regulates the concentration of various blood-borne molecules. There is a threshold concentration for each molecule. Below the threshold, the molecule is reabsorbed by active transport, but above it, the molecule remains in the filtrate and leaves the kidney in the urine. The threshold varies from one molecule to another. For urea, which is a waste, the threshold is very low, but for glucose, which is a nutrient, it is high. In a person with normal metabolism, the glucose

level in the blood is not exceeded, and no glucose appears in the urine. But diabetics, who have an unusually high level of glucose in the blood, exceed the threshold and excrete glucose in their urine (see Figure 14–8). Thus, the common laboratory analysis for diabetes tests the urine for sugar.

Summary

All cells must maintain certain water, salt, and ion concentrations in order to remain functional. This chapter concerns the mechanisms in some single-celled and a wide range of multicellular organisms living in salt and fresh water and on dry land.

In plants, morphology is a large factor in determining water loss. Plants that do not escape the driest periods of the year in arid regions by surviving in the seed or root stage have morphology and physiology adapted for water conservation and heat tolerance. The rigid cell walls of plants resist expansion and rupture in the presence of excess water.

Most animals that experience a wide range of salt and water concentrations physiologically regulate the salt, water, and ionic compositions of their bodies within relatively narrow limits. To some extent, the regulation is done by individual cells; but for the most part, the cells in multicellular animals are in equilibrium with the fluid medium that surrounds them in the body, and the content of this medium is regulated by such organs as gills and kidneys in vertebrates. We examined the function of the kidneys in detail and compared it with the function of several analogous organs found in invertebrate animals.

19

Temperature and Temperature Regulation

Heat, like water, imposes limits on life. Air temperatures on earth vary from about −70°C to more than 85°C, yet life as we know it is restricted to a fairly narrow band from about 0 to 40°C. With a few exceptions, like the algae and bacteria inhabiting hot springs with temperatures of 90°C, all functioning living systems must remain within this range.

These specific temperature demands depend in part on the chemical characteristics of carbon, oxygen, and water. At very low temperatures, these atoms cannot interact to produce the complex molecules of living things, and at high temperatures the molecules break up. Life is more than a structural phenomenon, however, and the rate of chemical reactions necessary to life also depends on temperature. In most physiological systems, biochemical reaction rates double with every temperature increase of 10°C (Figure 19–1). The rate of change as a function of a 10°C change in temperature is defined as the Q_{10}. A Q_{10} of 2 implies a doubling of rate. For example, suppose a protozoan moves its cilia at 10 beats per second at 10°C. If the organism has a Q_{10} of 2 throughout the temperature range of 10 to 40°C, it would do 20 beats a second at 20°C, 40 at 30°C, and 80 at 40°C. Beyond a certain point, however, increasing temperature decreases the rate of reaction, finally halting it altogether. The reason is that the enzymes catalyzing the reactions lose their three-dimensional structure (they may coagulate) and thus their ability to catalyze. Thus, although at high temperatures the reaction rate of individual enzyme molecules is often promoted, enzyme molecules are also deactivated, so that the overall reaction rates decline.

Living systems must be able either to maintain a temperature within some acceptable range or to adapt to extremes. In this chapter we will look at various temperature strategies, beginning with plants and then turning to animals.

Temperature and Plants

Tissue Temperatures Since plants are unable to move about, their temperatures, much more than animals', are at the mercy of the environment. Protoplasm in most plants dies if heated beyond 50 to 54°C. Although air temperatures in tropical and subtropical deserts seldom go over 50°C, high levels of radiant energy from sunlight can raise leaf and stem temperatures well above the lethal point, particularly if the leaves or stems touch the soil.

Some desert plants avoid the problem. Palms growing around an oasis have an ample water supply. During heat

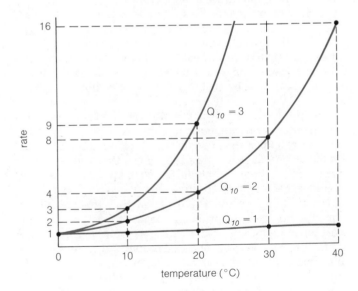

Figure 19–1 The Q_{10} of three hypothetical rates. Most biological systems correspond to the middle or the top curve. Few reactions are temperature-compensated ($Q_{10} = 1.0$).

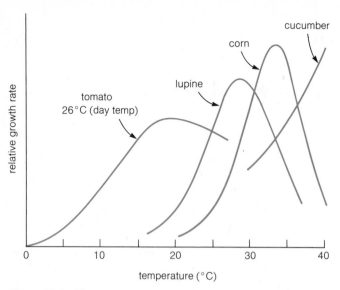

relative growth rate

tomato
26°C (day temp)

lupine

corn

cucumber

temperature (°C)

Figure 19–2 Plant growth curves as a function of temperature. The optimum temperatures for growth rate vary greatly among species.

spells, transpiration increases sharply, and evaporation cools the plant. However, most desert plants have to conserve water, retaining it within the leaves and stem. When they do not lose water evaporatively, plant and air temperatures are roughly equal. Since air temperatures rarely exceed the lethal point, this situation is tolerable. However, some tissues, such as those in the base of the stem touching the soil, may reach temperatures well above 50°C, yet they survive. How they do so remains a mystery.

As we will see shortly, most animals regulate their body temperatures to some extent. Plants, however, passively depend on environmental factors for their tissue temperatures. Why can't plants regulate their internal temperatures as animals do? Obviously, this question involves too much speculation to be answered with certainty, but it may be that the plants' ability to photosynthesize rules out temperature regulation. Unable to move about, plants cannot use large sources of food energy the way animals can. Instead, they are assured a steady supply of energy from the sun. Capturing that energy requires a large surface area. When the sun falls on this surface, internal temperatures build up, but when the sun is gone, that same surface allows much of the captured heat to radiate back into the atmosphere. Thus, wide swings of temperature are almost inevitable. Animals, however, have to move about quickly, necessitating a high and relatively constant rate of enzymatic reactions. Plants and animals have therefore been shaped by different selective pressures. Plants have evolved enzymatic systems that work fairly slowly but evenly over a wide

range of temperatures, whereas animals have enzymes with high reaction rates specialized to hit their peak only in the narrow range of regulated body temperature.

Further clouding this issue are plants that are somewhat independent of the environmental temperature. Such small-leaved desert species as the creosote bush, *Larrea tridentata* (see Figure 18–2), unload by convection most of the heat they absorb from sunshine. And at least one plant can raise its temperature above the environment's by internal heat production. At a specific time of the day, once in the lifetime of the flower, the metabolic rate of the inflorescences of some of the *Arum* flowers increases sharply, and internal temperature rises 10 degrees or more above air temperature. According to one hypothesis, the higher temperatures disperse volatile chemicals attractive to pollinating insects. However, this hypothesis raises more questions than it answers. Why hasn't the plant evolved volatile scents active at lower temperatures? No one knows.

Thermoperiodism Besides affecting a plant's chance for survival, temperature influences its rate of growth (Figure 19–2). For a given species, there is often a minimum temperature, below which the plant will not grow (for example, 0 to 5°C); an optimum temperature, at which the plant grows best (for example, 28 to 35°C); and a maximum temperature, above which the plant stops growing and may die (for example, 45 to 52°C). These *cardinal temperatures* vary from one species to another.

Since enzymes are critical to life, biologists long thought that the cardinal temperatures represented the range in which the enzymes characteristic of a particular plant could function. At the maximum temperature, for example, it was believed that the enzymes were deactivated, and cellular reactions slowed or stopped. However, it soon became apparent that such a simple explanation for plants' reactions to temperature does not fit all the facts.

In the 1940s and early 1950s, Frits Went carefully documented a phenomenon already suspected by people who were experienced in growing plants at controlled temperatures, as in greenhouses. From the simple explanation of cardinal temperatures by enzyme action, one would expect a plant to grow best if kept always at its optimum temperature. Went found that this is true of some plants, but not others. Instead, many species grow best when high (day) and low (night) temperatures differ, a phenomenon Went called **thermoperiodism.** For example, tomatoes grow best at 26°C in the day and 18°C at night, temperatures that vary somewhat with the age and variety of the plant. Went even discovered why some people have so much trouble raising African violets. These prized cultivated plants have thermoperiods just the opposite of most, growing best when temperatures at night are somewhat higher than during the day. People who turn their thermostats down at night and throw the windows open will have less success with these "temperamental" plants than those who leave the thermostat up and the windows down.

Went studied the thermal responses of many species, comparing them with the average annual temperature patterns of the localities in which they grow. Some of the species Went researched had unusually low maximum nighttime temperatures. One such plant is *Baeria charysostoma,* a small annual flower common in the spring in California mountain valleys, foothills, and occasionally the western Mojave Desert. Grown under short-day conditions, experimental plants die at nighttime temperatures over 26°C, which is the optimum nighttime temperature for African violets. At a nighttime temperature of 26°C, the plants live about two months, but at 20°C, they grow for at least 100 days. The thermoperiodicity of this plant is particularly marked. Although temperatures in excess of 26°C prove lethal at night, *Baeria charysostoma* flourishes at such daytime temperatures as long as the nighttime temperature is low enough. Went uncovered similar patterns in other species.

Temperature Regulation in Animals

Like plants, animals are limited in their growth and activity by temperature. The particular high and low temperatures at which enzymes—and life—cease to function vary from species to species. Animals are mobile, however, freed of the thermal captivity experienced by plants. In addition, their high metabolic rates generate considerable heat, making them better able to adapt to cold.

The body temperature of any organism is the net result of the heat produced as a by-product ("inefficiency") of metabolic events like muscle contraction and the heat obtained from the environment (such as solar radiation) minus heat lost to convection, radiation, conduction, and evaporation. Various combinations of these factors affect or control animals' body temperatures.

With regard to temperature regulation, animals can be divided into two basic groups: **poikilotherms,** whose body temperature is much the same as that of the environment, fluctuating as the environmental temperature changes, and **homeotherms,** which maintain a constant body temperature despite environmental fluctuations (Figure 19–3). Usually this means keeping body temperature higher than the environmental temperature, and the source of the extra heat varies. Animals that take up heat from the environment are called **ectotherms.** Others, that use heat produced by their own metabolism, are called **endotherms.** In some animals, thermal behavior changes

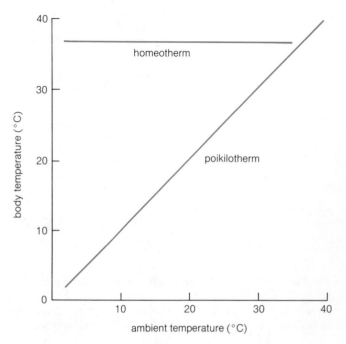

Figure 19–3 The relationship between body temperature and ambient temperature in a homeotherm and in a poikilotherm.

from endothermy during activity to ectothermy at rest. This behavior, particularly common among insects, is called **heterothermy.**

Few animals use any one thermal behavior exclusively. Humans, for example, are largely endothermic, but we often bask in the sun or move closer to a fire to get warm, behaviors that are typically ectothermic. The differences in thermoregulation among various groups of animals are less a matter of kind than of degree.

Poikilothermy Most small invertebrates, particularly those living in water, do not control their body temperatures. Sluggish at low temperatures, they move faster as ambient (environmental) temperatures rise. This is also true of some aquatic vertebrates as well. For this reason, fresh-water fishermen often notice pronounced differences in the behavior of game fish from one season to the next. In early spring, when the water is cold, a pike may be pokey, but later in the year, as the water warms, it becomes a fierce battler.

Poikilotherms may produce considerable heat during activity yet be poikilotherms because they also lose considerable heat. Most of the fishes fall into this category. Blood passing through a swimming fish's muscles picks up heat, which it then gives up to the cooler water through the thin walls of the gill capillaries. Water is such a good

Figure 19–4 A basking Andean lizard, *Liolaemus*.

conductor of heat—it draws heat away from the organism so easily—that endothermy is impossible in all but the largest fishes, which have smaller surface areas relative to volume and thus better retain the heat they produce.

Ectothermy: Temperature Regulation through Behavior The environment is a mosaic of different temperatures. The temperature of the desert air may be quite different from that of a rock sitting in the sun; in a still pond, the surface is warmer than the bottom. The smaller the animal, the more extreme the temperature mosaic available to it in the environment. A beetle may encounter lethal temperatures in sunshine on desert sand, but by moving a centimeter into the shade of a blade of grass, it can suddenly be in a suitable thermal environment. Many poikilotherms can vary their body temperatures simply by shuttling to and from areas of different temperature to pick up or unload heat as needed. The simplest form of this behavior is a trial-and-error search for the preferred temperature. Ciliate protozoa like *Paramecium,* for example, gather in regions of appropriate temperature because of such random movements.

Many animals that raise their body temperatures above the environmental temperature, including even the normally endothermic and homeothermic human, occasionally do so by behaviors fashioned to take up heat from the surroundings. Many lizards and butterflies take on heat from their surroundings regularly. Usually these mechanisms can be observed only in the animal's normal environment, not in the laboratory, simply because they are adapted to those specific surroundings. A lizard caged in the shade is poikilothermic. But in the field, active lizards may keep their body temperatures much higher than that of the surrounding air. Oliver P. Pearson of the University of California at Berkeley studied the Andean lizard, *Liolaemus* (Figure 19–4). In *Liolaemus'* moun-

tain environment at altitudes of 15,000 feet or higher in Chile, solar radiation in the daytime is intense, but the ambient temperature may not exceed 0°C. *Liolaemus* basks in the sun and warms up to 35°C even when temperatures are below freezing.

Maintaining a high body temperature allows the lizard to be active at low air temperatures. But a high body temperature may also have other functions, like combating disease organisms that are sensitive to high temperatures. It has recently been shown that lizards injected with pathogens, and then permitted to choose a temperature in a temperature gradient, chose to maintain a high body temperature. These warm animals survived their infection better than controls kept at a lower temperature. Fever in sickness may have a similar function.

The sidewinder rattlesnake *(Crotalus cerastes),* a successful desert dweller, maintains a fairly even body temperature around 30°C by seeking cool refuges in the heat and warm sunny places in the cold. Basking is common in insects. Many butterflies spread their wings and turn their bodies so that the sunlight falls on them perpendicularly, raising their body temperature several degrees per minute (Figure 19–5).

For the sun's rays to produce heat, they must be absorbed. An object's color indicates how well it absorbs sunlight. Dark objects look dark because they absorb most of the visible spectrum and reflect little; light-colored objects reflect more and absorb less. The roadrunner has black patches of skin on its body. In the early morning, when the desert air is still cold, the roadrunner lifts its feathers and exposes the black patches to the sun, warming itself and saving energy. Color differences have also been shown to be important to the rate of warm-up in grasshoppers, beetles, and butterflies in the sunshine.

Behaviors can be used to unload heat as well as to pick it up. When it is too hot, insects often simply retire to the shade. The antelope ground squirrel of the Mojave

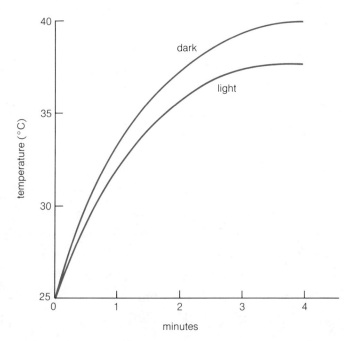

Figure 19–5 Heating curves of *Colias* butterflies of light and dark pigmentation while basking in sunshine.

Desert is active in the heat of the day. When it scurries outside to forage in the sunlight, the squirrel heats up to 43°C. Then it returns to its burrow, applies its belly to the cool sand within, and gives off heat until its body temperature returns to a more "normal" 36°C, readying the rodent for another foray to find food.

Endothermy and Thermoregulation at Low Temperatures Most small invertebrates have a relatively low metabolic rate, and the heat they produce is rapidly lost to the environment. Some insects, particularly those covered with insulating hairs, like the moths and bumblebees, can maintain an endothermic body temperature higher than that of birds or mammals (see Supplement 19–1).

Birds and mammals, whether moving or not, are usually endothermic (internal heat producers), making them ready to flee or fight at almost any time. Regardless of size, cooling rate, and environmental temperature, most mammals, including humans, maintain core body temperatures around 37 or 38°C. Birds tend to be 3 or 4°C warmer.

The remarkable thing about mammals and birds is their ability to maintain that internal temperature even when environmental temperatures are extremely low. In mid-winter, when air temperatures regularly approach −30°C, the Arctic fox and the ptarmigan, a kind of grouse, remain at their usual body temperatures. (Small mice and shrews also maintain the same high body temperatures in the Arctic at very low air temperatures, but they spend most of the winter under deep layers of snow where temperatures remain considerably higher than those of the air above the snow.)

As the air temperature drops and more heat is lost to the environment, the first physiological response is raising the hairs of the coat or fluffing the feathers, an action called **piloerection.** Piloerection creates a larger insulating air space under and between hairs and feathers and thus increases insulation. Humans have little body hair, but you can see a remnant of piloerection in the "goose pimples" we get when chilled. In addition to piloerection, constriction of the blood vessels reduces blood flow to such extremities as the feet, hands, and ears. In humans, this response cuts radiant heat loss by one-sixth to one-third. This is also the reason that frostbite affects the extremities first; deprived of blood, the fingers and toes are least able to resist freezing under severe conditions. However, when our bodies are warmly clothed, our hands and feet, although poorly insulated, can remain warm because heat is brought to these extremities from the body by blood circulation.

Marine mammals face a particularly acute problem with respect to heat loss from the limbs. Because water is such a good conductor, it draws heat rapidly away from the limbs, with their large surface-to-volume ratio. To act as paddles, the limbs cannot be covered with thick layers of insulation. Nevertheless, the mammal must maintain a high blood flow to the muscles of those limbs in order to swim. Seals and porpoises cut heat losses by reducing the temperature of their flippers with a *countercurrent heat exchanger.* Veins surround the major arteries carrying blood to the flippers (Figure 19–6). Cool blood returning from the flippers through these veins picks up heat from the warm blood flowing through the arteries into the flippers from the trunk. Thus, the arterial blood gives up its heat to the venous blood, not to the environment, and the venous blood is warmed before entering the trunk. If the seal or porpoise needs to unload heat, blood returning from the flippers is shunted away from the countercurrent system and into other veins near the surface of the skin, providing a thermal "window" in an otherwise well-insulated body.

If piloerection and circulatory control are not adequate to keep up the body temperature of a terrestrial animal, then shivering begins. Shivering exercises the muscles, causing them to produce heat through metabolism. Both cold receptors in the skin and a temperature regulation center in the hypothalamus of the brain (to be discussed in Chapter 20) can trigger shivering. The hypothalamus detects changes in the temperature of the blood vessels and acts something like a thermostat regulating a furnace. The greater the rate of heat loss, the

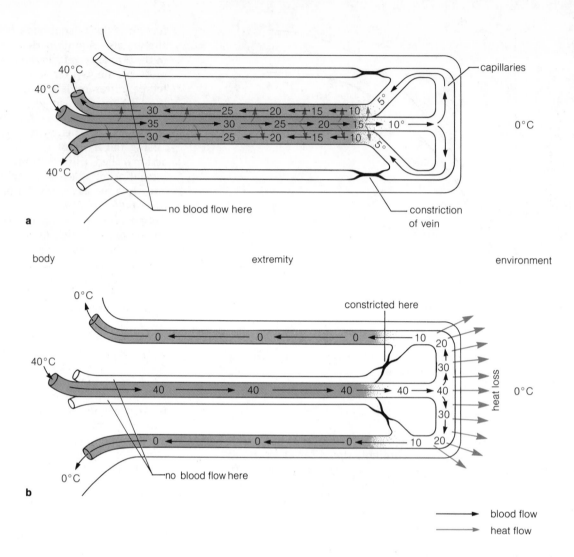

40°C

40°C

30 ← 25 ← 20 ← 15 ← 10
→ 35 → 30 → 25 → 20 → 15 → 10°
→ 30 → 25 → 20 → 15 → 10

5°

5°

10°

0°C

capillaries

constriction
of vein

40°C

no blood flow here

a

body   extremity   environment

0°C

constricted here

0 ←   0 ←   0 ← 10
20
30

40°C

40 → 40 → 40 → 40 → 40

30

20

heat loss

0°C

0 ←   0 ←   0 ← 10

0°C

no blood flow here

b

Figure 19–6 A countercurrent heat exchanger showing the regulation of (**a**) heat retention and (**b**) heat dissipation.

→ blood flow

→ heat flow

more pronounced the shivering and the more heat produced. When body temperature returns to normal, shivering ceases.

If, however, body temperature continues to decline, the endocrine system comes into play. The pituitary gland releases adrenocorticotropic hormone (ACTH) and thyroid-stimulating hormone (TSH). These hormones affect the adrenal cortex and the thyroid gland, respectively, triggering the release of corticosteroids from the former and thyroxine from the latter. Norepinephrine and thyroxine raise the metabolic rate, increasing heat production.

The importance of the adrenal gland and the thyroid in temperature regulation can be seen in surgical experimentation on rats. A normal white rat can successfully maintain its body temperature of 38°C at air temperatures down to −10°C. But a rat that lacks either the adrenal or the thyroid cannot thermoregulate below 2°C. With both glands gone, thermoregulation fails at 10°C.

The rate of heat loss is related to body size: The smaller the animal, the faster it loses heat and the higher the metabolic cost of maintaining an elevated body temperature by endothermy. The pygmy shrew, *Microsorex hoyi*, which weighs only 3 grams, has a weight-specific metabolic rate 55 times higher than a 68 kilogram (150-pound) human's. Thus, a shrew deprived of food for several hours starves to death, while a human, particularly if fat, can survive weeks without eating. It is very likely that body size has much to do with the activity patterns characteristic of endothermic species.

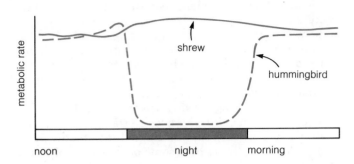

Figure 19–7 Some small animals, like hummingbirds, allow their body temperatures to decline to near ambient temperature when they are not active. This results in considerable energy economy. In contrast, the shrew of comparable size is homeothermic and, because of its continual high energy demands, is always within hours of starvation.

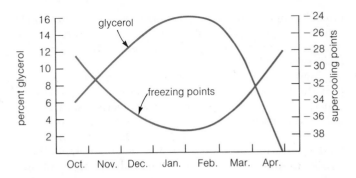

Figure 19–8 Glycerol acts as an antifreeze compound, and some insects that overwinter as larvae or as adults sequester large amounts of glycerol in their bodies before entering hibernation. Shown here is the percentage of glycerol in the blood of some moth larvae that overwinter as caterpillars in Canada. After coming out of hibernation in the spring, the animals cease to maintain glycerol in their blood.

One way of escaping the high metabolic costs of endothermy is to enter torpor—that is, to allow body temperature to drop to a point where the animal is inactive and its metabolism slow (Figure 19–7). Large animals like wolves, elephants, and humans never enter torpor, but many small desert rodents become torpid at night when temperatures drop and food is scarce. Hummingbirds, which weigh only 2 or 3 grams and have high metabolic rates, become torpid at night, and whippoorwills and swifts likewise enter torpor when food supplies are meager and temperatures low. In Chapter 25 we will investigate such behavior more closely.

Feathers and fur lower the rate of heat loss considerably, thus reducing the metabolic price paid for endothermy. In general, Arctic mammals have about nine times the insulation of tropical species. The northern fur seal is so well insulated that, if forced to march overland at an air temperature of 15°C, it will soon overheat and die. In the water, the fur of these seals, like the coat of the muskrat and the feathers of such diving birds as loons and grebes, entraps a layer of air, which is a good insulator. You may recall the old woodsman's bromide that in cold weather several light layers of clothing will keep you warmer than one heavy layer. The reason is that insulation depends more on the number of trapped air layers than on their thickness. Another superb insulator is fat. Seals, walruses, and whales have thick layers of blubber beneath their skin. A whale's skin is about the same temperature as the water, but its core temperature of 36 or 37°C falls within the typical mammalian range. In fact, whales can hardly enter water warmer than 18°C. The blubber holds in the heat of their muscular metabolism so effectively that they soon overheat and must retreat to colder water.

Acclimation: Getting Used to the Cold If an animal is kept for a long time at low temperatures, it *acclimates;* that is, it adjusts physiologically to the new conditions. In poikilotherms, acclimation involves actual cellular changes. Enzyme concentrations change; different enzymes with different optimum temperatures are induced; water concentrations shift; conditions affecting enzyme activity, like pH and ionic concentrations, are altered. As a result, the metabolic rate at any given temperature increases.

In homeotherms, cells remain at the same temperature no matter what the air temperature, and their enzyme systems do not change. Nevertheless, acclimation does occur. The fur grows thicker, circulation shifts to reduce heat losses to the environment, and higher levels of thyroxine flow in the blood to spur the metabolic rate.

Adaptations to Low Temperatures: Enduring the Cold Most animals are inactive at 0°C. But in the deep fjords off Newfoundland, where the water temperature all year long is almost constantly −1.7°C, many fish swim, feed, and breed normally. These fish have enzymatic machinery adapted to a narrow band of temperature; in water at 1 or 2°C, they quickly "overheat" and die.

Many Arctic and Antarctic fish are *supercooled*—that is, their body temperatures are actually below the freezing point of their body fluids. Nevertheless, they remain active, and their fluids do not freeze. However, these supercooled fish freeze solid almost instantly if touched with an ice crystal. The crystal provides a surface for other crystals to form, and the fish's body fluids quickly turn to ice. Many cold-water fish have glycoproteins in their blood besides the usual blood components. The additional solutes act like antifreeze, lowering the temperature at which the blood freezes, a phenomenon quite different from supercooling.

Similar adaptations are found in many insects, which pass the winter either by hibernating in a dry place and supercooling or by accumulating antifreeze (sugars and glycerol) in their blood before entering torpor. These solutes, like the fishes' glycoproteins, lower the freezing temperature of the blood (Figure 19–8), though other, largely unknown, factors are involved in supercooling.

Some animals and plants can survive freezing solid. Many Arctic and alpine plants and insects freeze solid in the winter and thaw out in the spring, seemingly none the worse for wear. Similarly, many intertidal invertebrates, such as mollusks and barnacles, freeze in the winter without harm. Small goldfish may freeze overnight and thaw out alive the next afternoon.

Surviving freezing requires special adaptations. Since water expands when it freezes, ice crystals forming inside a cell would normally rupture the cell membrane, often fatally. We don't know how some organisms avoid such

---

Essay 19–1

Pupfish, Temperature, and the Supreme Court

One of the lesser-known fishes, the desert pupfish, was the subject of a U.S. Supreme Court decision in 1976. The court's ruling is important, and the pupfish will surely go down in legal history as a major contributor to environmental conservation. Before we knew the outcome of the case, we had asked James H. Brown, now at the University of Arizona, to write about his confrontation with the pupfish on the scientific, rather than the legal, level. This is what he wrote:

The majority of students and fellow scientists who have heard of me at all probably associate my name with work on the physiological adaptations of desert pupfish. This always seems strange to me because my association with pupfish was only a short and passionate affair that interrupted a long and productive relationship with my real love, the ecology of desert and montane mammals. The affair began in the fall of 1969, when a UCLA graduate student, Bruce Turner, came to talk with me about his thesis project, a study of the genetic relationships among some populations of desert fishes. As he described the desert pupfish (genus *Cyprinodon*) and the physical characteristics of the springs and streams where they live, I began to get excited.

Most of us who are interested in the process of adaptation have a fascination for what we call "Oh my" organisms—those that have evolved particularly extreme or bizarre adaptations to unusual environments or ways of life. The pupfish that Turner described certainly has amazing physiological capabilities. Some populations live in shallow streams or marshes that freeze in winter and

heat to approximately 40°C in summer. Others occur in thermal springs of constant temperature, but some are as hot as 42°C. Some populations live in fresh water, others in waters more saline than the oceans. These populations all are found within a few miles of each other in isolated springs, streams, and marshes in the Death Valley region of California and adjacent Nevada. The present populations represent the remnants of a widely distributed ancestor that inhabited the area during the Pleistocene (ice ages), when Death Valley was a lake fed by permanent rivers. The present populations have been isolated for different lengths of time, varying from hundreds to tens of thousands of years.

The pupfish impressed me as a likely system to investigate a problem that had always interested me: How do organisms lose attributes that no longer have a function? It has long been known that cave animals evolve to lose their eyes and skin pigments, for example, but there is still no generally accepted mechanism to account for such losses. Pupfish that had evolved for thousands of years in springs of constant temperature had no need for the wide thermal tolerances of their ancestors or contemporary populations that inhabit thermally fluctuating streams and marshes. It should be possible to determine whether pupfish living at constant temperature have lost their ability to withstand extreme temperature fluctuations and, if so, how rapidly this loss had occurred. Unfortunately, I had never worked with fish or any other aquatic organisms, and I was reluctant to tackle the problem. However, when I mentioned my ideas to C. Robert Feldmeth, a colleague interested in the physiological adaptations of fresh-water animals, he greeted them with enthusiasm. Immediately we became collaborators, and Bob provided many of the ideas and most of the technical expertise for the subsequent research.

damage. However, there is good evidence that resistance to freezing and to dehydration go hand in hand. A good example of the correlation between dehydration resistance and temperature resistance is the fly *Polypedilum vanderplankii,* an inhabitant of temporary pools in arid regions of equatorial Africa. The larvae can lose up to 82 percent of their body fluids and survive for years, rehydrating like "instant insects" when water returns. The dehydrated larvae can withstand temperatures from slightly above the boiling point of water down to $-270°C$, about as close to the lowest temperature in the universe (absolute zero, $-273°C$) as human technology can reach. Perhaps the reason for this combination of resistances is that, as the animal dehydrates, solute concentration increases sharply and crystals large enough to rupture the cells cannot form.

Regulation at Higher Temperatures: Taking the Heat The muscular effort of running a mile produces considerable by-product heat. If you are running when the air temperature is low, that heat helps maintain your

---

At that point neither Bob nor I had ever seen a live pupfish, and I had never been in Death Valley. In November 1969 we made our first field trip to the area. Until then everything had been an intellectual exercise, interesting but not addictive. My love affair with pupfish really began when we waded through salt-encrusted marshes and stood on the shores of tiny, crystal-clear springs and watched the bright blue male pupfish defending their territories. My infatuation lasted for about two years. In 1971, when I left southern California, much of the passion had already been eroded by the disillusionment of seeing beautiful springs destroyed by greedy, thoughtless humans and by a rekindling of my interests in mammalian ecology. In the meantime, Bob Feldmeth and I showed that pupfish populations that had lived in constant-temperature springs for tens of thousands of years (long enough to become dwarfed, lose a set of fins, and evolve distinct color patterns) had not lost any of their amazing ability to tolerate fluctuating temperatures. We did other work on pupfish physiology and population ecology, and my wife, Astrid, began an extensive study of their territorial and breeding behavior. Perhaps the affair will start up again, because as I write this my wife is trying to persuade me to go to Mexico to study four species of pupfish that have recently been discovered living together in the same lake.

The action that Brown characterized as "greedy, thoughtless" was the basis of the suit brought before the Supreme Court. The case centered around a particular species known as the Devil's Hole pupfish. It is believed to have lived at least 30,000 years in a 200-foot deep limestone pool in Death Valley in Nevada. Some 200 to 800 fish of this species live in this pool and are found nowhere else on earth. Their food supply is blue-green
algae growing on a sloping ledge measuring 8 by 18 feet on one side of the pool, the pool measuring some 10 by 65 feet.

In 1968, Francis and Marilyn Cappaert, owners of a 12,000-acre cattle ranch adjacent to Death Valley National Monument, started pumping irrigation water from deep wells. The pumping reduced the water level in the pool so that, by 1972, about 60 percent of the sloping ledge was exposed.

In August 1971, the federal government sued the Cappaerts. It claimed the right to control pumping of the water under a proclamation by President Harry Truman in 1952 that set aside Devil's Hole as part of Death Valley National Monument. U.S. District Judge Roger Foley, of Las Vegas, ordered the Cappaerts to limit their pumping so that the water would not drop more than 3 feet below a copper washer placed on the wall of the pool in 1962 to mark the water level. If the water dropped more than 3 feet, Foley found, the survival of the pupfish would be endangered.

The Cappaerts appealed to the U.S. Court of Appeals and then to the Supreme Court. Briefs in support of the ranch owners were filed by California, Colorado, North Dakota, Washington, Arizona, Hawaii, Idaho, Kansas, Montana, Nebraska, South Dakota, Utah, and Wyoming. The officials of these states said the decision threatened the rights of states in arid portions of the West to control vital water supplies. They noted that Devil's Hole is part of a 4,500-square-mile ground-water system.

The Supreme Court, in a unanimous opinion written by Chief Justice Warren Burger, ruled that the federal government has the right to control water pumping that threatens a rare species with extinction. Thus, the pupfish survived its day in court, as well as its difficult desert environment.

high mammalian body temperature. But at high air temperatures, even though the by-product heat far exceeds what body temperature requires and heat losses to the environment are much reduced, body temperature remains close to its normal reading. How does the body rid itself of the unneeded heat?

First, mechanisms involved in retaining heat are altered so that they unload it instead. Mammals and birds dilate arterioles and capillaries and send more blood to the skin and extremities, carrying heat from the interior to the surface for loss by radiation. If this shunting of the blood is not sufficient to keep body temperature from rising, many mammals begin to sweat. The evaporation of water requires energy in the form of heat. When sweat lying on the skin evaporates, it draws that needed heat from the skin and lowers body temperature. Obviously, cooling by the evaporation of sweat works best when the outside air has a low relative humidity—that is, when the concentration of water vapor is low and the gradient for evaporation is high. This is the reason why a hot muggy day in New York City can be so much more uncomfortable than an even hotter day in the desert of central Sonora in Mexico. Desert air has a low humidity and sweat evaporates readily, while the muggy air of the city is already close to saturation and evaporation is slow.

In reptiles, birds, and some mammals, which do not sweat, the mouth and respiratory tract are the site of evaporative cooling. We are all familiar with the way a dog pants on a hot day. This shallow breathing is one way of moving air over moist respiratory surfaces to cause evaporative cooling.

A similar mechanism is found in birds. The pelican's prominent pouch is called a *gular pouch.* In the heat of the day, pelicans flutter the pouch, a behavior called *gular flutter,* which moves air over its surface and down into the throat to speed evaporation. Gular flutter is most dramatic in the pelican because it has the most dramatic pouch, but the same fluttering can be seen in quail, nighthawks, hummingbirds, and many others (Figure 19–9). A study by the late R. Lasiewski and W. Dawson, now at the University of Michigan, shows that at an ambient temperature of 45°C, the common nighthawk *(Chordeiles minor)* maintains its normal body temperature of 40°C by gular flutter, which increases the rate of water loss up to nine times.

Temperature Regulation and Immature Animals: Keeping the Young Warm Temperature influences the rate of biochemical reactions, including those controlling growth and reproduction. As a result, temperature can directly influence the reproductive rate of an animal. These effects are particularly pronounced in some poikilotherms. For example, the gestation period of the

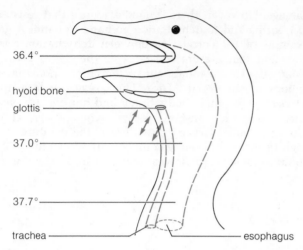

Figure 19–9 When heat-stressed, many birds open their beaks and rapidly vibrate their gular areas (arrows) in a behavior called gular flutter.

Northern Pacific rattlesnake *(Crotalus viridis oreganus),* which bears its young alive, varies with climate. In the area around Santa Barbara, California, the southern edge of the snake's range, winters are generally mild enough that the pregnant (gravid) snake can remain active and warm all year round. As a result, the young inside her grow rapidly, and mature within a year after she mates. In central British Columbia, the northern boundary of the snake's range, winters are harsh, and the gravid snake is torpid and cold for several months. It takes the unborn snakes two years to develop enough to be born. Obviously, temperature directly affects the rate of reproduction in this animal, since the California individuals can breed twice as often as their Canadian counterparts.

Evolution has produced some strange ways of raising the temperature of the eggs and the unborn so that they develop more quickly. Both alligators and *Mallee* fowl lay their eggs in heaps of decaying vegetation. The decay produces heat, which warms the eggs. After they hatch, the young dig themselves out. Snakes, most of which lay eggs rather than bear live young, usually deposit the eggs in some warm spot and abandon them. The African python, however, remains with the clutch and incubates it. The female wraps her coils around the eggs and contracts her powerful muscles, producing heat to warm the clutch.

The young of reptiles must fend for themselves almost immediately after birth, but young birds require

Figure 19–10 Altricial nestlings being brooded.

at least some parental care. Some birds, such as the ducks and the fowls (chickens, turkeys, quail), produce *precocial* or well-developed young. Covered with down, these chicks have open eyes and can leave their nest within minutes or hours of hatching. Precocial young can thermoregulate to some extent from birth. Other birds, including all of the songbirds, have *altricial* young, which are born naked, blind, and helpless. Altricial nestlings cannot thermoregulate, and the parent birds brood the young just as they do the eggs (Figure 19–10).

Summary

Temperature is a potent factor affecting almost all aspects of living organisms. In some plants, specific temperatures act as cues in synchronizing the life cycle with the seasons.

In most physiological systems, biochemical reaction rates double with every 10°C rise. Generally, the more active the animal, the more its tissues are specialized to operate within a narrow (and high) range of temperature. Survival, however, is possible at very low temperatures, particularly in some animals that do not regulate their body temperature.

Some animals, called poikilotherms, have body temperatures that are always close to environmental temperatures. Others, called ectotherms, achieve some elevation

of body temperature by behaviorally taking advantage of external heat sources. Endotherms warm up by their own metabolism, and homeotherms maintain stable body temperatures by regulating their heat production and heat loss.

In this chapter we discussed the comparative physiology and the ecological relevance of thermoregulatory mechanisms in a wide range of organisms.

Supplement 19–1
Hot-Bodied Bumblebees

We are all too easily deceived in our observations of the world about us by preconceived ideas. In particular, humans often have fixed ideas about insects and insect physiology. In this essay, one of the authors of this text, Bernd Heinrich, tells of some of his research with bees and the unusual facts he has found about them.

When an athlete races the mile, the muscles driving his legs contract about once per second. Each time they contract, they produce large quantities of heat. To be capable of maximal contraction rates and power output, human muscles must be at about 38°C. Similarly, the muscles driving the wings of a bumblebee during flight are about 36 to 40°C during flight, when the muscles contract about 150 times per second. Both the athlete and the bee produce heat as a by-product of the muscular effort, but the muscular effort and the rate of heat production of the flying bee are much greater than those of any human athlete.

Although it seems intuitively obvious that some insects would have a high body temperature during flight, our rigid concepts of invertebrates as "cold-blooded" has inhibited the study of the thermoregulatory physiology in some of these animals. However, some previously unsuspected phenomena have recently come to light. This is a personal account of how I see an area of investigation unfolding, of what has been found out, and of what remains unknown.

I started my study of temperature regulation in insects during work for a Ph.D. in zoology by examining the physiology and energetics of hummingbird moths (family Sphingidae). It had already been shown that these insects shivered (this can be observed by watching their wings vibrate) before flight until the temperature of their flight muscles was near 38°C, when they initiated flight. It was not surprising that the moths maintained the same muscle temperature during flight at a wide range of air tempera-

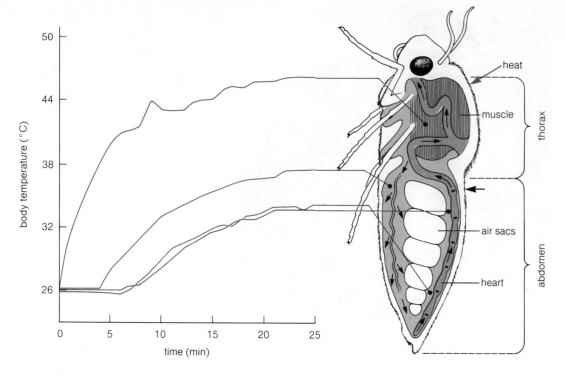

Figure 19–11 Longitudinal section of the sphinx moth *(Manduca sexta)* and body temperatures recorded during an experiment showing heat transfer from thorax to abdomen. Normally the moth produces heat with the muscles in the thorax only before and during flight. Here the moth was stationary but heated with a narrow beam of light in the thorax while body temperature was measured in four places. Heat is transferred to the abdomen when the heart begins to beat rapidly. If the heart is tied off at the black arrow, the moth soon overheats in flight and must stop to cool.

tures, but it was exciting, because it meant that the moths had some sort of unknown mechanism for regulating body temperature. And it *was* surprising to learn ultimately that the moths produced so much heat during flight that they actively dissipated heat from their flight muscles by blood circulation through the thorax into the relatively cold abdomen. The abdomen acted like the radiator in a car, dissipating excess heat from the "motor." (Figure 19–11 shows this process in the sphinx moth.)

These discoveries were interesting, but they provided little groundwork for further work in biology. Laying "groundwork" in biology, however, was furthest from my mind. Instead I began to measure the body temperature of bumblebees after picking them from flowers during one summer vacation. I was curious because I suspected they could be "hot" like the moths, but it was also possible that they had a low body temperature when they were not in flight. It turned out that *both* suppositions were true; some bees had a low muscle temperature, while others maintained their muscle temperature many degrees above air temperature even though they remained stationary for long periods and could not be seen to "shiver" (their wings did not vibrate). These results immediately suggested at least two new problems. First of all, how did they maintain such high rates of heat production while stationary and apparently not "working"?

Second, why were the bees sometimes "hot" and sometimes "cold"?

We may not have the final answers to these questions for some time, but partial answers usually follow quickly after the right questions have been posed. It was already known that some insects, particularly bees and beetles, can activate their flight muscles without *visible* wing vibrations. The muscles are "unclutched" from the flight motor by a delicate lever mechanism in the wing bases. Armed with this knowledge, and in collaboration with a colleague, Ann E. Kammer, who was familiar with electrical recordings from insect flight muscles, I examined the electrical activity of the flight muscles. We found that the bumblebees had an "invisible" shivering response that also produces great quantities of heat (Figure 19–12). We now believe that during this warm-up behavior the opposing flight muscles are in isometric (simultaneous) contraction, at least when they are activated at high frequency from the central nervous system.

To find out why bees are sometimes "hot" and sometimes "cold" we examined the environment of the bees and attempted to correlate it with their behavior. It was apparent that the bees appeared at some kinds of flowers early in the morning when air temperature was low. The bees foraging from these flowers were always "hot" and ready to fly from one floret to the next in rapid succession,

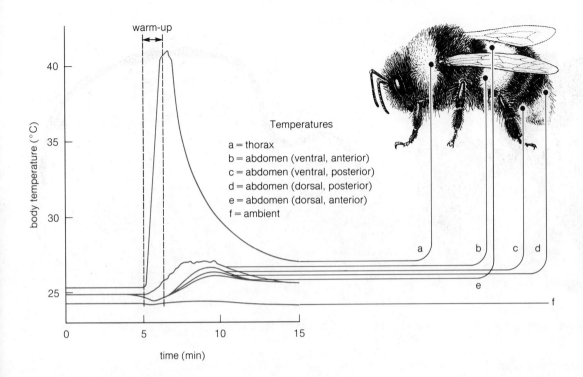

warm-up

Temperatures

a = thorax
b = abdomen (ventral, anterior)
c = abdomen (ventral, posterior)
d = abdomen (dorsal, posterior)
e = abdomen (dorsal, anterior)
f = ambient

body temperature (°C)

time (min)

Figure 19–12 Body temperatures during warm-up by "invisible" shivering in a bumblebee. Points on the diagrammatic sketch of the bee connect with the temperatures measured at that spot during cool-down of the thorax at the cessation of shivering, when heat was dumped into the abdomen.

usually visiting 20 to 30 flowers per minute. In contrast, they did not visit other kinds of flowers at all, or only when the air temperature was higher. While they were foraging from these other flowers, they were often sluggish and incapable of immediate flight because of low muscle temperature. Instead of flying from flower to flower, they often walked. These differences correlated with the amount of nectar in the flowers. The bees (in particular the drones) were unwilling to expend the large amounts of energy required for flight or for shivering while foraging for nectar when food rewards were low. It appeared that the bees were somehow able to "compute" the balance of the energy cost of foraging and the energy reward. I have attempted to calculate whether the bees are making the "correct" decisions. This has proved to be more difficult than anticipated, but the attempts clearly show that the bees could not make an energy profit on some flowers if they expended the energy necessary to maintain a high body temperature at low air temperatures. Similarly, these calculations show that the "strategy" of maximizing the speed of foraging, even at low air temperatures, by investing the extra energy for heat production is energetically justified for those flowers visited in cold weather. These results point to a *new* problem: How do bees "know" how to forage in order to make a profit? In addition, they point to problems relating

to the plants, which are pollinated by the bees and which must evolve to provide the right amount of food reward.

The energy balance that bumblebees have to maintain in the long run must do more than just keep them alive. Like all animals, they must have sufficient reserves for reproduction. In bumblebees, this does not involve just producing gametes. The bees also need fuel to heat their nests. Their nests are usually maintained near 30°C so that the eggs, larvae, and pupae continue to develop rapidly even at low air temperatures. The "fuel" for heat production in the nest is, of course, the sugar from the nectar of flowers. It is oxidized by the flight muscles of the adult bees during shivering.

The lone queen that initiates the annual colony in the spring perches on her brood clump containing eggs that develop into the first brood of workers, her first helpers in the seasonal colony build-up (Figure 19–13). She builds a small honeypot out of wax and places it beside the initial brood clump. She fills it with honey during the day, providing herself with the fuel that is used during heat production in the night, when she perches on the brood clump. Using tiny wire thermocouples and electric recording techniques, I measured the temperature of the brood clump continuously and found that indeed it was *incubated*. It was of particular interest that, like a bird, the bee incubates primarily with the

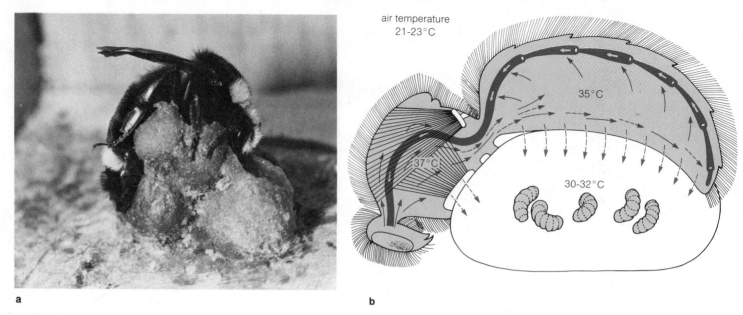

Figure 19–13 (**a**) A queen bumblebee *(Bombus vosnesenskii)* incubating a brood clump containing eggs, larvae, and pupae. Her head is pointed into her honeypot. (**b**) Diagrammatic representation of the bee while incubating, showing the source of the heat (the muscles in the thorax) and the direction of heat flow. The bee's internal temperatures were recorded at an air temperature of 21–23°C (room temperature).

abdomen, which is tightly wrapped about the brood clump. However, the heat is produced in the thorax with the "flight" muscles by shivering. In *flight,* when the thoracic temperature is above 35°C at a wide range of air temperatures almost to the freezing point, the temperature of the abdomen is always relatively low compared to the thorax. In contrast, during the *incubation* of the brood, it is close to the thoracic temperature. The heat transfer between thorax and abdomen involves blood circulation; tying off the heart (with a human hair) completely abolishes all heat transfer to the abdomen. The same mechanism allows bees to incubate as well as to stabilize the muscle temperature during flight at high air temperatures.

20

The Nervous System

For sheer grace and economy of motion, there are few sights in the natural world to surpass that of a lion hunting an antelope. Slowly the cat stalks its prey, moving in almost perfect silence on a course chosen to keep it hidden until the last possible moment. At that moment, when the cat is close enough to catch the antelope before it can escape, the predator bursts into a sprint, covers the distance in a series of great bounds, seizes the prey in its forepaws, and dispatches it with such speed that the human eye sees only a multicolored blur of fur and dust.

Such dramas of predator and prey are repeated so often that we easily lose sight of the almost unfathomable complexity of the cat's behavior. Every day millions of animals fall prey to millions of predators, and each of these predators—as well as the prey attempting to escape and usually succeeding—exhibits the same complex coordinated response that makes all the body's cells behave as one.

The lion's hunting behavior involves three fundamentally different processes: sensory, integrative, and motor. Specialized sensors in the cat's eyes, feet, whiskers, skin, and other exterior areas, as well as in such internal structures as the skeletal muscles, translate stimuli into electrical signals. One such stimulus is the first sight or smell of the antelope. The sensory signals follow nerve pathways to the brain, where the constant variety of incoming information is sorted and integrated. At this point, the cat's brain sets the goal of hunting the antelope and picks an appropriate strategy. Now begins the motor phase, as the muscles propel the stalking cat and produce the final lethal sprint. All the while, the sensors pick up new stimuli—say, sudden excited movements by the antelope—and relay them as electrical signals along the nerves to the integrative centers in the spinal cord and brain, which interpret them and settle on an appropriate motor response—say, holding stock-still until the antelope has relaxed and let down its guard again.

Such complex coordination of all the cells and tissues of the body is due largely to the unique physiology of the nerve cell. In this chapter, we will look at how the nerve cell works and at how billions of nerve cells are organized into the complex nervous systems characteristic of higher animals.

The Parts, Forms, and Workings of the Nerve Cell

Some kinds of coordinated behavior are possible without benefit of a nervous system. When a *Paramecium* moves, rows of external cilia beat in waves triggered by changes in electrical charge, or depolarizations, traveling along the cell membrane. This depolarization is similar to the nerve impulse we will discuss shortly and is found in many unicellular organisms besides *Paramecium*.

Plants generally respond to the environment with differential growth, yet electrical signals similar to nerve impulses do occur in certain species. The sundew *(Drosera)* found in most bogs in North America is an example. The hairy surfaces of the club-shaped tentacles growing from the upper sides of the leaves of this tiny carnivorous plant are covered with a sticky secretion. When an insect lands and is trapped by the sticky liquid (much like a fly on flypaper), its weight bends the hairs, producing a mechanical stimulus that sends an electrical signal to the base of the tentacle. The tentacle folds shut around the insect and secretes enzymes that digest it.

Although the sundew's insect-trapping is somewhat similar to a true nervous response, it is only 0.0001 as fast. Quick and highly coordinated behaviors are very much the province of the nerve cell.

Cellular Anatomy of the Neuron The nerve cell, or **neuron,** is the functional unit of the nervous system.

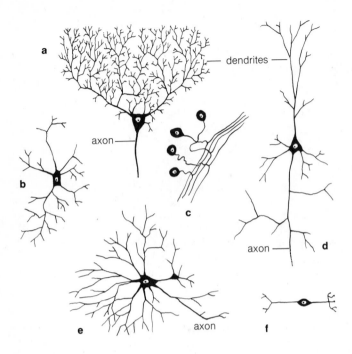

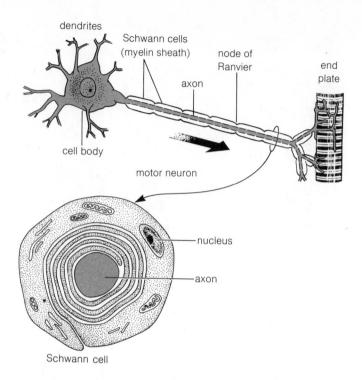

Figure 20–1 Some nerve cells: (**a**) purkinje cell from cerebellum; (**b**) interneuron from spinal cord; (**c**) sensory cells from spinal ganglion; (**d**) pyramidal motor cells from cerebral cortex; (**e**) motor cell from spinal cord; (**f**) bipolar cell from retina of eye.

Figure 20–2 Vertebrate axons are often (but not always) myelinated. The myelin sheath, interrupted by nodes of Ranvier, surrounds the axon. The cross section shows the layers of Schwann cell membrane surrounding the axon.

Neurons are the most morphologically varied cells in the body: The optic lobe of the fly has 30 different kinds; the human cerebral cortex, 44. Like other cells, the neuron contains a nucleus and cytoplasm, but the cytoplasm of the neuron extends out into long branches leading away from the cell body.

These cytoplasmic branches are of two types: dendrites and axons (Figure 20–1). *Dendrites* are generally numerous, tapered, and, particularly in higher vertebrates, often profusely branched. Usually they carry nerve impulses toward the cell body. *Axons*, by contrast, rarely number more than one per neuron, do not taper, and branch only at the terminal end. The axon generally bears nerve impulses away from the cell body. Vertebrate axons are often wrapped in a *myelin sheath*, which is the membrane of the Schwann cells that encircle the axon and act as insulators (Figure 20–2). Bare spots called the *nodes of Ranvier* interrupt the sheath at intervals of a few millimeters.

The exact length of particular dendrites or axons depends on the location and function of the neuron in question. Some are mere fractions of an inch long; others ex-

tend distances that, for a single cell, are enormous. For example, single axons and dendrites span the 2 or 3 meters from neural cell bodies in and around the giraffe's spine to motor and sensory cells in the animal's hooves.

Axons and dendrites traveling through the tissues are organized into discrete bundles called **nerves.** A nerve is much like a telephone cable in that it encloses a great number of individual nerve extensions, or fibers. The fibers in a given nerve may be all axons, all dendrites, or a mixture of both.

Neurons are grouped into three major types according to their functions. *Sensory* neurons, leading from sense organs to the central nervous system, make up the *afferent* portion of the nervous system. They collect information by converting external and internal stimuli to electrical signals. *Motor* neurons, leading from the central nervous system to effectors such as the muscles or other organs, make up the *efferent* system. They carry electrical signals to the organs and limbs. *Association* neurons, found in the *central nervous system* of the brain and spinal cord, interconnect sensory and motor neurons. They process the information brought in by the afferent system and, on the basis of those data, relay commands to the body over the nerves of the efferent system.

The Nerve Impulse When scientists first realized that the impulses traveling along nerves are electrical, they assumed that a nerve fiber worked much like a piece of copper wire passively carrying a current. Initial experimental work, however, ruled out this assumption. For example, physiologists found that the strength of a nerve impulse does not decrease as it travels along a nerve fiber, while ordinary electrical current flows through nerve fibers for only a few millimeters before dying out. Clearly, the nerve impulse was something different, and a new model was needed.

In 1902, Julius Bernstein of the University of Halle in Germany proposed that model, basing it on the neural membrane's selective permeability to various charged ions. Bernstein's model won wide acceptance almost from the start, but not until the late 1930s did experimental data confirm the theory. This research, performed by A. L. Hodgkin, A. F. Huxley, and B. Katz in England, used microelectrodes of drawn glass filled with a solution of a conducting salt, potassium chloride, and inserted into giant axons taken from squid. These giant fibers are as much as a millimeter in diameter, about 100 times thicker than the largest nerve fibers of the human body, and thus much easier to handle in the laboratory. This work won Hodgkin and Huxley a Nobel Prize in 1963 and showed that Bernstein's model, with a few corrections, did in fact account for neural conduction.

The key to the Bernstein model of nerve impulse transmission is the observation that the interior of the neuron is negatively charged relative to the exterior. This electrical difference, or polarity, results from variations in the kinds and concentrations of ions on the two sides of the membrane: the cytoplasm of the nerve cell on the one side, and the extracellular fluid on the other. The cell membrane acts like a selectively permeable barrier dividing inside from outside. The cell interior contains K^+, Cl^-, and negatively charged protein ions. The cell exterior contains more positive ions than does the inside. These ions are primarily Na^+. Since Na^+ concentration outside the membrane is 10 times greater than inside, and K^+ concentration inside is 30 times greater than outside, strong concentration gradients exist across the membrane. However, few ions leak through. Instead, the energy-consuming active transport mechanisms of a "sodium-potassium pump"—whose precise workings remain unknown—move sodium out of the cell and bring potassium in, maintaining the concentration gradients against the force of diffusion. The negative difference in electrical charges between outside and inside due to the expelled Na^+ can be measured at some 60 to 70 millivolts. This value is called the *resting potential.*

Polarity across the membrane is not unique to neurons. Whenever any cell is disturbed by a mechanical, chemical, or electrical stimulus, the membrane in the area of the disturbance depolarizes. That is, the ions that had been held back flow across the membrane, and the polarity of electrical charge is lost. In non-nerve cells, this depolarization is confined to the area of the disturbance. In the neuron, however, the disturbance travels from the point of origin through the length of the cell in a wave of depolarization. In essence, the wave of depolarization is the nerve impulse, or action potential.

Figure 20–3 outlines the basic chemical and electrical happenings of the nerve impulse. When the neural membrane is disturbed, it suddenly becomes very permeable to Na^+. An unknown mechanism opens the "gate" that holds out the sodium, and these positive ions rush through, drawn in by the concentration gradient and by the negative charge inside the cell. Na^+ ions continue to enter until the cell interior actually becomes positive. At this point, K^+ ions, repelled by the positive charge on the Na^+ ions, move through the membrane and out of the cell. Then, within a thousandth of a second, the sodium pump goes back to work, moving sodium out of the cell (and causing some potassium to move back in), so that the original negative charge of the resting potential is restored, and the membrane is again able to relay an impulse.

If we insert an electrode into a nerve and lay a reference electrode on the nerve's surface, an oscilloscope connected to the electrodes will first measure the negative resting potential and then display the nerve impulse as a positive disturbance. The disturbance, known as the *action potential,* is due to the rapid entrance and exit of Na^+.

As we noted earlier, the unique property of the neural membrane is that the disturbance moves throughout the cell's length. The depolarization of one segment of the membrane opens the sodium "gate" of the adjacent segment. Sodium rushes in, the charge of that segment of membrane changes, and the "gate" of the next segment of membrane opens. Thus, the impulse proceeds from one end of the cell to the other, much like a line of dominoes stood on end and sent tumbling one into the next.

The Synapse When a sensory impulse originates in the foot, it travels along an afferent nerve fiber to the cell body near the spine, moves on through association neurons in the spine and brain, and then follows an efferent fiber to some effector cell like a section of muscle. The nerve impulse must be able to move from one cell to another, either between neurons or between neurons and effectors.

The terminal end of an axon branches into a great many *processes,* each of which ends in a *synaptic knob*

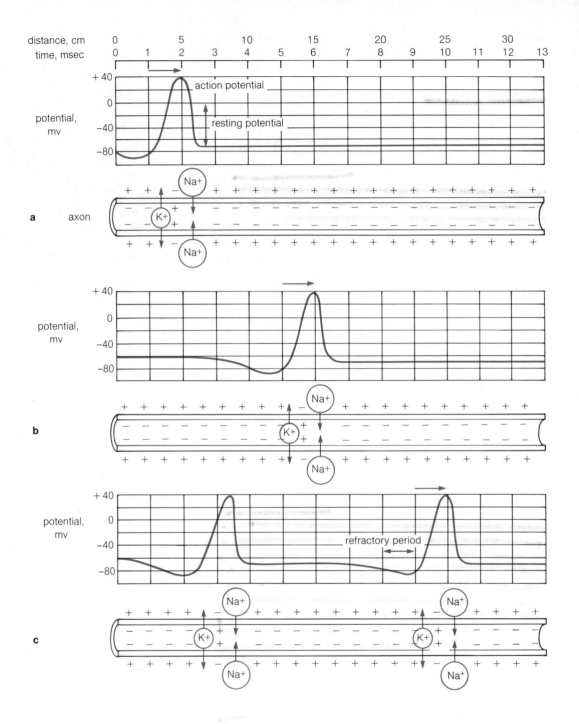

Figure 20–3 Stimulation of a nerve and conduction of the action potential.

that contains small round globules called *synaptic vesicles.* Since these processes are numerous, the cell the axon contacts may be quite liberally sprinkled with synaptic knobs, as shown in Figure 20–4a. Each of these junctions between an axon terminus and another cell is called a

synapse, a term first used by the English biologist Charles Sherrington.

Figure 20–4b takes a close-up look at the anatomy of a synapse. Notice that the membrane of the knob, or presynaptic membrane, and the membrane of the other

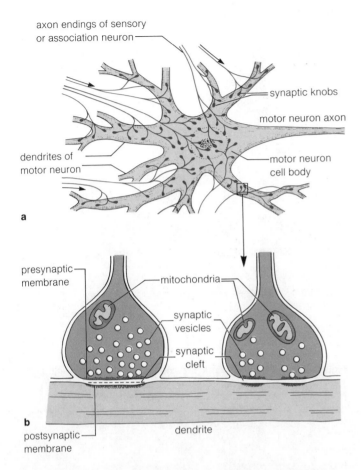

a

b

presynaptic membrane
mitochondria
synaptic vesicles
synaptic cleft
postsynaptic membrane
dendrite

axon endings of sensory or association neuron
synaptic knobs
motor neuron axon
dendrites of motor neuron
motor neuron cell body

Figure 20–4 (**a**) Motor neuron cell body and dendrites contacted by numerous axon terminals (synaptic knobs) from other neurons. (**b**) Enlargement of one of the synaptic knobs. The synaptic knobs deliver a chemical transmitter substance located in synaptic vesicles into the synaptic cleft. The stimulus for this release is membrane depolarization (action potentials).

cell, or postsynaptic membrane, do not quite touch. Instead, they are separated by a space, called the *synaptic cleft,* which is usually some 20 nanometers wide. Obviously, any nerve impulse that is to travel from one cell to another must somehow cross this gap.

The synapse is more than just a near junction between cells. It is the mechanism ensuring that nerve impulses travel in one direction only. If an animal is to coordinate the action of its muscles with the information it is receiving from its senses, the nerve impulses cannot be running up and down the same nerve, canceling one another out and interfering with one another. Thus we find that sensory impulses always travel to the spinal cord and the brain; motor impulses travel in the opposite direction.

If an axon is stimulated experimentally, however, the

nerve impulse can be seen to travel in both directions away from the point of the stimulus. Therefore, the one-way travel of nerve impulses is not a property of the axon itself. Instead, the synapse acts like a chute that allows neural impulses to pass in only one direction. This fact can be readily demonstrated experimentally. A recording electrode is placed on a cell body, and an axon that synapses with the cell body or its dendrites is given an electrical stimulus. The electrode will record an action potential in the cell body. But if the procedure is reversed, so that the electrode is on the axon and the electrical stimulus is given to the cell body, no action potential is recorded. The synapse guarantees one-way transmission of neural impulses from axon endings of one neuron to dendrites of a second neuron.

Much of the credit for determining how impulses cross the synapse goes to John Eccles of the Australian National University, who studied under Sherrington and shared a Nobel Prize in 1963 for his work on the synapse. In 1951, Eccles revolutionized the study of synaptic transmission by recording electrical activity within the nerve cell. He accomplished this feat by filling a glass pipette only 0.00002 inch in diameter with an electrically conductive salt solution and then inserting the pipette into the cell. Eccles discovered localized and seemingly random depolarizations of the postsynaptic membrane. He also found that, as more and more of the synaptic knobs on a cell depolarize, the resting potential of the postsynaptic membrane falls. When the depolarization of the postsynaptic membrane reaches some threshold value, the Na$^+$ "gate" of the membrane opens, and an action potential travels from the postsynaptic membrane through the cell. If many impulses pass over the synapse rapidly, the number of synaptic vesicles in the synaptic knobs decreases.

These data underlie our current model of synaptic transmission. Apparently, when a neural impulse depolarizes the synaptic knob, the synaptic vesicles release a *transmitter substance* into the synaptic cleft. At every depolarization, several thousand molecules of the substance are released into the cleft, and the number of vesicles decreases. In the frog, for example, at the point where a nerve joins a muscle—a site called a *neuromuscular junction* or *motor end plate* (Figure 20–5)—600,000 synaptic vesicles may be counted at each synapse before stimulation. After one minute of stimulation, however, the number drops to 400,000. The vesicles contain transmitter molecules which are released by electrical stimulation of the neuron and then diffuse across the cleft and attach to specific receptor sites on the postsynaptic membrane, opening "gates" to allow the passage of Na$^+$. This influx of ions depolarizes the membrane locally (Figure 20–6).

Typically, it takes a number of such local depolarizations added together to generate an action potential in the postsynaptic cell body. This accumulation toward the

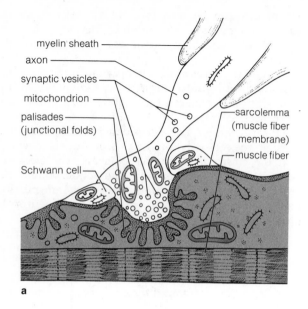

myelin sheath
axon
synaptic vesicles
mitochondrion
palisades (junctional folds)
Schwann cell
sarcolemma (muscle fiber membrane)
muscle fiber

a

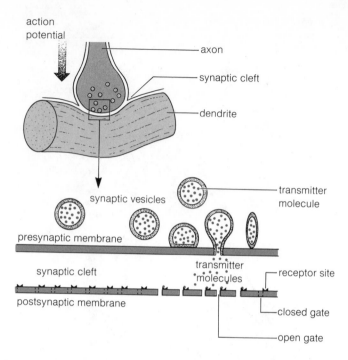

action potential
axon
synaptic cleft
dendrite

synaptic vesicles
transmitter molecule
presynaptic membrane
synaptic cleft
transmitter molecules
postsynaptic membrane
receptor site
closed gate
open gate

Figure 20–6 The transmitter substance molecules released into the synaptic cleft from the vesicles act on the surface of the nerve cell membrane below. In the excitatory synapse, the transmitter molecules diffuse across the cleft and open gates in the postsynaptic membrane upon contacting receptor sites. Ions may then pass through.

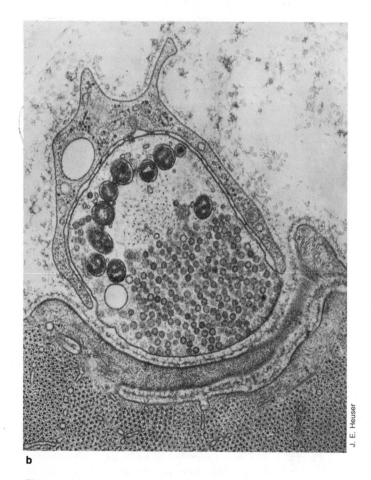

J. E. Heuser

b

Figure 20–5 Motor end plate: (**a**) diagram; (**b**) electron micrograph.

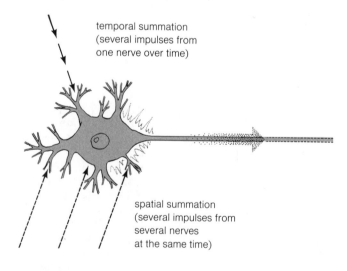

temporal summation (several impulses from one nerve over time)

spatial summation (several impulses from several nerves at the same time)

Figure 20–7 The two ways in which a neuron adds up the impulses it receives from other neurons.

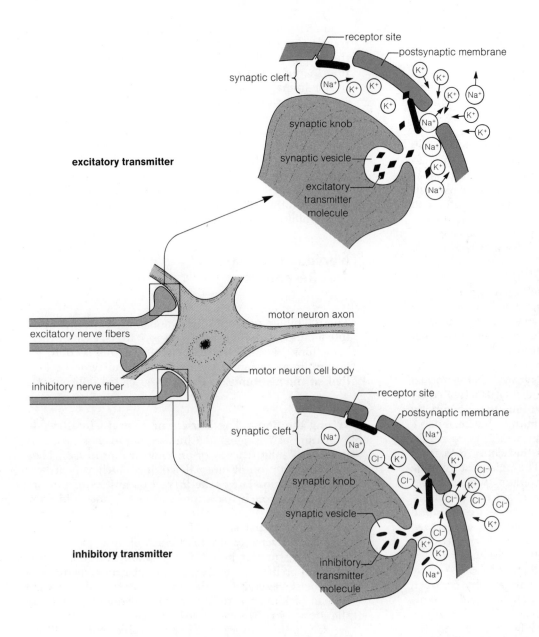

receptor site

postsynaptic membrane

synaptic cleft

synaptic knob

excitatory transmitter

synaptic vesicle

excitatory transmitter molecule

motor neuron axon

excitatory nerve fibers

motor neuron cell body

inhibitory nerve fiber

receptor site

postsynaptic membrane

synaptic cleft

synaptic knob

synaptic vesicle

inhibitory transmitter

inhibitory transmitter molecule

Figure 20–8 Excitation and inhibition of a nerve cell are accomplished by the activation of the nerve fibers that form synapses on its surface. Excitatory transmitters and inhibitory transmitters have different effects on the conduction of a postsynaptic membrane.

required stimulus threshold is known as *summation* (Figure 20–7). The threshold may be crossed either by several synaptic transmissions from one axon in a short period of time *(temporal summation)* or by nearly simultaneous transmissions from many axons *(spatial summation)*. The threshold varies from one neuron to another, depending on the particular location and function of the cell.

The synaptic transmission we have described thus far, in which stimulation by the presynaptic membrane of the knobs depolarized the postsynaptic membrane, is called an *excitatory synapse*. The depolarization, or low-

ered potential, of the postsynaptic membrane is called the *excitatory postsynaptic potential* (EPSP).

Some synapses have just the opposite effect (Figure 20–8). Such *inhibitory synapses* increase the negative charge of the interior of the postsynaptic cell and thus make the membrane more polarized. This has the effect of increasing the number of excitatory impulses needed to generate an action potential in the postsynaptic membrane. The increased resting potential caused by inhibitory synapses is called the *inhibitory postsynaptic potential* (IPSP).

The difference between excitatory and inhibitory synapses lies in the effects of the transmitter substances they release (see Figure 20–8). It is thought that the excitatory transmitters allow Na^+ to enter the postsynaptic membrane. Inhibitory transmitters, in contrast, probably open channels in the postsynaptic membrane that prevent the passage of Na^+, although the K^+ or Cl^- can probably penetrate.

Many of the transmitter substances of both varieties have been identified. In some insect and crustacean muscles, glutamate is an excitatory substance, and gamma-amino butyric acid (GABA) acts as an inhibitor. Acetylcholine acts as an excitatory transmitter in motor end plates (neuromuscular junctions) in vertebrates and crustaceans and in the vertebrate central nervous system. Norepinephrine (see Chapter 14) is also a transmitter of the autonomic nervous system of vertebrates.

There is some evidence that the activity of a transmitter resides less in the transmitter itself than in the receptor molecules of the postsynaptic membrane. In humans, for example, although acetylcholine is an excitatory transmitter in the central nervous system and in neuromuscular junctions, in the heart it acts as an inhibitor.

Certain poisons and drugs have shown us the practical importance of the chemistry of the synapse. In the normal synapse, released acetylcholine is quickly deactivated by the enzyme cholinesterase, preventing the potentially dangerous build-up of the transmitter. Various nerve gases keep the cholinesterase from acting. As a result, acetylcholine levels increase until the victim dies from violent spasms and convulsions. The poison strychnine and the toxin of the tetanus bacterium, which causes "lockjaw," block inhibitory synapses, but not excitatory ones. This produces a permanent, unrelaxed contraction of the muscles—the reason for the locked jaw of lockjaw. The nicotine in tobacco mimics the action of acetylcholine and serves as a stimulant. Another stimulant is the amphetamine Dexedrine, which induces excessive release of excitatory transmitters. Chlorpromazine, by contrast, blocks the receptor sites for acetylcholine and norepinephrine, producing the tranquilizing effect for which it is commonly prescribed. The toxin produced by the bacteria *Botulinus,* ounce for ounce the most lethal poison on earth, prevents the release of acetylcholine from the synaptic vesicles. Cobra venom paralyzes muscles by blocking their acetylcholine receptors, causing death when the paralysis reaches the respiratory muscles.

A great deal of research on synaptic transmission remains to be done. Presumably the receptor sites of postsynaptic membranes consist of specific molecules, but the identities of those molecules are unknown. Nor do we know how the channels in the membranes open, how synaptic vesicles move to the presynaptic membrane, or how they release their transmitters into the synaptic cleft.

What we do know is that nerve cells conduct impulses by a chain reaction of permeability to sodium ions that travels along their membrane. Started experimentally in the middle of the nerve fiber, the membrane depolarization (nerve impulse) travels both ways along the fiber. In the animal, impulse traffic from one nerve cell to the next is only one way, largely due to the synapse, the junction where a nerve cell contacts another nerve cell or a muscle cell. The membranes of these cells, in turn, are depolarized by chemical transmitter substances released at the cell-cell junction when an impulse travels to the synapse. Such "excitatory synapses" contrast to "inhibatory synapses" where the transmitter substance polarizes rather than depolarizes the cell at the receiving end, inhibiting rather than stimulating its response.

Nervous Pathways and Responses

The 100 billion neurons of the human body are organized into a multitude of working groups or units, each of them something like an electrical circuit. In this section, we will examine certain aspects of that circuitry and take a look at the neuromuscular responses called reflexes.

Wiring the Body: The Variety of Neural Circuitry In a creature as complex as a human, nerve cells take part in a nearly infinite array of physiological processes. They monitor and prompt internal activities, such as heartbeat and ventilation. They take in vast quantities of sensory information: touch, pain, pressure, heat, and cold from the skin; taste from the tongue; odors from the nose; all sorts of visual signals for color, texture, and form; sounds as diverse as the loud crash of an explosion and the delicate quarter-notes of a symphony. Nerves coordinate and orchestrate elaborate muscular movements, like a gymnast's exercise on the uneven bars. And that most human of human activities, rational thought, is entirely the property of the nerve cells of the brain.

Given this variety, one might assume that the differences among neural functions originate in the activity of the neuron itself—that, for example, a neuron in the retina of the eye responsible for picking up visual signals works differently from a neuron of the cerebral cortex specialized for linguistic function. Yet this is not the case. Neural impulses in all neurons involve the same set of chemical and electrical happenings just outlined. Nor can a single neuron give off signals of varying strength. Whether the stimulus is the light touch of a feather or the smashing drive block of a Big Ten tackle, a tactile nerve transmits precisely the same sort of action potential. This all-or-nothing character of the nerve impulse is true of every neuron.

How, then, do essentially similar neurons fill their varied roles? Because of the extreme complexity of nervous interconnections, we may never learn how all neural circuits are integrated to produce particular behaviors or how environmental or internal input modifies these circuits. But we do have some idea how the all-or-nothing response varies among neurons, how action potentials can be used to code different sorts of information, and how some of the simplest interconnections result in different patterns of output.

The speed at which a neuron impulse travels along the membrane varies among neurons, and this fact is often important to the arrangement of nervous circuits. When a squid needs to escape an enemy, it forces the water in its mantle through its forward siphon, creating a form of jet propulsion. For this action to work right, the muscles in the mantle must contract almost simultaneously, with the hindmost muscles contracting a little ahead of the rest to force the water forward and out through the siphon. If the speed of transmission along each axon were the same, then the forward muscles of the mantle, which lie closest to the origin of the nerve impulse in the squid's brain, would contract first, and the escaping squid would not travel very far or very fast. This apparent problem is solved by differences in the diameters of the axons leading from the brain to the mantle muscles. The thicker an axon is, the faster it conducts an impulse. In the squid, the axons leading to the rear muscles are far thicker than those leading to the forward muscles. These large axons, the giant axons of experimental fame, conduct a neural impulse at the rate of 100 meters per second, about 100 times faster than the thinnest fibers. Similar arrangements, based on varying thicknesses of efferent fibers, are found in insects and other invertebrates, where they are usually associated with quick-escape behavior much like the squid's.

Vertebrates have no giant axons, but another feature of the neuron speeds conduction of the impulse. Axons are often encased in a myelin sheath interrupted by the nodes of Ranvier (Figure 20–2). In these axons, the impulse jumps from node to node in so-called saltatory conduction (from the Latin *saltus*, meaning "leap") that achieves velocities of up to 120 meters per second. Although dendrites are the typical path for impulses heading toward the cell body, sensory fibers in humans and other vertebrates are of the myelinated axon type, an arrangement that greatly speeds the signals traveling along these fibers.

Another method for controlling the speed of conduction is to vary the number of neurons, and therefore of synapses, involved. Since synaptic transmission occurs by diffusion, it takes longer—at least 0.5 millisecond—than does a nerve impulse. Thus, the more neurons and the more synapses in a circuit, the slower an impulse can travel down that pathway. The fastest circuits have but one neuron; in humans, as in other vertebrates, single neurons span the distance from toes to spinal cord. Adding neurons and increasing the number of synapses decreases the speed of the conduction. In the segmental ganglia of annelid worms, which have numerous synapses, the speed of conduction is less than 0.05 meters per second. In the single fibers that extend the length of the worm, the speed is on the order of 10 to 25 meters per second. This principle can solve coordination problems similar to the squid's jet propulsion, in which effectors at various distances from the origin of the impulse have to be activated either simultaneously or in a precise sequence.

Simultaneous response of a number of effectors, all at a given distance from the point of origin, can be achieved by the branching of a single axon, either directly to the effectors or to other neurons synapsing with the effectors. For example, in many vertebrates, a single motor neuron provides the sole innervation for tens, hundreds, even thousands of muscle fibers. The muscle fibers controlled by a single axon constitute a *motor unit* (Figure 20–9), and each of the major muscles that move our limbs contains hundreds or thousands of motor units. Never are all the motor units in a muscle activated simultaneously. If they were, the bone the muscle is attached to would most likely break. Instead, nerve impulses activate the motor units asynchronously, so that the muscle contracts in its usual smooth and even manner.

Although the strength of nerve impulses does not vary, their frequency—that is, the number of impulses per unit of time—does vary, and information carried by a nerve can be coded in terms of frequency. Typically, the stronger or more intense the stimulus, the higher the frequency of nerve impulses in response. In our previous example of touch of a feather compared to the drive block of a football lineman, the violent collision of the block will elicit a far higher frequency of neural impulses than will the delicate brushing of the feather.

In many cases, nervous response depends on the reaction of a sensory receptor to a stimulus in the environment or within the body. That reaction may well vary from one instance to the next. Typically, continuous stimulation decreases the sensitivity of a receptor; that is, as time goes on, it takes a stronger and stronger stimulus to initiate an action potential in the receptor. Such a loss of sensitivity, called sensory adaptation, is particularly noticeable with the sense of smell. If you walk into a closed room where someone has just broken a bottle of ammonia, the piercing smell will practically overwhelm you. But within a few minutes, you will scarcely notice the smell at all. The reason is the sensory adaptation of the olfactory tissues in the nose.

In some cases, stimuli are ignored, not because the receptor has become adapted, but because the brain considers these stimuli unimportant and irrelevant. This is

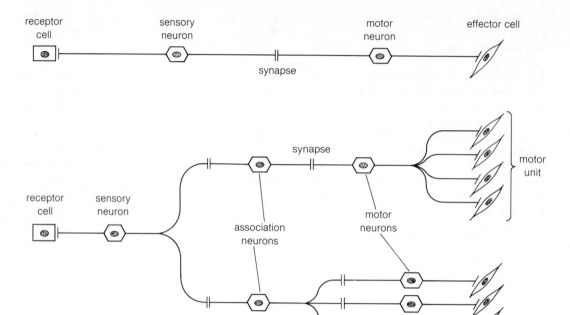

Figure 20–9 Stylized representations of nervous pathways. Most nervous pathways have one or more association neurons within the cells interposed between the sensory and motor neurons. The impulse can often follow numerous alternative routes. Inputs to the pathway from other neurons alter the probability of one pathway being used over the other. In addition, some pathways become "preferred" over others through use. A motor unit consists of all the effectors acting together under the control of a single motor neuron.

an example of *sensory gating,* a process in which nerve tracts in the thalamus of the brain filter incoming information, classifying it as relevant and worth bothering about or irrelevant and worth ignoring. A good example of sensory gating occurs every night while you sleep. Most sleep areas are actually relatively noisy—roommates tramping in and out, car noises in the street, tomcats fighting, sirens screaming—yet you sleep through the racket because the sensory gate classifies these noises as normal and insignificant. However, a very small sound—say, the rattling of a doorknob by an intruder—may awaken you in an instant. In this case, the sensory gate classifies the noise as significant and awakens you immediately to attend to it.

Lowered perception of stimuli, whether due to sensory gating or to sensory adaptation, is called *habituation.* If the nervous system were equally responsive to all stimuli, we would be constantly ajitter with all sorts of unnecessary responses. Because of habituation, we are aware less of the absolute level of stimuli than of the relative level, an arrangement that makes us sensitive to changes in the environment.

The Reflex Arc: Reaction in an Instant If you asked a human engineer to design the simplest possible nervous pathway, he or she would probably devise a single cell

connecting a receptor directly to an effector. However, the simplest pathway found thus far in nature, among both crustaceans and vertebrates, is a two-cell system, involving a receptor and an effector with a synapse between them. Usually one or more conductor cells are placed between the receptor and the effector. Such neuronal circuits linking receptor and effector are called **reflex arcs.**

Reflexes typically produce rapid and relatively automatic responses. For example, most of our involuntary body functions, such as breathing, heartbeat, blood pressure, coughing, swallowing, digestion, and body temperature, are monitored and controlled reflexively. Most reflexes are totally unconscious and do not need to be learned. However, many reflex arcs include association neurons, and their responses can be changed or inhibited by external input like learning. As a general rule, the fewer the association neurons in a reflex arc, the faster and more automatic the response. Still, it is impossible to draw a hard and fast line between reflexes and other behaviors.

The unconsciousness of reflexes serves many useful purposes. If we had to control consciously such housekeeping chores as breathing or swallowing, we would get little else done. Reflexes attend to these routine tasks and free higher nervous centers for other activities. In addition, conscious behavior is relatively slow, and that

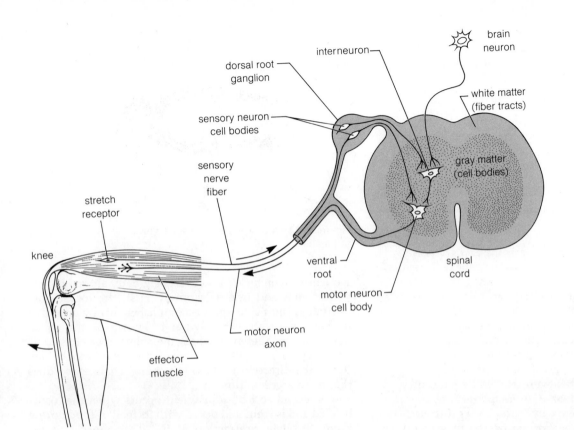

brain
neuron

interneuron

dorsal root
ganglion

white matter
(fiber tracts)

sensory neuron
cell bodies

gray matter
(cell bodies)

sensory
nerve
fiber

stretch
receptor

knee

ventral
root

spinal
cord

motor neuron
cell body

motor neuron
axon

effector
muscle

Figure 20–10 The neural diagram of the knee-jerk reflex arc. The motor neuron cell body also synapses with other neurons, from both the central and peripheral nervous systems. Some of these neurons inhibit and can override the stimulations from the stretch receptor.

slowness can be very costly. If you laid your hand on a hot stove and the message "hand burning" had to be relayed through the brain for a conscious choice to be made to withdraw the hand, you would probably be seriously burned. Instead, because reflexes allow such quick responses, the hand is pulled away well before any sensory message reaches the brain.

A classic example of a reflex, involving but one sensory neuron and one motor neuron, is the knee-jerk reflex arc (Figure 20–10). The sensory neuron is the stretch receptor in the tendon of the muscle that raises the leg. When a physician stimulates the stretch receptor by tapping the tendon with a hammer, impulses are sent from the receptor up the dendrite through the cell body, which is located just outside the spinal column in the *dorsal root ganglion,* and on into the spinal cord. There the axon terminals of the sensory neuron synapse with the dendrites of a motor neuron. The impulse follows these dendrites to the cell body, which is part of the gray matter of the spinal cord, leaves the spinal cord through the axon in the ventral root of the nerve, and follows that nerve back to muscle attached to the tendon. The nerve impulse causes the muscle to contract, and the lower leg jerks.

This reflex helps us maintain balance when standing. If you start to lean backwards, the stretch receptors are stimulated, and the reflex response contracts the muscle of the upper leg, drawing you upright again. Usually the neural information about the backward sway and its correction is not relayed to the brain, and you are unaware that anything has happened.

The knee-jerk reflex is inborn, but some reflexes are shaped by conditioning and are called *conditioned reflexes.* The classic work on conditioning was done by Ivan Pavlov, the famous Russian physiologist of the early twentieth century. Pavlov noted, as many had before him, that dogs salivate at the sight of food. Curious about the automatic nature of this response, Pavlov rang a bell every time he fed the dogs. Then he tried ringing the bell only, without presenting any food. The dogs responded by salivating. They had substituted a learned stimulus—the bell—for an inborn stimulus—food.

Obviously, a learned reflex like Pavlov's classic example involves some change in the nervous pathways. The exact nature of this change remains unknown. One theory maintains that when particular neurons or neural pathways are used, they are more easily depolarized in the future. Such *facilitation* has the effect of "engraving" certain pathways on the nervous system, preferring some to others. There is also evidence that increased use of nervous pathways causes nerve fibers to grow, suggesting that new nervous connections are created.

Reflexes and Behavior A wide chasm separates the automatic reflex of the knee jerk from the complexity of the neural events that produced *War and Peace* in the mind of Leo Tolstoy. Neural physiologists want to be able to understand the workings of Tolstoy's brain, but that is obviously too confusing a place to begin. Instead, they seek out simpler and more easily understood behaviors that better lend themselves to models relating nervous function and behavior.

In many of the simpler animals, learning and volition have little to do with behavior. Instead, reflexes are "wired" in sequence, with one reflex initiating the next in line and that reflex triggering the one following, so that seemingly complex behaviors actually consist of a series of simple steps. Although external input does modify these reflexes to some extent, the animal behaves in a relatively predictable, mechanical fashion.

A good example of this type of behavior is the way a fly regulates its feeding, a behavior studied by Vincent Dethier and his former student Alan Gelperin. The sequence of events begins when a fly lands on a substrate containing food. Hairs on the bottom of the fly's feet contain taste receptors. If they detect food, a reflexive response in the fly's central nervous system sends motor impulses to the muscles of the labellum, the fly's feeding structure (Figure 20–11), and the labellum is lowered. The labellum also contains hairs with taste receptors, and these receptors test the substrate a second time. If the food is found suitable, a reflex starts the muscles in a pump in the head. The fly sucks the food into its crop. The filling of the crop stimulates stretch receptors located in the net of nerves overlying the organ. Impulses from these receptors increase the stimulus threshold of the taste receptors in the labellum, so that the fly tastes the food less and less and finally stops feeding.

Thus, the fly's feeding behavior is controlled by a series of reflexes: the first triggering the second, the second triggering the third, and so forth. The automatic nature of this chain of events can be demonstrated by cutting the nerves leading from the stretch receptors of the crop. Deprived of the inhibitory stimuli, the fly continues to eat until bursting.

Reflexes in sequence trigger the precise muscular movements required for locomotion, particularly flight, in insects. A well-studied example is the migratory locust, a large grasshopper, first investigated in the mid-1960s by Donald Wilson, then at Stanford University. The locust's central nervous system activates the insect's upstroke and downstroke flight muscles in relatively fixed

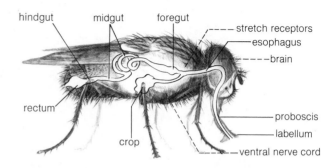

Figure 20–11 A blowfly's feeding apparatus.

sequences, which can, however, be greatly altered by environmental and internal stimuli. First, the locust normally flies right side up. It accomplishes this by orienting its body with respect to what is known as the *dorsal light reaction,* a reflex in which the animal seeks to stimulate its eyes with light from above rather than from below. But if the direction of the light source changes, so does the locust's orientation.

A second sort of adjustment occurs when the locust has fed and is weighted down with its load of food during flight. Stronger contraction of the flight muscles is required for the insect to remain airborne. This is accomplished by increasing the frequency of the impulses sent from the central nervous system to individual motor units of the flight muscles, as well as recruiting more of the available motor units. When the locust is unloaded, greater contractions would produce wingbeats of greater amplitude, which in turn would stimulate stretch receptors at the base of the wings and send inhibitory impulses back to the central nervous system, thus decreasing the activation of the flight muscles.

A third reflex keeps the locust flying in a selected direction. The locust's face is covered with beds of fine hairs that bend in the wind, thus indicating the direction of flight relative to the direction of the wind. By reflex, the flying locust orients itself so that the hairs bend in a constant direction. If the insect's flight direction changes, so does the bend of the hairs, and the locust alters its flight path until the initial stimulus from the hairs is restored.

Some researchers are carrying their work on the neural aspects of behavior down to the cellular level, seeking to associate specific neurons with particular functions. This work has progressed much further with invertebrates than with vertebrates. Of the nearly 2,000 neurons in the sea hare *Aplysia,* a snaillike marine animal, 30 have been identified by size, position, and dendritic and axonal processes, as revealed by staining. The functions of some

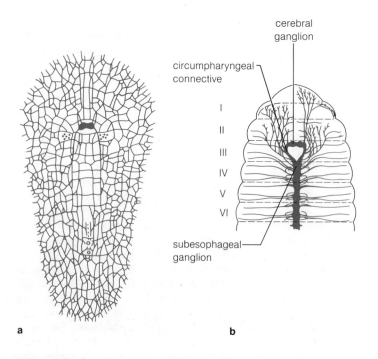

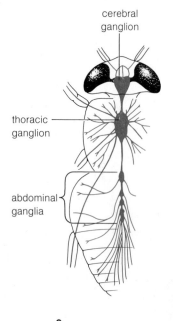

Figure 20–12 Beginning stages of cephalization. (**a**) Concentration of nervous tissue in the flatworm's dorsal area. (**b**) The development of a central nervous system in annelids. Shown here is a dorsal view of the ventral nerve cord of the common earthworm in segments I through VI. Minor ganglia occur in each segment. (**c**) Dorsal view of the central nervous system of an insect, showing the cerebral ganglion (subesophageal ganglion and circurnesophageal ganglion connectives are present but here hidden from view), thoracic ganglion, and abdominal ganglia.

of these neurons have also been specified. One such neuron, coded "L-15" by researchers, spontaneously discharges at regular 24-hour intervals. This neuron may serve as a pacemaker that initiates daily activity patterns (see Chapter 25). Other research has identified some 50 neurons in the cerebral mass of another marine snail, *Tritonia.* When this animal is allowed to move freely, nerve discharges among the cerebral neurons follow specific patterns apparently related to the animal's movement.

Still another research approach is to study inheritance. Since reflexes are inborn, they are inherited, just like eye color or wing shape. The extensive genetic work on *Drosophila,* the fruit fly, has isolated numerous behavioral mutants. Comparing these aberrant animals with normal individuals and following the inheritance of behavioral traits can provide insights into the neural mechanisms of behavior.

The gap between behavioral inheritance in *Drosophila* and Tolstoy's composition of *War and Peace* remains wide and deep, but careful research on these relatively simple behaviors and their nervous pathways promises an eventual understanding of our own complexities.

The Central Nervous System

One of the most pronounced trends noted in contemporary theories of animal evolution is the increased number of neurons and the concentration of those neurons in the central or forward part of the body. Here we will examine that trend and detail some of the anatomy and functions of the central nervous system it has produced.

Evolution and Cephalization The simplest kind of nervous system is the *nerve net* of such radially symmetrical animals as hydra and jellyfish and of some flatworms (Figure 20–12a). The neurons of the nerve net are distributed more or less uniformly throughout the animal's body. There is no central nervous area where information is processed and responses are coordinated. Axons cannot be distinguished from dendrites, and impulses simply spread out in all directions from the point of stimulation. The stronger the stimulus, the stronger the response and the farther it travels through the nerve net.

As animals evolved toward greater complexity and became better able to move about, their sensory capabilities increased. These capabilities were concentrated in the forward end of the body, since that was the part of the moving animal that first contacted the environment. As the sensory apparatus became more complex, so did the association centers where incoming stimuli were processed and responses were relayed to the effectors. These developments are known as cephalization and represent the beginning of the brain.

The beginnings of this trend toward the location of sense organs and nervous tissues in the head can be seen in the flatworms. Most of the species in this class of animals have nerve nets. But in some of the more advanced species, neurons accumulate in the forward end

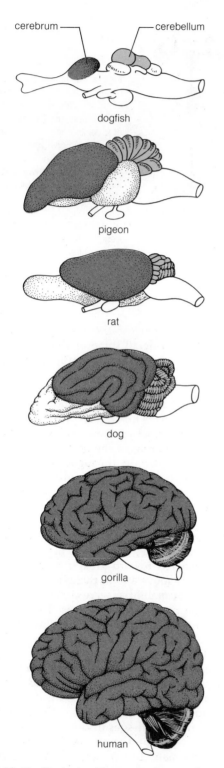

cerebrum — cerebellum

dogfish

pigeon

rat

dog

gorilla

human

Figure 20–13 The brain of humans and some other animals.

of the body and along several longitudinal tracts. By stretching the meaning of the words a bit, these accumulations can be referred to as a "brain" and as "nerve cords," respectively.

Both the annelid worms and the arthropods have a central nervous system consisting of a ladderlike longitudinal nerve cord, with one rung for each body segment (Figure 20–12b). In the annelids and in the primitive insects, two masses of nerve cells called *ganglia* occur in each segment at the point where the "rungs" of the ladder meet the nerve cord. The ganglia contain many of the neural cell bodies, and the nerve fibers radiate out into the body from them. In the higher insects, the two segmental ganglia fuse into one, so that the central nervous system looks more like a string of beads than a ladder (Figure 20–12c). The insects also exhibit a tendency for the abdominal ganglia to move forward and to fuse. In some flies, for example, all the abdominal and thoracic ganglia have fused into one large ganglion in the thorax. In the head, the central nervous system forms a ring around the esophagus. The portion of the ring above the esophagus is called the brain by convention, but it is nowhere near as important an organ in these animals as in vertebrates. Among insects with a highly developed sense of vision, from 30 to 80 percent of the brain is made up of the optic lobes. The integrative centers of the insect central nervous system are located not in the brain, but in the ganglion under the esophagus (the subesophageal ganglion) and in the thoracic ganglion or ganglia.

The vertebrate central nervous system differs in a number of ways from that of the annelids and the arthropods. Embryologically it develops as a tube that lies dorsally rather than ventrally. The longitudinal cord is single, not paired, and the brain replaces it as the master control center of the body. In addition, the entire system is encased in a protective cover of bone.

The Brain Many of the behaviors and physiological functions of invertebrates are controlled by reflex arcs centered in the longitudinal nerve cords. We have noted how sequential reflexes guide the flight of a locust and the feeding of a fly. If a cockroach is decapitated and thus deprived of both brain and subesophageal ganglion, it can still walk about, respond to tactile stimulation, and breathe. Since many of its integrative functions are handled in the thoracic ganglia, the cockroach can get along fairly well without its head.

On occasion, an insect actually performs some tasks better without its head than with it. The classic example is the praying mantis, a relative of the cockroach. Often, when a courting male mantis approaches a female, the female bites off his head. The headless male then mounts the female and copulates with her for hours. The subeso-

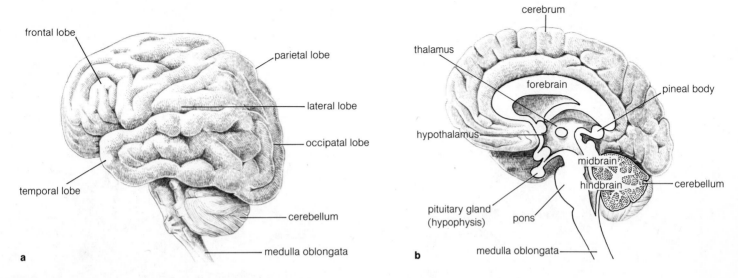

Figure 20–14 The human brain: (**a**) side view of entire brain; (**b**) the brain in longitudinal section.

phageal ganglion inhibits copulation, so the beheaded male is actually more successful at mating than the mantis with a head. In the end the female eats the male, who becomes protein for the female's developing eggs.

A headless human lover would be an unbelievable phenomenon. Since almost all vertebrate integrative functions have been moved into the brain, removal of that organ kills almost any vertebrate immediately. The evolution of the vertebrate brain has reached its contemporary peak in the birds and the mammals, whose brains are massive relative to their body weight. For example, a human brain is 300 times as big, relative to body weight, as the average snake brain.

Besides using the usual anatomical techniques to determine the structure of the brain, researchers have mapped the areas responsible for various functions by either stimulating or destroying various areas and observing the effects. All vertebrate brains can be divided into three general areas: the *hindbrain,* the *midbrain,* and the *forebrain.* As Figure 20–13 shows, the relative proportion of each of these areas varies greatly among the groups of animals.

Each of these major regions can be further divided. The hindbrain consists of the **medulla oblongata** and the **cerebellum.** The medulla, which is the enlarged end of the spinal cord, connects the spinal cord with the remainder of the brain. It regulates such involuntary functions as breathing, swallowing, vomiting, and the constriction of blood vessels and is an important part of the autonomic nervous system, which we will examine shortly. The cerebellum is an outgrowth of the medulla. It is the center for involuntary muscular coordination,

and it is most highly developed in the birds and mammals, creatures noted for their varied and complex movements.

The midbrain consists primarily of the optic lobes, which integrate visual information. The midbrain is prominent in lower vertebrates but much reduced in the mammals.

The distinctive structure of the mammalian brain is the forebrain. It consists of the **thalamus,** the **hypothalamus,** and the **cerebrum** with its olfactory lobes (Figure 20–14).

The thalamus of the human forebrain is completely concealed by the enlarged cerebrum. In humans, as in all vertebrates, the thalamus is the center where incoming sensory information is integrated. The human thalamus contains a network of neurons called the *reticular formation* that interconnects with most other areas of the brain. The reticular formation appears to regulate sleep and wakefulness, acting like an alarm clock that activates the brain when one is awake and a blindfold that shuts out disturbing stimuli when one is resting or asleep. The thalamus is the site of the sensory gating mentioned earlier. It suppresses some stimuli and magnifies others, sending them on to other brain centers for further action. The thalamus prevents us from reacting indiscriminately to each and every stimulus by separating the important from the unimportant. A person with a damaged thalamus often loses the ability to discriminate, reacting as strongly to an insignificant stimulus as to a significant one.

The hypothalamus, which lies just beneath the thalamus, is the brain center responsible for internal homeostasis. It monitors and controls such physical and chemical characteristics as body temperature, water balance, blood

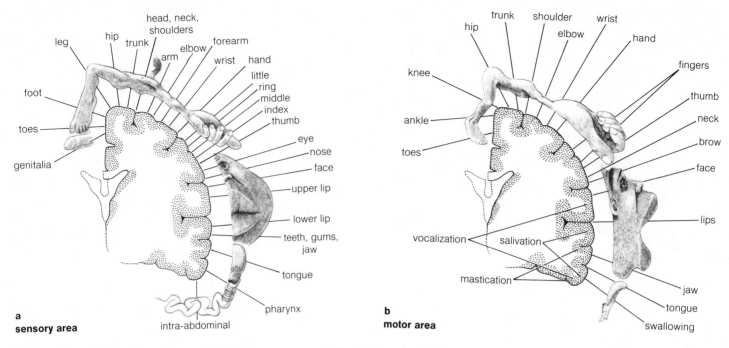

a
sensory area

b
motor area

Figure 20–15 (**a**) Map of the sensory area of the human cortex associated with touch. (**b**) Map of the motor area of the human cortex.

pressure, and carbohydrate and fat metabolism. It is also the seat of such basic moods and drives as hunger, thirst, fear, rage, sexual excitement, and pleasure. The mood of an experimental animal can be controlled by sending a current into an electrode implanted in the hypothalamus. For example, a tranquil cat with an electrode in the "rage center" of the hypothalamus can be turned into a snarling image of feline fury at the flick of a switch. James Olds at McGill University in Toronto implanted electrodes into the pleasure center of rats and then taught them to stimulate themselves by pressing a lever. The rats pressed the lever as often as 7,000 times an hour, ignoring hunger, thirst, and all other needs in their quest for pleasure. This is not, however, the technological ticket to Nirvana it seems. After a while, sensory adaptation sets in, and the animal gets progressively less pleasure from each stimulation.

The hypothalamus produces the hormones released by the posterior pituitary, directs the synthesis and release of the anterior pituitary, and directs the synthesis and release of the anterior pituitary hormones by secreting releasing factors into the portal blood (Chapter 14). The hypothalamus is the site where the nervous system connects with the endocrines. It is the structure that ensures the necessary synchrony between these two systems of coordination.

The cerebrum of fishes and amphibians is concerned primarily with smell and is not essential to immediate survival. A frog without its cerebrum cannot smell, but it functions fairly well otherwise. As the vertebrate line has evolved, however, the cerebrum has taken on more functions and grown in size, a trend clear in Figure 20–13. The cerebrum has reached its greatest relative size and complexity in humans, where it has grown so large that it completely covers the original midbrain and the rest of the forebrain.

Essentially the human cerebrum coordinates sensory stimuli and directs motor responses. It is the site of our conscious mental processes. Specific areas of the cerebrum are responsible for the sensory and motor functions of particular parts of the body. These areas have been mapped out by systematically stimulating with an electrical probe the exposed surface, or cortex, of the cerebrum, usually of a patient undergoing brain surgery, to test for unusual responses. If the probe touches a sensory area, the patient will feel a sensation in the corresponding part of the body. If a motor area is touched, the response is an involuntary movement. This routine procedure, which causes no pain to the patient, has given us a fairly clear idea of the location of the nerve centers in the cerebrum.

There are two points to be noted about the distribution of responsibility in the cerebrum. The first is that the size of a portion of the body in no way determines how much cerebral tissue it is assigned. Figure 20–15 shows the cortical areas devoted to various parts of the body. For example, the legs and the lips have about the same amount of sensory cortex. As for motor functions, the neurons devoted to the hand are the most numerous. Undoubtedly the extraordinary dexterity of the human hand is due not only to the unique mechanics of the apposable thumb but also to the large controlling nerve area that refines and coordinates its movements.

The cerebrum has two halves, or hemispheres, and each has somewhat different functions. It appears that the association cortex of the left hemisphere is specialized for language and logic, whereas the right hemisphere deals with such nonlinguistic functions as spatial orientation and body awareness. It may be that the intellectual differences between artists and sculptors, who have a highly developed sense of form and color, and writers, who are uniquely sensitive to words and language, are in part variations in the left and right association cerebrums.

The cerebrum is very much a physical part of what we call personality. In 1868, Phineas Gage, the foreman of a road gang, was severely hurt in a dynamite explosion. The blast rammed a crowbar through his upper jaw and out the top of his head, destroying a frontal lobe of his cerebrum. Surprisingly, Gage survived, but the damage to his brain altered his personality. He had the temper tantrums of a child, changed elaborately drawn plans on the least whim, and swore when and where he pleased. Along with his frontal lobe Gage lost his will and his sense of what was socially acceptable.

Not so long ago, some psychiatrists and neurosurgeons decided that the best way to calm and control patients with severe mental illness was to remove the frontal lobes (lobectomy) or to sever their connections with the rest of the brain (lobotomy). The effects of lobotomy varied from person to person, although the operation did generally calm patients down and make them easier to handle. It became apparent, however, that this was an outright assault on a personality, and lobotomy is no longer considered reputable surgery. Today mental patients are controlled with heavy doses of tranquilizers, a procedure that has also been questioned and may in time become no less repugnant than lobotomy.

The large amount of association cortex in the human brain probably has much to do with our behavioral versatility. New patterns of behavior are learned most easily in youth, when the brain is still developing, but the capacity to incorporate new patterns is never fully lost. You *can* teach an old dog new tricks, especially an old human dog. Versatility of response may decay with time, because we irrevocably lose some 10,000 brain neurons a day and because the interconnections between neurons deteriorate with age. However, the brain contains so many neurons that this loss is almost insignificant: If you live to be

100, only 0.4 percent of your brain neurons will die before you do. Experiments with rats have shown that ethyl alcohol increases the rate of mortality in neurons, a fact that may explain the severe brain debility common among chronic alcoholics.

The Autonomic Nervous System   The term **autonomic nervous system** refers to the portion of the nervous system that reflexively controls bodily functions generally considered outside conscious control. The autonomic nerves arise in the medulla oblongata and the spinal cord and terminate in the eyes, heart, lungs, stomach, kidneys, intestines, uterus, bladder, external genitalia, and some glands (Figure 20–16).

The autonomic system has two divisions, the **sympathetic** and the **parasympathetic.** Both systems innervate the same organs, but they oppose each other. The sympathetic nervous system, which contains so-called adrenergic fibers, releases norepinephrine as the transmitter substance that crosses the synaptic cleft between adrenergic fibers and target organ. Thus it is hardly surprising that sympathetic nervous stimulation has much the same effect as norepinephrine released by the adrenal medulla into the blood. In the case of the fight-or-flight response, the sympathetic system provides the first rush of readiness, and hormonal norepinephrine maintains the excited state for a longer period, serving as a sort of reserve stimulant. The stimulation or inhibition of target organs depends on the organ or structure in question. For example, the sympathetic system speeds up the heart but slows the gut.

The parasympathetic system inhibits the sympathetic response. The parasympathetic fibers, also known as cholinergic fibers, release acetylcholine as a transmitter, the same chemical found as an excitatory transmitter in the neuromuscular junctions of the voluntary muscles. If the lung has been stimulated by the sympathetic system and is ventilating rapidly, parasympathetic stimulation will slow it down. The largest single nerve of the parasympathetic system is the **vagus nerve,** or tenth cranial nerve. Approximately 90 percent of all the parasympathetic fibers leave the medulla oblongata through the vagus nerve.

Continued sympathetic stimulation would put the body in such a state of alarm that complete exhaustion would soon result. Continued parasympathetic stimulation, on the other hand, would produce trancelike lethargy. Obviously we spend most of our lives in some intermediate state, resulting from the combined influence of both systems.

Heart rate is a good example of the tandem control of the parasympathetic and sympathetic systems. In Chapter 17 we learned that the SA node of the heart acts as a pacemaker that will go on working even after the heart is removed from the body. However, heartbeat

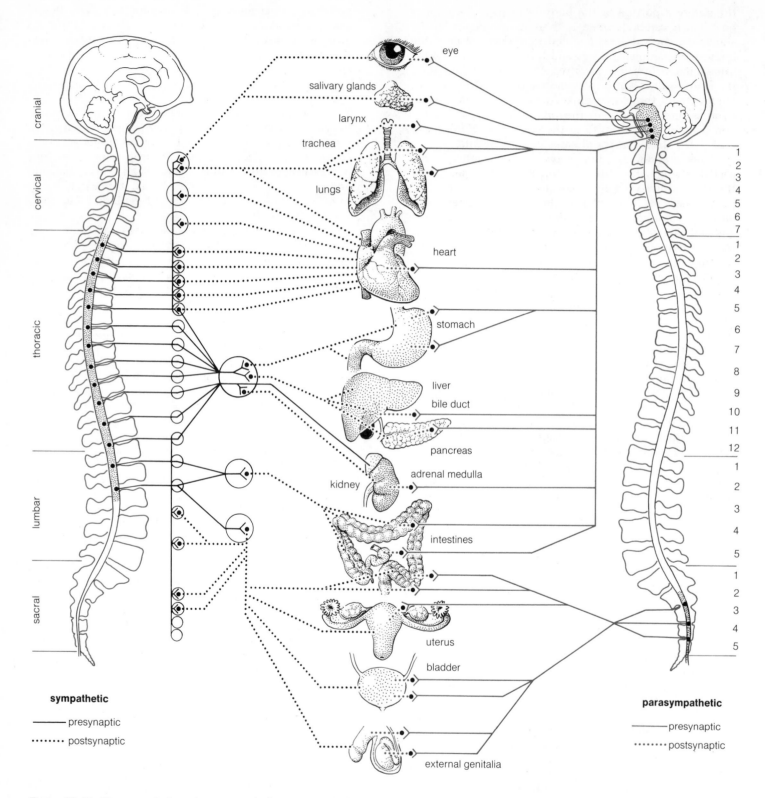

cranial

cervical

thoracic

lumbar

sacral

eye

salivary glands

larynx

trachea

lungs

heart

stomach

liver

bile duct

pancreas

adrenal medulla

kidney

intestines

uterus

bladder

external genitalia

1
2
3
4
5
6
7
1
2
3
4
5
6
7
8
9
10
11
12
1
2
3
4
5
1
2
3
4
5

sympathetic

——— presynaptic

········· postsynaptic

parasympathetic

——— presynaptic

········· postsynaptic

Figure 20–16 The sympathetic and parasympathetic components of the autonomic nervous system.

is continually monitored in the medulla oblongata and either prompted or slowed by nervous impulses. The SA node is innervated both by sympathetic fibers, which release norepinephrine and raise the pace, and by parasympathetic fibers from the vagus nerve, which release acetylcholine and slow it. The medulla receives information about the heart in the form of nerve impulses from stretch receptors in the right atrium and from sensors in the arteries of the neck that monitor the carbon dioxide concentration of the blood. If the atrium stretches or if carbon dioxide becomes more concentrated, the medulla speeds the heartbeat by relaying impulses along the sympathetic fibers to the SA node. This faster rate raises blood pressure, which stretch receptors in the aorta monitor. Nerve impulses travel from these receptors to an inhibitory center in the medulla, which relays nerve impulses along the vagus nerve to the SA node, slowing the heartbeat again and thus moderating the effect of sympathetic stimulation.

Although the autonomic nervous system functions unconsciously, it is not robotlike. The medulla is influenced by other brain centers. Emotional state, among other things, alters autonomic function. Excitement and anxiety raise the heart rate and blood pressure, and prolonged pressure, such as a student undergoes when preparing for finals, can disturb the workings of the intestines. In recent years, it has come to light that we can often control autonomic functions consciously, usually by learning special techniques. Women who use the Lamaze method of natural childbirth control the uterine contractions of labor with deep-breathing exercises. Laboratory rats have been trained to lower their blood pressure, and the techniques learned in this research are now being used experimentally with humans suffering from hypertension. But the most astounding example of conscious control of purportedly autonomic functions comes from certain masters of Indian yoga who can slow their hearts at will.

Taking these findings together with what we will learn in the rest of this section, we will be able to appreciate how physiological processes are often tightly interrelated with perceptions, and, in humans, perhaps with thought. The brain not only controls the various hormone concentrations that affect body function and development; it also affects the physiological functions directly. In humans, conscious control can affect bodily processes that ordinarily proceed without conscious involvement. This suggests the existence of possibly still unknown powers of the conscious mind to affect sickness and health. However, the actual role of the conscious mind in the functioning of the human body is so far largely unknown.

Summary

Like the endocrine system, the nervous system acts as a communication system that integrates functions in various portions of the body, while allowing the body to respond as a whole to external stimuli. However, unlike the endocrine system, the nervous system is adapted for immediate response. Also, its versatility and ability to direct specificity of responses does not reside in a multitude of chemical messengers, but mostly in the complexity of its physical interconnections.

The basic unit of the nervous system is the nerve cell, or neuron. All information transmitted via neurons involves the propagation of a depolarization along the cells' membranes. In this chapter we discussed how the neuron conducts an electrical impulse, and how some neurons are arranged to permit simple behavior patterns.

One major function of the nervous system is to integrate a variety of information and to provide an appropriate response that is translated to the effector organs. In vertebrates, most of the integration occurs in the brain. Various portions of the brain are specialized for various functions, and the relative area allotted for these functions varies among different animals.

21

Movement

In the seas of the ancient earth where life arose, the precursors of all organisms drifted about at the whim of the currents, consuming the abundant organic molecules dissolved in the water. But as heterotrophs evolved into larger and larger aggregations of cells, they needed more food, which became less abundant as life-forms multiplied. Organisms able to move about on their own to seek food and escape predators had an advantage, and natural selection favored the elaboration of cells, tissues, and organs specialized for movement. Each advance in locomotor abilities has uncovered novel methods of foraging and escape.

In this chapter, we will explore the variety of movement in the biotic world. We will look at animal movement in terms of muscles, skeletons, and the various modes of locomotion, and we will survey movement in plants.

Muscles: Their Variety, Structure, and Function

The Varieties of Muscle   In the division of cellular labor that characterizes the organization of complex animals, muscles provide the power for movement. Muscles are classified into three types: **skeletal,** also known as striated (striped), or voluntary; **smooth,** also known as visceral, or involuntary; and **cardiac,** or heart (Figure 21–1). The skeletal muscles, the most abundant tissue in the human body, move our arms, legs, jaws, eyeballs, and so on. These muscles are also most abundant in the bodies of other vertebrates, and if you buy a beef roast at the store, you will be buying the skeletal muscle of a steer. Each muscle, from the biceps that draws the forearm toward the body to the tiny stapedius that moves one of the bones in the middle ear, is made up of many tiny muscle fibers. Alternating light and dark bands, or *striations,* on the fibers give the whole muscle a striped appearance under the microscope and give the muscle one of its names. Each fiber is a single cell with many nuclei and a diameter of 10 to 100 micrometers. The fibers often run the whole length of the muscle.

Vertebrate skeletal muscle fibers are usually classified as *red* or *white,* although some muscle physiologists recognize a third variety, *intermediate.* Generally, red fibers are found in muscles in prolonged use, such as the shoulder muscles of a running animal like a wolf or the flight muscles of a bird that spends much time aloft, like a gull. These muscles are highly aerobic, and they resist fatigue well. Their dark-red color comes from the abundant myoglobin they contain. Myoglobin (Chapter 17) extracts oxygen from the blood and makes it available to cells. Also, red fibers contain many energy-releasing mitochondria and energy stores in the form of fat droplets.

White fibers are thicker than red, are clear in color, and have far fewer mitochondria and fat droplets. They contain only low levels of oxidative enzymes, but large amounts of glycogen, a carbohydrate that can be used anaerobically during brief bursts of activity. Since anaerobic respiration produces large amounts of lactic acid, white fibers have little endurance. The flight muscles of running birds like quails and partridges are composed primarily of white fibers ("white meat"), and they are used only for short flights, usually to escape a predator, while their leg muscles are composed of red fibers (hence "dark meat").

Intermediate fibers share the characteristics of both red and white, and any vertebrate skeletal muscle comprises some mixture of the three types. Relative proportions vary among different muscles in the same animal and among homologous muscles in different animals.

As we noted in Chapter 17, the oxygen demands of repeated exercise increase the amount of myoglobin in the red muscles. Although the relative proportion of red and white fibers is genetically determined, there is also some evidence that exercise causes white fibers to change

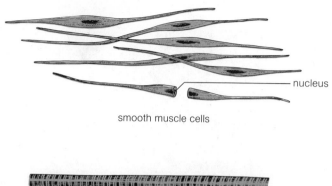

smooth muscle cells

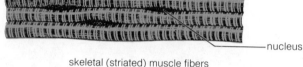

skeletal (striated) muscle fibers

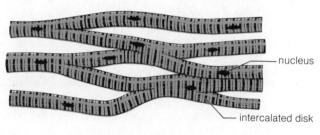

cardiac muscle fibers

Figure 21–1 The three main types of muscles in the body.

into red fibers. In a well-exercised guinea pig, the percentage of red fibers (determined by myoglobin content) in a large calf muscle increases from 38 to 53 percent, whereas the white fibers decrease from 52 to 40 percent. The type and number of nerve connections and the length of contractile proteins, characteristics that also vary from red to white fibers, probably do not change with exercise.

There is a wide range of difference in the intrinsic rates of contraction of various muscles. White fibers generally contract faster than red, but in some cases the reverse is true. Usually, the rate of contraction decreases as body size increases, so that elephant muscle contracts slower than rat muscle. However, the sloth and the cat are about the same size, but sloth muscles contract only one-third as fast as homologues from the cat. The sloth, as its name implies, is a slow-moving creature, largely because of its muscle physiology. Such variations are due to chemical differences in the activity of ATPase, the enzyme that affects the rate of ATP hydrolysis, and to differences in morphology and innervation. In mammals,

"slow" muscles are innervated by motor neurons that discharge about 10 to 20 times per second; in "fast" muscles the discharge rate is 30 to 60 times per second.

Smooth muscle is found in such hollow organs as the intestines, uterus, urinary bladder, and stomach, and in the wall linings of arteries and ducts. The individual fibers have only one nucleus, and the muscle lacks the striations characteristic of skeletal muscle. Smooth muscle fibers are interlaced into sheets that contract slowly. For example, the peristaltic action that moves food masses through the intestines results from the slow contraction of smooth muscles in the digestive tract.

The cardiac muscle of the heart has the striations and multiple nuclei of skeletal muscle. The fibers of cardiac muscle are so tightly pressed together that the junctions between cells, the intercalated disks, can be seen only under the electron microscope.

The Microanatomy of Muscle and the Mechanism of Contraction Muscles power movement, and muscles work by contracting. The contractile ability of muscle results from the arrangement of its microscopic components. The model that follows is drawn from work on skeletal muscle, which has been studied in the most detail, but it probably applies to smooth and cardiac muscle as well.

Each muscle fiber, or cell, contains many parallel elements called **myofibrils,** each about 1 micrometer in diameter. As you can see in Figure 21–2, the fibril is striped vertically at regular intervals, dark *A-bands* alternating with light *I-bands.* In the center of each A-band is a narrow, lighter region called the *H-zone.* Halving the I-band is a thin, dark *Z-line.* The region from one Z-line to the next is called a *sarcomere* (Figure 21–3c). The size of the sarcomere varies with the animal group, from 2 or 3 micrometers in vertebrates up to 40 in earthworms.

If you stop now and look back at the muscle anatomy presented so far, you can see that a muscle is actually a complex of repeatedly divided structures. A single muscle (Figure 21–3a), say the biceps of the arm or the gastrocnemius of the leg, comprises many muscle fibers, each of which is a cell. The fibers, in turn (Figure 21–3b), are made up of many myofibrils. The myofibrils (Figure 21–3c) contain sarcomeres, and the sarcomeres, which are the contractile units of the muscle, are composed of alternating A- and I-bands.

The sarcomere and its bands can be further subdivided as well (Figure 21–3d). They are made of two kinds of filaments, each containing a characteristic protein. The thin filaments are made primarily of **actin,** and the thick filaments, which are twice the diameter of the thin, contain only **myosin.** The overlapping of these fibers, as shown in Figure 21–3d, creates the light-and-dark banding effect in the myofibril. The myosin filaments extend

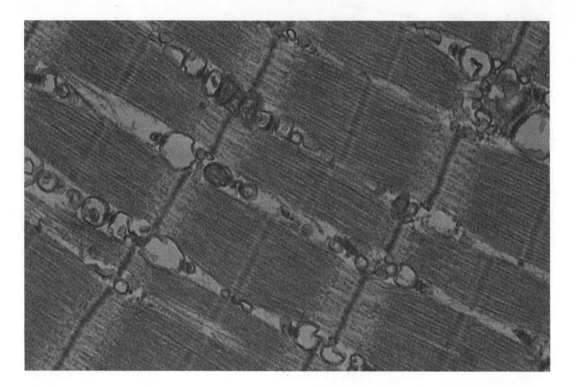

Figure 21–2 Striated muscle as seen with the electron microscope. Each of the diagonal ribbons is a thin section of a muscle fibril. Clearly visible are the dense A-bands, bisected by H-zones, and the lighter I-bands, bisected by Z-lines.

only the length of the A-band, overlapping with the actin fibers from the I-band to create a dense area that looks dark to the eye or the microscope. The H-zone contains only myosin fibers and the I-bands only actin fibers, making them appear lighter in color.

Anatomical studies of the muscles show that when a muscle contracts, the I-bands and the H-zone shorten, sometimes disappearing altogether, but the A-band remains the same length. This observation and others consistent with it led H. E. Huxley, A. F. Huxley, and J. Hansen to propose the *sliding-filament model* of muscle contraction, which is now generally accepted. According to this model, a muscle contracts when the actin filaments slide over the myosin filaments toward the center of the sarcomere, meeting and sometimes even overlapping in the H-zone. The individual actin and myosin filaments themselves do not change length. Instead, the entire sarcomere shortens, as one Z-line moves closer to the other, pulled along by the actin filaments attached to it. Figure 21–4 is a schematic diagram of the sliding-filament model.

To understand the mechanism by which the filaments slide to produce contraction, we need to examine their molecular structure. Each thick filament is composed of several hundred myosin molecules. Myosin molecules are thin and rod-shaped, with double globular heads at one end (Figure 21–5a). The myosin molecules are arranged within the thick filament to form a cigar-shaped sheaf, studded with the head ends of the myosin molecules pro-

jecting along the sides. There is a bare zone in the middle because the tail ends of the molecules are toward the center. The projections provide the cross-bridges that link thick and thin filaments.

The thin filaments contain three major proteins. We have already mentioned that the most abundant of these is actin. Actin molecules are small, roughly spherical molecules, strung together like a strand of beads. Two strands of actin molecules are twisted together to provide the backbone of the thin filaments (Figure 21–5b). The groove between the paired polyactin strands contains *tropomyosin,* long thin molecules each of which extends over seven actin molecules, as well as another protein called *troponin.* Troponin is globular and sits astride the tropomyosin molecule. There is one troponin molecule on each tropomyosin molecule.

Let us now examine how the proteins interact in muscle contraction (Figure 21–5c and d). The energy source for muscular contraction, as for all other biological work, is ATP. ATP combines with the myosin head, and this "charged" head then binds an actin molecule of the thin filament to produce an *active complex.* Removal of the phosphate group of the ATP releases the energy of the high-energy bond. The actin–myosin bond then breaks, and the myosin head (when later recharged with ATP) binds an actin molecule farther along the thin filament. Thus the actin slides along the thick filament as the muscle shortens.

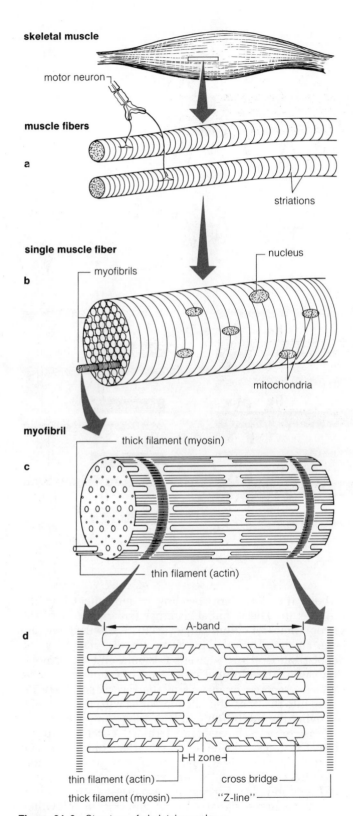

skeletal muscle

motor neuron

muscle fibers

a

striations

single muscle fiber

nucleus

b

myofibrils

mitochondria

myofibril

thick filament (myosin)

c

thin filament (actin)

d

A-band

H zone

thin filament (actin)

thick filament (myosin)

cross bridge

"Z-line"

Figure 21–3 Structure of skeletal muscle.

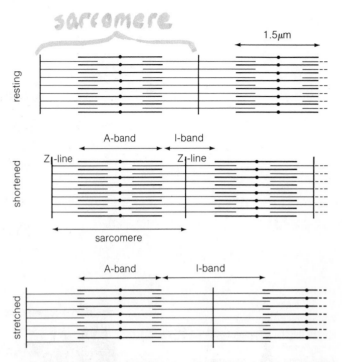

sarcomere

1.5μm

resting

A-band | I-band

Z-line | Z-line

shortened

sarcomere

A-band | I-band

stretched

Figure 21–4 Diagram showing the relationships of the A-band and the I-band as the muscle fiber contracts.

Release of Muscle Contraction in Fibers A muscle contracts in response to nervous signals. Like the neural membrane, the membrane of a muscle cell (the sarcolemma) is polarized: The interior is negative relative to the exterior. The release of a transmitter substance at the neuromuscular junction (Chapter 20) causes a local depolarization of the muscle membrane. If this depolarization exceeds a threshold value, a wave of depolarization travels the length of the fiber, much like an action potential in a nerve cell. The difference between the two mechanisms is that the ion flowing through the muscle membrane is calcium (Ca^{2+}), not sodium.

The other two proteins of the thin filament, troponin and tropomyosin, are involved in regulating contraction, which is mediated by Ca^{2+}. A nerve signal arriving at the muscle releases Ca^{2+} into the fluid surrounding the filament. The calcium binds to the troponin, changing the filament from an inactive to an active state. Without calcium, ATP can still combine with the myosin head, but this charged head cannot bind to the thin filament to form the active complex preceding ATP hydrolysis. The exact role of tropomyosin is not clear, but it is thought that it converts the independent actions of the seven actin molecules of a thin filament under it into a highly cooperative unit.

At first, many scientists would not accept Ca^{2+} as the trigger for contraction. All the myofibrils in a fiber

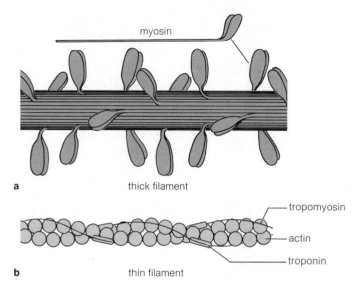

myosin

a   thick filament

tropomyosin

actin

troponin

b   thin filament

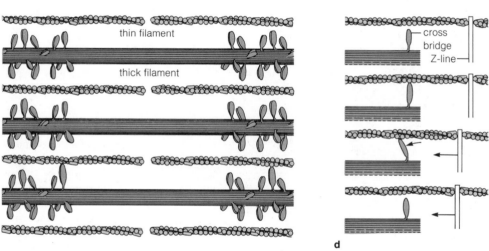

thin filament

thick filament

cross bridge

Z-line

c   **d**

Figure 21–5 (**a**) Molecular structure of the thick filaments consists of an assembly of myosin molecules—long rods with a double "head" at one end. The head has an active site that reacts with the actin molecules of the thin filaments (**b**). The thin filament has two additional proteins, tropomyosin and troponin, besides the two strands of actin molecules that are linked together like beads to form a double helix. (**c, d**) The muscle contracts when the myosin heads attach to actins, swivel, break contact, and reattach to another actin further along the chain.

must contract simultaneously; yet calculations showed that diffusing Ca^{2+} could not reach the centermost myofibrils in the short interval between stimulation of a fiber and contraction. Anatomical studies with the electron microscope finally solved the puzzle. In 1955, Stanley Bennett and Keith R. Porter of the Rockefeller Institute reexamined tubules previously seen in vertebrate muscle fibers. Their studies showed that these tubules are actually indentations of the cell membrane leading into the fiber near the Z-lines (Figure 21–6). The tubules are collectively called the *transverse system,* or the *T-system.* In intimate contact with the T-system near the Z-lines are sacs called *terminal cisternae* that are part of an elaborate endoplasmic reticulum known in muscles as the sarcoplasmic reticulum. A depolarization traveling along the membrane of the muscle cell apparently enters the fiber

through the T-system, speeding its movement into the cell interior. The terminal cisternae (enlargements of the sarcoplasmic reticulum adjacent to T-tubules) contain large amounts of calcium. They are stimulated by the depolarization of the adjacent T-tubule, and the stimulation causes the terminal cisternae to become permeable to Ca^{2+}. As great quantities of the Ca^{2+} are quickly released into the fiber, the calcium bonds with the troponin, removing the inhibition and triggering contraction.

Fast muscles have many terminal cisternae, apparently for quick take-up of Ca^{2+} after contraction and release before it. This arrangement allows the muscle fiber to contract more often in a given unit of time.

Some muscles contract in response to stretching as well as to nerve signals. While working on blowflies, J. W. S. Pringle of Oxford University noticed that their

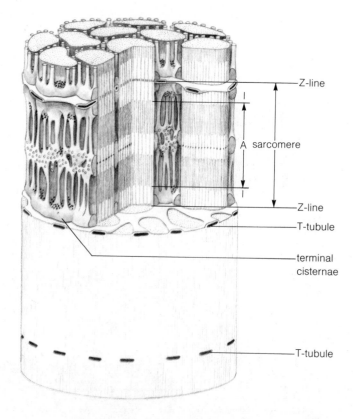

Z-line

A sarcomere

Z-line

T-tubule

terminal
cisternae

T-tubule

Figure 21–6 Diagrammatic representation of a muscle fiber showing the T-tubules and the terminal cisternae. The tubules of the T-system are invaginations of the plasma membrane. The terminal cisternae (which are part of the sarcoplasmic reticulum) are intimately associated with the T-system.

wing muscles contract at rates several times greater than the frequency of incoming motor-nerve impulses. He found that the muscles work somewhat on their own. At the top of the upstroke, the downstroke muscles are stretched. The stretching triggers a contraction in the muscles, pulling the wing down. At the bottom of the downstroke, the upstroke muscles are stretched, a contraction is triggered, and the wings are drawn up. These *asynchronous muscles*, as they are called, make the wing-beat essentially self-perpetuating. Only a periodic nervous impulse is required to keep the wings moving up and down.

Neural Control of Muscle Contraction The contraction of muscle fiber is an all-or-nothing proposition: The fiber contracts fully or not at all. There is no intermediate contraction. Yet we know from our everyday experience that the biceps, for example, exerts far less force lifting a feather than lifting a 100-pound barbell. How can such a wide range of control come from so limited a mechanism as muscle fiber contraction?

The answer is that a variable number of muscle fibers can be activated at one time. Several motor neurons innervate a large muscle, and each of these neurons branches repeatedly, activating only a limited number of muscle fibers. The muscle fibers that act together, being controlled by one motor neuron, are called a *motor unit*. In the less than a dozen major muscles controlling the movement of a mouse's front leg, there are about 2,600 motor units. To lift its leg gently, the mouse need activate only a few of the motor units in the muscles involved. But a strong action, like a jump, requires the repeated activation of many motor units. Thus, the strength and speed of a motion is controlled by the number of motor units activated and the timing of that activation.

A single muscle contraction, whether delicate or strong, is called a **twitch.** The strength of a twitch depends on the number of activated motor units. A twitch is divided into three phases: the *latent period,* which is the time lag between the stimulus and the beginning of the responding twitch; the *contraction phase,* in which the muscle contracts fully; and the *relaxation phase,* in which the muscle returns to its original resting state. Twitch durations vary greatly in different animals and in different muscles of the same animal, spanning a range from a millisecond to several seconds.

Laboratory studies of muscle contraction often use an apparatus similar to the one shown in Figure 21–7. When an electrical stimulus is applied to the muscle, it twitches, just as an intact muscle in the body twitches in response to a nerve impulse. If a second stimulus is applied to the muscle before it relaxes completely, it will contract again. If still more stimuli are applied rapidly, the muscle contracts again and again, as the same motor units are activated repeatedly or as more motor units come into play. As a result of the repeated stimulation, the overall strength of the contraction increases, a phenomenon called *summation.* If all the motor units are activated and the muscle remains contracted, it is in a state of *tetanus.* Tetanus doesn't last long. Soon fatigue sets in, a result of such chemical changes in the muscle as the accumulation of lactic acid, and the muscle relaxes.

The disease tetanus takes its name from this technical term for an unrelaxed muscle contraction. The disease is caused by the bacillus *Clostridium tetani,* whose metabolism produces a toxin that causes spasmodic contractions of the voluntary muscles, often with lethal consequences.

In sum, the mechanics of the twitch in a vertebrate muscle allows for contractions of immensely varying strength. A single stimulation of a few motor units yields a slight contraction, but repeated stimulation of many motor units, whose effect is increased by summation, produces movements of remarkable strength.

But what of smaller animals lacking the internal space for elaborate muscle systems containing many motor units? In the invertebrates, miniaturization requires that there be fewer motor units, but many of these animals are capable of precise movements of varying strength. In part, this range of contracting strength is achieved by two different kinds of axons leading to the muscles, the "fast" and the "slow." The slow axon of a given muscle triggers a slow rate of contraction; the fast axon stimulates a quicker reaction. Activating different axons in various muscle groups results in graded muscle contractions.

Skeletons: Support and Muscle Attachment

The basic physiology of muscle contraction varies relatively little from animal to animal: When a clam closes its shell and when a human sprinter bursts out of the blocks at the gun, the chemical events are parallel. However, muscles can be arranged in a nearly infinite variety of ways, and the particular arrangement determines any animal's locomotive abilities. Skeletons, which we tend to think of as passive supports, are the frame on which muscles are arranged and therefore have much to do with how animals get around.

Hydrostatic Skeletons: Fluid Support Water resists compression; that is, it takes a tremendous pressure to force a given volume of water into a smaller volume. Thus, body fluids enclosed within a cavity can offer a firm structure against which a muscle can contract. In the simplest movement allowed by such a *hydrostatic skeleton,* the contraction of the muscles against the body fluids extends the body in the uncontracted region something like a moving piston (Figure 21–8). More complex movements are also possible. Flatworms, annelid worms, and caterpillars contain two sets of muscles, one longitudinal and the other circular. When the circular muscles contract, the body extends; when the longitudinal muscles contract, the body shortens. Each segment in the common earthworm contains its own set of muscles and its own compartmented body fluids, allowing each segment to act independently of the others. The earthworm moves by lengthening some segments and shortening others. Circular muscles in forward segments contract, lengthening the front of the worm and extending it. At the same time, the longitudinal muscles in the segment to the rear of the extending segment contract, shortening that seg-

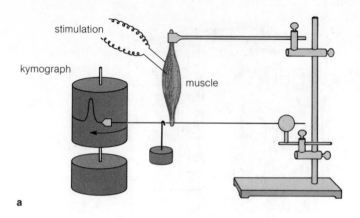

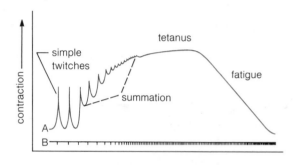

Figure 21–7 (**a**) Kymograph apparatus used to examine muscle contraction. The muscle is mounted to suspend a weight on a marker pushed against a revolving smoked drum. The muscle can be stimulated to contract by applying current with a stimulator. As the muscle contracts, the pen leaves a record of the contraction on the smoked paper on the revolving drum. (**b**) A record of a muscle contraction obtained by applying electrical stimuli at greater and greater frequency. At low stimulus frequency, the muscle has time to relax after contracting. At higher stimulus frequency, the muscle may not have fully relaxed before it is stimulated to contract again. As a result, contractions are stronger. At very high stimulus frequency there is no relaxation of the muscle at all, until fatigue sets in and it is no longer able to remain in the contracted state.

ment and pulling the posterior segments forward. Bristles *(setae)* on the worm's underside provide traction. Caterpillars lack the compartmented body fluids of the annelid worms, and they do not move with the whole body surface touching the substrate. Instead, they hunch along, a motion most exaggerated in the loopers, or inchworms, which extend the forward end, anchor it, and then loop the rear up to the front. Forelegs and fleshy "claspers" in the rear anchor the caterpillar. When the caterpillar is punctured it is unable to move forward because the

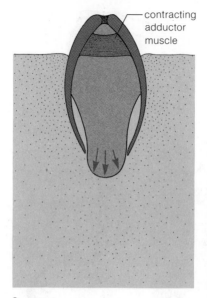
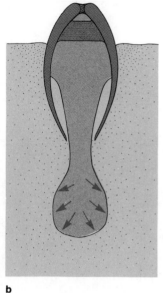
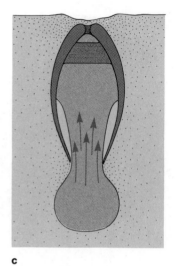

contracting
adductor
muscle

a   b   c

Figure 21–8 Mechanism of burrowing using the hydrostatic skeleton in a bivalve mollusk, the clam. (**a**) Clamping down the shell adductor muscles forces fluid into the foot; (**b**) the foot is pushed into the mud and expands, providing an anchor; (**c**) the clam then contracts the muscles to the foot and is thus pulled into the mud. The mechanism is used to move down as well as laterally. In the snail, with an incompressible shell, the foot is extended by inspiring air—by exaggerating the respiratory cycle.

fluid can no longer be compressed and thus it cannot act as a piston.

Some animals lacking a true hydrostatic skeleton use a hydraulic system for locomotion. One such animal is the starfish, which moves by projecting tube feet from its rigid shell. A system of tubes that terminate in numerous sacs (ampullae) contains the fluid (Figure 21–9). Muscles in the ampullae contract, forcing fluid down into the "feet" and extending them in a direction determined by additional muscles in the feet. Suckers on the bottom of the feet take hold of the substrate, and muscular action forces the fluid back into the ampullae, shortening the feet and pulling the starfish forward. The starfish also uses its tube feet to pull open the shellfish it preys on, and a hydraulic mechanism extrudes its stomach into the prey's shell to digest the soft body.

Exoskeletons: Outside Support Some animals are covered with a relatively hard exterior skeleton, or *exoskeleton.* This kind of skeleton is characteristic of adult arthropods, which have a covering, or cuticle, not unlike the iron armor worn by medieval knights. Locomotion comes from legs that extend from the body and act as levers. The arthropod exoskeleton is not completely rigid. Flexible membranous areas between the body plates and in the joints of the legs allow movement (Figure 21–10). Besides protecting the arthropod's internal organs, the exoskeleton acts as the attachment for the muscles (Figure 21–11).

The cuticle is remarkably light and durable. It is composed primarily of a polymer of the polysaccharide

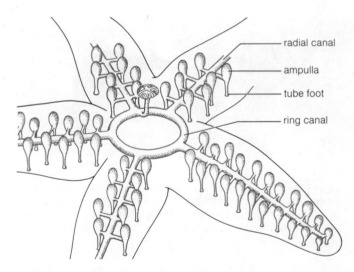

radial canal

ampulla

tube foot

ring canal

Figure 21–9 Water-vascular system of the starfish showing essential structures for locomotion using hydrostatic pressure.

N-acetylglucoseamine—commonly called chitin—cross-linked with protein. The crustacean cuticle is reinforced with calcium carbonate crystals. The flexibility of the cuticle varies with the species, from elastic to extremely rigid. One reason insects can inhabit the land successfully is that their exoskeletons are impermeable to water. The exterior of the cuticle is lined with a single layer of polar lipid molecules whose water-binding ends point inward

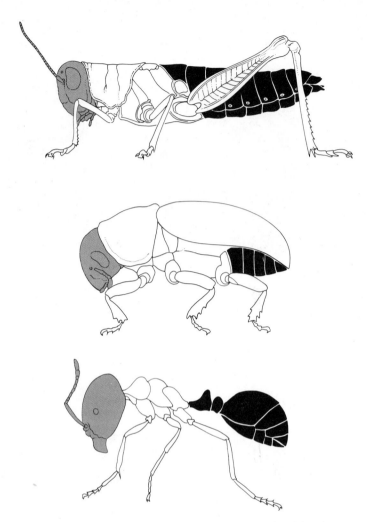

Figure 21–11 Insects muscles are attached to their exoskeletons. Shown here is a typical insect leg of 10 joints and the associated muscles and tendons used to control its movements.

Figure 21–10 The major body areas of an insect are head (gray), thorax (white), and abdomen (black). Each area is covered by hard plates of cuticle. In the adult with a rigid exoskeleton, movement is possible only along membranous areas beneath the overlapping armored plates.

(Figure 21–12). This feature makes the cuticle impermeable to water and greatly reduces water losses to evaporation, permitting insects to succeed in dry terrestrial habitats. It also acts as an osmotic barrier in aquatic insects. If this lipid layer is destroyed experimentally with abrasive dust or high temperatures, insects dehydrate in the terrestrial environment.

Since the cuticle is relatively rigid, it does not grow with the animal. Arthropods must periodically shed a cramping cuticle for one of the right size, a process called molting, or **ecdysis.** When ecdysis begins, cells in the epidermis secrete enzymes that digest the lower layers

of the cuticle, leaving the outer, less digestible portions. At the same time, the epidermal cells secrete a new cuticle. When most of the new cuticle has been laid down, the old one ruptures along the *ecdysial lines* and is shed. The new cuticle stretches under hydrostatic pressure as the animal gulps air or water—essentially "blowing up" the new exoskeleton. Chemicals released by the epidermis cause the protein molecules in the new cuticle to make extensive cross-linkages with each other and with the chitin, hardening the exoskeleton. Until the cuticle hardens completely, the arthropod is vulnerable to predators, lacking the armor that protects it the rest of the time. The so-called soft-shell crabs that are a gourmet's delight in East Coast markets and restaurants are edible crabs that have recently molted.

It seems likely that molting has limited the possible size of insects. The bigger the insect, the more difficult it would be to maintain body shape while the cuticle was soft. Also, the bigger the insect, the less likely it would be able to find a safe place to molt and escape predators.

Endoskeletons: Inside Support Vertebrate skeletons are systems of rods or tubes contained within the body to make an *endoskeleton.* An endoskeleton allows continuous growth, making periodic molts unnecessary. The endoskeleton provides less protection than the exoskeleton, but the most sensitive tissues of the body are enclosed within the skull, the spinal cord, and the rib cage.

In the elasmobranch and agnath (lampreys and hagfish) fishes, the skeleton is composed entirely of cartilage, a relatively soft and flexible bonelike material. In vertebrates, cartilage is the primary skeletal component only in the young. As the juvenile animal grows into an adult, bone replaces cartilage in all but a few places, like the joint of ribs and breastbone, the tip of the nose, and the linings of the knees.

Bone is a living tissue, made up largely of protein and the inorganic salt calcium phosphate, which hardens bone and does not occur in cartilage. The structural unit of bone is the microscopic **Haversian system.** Each of these repeating, roughly cylindrical units is composed of a central canal, called the Haversian canal, enclosed within concentric layers of calcium phosphate salts. Blood vessels and nerves pass through the canals. Between the concentric layers are small cavities, or lacunae, in which the true bone cells are found. Thin tubules called *canaliculi* connect these cells to the Haversian canal and its blood vessels and nerves (Figure 21–13).

The human endoskeleton contains 206 bones, over half of which are found in the hands and feet (Figure 21–14). The bones articulate, or meet, at joints. The bones of the human skull meet in rigid joints, or *sutures* (Figure 21–15). All the other joints are movable. Bands of tough connective tissue called **ligaments** span the joint and bind its elements together, much as a hinge holds a door to the jamb. Sprains are actually injuries to the ligaments, usually caused by a violent jerking of the joint that pulls the ligament farther than it can go, rupturing blood vessels and causing local swelling. Since damaged ligaments are soft and can easily be pulled out of their proper shape, a sprained joint must be immobilized until it heals. Otherwise, the ligaments will be distorted, and the joint will be loose and easily damaged again.

Like the exoskeleton and the hydrostatic skeleton, the endoskeleton provides the attachment for the muscles. White, inelastic **tendons,** which are connective tissue like the ligaments, attach the muscles to the bones. Typically, a muscle attaches to a relatively immobile bone at one end, spans a joint, and attaches to another bone in a movable body part. The immobile attachment is called the *origin;* the movable one, the *insertion.* A given muscle may have origins or insertions on more than one bone. For example, the triceps muscle on the back of the upper arm has origins on both the humerus (the upper arm bone) and the scapula (the shoulder blade).

In humans, as in the other vertebrates, muscles do their work by pulling, never by pushing. If you hold your arm out straight and bring your hand to your shoulder, the action is powered by the contraction of the biceps,

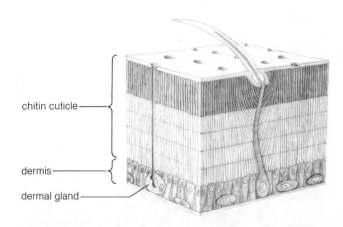

chitin cuticle

dermis

dermal gland

Figure 21–12 (**a**) Diagram of a portion of generalized insect exoskeleton. The dermal glands produce secretions that travel by ducts to the surface of the cuticle. The cuticle is perforated by thousands of pore canals per square millimeter. The pore canals carry lipid from the dermal cells to the top of the cuticle. (**b**) The waterproofing of the cuticle is due primarily to a monolayer of lipid.

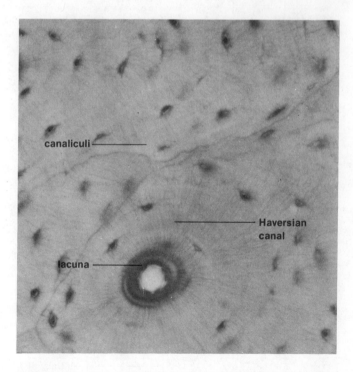

canaliculi

Haversian canal

lacuna

Figure 21–13 The cellular structure of bone. The cell bodies are located in the lacunae, from which numerous cytoplasmic extensions radiate out through the canaliculi.

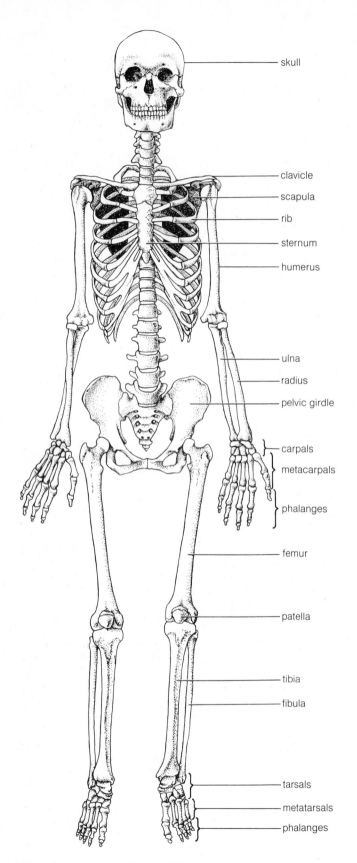

skull

clavicle
scapula
rib
sternum
humerus

ulna
radius
pelvic girdle

carpals
metacarpals

phalanges

femur

patella

tibia
fibula

tarsals
metatarsals
phalanges

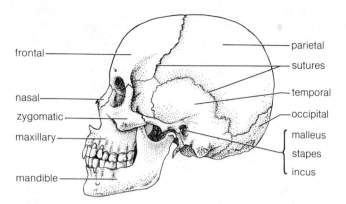

frontal
nasal
zygomatic
maxillary

mandible

parietal
sutures
temporal
occipital
malleus
stapes
incus

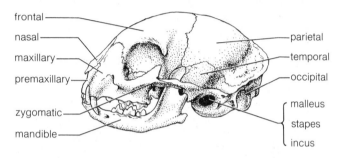

frontal
nasal
maxillary
premaxillary

zygomatic
mandible

parietal
temporal
occipital
malleus
stapes
incus

Figure 21–15 Human and cat skulls.

pulling the lower arm toward the body (Figure 21–16). To reextend the arm, the triceps contracts, pulling the lower arm back. This simple example also illustrates another important aspect of the skeletal arrangement of muscles: They are grouped in opposing or antagonistic pairs. The biceps and the triceps, for example, pull the lower arm in opposite directions.

Varieties of Locomotion

Natural selection through the eons of life's existence has produced many ways for animals to get around, each with its own muscular and skeletal adaptations.

Creeping Locomotion without limbs, or creeping, is found in many animal groups. We have already seen the role of the hydrostatic skeleton in the creeping locomotion

Figure 21–14 Human skeleton.

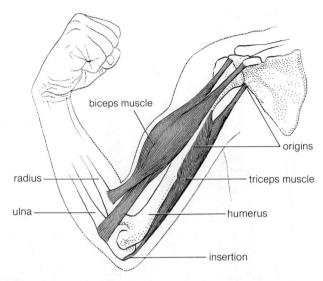

Figure 21–16 Opposing muscles allow the arm to be flexed or extended.

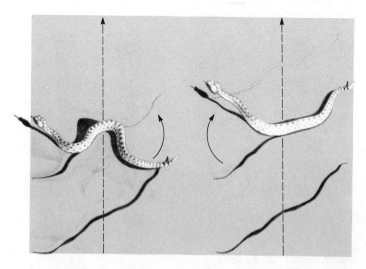

Figure 21–17 Locomotion by sidewinding. The thick arrows indicate general direction of movement of two snakes, which are shown at different stages in the movement sequence.

of the earthworm and the caterpillar. The clam, depicted in Figure 21–8, creeps in its own curious fashion, by thrusting an expandable "foot" into the sand as an anchor and then pulling itself along behind the foot. The muscular underside of snails and sea anemones undulates in waves that move from head to tail and carry the animal forward. A layer of sticky mucus secreted by the animal's ventral surface provides traction.

Snakes, which evolved from lizards with limbs, creep in several different ways. Some loop their bodies forward, pushing off against irregularities in the surface. Others alternately contract and relax their rib muscles. Some snakes raise the forward section of their body, which reduces friction. Traction with the surface comes from the belly scales, which point backward and anchor the animal the way the setae hold an earthworm.

Desert snakes face the locomotive problem of moving across sand, which slides easily and is hard to push against. To understand the snake's problem, try running on a beach. A number of snakes, including one American rattlesnake, use sidewinding as their mode of locomotion. Sidewinding is energetically expensive, but a fast and effective way of moving in sand. As Figure 21–17 shows, the sidewinding snake begins moving with its body parallel to the direction of travel and its head pointed almost backwards. The snake lifts its head and places it on the sand in the direction of travel. It then raises the rest of its body and thrusts it forward and then to the side in a loop. Again the snake raises its head, moves it forward, and then brings up the remainder of the body. The movement is like that of the inchworm, but the whole body

is used lengthwise to anchor the animal, rather than only the ends of the body.

Walking   To walk, an animal must have rigid levers that act as stilts to hold its body up off the ground and as organs of propulsion to move it forward. The necessary rigidity can be provided by any of the three skeletal arrangements examined earlier. The starfish and other echinoderms such as the sea urchin use the hydraulic tubefoot system; arthropods and vertebrates have exoskeletal and endoskeletal legs, respectively.

Running   The evolution of running from walking provided tremendous selective advantages. Running animals can forage over wide areas—African hunting dogs may range over 1,500 square miles—and they can catch prey or escape predators more successfully. The champion sprinter of the animal world is said to be the cheetah, which hits 70 miles per hour for short distances. A riderless horse cannot beat 50 mph, but it can maintain a constant speed of 35 mph for as long as 15 miles. Humans are far slower. The record for humans at 100 yards is 22.73 mph, (9.0 seconds, set by Ivory Crockett, 1974); and at the longer distance of the mile, the speed falls to 15.57 mph (John Walker, 3:49.4, 1975).

Running requires an array of skeletal and muscular specializations. Almost all fast runners have long legs relative to the rest of the body. Usually, this added length appears in the parts of the leg farthest from the body.

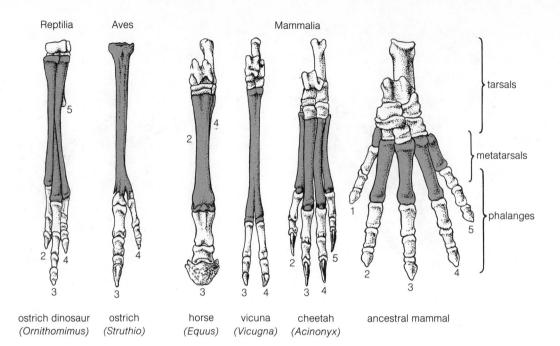

Reptilia   Aves   Mammalia

tarsals

metatarsals

phalanges

Figure 21–18 Adaptations for running in the feet. The adaptations involve lengthening, compaction, or fusion of metatarsals, loss of lateral digits, and squaring or fusion of tarsals.

ostrich dinosaur
(*Ornithomimus*)

ostrich
(*Struthio*)

horse
(*Equus*)

vicuna
(*Vicugna*)

cheetah
(*Acinonyx*)

ancestral mammal

In running animals, the metacarpals and metatarsals, the bones homologous to the palm and instep in the human, are greatly elongated (Figure 21–18). As a result, the carnivorous mammals and the birds run on what corresponds to the ball of the human foot. The now extinct running dinosaurs ran the same way. The hoofed mammals (ungulates) have even longer legs and run on their toes, as a ballet dancer might. You may have noticed that human sprinters exploit this same principle of greater speed through longer limbs by running on the balls of their feet. When several regions of the lower leg are lengthened rather than just one, the set speed of movement is increased. In other words, adding several joints, all of them moving in the same direction, increases the overall velocity at the far end of the limb.

The length of a running animal's stride is also increased by modifications of the shoulder or by added suppleness in the spine. When the forelimb is attached to a collarbone it has rigid support, but its movement is restricted. Carnivores have only a vestigial collarbone, and ungulates lack the bone entirely. The forelimbs are attached loosely to a narrow and deep chest, allowing them to swing back and forth and adding inches to their stride. Running carnivores like cats and dogs have supple spines that further lengthen their stride. In midstride, when the hindlegs are airborne, the spine bends under. Then, when the feet strike the ground, the spine straightens out, lengthening the stride. The weasel presents an extreme example of this adaptation. It has short legs, but

a highly flexible spine, and thus can reach a good speed over short distances.

Speed is a matter not only of stride length but also of stride frequency—that is, the number of times the limb can move through its arc in a given unit of time. In part, stride frequency depends on the speed at which the muscles move the limb. High-speed motion requires a long, thin muscle. For example, a series of five rubber bands, when maximally stretched, will shoot a paperclip five times as fast as a clip shot from just one rubber band. The units are additive. If they are in parallel, on the other hand, they would have five times the power, but the speed of one. That is, five in parallel will be no faster than one, but they will be able to propel a five times heavier weight. The same principle applies to muscle arrangements. In addition, the closer to the joint the insertion of the muscle, the faster the stride. The reason for this is a principle of physics: The limb works like a lever on a pivot, and attaching the muscle close to the pivot (joint) lengthens the lever (limb) and increases the speed produced by a single muscle contraction. Thus, in the horse, most of the running muscles are close to the shoulder. However, inserting the muscle close to the pivot sacrifices power, again because of physical principles, making any extra weight in the limbs an added burden. As you can see in Figure 20–20, many cursorial (long-distance running) animals, particularly ungulates, have reduced digits (fingers and toes), a development that decreases the weight of the limb. The joints are simplified,

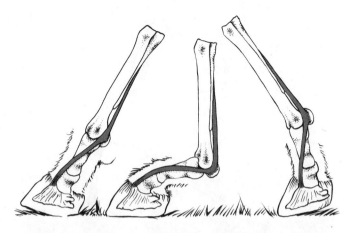

Figure 21–19 Springing ligaments in the leg of a horse. Impact of the foot against the ground bends the fetlock joint (center) and stretches an elastic ligament, storing energy that would otherwise be lost. As the foot bends back the stored energy is released, helping to give the leg an upward impetus and reducing the need for heavy muscles.

allowing movement in only a limited plane. And the musculature is reduced and modified. The horse, as one example, has an elastic ligament in its foot that stretches when the hoof strikes the ground, then springs back when the hoof lifts off, adding an extra push to an already fast stride (Figure 21–19).

The power/speed trade-off in limb anatomy can be seen in human athletes. A discus thrower or weight lifter needs strong arms and legs, and these athletes are powerfully muscled people with highly developed limbs. Distance runners, particularly those who compete at such distances as 10,000 meters and the marathon (26 miles, 385 yards) require high stride frequency. A weight lifter's legs would be a hopeless liability for a distance runner, and successful runners have thin, small-boned legs with long, supple muscles. Supplement 21–1 discusses factors that limit the endurance of distance runners.

Jumping Kangaroos, jumping rodents like the jerboa and the kangaroo rat, and many birds travel on the ground by jumping. In principle, jumping in a vertebrate is little different from walking, except that both legs exert their force simultaneously, increasing the power of the stride and lifting the animal clear of the substrate.

Certain jumping insects, like the prodigiously leaping flea and the leafhoppers, have a more refined jumping mechanism. A jumping insect has short limbs. For an insect to jump like a long-limbed kangaroo, its muscles would have to contract much faster. However, the intrinsic rate of contraction in insect muscle is roughly the same as in vertebrate muscle. To amplify the rate of contraction, the flea and other jumping insects have a special energy-storage mechanism. The insect gradually contracts its jumping muscle, and the muscle's mechanical energy, rather than acting directly on the leg, is stored by the stretching of a rubberlike protein called resilin. In the case of the flea, a second muscle must act as a "trigger" to release the power. When a "releaser muscle" pulls the tendon of the jumping muscle into a position where it can act on the leg, the energy stored in the resilin is released suddenly, and the flea leaps.

Jet Propulsion Some aquatic organisms move by propelling themselves with a jet of water. The scallop on the familiar Shell Oil gas station sign is a swimming clam that can escape the starfish, its principal natural enemy, by closing its twin shell rapidly and forcing out a water jet. The squid squeezes water in its mantle out through its siphon with great force. A similar mechanism is seen in dragonfly larvae, which take water into the rectum and expel it with muscular contractions of the body wall.

Swimming Fish have massive muscles running along their sides and attached to their backbones by the dorsal and lateral spines and the ribs. When they swim, fish contract the muscles on one side of the body, bending the spine. These muscles then relax, and those on the other side contract, bending the spine back. Waves of muscular contractions and relaxations pass down the body, whipping the lever of the tail back and forth on the fulcrum of the head.

Whales, dolphins, and porpoises swim much the same way, except that their tails move up and down instead of side to side. Human swimmers use the same motion in the so-called dolphin kick of the butterfly stroke. Whales can swim at remarkably high speeds for great distances without expending large amounts of energy. Their body shapes are adapted to reduce turbulence to a minimum, leading naval architects to imitate the whale shape in designing submarines. Also, when a whale surfaces to blow, it leaves a small slick on the surface. The slick comes from an oily substance on the skin that "greases" the whale through the water, reducing friction. It is thought that fish slime performs the same function.

Flying In energy expenditure per given unit of time, swimming is the least demanding mode of travel, running is somewhat more costly, and flying is the most demanding of all. However, flying is also the fastest, and a flying animal can travel farther than a swimmer or a runner, with the same expenditure of energy. A field mouse covering 1 mile uses about as much energy as a bird of similar size flying 20 miles.

Flight has evolved in prehistoric reptiles, birds, mammals (bats), and insects, and each group has its own locomotive mechanisms. Birds and bats have skeletal components similar to those in walking animals. The humerus joins the pectoral girdle at the shoulder, and the radius and the ulna join to the humerus (Figure 21–20). In bats and birds, these bones are relatively thin and light, and the muscles are rather undeveloped, used only to control the wingstroke and not to generate it. Most of the power for the wingbeat comes from large chest muscles. The pectoralis, which is directly under the mamary gland in humans, is the main downstroke muscle, and the supracoracoides directly under the pectoralis is the upstroke muscle. In birds, these muscles originate in the center of the breast on the protruding keel bone, and they insert on the upper and lower surfaces of the humerus. Anyone who has eaten a chicken breast is familiar with these large flight muscles.

The major adaptation to flight in the anatomy of the forelimb involves the digits and the limb covering. In bats and prehistoric flying reptiles, the digits are very long, as is shown in Figure 21–20. The webbing between them reaches the sides of the body and the legs, providing the flight surface. Birds, on the other hand, have rudimentary digits. The surface area of the wing comes from large overlapping feathers attached to the following (posterior) edge of the wing. Feathers are lightweight, but they wear out and have to be replaced periodically. Ducks and geese shed (molt) their flight feathers all at once, leaving them flightless and vulnerable for two or three weeks in the late summer. Most other birds replace their feathers a few at a time, so that they are never flightless because of the loss of wing feathers.

Precisely how wings create the lift and thrust needed for flight is an inordinately complex question. Essentially, though, the wings follow a cycle similar to the movement of the arms in a swimming human's breast stroke. The power comes from the down and back movement. The reverse stroke returns the wing to its beginning position for another power stroke. An advantage of feathers is that they reduce the resistance of the recovery stroke, allowing the air to flow between them as if through the slats of a venetian blind.

Wing size relative to body size varies greatly. Generally, smaller wings are aerodynamically less efficient because of the turbulence they create, but they are much more maneuverable. Although flying animals with small rapidly beating wings, like the hummingbirds, bees, and sphinx moths, expend large amounts of energy relative to their weight, their precisely controlled short wings allow them to exploit rich energy sources like flower nectar. Of all the birds, only the hummingbird can fly backwards.

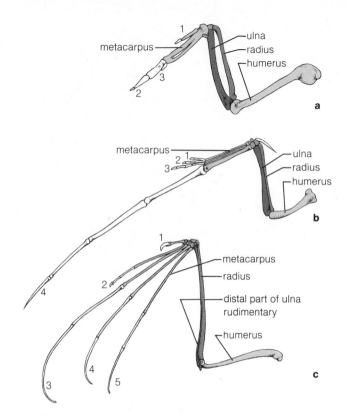

Figure 21–20 Wing skeletons of (**a**) a bird, (**b**) a pterosaur, and (**c**) a bat.

At the other end of the scale of relative wing size are the large birds like the albatrosses, vultures, and storks. These birds have wings so long that they can barely stay aloft under their own muscular power. They fly by extracting energy from the atmosphere in the form of thermals and other updrafts. The large wings catch the updraft, and the bird spirals upward, gliding away from a high altitude and dropping down to another updraft. Since we humans are so used to moving under our own power, this method sounds too passive to be effective, but African vultures have been known to rise as high as 3,500 meters in 10 minutes and then glide 18 kilometers in 15 minutes. The advantage of gliding on long wings over flying on short ones is that the animal can forage over a wide area while expending little energy. Six common species of African vulture regularly make round trips of 200 kilometers or more from their nests to the migrating game herds whose dead they eat.

The wings of some extinct reptiles dwarf those of present-day birds. Recently Douglas A. Lawson, a paleontology graduate student at the University of California,

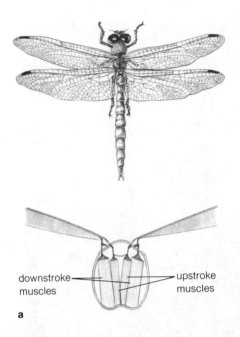
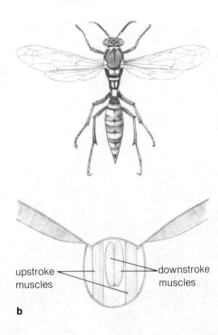

downstroke muscles → ← upstroke muscles

a

upstroke muscles → ← downstroke muscles

b

Figure 21–21 Relationship between the main power muscles of the thorax that drive the wings in (**a**) a dragonfly with direct and (**b**) a wasp with indirect modes of operating the wings.

Berkeley, found partial skeletons of a late-Cretaceous pterosaur (flying dinosaur) in west Texas. From tip to tip, the wings of this reptile measure an estimated 15.5 meters, about the same size as the wings of some modern-day fighter planes. These animals had long necks and toothless jaws, indicating that they may have fed on the carrion of the giant dinosaurs, occupying an ecological niche similar to that of the modern vulture.

Flight evolved in insects more than 300 million years ago, for at that time forests of the earth were already inhabited by dragonflies with two-foot wingspreads. Insect flight mechanisms vary greatly from those of the vertebrates. The wings are not adapted forelimbs. Instead they are thought to have evolved from gliding "fins" protruding from the cuticle or from structures used originally in sexual display. Insect wings are rigid, and they contain no muscle. The power for flight comes from large muscles enclosed in the thorax. Also, most flying insects, with the exception of the flies and beetles, have four wings rather than two. The wings of bees and butterflies are normally hooked together in flight so that they act as one; the dragonflies and grasshoppers beat theirs independently, but in phase and at the same frequency. The forward wings of the beetles have evolved into protective wing covers, and the rear wings of the flies have become balance organs that work like gyroscopes.

Unlike the wings of most birds and bats, insect wings produce thrust during both the up- and the downstroke. Figure 21–21b shows how the flight muscles are arranged inside the thorax of some insects (flies, bees, beetles, but-

terflies). You will notice that the muscles do not attach directly to the wings. Instead, they move the wings by deforming the thorax and are called *indirect muscles.* The downstroke muscles run the length of the thorax, and the upstroke muscles are at a right angle to them, spanning the thorax from front to back. When the downstroke muscles contract, the rigid thoracic box is deformed and, because of complex articulations at the base of the wing, the wing moves down. When the upstroke muscles contract, the dorsal surface of the thorax is pulled down, and the wing flips up again. Because of the complex wing joints in certain insects, a 1 percent shortening of the muscles yields a 180° wingstroke. This adaptation, along with the asynchronous muscles discussed earlier, permits high-speed wing oscillations in the insects. For example, mosquitoes fly with wingbeat frequencies of from 400 to 600 cycles per second. Dragonflies have a different arrangement. The main power-producing muscles of their wings are attached directly to the wings. The muscles are therefore called *direct muscles* (Figure 21–21a).

Movement in Plants

Plants are sedentary organisms incapable of locomotion, and their movements are easy to overlook; but they do in fact move. Growing shoots twist and bend toward the light, flowers open in the morning and close in the evening, the tendrils of vines touch a branch and then wind

Figure 21–22 Response of the mimosa, or sensitive plant. The tip of one leaf is stimulated in such a manner (in this case, by using a flame) that the entire plant is not jarred. After 14 seconds the petiole of that leaf has collapsed, and many of the leaflets have folded up. As the stimulus is transmitted back along the stem, other leaves collapse, and their leaflets fold up. Some of this activity can be observed even after 90 seconds.

around it. Some plants can move rapidly. The leaves of the mimosa plant fold within a few seconds when touched (Figure 21–22), and the leaf of a sundew plant closes within minutes around an insect stuck to its surface.

Charles Darwin was one of the first biologists to study plant movement, and he wrote a book titled *The Power of Movement in Plants.* The research he began has led to our present knowledge that plant movement results from two mechanisms. The slower movements involve differential growth mediated by the growth hormones, as discussed in Chapter 11. The fast movements, such as those of mimosa and the sundew and the daily opening and closing of flowers, result from turgor changes in some of the cells, as shown in Chapter 3. Fluid loss makes the cells flexible, and adjoining cells with normal turgor exert a force against the now flexible regions, bending the plant.

Summary

In all animals, movement is due to muscle contractions. In this chapter we have examined the structure of muscle

and the mechanism of its contraction. The differences among various morphological types of muscles are relatively slight in different animals. But the ways they are attached to supporting structures and innervated by the nervous system largely account for the varied types of locomotion. Movements in plants occur by way of differential growth and by way of changes in turgor pressure.

Supplement 21–1

On Human Endurance: Factors Limiting the Marathon Runner

In 490 B.C., the Athenians defeated their ancient enemies, the Persians, near the village of Marathon. A Greek messenger bore the news of the Greek victory to Athens, a little over 26 miles away. The messenger ran the distance, but collapsed and died after delivering the message. As a sport, running the distance of 26 miles and 385 yards, the "Marathon," has persisted through the ages, and has at no time been as popular as it is today—at least if the 2,362 starters at Hopkinton, Massachusetts, and the 2,040 finishers at the Prudential Center in the seventy-ninth annual Boston Marathon in April 1975, are any indication. And about double that number competed in the 1978 marathon.

The marathon is in many ways the ultimate test of a runner, and to some it is a test of much more than athletic ability. After 20 miles, most runners find each lifting of the foot an effort of will. Release from the agony becomes an overpowering urge. Some submit. Some lose against themselves.

Ken Moore, a world-famous marathoner, states:

Men (and women) submit to the ordeal not in spite of pain but because of it. Competitive urges can carry you for 10 or 15 miles, but then the distance and discomfort already endured scream that this must not be for nothing. So you go on.

Afterward, in the dressing room, men hang stiffly on one another, too exhausted to untie their shoes—and jabber uncontrollably. . . . The pain has made everything suffered so extraordinary that it *has* to be expressed.

Perhaps when one stops, when one comes to accept that this is merely the way pain works on minds, that there is no inherent meaning in repeating this sort of suffering, one will have died a little.

The enduring satisfaction of distance running is . . . in knowing that you learned to be brave and to do something better than you first thought you could.

The will to persist, which ultimately determines success or failure, and hence shapes reality, is mental. No human would voluntarily submit to such rigors for the sake of experiments on the limits of human physical performance. But men, women, and children run the marathon—giving the physiologist ample and willing subjects in which to study the effects of the ultimate in endurance performance.

David L. Costill and his colleagues at the Human Performance Laboratory, Ball State University, Muncie, Indiana, have for years been taking blood samples and muscle biopsies of marathon runners. They have measured the performance of the heart and the lungs and their interactions in oxygen transport, both during training and in the race. These studies have shown us how the normal, healthy organism operates—or how it can *operate. The following essay summarizes some of their results and conclusions.*

Few human activities place greater demands on the endurance capacity of human physiology than the sport of marathon running. To the physiologist, the marathon runner offers an ideal model for studying the limitations of human endurance. Many theories have been proposed to explain the various stages of exhaustion experienced by the runner during the 26.2 miles (42.2 kilometers) of a marathon race. In general, these theories can be listed under the following headings: *hypoglycemia* (low blood sugar concentration), *muscle glycogen depletion, dehydration* (loss of body water), and *hyperthermia* (overheating).

Hypoglycemia

During prolonged exercise, the liver plays a crucial role in maintaining a constant supply of blood glucose for the active tissues such as nerves and muscles. By releasing glucose into the blood, the liver is able to keep pace with the muscles' demands for blood-borne carbohydrates. Unfortunately, the liver's supply of stored carbohydrate (glycogen) is limited. Consequently, during very long exercise bouts, the liver may be unable to mobilize glucose as fast as it is being used by the running musculature. As a result, blood glucose concentrations decrease and the muscle fibers are deprived of an essential source of fuel.

Most marathon runners attempt to supplement the liver's glucose supply by drinking sugar solutions during the race. Although the runners claim a nearly immediate energy lift from this practice, research has shown that the real physiological benefits of oral sugar are not realized for 15 to 20 minutes after drinking the solution. This is not to say that such feedings do not benefit the runner's performance. To the contrary, although delayed, the sugar that reaches the blood from the intestines lessens the demands on the liver. In that way, liver glycogen stores are less likely to be exhausted and blood sugar levels are more easily maintained.

Muscle Glycogen Depletion

Studies have shown that when the muscles' carbohydrate stores (glycogen) are depleted, the runner is exhausted and unable to perform even moderate exercise. Muscle tissue samples obtained from the leg muscles (vastus lateralis and gastrocnemius) of runners before and after a marathon race reveal that many muscle fibers are exhausted of all glycogen. Microscopic examination of these muscle samples shows that the depleted fibers are predominantly one type, slow twitch (ST) fibers, which are characterized by their endurance qualities. At the same time, the fast twitch (FT) fibers (sprint type) may still contain considerable amounts of glycogen. These findings suggest that during distance running the heaviest energy demands are placed on the ST muscle fibers. When these have consumed their glycogen stores, the FT fibers are unable to generate enough tension, since they are recruited principally for sprint running. As a result, the runner finds each stride more difficult and eventually becomes exhausted when the ST fibers fail to contract. This selective depletion of muscle glycogen is undoubtedly the cause of the muscle distress of the final stage of marathon competition often described by the runner.

In recent years, considerable attention has been given to the importance of stockpiling large quantities of glycogen in the muscles before endurance events. This is accomplished by a combination of exhaustive exercise followed by three to four days of a high-carbohydrate diet (such as spaghetti). This regimen has been shown to be effective in preventing premature exhaustion and has been credited with helping many outstanding marathon performances.

Dehydration

Body water plays a crucial role in satisfying the muscles' need to maintain a constant internal environment. This means that nutrients (glucose, fat, and so on) must be delivered to the muscles and, at the same time, the by-products of metabolism (including carbon dioxide and heat) must be carried away. This is the major responsibility of circulating blood, which is roughly 80 percent water. Marathon runners may lose 8 to 13 pounds of water, principally as sweat, during the 2.5 to 4 hours of competition. As a result, undue stress is placed on the circulatory system and other body water compartments. Dehydration to this degree may cause a 13 to 21 percent loss of plasma water, and more than an 11 percent decrease in muscle water content.

Besides a water deficit, heavy sweating during distance running can cause a substantial loss of body salts. Theoretically, any great disturbance in the balance of electrolytes (such as sodium and potassium) in body fluids could interfere with the contractile qualities of the muscle, thereby limiting performance. This is usually not the case, however. Although the concentrations of muscle and blood electrolytes are markedly altered after two hours of running, the muscle cell membrane is no less sensitive to stimulation. Consequently, these disturbances in electrolyte balance cannot be held responsible for the fatigue experienced by the runner.

Hyperthermia

One of the primary responsibilities of the circulatory system is to transport heat generated by the muscles to the surface of the body, where it can be transferred to the environment. A major fraction of the blood pumped from the heart each minute (cardiac output) must be shared between the skin and working muscles. As a result, running at a fast pace on a warm day will overload the circulatory system, impairing heat dissipation. Since the runner produces such large amounts of muscle heat, moderate air temperatures, high humidity, and/or bright sun can result in a critical accumulation of body heat. Body temperatures (rectal) of 106.5°F (41.4°C) have been reported for runners at the finish of marathon races. It is not unusual to observe rectal temperatures above 105°F (40.6°C) in competitors after distance races of 6 to 26 miles. Certainly, part of the distress experienced during the final phase of a distance race might well be attributed to the stress of excessive body heat on the nervous system.

Since there is little runners can do about a warm environment, it is obvious that they must slow their running pace to reduce the rate of muscle heat production. Drinking fluids at frequent intervals (10 to 15 minutes) alleviates, in part, the problem of overheating and minimizes the threat of circulatory collapse.

Summary

We can see that, although the factors that limit prolonged endurance exercise are varied, they are in some ways related. The muscles' ability to sustain a high energy output and at the same time maintain homeostasis depends on the capacity of the circulatory system. If the environmental heat stress or dehydration is such that the circulatory system cannot transport nutrients and metabolic waste materials, the muscles are subjected to less than optimal conditions.

It is not surprising, therefore, to find that successful endurance athletes are characterized by a highly developed circulatory system. Although it is often difficult to determine how much of the athlete's capacity is dependent on adaptations to training, champion distance runners are the product of genetics. However, it is repeated exposure to prolonged exercise that enables the runner to resist the stresses that tend to limit endurance.

22

Windows to the Environment

If you took your dog for a leisurely walk along a path in the woods and then afterwards, through some miracle of technology, interviewed the animal, you would find that its experiences differed significantly from yours. You would probably have paid particular attention to the shape and color of the leaves and flowers, noticed the fragrance of the more pungent blossoms, and picked up such sounds as the chattering of a squirrel or the sighing of a strong wind in the treetops. The dog, although unable to see color, would have registered so many smells in such detail that it would know who and what had come down that path in the last day or two. You would have walked right past the deer in a thicket 25 yards off the path, but the dog's acute hearing would have detected it there.

But even a combined inventory of human and canine perceptions comes nowhere near exhausting the full range of stimuli in our earthly environment. Sensitive recording instruments set in any local environment for a period of time will register changes in a variety of physical and chemical components of that environment—X rays, light waves, radio waves, heat, mechanical vibrations of atmosphere and substrate, and magnetic and electrical fields, among others. All the data from these changes provide information about the environment, and any animal possessing a sensor for the changes would also be taking them in and recording them. But the survival value, and thus the relevance, of any particular bit of data depend on the perceiving organism's mode of life.

Each species has specialized sensory capabilities—particular windows to the environment—that admit the relevant information and relay it to the central nervous system to be translated into behavior. In this chapter, we will explore the variety and function of these sensory windows.

The Nature of Sensation and Perception

For all the many variations of sensory reception found in the animal world, the basics remain the same: Some form of energy in the environment is translated into the electrochemical energy of the nerve impulse by some specialized receptor.

Sensory Receptors as Transducers In electronics, the word *transducer* refers to a part of a circuit that takes in one kind of energy and gives off another. A good example is the photoelectric cell, which absorbs light and emits an electric current. Sensory receptors work in an analogous fashion. The simplest receptors are free nerve endings (those lacking a myelin sheath). More commonly, the endings of the sensory nerve fibers are modified so as to be sensitive only to specific stimuli, or they connect with specialized receptor cells that pick up the stimulus and transfer it to the fiber.

Biologists have long been interested in how specific stimuli are transduced into nerve impulses. Although we have traced the complete sequence of events for only a few kinds of sensory reception, we know that all such reception involves a stimulus contacting the membrane of the transducer neuron or receptor cell and causing a change in the membrane's permeability. Commonly, the transducer becomes more leaky to sodium ions, a change that reduces the resting potential (see Chapter 21). This initial depolarization, or *generator stimulus,* is directly proportional to the intensity of the stimulus (Figure 22–1). Progressively stronger stimuli and repetitive small stimuli add to produce larger generator potentials (weak,

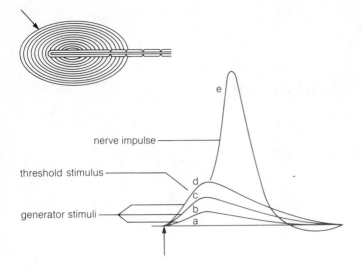

Figure 22–1 Diagrammatic section through a Pacinian corpuscle (receptor cell) showing unmyelinated nerve terminal enclosed in onionlike lamellae. Four nodes of Ranvier are shown. A weak stimulus (applied at arrow) produces a weak generator current (*a*); progressively stronger stimuli produce correspondingly stronger generator currents (*b, c*) until at the threshold stimulus (*d*) an all-or-nothing nerve impulse (*e*) is induced.

nonpropagated membrane depolarizations). Once the potential exceeds a critical level called the *threshold stimulus,* an action potential (a propagated wave of membrane depolarization of the neuron) is triggered.

An action potential is an all-or-nothing phenomenon (Chapter 21). Although the measured value of the generator potential varies with the intensity of the stimulus, there is only one value for the action potential of a neuron. Thus, stimulus intensity must be coded in some other way, typically in the frequency or pattern of the nerve impulses. The more intensely the sensor is stimulated, the more action potentials are produced per unit of time.

The Perception of Sensation All the action potentials traveling along sensory axons to the central nervous system are qualitatively the same, no matter what stimulus triggers them. The nerve impulses generated by a massive, mangling injury to the arm are electrochemically identical to the taste and smell impulses of a slowly savored sip of vintage wine. To be distinguished, the incoming action potentials are processed and integrated in the central nervous system, and it is there that sensation, whether the pain of injury or the delicate bouquet of wine, is perceived.

However, most sensory action potentials never reach

the sensory areas of the cerebral cortex (Chapter 21). They are stopped at the lower brain, where information about the stretching of the muscles, the position of the joints, the carbon dioxide concentration in the blood, the blood pressure, and so forth is acted upon but not felt as sensation. We are conscious of only a small fraction of our bodies' millions of sensors at work. Automatic processes, all those "subconscious" basic housekeeping chores of the body, are governed by sensory input that we do not actually perceive as sensation.

The so-called lower animals are probably as unconscious in their intake of all sensory data as we are about this large fraction of ours. But if these animals actually have conscious perceptions, they differ greatly from our own. When we put a worm on a fishhook, for instance, it writhes and thrashes about, but we have no way of knowing whether it feels pain. Its muscular contractions are just as likely to be innate responses favored by natural selection because they aid the animal in escaping predators and mechanical obstructions.

Receptors of the Skin and Skeleton

The five familiar senses in humans are vision, hearing, taste, smell, and touch. However, these five barely begin to include all the senses we actually have.

Cutaneous Receptors The skin alone contains specific receptors for touch, pain, heat, cold, and pressure. Figure 22–2 shows the location of these receptors in a section of mammalian skin. Each of these types of receptors is morphologically distinct from the others. Those adapted to perceive mechanical stimuli such as touch and pressure are called mechanoreceptors. The simplest are unmyelinated nerve endings that serve as pain receptors. Others are enclosed in bulbs or corpuscles and are often closely associated with hairs or other structures that magnify the stimulus.

Pacinian corpuscles (Figure 22–2f), for example, provide the sensation of deep touch or vibration. These relatively large receptors (0.5 millimeters in diameter) are located in the underlayers of the skin and in surrounding muscles, tendons, and joints, as well as in the structures supporting internal organs. Each Pacinian corpuscle consists of a sensory nerve terminal enclosed inside 20 to 50 onionlike lamellae of connective tissue. When something touches the skin and deforms this multilayered capsule, the nerve terminal is distorted and its permeability increases. If the resulting depolarization reaches the threshold level, an action potential is conducted along the axon.

Proprioceptors If you were asked how you know where your arms and legs are at any given moment, you might say that you keep track of them with your eyes. However, even blindfolded, you still know precisely the location of your limbs. The reason is that internal receptors called **proprioceptors** keep you informed about the position and status of the various body parts. They sense not only relative position but also the angle formed at each joint and the degree of shortening, or tension, in the muscles. There are three different kinds of proprioceptors.

Joint receptors are complex nerve terminals located at a joint that can sense movement within it. They detect the onset of a limb movement and the speed and direction of that movement. Once the limb comes to rest, some of the joint receptors put out a sustained discharge of action potentials to indicate the steady position of the limb.

The **Golgi tendon organs,** named after the same Italian histologist who discovered the Golgi apparatus of the cell, are much like the joint receptors. They are associated with tendons, which attach muscles to bones, and they register changes in the tension of the muscle.

Muscle spindle receptors measure the shortening, or contraction, of a muscle. When the length of a muscle fiber increases, the firing rate of the sensory axon of the spindle receptors increases as well; when the muscle contracts and shortens, the frequency of action potentials drops.

Insects by and large lack the internal proprioceptors of the vertebrates. Most of their proprioceptors are located in the external skeleton, some of them consisting of pads of hairs or minute dome-shaped organs at the joints. One such set of hairs keeps the insect up off the ground. Sensory neurons at the base of the hairs are triggered when the insect's sinking weight pushes the joints together. The incoming action potentials trigger a reflex in the insect's central nervous system that causes the muscle that extends the leg to contract, thus raising the animal up off the substrate. If these hair patches at the joints are removed, the animal can no longer sense the pull of gravity on its body and sinks limply to the ground.

Vibration Receptors

When a tuning fork is struck and gives off a sound, its prongs are vibrating back and forth at perhaps hundreds of oscillations per second. Each of the fork's vibrations produces a pressure wave in the air that travels out in all directions. The energy of that wave may be gradually dissipated as heat, or the wave may strike some other object and cause it to vibrate. The detection of such mechanical stimuli is the basis for the sense of hearing.

Sound travels not only through the air, as we are

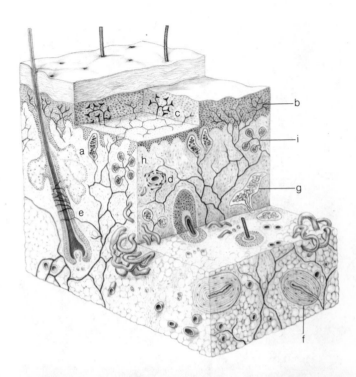

Figure 22-2 Three-dimensional construction of mammalian skin showing a variety of cutaneous sensory receptors. *a* = Meissner's corpuscles (touch); *b* = net of free endings (perhaps pain); *c* = Merkel's disks (touch); *d* = nerve net around blood vessel (probably pain); *e* = nerve terminals around hair sheath (hair movement); *f* = Pacinian corpuscle (pressure); *g* = Ruffini endings (warmth); *h* and *i* = Krause's end bulbs (cold).

used to thinking of it, but also through the ground and on and in the water. Loud noises like an explosion or the rumble of an approaching train produce perceptible ground vibrations, which often precede the sound itself. Addicts of Western movies are familiar with the scene of an ace tracker in fringed buckskins putting his ear to the dust of the prairie and announcing that seven Cheyenne on pinto ponies are pursuing 143 panicked buffalo and 2 pronghorn antelope 8 miles to the northeast.

In truth, perception of ground vibration is of limited value for a creature as large as a human, since it takes a tremendously powerful sound pulse to set the surface of the skin in motion and to stimulate stretch receptors. Some of the small-bodied insects, however, are quite sensitive to substrate vibration, and they have evolved intriguing mechanisms for orienting themselves to various types of such stimuli.

For example, the water strider, which actually walks on the water as its name suggests, can use vibrations in the surface of a still pond to locate its prey. The water

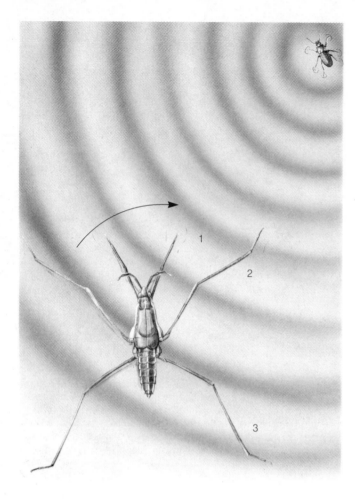

Figure 22–3 A water strider homing in on its prey.

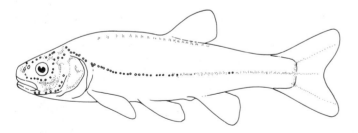

Figure 22–4 Lateral line system in side view.

strider detects such vibrations with special receptors in its leg joints that consist of hairs with neurons at their bases. Like the proprioceptors just described, they fire when the hairs bend. If an insect falls into the water and its struggles make a wave 90° to the left of the strider,

its three left legs will be lifted by the wave slightly before the right three, and their receptors stimulated simultaneously. If the wave originates 45° to the right, the right front leg will pick it up slightly before the others. The water strider can detect these slight time differences and orient itself in the direction of the wave's origin (Figure 22–3).

Besides the organs of hearing, all fishes and some amphibians have a lateral line receptor organ, which is visible as a series of pits along the sides of head and body (Figure 22–4). The lateral line organ is innervated by the eighth cranial, or acoustic, nerve. The nerve endings terminate in groups of hair cells within lymph-filled canals. When the lymph is set in motion, the hairs bend, and the nerves fire. One of the several roles of the lateral line organ is to pick up water-borne vibrations, probably as a way of detecting prey or predators. Sharks, for example, are known to be attracted to unusual vibrations, particularly those of a fish in distress. Nonfishermen are often surprised that lures like spinners and wobbling spoons, which bear little resemblance to anything alive, catch fish. When these lures are pulled through the water, they give off vibrations similar to those of a struggling baitfish. Hungry predators like bass and trout are drawn to the vibration and may strike the lure.

Snakes are deaf; that is, they have no organs for picking up airborne vibrations. Instead, the acoustic nerve is attached indirectly to the lower jaw, which rests on the ground in the snake's normal posture. Ground vibrations are transferred to the jaw and thence through the acoustic nerve to the brain.

Sound Receptors in Insects In 1879, Hiram Maxim, a Maine handyman (who later invented the machine gun and was knighted in England after becoming a British citizen), was installing electric lights at the Grand Union Hotel in Saratoga Springs, New York. An astute observer, Maxim noticed mosquitoes buzzing around the electric lines. Putting his inventor's mind to work, Maxim wondered whether the insects might be attracted by the humming of the lines. He tested this idea by striking a tuning fork that hummed at the same frequency. Sure enough, male mosquitoes gathered around the fork, too. Maxim proposed that the male mosquitoes mistook the buzz of the electric lines for the flight noise of female mosquitoes, and that the insects' feathery antennae acted as "ears." In fact, the male's antennae are so constructed that they vibrate most vigorously at the frequency of the female's wingbeat. These vibrations stimulate rows of mechanoreceptors located at the base of each antenna.

Such antennal organs are by no means the only mode of hearing in insects. A design common in insects, as well as in vertebrates, involves the reverse-drum principle. A drum, when struck with the flat of the hand or

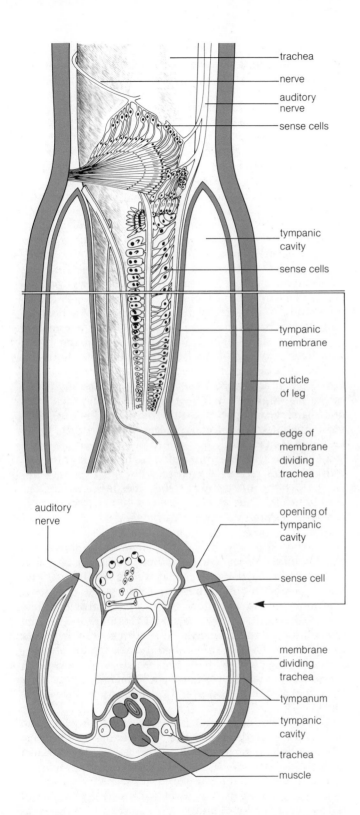

Figure 22-5 Longitudinal and transverse sections through the foreleg of a katydid, showing the structure of the tympanic organs.

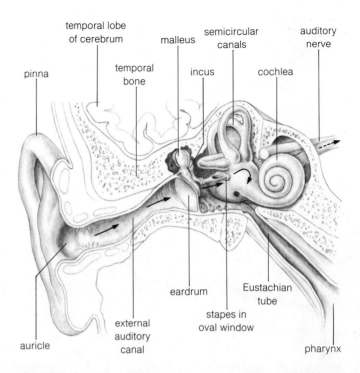

Figure 22-6 Mammalian ear. Ear structures in humans. Arrows indicate the path of sound waves.

a drumstick, vibrates and gives off sound. The principle can be reversed: Sound waves hitting a drum cause it to vibrate. A drum set up to transfer that vibration to mechanoreceptors makes for a serviceable organ of hearing. These hearing receptors, called *tympanic organs,* are found in various locations in various insects—on the thorax, on the abdomen, or even in the front legs (Figure 22–5).

The tympanic organs of most insects are tuned to a specific frequency, and sounds of other frequencies are not perceived. Thus, it is hardly surprising that the mating songs of most species are monotonous. The significant character of an insect song is less the melody than the rhythm, the sequential pattern of pulses.

The Mammalian Ear The mammalian ear, like the insect tympanic organ, is basically a drum that transfers vibrations to mechanoreceptors. But there the similarity ends. Whereas the tympanic organ is specialized to receive only a few pitches, the mammalian ear has evolved into an instrument sensitive to a wide range of sounds.

The external part of the ear, the *pinna,* directs sound into the *auditory canal* (Figure 22–6). Bats, rabbits, and horses have large pinnae that are controlled by voluntary muscles and can be directed at a sound. Humans also

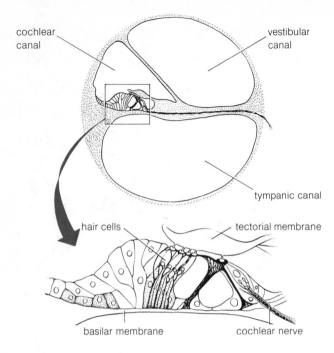

cochlear canal

vestibular canal

tympanic canal

hair cells

tectorial membrane

basilar membrane

cochlear nerve

Figure 22–7 Structure of the cochlea in cross section, showing the structure of the organ of Corti.

have fairly large outer ears, and a few can wiggle their ears at will. Hearing, it appears, was once far more important to our survival than it is now. The pinnae are greatly reduced in subterranean mammals such as gophers and moles, who depend more on smell than on hearing, and in submarine mammals such as seals and whales, whose swimming would be slowed by the drag of large external ears.

Sound traveling down the auditory canal strikes the **tympanic membrane,** or eardrum, and sets it in motion. On the other side of the tympanic membrane lies the cavity of the *middle ear.* Three small bones—the *malleus,* the *incus,* and the *stapes,* or hammer, anvil, and stirrup—bridge this cavity and connect the tympanic membrane to the *cochlea.* The cochlea is a coiled, bony structure partitioned by membranes into three fluid-filled canals, the sense canals. The upper and lower canals are connected at the far end of the cochlea, but the central canal remains separate. The cochlea, together with the *semicircular canals,* whose function we will examine shortly, constitute the *inner ear.*

Sound is transduced to nerve impulses in the central canal of the cochlea. The bones of the middle ear amplify the vibrations of the tympanic membrane and transfer them to a movable membrane in the cochlea known as the oval window. The oval window opens onto the upper canal of the cochlea. The sound waves follow this canal to the end of the cochlea, then enter the lower canal

and move along it to a second membrane, the round window, which bulges out and keeps the pressure inside the cochlea equal. The movement of fluid sets the *basilar membrane* of the central canal in motion. Hair cells attached to the basilar membrane scrape against the overhanging *tectorial membrane,* and the stimulation fires neurons at the base of the hair cells. These action potentials then follow the eighth cranial nerve to the brain, where they are processed and interpreted in the auditory portions of the cerebral cortex. The basilar membrane, the hair cells, and the tectorial membrane make up the *organ of Corti* (Figure 22–7).

Our ability to hear various tones and to distinguish one sound from another originates in the basilar membrane. The membrane narrows as it travels the length of the cochlea and therefore vibrates at different frequencies from one portion to the next. Since different frequencies stimulate different parts of the membrane—and thus different hair cells—many different patterns of action potentials can be produced in response to the sounds we hear. When these patterns are integrated in the brain, we perceive complex sounds, whether they be chords by Brahms or a cat's meow.

The higher the frequency of a sound, the higher its pitch. The highest pitch that can be heard varies from one mammalian species to another. The highest pitch most humans can detect is in the vicinity of 16 to 20 kilohertz (16,000 to 20,000 cycles per second), although some children can hear pitches as high as 40 kHz. Sensitivity to high pitches decreases with age, probably because the tympanic membrane becomes less elastic. Dogs readily hear sounds as high as 30 to 40 kHz; "noiseless" dog whistles are pitched in this range. Bats can hear still higher frequencies, as high as 100 kHz.

Sounds also vary in *intensity,* or what we commonly call loudness. When we turn the volume up or down on a radio or television set, what we are actually varying is the force with which the sound waves strike our ears. The more intense a sound, the more forceful the flow of fluid in the cochlear canals, and the more intense the stimulation of the hair cells. The softest sound the ear can detect—what might be called the hearing threshold—varies among species. It is because dogs, for example, can detect far fainter sounds than humans can that they have watched our flocks and homes for thousands of years.

By convention, sound intensity is measured on the decibel scale. A decibel is the least change in intensity that the human ear can detect. The scale is logarithmic, each 10-decibel increase signifying twice as intense a sound. Thus, a sound measured at 30 decibels is twice as loud as one measured at 20 decibels; a 40-decibel sound is four times as loud; and so on. Leaves rustle softly at 12 decibels; conversation proceeds normally at 45 to 50; noisy offices clatter on at about 65; a heavy truck startles you with 100 decibels; and a sonic boom rattles your teeth at 130. At 140 or above, the ear perceives less noise than pain; and if sound becomes intense enough, it can even kill, as Dorothy Sayers knew when she wrote *The Nine Tailors.*

It has long been known that prolonged exposure to loud sound or noise causes deafness, first evidenced by an inability to hear high-pitched sounds. This condition, commonly called "boilermaker's ear" because of its prevalence among workers in noisy foundries, is also found among jet airplane mechanics, trap and skeet shooters who do not protect their ears, and rock musicians who turn up their amplifiers full blast. Besides its specific effects on the ear, noise appears to disturb the entire functioning of the body. Various studies have implicated prolonged loud noise in reduced learning in children, ulcers, decreased sexual function, hypertension, weight loss, and possibly neurotic behavior and nervous breakdowns. Data of this sort have led the U.S. Environmental Protection Agency to treat excess noise as a pollutant.

Another health problem arises from the requirement that the middle ear have the same air pressure as the auditory canal. If the pressure varies from one side to the other, the tympanic membrane bulges and cannot function properly. To allow for pressure changes, the middle ear is connected to the pharynx by a duct called the **Eustachian tube.** That well-known sensation of popping ears when you change altitude, as in an airplane, a car on a mountain road, or an elevator in a tall building, is caused by an increase or decrease in air pressure. Swallowing allows air to move in or out of the middle ear through the Eustachian tube until the pressure in the middle ear equals the outside atmospheric pressure. Unfortunately, the Eustachian tube also gives disease-causing microbes easy access to the middle ear. The painful earaches common in childhood usually begin with infections of the throat, particularly the tonsils and adenoids.

The Ear and the Sense of Balance The inner ear functions to maintain balance as well as to receive sound. Two chambers connect the cochlea to the semicircular canals (see Figure 22–6). These chambers contain *otoliths* (literally, "earstones"), which are crystals of calcium carbonate resting on beds of sensory hairs. Gravity exerts an even downward pull on the otoliths, and a movement of the head moves them over the sensory hairs. The hairs bend. The bending triggers action potentials in associated neurons, and the cerebellum initiates muscle reflexes and reactions that keep the body upright and balanced. Since we can bend our heads without being thrown off balance by sudden muscular reflexes, it is apparent that visual input can override information from the otoliths. Typically, the senses of vision and balance work together. People with normal ears can keep their balance fairly

easily even when blindfolded, but those with damage to the otolithic organs can remain upright only as long as they can see, and cannot stand when blindfolded.

If you spin about in a tight circle for several seconds and then stop suddenly, you will feel as if you are still spinning. This demonstrates something of the working of the semicircular canals. The three canals lie at angles to one another, representing the different dimensions of our three-dimensional world. The canals contain clusters of sensory hairs at their lower, enlarged ends. When the head and body move, fluid in the canals is set in motion, and the movement stimulates the hair cells, which trigger neurons that send action potentials into the brain. The brain integrates the varying signals from the canals and constructs an overall picture of the body's motion.

The physics of the canals can be visualized by holding a sealed bottle half full of water at arm's length and then jerking your arm quickly. The water moves in the direction of the jerk, and that motion continues even after the arm stops. Likewise, after you spin in a circle, the fluid in your semicircular canals continues to flow and to indicate movement to the brain even after your body is still.

The semicircular canals serve largely to indicate the speed and direction of movement. When you watch a football player slide through a hole in the line, fake a linebacker inside and cut back out, then dodge through the secondary, you are witnessing the trained integration of three modalities of balance: vision, the otolithic sense of equilibrium, and the accelerational and directional sense of the semicircular canals.

Some of the invertebrates use organs of balance remarkably similar to ours. In crabs and some other crustaceans, an organ known as a *statolith* is found at the base of each antenna. The statoliths work much as otoliths do, with sand grains instead of calcium carbonate crystals. However, every time a crustacean molts, it must replace the sand grains.

The nineteenth-century Austrian zoologist Alois Kreidl designed an experiment to show the function of the statoliths. Putting molting shrimps into an aquarium that contained iron filings but no sand, Kreidl found that the shrimps renewed their statoliths with the iron filings. Kreidl then held a strong magnet above the aquarium, and the animals swam upside down. Magnetic attraction had taken the place of the pull of gravity and told the shrimps that "down" was in fact "up."

Insects have neither otoliths nor statoliths, yet they maintain their balance. In part they accomplish this task visually. However, another mechanism is more important for many species. In most insects, the head is attached flexibly to the thorax, and beds of hairy mechanoreceptors surround this joint. Gravity pulls the head down and bends the mechanoreceptors, stimulating the associated neurons. This stimulation triggers reflexes that orient the

Figure 22–8 Pinnae of a bat.

the navy's electronic system for echolocation of submarines.)

The best-known echolocation systems are found in the bats. These night-flying mammals feed on insects as small as fruit flies and mosquitoes, which they echolocate and catch on the wing. They also use sonar to avoid obstacles and can thread their way through an obstacle course of thin piano wires in a totally darkened room. Precisely how bats accomplish these feats was unknown until 1938, when the first clue appeared. Donald Griffin, then a graduate student at Harvard, carried a cage of bats into a physics laboratory equipped with instruments capable of picking up high-frequency sound and discovered that his seemingly silent bats were actually emitting sounds, but above the range of human hearing. Further research showed that flying bats emit a continuous sequence of high-frequency cries at the rate of about 10 per second. Some species release the cries through their mouths, but others use the nose, which aims the sound in a particular direction much the way talking through a tube casts the voice. Bats also have large maneuverable pinnae (Figure 22–8) and inner-ear mechanisms sensitive to high-frequency noises.

The success of echolocation as a hunting technique depends not only on the bat's sound system but also on its intricate and quick reflexes. A bat can detect and intercept a moth 3 to 5 yards away and catch the much smaller fruit flies and mosquitoes at about one-third of that distance. At such short distances, 0.0037 second (4 milliseconds) elapses between the emission of a sound and the reception of an echo. The bat can locate the insect by differences between the intensity of sound received from the echo in one ear and that in the other; but intercepting an insect means more than knowing where it is—it also means predicting where it is going to be. The bat accomplishes this readily, evidencing highly specialized neural circuitry.

The porpoises, dolphins, and whales use echolocation to avoid collisions with underwater objects and the ocean floor in their deep dives and to find prey in dark or poorly lit water. This sonar is far more sophisticated than the electronic gear used by the navy to look for submarines and by fishermen to find schools of fish. A blindfolded porpoise confined to a pool can immediately locate fish tossed into the water and can distinguish the fish from inedible imitations of the same size and weight.

Chemoreceptors

The chemoreceptors are sensors designed to recognize specific chemicals in the environment. By convention, we use the word *smell* to refer to chemoreception of gaseous substances and *taste* for those dissolved in liquid. In fact,

animal correctly to up and down. The exact reflex depends on which hairs are stimulated. For example, if the insect drifts to the right, one reflex draws it back; if it drifts left, the opposite reflex occurs. If a dragonfly's head is glued so that it is held stationary against the force of gravity, and if the insect is put in a dark room, where there can be no dorsal light reaction (another balancing mechanism), the dragonfly cannot tell up from down and will spend as much time flying upside down as right side up.

Sonar: "Seeing" with Sound   Most sensory receptors are designed to pick up signals originating in the environment. But some environments, such as the dark of night or the depths of the ocean, produce too few stimuli to guide a moving organism. Animals that inhabit such featureless places have evolved ways of making the environment give off intelligible signals.

Certain animals are equipped with sophisticated systems for emitting specialized sounds and processing the information conveyed by the echoes. Such a sensory system is called *echolocation,* or *animal sonar.* (The word *sonar* is a military acronym for *so*und *na*vigation *r*anging,

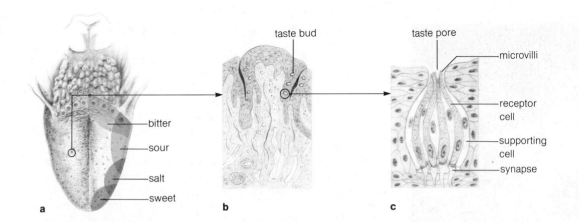

taste bud

taste pore

microvilli

receptor cell

supporting cell

synapse

bitter

sour

salt

sweet

a

b

c

Figure 22–9 Taste receptors. (**a**) Upper surface of the tongue, showing areas of maximum sensibility to the four primary taste sensations. (**b**) Enlarged region containing taste buds. (**c**) Longitudinal section of taste bud.

the two senses are often intertwined anatomically, and they generally work together.

In vertebrates, taste receptors occur most commonly on the tongue. The human tongue contains collections of sensory cells and associated nerve fibers in depressions called *taste buds.* Taste buds throughout the tongue are morphologically similar, but different regions are especially sensitive to particular tastes. Taste buds near the front of the tongue are receptors for "sweet." "Sour" and "salty" are perceived at the sides and "bitter" at the back (Figure 22–9). Few substances have a pure taste, and the perceived taste of a particular item is usually a blend of these four different sensory messages.

In addition, much of what we think of as taste is, in truth, smell. As food is chewed, volatile chemicals enter the nose and stimulate the olfactory receptors located in the olfactory epithelium in the rear nasal passages. This is why you cannot taste much when you have a cold: Excess mucus plugs the nose and blocks the olfactory receptors. Hot food seems more flavorable than lukewarm food because the heat volatilizes more molecules to be sensed in the nose.

Why do certain chemicals taste the way they do? There is no clear answer to this question. In general, "sweet" is a pleasant taste that indicates high-energy organic compounds such as carbohydrates, some alcohols, glycols, and amino acids.

The pure taste of "salt" comes only from sodium chloride (NaCl), common table salt, but other salts, such as potassium chloride (KCl), sodium fluoride (NaF), and calcium chloride (CaCl$_2$) all taste somewhat salty. People with known heart disease often have to eliminate sodium from their diet, and they use salt substitutes like KCl, which has almost the right taste and none of the culprit sodium.

"Sour" appears to be a function of hydrogen ion concentration, so that most acidic foods, like citrus fruits, taste sour.

"Bitter" comes from poisonous plant alkaloids like

nicotine and from many compounds with nitrate groups and sulfide bonds. Rejecting bitter food may once have had survival value for our species and, if so, it is no accident that coffee and other bitter-tasting beverages are usually acquired tastes.

In order to register taste, the sensory cells must be able to distinguish particular chemicals. But how do they do this? The tongue cannot classify substances as a chemist does, since completely unrelated compounds often taste much the same. And, as if to confuse the issue completely, the compound phenylcarbamide tastes sweet to some people, bitter to others, salty or sour to still others— and is tasteless to the rest. The question becomes all the more complex when one considers smell. Trained odor judges are said to be able to detect as many as 15,000 different odors. Of all the theories proposed to account for the mechanism of chemical discrimination, the most widely accepted is the one proposed by John E. Amoore of the U.S. Department of Agriculture. In his model, the key to smell is the shape of the molecule. Amoore examined hundreds of molecules that humans can smell and found that molecules with similar smells have similar shapes. Amoore found seven "primary odors." Presumably the membranes of the olfactory receptors contain binding sites, each specific for one of the seven basic shapes. When a molecule binds with the receptor site, the cell depolarizes, and an action potential is generated. The brain interprets the incoming action potentials and creates a single odor from a blend of the seven primary odors.

Mammals have a highly developed sense of smell, by and large considered an evolutionary adaptation to nocturnal activity. Humans have some 10 to 20 million olfactory nerve endings covering the 5 square centimeters of olfactory epithelium. Dogs, which have a much more acute sense of smell than humans, have an estimated 10 times as many nerve endings and a larger olfactory epithelium. To increase their sensitivity to odors, most wild animals sniff continuously, swirling new samples of air

Figure 22–10 Antennae of a male Polyphemus silkworm moth. A close-up of the antenna shows the sensory hairs.

over the olfactory sensors. Mammals use scent for a wide variety of tasks: telling friend from foe, locating territorial boundaries, finding food, distinguishing the sexes, and sensing nearby predators.

As far as we know, birds—with the notable exception of vultures and ravens—rely little on smell. Most odor-bearing molecules are heavy and hang close to the ground, where they are inaccessible to an aerial creature. Reptiles, the immediate evolutionary precursors of the birds, have a well-developed sense of smell. When a snake or lizard flicks its tongue, it is actually smelling its surroundings. The roof of the reptilian mouth contains a pair of small, richly innervated pits called *Jacobson's organs*. The tongue darts out, picks up airborne molecules, and brings them into Jacobson's organs for analysis. Snakes swallow their

prey whole and have no sense of taste as we think of it.

In the water, the distinction between taste and smell blurs. Thus, for fish, "taste" refers to chemoreception of matter the fish touches, whereas "smell" applies to chemoreception over a distance. Some fish have a very acute sense of smell. Apparently smell guides salmon back to their birthplace to spawn, since fish with plugged nostrils are unable to find their way.

Many insects are sensitive to smell, particularly to scents that act as pheromones (chemical messengers acting between individuals of a species). The olfactory sensors are located in the antennae (Figure 22–10). The olfactory abilities of some species are highly developed. Dietrich Schneider of Germany found that the male silk moth *Bombyx mori* can detect the female sex pheromone, bombykol, when only about 40 of his 40,000 olfactory receptors are stimulated. Males can locate an unmated female that is "calling" by releasing pheromone over a distance of a mile or more, provided that she is upwind.

Insects also have a good sense of "taste," but the development of this sense has followed a far different evolutionary path in insects than in vertebrates. For example, a fly's taste receptors are hairs on its feet. This observation was first made by D. E. Minnich at the University of Minnesota, when he noted that flies and butterflies extend their mouthparts to feed when a sugar solution is applied to their feet.

Light Receptors and the Sense of Vision

A great many organisms, from the one-celled algae to the higher plants and most of the animal kingdom, perceive and respond to light. Although the responses are diverse, every known case of light reception depends on the pigment carotene or one of its derivatives. Plants alone produce carotene, and animals require it in minute amounts in the diet as vitamin A. A single molecule of carotene splits to produce two identical molecules of vitamin A. For use in the animal visual pigments, the vitamin A is then converted to an aldehyde, retinene or retinaldehyde, and combined with one of the proteins known collectively as opsins.

Carotene's usefulness to the perception of light lies in the molecule's two different possible geometric conformations, or isomers. Normally in the so-called *trans* form, the molecule's shape changes to the *cis* form when struck by a photon of light. It then quickly reverts to the original *trans* form. This isomeric reaction in the presence of light is the key to the perception of, and the reaction to, light.

The specialized light receptors we call eyes in animals may have begun their evolution as concentrations of carotenoid pigments. The flagellate *Euglena*, which swims

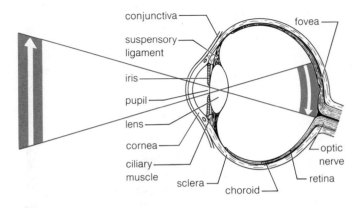

Figure 22–11 Diagrammatic cross section of the human eye, showing an inverted image projected onto the retina.

conjunctiva
fovea
suspensory ligament
iris
pupil
lens
cornea
ciliary muscle
sclera
choroid
optic nerve
retina

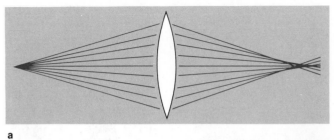

a

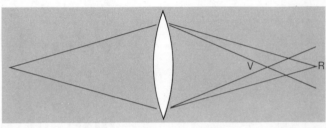

b

Figure 22–12 Spherical and chromatic aberrations. (**a**) Spherical aberration occurs when light is refracted by a lens with spherical surfaces. The light that passes through the edge of the lens is brought to a shorter focus than that which passes through the center. The result of this is that the image of a point is not a point but a "blur circle." (**b**) Chromatic aberration occurs when light of various colors is refracted by a lens made of one material. The light of shorter wavelength is refracted more than that of longer wavelength; that is, violet (V) is brought to a shorter focus than red (R). The image of a white point is a colored blurry circle.

toward a light source, perceives that light with a spot of carotene near the base of its flagellum. Many lower animals have similar eyespots capable of telling light from dark. In some cases, a thickened region of transparent epidermis over the pigment concentrates light and sharpens sensitivity. From these beginnings have evolved two basically different types of image-forming organs: the camera-eye of the vertebrates and higher mollusks, and the compound eyes of the arthropods.

The Vertebrate Eye The eye of the vertebrates and of such highly evolved mollusks as the squids and octopi is called the camera-eye because these eyes and the camera share the same basic optical mechanism: A single lens focuses a precise image onto a light-sensitive surface (Figure 22–11).

The human eyeball is more or less spherical and surrounded by a tough, elastic shell of connective tissue, the *sclera*. Light enters through the **cornea,** a transparent portion of the sclera, which forms the outside wall of a chamber containing the liquid *aqueous humor.* Beyond the aqueous humor lies the *iris,* the region that gives the eye its characteristic color. The iris is actually an extension of the *choroid,* a darkly pigmented tissue that lines the whole of the sclera except at the cornea and that absorbs scattered and reflected light inside the eye. Smooth muscle fibers in the iris control the size of the central opening, called the *pupil.* The pupil admits light to the crystalline *lens.* Now the light enters the second half of the eye, a large chamber containing the transparent fluid called **vitreous humor.** At the back of the eye lies the **retina,** which contains the sensory receptors. The nerve fibers associated with the retina leave the eye through the **optic nerve** and make their way to the visual portion of the cerebral cortex. Just as in a camera, the image focused on the retina in inverted; the brain reverses the image, so that we perceive it right side up.

The gross optics of the human eye was first worked out by Johannes Kepler, the famed German astronomer and physicist, in 1611 and by the French mathematician and philosopher René Descartes in 1664. Any object under view reflects light rays that scatter in all directions. To create an image, some of these rays have to be made to reconverge on a single point. If light is admitted through only a very small opening, then most of the reflected rays are eliminated, and reconvergence is little problem. This is precisely the principle employed in a "pinhole" camera, which is simply a light-tight shoebox with a piece of film at one end and a tiny hole punched in the other. However, such cameras are slow; even in bright daylight, it takes minutes to expose the film. The wider the opening, the more light is admitted and the less time is needed for exposure, but reconvergence of the rays, or focusing, becomes all the more crucial to prevent a fuzzy image or aberration (Figure 22–12).

A difficulty encountered in focusing light with its many wavelengths is that any given material, like a lens, does not bend every wavelength equally. The shorter wavelengths are bent (refracted) more than the longer

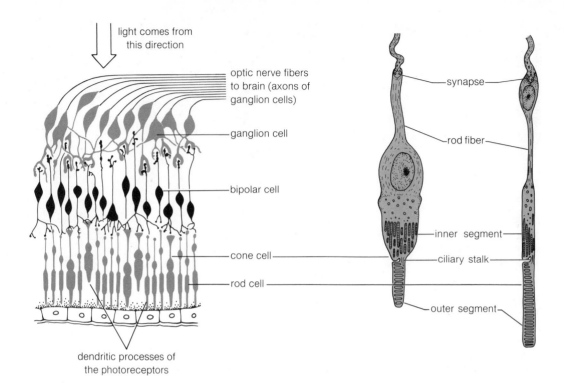

light comes from
this direction

optic nerve fibers
to brain (axons of
ganglion cells)

ganglion cell

bipolar cell

cone cell

rod cell

synapse

rod fiber

inner segment

ciliary stalk

outer segment

Figure 22–13 Nerve cells of the retina.

dendritic processes of
the photoreceptors

ones. The result is perceived as a distortion of the image (Figure 22–12). The human eye partially solves this problem with yellow pigments in the lens that filter out the shorter wavelengths.

In the human eye, the cornea and the lens focus the numerous rays of reflected light from a single point in the environment to a single point on the retina. Rays of light from a distant object are nearly parallel, and what little focusing they require is accomplished primarily by the cornea. Light from nearby objects requires additional refraction, a task accomplished by the lens.

The elastic lens is held in place by the *suspensory ligament* attached to the *ciliary muscles.* The key to the lens's ability to focus light is its changing shape: The more convex the lens, the more it refracts light. Tension on the suspensory ligament flattens the lens, and when that tension is relaxed, the lens again becomes convex. When the eye looks at something, reflexes adjust the tension of the smooth ciliary muscles so that the lens refracts the light just enough to produce a sharp image on the retina. This reflexive adjustment is called *accommodation.* Accommodation for close-up vision involves contraction of the ciliary muscles, while distance vision requires relatively little refraction and leaves the ciliary muscles more or less relaxed. This is why close work like reading a biology text for hours on end leaves your eyes feeling strained. The ciliary muscles have been contracted for a prolonged period and have grown fatigued.

A number of common structural defects in the eye result in improper focusing. A short eyeball places the focused image at a point beyond the retina. As a result, the ciliary muscles must contract even to focus on a distant object, and the eye may be unable to focus sharply on nearby points. Such an eye is said to be farsighted (hyperopic). Nearsightedness (myopia) involves the opposite defect, an overlong eyeball, and the resulting inability to see clearly at a distance. Astigmatism is an unequal curvature of the cornea that results in a distorted image. All these conditions can be corrected by artificial lenses that compensate for the defect and allow the eye to focus properly.

The eye has to control not only the degree of refraction but also the amount of entering light. The eye need admit only a small portion of the light available at midday in order to see, but in the poor light of evening, much more is required. The amount of light is controlled by reflex. Intense light causes the circularly arranged muscles of the iris to contract and thus decrease the size of the pupil. In low light, other muscles contract and open the pupil.

The retina contains two kinds of sensory cells (Figure 22–13). The longer and narrower receptors are called **rods,** and the thicker and shorter cells are known as **cones.** Rods outnumber cones by 125 million to 6 million, and they are two orders of magnitude more sensitive to light than the cones. The cones are sensitive only to bright light, but they are responsible for sharpness and for color.

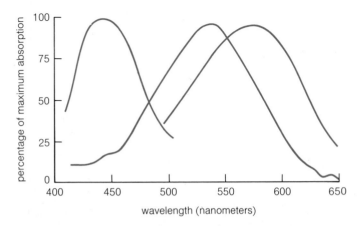

Figure 22–14 Spectral sensitivity curves of the three receptor types responsible for color vision in primates. The blue-violet receptor has peak sensitivity at 447 nanometers, the green receptor peaks at 540, and the yellow receptors have peak sensitivity at 577 nanometers. Sensitivities are widely overlapping. For example, the yellow sensors also absorb in the green region of the spectrum.

The cones are concentrated in a region near the center of the retina called the *fovea* (see Figure 22–11).

The light reactions of the rods and cones were first mapped by George Wald of Harvard University, work that earned him a Nobel Prize in 1967. Both rods and cones contain retinene combined with an opsin to form the highly unstable **rhodopsin,** or visual purple, which is the light-absorbing pigment. When light strikes rhodopsin, the *cis-trans* isomeric reaction of the carotene portion of the pigment triggers, in some unknown way, a change in the sodium permeability of the membranes of the associated optic nerve fibers. The rhodopsin in the rods, which are responsible for black-and-white vision, absorbs light maximally at about 500 nanometers, in the blue-green portion of the spectrum (see Figure 22–14). Whether a given rod initiates an action potential depends on the intensity of the light stimulus at this particular wavelength.

But how does the eye distinguish colors? In 1802, making one of those remarkable prophecies that dot the history of science, the English physicist Thomas Young proposed that color vision involves three different types of receptors. Young's hypothesis has since been confirmed by experimental work on the cone cells of monkeys, humans, and goldfish, as well as on the visual cells in the eyes of many insects. Each of the three kinds of cone has a range of greatest sensitivity: one in blue, one in green, and one in red (Figure 22–14). The differing sensitivities are due to variations in the opsin portions of the rhodopsin molecules. The perception of color in a viewed object results from the combination of different cones stimulated, so that the resulting image is something like a full-color picture made up of varying proportions of blue, green, and red dots.

Although the physics and chemistry of vision are relatively uniform throughout the various vertebrate classes, many adaptations do fit particular animals to their modes of life. For example, most reptiles, fish, and birds see colors; such perception is important in their mating and food-gathering. Mammals, with the exception of the primates, generally do not—most likely, as noted earlier, in part because they are largely nocturnal. The opsin part of the rhodopsin molecules varies from one vertebrate group to another as well. In the marine fishes, the visual pigments absorb best at precisely the wavelengths that penetrate water to the greatest depth. Many birds active only during the day (diurnal) have retinas that contain almost entirely cones, giving them extremely acute daytime sight but poor vision in dim light. Nocturnal birds, like many of the owls, have essentially all-rod eyes that give them particularly good vision at night.

The Camera Eye of the Mollusks Cephalopods, the most highly developed mollusks, have eyes similar to those in vertebrates, although the molluscan eye and the vertebrate eye have evolved independently. The eye of *Nautilus* forms images, but it lacks a lens and works much like a pinhole camera (Figure 22–15a). The cuttlefish *(Sepia),* the squids, and the octopi have eyes with true lenses (Figure 22–15b). The lens is, however, fixed in place, and accommodation occurs by changes in the shape of the eyeball rather than of the lens. Fish focus their eyes by moving the lens back and forth, not by changing its shape.

We have made the traditional comparison between cameras and eyes, but the analogy should not be overworked. The camera "sees" everything in front of it with equal emphasis. The eye, by contrast, has a limited area of great sensitivity, the fovea (see Figure 22–11). Furthermore, this varying sensitivity is amplified by the selective processing of visual information in the retina. Because of complex neural connections between the retina and the brain, the visual stimuli are sorted, with the irrelevant set aside and the relevant magnified. The nervous portion of the retina has been called a bit of the brain moved forward to run a preliminary analysis on visual signals before passing them on to the visual centers of the brain for more sophisticated processing.

To understand the nervous functions of the retina, J. Y. Lettvin and his co-workers at the Massachusetts Institute of Technology have been studying the optic nerves of the frog. They have found at least four different kinds of fibers. One sort, called the "sustained-contrast detectors," is stimulated only by the edges of the image. Presumably they give the frog an outline of his environment. A second type, the "moving-edge detectors," also

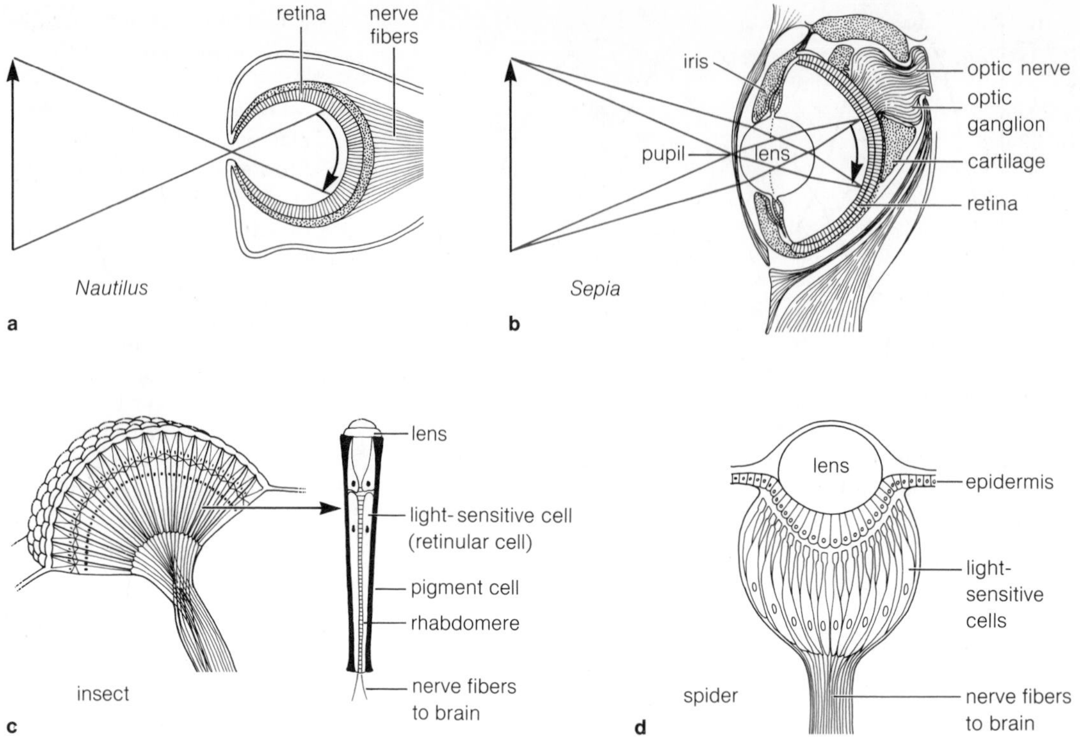

Figure 22–15 (**a**) The eye of *Nautilus* lacks a lens and acts as a pinhole camera. (**b**) The eye of the squid *Sepia* is like that of the human, built on the same principle as the camera. The image is formed by a cornea and a lens, and the lens can be changed to focus at different distances (a camera is focused by changing the distance between the lens and the film). (**c**) The compound eye of an insect is composed of ommatidia. Each ommatidium provides a point-source of light, and collectively a mosaic image is produced. The rhabdom is a group of fused rhabdomeres (retinular cells) that contains the visual pigment. (**d**) The simple eye of a spider.

responds to edges, but only those in motion. The third kind, "the net-dimming detectors," respond to suddenly reduced illumination in a large portion of the visual field. Most likely, these warn the frog that a large and overshadowing predator is approaching. The last type, "the net-convexity detectors," respond only to small dark objects. These detect the small dark objects of greatest interest to the frog: the insects it feeds on.

An analogous nervous arrangement has been found in certain insects. Their optic lobes contain different classes of neurons, each of which responds preferentially to stimuli from specific directions.

It seems probable that humans, who attend to many more aspects of their environment than do insects and frogs, take in more visual information and filter out less. However, the complexity of the several layers of nerve tissue in the retina and of the connections between the retina and the brain suggest strongly that much processing occurs. Little is yet known of its mechanics.

The Compound Eyes of the Arthropods　Many arthropods—insects and crustaceans—have eyes very different from the cephalopods and the vertebrates. The arthropod eye is called a **compound eye** because it is made up of many separate optical units, the *ommatidia.* Each omma-

tidium samples incoming light only from a restricted angle; its sensory cells are stimulated when the light exceeds a certain threshold. A dragonfly has up to 28,000 ommatidia per eye, each registering the light coming from a limited portion of the visual field.

The structure of the ommatidium is shown in Figure 22–15c. Essentially, the ommatidium is a small cylinder that contains from six to eight light-sensitive cells, the *retinular cells,* surrounded by pigment cells. The pigment serves as an insulator, keeping the light entering one ommatidium from stimulating the next. The top of the ommatidium is covered by a transparent cuticle (the cornea) and a crystalline cone, which focuses light onto the retinular cells. A specialized projection, the *rhabdomere,* extends along the length of each retinular cell and contains the visual pigment. All the rhabdomeres in the ommatidium lie in the center of the cylinder, where the crystalline cone directs the light. Stimulation of the rhabdomeres triggers action potentials in the axon at the base of the ommatidium, and these impulses make their way to the optic lobes of the brain.

Obviously, the mechanism of image-making in the compound eye differs much from that of the camera eye. According to the *mosaic theory* of insect vision proposed by Johannes Müller in 1829, each ommatidium acts as a unit, sensitive only to light entering along a small angle to its axis. Vision consists of an erect (right side up) image created by the juxtaposition of all the many small points of varying intensity and color. The picture that emerges is something like a newspaper photograph, which is a pattern of tiny individual dots perceived as one image.

But insect vision may be more complex than Müller's theory allows. For example, electrical recordings from the optic neurons indicate that insects have more resolving power than would be predicted from the visual angles of the ommatidia alone. This increased resolution must come from the neural processing of the information, but as yet no one knows how.

As a group, insects have much less visual acuity than humans. For example, a locust cannot distinguish two objects separated by an angle of less than 1°, but humans can distinguish two objects at less than 0.011°, approximately 60 times closer together. However, what many insects lack in acuity, they make up in speed. The ability to detect rapid movement depends on the speed with which the sensory receptors recover after stimulation. If a stimulus triggers a receptor and then a second stimulus follows the first before the receptor has chemically recovered, then the second stimulus is perceived not as a separate entity but as a continuation of the first stimulus. Fluorescent lights actually turn on and off about 60 times a second, but the human eye perceives the light as continuous, because it cannot recover fast enough to distinguish more than 40 repetitive stimuli a second. The speed at which repeated stimuli merge into a continuous stream is called the *flicker-fusion frequency.* Some of the fast-flying insects have flicker-fusion frequencies in the vicinity of 300 per second. Not only do they see fluorescent lights as a fast-moving strobe, but a movie, which to us looks like a continuous stream of movement, would look like a jerky, high-speed slide show to a theaterful of dragonflies. Because of their "fast" eyes, flying insects see objects in their moving environment as discrete objects rather than as indistinct blurs.

Insects perceive color. They can see in the ultraviolet region of the spectrum, which is invisible to us, but most of them do not see in the longer wavelengths. Except for certain butterflies known to respond to red colors in the mate's wings, insects perceive a room illuminated with red light as total darkness.

Simple Eyes   Besides their compound eyes, many insects have three *dorsal ocelli,* or *simple eyes,* on the front of the head. These eyes are little more than light-sensitive spots covered by a transparent cuticle, and individually they form no image (Figure 22–15d). Spiders, which lack compound eyes, may have as many as eight simple eyes. Although one eye gives only one bit of information, several light-sensitive spots (eyes) together can produce a crude "image" of the light environment.

Moth and butterfly larvae lack the compound eyes of the adult. Each side of the larval head capsule contains 6 relatively widely separated simple eyes called *stemmata.* Each of the 12 stemmata provides one point of light, and the resulting mosaic image is very coarse. The larvae may be seen sampling their visual environment by moving their heads slowly from side to side. Behavioral experiments have shown that, although the individual stemmatum cannot produce an image, the larva perceives form in a rough way and sees color.

The "Minority" Senses

The traditional list of five senses contains the sensory systems most important to humans and, in general, to other vertebrates. But other, less common, sensory receptors are significant for certain animals.

Temperature Perception: Seeing with Heat   Objects emit electromagnetic radiation in the infrared portion of the spectrum in proportion to their temperatures. As far as we know, no animal can see infrared waves, probably because of chemical limitations of the carotenoid pigments. However, when infrared radiation falls on an object, it agitates the molecules of that object. This agitation can be perceived as heat, or warmth. Humans have heat receptors in the skin (Figure 22–2g), but rather crude ones. Normally we can sense a warm object only on contact, or an extremely hot one at very close range.

Some animals have a highly developed sense of temperature perception. Bedbugs and some other blood-sucking insects locate a warm-blooded host with heat sensors in their antennae. Ticks, which are not insects but relatives of the spiders, respond most vigorously to hosts with a body temperature of 101°F.

The most elaborate heat receptors are found in the pit vipers, the New World family of poisonous snakes that includes the cottonmouth, the copperhead, and the 30 species of rattlesnake. The word *pit* in the family name of this group comes from the pair of heat-sensing *pit organs,* one on each side of the head midway between the eye and the nostril. Each pit organ contains a membrane with a surface area of 3 or 4 square millimeters studded with nerve endings. By way of comparison, these membranes contain about 100,000 more nerve endings than an equal area of human skin. The pit organs are so sensitive that at a distance of 30 centimeters a blindfolded rattlesnake can distinguish readily between two objects whose temperatures vary by only 1°C. The same blindfolded rattlesnake will strike repeatedly and accurately at a warm lightbulb dragged across its cage. The two facial pits give the rattlesnake binocular heat perception, allowing it to determine the direction and distance of the warm object. Pit vipers feed largely on warm-blooded prey, and they often hunt in cool surroundings, such as at nighttime or in underground burrows, where the hunted rodents' higher body temperatures contrast with the environment and make them stand out like the proverbial sore thumb.

Electricity: Location and Armament The continual flow of electrical potentials in the nerves and muscles of a living animal create a small electric field. Some sharks sensitive to electricity hunt prey buried in the sand by searching for these fields. Other fish have evolved ways of using their own electrical fields for a sort of "electrolocation."

Like echolocation, electrolocation involves a system for emitting signals and a system for detecting them. The African fresh-water fish *Gymnarchus niloticus* has tail muscles that do not contract, but emit weak electrical discharges at the rate of 300 per second. The electrical sense organs consist of skin pores on the head. The pores open onto jelly-filled channels containing nerve cells capable of registering electrical discharges from the tail. Placing an object between the head and tail disrupts the normal electrical flow, and the sense organs perceive the change. From the type of disruption, the fish can determine the location and size of the object. According to studies by H. W. Lissman, the various electrical fishes of Africa and South America feed at night in murky water, finding their way and their prey with the electrical sense,

thus exploiting a food resource unavailable to other, more conventionally equipped fish.

Some fish have evolved electrical fields, not as a method of detection, but as a weapon for stunning prey and driving off enemies. The amount of electrical power needed for this function far exceeds that required for electrolocation. The large electric fresh-water eel *Electrophorus* can generate 600 volts at a current of 1 ampere; the torpedo fish *Torpedo occidentalis,* a salt-water ray, can knock a human senseless. These electric organs are either modified muscles or motor end plates (Chapter 21) connected both in series and in parallel, like the cells of an electric battery. They are discharged in bursts on command from the central nervous system. Electric organs occur in taxonomically unrelated fishes of both fresh and salt water.

Summary

Each animal's behavior is geared to specific stimuli. Although there are potentially innumerable stimuli in the environment, animals have evolved to be sensitive to only a narrow range of information—the information most relevant to their survival. Other information is excluded. Exclusion of information is accomplished first by way of the morphology and physiology of the "windows" to the environment—the sense receptors. Additional filtering of information occurs in the central nervous system.

23

Communication

The bright colors of flowers, birds, and insects, the buzzing, shrieking, chirping, and singing of insects, birds, frogs, and mammals, the dance of the prairie chicken, the breath-taking aerial acrobatics of the raven and the falcon—these are but a few of the reminders that communication is an important aspect of the living world.

Let us begin to explore communication in organisms by transporting ourselves temporarily to one particular place, a marsh in the northeastern United States. At first the sights, sounds, and smells, only some of which we may notice, make little sense. The meanings only become clear from a study of behavior and sensory physiology. We are for the most part visually and acoustically oriented animals like birds, and to us birds are the most conspicuous animals in the marsh. The male red-winged blackbird on the low bushes in front of us is puffing out its brilliant red shoulder patches. The Wilson's snipe is scarcely visible on the ground because its colors blend into the background, but in flight it is conspicuous because of its song and aerial display high above the marsh. Similarly, the bittern is scarcely visible, but its voice can be heard for a half-mile or more at nesting time. In the hardwoods at the edge of the marsh we may see the brilliant scarlet of the male tanager against the foliage, into which its mate blends and is scarcely visible. The barred owl is rarely seen day or night, but in the evening its voice carries for nearly a mile.

There are similar patterns among the insects. Dragonflies in open areas along the water are conspicuous because of black, white, and blue bands across their wings and bodies. Katydids are hidden in the long grass, but their presence is obvious because of their voluminous and strident chirping.

There are other stimuli about us besides color and sound. If we had a keen sense of smell we would detect the odors emitted by the cecropia moths and hundreds of other moths and other insects, which are silent and remain hidden. It is clear that many animals have evolved features that make them conspicuous, even though there

has also presumably been great selective pressure to remain hidden. The various stimuli, some of which are aesthetically pleasing to us, apparently convey specific messages. These include sexual advertisement at the time when both sexes are ready for mating. Communication among different species includes attracting symbiotic organisms and repelling or confusing predators. Involved are acoustical, chemical, and visual stimuli, each fitted for specific occasions and situations.

Sound Communication

Sound Signals among Insects As we saw in Chapter 22, male mosquitoes are attracted by the sound of the female in flight. This is the simplest sort of acoustical communication, since it involves no special apparatus. Many insects, however, have organs specialized for sound production. The most common method, called *stridulation,* entails rubbing one roughened portion of the body (a file) over another (a scraper) to produce one pulse of sound. File and scraper are generally on body parts that rub together during normal activity such as walking or flying. In the insect order Orthoptera, which includes the crickets, grasshoppers, and katydids, the highly developed stridulation mechanisms are often located on the wings, or on the hind legs and the abdomen. The mechanism has evolved to include a membrane attached to the file to amplify the sound. Figure 23–1 shows the stridulatory mechanisms of one species of grasshopper.

With the stridulatory mechanism, the insect can produce specific sounds or series of sounds carrying precise messages. Male crickets and katydids chirp *calling songs* to attract females (Figure 23–2). The silent female seeks out the singing male. The male also has songs for confronting an intruder in his territory. If the intruder is male, the cricket sings an *aggression song,* but the appearance of a female triggers the *courtship song* instead.

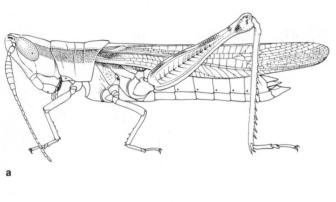

a

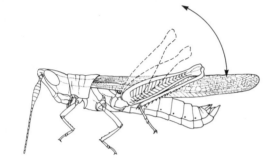

b

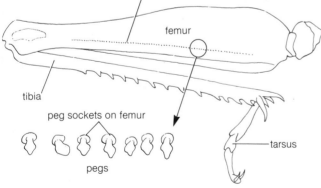

row of stridulatory pegs of scraper

femur

tibia

peg sockets on femur

pegs

tarsus

c

Figure 23–1 Stridulatory (sound-producing) apparatus of a male Acridid grasshopper (**a**). The grasshopper rubs his femur, bearing a row of pegs, along a prominent vein of the wing (**b**). The vibration of the wing produces the sound. (**c**) An enlarged view of the pegs, showing their position on the femur.

Each species—there are 2,500 cricket species and even more katydids—has its own songs. In any one environment, only members of a species respond to their specific songs.

Sound Signals and Birds We do not know whether birds find their own songs as beautiful as we do, but there can be no doubting their importance to the birds. There is evidence that communication by sound begins even before birds hatch out of the egg. Clutches of quail, turkey, and chicken eggs hatch at the same time, but this synchronous hatching occurs only if the eggs physically touch one another. Adrienne B. Orcutt of Cornell University has reported that quail embryos begin making different sounds before they hatch, perhaps somehow signaling the time to emerge from the egg. However it is achieved, this coordination of the clutch is obviously adaptive among precocial birds. Since both hatchlings and mother bird leave the nest shortly after the group hatch, any late-emerging young would find themselves orphaned.

What we think of as a bird's song is primarily the vocalization of the male in his breeding area. But this is actually only part of the avian sound vocabulary. Birds have "begging notes," which young birds or incubating adults use to get food from the parent or mate. The cry of baby mammals, including humans, induces solicitude from the parents. Flocking birds like quail and geese have "calling notes" that either assemble the group or keep its members in touch when they are out of sight of one another, as when flying in fog or foraging in dense cover. The "warning note" of birds sounds the alarm that a predator is lurking nearby. Mammals also have calls that serve as alarms and other signals.

The following example will show, however, that the meaning we should give to a particular animal sound is not always obvious. The true meaning to the animal can be assessed only through experimentation. Much recent attention has centered on the role of bird song. The male's song attracts females and may also warn other males of the boundaries of the territory. Clive K. Catchpole of the University of London investigated the precise roles of song in two European warblers (family Sylvidae, as opposed to New World warblers, Parulidae), the reed warbler *(Acrocephalus schoenobaenus)* and the sedge warbler *(A. scirpaceus).* Catchpole found that the male sedge warbler, which lives in a relatively open habitat, stopped singing once a female joined him. If the female left, the male began singing again. Catchpole concluded that the sedge warbler's song serves only to attract females. Since this species lives in an open area where individuals can see one another, the sight of the male warns other males that the territory is occupied. The reed warbler, however, lives in a dense habitat, and it continues to sing intermit-

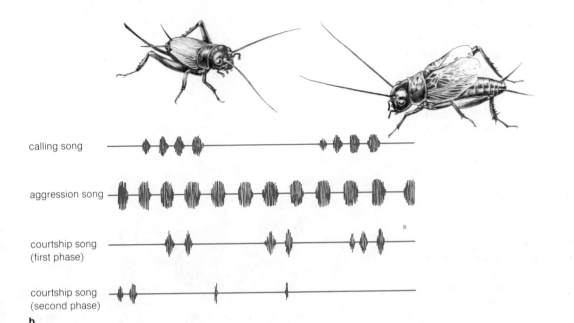

calling song

aggression song

courtship song
(first phase)

courtship song
(second phase)

b

a

Figure 23–2 (a) Posture of male
cricket while calling. Note the
raised wings (which are rubbed together to produce sound) of the
male. (b) Oscillograms of different
songs of *Gryllus campestris*.

tently even after it pairs with a female. Because the males
cannot see one another, their songs advertise their territories. This pattern of song and habitat seems to hold for
many species: Singing serves a territorial function in forest
birds; visual cues, particularly brilliant plumages, play
the same role in open-country species.

Birds respond not only to the quality or melody of
the song but also to its volume. William C. Dilger placed
a dummy male wood thrush and a loudspeaker inside
the territory of a male thrush. When Dilger broadcast
the song of the wood thrush through the loudspeaker
at low volume, the resident male attacked the dummy.
But when the volume was high, the resident male retreated. Dilger discovered a precise threshold of volume
below which the male attacked and above which he
retreated.

Sexual and territorial songs are intended only for the
ears of one's own species, and each species has its own
particular songs. Even closely related species inhabiting
the same range have readily distinguishable songs, which
prevents interbreeding. Calls other than the sexual and
territorial songs may be recognized by other species. Various species of finch often flock together, and each species
responds to the similar flocking calls of the others. Particularly fascinating are the two predator-warning calls of
the chaffinch, each of which elicits a different behavior
from nearby birds. If a hawk or owl is perched and not
immediately dangerous, the chaffinch gives out a "chink"
note. Other birds nearby look up attentively and locate
the predator. But if the hawk or owl is on the wing and

hunting, the chaffinch gives off a "seet" note that sends
the other birds scurrying for cover.

Mammalian Sound Signals and Human Speech
Mammals, like birds and insects, communicate with a
variety of sounds. However, we are often less sure of
the precise meaning of these sounds than we are of insect
or bird songs. Wolves sing in elaborate choruses that mark
the pack's territory but also apparently play a social role
in maintaining the group. Whales are a tremendously vocal group. The beluga, or white whale, of the Arctic produces such a wide variety of songlike sounds that sailors
call it the "canary of the sea."

All the acoustical communications we have investigated thus far code bits of information with specific
sounds. Typically, the ability to produce and interpret
the sounds is inherited and does not vary among the
individuals of a given population—one male robin sounds
the same as any other male robin. For the communication
to work, an animal need not know what information a
sound conveys; it need only respond correctly, even by
reflex. Humans certainly have unlearned vocal signals
common to all cultures, signals that indicate pain (cry),
sadness (sigh or groan), happiness (whoop for joy), and
humor (laugh). However, the unique human ability to
speak a language overshadows other forms of acoustical
communication.

We can speak because of our intellectual capacity
and because of our highly developed pharynx, the area

above the voice box (larynx) that produces the range of sounds needed for speech. Chimpanzees lack the neuro-muscular apparatus to produce varied sounds, and for a long time it was a moot question whether the chimpanzee brain had any capacity for language. The stumbling block to answering that question was the chimp's inability to talk. But in the late 1960s, Allan and Beatrice Gardner of the University of Nevada succeeded in teaching American Sign Language, used by the deaf, to a year-old female chimp named Washoe. In four years, Washoe's vocabulary numbered 160 words, which she used regularly in sign-language conversations:

Washoe: "Gimme, gimme."
Human: "What do you want?"
Washoe: "Sweet."

Washoe's ability went far beyond the instrumental conditioning by which dogs learn specific words associated with rewards or punishments. She was able to combine previously uncombined words to describe some new phenomenon or feeling. The first time Washoe saw a swan, she put together two signs to say "water bird." Washoe used "monkey" to refer to squirrel monkeys and gibbons, but she called a particularly nasty rhesus monkey "dirty monkey." Before that, she had used "dirty" only for feces and soiled objects.

Other researchers are using alternative ways of communicating with chimpanzees, such as plastic word symbols and symbol-coded computer consoles. No doubt remains that chimpanzees have enough language ability to construct sentences with complex grammar. This avenue of research promises continuing insights into the linguistic side of human nature.

Visual Communication

Of all the forms of animal communication, visual signals require the least energy. However, visual communication necessitates an unobstructed view, a requirement that may bring the signaling animal to the attention of a hungry predator as well as a willing mate. Thus, visual signals are often displayed only at restricted times and in safe places, and by animals equipped to escape predators rapidly.

Sight Cues and Signals   Although most nocturnal insects attract mates by sound or pheromone, a few use light signals produced by biochemical reactions. The lantern beetle *(Photinus),* colloquially called "firefly," gets its luminescence from a pair of light organs, or lanterns, on the back of two abdominal segments. Each lantern contains several hundred photocytes, cells that produce

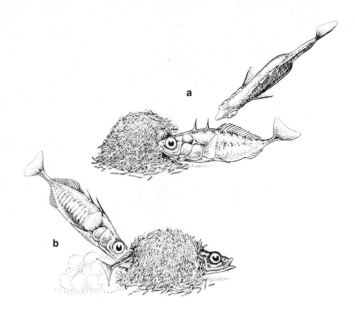

Figure 23–3   Some examples of visual and other signals used to coordinate a behavioral sequence. Movement: (**a**) male stickleback showing nest entrance to female. Touch: (**b**) female stickleback spawning in response to a male's "quivering."

light in an enzymatic reaction when stimulated by the central nervous system. The nervous control allows the beetle to produce specific flash codes by turning the lanterns on and off at different intervals.

A meadow on a summer evening may sparkle with flashes from several different species of lantern beetle. The males fly over the flightless females, emitting the sequence of flashes specific to their species. If the male passes over a female of the same species, she answers after a time delay of a particular number of milliseconds. The male recognizes the correct response and descends to the waiting female.

Among the vertebrates, fishes and birds in particular use visual mating cues. The breeding-season red of the male stickleback's belly releases aggression in other males. The female, on the other hand, is attracted to the red belly. She follows the male into the nest he has prepared and, responding to his tactile cues, deposits her eggs (Figure 23–3).

Although fish can change colors in minutes and thus communicate specific information about their behavior or mood, birds change color seasonally. As breeding time approaches and the secretion of reproductive hormones in the body increases sharply, many species molt into brilliant nuptial plumage that indicates their readiness to reproduce. This plumage advertises the male, function-

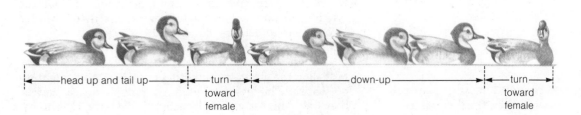

Figure 23–4 Gadwall duck courtship, showing the sequence of poses and movements. Similar poses are also part of the genetic heritage of other surface-feeding ducks (such as the mallard and the teal), but in the other species they occur in different sequences.

ing to drive off other males or to attract females, or both; and it may also help the bird to recognize its own kind. These functions overlap considerably, and their relative importance varies from one species to another.

Humans, too, respond to visual cues, particularly to facial expressions and body posture. In fact, some visual cues are so subtle that we are not conscious of them. In one experiment, male subjects were shown two pictures of the same woman and told to choose the more appealing. The only difference between the pictures was the size of the woman's pupils. The subjects chose the picture with the larger pupils, even though they were not conscious of the difference and could not tell why they preferred one picture to the other. Further tests showed that when subjects reacted favorably to a variety of different stimuli, their pupils enlarged. If the reaction was unfavorable, their pupils decreased in size. These data and others of the same sort indicate that humans send and receive visual information without being aware of the exchange.

Visual Displays Often visual signals are combined with behavioral cues called *displays.* Displays are most diverse and elaborate in the courtship of birds. The woodcock, a snipelike bird inhabiting woodlands of the eastern United States, puts on a dazzling aerial display every

spring. The male climbs into the sky until he is no more than a tiny dot. On the way up he emits a whistle that is produced by modified feathers in his wings. Then he dives toward earth, adding a vocal component to his song that increases in frenzy and pitch as he plummets. He continues spiraling and plummeting earthward for about a minute, then lands, wings fluttering loudly, within a few feet of the spot where he first took off.

Displays evolve gradually, arising from behaviors that originally had no role in communication. This modification of a behavior into a mode of communication is called *ritualization.* The male woodcock's display is a ritualization of flight behavior. Bathing movements have been exaggerated into courtship rituals in some ducks (Figure 23–4). The take-off leap, in which the bird bends its legs and spreads its tail feathers and wings, can be found in one ritualized version or another in the courtship displays of sandhill cranes and some herons and ducks.

Displays are intrinsic not only to courtship rituals but also to *agonistic encounters,* such as the confrontation between wolves shown in Figure 23–5. (Ethologists use the word *agonistic* in place of *hostile,* since the latter term is anthropomorphic and implies feelings we cannot prove exist in the animal.) Agonistic displays signal an animal's intentions and often prevent combat. A wolf pack is organized in two dominance hierarchies, one for

Figure 23-5 Displays signifying dominance (left) and submission (right) in a pair of wolves during an agonistic encounter.

males and one for females, and each wolf knows its place in the social order. When two wolves meet, the dominant wolf stares fixedly, raises hackles and tail, and walks on stiff legs. The subordinate wolf lowers head and tail, much as a dog does, and may roll over on its back to expose its belly, thus appeasing the dominant animal. The ritual prevents any actual fight between the two animals and is crucial to maintaining the pack. An outbreak of fighting among animals as powerful as wolves could kill or cripple many of them, reducing the pack's ability to hunt and probably leading to starvation for all. Thus, agonistic displays have a high adaptive value for social animals.

Chemical Communication

Chemicals in the form of scents are used as agents in communication among species. Equally important is the employment of chemicals to communicate within a species. These chemicals are the pheromones discussed in Chapter 14. Pheromones are chemicals released by one individual that act on other individuals of the same species. Chemical communication is largely the province of insects, but other animals, including mammals, use pheromones. Pheromones are classified in two groups. *Releaser pheromones* trigger an immediate behavioral response, while *primer pheromones* affect the endocrine system of the recipient and produce a delayed physiological effect. Usually releaser hormones are perceived by smell; some of the primers are passed by mouth.

In mammals, releaser pheromones attract mates and mark boundaries. When females of the dog family (Canidae) go into heat and are ready to mate, releaser pheromones in their urine and estrous flow attract males, sometimes over long distances. Male dogs mark their territorial boundaries with urine.

In insects, releaser pheromones are so numerous and varied that only a few can be mentioned here. As in the mammals, pheromones attract mates. Sex attractants have been most carefully studied in the nocturnal moths. The female's scent glands are modified intersegmental membranes that emit a highly volatile scent specific to the species. The scent drifts downwind. Males who perceive the scent with the olfactory sensors of their antennae fly upwind. If they lose the scent, they zigzag until they find it again, continuing upwind all the while (Figure 23–6). It is reported that some moths can locate a female 2 miles away. Scent has definite advantages as a sex attractant: It is specific, and it can be emitted without attracting predators. The major drawback is that it works in only one direction—downwind.

In recent years, scientists have synthesized many pheromones. These chemicals can help to control insect pests like the gypsy moth without resorting to dangerous pesticides. The problem with pheromones as a bait is that only the males are attracted. But Harry H. Shorey and his co-workers at the University of California at Riverside added to the ways of fighting with pheromones when they found that releasing large amounts of sex attractant in the air continuously stimulates the males. The abundant pheromone confuses the males by masking the female's scent; they continue to fly upwind right past the females actually emitting scent. In 1972, Shorey and his team used this method in the agricultural Coachella Valley of California to almost totally prevent the males of the fruit-destroying codling moths *(Laspeyresia pomonella)* from finding the females. Only isolated females were mated. The same methods have succeeded with other pests.

The greatest array of releaser pheromones is found in the social insects, among whom they coordinate the actions of the nest or hive (see Supplement 23–1). Ants and bees have several pheromone-producing glands (Figure 23–7). The effect of the secretions of each of these glands may vary with the context. Some pheromones induce the workers to attend to the brood. A pheromone released by dead ants causes living workers to carry the corpses out to the refuse heap for disposal. If a live ant is covered with this dead-ant pheromone, other workers will cart it away despite all its struggles, showing that the releaser for corpse-disposal behavior is the pheromone and not a lack of movement. Alarm pheromones alert the hive or nest to danger. These pheromones are extremely volatile chemicals that diffuse rapidly through the nest and dissipate soon after the threat is gone, lest

wind →

♀
calling

pheromone plume

flight path of ♂

a

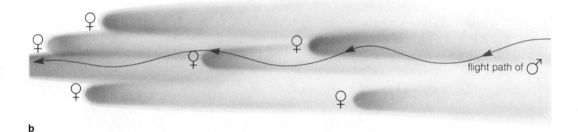

flight path of ♂

b

Figure 23–6 Pheromone plumes of a single female (**a**) and of many in the same area (**b**).

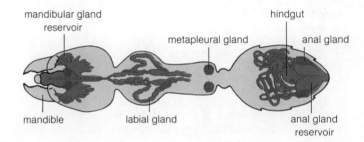

mandibular gland reservoir

metapleural gland

hindgut

anal gland

mandible

labial gland

anal gland reservoir

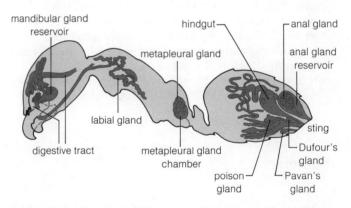

mandibular gland reservoir

hindgut

anal gland

metapleural gland

anal gland reservoir

labial gland

digestive tract

metapleural gland chamber

sting

Dufour's gland

poison gland

Pavan's gland

Figure 23–7 Some of the sources of pheromones used in chemical communication in ants. Some pheromones are stored in reservoirs and released as needed, while others are secreted continuously. The same gland does not produce the same functional pheromone in all species. For example, trail substances are produced in Dufour's gland, Pavan's gland, or poison glands. In some bees, chemicals acting as trail pheromones are produced by the mandibular glands. But in ants, the mandibular glands, as well as the anal glands in some species, produce alarm substances.

the community remain agitated longer than necessary.

Ants and some bees communicate the amount and source of food at a foraging site with trail pheromones laid down by the returning workers (Figure 23–8). The more desirable the food, the more pheromone they lay down. The colony's response depends on the amount of pheromone present; an entire colony can be evacuated if enough trail pheromone is deposited at the entrance of the nest. Like the alarm pheromones, trail pheromones dissipate quickly, usually in about 100 seconds, saving the insects from following old trails to an exhausted food source.

Primer pheromones, like releasers, are found in both mammals and insects. The scent of the male mouse affects reproduction in the female. For up to four days after coitus, the smell of a male other than the mate will cause a pregnant female to abort. When female mice are crowded close together, their estrous cycles cease, and many of the mice suffer pseudopregnancies. If the sense of smell is destroyed, normal estrous cycles resume in the still crowded females. In both cases, smell triggers changes in the hypothalamus, which affects the production of gonadotropin in the anterior pituitary and thus influences the ovaries and uterus. These pheromones act as population regulators. If mouse colonies are populated too densely, all pregnancies fail until mortality reduces the population to its proper balance.

Primer pheromones affect the development of the reproductive caste in social insects. The queen honey bee produces a pheromone that the workers lick off her body

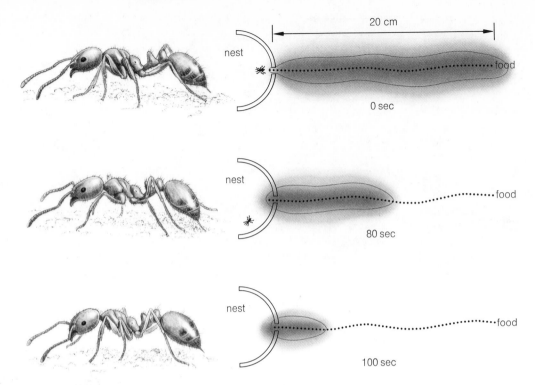

Figure 23–8 The fire ant *Solenopsis* lays down an odor trail from its extruded sting (top). The pheromone is present in levels above threshold concentrations in a semiellipsoidal active space as it diffuses from the points of application. The entire signal, and hence the trail, fades in about 100 seconds, but as long as many foraging ants returning along the trail have been successful, the signal will be maintained or amplified.

and feed to each other. If the pheromone—and thus, presumably, a queen—is lacking, the workers set about rearing new queens.

Termite societies comprise three castes—king and queen, soldiers, and workers—each of which produces primer pheromones. When each caste is at its proper strength, the level of primer pheromone produced by all the members combined prevents the development of any more individuals. If the population of one caste falls too low, the relative lack of that caste's hormones allows new individuals to develop to fill the depleted ranks.

Summary

In this chapter, we have seen some of the ways in which organisms communicate with each other. The range of such communication is great. The different modes of communication that have evolved are fitted for specific informational "messages" in specific environmental contexts. Features used for communication usually are evolved from behaviors and structures that originally had other functions. Communication is essential for the survival of the species.

Supplement 23–1
Honeybee Communication

Communication within most species is a mixture of visual, vocal, and chemical modes. To see how this mixture works in a single species, we will look more closely at communication in the honeybee *(Apis mellifera)*.

Food Sources and
the Dancing Bee

When Karl von Frisch was conducting color-discimination tests among bees, he noticed that shortly after a worker found a food dish and returned to the hive, many bees came to the dish to feed. Curious about how so many bees came to the location of the food so quickly, von Frisch designed experiments to test if they could communicate. In one test, a few bees were fed at one dish and then allowed to return to the hive. Several other feeding dishes were placed about the hive at the same distance as the first dish, but in different directions (the "fan" experiment). Counting the bees that came to each dish, von Frisch's observers found that the greatest number of bees came to the location of the first dish. In a second experiment, bees were again fed at a single food dish, and other food dishes were set out. This time, the dishes were all in the same direction from the hive as the first dish, but at different distances (the "step" experiment). Still, more bees arrived at the location of the initial dish. These experiments suggested that the forager bees could be conveying both the distance and the direction of the food source to recruits in the hive.

By what mechanisms might the bees be communicating direction and distance of food? To find out, von Frisch built a hive with glass sides through which he could watch the returning foragers. When the forager returned, other bees crowded around, and the forager fed them. Then it performed one of two dances. If the food source was close to the hive, the bee did the *circle dance,* running about rapidly in a tight little circle, first in one direction and then in the other. If the food source lay some distance away, the bee did the *waggle dance* (Figure 23–9)—running straight ahead, waggling its abdomen; then looping in a semicircle, repeating the waggling, and looping in a semicircle in the opposite direction. Von Frisch found that the greater the distance, the fewer runs per minute. When the food was 300 meters from the hive, the forager made 30 runs per minute; but at 600 meters, the number of runs dropped to 22. The waggle dance contained information about how far the food was from the hive and in what direction.

Then von Frisch noticed that, as the day passed, the straight part of the waggle dance changed direction. When the food dish lay in a direct line with the sun, the returned bees went straight up the comb during their waggle run. If the food was directly opposite the sun, the waggle run pointed straight down. If the food lay at 20° to the right of the sun, then the dance was oriented 20° to the right of vertical. The direction of the bee's dance reveals the direction of the food source relative to the sun. Up and down on the comb signify "toward" and "away from" the sun, respectively. Von Frisch concluded that the waggle dance is the mode by which the returning forager tells recruits in the hive where to find food.

The Bee Dance Language Controversy

As with almost every finding in biology, von Frisch's experiments and their interpretation were carefully studied, and attempts were made to duplicate the experiments and the results. This is, of course, a necessary part of the advancement of scientific knowledge, and through such attempts we both verify and extend our knowledge.

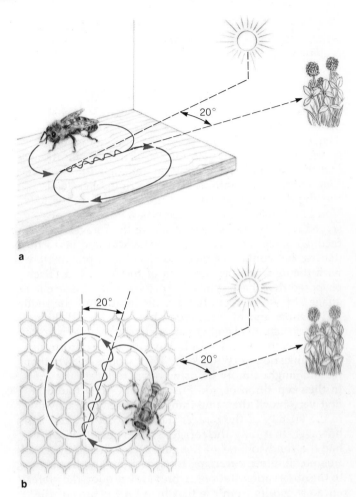

Figure 23–9 How the waggle dance of the honeybee indicates the direction to a food source. When dancing on a flat surface (**a**), the bee points directly toward the food source while rapidly vibrating her abdomen. After one straight run the bee turns around and repeats the behavior. In this case the food, and the direction of the straight run, are 20° to the right of the direction to the sun. However, the bees perform the dance primarily on the vertical combs (**b**), where they orient by the sense of gravity. A run straight up the comb signifies food directly toward the sun, and a run straight down indicates food directly away from the sun. In this case, the waggle dance is 20° to the right of vertical, indicating food 20° to the right of the direction of the sun.

It is also natural that controversy should arise in this process, and so it was with von Frisch's experiments. When they were repeated by other researchers, other interpretations of the data were put forth, and a controversy arose that only recently has appeared to be near resolution because of the work of James L. Gould of the Department of Biology, Princeton University. We asked Gould to write us of his experiments and results, and here are his remarks:

The controversy arose in 1967, when Adrian Wenner and Dennis Johnson (later joined by Patrick Wells) repeated Karl von Frisch's classic step and fan experiments. This style of experiment involves training a group of forager bees on sugar solution to a particular location, and then setting out an array of stations with a certain food odor. These stations attract recruited bees and thereby determine where they are searching in the field. Von Frisch had found that recruits were mainly observed at the stations nearest where the foragers were feeding. Since foragers return to the hive and perform dances that contain elements correlating approximately with the distance and direction of the food, von Frisch proposed that these symbolic cues serve as a "dance language" by which recruits learn the location of the food.

Wenner and his colleagues, however, found the notion that such small and phylogenetically remote creatures from humans as bees should have an abstract form of communication very unlikely. They wondered whether recruits might simply be using odors to locate the food. In their repetition of von Frisch's work, they found at first very much the same thing—most recruits arrived in the vicinity of the forager station. It occurred to them, however, that only the forager station and its vicinity had the "body odors" of foraging bees. When they tested this possibility by training foragers from another hive to the observation stations to provide bee odor, the preference of recruits from the first hive for the area of the forager station disappeared.

Michael Henerey, Michael MacLeod, and I put the bee odor hypothesis to the test in 1969 by training two groups of foragers from the same hive to two widely separated stations. We fed one group a concentrated sugar solution which caused them to dance vigorously in the hive. The other group was fed a solution so dilute that they did not dance. Recruits strongly favored the station being visited by the dancing foragers even when bee odor was present at both stations. Martin Lindauer obtained the same results using the same approach. Recruits were using either the dance information or some other cue for which we did not control (or both).

Meanwhile, Wenner and his colleagues discovered an uncontrolled cue that might have explained everyone's results: unique site odors. According to this hypothesis, foragers return to the hive bearing odors both of the food and of the locale of the forager station. Recruits, familiar with the "olfactory landscape" (the arrangement of odor landmarks in the area), determine the approximate location of the food from the locale odor carried back by the forager.

Unsurprisingly, both the dance language and odor hypotheses (including the more recent version) can predict the observed phenomena. Adherents of each are able to explain to their own satisfaction the results of the other camp and firmly disagree about which controls are important. To settle this controversy, one must cause foragers to "lie" about where the food is located. If recruits use the dance correlations, they would then be misdirected to specific locations well away from the forager station. If recruits use any other cues, they would not be deceived, and would go to the forager station, where all other cues would be present.

Recruits *can* be made to lie about either distance or direction. When, in 1974, Burkhard Schricker fed the insecticide Parathion to foragers, he found that their dances became abbreviated, as if the food they had found were located closer to the hive than it actually was. He found that recruits began to favor stations closer to the hive after the foragers were treated. Recruits must have been following the misleading distance information in the forager dance.

I was able to cause foragers to "lie" about the *direction* of the food. Dancers normally perform their dances on the vertical sheets of comb in the dark hive. The angle of the dance with respect to vertical (gravity) corresponds to the angle of the food with respect to the sun (a transposition common among arthropods). When a bright light is shone upon the dancers, they reorient their dances to this "artificial sun" and ignore gravity. By itself this phenomenon is not useful, since the recruits are similarly reoriented. However, when the foragers' ocelli (three simple eyes between the bees' compound eyes) are painted over, they forage normally but fail to respond to the light. Hence, *they* dance with respect to vertical, while *recruits* would be expected to interpret the dances with respect to the light. Recruits, *if* they use the direction correlation of the dance, should misinterpret the direction of the food by just the angle between the light and vertical in the hive. Of course, if the correlation is not used, there should be no effect on recruits: They should go to the area of the forager station.

In a series of 18 experiments in 1974, I found that recruits go reliably to the location specified by the dance rather than to the forager station. Whenever I altered the angle of the light, recruits began arriving in full strength at the newly specified location after a delay of about 6 minutes.

My control experiments showed that while von Frisch's techniques may have exaggerated recruit accuracy somewhat, the training techniques used by Wenner and his colleagues virtually abolished the dance communication.

Wenner's technique results in a substantial accumulation of the experimental odor in the hive before the experiments begin, while von Frisch's bees were never exposed to the odor until the experiment began. This result leads me to wonder whether recruitment might not be a two-stage process: Perhaps when a food source is new, recruits use the dance language, but as the source continues to be available and its odor builds up in the stores of nectar in the hive, recruits begin to use odor to locate the food source. If this is the case, then von Frisch and the Wenner group may have been studying two entirely different phases of a single process.

Although clearly wrong in their contention that honeybees have no dance language, Wenner and Wells have created much new interest in honeybees and have reminded us of the importance of odor information, so long neglected in favor of the more spectacular idea of a dance communication. The controversy has also been useful in forcing a reexamination of the relevant evidence and the adequacy of present techniques. As a result, I, for one, feel that the precise limits of our knowledge about honeybees are now much more clear. I hope that the controversy will encourage and focus new research on such crucial questions as how recruits receive, process, and then use the dance information in the field.

24

Behavior

A keen observer sitting quietly near the edge of a pond on a spring morning sees a variety of animals doing different things. A dragonfly nymph comes out of the water and crawls onto pickerel weed. A dragonfly emerges from the mud-covered armor of chitin, gracefully flies off, and patrols the shore, scooping small gnats out of the air. A bumblebee alights on the purple rhododendron flowers along the shore, visiting many thousands of the blossoms yet bypassing other kinds of flowers. A wasp catches a spider lurking under a flower, stings it, and carries it to a burrow it has dug in the sand. Another species of wasp hunts only caterpillars. A loon gliding on the water's surface silently submerges, swims a hundred meters under water, and catches a small perch. An osprey catches another fish in its talons, but by diving from high in the air. A heron walks slowly along the shore and spears a minnow, using its long neck and sharp beak. The animals are feeding, each species in its own specific way.

Sometimes the significance of behavior is not so obvious. For example, a male duck bobs up and down as if seized by strange paroxysms. A male Maryland yellowthroat warbler drives another male from the area, returning to sing loudly from a perch above the marsh grass. Its hidden mate flits with bits of grass to a tussock of sedge where she is shaping the grasses into a nest-cup. The kingbird is building a nest out of small sticks and lichen on a dry branch overhanging the water. The ancestors of these animals have behaved nearly the same way for thousands of years. Their offspring will continue to do so in the future.

A beaver slaps its tail on the water and dives, ending our reverie. We have enjoyed the spectacle and wonder why each organism behaves in one way and not another at any one instant.

Behavior is a sequence of nerve impulses and muscular contractions, and a large branch of the biological sciences concerns itself with trying to understand the underlying basis of the resulting movements. These various movements can be related to feeding, mating, and other aspects of survival. They are an integral part of the animals' equipment for survival that draws heavily on the sensory processes and on the central nervous system. An animal's behavior is often as characteristic of the species as external appearance—and sometimes more so. Like claws, teeth, and digestive processes, behavior has evolved to adapt animals to survive in their environments.

The Pitfall of Teleology

One problem all behaviorists face—and never completely solve—is how to describe an observed behavior accurately and objectively. Our language is human language, and it sees the world from the human point of view. Thus, the very words we use to describe a particular behavior may say things about it that are true of humans but not of the organism in question. For example, if you observe a chicken sitting on a clutch of eggs until they hatch, you may well report that the chicken sat on the eggs *in order to hatch out the young.* But if you do, you will be guilty of anthropomorphic (human-structured or human-biased) thinking. Specifically, you will have fallen into the trap of **teleology,** by assuming that the bird *purposefully* sits on the eggs to bring the young forth. The assumption is unwarranted. Obviously eggs hatch because the hen incubates them, and they will not hatch if she does not; but no one has proved that the chicken has this purpose in mind. If we want to stick to facts, then, we have to say that any "purpose" thus far shown for incubation comes not from the individual hen but from the course of evolution. Hens that do not incubate their eggs do not reproduce, and the nonincubator genes become less frequent in the population.

Careful behaviorists describe only what they see and try to avoid any statement implying insight into the mental state of an animal. However, since words with such implications are so abundant in our language, these statements inevitably appear.

The Two Schools of Behavioral Thought

Until quite recently, scientific students of behavior belonged to two major schools of thought, the one largely American and the other largely European, each with its own theoretical assumptions and research techniques.

The American school, an offshoot of psychology, is tied to **learning theory.** Learning theorists maintain that the brain at birth is a *tabula rasa,* or blank slate, and that behavior is a result of learning based on rewards and punishments. (We will examine this mechanism in more detail shortly.) Learning theorists work under controlled conditions in a laboratory, often using the white laboratory rat *Rattus norvegicus.* Monkeys, pigeons, and college students have also been used.

With considerable success, learning theorists have tended to extrapolate from rats to humans, assuming that what they find true for rats applies to us as well. Behavioral methods drawn from learning theory have been incorporated into teaching machines and are used in treating emotional and mental disturbances. The most prominent American learning theorist is B. F. Skinner, a psychologist at Harvard University. In addition to his professional scientific work, Skinner has written such popular books as *Walden Two* (1948) and *Beyond Freedom and Dignity* (1971), advocating learning theory as the cure for the world's ills and the only hope for its future.

Those who subscribe to **ethology** begin differently. Interested more in innate behaviors than in learned responses, they study animals under natural conditions in the field. Typically, they describe exhaustively what they see and try to determine the precise structure and meaning of the observed behaviors. Ethologists are intrigued primarily by the evolution of behavior and its adaptive significance. Since behavior leaves no fossils, the only way to learn how a behavior has evolved is to examine equivalent behaviors in closely related living species and try to deduce the original behavior and its probable line of evolution. A researcher interested in the evolution of mating behavior might select a closely related group such as the ducks, observe and record the courtship rituals of each species, deduce the likely evolution of the differing rituals, and assess how each ritual has helped that species adapt to its ecological role. Ethologists have studied a wide variety of animals—geese, ducks, hyenas, bees, fish, dogs, wolves, parrots, chimpanzees, and gulls, to name a few.

Ethology has sprung largely from the work of Niko Tinbergen, an Oxford biologist, and the two Austrians Konrad Lorenz and Karl von Frisch. These three men shared the first Nobel Prize ever given for work in behavioral science (1973).

The two schools of behavioral thought embody the so-called nature-nurture controversy: Is behavior inborn or is it learned? The many debates on this question have engendered far more heat than light. Research from both sides now shows that this controversy is generally a waste of time and that learning and innate capacity complement each other in behavioral development.

Simple Behaviors

Whatever their theoretical disagreements, ethologists and learning theorists agree on the nature of behavior: an action directed toward achieving some goal, or "target value." This does not mean that the animal makes a conscious choice and then pursues its end rationally. Instead, behavior fulfills some biological need, like finding a mate or a nesting site or avoiding a hungry predator. Behavior involves, first, a response to an external or internal stimulus related to the goal and, second, a way of monitoring progress, so that once the goal is achieved, the behavior stops.

In some cases, the relationship between a behavior and its goal is obvious, as when a fly alights on spilled maple syrup, feeds until it is full, and stops feeding (see Chapter 20). In other cases, the behavior is so complex and drawn out that the one major goal is probably broken down into shorter-term, or interim, goals. The classic example is a migratory bird like the Arctic tern that journeys thousands of miles between its winter and summer ranges. In this case, the single goal of migration is composed of a series of target values like finding food along the way.

As long as an organism is alive, it is simultaneously responding to many different target values with different "due dates." The migrating tern, for example, has the long-term goal of reaching its destination, the daily need of orienting to stimuli such as landmarks, and the present needs of adjusting to air currents and neighboring birds in its path. Obviously, animals continuously monitor their progress and adjust their goals and their behavior with regard to various internal and external stimuli.

Trying to understand the whole complex of behaviors that guide the daily activities of a mammal or a bird leads quickly to confusion. Therefore, we will begin by analyzing the mechanisms of behavior of simple organisms.

Tropisms and Taxes The behavior of a plant shoot in growing toward the light or away from the pull of gravity

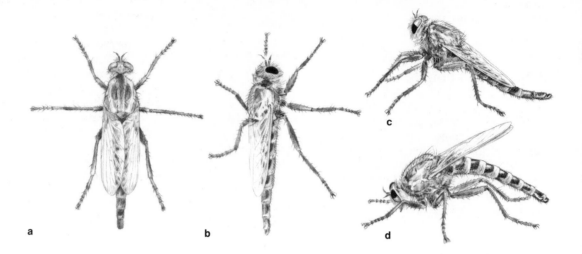

Figure 24–1 The role of positive phototaxis in maintaining normal resting posture (**a**) in the robber fly. (**b**) The fly's right eye has been painted so that it no longer receives light stimuli. The fly now twists to the left, "toward the light." (**c**) The ventral portions of both eyes have been blackened, and the animal compensates for the reduced ventral stimulation by rearing up. (**d**) The dorsal portions of the eyes have been blackened, and the fly now bends down, attempting to increase light stimulation by moving "away from" the reduced dorsal stimulation.

a

b

c

d

is called a *tropism* (from the Greek *tropos,* meaning "turn"), and a combining form is added to the word to indicate the factor affecting that particular growth. Thus, *phototropism* is growth influenced by light, and *geotropism* is growth influenced by gravity (Chapter 11).

At the turn of the century, the physiologist Jacques Loeb, a German who also worked in the United States, borrowed the word *tropism* from the botanists and applied it to reactions like the apparent heat seeking behavior of swimming protozoa. In fact, Loeb thought that the tropism was the key to behavior and that all behaviors, even the most complex, would one day be understood as nothing more than sequences of tropisms. Now, however, the word is again restricted to plants, and the term **taxis** (Greek, meaning "arrangement") is applied to animals' apparently innate directed movements toward or away from a stimulus. Among the possible stimuli are light, temperature, chemical properties, and gravity; the corresponding taxes are called phototaxis, thermotaxis, chemotaxis, and geotaxis. In addition, depending on whether the direction of movement is to or from the stimulus, a taxis can be described as positive or negative. Thus, the *Paramecium*'s movement toward light is a positive phototaxis. Figure 24–1 illustrates positive phototaxis in a fly.

Light and Compass Reactions: "Like a Moth to the Flame"   Taxes are undeniably simple behaviors. Shine a light into one side of a darkened tank containing a flatworm, and the flatworm will swim toward the light. Now move the light to the opposite side. The flatworm will turn and swim toward the new light. Shine a light from both sides, and the flatworm will swim to a point midway between them.

Modified taxes can produce behavior more complex than swimming toward a light. For example, in small animals like the insects, the pull of gravity does not reliably indicate up and down. These animals maintain their orientation and stability in flight by a modified phototaxis that keeps the light source above them. This is the *dorsal light reaction* mentioned in Chapter 20. A locust, for example, orients itself so that the upper parts of its compound eyes are illuminated. Fish orient themselves partly by gravity and partly by light in a dorsal light reaction (Figure 24–2).

The water boatman *(Notonecta),* a common insect of stagnant ponds and backwaters, swims on its back. The boatman has a *ventral light reaction,* which orients the insect with its belly toward the light. If a water boatman is placed in a tank illuminated only from below, the insect will turn itself over and swim with its dorsal side rather than its ventral side up.

In the ventral and dorsal light reactions, the taxis orients the animal toward a constant area of greatest illumination. A similar reaction, modified to take into account a varying light source, allows social insects like ants and bees to journey long distances from their nests, forage in different areas, and then return home unerringly.

In the early twentieth century (1911), the Frenchman F. Santschi observed that if he picked up an ant traveling in one direction, turned it around, and set it down somewhere else, the ant would immediately turn itself toward the original direction and continue its travels. Apparently the ant was not orienting itself by familiar landmarks. Santschi wondered whether the ant was taking a fix on the sun, much the way a navigator does to determine his ship's location. He noted the angle to the sun of an ant's path of travel, hid the sun from the ant, moved the animal, and showed it the sun in a mirror. The ant

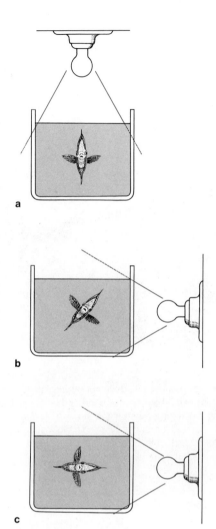

Figure 24–2 A fish maintains its vertical position in water by keeping the light overhead (**a**), and by reacting to the pull of gravity on the otoliths in its inner ear. With the otoliths present, the fish will tilt toward a laterally presented light (**b**). When the otoliths are removed, the fish will maintain the light dorsally regardless of the direction from which it comes (**c**).

Instincts The light reactions of the flatworm, the ant, the bee, and the moth are all relatively simple, automatic, and innate (inborn) responses—every normal night-flying moth can navigate by the moon without having to learn how. This sort of behavior fits the definition of a reflex (see Chapter 20).

One of the continuing problems in the study of behavior is that the terms used to describe particular responses are not mutually exclusive. Thus, a flatworm's orientation to light is both a phototaxis and a reflex. The term *reflex* refers to a particularly wide range of behaviors, from the single reflex arc of the knee jerk to the much more complex light-compass reaction of the moth, which involves thousands of wingbeats coordinated by their own reflexes as well as stabilizing reflexes incorporating sensory input from the insect's eyes and antennae. Although we could describe still more complicated behaviors in terms of hierarchies of reflexes, we often call these complex reactions *instincts*.

Instinct means so many different things to different people that some behavioral scientists have abandoned the term as useless. In a general way, **instincts** are rigid stereotyped behaviors that are inborn—that is, built into the neural circuitry in the course of growth and development—and not modified by learning. However, as we will see, some behaviors thought to be instinctive do involve a measure of learning.

The instinct concept probably originated as a semantic device for distinguishing humans from all other organisms. "Reason" governed human actions, while base "instinct" governed the lowly behavior of the animals. This conception elevated the human sense of self and no doubt justified our sometimes cruel use of other animals for our own purposes. However, the study of behavior has shown that using words like *reason* and *instinct* to distinguish humans sharply from other animals simply does not work. Instead, there is a continuum of behavior ranging from the simplest automatic taxes to the most complex and flexible types of learning. Many insects, particularly bees, are well known for their ability to learn certain things while humans, like other animals, also have a complement of reflexes and instincts.

Much of the contemporary work on the mechanisms underlying instincts has been done by Konrad Lorenz. As Lorenz sees it, specific behaviors are triggered by definite environmental stimuli called *releasers* (also known as *sign stimuli,* or *releasing stimuli).* Although response to a releaser is often relatively automatic, a releaser is not just an "on-off switch" as on an electrical appliance. The releaser's effectiveness may depend on the hormonal state of the animal, its previous experience, or both. For example, a hen will not sit on eggs and incubate them unless she is hormonally primed and in breeding condition.

Let us look more closely at incubating behavior. Birds

did in fact change directions, but the angle of the new path to the mirrored sun was precisely the same as the angle of the former path to the unhidden sun. This experiment showed that the sun served as a fixed point no matter what the direction of travel. This behavior is called the *sun-compass orientation.*

Bees, too, use the sun as a reference. Since the sun moves, any insect that orients itself by the sun over prolonged periods must have a sense of time. Bees, for example, rapidly learn the sun's varying positions through the day. On cloudy days, both bees and ants use landmarks to get around.

that are hormonally primed sit on their eggs, behavior apparently triggered by the sight of the clutch. But precisely what visual stimuli do the birds respond to? By varying the shape, color, and size of the eggs in the clutch, ethologists have been able to determine which stimuli are pertinent—and act as releasers—and which are irrelevant.

These releasers vary from one bird species to another. For some, the releasers stimulating incubation behavior are relatively broad. Herring gulls, for example, will accept almost any egglike item in their nests regardless of its shape or color. They will even incubate a potato! But size is clearly important in some way. The herring gull (Figure 24–3) will incubate wooden eggs in preference to its own if the wooden surrogate is bigger than the natural egg. The oversized egg serves as a *supernormal stimulus.*

In contrast to the gulls and oyster catchers, the murres (the Northern hemisphere's counterpart of the penguin) will not accept substitute eggs. Each murre can, in fact, distinguish its own eggs from those laid by another murre and will not incubate another murre's clutch. Murre eggs vary greatly in color and pattern, and the unique design of each egg probably serves as a releaser enabling the bird to distinguish its eggs from the many near it.

Why can one species of bird instinctively recognize its eggs while another cannot? Ethologists would immediately look for an evolutionary answer to this question, assuming that the observed behavior helped each kind of bird (or at least did not hinder it) in adapting to its particular environment. Virtually all incubating birds leave the nest from time to time, if only to feed. Birds that find their way back to the right nest will reproduce more successfully than those that do not. Murres nest with many other murres on the naked rock ledges of seaside cliffs. The ledges are so much the same in shape and color that often only the eggs themselves can reliably differentiate one nest site from another. Thus, the murre's highly developed ability to recognize the color and pattern of its own eggs is a reproductive advantage in its drab environment.

Herring gulls, on the other hand, nest singly among low-growing plants and locate their own nests by the topography and vegetation around the nest. Thus, the color, shape, and pattern of the eggs in the clutch are irrelevant for this behavior. All we can say, then, is that the herring gull's inability to discriminate is *not* a reproductive disadvantage.

Releasers continue to be important to birds' survival after incubation. After gull eggs have hatched, for instance, the feeding of the young is triggered by specific releasers. The chicks peck instinctively at a red spot on the adult's yellow lower mandible. This stimulus releases the instinctive regurgitation in the adult. The red spot is known to be the releaser for the chicks' pecking since

Figure 24–3 The herring gull does not distinguish its own eggs from artificial eggs placed in its nest. Here the bird is attempting to incubate a dummy egg much larger than any of its own. Apparently egg size is an important cue, and the bird is responding to a supernormal stimulus. The response to large objects in the ordinary context would prevent the bird from incubating pebbles near its nest.

an unspotted model bill elicits only 25 percent as many pecks as does the same model with the spot.

In some instances, releasers change with time and the development of the animal. Figure 24–4 details the changes in sequence in the feeding responses of young perching birds like thrushes or sparrows. The adults' feeding of young is released by the gaping of the naked and sightless young. That gaping is in turn released by the nest vibrations set up as the adults land. A human finger vibrating the nest will do just as well at this stage. But as the birds grow older, they need more and more specific stimuli to initiate gaping. After they can see, they learn to associate feeding with whoever or whatever happens to be the provider, and they gape in response to the learned visual stimulus.

Ethologists are often surprised when they learn what releases a particular behavior. During the breeding season, the male European robin (unrelated to the American robin) establishes a territory and drives out any male invaders. In the 1930s, David Lack of England studied

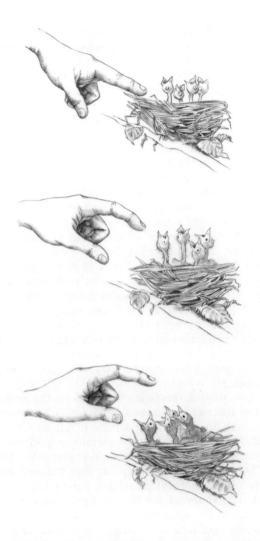

Figure 24–4 Young songbirds innately rear up their heads and gape when hungry in response to mechanical stimuli. When still blind they stick their heads only straight up, but after their eyes open (about a week after hatching) they respond to the sight of a human hand, or the parent bird, and orient toward the stimuli normally associated with feeding.

Figure 24–5 Adult male European robin threatening a red-breasted bundle of feathers.

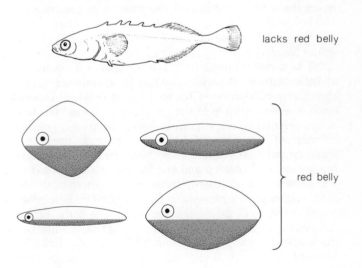

lacks red belly

red belly

Figure 24–6 When these models were tested on male sticklebacks in their territory, the realistic model lacking the red belly was attacked the least frequently, while the oddly shaped models were attacked most frequently as long as they were red ventrally.

this bird intensively and came to wonder what specific stimulus triggered the male robin's aggression. Lack made different model robins, introduced them into a male's territory, and observed the defending bird's response (Figure 24–5). He found that a resident male paid little attention to a mounted immature robin—which is robinlike in all details but has no red breast—but violently set upon a simple tuft or ball of red feathers. The red breast—rather than a robin silhouette, for instance—serves as the releaser.

Much the same reaction has been observed in red-winged blackbirds (in which the scarlet epaulets trigger the male bird's breeding-season aggression) and in the red-bellied stickleback, a small fish. Lifelike stickleback models without the red belly are ignored (Figure 24–6).

Insects are truly creatures of instinct, with inborn behaviors triggered by a wide variety of visual, acoustical, and olfactory releasers. Tinbergen found that male grayling butterflies chase paper dummies shaped like female butterflies and dangled from the end of a fishing pole. In fact, the butterflies will pursue almost any moving object, including falling leaves; but the larger and darker the model, and the faster it jiggles up and down, the more likely the males are to give chase. Like the herring gull in its reaction to oversized eggs, the butterfly responds preferentially to exaggerated stimuli.

Sound triggers the breeding behavior of male mosquitoes, which will fly to a tuning fork vibrating at 450 to 600 cycles per second, a tone corresponding to the flight sound of the female mosquito. If some suitable surface, like a piece of gauze, is provided, the male mosquitoes will actually try to mate with the tuning fork. Acoustical releasers are also found in many other animals. Hens, for example, quickly respond to distress calls from their chicks even when they are out of sight but will ignore the distress of a struggling chick in full view if they cannot hear the young bird's calls.

Odor serves as the releaser in many leaf-eating insects. The larvae of these insects typically feed on leaves from only one or two species of plant, and feeding behavior is triggered by an odor associated with the food plant. In some such insects this odor will initiate feeding even when the material giving off the smell is not nutritious. The larvae will feed just as readily on a piece of moistened filter paper impregnated with the releaser odor as on the actual leaves.

The same stimulus may release different responses at different times, depending on hormonal state or internal chemical condition (see Chapter 14) and previous experience. A good example of the influence of hormonal state is the orientation of migrating indigo buntings, studied by Stephen T. Emlen of Cornell University. In the spring these brightly colored finches fly north from their wintering grounds in Mexico and the West Indies to breed in the eastern and southern United States. In the fall the birds retrace their journey and fly south. After first proving that the birds orient themselves by the stars, Emlen divided his experimental population into two groups. Using artificial photoperiods (periods of light and dark), he brought one group into "fall" hormonal readiness and the other into "spring" hormonal readiness. When released in a planetarium, the fall birds oriented themselves toward planetarium south, whereas the spring birds oriented themselves in the opposite direction, planetarium north.

Our feeding fly of Chapter 20 showed how a series of relatively simple reflexes can be wired in a sequence to produce complicated behavior. A similar phenomenon occurs with the higher-order instinctive behaviors released by sign stimuli. Many species of birds have elaborate, even spectacular, courtship rituals (Figure 24–7). These rituals are not so much single inborn behaviors as a series of instinctive responses, each of which is released by its own stimulus. Typically that releaser is the behavior of the other bird. Thus, the male's initial behavior releases the female's response; that behavior triggers the male's next behavior; his action initiates the female's following response; and so the pattern goes, an ornate choreography of releasers and behaviors.

The reciprocal nature of its courtship behavior has proven a point of vulnerability in the survival of the marsh hawk. The marsh hawk's breeding ritual comprises an intricate series of aerial displays, in which the male and female pass killed prey, usually a snake, back and forth. If DDT accumulates in the male's body, some portion of his central nervous system is affected, and one of the female's behavior patterns fails to trigger the appropriate response. As a result, the female loses interest, the pair does not mate, and marsh hawks become rarer in polluted areas.

What can all this research tell us about the role of releasers and instincts in human behavior? Chances are that sign stimuli do release some human behavior. Certain facial expressions may well elicit certain responses—in other words, there may really be a recognizable "look of love." There has been speculation that the pug nose and rounded face of an infant trigger parental behavior in humans. The shape of the human breast may arouse sexual feelings in the adult male. Although one can easily be led astray by applying information about other animals directly and unquestioningly to humans, with proper caution, such data can prove useful.

Behavior and Its Inheritance

The behaviors we have discussed—taxes, reflexes, and instincts—are inherited, passed on from one generation to the next as part of the genotype. Many insects have provided the clearest evidence so far that specific behavior can be coded on genes like any other characteristic. The calling songs of male crickets, for example, consist of highly specific sequences of trills. Each species has its own song, and a male cricket sings the proper one whether he has heard it before or not. In research published in 1974, David Bentley of the University of California and Ronald R. Hoy of Cornell University reported that interspecies hybrids produce intermediate songs. They also found that the song of the hybrid resembled more closely the song of the male parent, leading to the conclusion that the song pattern is encoded on the X chromosome of the male. (In grasshoppers and crickets, XO is a male, and XX is a female.)

Another clearly inherited insect behavior is the nest-cleaning of honeybees. Honeybee colonies are sometimes attacked by foulbrood bacteria that infect and kill the larvae and pupae in their wax cells. Some, but not all, strains of bees uncap the infected cells and remove the dead young from the nest, disposing of some of the bacteria in the process and losing fewer young in the long run. This behavior is clearly adaptive. In 1964, Walter C. Rothenbuhler of Ohio State University bred honeybees for differences in nest-cleaning behavior and created four distinct lines: (1) uncappers only; (2) uncappers and removers; (3) removers after uncapping by the researcher; (4) neither uncappers nor removers. Rothenbuhler's data suggest that nest-cleaning behavior depends on two genes, one for uncapping and one for removing, both of them recessive. Thus, nest-cleaning honeybees must be homozygous at two gene loci.

These examples of insect behavior could easily lull one into the false security of a "one gene–one behavior" hypothesis. In fact, many behaviors, particularly in higher organisms, result from the interplay of many genes—that is, they are *polygenic*—and must be studied statistically. In addition, some genes are switched on only under specific circumstances. Thus, although the presence of a particular behavior definitely means that the underlying genes are present, the absence of the behavior does not necessarily mean that the genes are absent. It may merely mean that they are not expressed.

The following example showed that genetically determined behavior can be modified. When gathering material to build its nest, the peach-faced lovebird tucks the material under its tail, closes the feathers down like a clamp, and flies back to the nest site with its beak free. Fisher's lovebird carries nest material in the more usual way, in its beak. To see whether these behaviors are inherited, William C. Dilger of Cornell University crossbred the two species in the laboratory. When the hybrids started to build their first nest, they acted confused. They tried to tuck nesting material into their tails, but their movements were so uncoordinated that they failed. After three such unsuccessful attempts, however, the hybrid birds started carrying the nesting material in their bills. Clearly the material-carrying behavior is inherited, but the birds have some behavioral flexibility.

Inherited differences in behavior occur not only between species or strains of the same species but also between sexes. Young roosters crow; immature hens do not. Until about six or seven months of age, dogs of both sexes squat to urinate. After that age, the females continue to squat, but the males elevate one leg in the familiar way. In his classic studies of the rhesus monkey, Harry

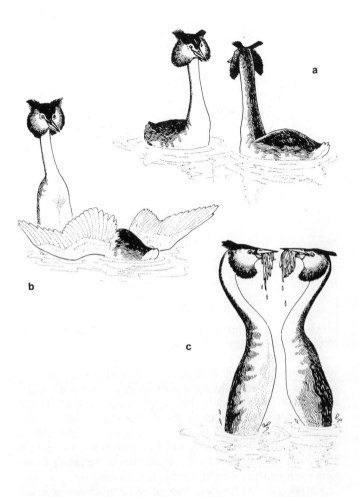

Figure 24–7 Courtship in the great crested grebe, as studied by Julian Huxley in 1914. (**a**) Head-shaking: Facing each other, the birds display their head feathers and shake their heads from side to side. (**b**) Male dive: The male approaches the female with his head submerged and suddenly shoots high out of the water just in front of her. (**c**) Mutual presentation: Each presents water weeds to the other.

F. Harlow observed behavioral differences between males and females from the second month of life onward.

Intelligence A fascinating and troubling area of polygenic inheritance is intelligence. The troubles arise first because we humans have long considered ourselves the most "intelligent" creatures on earth. This prejudice makes our interest in the subject more vested than scientific. Second, intelligence, whether for humans or for animals, has never been precisely defined. Criteria used to determine animal intelligence are generally too anthropocentric to be valid. Moreover, genetic effects are modified by learning and probably by diet. Finally, the polygenic

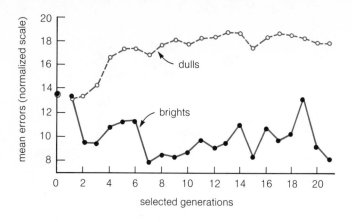

Figure 24-8 The results of selective breeding experiments in rats for maze brightness and maze dullness.

factors affecting intelligence can be sorted out only with extensive controlled breeding experiments.

Some of these difficulties can be surmounted by using rats as experimental animals. One way of testing the rat's abilities is to put a hungry animal in a maze with many blind alleys and a reward of food at the far end. To get the food, the rat has to negotiate the maze successfully, sorting its way through the blind alleys. Once the rat learns the food is there, it will be motivated to negotiate the maze more quickly in future trials, entering fewer blind alleys than on the first run-through.

If the ability to negotiate a maze without errors is an indicator of rat intelligence, then intelligence can be shown to be inherited. R. C. Tryon found that some rats negotiated a maze faster (with fewer errors) than others. Separating these "bright" rats from the "dulls," he allowed each group to breed only within itself, creating two divergent lines. After eight generations, the "brights" averaged only 9 errors per trial, whereas the "dulls" averaged 18. There was no overlap between the scores of the pure lines (Figure 24-8), but hybrids between them showed intermediate performance. Clearly, different levels of intelligence (or at least of maze-learning ability) were passed down through the two lines.

But intelligence is not so neatly heritable as the tidy data make it seem. When Tryon changed his intelligence test from maze-learning to a water escape, the "dulls" surpassed the "brights."

Flexible Behavior   Both the rigid inherited behaviors we have discussed and behaviors that can be modified by experience have their adaptive advantages. The innate stereotyped behaviors of the insects benefit organisms with such short life-spans and small nervous systems.

They have neither the time nor the equipment to learn all they need to know to survive. Likewise, the Arctic tern and the indigo bunting are well served by instincts that direct their lengthy migrations so unerringly. However, rigid behavior quickly becomes maladaptive if the environment changes in a critical way. Suppose, for example, that the Arctic tern arrives at the end of its grueling journey to find its nesting grounds under water or its food species vanished. If it cannot go beyond this pattern—fly a little farther to the new shoreline, learn a new food source—it will starve to death, and no flock of terns will head south from that area at the end of summer.

Learning

Behaviorists define **learning**—which is what our hypothetical tern needed at its crucial moment—as the modification of behavior by experience. We will need to keep this definition in mind, because we usually use the word in a narrower sense (as in "she is learning Greek").

Learning increases the organism's ability to respond to changes in the environment. The key word is *flexibility*. Flexibility, however, exacts its own price. The more of an animal's total behavior that has to be learned, the longer the animal tends to remain immature. This immature state, in which learning is so rapid, is also a vulnerable state, when many of the young may be killed by predators or may starve. Newborn rattlesnakes crawl away from their mother's den an hour or two after birth; humans may not leave home for 20 years.

Learning becomes more and more important to behavior as we move toward more complex organisms, but in any given animal, the line between innate and learned behavior is imprecise at best. The behavior of young wolf pups avidly chasing a moving animal, such as a porcupine, is most likely instinctive. But this instinct can be shaped by learning, so that the pups learn it is "wrong" to contact the porcupine but "right" to bite into a rabbit. The picture is further confused by the many kinds of learning, which we will now examine.

Respondent Conditioning   In Chapter 20, we outlined briefly the nature of *conditioning*. During his work on digestive physiology, Ivan Pavlov noted that his experimental dogs salivated before any food was given to them. The mere sight of food—or even the appearance of the person who always brought it—was enough to trigger salivation. Intrigued by this apparent association of stimuli, Pavlov experimented further. For a while, he rang a bell at every feeding time. The dogs were soon salivating in response to the bell. Then he rang the bell without presenting any food. The dogs salivated anyway. They had learned to associate another stimulus—the sound of the bell—in addition to the former ones—sight and smell—with the food.

This kind of learning is called *Pavlovian conditioning* or *respondent conditioning,* the term used increasingly by psychologists. The important thing to note about conditioning is that it is the stimulus, not the response, that is learned. Salivation, for example, is an involuntary response. The dog learns to associate one stimulus with another because the two are presented close together.

In a later series of experiments, Pavlov tested the effects of conflicting stimuli. Dogs were conditioned to expect food if they saw a circle and to expect an electric shock if they saw an ellipse. When this conditioning was well established, he started making the ellipse more and more nearly circular and the circle more and more elliptical. The dogs became more and more agitated until in the end, when ellipse and circle were indistinguishable, they appeared to behave neurotically.

Instrumental Conditioning From the moment of birth, an animal responds in functional and nonfunctional ways to the various stimuli it encounters. Some of these responses have no result. Others may gratify a need such as hunger and produce a reward. Still others may have an unpleasant outcome—what we call a punishment. Over time, the responses that are rewarded become more frequent, while those that are punished become less frequent. This is *instrumental conditioning, operant conditioning,* or *trial-and-error learning.*

Instrumental and respondent conditioning should not be confused. First, in instrumental conditioning, it is the response, not the stimulus, that is learned. Second, in respondent conditioning, the stimulus precedes the response. This order is reversed in instrumental conditioning: The response occurs first, then the stimulus. A stimulus that increases the frequency of a response is a *reinforcement;* one that decreases frequency is a *punishment.*

Instrumental conditioning is the basis of learning theory. B. F. Skinner maintains that the entire behavioral repertoire of an organism can be understood as a body of responses shaped and controlled by reinforcements.

Although learning theorists can properly be criticized for building a grand theory of behavior around a single type of learning, our everyday life is full of instrumental conditioning. If you train your dog to roll over by giving it a treat for every success, you are practicing instrumental conditioning.

In nature, instrumental conditioning can be a means of rapid learning. Lincoln and Jane Brower have shown that toads need only one or two trials to learn to avoid a bad-tasting food item. Theodore C. Schneirla of New

York's American Museum of Natural History has found that ants can quickly negotiate a maze where they have previously found food, even without odor trails.

Learning speed may depend on the nature of an animal's surroundings as much as on its inheritance. David Krech and his associates at the University of California found that rats raised in a complex environment (in which they were in contact with other rats and had access to a maze and to wooden "toys") made fewer errors in maze-learning than littermates raised in isolation in an empty or stark environment. Abundant and varied stimulation apparently increases learning speed in rats.

Habituation Many features in the environment are of little or no immediate relevance. Although an animal may react to such irrelevant features on first meeting, it will eventually cease to respond to them. This loss of response is called *habituation.*

Newborn animals are typically shy, but in time they accept familiar things, retreating only from unfamiliar ones. A human child gradually accepts family friends, for example, but fears strangers.

In an experiment conducted in 1937, Tinbergen and Lorenz uncovered a similar habituation in young shorebirds. They built a large bird model that looked something like an elongated ping-pong paddle and had a crosspiece for wings. The model was suspended from a line that ran above the young birds. Pulled along in one direction, the model approximated the silhouette of a goose with its long protruding neck. Drawn the other way, so that the long projection was in the back, it looked more like the long-tailed accipiter or "chicken" hawks, which prey on these birds. When the young birds fled the hawk model but ignored the goose model, Tinbergen and Lorenz first thought that the birds were responding to two different inborn releasers. Later research showed that the newly hatched chicks crouch down at the sight of anything overhead, even falling leaves. But gradually, since geese are numerous and hawks rare in their environment, the shorebirds become used to goose shapes and ignore them, but continue to fear the uncommon and dangerous shape of the hawk.

Habituation is what kills most animals taken from the wild when young and returned there at maturity. In captivity, the animal has lost its protective wariness; when released, it falls victim to wild or domestic predators, vehicles, or hunters, who are amazed to find game standing so still.

Imprinting Imprinting means just what it says—a permanent impression made on an animal's brain by a stimulus or a group of stimuli, perhaps forming a visual or acoustical image. This rapid kind of learning can generally

Figure 24–9 A flock of graylag goslings following Konrad Lorenz, on whom they have been imprinted.

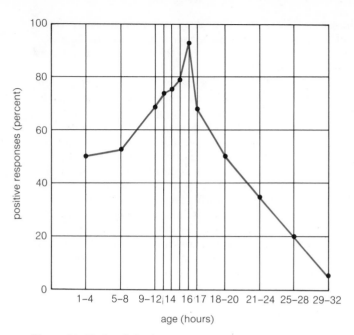

Figure 24–10 Imprinting in young precocial birds occurs within a narrow time range, the *critical time*. The curve shows that baby ducklings imprint most strongly at about 16 hours after hatching. At 32 hours after hatching they no longer respond to new stimuli by following.

take place only during brief, genetically determined periods of the life cycle.

More than any other kind of learning, imprinting illustrates the difficulty of separating instinct from learning. People have long noticed that goslings follow their mother with remarkable persistence. Since they do this within hours after hatching, their behavior might seem purely instinctive. Figure 24–9, which shows a line of goslings trailing a gray-haired man, proves otherwise. Konrad Lorenz, the man in the photograph, showed that newly hatched goslings do not know innately who their mother is; that is, they are not rigidly programmed to

attach themselves to the adult female goose who incubated them. What they *are* rigidly programmed to do is to attach themselves to and follow after the first moving thing of approximately the right size that they see when they leave the shell. It just happens that this usually *is* their mother. As behaviorists would put it, the inherited behavior must be completed by the learned behavior.

Lorenz demonstrated this by removing mother geese from the nest just as their clutches were starting to hatch. To the young, who therefore did not see their mothers, he presented various living and nonliving things as their first visual experience. He found that the hatchlings took to people, other kinds of birds, dogs, cats, toy trains, and even ticking alarm clocks as readily as to an adult goose. The surrogate mothers need only move (or be moved by the experimenter), make some sounds, and conform to certain size limits.

It was also Lorenz who found that this kind of learning is limited to a short time now called the *critical period*. Goslings, for example, imprint to a "parent" image before they are 36 hours old (Figure 24–10 shows the curve for ducklings). During this critical period, they actively seek out a moving object that fulfills their inborn need to follow it, and once the choice is made, they will avoid everything and everyone else. Thus, goslings who are imprinted on a certain dog will flee to their "mother" when any other dog enters the barnyard. They have not learned to accept dogs in general, but only to accept one particular dog.

Imprinting on the mother in geese—as in ducks, swans, quail, partridge, and turkeys—has obvious survival, or adaptive, value. The normally imprinted young of these species stay close to their mother, who protects them and leads them around in areas where there is food. When they no longer need protection, they lose the following response; but apparently they will carry with them an indelible image of what is safe and familiar—a reference image of their own kind, so to speak. This image itself has survival value if it is normal (if it is the image of its own species), for it ensures that newly matured birds will choose members of their own species as mates.

Perhaps because Lorenz's work is so widely known, many people think of imprinting as restricted to birds, and specifically to the kind of learning that leads to the following response in hatchlings. However, critical-period learning has been identified in a wide variety of animals, and in other periods of birds' lives as well. Peter Klopfer of Duke University has shown that goats learn to accept their young only during a few hours after birth. If a mother goat is separated from her newborn kid until after this critical period, she will reject it.

Imprinting also plays a role in song-learning in some songbirds. Bird-song studies have taught us that not only the learning period but the substance of the lesson is limited in imprinting. The degree of learning varies greatly from species to species; the whole spectrum from completely innate song to learning by imitation is observed. W. H. Thorpe of Cambridge University found that European chaffinches raised in isolation do not sing the normal chaffinch song. However, if a recording of the song is played to isolated birds at about six months of age, they will sing the normal song when they come into breeding condition the following year.

One might suppose that, just as the goslings follow any suitable "mother," the chaffinches would sing any song played to them at the critical age. Such is not the case. The birds ordinarily learn only their own song. They are so well attuned to it that they can respond even when it is played backwards; but they cannot learn the song of another species. The European bullfinch, however, will learn practically any song taught to it during its critical period. One German behaviorist has taught a bullfinch a popular German drinking song, which the bird warbles as if it were its rightful song. Bullfinches raised by canaries sing a nearly perfect canary song, and as long as the birds are isolated from other bullfinches, their offspring will also sing like canaries.

In general, the less restrictive the learning in the critical period—the more like the bullfinch—the more "intelligent" an animal is thought to be. In humans, for exam-

ple, the ability to learn language is almost certainly inborn, but there is no inherited predisposition to learn the language spoken by one's parents as opposed to any other human tongue.

Latent Learning   Through repeated experience, an animal may learn things about its environment that have little or no value at the time they are learned. In the course of its daily forays for mice, for example, a fox becomes familiar with its home area. Eventually the animal may get some value or reward from its previous learning, as when it needs a den or an escape route from a pack of dogs.

Such *latent learning* is probably most important to the higher animals, which can store and process much more information about their environment than immediate survival requires. Probably most latent learning occurs before an animal reaches maturity. The play of kittens and puppies teaches the growing animals motor skills that will help them survive as predatory adults.

Insight Learning   *Insight learning* or *reasoning*, enables an animal to respond correctly on or near the first try in a new situation. Usually insight learning involves applying previous learning to the present situation. It may also involve, consciously or unconsciously, visualizing a number of possible outcomes or solutions. The unworkable ones are dismissed without being tried. Insight learning is most highly developed in the primates, particularly humans.

Indeed, we humans are so focused on reasoning that we tend to see it in other animals' behavior whether it is there or not. To evaluate animal behavior accurately, however, we must distinguish reasoning from trial-and-error learning (instrumental conditioning) and from generalization. A rat running a maze in search of food will try one behavior after another until it chances upon one that works. The rat's second trip through that same maze will be much faster.

In *generalization*, an animal sees the similarity between a new stimulus and one to which it has already been conditioned, and responds in the same way. For example, a rat that has run mazes before will get through a completely new maze faster than rats encountering a maze for the first time. The experienced rat recognizes similarities between the general class of stimuli that we call mazes: It has generalized from one to the other.

True insight learning (detecting relationships not merely through direct experience) may involve both generalization and trial-and-error learning, but it goes beyond both of them. It is the most flexible of all the kinds of learning. To qualify as capable of insight or reasoning, an animal must detect not merely similarities between

Figure 24–11 Improvement of performance with time can be due to maturation rather than learning. Prisms in the goggles of the chick deflect its vision to the right. Peck marks in clay (right) show an improvement of aim toward a seed over a three-day period. Although the aim of the goggled chick improves with age, it is still to the right of the mark. Most higher animals would learn to adjust.

Wallace Kirkland/*Life* Magazine © 1954 Time, Inc.

partridges and chickens or among mazes (generalization), but differences (discrimination) as well. Insight learning involves perceiving similarities (and differences) far more subtle than those usually involved in generalization. Information and experience gained in several different contexts may be combined in this kind of learning.

Possible reasoning has repeatedly been demonstrated among primates. An orangutan given a length of broomstick and a hollow tube with a piece of candy wedged firmly inside will soon shove the broomstick through the tube and knock the candy free.

Using tools, as the orangutan does, often indicates

reasoning, but not always. After filling in its burrow to protect its eggs and larvae from parasites, the sand wasp *Ammophila* uses a small stone held in its mandibles to tamp down the loose earth. This behavior is certainly innate. Among social animals, tool-using may also be transmitted culturally—that is, by imitation. Chimpanzees "fish" for termites by running a long, thin stick into a mound. They leave the stick inside for a while, then pull it out and lick off the clinging insects. This behavior may have originated with one particularly insightful chimpanzee—or with an accident that happened in play. Other chimpanzees, seeing how easily food could be ob-

a   b   c

Figure 24–12 A young monkey brought up without a mother or other young monkeys is fearful of others (**a**) and is easily dominated by a youngster its own age (**b**). The unmothered infant when grown up is a poor mother herself, ignoring her baby (**c**).

tained, imitated the rewarding technique. Today each new generation of young chimpanzees learns how to catch termites by mimicking their mothers' actions.

The Development of Behavior: Maturation

Among the more complex animals, the newborn possesses only a fraction of the behaviors of the adult. As the organism grows, its nervous, sensory, and muscular systems develop, and it gains new behavioral capabilities. This behavioral change due to growth and development, as opposed to learning, is called *maturation* (Figure 24–11).

Some instances of maturation are obvious. No animal can reproduce until its sexual organs are developed, and breeding behavior begins largely as the product of maturation. But in other cases, simple observation cannot reveal whether learning or maturation produces a given behavior, and we must resort to experiment.

J. Grohman of Germany devised an experiment to determine whether flight in birds is influenced by learning or simply a function of maturation. As flight feathers appear and young birds approach the time when they are to leave the nest, they often spend hours flapping their wings vigorously. Grohman reasoned that if flight was in part learned, then birds kept from beating their wings could not fly as soon or as well as unrestrained birds. He raised young pigeons in tubes too narrow for the birds to exercise their wings. But when these prisoners were released, they flew as well as a control group that had been allowed to beat their wings at will.

Likewise, human infants probably do not "learn" to walk. Babies that never exercise, such as some American Indian infants bound tightly in the traditional packframe, can walk as early as unrestrained babies.

The juvenile environment may have subtle effects on overall behavior through maturation. Harry F. Harlow of the University of Wisconsin's Primate Research Laboratory showed how important the normal mother-infant relationship is to adult behavior in the rhesus monkey and, presumably, in apes and humans as well. Taking newborn monkeys away from their mothers, Harlow raised them with wire-and-cloth substitutes that held a bottle. As adults, these experimental monkeys were what we would call sociopathic: They stared mutely into space, were indifferent to others, and often burst into a frenzy of violence when approached (Figure 24–12). Generally they would not copulate, fighting viciously with prospective mates. However, some exceptionally persistent males impregnated some of the sociopathic females. When these unmothered monkeys became mothers themselves, they were "helpless, hopeless, heartless mothers devoid, or almost devoid, of any maternal feelings." Harlow concluded that maturing monkeys require the security of a mother if they are to develop into normal adults. Without it, they are too fearful to experience the outside world, fail to develop cooperative bonds with their peers, and thus become either aggressive or indifferent.

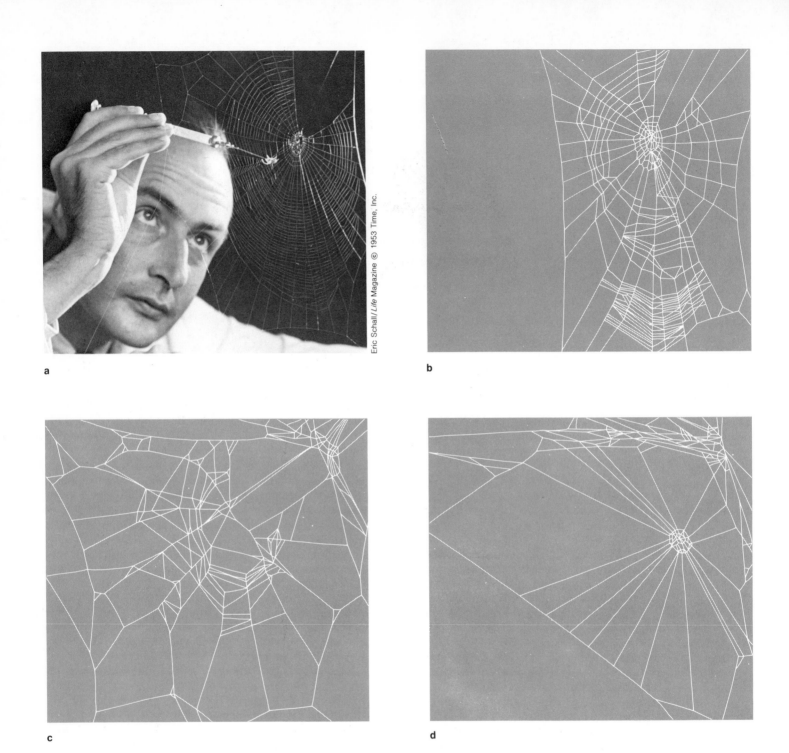

a
b
c
d

Figure 24–13 The webs of spiders are a record of their behavior, and this behavior can be influenced by drugs. A spider is being given a drug at top left. Pervitin, a Benzedrinelike stimulant, makes the spider spin an incomplete and disorganized web (top right). Caffeine causes the spider to produce a haphazard tangle of threads (lower left). Under the influence of chloral hydrate, the spider spins only a partial web (lower right).

Eric Schall/*Life* Magazine © 1953 Time, Inc.

The Physiological Context of Behavior: Motivation

A rat in a maze left unfed for 24 hours has an internal state that prompts behavior aimed at finding food. This internal state, the immediate cause of behavior, is called *motivation*.

Motivation may be hard to quantify. With our hungry rat as an example, we can summarize the events in physiological motivation. Dropping blood sugar levels trigger hormonal and neural events that the rat will experience as a need for food. These events in turn prompt an action, or behavior, directed toward the goal of meeting this need. If successful, this behavior uncovers something edible, and the rat eats. The food then produces internal changes, among them an increase in blood sugar, that return the rat's physiological state to a balanced condition and decrease its motivation to continue seeking food.

One way to study motivation is to introduce various physiologically active chemicals into experimental animals and observe the behavioral effects. Many hormones produce immediate behavioral reactions. Injecting a female canary with estrogens will cause her to build a nest even when it is not the breeding season. Turkey chicks of both sexes injected with androgens will strut like toms and try to mount other chicks. Nonhormonal drugs also affect behavior. There is a gas, whose formula is a military secret, that will cause a cat to flee at the sight of a mouse. One can imagine what an effective weapon this secret gas might be.

It is generally difficult to identify how a drug affects a lower animal, but spiders leave a visible record of their behavior: the web. Typically, each species of web-spinning spider builds a characteristic net. Peter Witt, a pharmacologist, has found that drugs change normal web-spinning and that each drug he used produces an identifiable pattern (Figure 24–13).

An important goal of research into motivation is to locate the exact cells or structures that direct or influence particular behaviors. Recent research has located centers in the vertebrate hypothalamus that respond to changes in temperature and in the concentration of certain ions and thus control the motivations to seek warmth, cold, or water. Motivational researchers have had more success with insects than with vertebrates, though. Vincent De-thier and his associates at Princeton University, for example, using the fly as a model system, have mapped the series of events involved in hunger motivation, from the firing of sensors to the final feeding behavior.

The External Context of Behavior: Displacement

In our discussion thus far, we have assumed that the releasers or stimuli in a particular situation are obvious and that an animal responds without confusion. In truth, the real world is somewhat more complex than this assumption allows, and, like Pavlov's dogs and their nearly circular ellipses, animals may be unsure how to respond.

Often an animal could respond to a particular situation in two opposite ways. For example, a courting male herring gull faced by a rival for a mate, food, or space may be equally motivated to fight and to flee. Instead, the gull tears up grass, a response that has nothing to do with either fight or flight but only with nest-building. An oyster catcher in a similar situation may tuck its head under its wing, as it does when it sleeps, whereas a stickleback stands on its head. This kind of behavior, in which an animal cannot choose among alternatives and instead behaves in an irrelevant way, is called *displacement activity*.

Summary

Behavior is a sequence of nerve impulses and associated muscle contractions in response to external and/or internal stimuli. Specific behaviors in many animals are triggered by specific stimuli. The science of behavior concerns the physiological mechanisms of different behavior patterns and the relevance of these patterns to the animal.

Behaviors include a continuum of responses from innate reflexes, to hierarchies of reflexes, or instincts, to highly variable responses that are shaped or modified by experience. In this chapter we examined the mechanisms, the relevance, and the genetics of some relatively "simple" behaviors. We also discussed the modification of behavior through maturation, motivation, and various types of learning.

25

Synchronization to the Environment

Although we often write and speak of the "environment" as if it were an eternally constant set of physical circumstances, in fact local environments change, sometimes drastically. In the daytime, sunlight is readily available, temperatures fairly high, and humidity low. When night falls, the sun disappears, temperatures drop, and humidity rises. These daily differences pale before seasonal changes. In the summer, a small pond in Minnesota is green and lush with trees and plants, alive with the sounds of frogs and birds. But when winter comes, with its freezing temperatures, and storms blow down the prairie, the pond becomes a snow-and-ice-covered expanse, seemingly lifeless, rimmed by naked trees.

To survive, an organism must have some way of handling these changes. We have already seen how some organisms adjust. For example, the photoperiodicity of flowering plants (Chapter 11) ensures a reproductive cycle in tune with the seasons. In this chapter, we will look further at the mechanisms that synchronize an organism's life cycle to the environment. We will look first at rhythmic activities, then at strategies for avoiding times of pronounced environmental stress.

Rhythmic Activities and the Sense of Time

Photoperiodicity in plants exemplifies one way of handling environmental change: coordinate activities with changes in the day and the year.

Circadian   Rhythms   Apparently   self-contained rhythms of behavior and physiology have been observed in many species (Figure 25–1). Photosynthesis in the one-celled alga *Gonyaulax polyedra* follows a daily cycle when it is kept in constant light. Cockroaches and flying squirrels, which are nocturnal animals, become active in the

evening, even when they are kept in experimental conditions of constant light. Fiddler crabs are most active at low tide, which comes at a different time each day. Even in a laboratory far from their habitat, these crabs retain activity patterns set to the tidal pattern on their home beach. Humans, too, show daily changes. Diabetics are more sensitve to insulin at night than during the day, and human tolerance for alcohol is highest at 5 P.M., the cocktail hour.

In 1959, Franz Halberg of the University of Minnesota coined the name **circadian rhythm** (from the Latin phrase *circa diem,* "about the day") for these daily changes. Circadian rhythms have been the subject of serious biological investigation for the past 30 years.

The existence of circadian rhythms is obvious. The important question is what triggers them. Are circadian rhythms released by environmental cues, much as we might awaken at the sound of an alarm clock, or is there an internal timer, something like a physiological alarm clock?

Many animals, including humans, respond to the daily light-dark cycle. When such animals are brought into the laboratory and subjected to an experimental light-dark cycle different from the one in their natural environment, they usually adjust quickly and set their rhythm to the new intervals. Once the new rhythm is set, it persists, even when the organisms are subjected to constant light or constant dark. However, the new rhythm rarely hits 24 hours exactly, running instead a little longer or a little shorter. For example, a sparrow set to an experimental light-and-dark cycle may begin moving about 35 minutes before the lights-on time every 24 hours. When subjected to total darkness, the sparrow might continue to begin its movements rhythmically, but on a cycle of 25 hours instead of 24. Thus, on the sixth day of total darkness, the sparrow begins hopping about 6 hours later than it did on the first day (Figure 25–2). Erwin Bünning of the University of Tübingen, Germany, has noted a similar effect in plants. Experimental

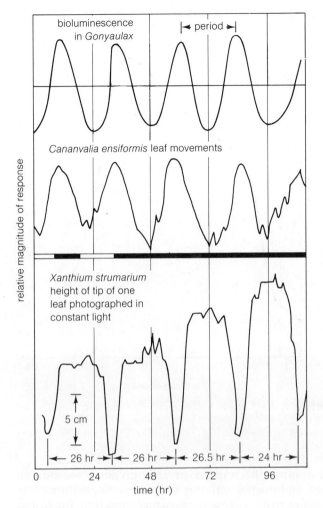

Figure 25–1 Data for various circadian rhythms. Top: bioluminescence in *Gonyaulax*, as measured for plants kept under constant conditions of dim light. Middle: a leaf movement record from *Cananvalia ensiformis*. Light and dark conditions are indicated by the bar. Note the gradual shift of the peak as the cycles progress during constant darkness. Bottom: leaf movements of cocklebur *(Xanthium strumarium)*, as recorded by time-lapse photography. Period lengths between the troughs are indicated. The increase in height of the waves results mostly from growth of the stem during the course of the experiment, but the range of leaf movement also increased. Light came entirely from fluorescent lamps.

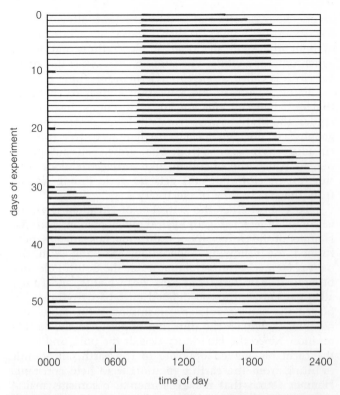

Figure 25–2 Circadian rhythm of activity in a house sparrow. Each of the 55 horizontal lines represents one day of the experiment. The hopping activity was recorded on the chart moving at constant speed each time the bird landed on a perch which, when depressed, completed a circuit activating the pen. From day 1 to day 20, a light cycle was imposed (lights on at 0800 hr, or 8 A.M.), and the bird initiated hopping each day 35 minutes before the lights went on. When the bird was placed in constant darkness (after day 20), it began hopping about an hour later each day. That is, it had a free-running circadian rhythm of about 25 hours.

bean plants kept in total darkness have rhythms of leaf movement set to cycles ranging from 22 to 28 hours.

Experimental data like these indicate that circadian rhythms, although affected by environmental stimuli, are internally determined, or *endogenous.* Probably the most convincing evidence for endogeny is that the cycles prevailing in total darkness or total light do not follow a perfect 24-hour pattern. If they did, it would be possible

that the experimental organisms were responding to a daily geophysical cue to which we humans are insensitive. The rhythm itself is part of the organism's makeup; its precise timing depends on environmental conditions. In the experimental condition of total light or total dark, the rhythm is said to be *free-running;* when set by environmental conditions, the rhythm is called *entrained.* In nature, the light-dark cycle entrains most circadian rhythms so that their periodicity never drifts far from the 24-hour day.

Entraining can be seen clearly in an experiment in which a number of organisms of the same species are confined in total darkness long enough for the variations in their rhythms to appear. Considerable individual differences may arise, so that one animal exhibits an activity at one hour, another at a different time altogether. If this group of animals is then exposed to a brief burst of light, they will all "reset" their rhythms and act in synchrony for a while. The stimulus that resets the internal clock is called a *zeitgeber* (German for "time giver").

Frank Brown of Northwestern University believes that circadian rhythms are not endogenous. Instead, he argues, the rhythms arise from subtle zeitgebers, perhaps geophysical phenomena. The oscillations of free-running circadian rhythms arise from what Brown calls autophasing, in which the organism, like an electric clock, can run fast or slow but still gets its basic timing signals from the steady flow of electricity.

To test whether the zeitgeber of circadian rhythms is some undetected geophysical phenomenon, Karl C. Hamner of the University of California, Los Angeles, performed the following experiment. Working with the navy, Hamner set up a laboratory at the South Pole and brought in hamsters, fruit flies, bean plants, and fungi, all organisms with well-defined circadian rhythms. He placed the experimental organisms on a turntable that revolved slowly in the direction opposite to the earth's rotation. Since the laboratory was at the pole, and since the turntable moved opposite to the earth, all stimuli resulting from the earth's rotation were held constant. Hamner found that his experimental organisms maintained their circadian rhythms despite the lack of geophysical cues, adding to the evidence that they are indeed endogenous.

Jürgen Aschoff and his associates at the Max Planck Institute for Physiology and Behavior at Erling-Andechs, Germany, have been studying circadian rhythms in humans intensively. For this work, the institute has an isolation facility where subjects are removed from all environmental cues and special monitoring equipment records their activities. The subjects, who have no watches or clocks, are isolated for three or four weeks at a time. Their only contact with the outside is the experimenter, who comes in at random times to bring such necessities as food and one bottle of Andechs beer per day. Subjects adjust their own light-and-dark cycle, turning the lights off when they go to bed and turning them on again when they get up.

All 26 subjects observed in the isolation facility up to 1965 exhibited circadian rhythms. Twenty-two of the subjects had free-running rhythms longer than 24 hours, usually about 25 hours. The period of the rhythms, however, is not absolute in any subject. In day-active animals, an increase in light intensity decreases the period of the activity cycle; in night-active animals, the effect is just the opposite, but the significance of this difference is not known.

Aschoff has found that more than 100 functions of the human body oscillate daily between a maximum and a minimum. Among them are urine production, body

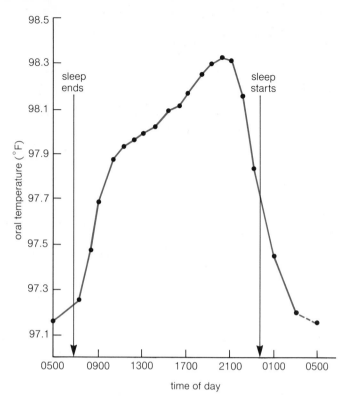

Figure 25–3 Daily rhythm of oral temperature in humans.

temperature, the number of cells dividing, reaction to drugs, and mental activity. The body-temperature cycle correlates with cycles of vigilance, reaction time, and other mental functions (Figures 25–3 and 25–4).

A practical reason for the interest in circadian rhythms is that peculiar modern problem called jet lag. Jet lag results from traveling rapidly between time zones. For example, San Franciscans who fly to New York find themselves three hours behind people on the East Coast. Their endogenous rhythms say one thing, but the environmental cues to which they are entrained say another. Jet lag can cause a range of physical and psychological disturbances, such as loss of sleep, constipation, depression or irritability, and reduced intellectual function. Already there is some concern that diplomats who jet over the globe from one high-powered meeting to another, making decisions that affect the lives of millions, too often do so without regard to the effects of jet lag on their bodies and minds. An even more severe problem arising from disrupted circadian rhythms could appear in long-term space travel.

The self-perpetuation of circadian rhythms means that the organism must have some way of telling time. To keep a rhythm on a 24-hour cycle, the organism has

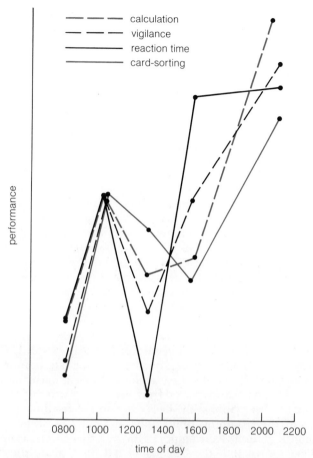

Figure 25–4 Relative performance scores in calculation, vigilance, reaction time, and card-sorting of human subjects as a function of time of day. Lowest performances were observed early in the morning and near noon. The highest performances were near 8:00 P.M. (2000 on the 24-hour clock). The rhythms continue without the intervention of external stimuli.

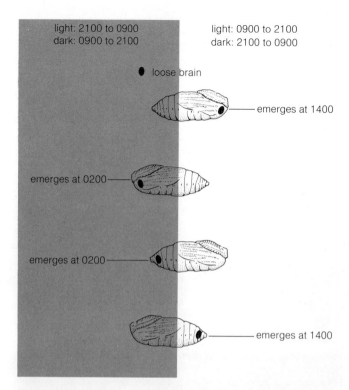

Figure 25–5 The response of silkmoth pupae to various lengths of light depends on the photoperiod to which its brain is exposed. In this experiment, the pupae were inserted into a light-opaque position between two chambers. The chamber at the right had lights on from 0900 to 2100 hr, and the chamber at the left had lights on at 2100 to 0900 hr. The brains of the moths had been excised (loose brain) and reinserted into the front end of the pupa or into the abdomen. The moths normally emerge about five hours after lights-on, and this timing is here shown to depend on the photoperiod impinging on the brain.

to know when 24 hours have passed. However, no one has yet discovered this internal clock. Karl Hamner, who has been studying time measurement in plants for over 20 years, says, "Our understanding of rhythms really stands about where gravitation stood before Isaac Newton. What we need is another Newton."

Almost certainly, the circadian clock is complex. For example, it remains accurate through a wide range of temperatures. Since most biochemical reactions double in rate for every 10°C increase in temperature, one would expect the clock to run faster at high temperatures and slower at low ones. Instead, it remains precise, except at extremes of heat and cold. Certainly the circadian clock must involve a complex series of compensating reactions.

That complexity, of course, makes the clock all the harder to identify.

It has proved easier to find the clock's location than its identity or mechanism. The insect circadian clock resides in the nervous system, specifically the brain (Figure 25–5). James W. Truman, now at the University of Washington, studied the internal clock in giant silk moths. The Cecropia silk moth emerges from the pupal case in the morning, whereas the closely related Pernyi moth emerges in the afternoon. Apparently the moths can tell the time of day. In the first experiment, Truman moved the moth's brain from the head to the abdomen. This did not affect the time of emergence. Then Truman transplanted the brain of a Pernyi moth into a Cecropia and

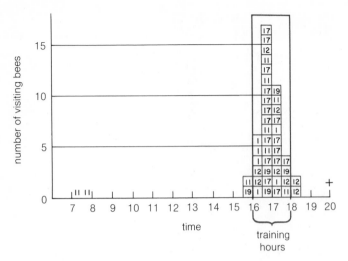

Figure 25–6 Time sense in honeybees. Five individually marked (#1, 11, 12, 17, and 19) bees were fed at a food dish for several days, but only between 1600 and 1800 hr. On the test day, when no food was provided, an observer recorded the time at which the bees came to the food dish. Except for bee #11, which came twice between 0700 and 0800 hr, most of the visits occurred at the time when the bees had originally been fed at the dish.

the brain of a Cecropia into a Pernyi. These experimental animals emerged at the time of the brain-species, not that of the body-species. A hormone has to be released for the moths to emerge, and Truman's experiments show that the clock timing that release is controlled by the insect's brain.

However, circadian rhythms also appear in animals without brains. Plants, of course, lack brains, but they are complex organisms. A remarkable case of a circadian clock in a simple organism has been uncovered by the research work of Beatrice M. Sweeney at the University of California at Santa Barbara, who has detected circadian rhythms in the one-celled green alga *Gonyaulax polyedra* (Figure 25–1). This marine flagellate is only 0.05 millimeter in diameter, yet it shows rhythms in photosynthesis, capacity for luminescence, and cell division. Exposure to six hours of darkness shifts all the rhythms similarly, suggesting that they are controlled by a single timer.

The emergence time of a moth and the daily patterns of a *Gonyaulax* or a human are expressed at a *specific* time in relation to a zeitgeber. In some cases, however, the output of the circadian clocks is indefinite, being closely related to stimuli other than the zeitgeber. The best-known case is the behavior related to the circadian-controlled time sense of the honeybee, which was first observed by serendipity.

The story begins with the domestic habits of Auguste Forel, an early-twentieth-century Swiss psychiatrist whose reputation rests on his studies of ants. Forel was a man of regular habits, who liked to have breakfast with his family on the terrace near the garden each morning at the same time. Forel liked sweet jams, of which the bees were also fond. Forel noticed that the bees came only at the regular breakfast time, not at other hours. They came at that hour even when breakfast was served inside and there were no olfactory cues to summon them. Forel wondered whether the bees could tell time.

A few years later, a German biologist named von Buttel-Reepen observed that honeybees promptly left fields of buckwheat in midmorning, when the flow of nectar ceased. The buckwheat smelled just as sweet in the afternoon as in the morning, yet the bees did not return. Like Forel, von Buttel-Reepen wondered whether they could tell the time of day and knew that afternoon visits would be futile. He guessed that they had an internal clock, for which he coined the term *Zeitgedachtnis* (German for "time memory").

The time memory was not investigated until 1929, when Ingeborg Beling, a graduate student working under Karl von Frisch at the University of Munich, began her experiments. She marked groups of bees, then trained each group to feed only at particular stations at particular times of the day. As expected from Forel's and von Buttel-Reepen's observations, the bees came to the feeding stations at their training times only, and they came even when the stations were empty. However, this observation did not prove that the bees could tell time, since they might be responding to an environmental cue like temperature, humidity, or sunshine. Next Beling trained bees in the laboratory under constant conditions of light, temperature, and humidity. The bees remained as regular as before, coming to the feeding dishes every 24 hours (Figure 25–6). Then Beling imposed a 19-hour day—for example, 10 hours of light and 9 hours of dark—on the bees. Still they kept their 24-hour day. Beling assumed that they regulate their coming and going by some external factor coinciding with the earth's 24-hour rhythm.

The obvious candidate was the sun. Karl von Frisch had already shown that bees use the sun as a compass. He had also found that bees shifted the direction of their waggle dance with the sun's movement, even in a darkened hive. The bees seemed to know where the sun was without seeing it. The question now was whether they perceived the sun's position from an endogenous time sense or from subtle environmental cues.

Von Frisch and Max Renner designed an experiment to answer this question. In the summer of 1955, they trained 40 bees in a room in Paris, feeding them only between 8:15 and 10:15 A.M. French Summer Time. A room identical to the one in Paris was prepared at the American Museum of Natural History in New York City.

After the bees were trained, the hive was sealed and flown to New York. The flight involved a 5-hour change in time zones. If the bees were responding to an internal time sense, they would feed 24 hours after their last feeding in Paris, but if they told time from environmental cues, they would feed 29 hours later. The answer was clear. When the hive was opened, the bees fed at 3:15 P.M. Eastern Daylight Time, 24 hours after their last Paris feeding. Time sense in the honeybee is endogenous, being genetically determined on a 24-hour "clock." The sun is the environmental cue that sets it; but, as Beling's experiment showed, the "clock" cannot be made to run at a different speed.

Why are there circadian rhythms? The adaptiveness of some circadian rhythms is obvious. If mice, flying squirrels, cockroaches, and other nocturnal animals lacked an endogenous alarm clock, the only way they could tell whether night had fallen would be to get up and take a look. That strategy is energetically wasteful and possibly dangerous, since the animal would expose itself to day-active predators. Another profitable adaptation of a circadian rhythm can be seen in the emergence rhythm of the desert fruit fly *Drosophila melanogaster*. The fly's emergence coincides with dawn (Figure 25–7). This timing allows the flies to escape the pupal case while the air is still moist and cool, avoiding the risk of desiccation present later in the day.

In other cases, however, the precise adaptive significance of a rhythm is a mystery. For example, the rhythmic leaf movements of some plants serve no known purpose. Bünning suggests that the rhythm prevents photostimulation by moonlight, ensuring that the accounting of daylight hours needed for photoperiodicity is not thrown off. Most likely, the significance of these mysterious rhythms lies in unique ecological and physiological demands on the organism in question, demands that biology still understands poorly.

Longer Rhythms Certain organisms have rhythmic activities following intervals longer than a day. One example is the marine midge *Clunio marinus*, which responds to the cycle of the moon. The reproducing midges lay their eggs on tidal flats exposed by the lowest tides, which occur twice a month. The developing midges must emerge from the pupae at the same low tides, or they will drown. The curious thing about this rhythm is that it occurs not once a month, roughly equal to the cycle of the moon, but twice a month, equal to half the moon's cycle. This cycle is therefore called a *semilunar rhythm.*

Klaus Neuman in Germany found that a short exposure to moonlight entrains the semilunar cycle in the midges. His study of the midges' emergence rhythm showed that adding four nights of "moonlight"—actually

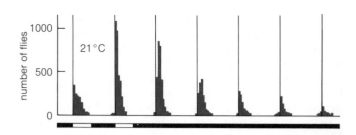

Figure 25–7 Circadian rhythm of emergence of fruit flies, *Drosophila*, from their pupae. Lower broad line indicates darkness interrupted by two periods of light each of 12 hours. The figure shows that when flies of the population are ready to emerge they do not do so at random. Most of them emerge near lights-on, or at dawn. In the laboratory in continuous darkness, different individuals of the population continue to emerge at approximately 24-hour intervals (indicated by thin vertical lines), which correspond to the times of dawn if the "night" were not continuous.

faint light—to the regular light-dark cycle of the day resulted in peak emergences every 15 days.

Generally, researchers have thought that annual activities like hibernation, molting, and migration were due to photoperiodicity. However, some animals retain their annual rhythms even when the photoperiod is held constant. Eberhard Gwinner of the Max Planck Institute took a number of juvenile European warblers (family Sylviidae) from the nest in northern Europe. He divided the birds into four groups. Two groups were kept in Europe, and the other two were taken to Central Africa, the warblers' usual winter range. In both sites, one group of birds was kept in constant laboratory conditions of 21°C and a day divided into 12 hours of light and 12 hours of dark, and the other group was kept outside, where temperature and photoperiod changed normally. All four groups displayed the same migratory restlessness and physical changes on the same annual cycle. Gwinner has kept some of these hand-reared warblers at a constant photoperiod for seven years, and they still molt, put on weight, and grow restless on an annual cycle. He concludes that these changes are annual rhythms, not behaviors prompted by changes in photoperiod.

In the early 1960s, Eric T. Pengelley of the University of California at Riverside and Kenneth C. Fisher of the University of Toronto observed a similar phenomenon while researching hibernation in the ground squirrel *(Citellus lateralis).* Normally these animals put on weight, mostly fat, before hibernation, then lose the weight as it is consumed by body metabolism during hibernation. Pengelley and Fisher kept squirrels under conditions of ample food and constant light and temperature for periods up to four years, and the animals still exhibited rhythmic weight gains and losses. These cycles—called "circannual"—were independent of temperature. At 35°C (95°F), the squirrels do not hibernate, but they still gain and lose weight as if they did.

Humans may have these annual cycles, too. Franz Halberg and his associates have observed an annual variation in the level of 17 ketosteroids excreted in the urine. The lowest level is in May, the highest in fall and winter. Manic-depressive psychosis, in which euphoria alternates with crushing depression, generally follows a precise annual cycle in each psychotic. The stumbling block to finding out more about annual rhythms in humans is that subjects must be isolated for several years from all environmental clues—such as sunlight or temperature cycles—that might give clues of the season. Thus, volunteers are few.

On Escaping Stress

The first frosts sparkle in the forests of the eastern United States, and the pattern of life changes dramatically. In the summer, with its lush annual plant growth and myriad insects, food has been abundant. But as the cold deepens toward winter, the insects disappear and the plants die back, their reproductive cycles at an end. The crisis worsens with the first snows. Bark and bud foragers like the porcupine and the ruffed grouse can still find food, as can the nuthatches, chickadees, and other birds that prey on insect eggs and pupae. But for many forest dwellers, survival requires special measures. Many of the birds leave, migrating to warmer climates. As for the chipmunks and woodchucks, they find a suitable den and settle down to a winter's hibernation.

Going Underground: Hibernation *Hibernation* is a physiological state in which an animal lowers its body temperature for a prolonged period, becomes inactive, and conserves energy. No clear-cut line separates hibernators from nonhibernators. Instead, there is a continuum of physiological variations.

At one end of the spectrum is a typical nonhibernator, for example, a pine siskin or some other northern finch. William Dawson at the University of Michigan has shown that these birds can maintain a high body temperature even at −60°C. However, even though these birds have a high capacity for endothermy, their body temperature drops slightly at night. This pattern, as we noted earlier, is also true of humans. Keeping body temperature high consumes large amounts of energy, and on really cold nights, siskins will use up half of their body fat.

Some birds enter a much deeper nocturnal torpor, thus saving their fat stores. In 1954, Oliver P. Pearson

of the University of California at Berkeley showed that the nighttime temperature of the Anna's hummingbird *(Calypte anna)* drops so low that the bird can be plucked from its roost like a plum from a branch. This nocturnal torpor allows great savings in the hummingbird's energy economy. George A. Bartholomew and his associates at the University of California at Los Angeles have studied similar relationships between food supply, torpidity, and energy expenditure in small desert rodents. Vance Tucker from Duke University discovered that the tiny pocket mouse *(Perognathus californicus)* becomes torpid for days or weeks when food supplies are low. If food is abundant, the pocket mouse remains active, no matter what the temperature. Curiously, the rodent stores food in its den and enters torpor long before these stores are exhausted, suggesting that psychological factors are involved.

True hibernation extends this kind of torpidity over a longer period. As winter approaches, the nighttime body-temperature drop increases each night and lasts longer into the day, until the torpidity of one night overlaps into the next. This kind of hibernation is found in bats, ground squirrels, chipmunks, woodchucks (of particular significance on Groundhog Day), and the European dormouse, which may hibernate for as long as nine months, although not continuously. There is some truth to the popular Groundhog Day legend that the woodchuck emerges from its den to take a look around. All the hibernators rouse for short times.

Bears do not hibernate. Like skunks and raccoons, they exhibit only a small drop in body temperature, although they appear to sleep through the winter. During warm spells they awaken and leave their dens. It may take an hour or two for a hibernating woodchuck to become active, but a sleeping bear may be fully roused in a matter of minutes.

During hibernation, the animal's body temperature usually drops close to that of the surrounding air. If, however, the air temperature falls below freezing, then the animal's body temperature holds steady at some set point above freezing. Body-temperature control is thus not relinquished. The hibernating metabolic rate is but a fraction of the nontorpid rate, and life proceeds slowly.

Before entering the winter torpor, most hibernators build up large deposits of fat. This means that the animal enters torpor at the very time when its fat stores could best provide energy to keep body temperature high. The relationship between torpidity and food economy is indirect, complex, and poorly understood. If the woodchuck waited until the supply of clover fell off before physiologically "planning" for winter by storing fat, it would be too late, and the animal would probably die. Instead, the response is indirect, through either a photoperiodic response or an annual rhythm, making sure that fat is stored before it is needed.

What mechanism triggers the deposition of fat? Recently, Albert M. Meier of Louisiana State University has found that the release of an adrenal cortical hormone and prolactin affect fat build-up. Probably these hormones are released in response to changes in the usual function of the brain. N. Mrosovsky of the University of Toronto has shown that destroying certain portions of the hypothalamus (Chapter 20) makes rats, which are nonhibernators, act like hibernators nearing winter. The rats tend to gain weight, and although sluggish at hunting for food, they gorge themselves when they find it. This research on brain function, hormones, and fat deposition may eventually show why some humans are obese and remain so despite many frustrating attempts to lose weight.

Leaving Trouble Behind: Migration   Rather than become dormant and wait winter out, some animals abandon one area and travel, or *migrate,* to a more hospitable range. Migratory behavior occurs primarily in animals that can travel long distances with comparative ease, like the ungulate mammals and the birds. Migration in these animals probably evolved more as a response to reduced food supplies than as a response to temperature extremes. For example, cardinals, which normally migrate south in winter, will remain in the northern states if they can find a steady food source like a backyard feeder; and bluejays, which normally do not migrate, will leave during particularly harsh winters. In most cases, however, migration behavior is fixed genetically, and the animals travel no matter what the local food conditions.

Migration may also have an adaptive significance other than ensuring food supplies. Food supplies in the ocean are fairly constant, yet many wide-ranging sea animals, like whales, seals, and many marine birds, congregate at specific breeding grounds at specific times. Probably these migrations aid successful reproduction by congregating all the individuals of a widely dispersed species in one place.

Humans have long been fascinated by the migratory virtuosity of birds. Most of the songbirds of the United States fly 1,000 miles or more twice a year, from Central or South America to North America and back again. The greatest migrator of them all is the Arctic tern. This bird breeds within a few hundred miles of the North Pole. As winter approaches, the tern flies over the Arctic ice, crosses the Atlantic, and follows the European and African or South American coasts to summer grounds close to the Antarctic (Figure 25–8).

In many bird species, the young of the year fly on their own, without guidance, yet they still find their way. Other birds fly at night, when landmarks are of little value. How do the birds tell where to go?

It has long been known that caged migratory birds become restless in the spring and fall, when but for the

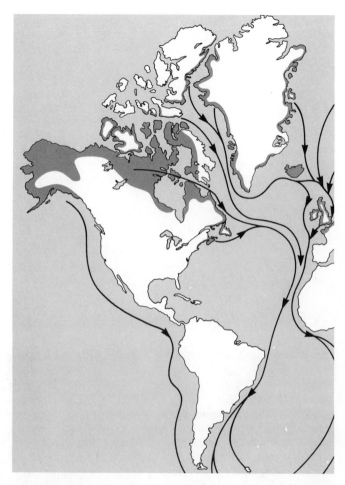

Figure 25–8   Migratory route of the Arctic tern *(Sterna paradisea)* from its breeding grounds (in black) to its wintering grounds off the coasts of South America, Antarctica, and Africa.

cage they would be migrating. In the early 1950s, the late Gustav Krämer at the Institute for Marine Biology in Wilhelmshaven, Germany, designed a series of pioneering experiments around migratory restlessness. For his subjects, Krämer chose starlings, which migrate during the day. He confined the starlings to cages with transparent bottoms and spent many days watching from underneath, recording the birds' positions inside the cage every 10 seconds. In October, the birds became restless, as expected, and they gathered in the corner of the cage to the southwest, the direction of their normal migration to Africa. In the spring, the birds faced northeast, the direction back to Europe. These observations held true, however, only when the sun was shining; on overcast days, the birds' movements were random. Krämer blocked

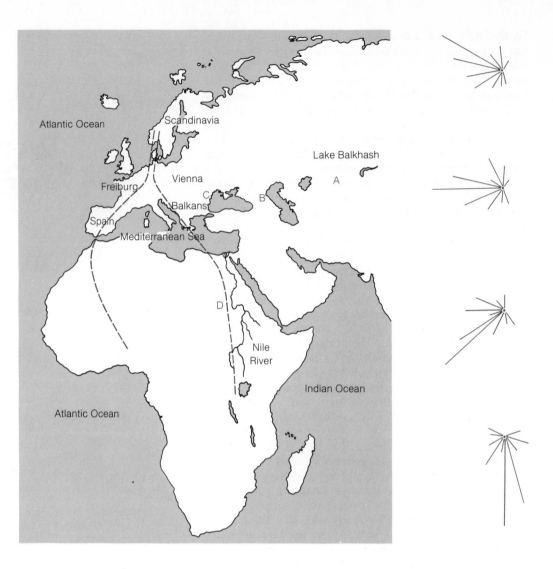

Figure 25–9 Paths of migration (broken lines) of two species of European warblers are not straight. When the planetarium sky was adjusted to positions lateral to the birds' origins (C, B, A), the animals adjusted their "flight direction" on a course that should take them to their usual migratory route.

out the sun and showed it to the birds from another direction in a mirror. The starlings changed their orientation to conform to the new position of the sun.

Krämer's experiments were simple, but they brilliantly highlighted certain facts. First, the starlings had no innate sense of direction; they took their bearings from the sun. Second, the starlings did have an innate sense of time. The sun moves throughout the day, yet the starlings keep the same orientation. Without a time sense by which the birds could compensate for the sun's movement, their orientation would also change.

Schmidt-Köenig experimented further on the internal clock. He showed that the clock can be "reset" either back or ahead by keeping the birds in artificial days out of phase with the natural 24-hour day. For example, 6 hours is to 24 as 90° movement of the sun is to 360°.

Turning lights on 6 hours before dawn and turning them off 6 hours before sunset advances the internal clock 6 hours. Birds so treated orient themselves 90° to the left of the usual direction of migration. Setting the clock back by 6 hours shifts the orientation 90° to the right of the correct direction.

Many birds migrate at night, when the sun is not visible. How do these species find their way? The moon, often the most conspicuous celestial object, changes its appearance and position nightly. Thus, lunar navigation requires a more sophisticated memory and time sense than solar. The stars seem an even more remote and difficult guide. Yet, surprisingly enough, night migrators do steer by the stars. Franz and Eleanore Sauer of the University of Freiburg, Germany, performed experiments similar to Krämer's with night-migrating Old World warblers.

Their caged birds preferred the northeast in April and the southwest in late August, the correct migratory directions. On overcast nights, the birds moved about randomly. Following Krämer's example of the sun in the mirror, the Sauers took their warblers to a planetarium, where the star patterns of any time and place can be projected onto the dome. As predicted from their previous experiments, rotating the star pattern shifted the birds' orientation. Some of the warblers do not follow straight courses, but shift their direction of travel at particular points on the journey. The Sauers found they could change these warblers' orientations by rotating the star patterns characteristic of these points (Figure 25–9).

The Sauers showed that the birds oriented themselves by the stars, but they did not discover which stars in particular gave the birds their bearings. In his research on indigo buntings, Stephen Emlen of Cornell University found that his experimental birds could orient themselves correctly if only a small area of night sky surrounding the North Star was left visible. But if that area was erased and the rest of the sky shown, the birds could not get their bearings. The buntings apparently use the North Star as human navigators have done for untold centuries. The significance of the North Star is that it seems to remain fixed all night long, while the other stars rotate about it. The magnetic sense found in some birds may also aid orientation.

No one knows why some birds migrate at night and others during the day. Generally, the night migrators are small species, whereas the day migrators are large (ducks, geese, and shorebirds). Perhaps the reason for the difference lies in energy demands. The small birds put on massive amounts of fat—up to 50 percent of body weight in some species—before migration, but at the end of the journey, in which they may cover up to 600 miles in a day, they are emaciated. Flying at night allows these little birds to stop during the day and feed. Daytime flying would rule out such possibly crucial snacks and greatly reduce the distance the birds could cover.

Birds sometimes migrate to specific locations, using the same small wintering ground or nesting site again and again. Salamanders, turtles, mammals, and fish may exhibit a similar strong attraction to a home site. Perhaps the most famous examples are the various species of salmon. After years at sea, salmon return to breed in the very fresh-water stream where they hatched. The salmon appear to follow olfactory stimuli, "smelling" their way upstream, since fish with plugged nostrils cannot find the original breeding pool. The eel, another migratory fish, has a sense of smell so acute that it can detect one-half teaspoon of β-phenylethyl alcohol diluted in a large lake. American (Anguilla rostrata) and European (A. anguilla) eels spawn in the Sargasso Sea of the Atlantic Ocean. The young eels return to the lakes and rivers of their respective continents. The journey takes 1 year for the American eels and 2.5 years for the European species. They remain upstream for 8 to 13 years, then become sexually mature and return to the Sargasso Sea to reproduce.

Summary

Environment relevant to different animals changes on a daily, seasonal, and occasionally fortnightly basis. Celestial cues—sun, moon, stars—are reliable indicators not only of time, but also of location for animals that move over short and long distances. But the celestial cues are useful for fixing location in seasonal time and in space *only* when coupled to a time sense. The time sense can be considered analogous to a clock that can be set and reset by the celestial and other cues. But its speed is genetically determined and unvarying with temperature and with all other environmental stimuli. By use of the genetically determined chronometer, set by environmental stimuli, animals synchronize their physiology and their movement to exploit appropriately the temporal changes that are linked with celestial events.

Suggested Readings

Adolph, E. F. 1967. The heart's pacemaker. Scientific American 216(3):32–37.

Axelrod, J. 1974. Neurotransmitters. Scientific American 230(6):58–66.

Baker, P. F. 1966. The nerve axon. Scientific American 214(3):74–82.

Bentley, D., and R. R. Hoy. 1974. The neurobiology of cricket song. Scientific American 231(2):34–44.

Bullock, T. H. 1973. Seeing the world through a new sense: Electroreception in fish. American Scientist 61:316–325.

Casey, F. G. 1973. Fishes with warm bodies. Scientific American 228(2):36–44.

Chapman, G. 1958. The hydrostatic skeleton of invertebrates. Biological Review 33:338–371.

Cohen, C. 1975. The protein switch of muscle contraction. Scientific American 223(5):36–45.

Davenport, H. W. 1972. Why the stomach does not digest itself. Scientific American 226(1):86–92.

Davidson, E. H. 1965. Hormones and genes. Scientific American 212(6):36–45.

Easton, T. A. 1972. On the normal arc of reflexes. American Scientist 60:590–599.

Eccles, J. 1965. The synapse. Scientific American 212(1):56–65.

Edelman, G. M. 1970. The structure and function of antibodies. Scientific American 223(2):34–42.

Emlen, S. T. 1975. The stellar-orientation system of a migratory bird. Scientific American 233(2):102–111.

Gamow, R. I., and J. R. Harris. 1973. The infrared receptors of snakes. Scientific American 228(5):94–100.

Gould, J. L. 1975. Honeybee recruitment: The dance-language controversy. Science 189:685–693.

Guillemin, R., and R. Burgus. 1972. The hormones of the hypothalamus. Scientific American 227(5):24–33.

Hartman, J. P. 1969. How an instinct is learned. Scientific American 221(6):98–106.

Heinrich, B. 1973. The energetics of the bumblebee. Scientific American 228(4):96–102.

Heinrich, B., and G. A. Bartholomew. 1972. Temperature control in flying moths. Scientific American 226(6):70–77.

Hess, E. H. 1958. "Imprinting" in animals. Scientific American 198(3):81–90.

Hildebrand, M. 1960. How animals run. Scientific American 202(5):148–157.

Hodgson, E. S. 1961. Taste receptors. Scientific American 204(5):135–144.

Horridge, G. A. 1977. The compound eye of insects. Scientific American 237(7):108.

Hoyle, G. 1970. How is muscle turned on and off? Scientific American 222(4):84–93.

Hubel, D. H. 1963. The visual cortex of the brain. Scientific American 209(5):54–62.

Huxley, H. E. 1969. The mechanism of muscular contraction. Science 164:1356–1366.

Jacobson, M., and M. Beroza. 1964. Insect attractants. Scientific American 211(2):20–27.

Jones, J. C. 1978. The feeding behavior of mosquitoes. Scientific American 238(6):138–148.

Kandel, E. R. 1970. Nerve cells and behavior. Scientific American 223(1):57–70.

Kimura, D. 1973. The asymmetry of the human brain. Scientific American 228(3):70–79.

Lester, H. A. 1977. The response to acetylcholine. Scientific American 236(2):106.

Levine, S. 1971. Stress and behavior. Scientific American 224(1):26–31.

Lorenz, K. Z. 1958. The evolution of behavior. Scientific American 199(6):67–78.

Luria, A. R. 1970. The functional organization of the brain. Scientific American 222(3):66–78.

Margaria, R. 1972. The sources of muscular energy. Scientific American 226(3):84–91.

Mayerson, H. S. 1963. The lymphatic system. Scientific American 208(6):80–90.

McEwen, B. S. 1976. Interactions between hormones and nerve tissues. Scientific American 235(7):48.

Michael, C. R. 1969. Retinal processing of visual images. Scientific American 220(5):104–114.

Milne, L. J., and M. Milne. 1978. Insects of the water surface. Scientific American 238(4):134–142.

Mulroy, T. W., and P. W. Rundel. 1977. Annual plants: Adaptations to desert environments. Biological Science 27(2):109.

Murray, J. M., and A. Weber. 1974. The cooperative action of muscle proteins. Scientific American 230(2):58–72.

Nathanson, J. A., and P. Greengard. 1977. "Second messengers" in the brain. American Scientist 237(2):108–118.

O'Malley, B. W., and W. T. Schrader. 1976. The receptors of steroid hormones. Scientific American 234(2):32–43.

Palmer, J. D. 1975. Biological clocks of the tidal zone. Scientific American 232(2):70.

Pastan, I. 1972. Cyclic AMP. Scientific American 227(2):97–105.

Pengelley, T., and S. J. Asmundson. 1971. Annual biological clocks. Scientific American 224(4):72–79.

Perutz, M. F. 1964. The hemoglobin molecule. Scientific American 211(5):64–76.

Pike, J. E. 1971. Prostaglandins. Scientific American 225(5):84–92.

Premack, A. J., and D. Premack. 1972. Teaching language to an ape. Scientific American 227(4):92–99.

Rushton, W. A. H. 1975. Visual pigments and color blindness. Scientific American 232(3):64.

Schneider, D. 1974. The sex-attractant receptor of moths. Scientific American 231(1):28–35.

Shepherd, G. 1978. Microcircuits in the nervous system. Scientific American 238(2):92–102.

Smith, D. S. 1965. The flight muscles of insects. Scientific American 212(6):76–88.

Tucker, V. A. 1975. The energetic cost of moving about. American Scientist 69:413–419.

von Frisch, K. 1962. Dialect in the language of bees. Scientific American 207(2):78–87.

von Overbeck, J. 1968. The control of plant growth. Scientific American 219(1):75–81.

Wald, G. 1959. Life and light. Scientific American 201(4):92–108.

Wilson, E. O. 1972. Animal communication. Scientific American 227(3):52–60.

Wood, J. E. 1968. The venous system. Scientific American 218(1):86–96.

Wurtman, R. J., and J. Axelrod 1965. The pineal gland. Scientific American 213(1):50–60.

Zimmermann, M. H. 1963. How sap moves in trees. Scientific American 208(3):132–142.

SIX

Ecology, Population Biology, and Evolution

26

The Ecosystem: Structure and Function

Since the time of Columbus and Copernicus, we have known that we live on a spherical mass spinning around the sun, but it took photographs from space satellites to show the earth as the shimmering blue-and-white jewel it is. These same photographs make it graphically clear that the earth is an isolated, self-contained unit traveling in the dark void of space. Enormous as the earth seems to us who live on its surface, it is at best a medium-sized planet, a speck of dust in the universe. Yet the earth contains a wealth of living things in an intricate web of interrelationships that binds and supports us all. We will study that web in this section.

All earthly life evolved and exists in a film of air, water, and soil only about 9 miles thick. This film is known as the *ecosphere,* or *biosphere.* It can be divided into three major spherical layers: the *atmosphere* (air), the *hydrosphere* (water), and the *lithosphere* (earth), as shown in Figure 26–1. Above us is a thin layer of usable atmosphere no more than 7 miles (about 12 kilometers) high; around us is a limited supply of water in rivers, glaciers, lakes, oceans, and underground deposits; and below us lies a thin crust of soil, minerals, and rocks extending only a few thousand feet into the earth's interior. Here are all the water, minerals, oxygen, nitrogen, carbon, phosphorus, and other chemicals necessary for life. Because we live in a closed system (that is, little matter enters or leaves the earth), these vital chemicals must be recycled again and again for life to continue.

The ecosphere is too large a unit to be studied in detail, so ecologists usually study smaller units, called **ecosystems,** that include the organisms living in the system and the environmental factors related to them. An ecosystem has three main parts: energy flow; an *abiotic,* or nonliving, component of both inorganic matter and nonliving organic matter; and a *biotic,* or living, component. The boundaries of an ecosystem are arbitrary, set as a matter of convenience to the ecologist, but they usually enclose manageable and meaningful bodies such as lakes, streams, ponds, fields, or woodlots.

The Energy Component of Ecosystems

The energy powering the activities of ecosystems flows essentially one way—through and out. As chemical or biological change takes place, some energy is always converted to heat and lost to the system. For this reason, the activities of ecosystems can continue only if usable energy constantly streams in from outside. The outside energy source for the largest ecosystem, the earth itself (the ecosphere), is the sun. Similarly, within the ecosphere, ecosystems depend either directly or indirectly on sunlight for their energy supply. Most ecosystems depend directly on sunlight; they are based on plants that absorb the energy of sunlight and convert it to chemical energy through photosynthesis. Ecosystems that do not depend directly on sunlight use organic matter built up in directly dependent ecosystems as an energy source. For example, in a forest litter ecosystem, fungi, animals, and bacteria oxidize the organic matter of fallen leaves as an energy source. If traced back to ultimate origins, however, the energy for all ecosystems comes from the sun.

Energy flow in ecosystems obeys the same two basic thermodynamic laws as energy changes in any physical or chemical system (see Chapter 2). According to the first law, energy can neither be created nor destroyed. In an ecosystem, this means that the energy flowing into the ecosystem must eventually flow out to its surroundings in exactly the same quantity. For the largest ecosystem, the ecosphere, the first law says that all the energy flowing to the earth from the sun ultimately flows out toward outer space, either as heat or as reflected light.

The second law of thermodynamics states that chemical or physical changes can take place spontaneously only if the total entropy, or disorder, of a system increases. For ecosystems to run, this law exacts a price in the form of entropy for each spontaneous change, activity, or reac-

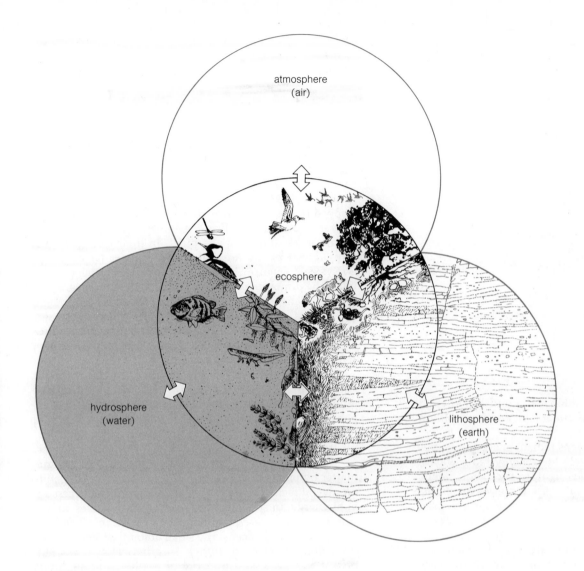

atmosphere
(air)

ecosphere

hydrosphere
(water)

lithosphere
(earth)

Figure 26–1 The ecosphere.

tion anywhere in the system. This entropy toll is eventually paid out as heat, which radiates from the ecosystem and is lost as a potential energy source. To make up for the energy lost to entropy, and to keep the system operating spontaneously, energy must continually be supplied from outside. Payment of the entropy toll is the primary reason why energy flows only one way through ecosystems and does not cycle to be used again.

Solar Energy The sun is a medium-sized star made mostly of hydrogen. At its center, the sun is so hot that a pinhead of heat at this temperature could kill a person 100 miles (166 kilometers) away. At such temperatures (and pressures), light hydrogen nuclei can fuse to form

heavier helium nuclei. In this nuclear fusion, some of the mass of the nuclei is converted to energy. The amount of energy released is enormous.

The sun, then, is really a gigantic thermonuclear reactor some 93 million miles (155 million kilometers) away, liberating about 100,000,000,000,000,000,000,000,000 (10^{26}) calories of energy per second. If we could harness all this energy, each person on earth would have for his or her own personal use, each second, over 70,000 times the annual power consumption of the United States. The sun uses up about 4.2 million tons (4.1 billion kilograms) of its mass every second in producing this enormous amount of energy. However, we need not worry about the sun running out of fuel. In the normal life cycle of stars, the sun is entering middle age. It has probably been

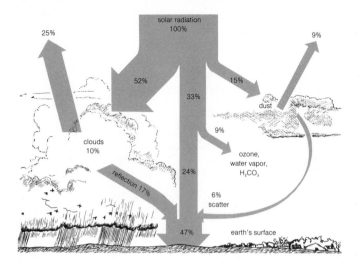

Figure 26–2 The fate of solar radiation as it passes through the earth's outer atmosphere.

in existence for at least 6 billion years and has enough hydrogen left to keep it going for at least another 8 billion years.

The sun's energy leaves it as radiant energy, at wavelengths ranging from relatively low-energy radio waves, through visible and ultraviolet light, to very short, high-energy X rays and gamma rays. Of these wavelengths, only light in the visible and infrared wavelengths is used directly as an energy source by the earth's ecosystems.

Only one-fifty-millionth of the total light energy radiated by the sun falls on the earth's atmosphere. Of this fraction, more than half is reflected back to space by clouds and dust particles or absorbed by the gases of the atmosphere (Figure 26–2). The remaining light, slightly less than half the energy entering the atmosphere, reaches the earth's surface. Some of this light is reflected back into the atmosphere and outer space by brighter regions of the earth's surface, such as sand, snow, rock, and bodies of water. Water also absorbs light, with the result that deeper levels of lakes and oceans receive less light than the surface; no light penetrates to levels deeper than about 100 meters.

Most of the light absorbed in the atmosphere, at the surface of the earth, and within bodies of water is converted directly into heat at this stage and becomes unavailable as an energy source for ecosystems. Even so, this heat energy contributes to ecosystems by creating the temperature necessary for life to exist. In addition, solar heating of land and air masses causes water to evaporate and produces air currents that interact to create the hu-

midity, cloud cover, and precipitation typical of different regions of the earth.

Capture of Light Energy in Ecosystems Green plants capture some of the light falling on the earth's surface and make it available to power the activities of ecosystems. In photosynthesis, light in the visible wavelengths is absorbed by pigmented molecules of the plants, primarily chlorophylls and carotenoids. In the process, the absorbed light is transformed into chemical energy in the form of high-energy electrons (see Chapter 4). The energy trapped in the electrons is then used to synthesize carbohydrates, amino acids, and eventually all the complex chemical substances of the plant. In their role as energy traps, green plants are called the *primary producers* of ecosystems (Figure 26–3). In various terrestrial ecosystems, the primary producers may be trees, shrubs, grasses, mosses, lichens, or photosynthetic bacteria. In aquatic ecosystems, aquatic plants or algae are usually the major primary producers. Often, these aquatic primary producers are floating microscopic algae called *phytoplankton* (*phyto* = plant, *plankton* = drifting).

The green primary producers actually absorb about half the light energy falling on them and transmit or reflect the rest. Plants manage to convert about 1 to 5 percent of the absorbed light into chemical energy, "fixed" into the structure of complex carbohydrates, lipids (fats), proteins, and nucleic acids of the plant tissues.

The total amount of energy converted into organic matter by plants in different regions of the earth depends on several factors. One is the availability of the necessary light energy. In some regions, such as land masses almost continuously covered by clouds, or the deeper levels of lakes, rivers, and oceans, the small amount of light reaching plants limits productivity. In other regions, scarcity of water, a basic raw material of photosynthesis, limits productivity even though light intensity is high. For these reasons, ecosystems in the deeper regions of oceans, where light intensity is low, or in deserts, where water is scarce, have relatively low productivity. Figure 26–4 summarizes the relative primary productivity of different major regions of the earth.

After light energy has been converted to chemical energy, we can trace it in the fate of the complex chemical substances manufactured by the primary producers (Figure 26–5). Part of the material produced, primarily carbohydrates and lipids, is used by the plants themselves for respiration and is lost to the ecosystem as entropy or unusable heat energy. Depending on the ecosystem and the particular kinds of plants present, 20 to 50 percent of the captured light energy is degraded (changed to heat) in this way. The remaining chemical energy, which forms the *net primary production* of the plants (the total production before losses to respiration is the *gross primary*

Figure 26–3 A greatly simplified version of a forest ecosystem.

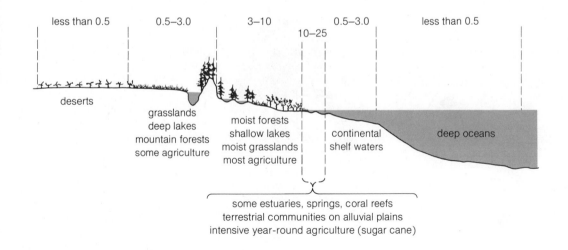

| less than 0.5 | 0.5–3.0 | 3–10 | 10–25 | 0.5–3.0 | less than 0.5 |

deserts

grasslands
deep lakes
mountain forests
some agriculture

moist forests
shallow lakes
moist grasslands
most agriculture

continental
shelf waters

deep oceans

some estuaries, springs, coral reefs
terrestrial communities on alluvial plains
intensive year-round agriculture (sugar cane)

Figure 26–4 Relative primary productivity, in grams of dry matter per square meter per day, of different major ecosystems.

production), is stored in the plant tissues, some to be consumed by herbivores and some to be added to the decaying plant matter forming the upper layers of the soil. These paths of energy on its way through the ecosystem are summarized in Figure 26–6. For the present, we will follow the path supplying chemical energy to the herbivores.

Energy Flow to the Primary Consumers Herbivores get the energy they need for their activities by feeding directly on the tissues of the primary producers. In doing so, they become the *primary consumers* of the ecosystem. In land ecosystems, the main herbivores are grazing animals such as rabbits, deer, cattle, insects, or humans, feeding on leafy plants, shrubs, grass, or mosses. In water

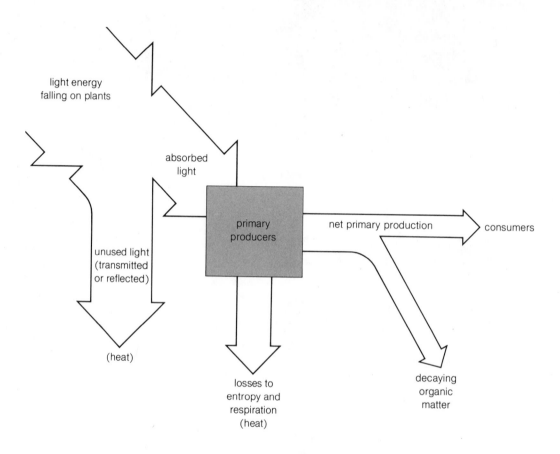

Figure 26–5 Energy flow through the primary producers.

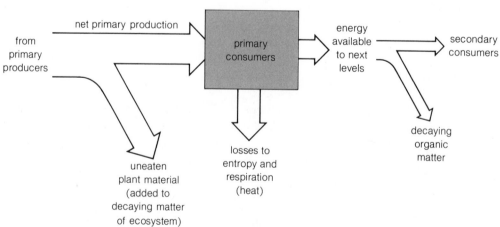

Figure 26–6 Energy flow through the primary consumers.

ecosystems, the primary consumers are usually micro-scopic floating animals and larvae called *zooplankton* (Figure 26–7), feeding on the phytoplankton.

Depending on the type of ecosystem, primary consumers consume and use 5 to 20 percent of the energy stored in the primary producers (see Figure 26–6). Within the primary consumers, about 30 to 60 percent of the energy taken up is used in respiration and lost from the ecosystem as heat; the rest is stored in their tissues. This energy, stored as complex organic molecules in the primary consumers, has the same possible fates as the energy stored in the plant tissues of the primary producers. A part is consumed by carnivorous animals feeding on the primary consumers, and a part is added to the decaying organic matter in the soil. These paths are summarized in Figure 26–8.

The Next Levels in Energy Flow: Secondary and Tertiary Consumers The carnivores of the ecosystem, called the *secondary consumers,* live by capturing and eating the herbivorous primary consumers. Secondary consumers in land ecosystems include animals such as spiders, snakes, weasels, insectivorous birds, and, to some extent, humans. In watery environments, animals such as whales, trout, bass, and cod fill the same role. Microscopic forms, including the protozoa, are also important secondary consumers in some aquatic ecosystems. These carnivores manage to capture about 25 to 30 percent of the energy stored in the primary consumers; the rest is added to the decaying matter of the environment (see Figure 26–8). Secondary consumers pay a high price in respiration and entropy for their hunting and capturing role as carnivores: As much as 60 to 70 percent of the food they capture goes to provide energy for their activities and eventually leaks away to the ecosystem as heat.

The secondary consumers may themselves be consumed by yet another level of larger carnivores such as wolves, lions, and cougars on land, or sharks, tuna, and barracuda in the water. These animals act as *tertiary consumers;* that is, they feed on secondary consumers. Or, as in all the other levels of the ecosystem's producer-consumer chain, the secondary consumers may simply die and add their organic matter to the substrate.

Consumers that specialize on prey past the secondary or tertiary levels are rare, because too much energy is lost at each successive level. As a rule of thumb, ecologists estimate that only 10 percent of the energy captured at one level is actually passed on to the next level in the

Figure 26–7 Zooplankton, the small, floating or swimming primary consumers of many aquatic ecosystems (×16).

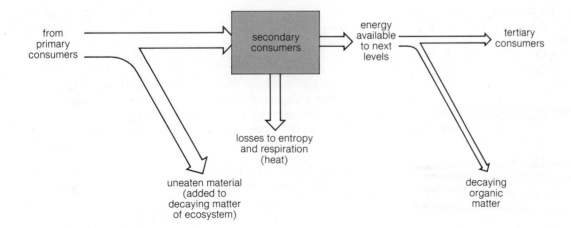

Figure 26–8 Energy flow through the secondary consumers.

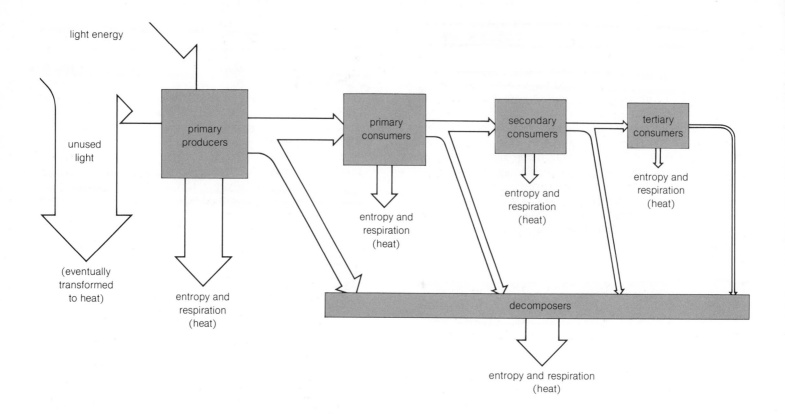

Figure 26–9 Summary diagram showing energy flow through producers, consumers, and decomposers. The size of the boxes gives a relative estimate of the biomass (dry weight of organisms) at each level.

primary producer → primary consumer → secondary consumer → tertiary consumer pathway. That is, for every 100 calories of light energy converted to chemical energy by the primary producers, only 10 calories are likely to be taken up and stored as a part of the chemical constituents of primary consumers. Only 1 calorie of this energy, on the average, finds its way to the secondary consumers to form the complex molecules of the organisms at this level. This leaves only 0.1 calorie for tertiary consumers at the next level, and so on. At this point, little energy is left for further levels of consumers.

The Final Users of Energy: The Decomposers Because herbivores and carnivores are rarely 100 percent effective in eating the level just before them in the energy chain, many plants, animals, and microorganisms simply die. Thus, they add their energy, in the form of complex organic molecules, to the decaying matter of the substrate. This organic matter, still rich in stored energy, is called *detritus*. The organisms feeding on it, such as bacteria,

protozoa, and small aquatic and terrestrial worms, are called *detritus feeders,* or *decomposers.* At this level (Figure 26–9), most of the remaining energy of the organic matter is used for respiration and lost to entropy in the form of heat. Some decaying organic matter is simply oxidized by the molecular oxygen of the atmosphere, either slowly, over the ages, or rapidly, as when fire sweeps through the litter of a forest floor. Eventually, all the chemical energy of the organic molecules remaining in the ecosystem as detritus is converted to heat by one of these two paths.

How much of the energy captured by the primary producers goes directly to the decomposers depends on the ecosystem. In some land environments, such as a mature forest (Figure 26–10), herbivores eat comparatively little of the vegetation. In this situation, most of the organic matter produced by the photosynthetic plants falls to the forest floor as detritus, to be gradually consumed by decomposers or oxidized by molecular oxygen. In forest ecosystems, the total energy flowing through this route may be 10 times as great as in the primary producer →

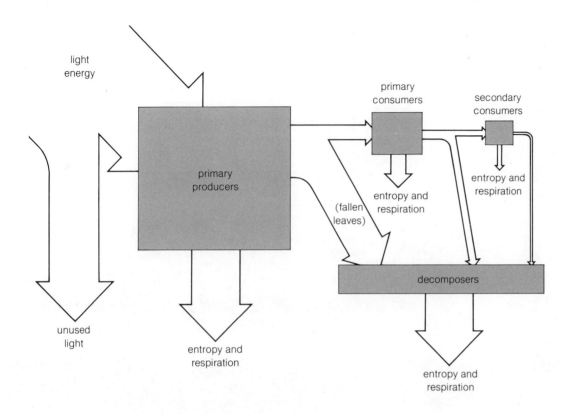

Figure 26–10 Energy flow in a mature forest ecosystem. Most of the energy flows directly from producers to decomposers; relatively little travels through the primary consumer–secondary consumer path.

Diagram labels:
light energy

primary producers

unused light

entropy and respiration

(fallen leaves)

primary consumers

entropy and respiration

secondary consumers

entropy and respiration

decomposers

entropy and respiration

primary consumer → secondary consumer route. In other ecosystems, such as the phytoplankton → zooplankton → whale ecosystem of marine environments (Figure 26–11), most of the energy flows from producers to consumers and comparatively little is left to the decomposers.

Trophic Levels, Food Chains, Food Webs, and Biomass The various levels of producers and consumers in the energy chain of an ecosystem are called *trophic levels* (from the Greek *trophikos,* "nourishment"). The primary producers form one trophic level, the primary consumers the next trophic level, and so on through the decomposers, the last trophic level. The entire chain of producers, consumers, and decomposers is called a *food chain.*

Food chains in natural ecosystems rarely run in the straight pathways we have described. Some organisms are omnivorous and eat plants, animals, and even detritus. For this reason, energy pathways in natural ecosystems often look more like a network than a straight line. Thus,

ecologists coined the name *food web.* Figure 26–12 shows a greatly simplified food web.

At any trophic level of an ecosystem, the amount of energy available is one of the primary factors limiting the number of organisms present. Ecologists often express the number of organisms living at any trophic level, or in the entire ecosystem, in terms of their weight or *biomass.* Usually, they use the dry weight of organic matter to express biomass, making direct comparisons between different trophic levels easier. In energy flow diagrams such as Figures 26–10 and 26–11, the size of the box enclosing the organisms at each trophic level indicates the relative amount of biomass.

Obviously, the energy potentially available to organisms in the food chain depends on how close their trophic level is to the primary producers. At the level of herbivores, energy has suffered only one loss to respiration, in the primary producers, and comparatively large amounts of energy persist in the form of organic molecules. By the time energy reaches the secondary consumers, much of it has been lost through respiration and

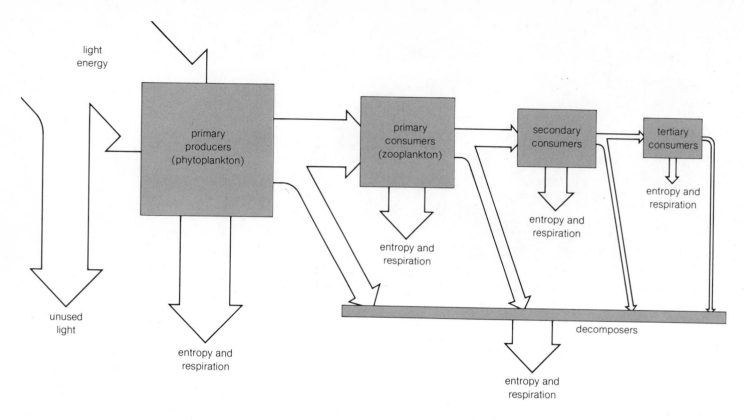

Figure 26–11 Energy flow in a marine ecosystem. Most of the energy flows through producers to consumers and relatively little through the producer-decomposer path as in forests.

inefficiency in feeding at intervening levels. This lesson is one our own species needs to learn, since we can enter ecosystems as either primary or secondary consumers. Vegetarians (primary consumers) use the energy captured by plants much more efficiently than meateaters (secondary or tertiary consumers). As the world's population soars and food energy grows scarcer, survival may depend on more of us becoming accustomed to a meatless diet, to being primary rather than secondary consumers.

The Abiotic Components of Ecosystems: Cycles of Matter

The nonliving chemical elements and compounds of the environment make up the abiotic component of ecosystems. Within smaller ecosystems of the ecosphere, inanimate matter may either flow or cycle. In a river or stream ecosystem, for example, some matter may enter from upstream, some may stay to cycle among atmosphere, water, and bottom sediment, and some may be lost with the outflow. However, the total matter of the ecosphere re-

mains intact and cycles in constant quantities, rather than flowing through as energy does. There is no net input of matter from outer space (except for the negligible quantities entering in meteorites), and almost no loss. Ecologists call these major, perfect cycles of the ecosphere *biogeochemical cycles (bios* = life, *geo* = earth).

More than 40 chemical elements, some in pure form and some in combination with other elements, are important in the major cycles of the ecosphere. Some of these elements, called *macronutrients,* cycle in relatively large quantities. These include the elements found in highest concentrations in living matter: carbon, oxygen, hydrogen, nitrogen, phosphorus, potassium, calcium, magnesium, and sulfur. Others, the *micronutrients,* shuttle in much smaller quantities or in trace amounts. Among these are the elements iron, sodium, chlorine, copper, zinc, boron, vanadium, and cobalt. (For a complete discussion of mineral nutrition see Chapter 15.)

In all cases, these elements cycle between large natural reservoirs and temporary lodgings within living organisms. There are three of these natural reservoirs: the earth's atmosphere, its bodies of water, and its sedimentary rocks. Most of the gaseous macronutrients, including

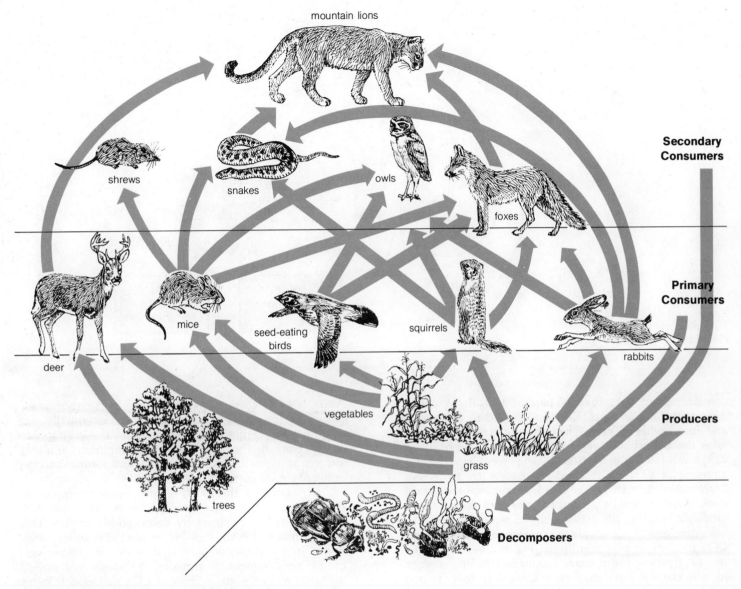

Figure 26–12 A greatly simplified terrestrial food web.

hydrogen, oxygen, nitrogen, and carbon in the form of carbon dioxide, are pooled in the atmosphere and (in solution) in bodies of water. Other elements, including the macronutrients sulfur, potassium, phosphorus, calcium, and magnesium, and all of the micronutrients, are pooled in sedimentary rocks of the earth's crust, from which they are slowly released by weathering and erosion, and to which they slowly return through the gradual depositing of new rock.

This section describes three vital biogeochemical cycles: the *carbon cycle,* the *phosphorus cycle,* and the *water cycle.* The nitrogen cycle, also important to ecosystems,

was described in Chapter 15. The nitrogen and carbon cycles are based on reservoirs in the atmosphere and bodies of water. The remaining biogeochemical cycle, the phosphorus cycle, is based on a sedimentary rock reservoir.

The Carbon Cycle   All organic compounds have carbon backbones. In addition to this structural role, carbon is the immediate energy source for almost all organisms. The major reservoirs of carbon (CO_2), as already mentioned, are the atmosphere and the sea. The atmosphere

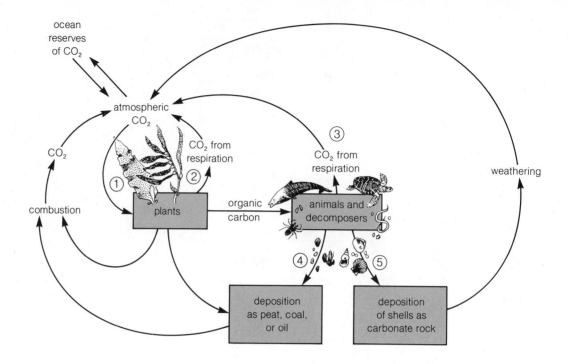

Figure 26–13 The carbon cycle.

contains about 0.03 to 0.04 percent CO_2 by volume. About 50 times as much CO_2 is dissolved in the oceans and other bodies of water. The larger ocean reservoir regulates the atmospheric reservoir. That is, if atmospheric CO_2 levels fall below the 0.03 to 0.04 percent level, greater amounts of CO_2 are released from seawater to restore the balance. On the other hand, if atmospheric levels rise above 0.03 to 0.04 percent because of increased respiration or burning, the ocean tends to absorb more CO_2.

Atmospheric CO_2 enters the carbon cycle through CO_2 fixation in photosynthesis (Figure 26–13, step 1). In CO_2 fixation, the primary producers use light energy to add electrons and hydrogen molecules to CO_2 and produce carbohydrate units (CH_2O; see Chapter 4):

$$CO_2 + H_2O \xrightarrow{\text{light}} (CH_2O) + O_2$$

Plants then use carbohydrate units to synthesize carbohydrates and lipids, or, after adding nitrogen, to synthesize amino acids, proteins, nucleotides, and nucleic acids.

The carbon cycle is completed through respiration and burning of the organic compounds synthesized by plants. The plants themselves use some of this material as an energy source (step 2, Figure 26–13), in a reaction that releases free energy and returns carbon to the atmosphere as CO_2:

$$(CH_2O) + O_2 \longrightarrow CO_2 + H_2O + \text{energy}$$

Animals and decomposers use the organic carbon compounds of the plants as an energy source, eventually also returning the carbon fixed by plants to the atmosphere as CO_2 (step 3). The complex carbon compounds of plants are also converted directly into CO_2 and returned to the atmosphere by forest or grassland fires.

A fraction of the fixed, organic carbon compounds of plants and (to a lesser extent) animals enters the sediment but is not oxidized by decomposers or fire. This organic material (step 4 of Figure 26–13) is slowly converted into deposits of peat, coal, or oil, the fossil fuels of our environment. Eventually, this deposited carbon also returns to the atmosphere as CO_2, either as it burns in natural fires or volcanic activity, or as humans use it for heating, lighting, or powering engines.

Some carbon is deposited as sediment in the form of carbonates derived from the shells of marine and freshwater animals (step 5, Figure 26–13). This material, gradually compacted into carbonate rock, eventually also gives up its carbon to the atmosphere as CO_2, through uplift and weathering.

The carbon cycle is essentially perfect in that almost all the atmospheric carbon dioxide fixed by plants returns to the atmosphere through respiration, natural fires, weathering of carbonate rocks, and burning of wood and fossil fuels by humans.

The part of the carbon cycle of greatest concern to humanity at present is the sedimentary part, the deposition and gradual conversion of organic carbon com-

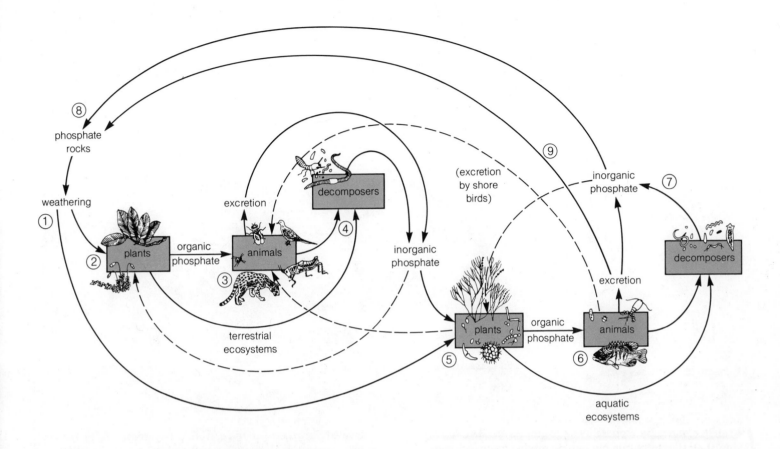

Figure 26–14 The phosphorus cycle. The greatest quantities of phosphorus in most ecosystems follow the paths indicated by the solid lines.

pounds to coal and oil. Only a small fraction of organic matter actually enters this part of the cycle, which takes millions of years to run its course. Since the beginning of the Industrial Revolution, we have been burning these fossil fuels much faster than they form—so fast, in fact, that the available reserves may be exhausted in 50 to 100 years or even less. (Actual estimates, depending on the politics and philosophy of the predictor, run from 10 years to inexhaustible.) Severe oil shortages already exist, as we are all aware, and both the world economy and the economic health of individual nations depend more and more on access to the earth's remaining oil resources. When oil and coal reserves run out, we will have to convert to other energy sources, such as solar energy, if our present material living standards are to be preserved and extended to underdeveloped parts of the world.

The Phosphorus Cycle   Phosphorus is an essential element in nucleic acids, phospholipids, and many proteins.

Because it is essential for life, a lack or a shortage of it in an area can limit plant growth. Countries with highly developed agriculture add millions of tons of fertilizer containing phosphorus (as phosphates) to their soils yearly to correct such lacks.

In contrast to nitrogen and carbon, phosphorus occurs mostly in phosphates deposited in sedimentary rocks of the earth's crust. These phosphates are slowly released by erosion and weathering to enter the phosphorus cycle (step 1, Figure 26–14). Carried to the soil by water runoff, they are taken up and used directly by plants to synthesize their organic phosphate compounds (step 2). These organic phosphates are used by the primary and secondary consumers of ecosystems (step 3), who can get their phosphates only in this way (animals cannot absorb inorganic phosphate directly from their surroundings). The terrestrial cycle is closed by decomposers (step 4), which break down the organic phosphate compounds of dead plants and animals and release the phosphorus to the environment as inorganic phosphate. Many animals also regularly excrete large quantities of inorganic phosphate.

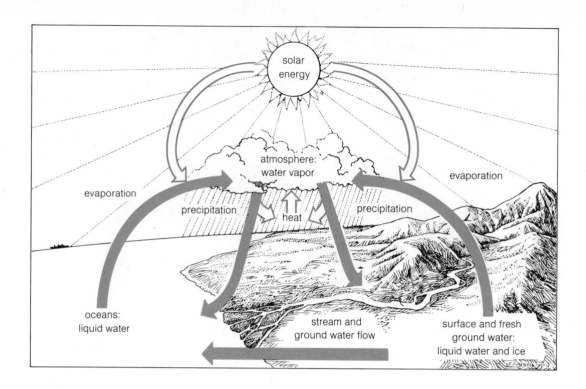

Figure 26–15 The water cycle.

The aquatic phosphate cycle follows a similar course. Almost all the inorganic phosphate eroded from sedimentary rocks drains into streams, lakes, and seas, either directly or after moving through the terrestrial phosphate cycle. The phosphorus entering the aquatic environments is used there by primary producers (step 5, Figure 26–14) and flows in turn as organic phosphates to the consumers of aquatic ecosystems (step 6). After death and decomposition of aquatic plants and animals, phosphorus is released as inorganic phosphate that settles to the bottom as sediment (step 7). In marine environments, which eventually receive most of the phosphate eroded from terrestrial rocks, the phosphate settles to the ocean floor in large quantities. Gradually, over millions of years, the phosphate sediments harden into rock, completing the phosphate cycle (step 8). This phosphate rock of the ocean floor may again enter terrestrial and aquatic phosphate cycles if it is raised and exposed to the air and weather by movements of the earth's crust.

Some of the phosphate entering marine environments returns to the land in a side cycle through shorebirds that feed on marine organisms (step 9, Figure 26–14). These birds, among them pelicans, cormorants, and gannets, excrete phosphate wastes (from digesting fish) on their shoreside nesting grounds. The excreted material, called guano, piles up in surprising quantities—between 300,000 and 400,000 tons a year. We mine the deposits for fertilizers and other chemical needs. Some phosphate

is also returned to terrestrial ecosystems by terrestrial organisms, such as humans, that feed on marine life.

Many ecologists are concerned that we are upsetting the balance of the phosphate cycle by our increased use of phosphates for fertilizers and detergents (see Supplement 26–1). Every year we add millions of tons of mined phosphate to the natural flow reaching lakes and seas, as runoff from fertilized fields and detergent wastes enter streams and rivers. Ocean phosphate sediments are now building up much faster than they can return to land through geological uplifting of the ocean floor, a process that requires millions of years. As a result, the natural phosphate cycle has become almost a one-way flow of phosphorus from land to sea. Although our terrestrial reserves of phosphate rock are extensive, a continued one-way flow may so deplete them that we will have to mine the oceans for phosphate.

The increased phosphate flow into bodies of water has also had some undesirable side effects on fresh-water lakes and ponds. The phosphate entering these lakes from fertilizer runoff and sewage discharge greatly improves conditions for algal growth, causing "blooms" of algae so dense that the water turns a murky green. Although the increased photosynthesis and oxygen production might be expected to improve conditions in the lake, the opposite actually happens. The algae foul the shorelines with green scum. At the end of the growing season, they die and sink in rotting masses to the bottom, where bac-

precipitation

recharge

percolation

percolation

well

water table

artesian well

impervious stratum

zone of saturation

impervious rock

aquifer

Figure 26–16 The ground-water system.

terial decomposition and respiration soar until they deplete the oxygen of the lake. This condition may kill most of the lake's animal life, including both insect and crustacean larvae and the game and commercial fish that feed on them.

We call this rapid accumulation of organic matter (and the resulting oxygen depletion) *eutrophication.* It has turned countless lakes and ponds into dead or dying bodies of water, where only bacteria, a few species of worms, and such fish as carp can survive.

The Water or Hydrologic Cycle Water forms a dynamic, continuous web that maintains and unites all life by cycling throughout the ecosphere. In this vast cycle (Figure 26–15), our fixed supply of water is renewed again and again. Although local supplies may be temporarily exhausted, the total supply of water on earth is relatively

constant. A few molecules of the water you drank or washed in today may once have been used by Cleopatra or Plato.

Solar energy drives water upward from the oceans, lakes, rivers, and continents by evaporation; gravity pulls it down again as rain, snow, or hail. Some of the water that falls on the land sinks, or percolates, downward into the soil and subsoil to form the ground-water system (Figure 26–15). But the soil, like a sponge, can hold only so much water. If the rain falls faster than the water percolates downward, water begins to collect in puddles and ditches and to run off into nearby streams, rivers, and lakes. This runoff causes erosion and sometimes water pollution in various forms.

The water that is not picked up by plant roots sinks on down. Eventually the water stops its downward movement and fills all the cracks and crevices between soil particles and bedrock. The top of this zone of saturation is the *water table* (Figure 26–16). The bottom of this zone contains the *aquifer,* a porous rock layer that traps flowing water between two impervious rock layers. In areas that have heavy rainfall, the water table may be only a few feet below the land surface; but in dry areas, it may be hundreds or thousands of feet down—or may not exist at all. In some places, water that collected in prehistoric times has been buried deep underground by geological processes. This water can be tapped and withdrawn for use—and is being tapped in north Africa today—but since it is replaced very slowly if at all, it should be considered an irreplaceable resource.

Because of the water cycle, the hundred or so pounds of water in your body are replaced a number of times each year. Water in a tree is replenished thousands of times as it grows. The water in the air falls and is replaced about every 12 days. River water is replenished within weeks, while lakes take from 10 to 100 years, depending on their depth. The ocean's water is renewed by rivers and streams about once every 3,600 years, and polar ice takes about 15,000 years.

Limiting Factors in Ecosystems: Climate and Soil

What determines the particular combination of plant and animal species found in a given ecosystem? Why is one area a desert, another a forest, and another a grassland? The particular type of *biome,* or major life zone, found in a given place at a given time results from the particular combination of temperature, precipitation (rainfall, snowfall, fog, and dew), and soil structure and chemistry found there.

As a general rule, climate determines which plants can live in an area. Climate and plants together determine the soil type; and climate, plants, and soil together determine the types and distribution of animals. This pattern has exceptions, but it is a useful generalization.

The particular associations of plant species (forest, grassland, and desert) found at any point on the earth's surface are determined primarily by three climatic factors: temperature, precipitation, and—most crucial—moisture. Moisture is determined by the amount and seasonal variation of precipitation and the water loss in the region through evaporation (transpiration) from the leaves and other exposed surfaces of plants. This water loss is greater at higher temperatures. Temperature also affects precipitation. Thus, the three basic climatic factors are all bound up together.

The soil is a dynamic body, always changing in response to the prevailing climate, types of plant life, and use or abuse by humans. Theoretically, it is a renewable resource, but in practice it comes back so slowly that it should be considered essentially nonrenewable. Soil is formed by interaction among climate, topography, parent rock material, deep-rooted plants, decay organisms, and time. It is produced and modified by the weathering of rocks and by the plant and animal organisms that live and die in it. Myriads of bacteria, fungi, protozoa, worms, mites, and small insects live in the soil. They burrow into it and help maintain its porosity, but their main function is to break down organic material such as leaves and dead animal bodies that fall to the soil surface. This decomposition returns vital nutrients to the soil in a form that plants can use for their growth.

However, a single factor can be crucial. Rainfall, humidity, temperature, or the concentration of an element such as manganese, iron, or phosphorus in the soil can change this combination, or by itself exclude some organisms, influencing the species census in an area and in turn determining the biome or ecosystem type.

This generalization is usually known as the *law of the minimum,* or the *limiting-factor principle:* The single factor that is most deficient in a biome or ecosystem is the one that determines the presence or absence of particular plant and animal species. For example, no matter how warm, sunny, or well supplied with minerals an area is, it can still be a desert if rainfall is only 1 inch per year.

Biotic Factors in the Ecosystem

The final factor in an ecosystem is its living organisms. We have talked of them in this chapter as producers and consumers, as users of energy, and as agents in the cycling of chemical elements, but they play a far greater role than this. Indeed, their interactions with us and with one another are so important to an understanding of the world around us that we devote the following chapter to them.

Summary

In this chapter we considered the basic concepts of ecology. We saw the importance of the second law of thermodynamics in understanding energy flow through the ecosystem. Plants are the only means of trapping solar energy and converting it to a form that can be used by other organisms. Plants are the producers, and all other organisms are consumers. Some of these are primary consumers feeding directly on plants; others are secondary and tertiary consumers feeding on organisms lower in the chain. The many intricate webs of consumers in any ecosystem are all based on the plants as producers. We also saw that ecosystems have both abiotic (nonliving) and biotic (living) components. Important in the abiotic component are the cycles of matter, such as carbon, phosphorus, and water. Biotic factors are also of great importance in the ecosystem, and we will treat them at greater length in the next chapter.

Supplement 26–1
Pollution and Its Control

Ecosystem stability is maintained by three major mechanisms: controlling their rate of chemical cycling, controlling their rate of energy flow, and maintaining a diversity of species and food webs so that stability is not threatened by the loss of some species or food web link. These principles of chemical cycling, energy flow, and species diversity can help us understand some of the major changes and problems that can arise in local, regional, or global ecosystems.

Disrupting the Rate of Chemical Cycling Natural volcanic eruptions spew clouds of dust and sulfur dioxide (SO_2) into the atmosphere. The eruption of Krakatoa in 1883 blew away several cubic miles of land. The resulting blast of dust into the atmosphere affected sunsets and world weather patterns for several years. But we should take no comfort from the fact that nature is also guilty, especially when human pollution approaches or exceeds natural injections. Herein lies the problem. We know that there are limits to nature's absorbing and recycling rates, but we are not certain what these limits are.

One example of the disruption of a chemical cycle is the addition of excess nitrates and phosphates to some lakes from fertilizer runoff, effluents from sewage treatment plants, runoff of livestock manure and phosphates in detergents, and erosion and runoff of soil from natural sources, farming, construction, and mining. This overloading of the nitrogen and phosphorus cycles can lead to eutrophication and the death of fish and other animals in the lakes.

Introducing Synthetic Chemicals: DDT Backlash Chemists have created many new organic compounds, some of which have saved or prolonged millions of lives. Unfortunately, the widespread use of some of these beneficial chemicals, in particular some pesticides (such as DDT) and herbicides, has had unforeseen and undesirable side effects. Especially worrisome are chemicals that find their way into the food chain or other natural chemical cycles.

For instance, malaria once infected 90 percent of the population of Borneo. In 1955, the World Health Organization sprayed the island with DDT and all but eliminated this dread disease. But besides killing the malaria-carrying mosquitoes, the DDT killed other insects, including flies and cockroaches that lived in houses. These insects were the favorite food of house lizards called geckos, which gorged themselves on dead insects and died from the DDT. These lizards, along with dead DDT-laden cockroaches, were then eaten by house cats. As the cats died, the rat population soared. The inhabitants of Borneo were then threatened by a new disease—sylvatic plague carried by fleas on the rats. Cats had to be parachuted into remote regions by the Royal Air Force in "Operation Cat Drop"!

As if this wasn't enough, the thatched roofs of some of the natives' houses began to fall in. The DDT had killed certain wasps and other insects that fed on a particular type of caterpillar that somehow avoided the DDT. With most of their predators eliminated, the caterpillars multiplied and began munching their way through one of their favorite foods, the leaves used as roof-thatching. On balance, the Borneo episode was a success story in that malaria and the unexpected side effects were brought under control. However, it is an excellent example of what can happen when we tamper with an ecosystem, even with the best of intentions.

Some Principles of Pollution A *pollutant* is any material or set of conditions that imposes a stress on or unfavorably alters an individual organism, population, community, or ecosystem beyond normal environmental limits. The range of tolerance to stress varies considerably with the type of organism and the type of pollutant. Thus, determining whether an effect is "unfavorable" may be difficult and highly subjective. To some, a given human alteration of the environment may be unfavorable, but to those whose livelihood depends on the activity it may be favorable or at least an acceptable risk in comparison to the benefits. The nature of tragedy, as Hegel pointed out long ago, is not the conflict between right and wrong, but the conflict between right and right.

On a global basis, most air pollutants come from natural sources. For example, hydrocarbons emitted by trees and plants throughout the world greatly exceed the hydrocarbons from automobiles, refineries, and other industries. Occasionally, this fact is used to argue that, since nature is the major polluter, we should not get so excited about air pollution from human activities. The fallacy in this reasoning is that natural air pollutants are produced and diluted throughout the entire globe. In contrast, human activities concentrate pollutants to potentially dangerous levels in urban and industrial areas.

Materials Balance: Everything Must Go Somewhere To exist is to pollute in some way or another, and to exist at a high level of affluence is to be a superpolluter. Strictly speaking, talk of "cleaning up the environment" and of "pollution-free" cars, products, or industries is a scientific absurdity. According to the law of conservation of matter, matter can neither be created nor destroyed, but only changed from one form to another. Everything must go somewhere. We can never avoid pollution completely; we can only shift its form and location to reduce undesirable effects. We are discovering that there is no such thing as a "throwaway society" and no such thing as a consumer. We do not consume anything. We only borrow some of the earth's resources for awhile—extract them from the earth, carry them to another part of the globe, process them into a new material, and then discard, reuse, or reformulate them. There is no "away"!

In scientific terms, there must be a *materials balance* in which matter put in equals matter put out. We can collect dust and soot from the smokestacks of industrial plants, but these solid wastes must then go into our water or soil. Cleaning up the smoke can even be misleading, if we remove only the visible parts of it, because the invisible pollutants (such as carbon monoxide, ozone, and hydrocarbons) are usually more damaging. They are also harder and costlier to remove. We can collect garbage and remove solid wastes from sewage, but they must either be burned (creating air pollution), dumped into our rivers, lakes, and oceans (creating water pollution), deposited on the land (creating soil pollution—and water pollution if they wash away), or recycled.

We could reduce air pollution from automobiles by using electric cars, but producing more electricity to recharge their batteries would increase air pollution (sulfur oxides, nitrogen oxides, and particulates), water pollution (heat and oil spills), and land pollution (strip mines and uglification). We could shift to nuclear fission power plants not dependent on fossil fuels, but this increases the possibility of releasing highly radioactive substances into the environment either at the plant site or at waste burial sites. We are faced with a problem of trade-offs, which in turn involves subjective and often controversial value judgments. (by G. Tyler Miller, Jr.)

27

Communities and Biomes

Look out your window or walk out your front door. You will find that no matter where you live you are in the presence of plants and animals. Not just any plants and animals, but those characteristic of the region in which you live. You know where you are as much by the plants and animals you see as by the climate. The maples and cardinals of the northeastern United States and the Douglas firs and Steller jays of the Pacific Northwest are as typical of their regions as hard or mild winters and warm or cool summers.

In the last chapter, we examined some of the basic features of the ecosystem and the physical factors that influence it. In this chapter we will look at the biological side of the picture. We will start by looking at communities: how to define them, factors influencing the diversity of species within them, and their distribution. We will conclude with a discussion of the larger picture of plant and animal distribution represented by biomes.

Community Structure

A **community** is a collection of plants and animals occupying a particular space. Biological communities are complex, and it is hard enough to define one, let alone describe it. However, even a casual inspection of a community reveals some aspects of its structure. Certain plants typically dominate: the trees of the forest, the grasses of the prairies, the shrubs and succulents of the deserts, and the herbs of the tundra. Often, we name the community after these dominant plants—grassland, deciduous forest, chaparral. We also could name communities after animals—and we do in the case of coral reefs—but animals are harder to analyze over vast areas. Where plants are not conspicuous, we may name communities after the landforms—such as tundra and taiga (from the Russian words for treeless plains and swampy forests).

In using a set of dominant plants to designate a community, we must not ignore the inconspicuous plants or the animals also present. We must always be aware of how severely we are simplifying. We will return to this point shortly, but for now we will take advantage of the convenience provided by naming communities after their dominant plants.

Recognizing that a community is a collection of plants and animals occupying a particular space, we can see that most have both a *horizontal* and a *vertical structure* in the way individuals of these species are arranged in relation to one another. Horizontal structure is perhaps easiest to see in desert communities, where water supply determines how close together plants can grow and animals can live. Vertical structure is most obvious in forests, where it is often called *stratification* (Figure 27–1). In a tropical forest, for example, there may be several layers of trees, with a few reaching great heights. Below these tree layers may be a level of shrubs, and below that a level of herbs.

Another easily observed aspect of community structure is that it may change from one time of the year or season to the next. This is true not only in temperate-zone forests, where deciduous trees are bare part of the year, but also in grassland and other communities, where one species may be in full flower and conspicuous during a part of the year and inconspicuous the rest of the year. Greater changes, of course, occur over longer periods of time.

If we start with an act that clears a particular piece of ground—say a fire, a retreating glacier, or the abandonment of a cultivated field—and observe the plants that appear over time, we will find a *succession* of communities through time. Each responds to the environmental conditions present during its establishment but, in turn, changes the environment so that a different community can become established. For example, in a typical pattern of succession in the northeastern United States, in the

coniferous forest profile

deciduous forest profile

tropical rain forest profile

Figure 27–1 Stratification in coniferous, deciduous, and tropical rain forests.

case of an abandoned field, small herbs first become established (Figure 27–2). Then shrubs appear and grow, shading and gradually eliminating the herbs. The shrubs are eventually replaced by pines that compete for light and water with the shrubs. In dense pine forests, however, the pine seedlings get too little light to become established. The seedlings of various broad-leafed trees, such as maples and beeches, can survive in the reduced light and will replace the pines. As the maple and beech seedlings can survive in the reduced light on the floor of a mature maple-beech forest, this type of forest is the one that becomes established as the **climax community,** a nearly stable mix of plants and animals that will not be replaced by another group.

There are many types of climax communities, and they may consist of various associations of plants and animals. For example, the deciduous forest may consist of a maple-beech association, an oak-chestnut association, and an oak-hickory association. These various associations blend into one another at their boundaries, which we call *ecotones.*

The Continuum Although it is convenient to describe communities by the dominant plants present, is that really the way organisms are found in nature? Do they actually cluster in discrete associations always found together under the same environmental conditions? Ecologists thought so at first, but now doubt that they do (see Supplement 27–1).

Instead, when environments can be examined carefully enough (plant distributions exactly measured over long distances or as climate changes, for instance), each individual species seems to show a unique pattern. If associations were really discrete, the distribution of species involved in such an association should look like Figure 27–3a; if they formed a continuum they should look like b. Actually, however, as you can see from Figure 27–3c and d, the various species overlap. Thus, as one moves along the earth's surface, there is a continuum of stands of plants, each somewhat different from the others in species composition. As the environment changes in a continuum, so do the species making up the stands. Such data have forced ecologists to modify the discrete-association concept to recognize the fundamental continuum in the distribution of plants and animals related to an environment.

In many areas, several environmental gradients must be considered at once in order to get a reasonable picture of community structure. Going up a mountain slope, for example, altitude increases, mean temperatures decrease, rainfall increases, wind speed increases, growing seasons shorten, and soils change.

From studies of a wide variety of environmental gradients, ecologists have drawn some general conclusions:

newly abandoned

1 year later

10–15 years later

100–200 years later

Figure 27–2 Plant succession in an abandoned field in eastern North America.

1. Ordinarily, productivity and biomass of living organisms decrease along gradients from favorable to unfavorable conditions. In the mountains of the American Southwest, for example, one goes from highly productive and massive forest communities at mid-elevations to low-mass desert communities in one direction and low-mass timberline communities in the other.

2. Community structure as reflected in either vertical or horizontal stratification is simpler at unfavorable extremes of environmental gradients.

3. There is a general correlation between the number of species present and the complexity of interactions between species in communities.

4. Certain growth forms occur in characteristic places along environmental gradients. For example, short, woody shrubs typically occur under specific desert and near-desert conditions, and tree ferns occur in areas of high and regular atmospheric moisture. This makes clear why a general kind of *convergence* (similar structures in similar environments) is common in plant communities that have evolved independently in different parts of the world. For example, plant communities dominated by woody shrubs with small, hard leaves (often evergreen) have evolved in areas of Mediterranean climate (hot, dry summers; cool, wet winters) in the Mediterranean region itself, and in California, Chile, South Africa, and Australia (Figure 27–4). Similarly, temperate rain forests in the Pacific Northwest of the United States and in southern Chile have structural features in common.

Patterns of Species Diversity

In their study of community structure, ecologists are interested in the diversity of species present. What exactly does *species diversity* mean? Suppose we sample two different acres of prairie land and find 20 species on Plot A and 18 species on Plot B. Plot A obviously has a greater species richness. However, suppose 1,000 individual plants were sampled in each plot, and in Plot A 900 of them belonged to a single species, 50 to a second species, and single individuals to several other species. In Plot B, by contrast, no species had more than 100 or fewer than 20 representatives. The distribution of species is more even in Plot B, and although it has fewer species ecologists would say it has greater diversity. Diversity thus measures the *number of organisms* in each species as well as the *number of species* present. In practice, the number of species present is usually a good indicator of species diversity.

Species diversity is a familiar thing. You can hardly walk out of doors without noticing it. You have probably also noticed that some areas have far more species than others. But what are the bases of this species diversity and its local variations?

Resources In general, the more resources a particular environment has—such as food, space, water, sunlight, shelter, and nesting sites—the greater its diversity of species. Consider a community with five species of insect-eating lizards. The lizard species, of different size, can each eat insects of a particular size range. Think of the total array of insects in the community as a **resource axis** (in this case, food is the resource) relative to the lizards. Each species of lizard uses a certain segment of the axis (Figure 27–5), and each species has part of the axis to themselves. However, each species overlaps with at least one other species in its use of this resource axis.

In areas of overlap, competition for food can occur. If the competition grew too fierce, some species might disappear, reducing the community's diversity. If we examine another resource axis, say that of space, we can see how these species might avoid competition. For example, assume that species 1, 4, and 5 live on the ground and 2 and 3 live in trees. Further, species 2 uses small twigs and branches in the crown of trees, while the larger species 3 uses the trunk and main branches within the tree. In this simplified example, we see how species diversity is related not only to the amount but to the diversity of resources.

Equilibrium Many communities with nearly constant supplies of resources show a dynamic balance in the num-

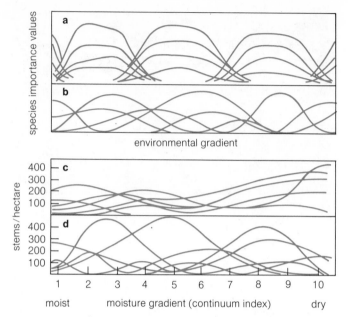

Figure 27–3 Expected and observed plant distributions according to the concepts of discrete communities or continua. (**a**) Curves for discrete communities. Species-importance values are plotted as a function of distance along some environmental gradient. Species occur in discrete units that overlap at their boundaries. (**b**) Curves for continuum. Each species is distributed according to its environmental requirements. (**c**) An actual example. Trees in the forest are plotted as the number larger than 2 cm in diameter per hectare (10^4 m) as a function of position along a moisture gradient. The gradient extends from moist ravines to dry southwest-facing slopes in the Siskiyou Mountains (760–1,070 m elevation) of northwestern California. (**d**) Similar data from the Santa Catalina Mountains (1,830–2,140 m elevation) of Arizona.

ber of species present. That is, they gain species at about the same rate as they lose them. An ecologist would say that the *colonization rate,* or the number of species that arrive and become established per unit of time, and the *extinction rate,* or the number of species that become extinct or disappear per unit of time, are about equal (Figure 27–6).

Obviously, the colonization and extinction rates together control not only species diversity but the stability of the whole system. The interaction between them provides a framework for analyzing community structure and evolution. Using island models at first, we can then extend the idea of species balance to other areas. For example, if on a hypothetical island we find a certain number of species at equilibrium, and the environment of that island suddenly becomes less favorable—grows drier, for instance—the extinction rate should increase (Figure 27–6). If the colonization rate does not also in-

Figure 27–4 The chaparral plants from southern California, shown here, are similar in appearance to chaparral plants found in Chile, South Africa, and Australia.

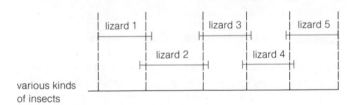

Figure 27–5 Resource axis (insects of increasing size) and five species of insect-eating lizards.

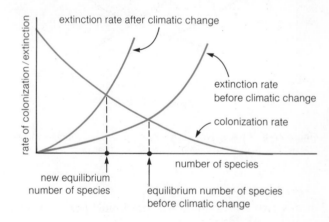

Figure 27–6 Extinction rates, colonization rates, and their role in species diversity.

crease, the new equilibrium number of species on the island will be lower. You can see how the species number at equilibrium will shift, either down or up, depending on conditions. The theory is useful in predicting relations between the total area occupied by a community and the number of closely related species (for example, the number of different types of birds such as finches) present.

Not surprisingly, islands lying off continents usually have fewer species than the nearby mainland. In part, this is because islands generally have fewer resources. But the main reason is that extinction rates are higher

and colonization rates lower than on the mainland. Island populations generally are smaller (and thus more likely to become extinct) than mainland ones, and the water barrier keeps most potential colonists from reaching the island. Extending this reasoning, we expect that the larger

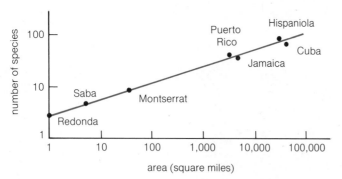

Figure 27–7 The area-species curve for the total number of species of amphibians and reptiles found on West Indian islands of increasing size.

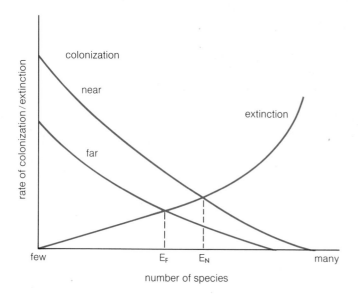

Figure 27–8 An island distant from the mainland equilibrates with fewer species, E_F, than an otherwise identical one near the mainland, E_N.

of any two such islands will have more species than the smaller, and this is in fact the case.

Thus, when we look at a group of different-sized islands, we see a characteristic relationship between area and number of species from any one taxonomic group—perching birds or rodents, for example (Figure 27–7). However, a tenfold increase in island size makes room not for 10 times as many such species, as you might expect, but for only about twice as many. These area-species curves tell us something important about community structure: Each population needs a certain amount of space (including, of course, many kinds of resources),

and species cannot endlessly be added to a community simply by subdividing resources through specialization. There is a practical limit to the number of close relatives that can be supported in an area of limited size, and that number can be predicted for an unknown or unexplored island from data for several known islands, assuming all are at or near equilibrium.

If islands deviate from the area-species curve, we should find out why, for such studies teach us much we can apply to community structure on continents. An island might deviate simply because it is too new to have reached equilibrium. In other words, colonization is still in progress. Or some of the factors normally leading to extinction of related groups—random population fluctuations, competition, and predation (including disease)—might have been upset. Small populations are obviously more vulnerable to extinction by random fluctuations than larger ones. Perhaps a particular colonizing species that has reached a deviating island tends to have large populations. Or predators might be more numerous or competitors more severe on one island than on another. In the case of birds, perhaps some unrelated group had reached the island earlier and reduced the resources available to the group of birds being considered.

However, the most important factor in an island's deviation from area-species curves should be its distance from a source of potential colonizing species (Figure 27–8). The common-sense reason for this, of course, is that water is a barrier to movement of land species; and the more water, the greater the barrier.

History In addition to equilibrium theory, we should look at geological history when we are trying to understand why an area has the diversity of species it does. In Central America, for example, the number of tree frogs (a successful tropical group) and salamanders (a successful north-temperate group) is about equal. We think both groups are approximately in equilibrium. Moving southward into South America, the number of species of salamanders drops greatly, whereas the number of species of tree frogs per unit of area remains about the same or increases. In the vast Amazonian lowlands, where there are nearly 400 tree frog species, there is only a single salamander species.

What causes this drastic decline in salamander species as one travels south? The most reasonable explanation seems to be partly evolutionary and partly geological. Not until Cenozoic times (some 65 million years ago) did the North American salamanders evolve a form suited for life in the tropics. These creatures apparently spread slowly southward. However, until about 5 million years ago, a broad water barrier separated Central from South America. When the barrier was replaced by a land bridge, a period of great change followed. Many North American mammals moved south and contributed to the extinction of certain South American mammals. Some South American groups—the opossum, sloths, and armadillos—moved north. Among the mammals, a near equilibrium seems to have been established. But salamanders travel very slowly; and their colonization of the area is probably only beginning, with equilibrium far in the future.

Habitat Structure Some environments are more complex than others, and this structural element, too, can influence species diversity. For example, forests, with their complex vertical arrangement, contain far more potential habitats than do grassland communities. Thus, they usually contain more species, especially of such groups as birds, which can use these habitats.

Some forests are far more complex than others. For example, two forest environments and their species of salamanders were studied, one in Guatemala and one in western North America. The Guatemalan environment, consisting only of woodland, supported about as many species as the North American one, consisting of both stream and woodland, because the Guatemalan woodland was more complex. It had been enriched by plants called bromeliads living in the trees and forming suitable habitats for some salamanders.

Forests usually become more complex, however, by becoming more highly stratified—divided into more layers—with respect to the available light. The uppermost level, or canopy, consists of trees that typically absorb or scatter about half the energy from the sun. Lower levels get less and less light, as each in turn absorbs and deflects available light. Finally, the herbs at ground level may get only a fraction of 1 percent of the sunlight striking the area.

Species diversity typically increases with stratification. More stratification, to begin with, reflects greater diversity of plant species. It also leads to increased complexity in the habitat, which in turn leads to more diversity of animal species. It should come as no surprise, then, that wet, tropical lowland forests show the greatest stratification and the greatest diversity of species of any in the world.

Climatic Stability There is a theory that a stable climate contributes to diversity of species. For example, in the relatively stable tropical forests, birds can breed throughout the year; in the less stable temperate zones, breeding is restricted to the most productive periods of spring and early summer. Since all the birds in the temperate areas need more food at the same time, the number of different species the area can support is limited. Tropical birds often have distinct breeding seasons, but these seasons vary for different species, which should spread the heavy

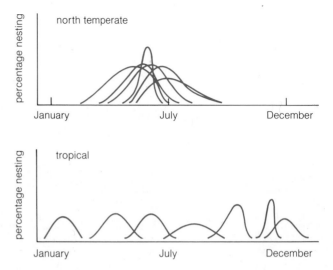

Figure 27–9 Breeding patterns of north temperate and tropical birds plotted against months of the year.

need for food and breeding sites throughout the year (Figure 27–9). Frogs, insects, and other groups show similar patterns. By spreading the time resource, organisms living in areas of stable climate have also divided other resources and have become more diverse than related groups in less stable areas.

Habitat Productivity It seems only common sense that where resources are more plentiful and more diverse, more kinds of organisms should be found. In the polar regions, as in the deep forest, light can be treated as a resource. During much of the year there is little light, and then, for several months, light is superabundant. Photosynthesis—the basis for all productivity—is limited in such areas, and diversity of plant and animal species is low. A few groups, such as birds, have evolved to take advantage of the plants' short summer bursts of productivity by nesting at this time, then migrating to other areas during nonproductive periods. In the tropics, light is not a problem; it is available in regular and more or less uniform quantities. There, both on land and in the sea, species diversity is usually high.

We can see similar correlations between productivity and species diversity if we contrast tide pools at similar latitudes in eastern and western North America. Western tide pools have much higher diversity. This may be the result of the higher productivity of these environments, which results from a phenomenon known as *upwelling* (current patterns bringing water from a lower to a higher

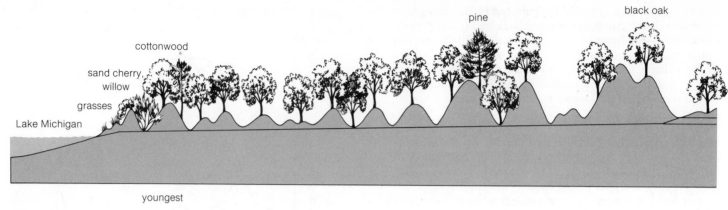

Figure 27–10 Diagrammatic profiles across Indiana sand dunes at the southern end of Lake Michigan. Successively older dune systems originated along earlier and higher beaches.

level). It cools the surface, but, most important, it brings abundant nutrients into shallow water areas. This increases the amount of plant growth, which in turn means more resources for the animals; hence the greater number of species.

Despite the seemingly obvious relationship between productivity and diversity, convincing empirical studies are too few to let us generalize.

Predation   In the last examples we have been concerned with resources. But if some of the species under consideration are preyed upon extensively, the way resources are subdivided is unimportant. The heavily hunted species will be so far below that particular environment's **carrying capacity** (the number of individuals of a species a given environment can support) that resources are not limiting. There can then be extensive overlap in usage patterns so that, theoretically, more species can be accommodated.

Under certain conditions, however, predation can actually increase species diversity. Starfish are formidable predators that can eat large numbers of mussels and barnacles. These crustaceans, attached to rocks in the intertidal zone, compete tightly for space. With these animals reduced in numbers by foraging starfish, there is space for other organisms. When the starfish were removed from one region in the San Juan Islands of Washington, the number of other species in the experimental habitat fell from 15 to 8. The mussels, in particular, came to dominate the community by literally crowding out other species.

Succession

Earlier in this chapter, we showed how species replace others through time in a particular place—in our example, an abandoned field. It is convenient to distinguish two patterns of succession. *Primary succession* begins on biotically virgin material, in either terrestrial or aquatic environments. For instance, a volcanic eruption creates a sterile area. No life appears in the solidified lava for a time. Usually the first living things are lichens, which become established on the barren rock. As the lichens grow, they begin to break the rock down into its mineral parts. Thus soil forms, and rooted vegetation becomes established. Once plants arrive, animals follow, and a succession is well begun.

Secondary succession occurs when a place previously modified by the activities of plants and animals is denuded (by plowing, fire, wind erosion, glaciation, or other activity) and then left undisturbed. Here, succession is generally much faster. For example, a freshly plowed patch of soil surrounded on all sides by virgin prairie will return to prairie vegetation within a few years.

One special kind of secondary succession occurs every spring in the vernal pools that form in temperate climates as the winter snows melt. Eggs, seeds, and spores of animals and plants, dormant since the previous summer, produce new life. *Daphnia* and fairy shrimp fill the pond, along with the protozoa and one-celled algae that they eat. Salamanders and frogs come to the pond to breed. The salamander embryos develop into larvae, which gobble up the *Daphnia* and fairy shrimp. Tadpoles chew the filamentous algae, and they filter microorganisms from

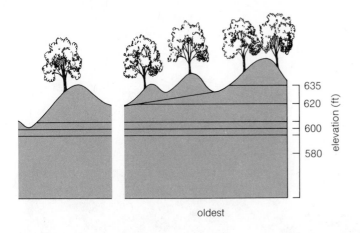

elevation (ft)

635
620
600
580

oldest

the water. Night herons wade the margins of the pond, stalking the salamander larvae and tadpoles. As summer arrives and the pond begins to dry up, the amphibians metamorphose and move onto dry land. Now garter snakes and birds come to prey on these numerous but relatively helpless amphibians. The pond dries, and the new crop of seeds, spores, and eggs are locked in drying mud, where they will lie dormant until the next spring.

The pattern of succession must be analyzed in each community, for no single pattern characterizes all places. Understanding long-term succession, of course, is fundamental to understanding the structure of communities.

The classic and still basically valid study of succession was done by H. C. Cowles of the University of Chicago on the sand dunes of the southern end of Lake Michigan. During the last glacial epoch, Lake Michigan contained much more water than it does now. As the glaciers receded, the water level fell, and benches of sand were left above water level. Winds blowing predominantly from the northwest produced a system of sand dunes (Figure 27–10).

The dunes offer an excellent opportunity for studying succession, for we know that the lake has been receding. Thus, the farther from the lake, the older the dune. The substrate must have been similar for the entire area, basically wind-blown sand. The first colonizer is manam grass, which spreads by means of underground stems and can stabilize a barren area of shifting sands in six years. Once the sand dune is stabilized, the sand reed grass and little bluestem grass come in. Sand cherry and willow also play a role in stabilizing the dune and in its species succession. The first sizable trees to appear are cottonwood. Within 50 to 100 years, jack pine and white pine

become established on the dunes, and distinctive plants and animals begin to use the low-lying swampy areas between successive, stabilized dunes. After 100 to 150 years, black oak replaces the pines.

All communities change, but in some, such as the dunes and the vernal ponds, changes occur over and over as a part of the internal dynamics of the community. The dominant factor that determines the changes in a community is the life history pattern of the dominant plant. Often, dominant plants decline in vigor as they get old, and they are replaced by competitively superior plants.

Succession is a dynamic process, resulting from climatic and biotic interactions, and we must understand it if we are to manage the resources of the earth intelligently. For example, we constantly interfere with normal succession by producing repetitious monocultures of corn, wheat, and other crops. Although these are often necessary and, when well managed, can be highly productive for long periods, they drastically reduce species diversity. The great tragedy is that often, especially in tropical environments, a great environmental modification yields only a year or two of productivity before soil nutrients are exhausted.

Humans also interfere with succession in other significant ways. The lowland wet forests of Vietnam were some of the most diverse species communities known. Repeated spraying of herbicides during the Vietnam war drastically disturbed communities in which a delicate balance had evolved through successional processes. What was once a relatively stable, highly diverse community is now a relatively unproductive, unstable, monotonous monoculture of bamboo thickets. The original vegetation—a highly productive mangrove environment that sheltered many animals—will take a hundred years to grow back.

Community Patterns

Although ecologists disagree about the definition of communities, or even whether they exist as meaningful units in nature, regions having similar physical and climatic characteristics do often display common patterns of community organization at a very general level. We will briefly consider some of the patterns that develop in response to different kinds of physical and climatic factors before considering the patterns that give visible and meaningful organization to our biotic world.

Aquatic Communities Aquatic ecosystems seem to divide into zones in response to different kinds of environments. Before looking at ocean communities, we can

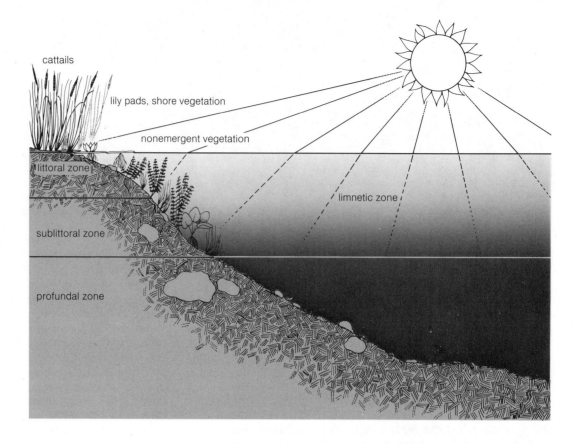

cattails

lily pads, shore vegetation

nonemergent vegetation

littoral zone

limnetic zone

sublittoral zone

profundal zone

Figure 27–11 Vertical organization of a lake.

briefly consider three kinds of fresh-water habitats: lakes, streams, and springs.

Lakes have a relatively simple vertical organization (Figure 27–11). The entire surface layer of the water is called the *limnetic* zone. This is the warm, circulating part of the lake, where productivity is highest.

A *littoral* zone extends from the edge of the water toward the center of the lake as far as rooted vegetation reaches to the surface. This is the zone of most diversity, where the great majority of animals and plants occur. It is also the zone of greatest light penetration.

Beyond the littoral zone is the *sublittoral* zone, characterized by rooted vegetation that does not reach the surface. Species diversity falls dramatically from the littoral to the sublittoral, and in most fresh waters there is a sharp decline in productivity as well. Below the sublittoral is the *profundal* zone, an essentially lifeless region in most lakes, but perhaps visited by some fishes. The waters below the limnetic are cold and usually uncirculating. So little light penetrates these waters that respiration is greater than photosynthesis, and there is negative net productivity.

Vertical zones are most important in lakes, but horizontal zones interest us most in flowing waters. Streams arise from snow melt in mountains, from springs, or from

drainage of surface water. They often have high productivity as a result of nutrients in runoff waters. Their course depends greatly on the geological structure of the region, but streams are generally shallow and torrential in their upper courses and become progressively broader, deeper, and slower as they reach low elevations. Suspended matter, especially one-celled plants, is much more abundant and diverse in the warmer, lower reaches than in the upper, colder areas.

Patterns of replacement are found in group after group of organisms. For instance, in the mountains of western North Carolina, the high mountain streams are populated by larvae of dusky salamanders (genus *Desmognathus*) (Figure 27–12). Because the low tail fins and small gills of these larvae reduce their resistance to the flowing water, the larvae can maintain their position in the torrential streams by using their strong limbs and horny toe tips. At lower elevations, where the streams broaden and water is deep and slow moving, mudpuppies *(Necturus)* are common. They are much larger than *Desmognathus* and have high, flat tail fins and large, feathery gills. Upon reaching the broad coastal plain, the rivers spread out into an extensive series of marshes. Here lives another salamander *(Stereochilus)*. Its larvae have high tail and body fins and long gills, and it is well

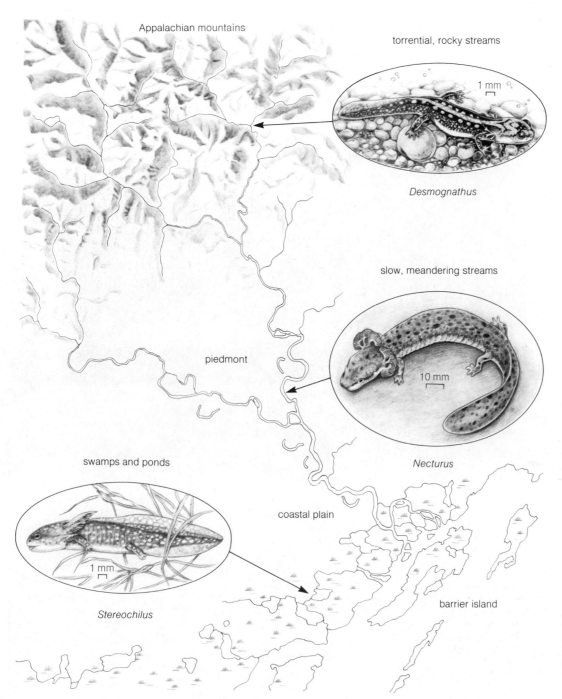

Appalachian mountains

torrential, rocky streams

1 mm

Desmognathus

slow, meandering streams

piedmont

10 mm

Necturus

swamps and ponds

coastal plain

1 mm

Stereochilus

barrier island

Figure 27–12 A bird's-eye view of North Carolina, showing the relation between larval forms of various species of salamanders and the type of stream or river in which they live.

equipped to move through relatively still waters and to extract oxygen from the warm, relatively stagnant water.

Springs show other zonal patterns. One illustration is the temperature gradient seen in a series of springs, or in a spring whose waters join a stream. For example, there are springs whose waters literally boil as they reach

the surface of the ground, and others that are near 4°C. At these extremes there is virtually no life, but just a short distance from either extreme along a temperature gradient, life is found, and plants and animals have evolved to exploit these extremes (see Essay 19–1). Certain species of bacteria and blue-green algae are found

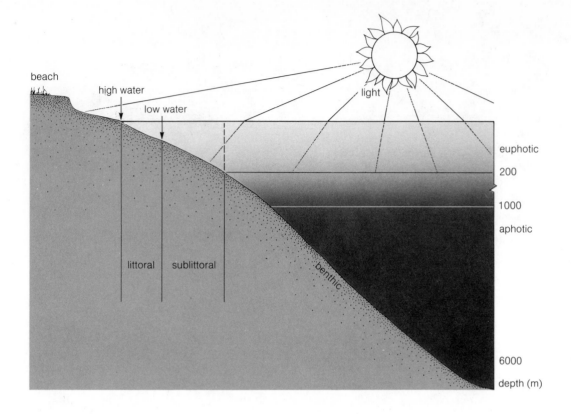

Figure 27–13 Marine environmental zones.

beach

high water

low water

light

euphotic

200

1000

aphotic

littoral sublittoral

benthic

6000

depth (m)

Figure 27–14 A Pacific Coast tide pool.

1 bushy red algae, *Endocladia*
2 sea lettuce, green algae, *Ulva*
3 rockweed, brown algae, *Fucus*
4 iridescent red algae, *Iridea*
5 encrusting green algae, *Codium*
6 bladder-like red algae, *Halosaccion*
7 kelp, brown algae, *Laminaria*
8 Western gull, *Larus*
9 intrepid marine biologist, *Homo*
10 California mussels, *Mytilus*
11 acorn barnacles, *Balanus*
12 red barnacles, *Tetraclita*
13 goose barnacles, *Mitella*
14 fixed snails, *Aletes*
15 periwinkles, *Littorina*
16 black turban snails, *Tegula*
17 lined chiton, *Tonicella*
18 shield limpets, *Collisella pelta*

19 ribbed limpet, *Collisella scabra*
20 volcano shell limpet, *Fissurella*
21 black abalone, *Haliotus*
22 nudibranch, *Diaulula*
23 solitary coral, *Balanophyllia*
24 giant green anemones, *Anthopleura*
25 coralline algae, *Corallina*
26 red encrusting sponges, *Plocamia*
27 brittle star, *Amphiodia*
28 common starfish, *Pisaster*
29 purple *sea urchins, Strongylocentrotus*
30 purple shore crab, *Hemigrapsus*
31 isopod or pill bug, *Ligidium*
32 transparent shrimp, *Spirontocaris*
33 hermit crab, *Pagurus*, in turban snail shell
34 tidepool sculpin, *Clinocottus*

only in hot springs. Between the extremes of temperature are environments favorable to a vast diversity of life.

The oceans contain several kinds of gradients. Figure 27–13 combines four of them in one diagram: depth, light, water movement, and relation to bottom. Many of the zones of the marine environment can be subdivided to show further kinds of gradients. For example, the littoral zone (defined for oceans as the region between high and low tide) of a rocky coast ranges from an uppermost splash region, which is exposed except at the highest daily tide, to a region that is exposed only during the lowest daily tide, and finally to an area that is exposed only during unusually low tides. Along this exposure gradient there are also temperature and force gradients. The fauna and flora change markedly from high to low littoral, as seen in tide pools (Figure 27–14). At the highest levels only barnacles, mussels, and some kinds of algae can maintain permanent homes, although snails, limpets, and crabs will venture rather high. Abalone and brittle stars are rarely exposed for long.

Major kinds of littoral communities are those associated with sandy beaches, rocky shores and tide pools, and muddy bays. The first two of these are high-energy environments, buffeted by waves, and animals and plants need specializations for keeping their place. Clams dig into the sand, and small worms and mollusks live among the sand grains themselves. On rocky shores, mechanisms for attachment are found: Mussels have byssal threads, barnacles have stalks, and algae have holdfasts. The littoral zone is a rich and complex region of interacting life.

The ocean sublittoral extends from the low-tide mark to the edge of the continental shelf, or to a depth of about 200 meters. It is much more stable than the littoral in terms of water movement and temperature. The sublittoral zone is one of the most highly productive regions on earth, often combining abundant light with adequate nutrients and warm temperatures.

Coral reef communities are essentially sublittoral, although they rise into the littoral. Coral reefs consist mainly of the skeletons of dead corals, colonial coelenterates that take minerals from the water to produce a stony structure persisting after the animals have died. Algae are also extremely important parts of coral reefs. Some, like the calcareous red algae, also form stony skeletons and can withstand the battering surf on the outer edge of the reef even better than the corals. Other algae (*zooxanthellae*) actually inhabit the living tissues of the corals in a mutually beneficial relationship. In essence, coral reefs are stable, well-integrated ecosystems characterized by high species diversity and much internal symbiosis.

Above the sublittoral is an open water zone of relatively high productivity, the *euphotic* zone, where sunlight penetrates (Figures 27–13, 27–15). Depth of the euphotic zone varies with many factors, from sky conditions to

Figure 27–15   The open ocean—plankton and nekton.

1 dolphins, *Delphinus*
2 tropicbirds, *Phaethon*
3 paper nautilus, *Argonauta*
4 anchovies, *Engraulis*
5 mackerel, *Pneumatophorus*, and sardines, *Sardinops*
6 squid, *Onykia*
7 *Sargassum*
8 sargassum fish
9 pilotfish
10 white-tipped shark
11 pompano, *Palometa*
12 ocean sunfish, *Mola mola*
13 squids, *Loligo*
14 rabbitfish, *Chimaera*
15 eel larva, leptocephalus
16 deep sea fish
17 deep sea angler, *Melanocetus*
18 lanternfish, *Diaphus*
19 hatchetfish, *Polyipnus*
20 "widemouth," *Malacosteus*
21 euphausid shrimp, *Nematoscelis*
22 arrowworm, *Sagitta*
23 amphipod, *Hyperoche*
24 sole larva, *Solea*
25 sunfish larva, *Mola mola*
26 mullet larva, *Mullus*

27 sea butterfly, *Clione*
28 copepods, *Calanus*
29 assorted fish eggs
30 stomatopod larva
31 hydromedusa, *Hybocodon*
32 hydromedusa, *Bougainvillia*
33 salp (pelagic tunicate), *Dolielum*
34 brittle star larva
35 copepod, *Calocalanus*
36 cladoceran, *Podon*
37 foraminifer, *Hastigerina*
38 luminescent dinoflagellates, *Noctiluca*
39 dinoflagellates, *Ceratium*
40 diatom, *Coscinodiscus*
41 diatoms, *Chaetoceras*
42 diatoms, *Cerautulus*
43 diatom, *Fragilaria*
44 diatom, *Melosira*
45 dinoflagellate, *Dinophysis*
46 diatoms, *Biddulphia regia*
47 diatoms, *B. arctica*
48 dinoflagellate, *Gonyaulax*
49 diatom, *Thalassiosira*
50 diatom, *Eucampia*
51 diatom, *B. vesiculosa*

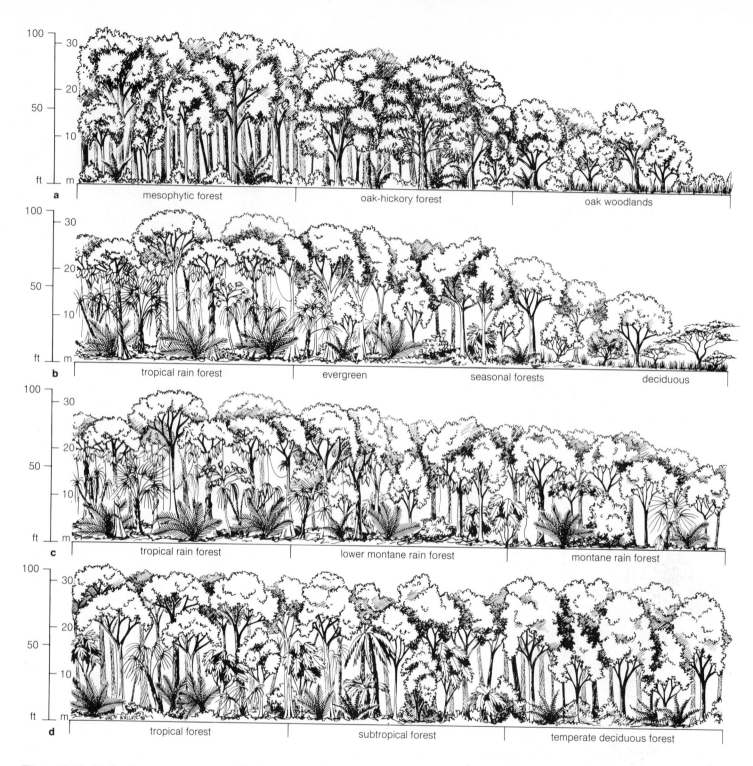

Figure 27–16 Profile diagrams for four ecoclines: (**a**) along a gradient of increasing aridity from moist forest in the Appalachian Mountains westward to desert in the southwestern United States; (**b**) along a gradient of increasing aridity from rain forest to desert in South America; (**c**) along an elevation gradient up tropical mountains in South America from tropical rain forest to paramo or alpine meadow; (**d**) along a temperature gradient from tropical seasonal forest northward in forest climates to the Arctic tundra.

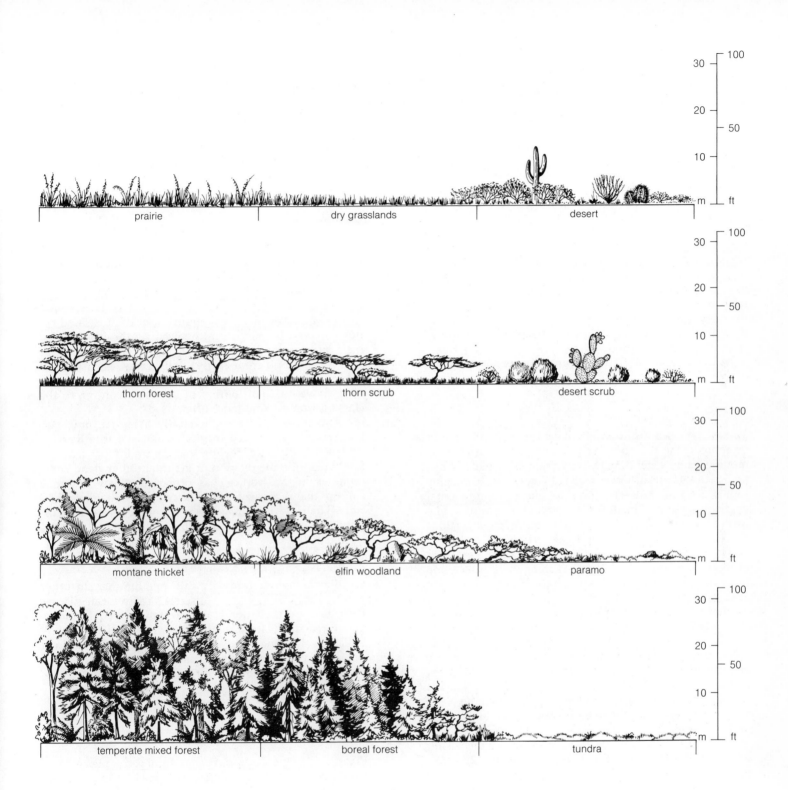

prairie   dry grasslands   desert

thorn forest   thorn scrub   desert scrub

montane thicket   elfin woodland   paramo

temperate mixed forest   boreal forest   tundra

the amount of suspended materials and even the density of organisms, but generally the zone is considered to extend to a depth of about 200 meters. Below the euphotic is the *aphotic* zone, where light does not penetrate. This region is unproductive and depends on the life above it for nutrients. Nevertheless, life is diverse in many parts of these cold, permanently dark waters (Figure 27–15).

The euphotic zone is the region of *plankton* and *nekton*. Phytoplankton, as mentioned earlier, are the producers of the open ocean, and zooplankton are the main consumers. They are floating organisms, powerless to control their horizontal position in the sea. Thus, they are at the mercy of current systems (Figure 27–15). The larger consumers are the *nekton*, swimming organisms such as fish that can navigate at will, and the *benthos*, organisms such as barnacles and worms that are attached to or rest on the bottom, or live in the marine sediment. Both nekton and benthos can feed on organisms from the very small (fed upon by filtration mechanisms) to the very large. In the aphotic zone, organisms are specialized either to feed on detritus or to prey on other relatively large organisms. The latter group includes some bizarre deep-sea fishes with huge mouths and long teeth (Figure 27–15).

Terrestrial Communities   Perhaps the most direct way of approaching the complex patterns of terrestrial communities is by extending the environmental gradient approach discussed earlier. Looking at Figure 27–16, imagine four routes, or transects, traveling along environmental gradients—**clines**—in different parts of the world. The first of these (Figure 27–16a) starts in the forested portions of eastern North America and goes westward, ending in the southwestern deserts. This cline is mainly a gradient of increasing aridity. At all points along the transect, one finds complex food webs that merge smoothly from one area to another.

In the East, the mesophytic forest is tall and complex. As the oak-hickory forest is reached, the trees are smaller and the woods more open. Trees get shorter and farther apart, until grassy areas are extensive. Trees disappear entirely in the prairie, which changes from tall grasses in the east to short grasses in the west. Finally, grasses blend into a scrub vegetation that culminates in the creosote bush desert of the interior Southwest in Arizona and California.

Similar trends are seen along gradients of increasing aridity in the tropics of South America (Figure 27–16b). Grasslands are absent from this particular transect, but depending on soil and other conditions, they, too, could be present. One of the major shifts along this transect is from the evergreen forests of the continually moist regions to the deciduous forests of the seasonally wet-dry regions.

The final two gradients involve temperature. From the tropical lowland forests to the nearby mountain tops, in climates of abundant rainfall, communities change as Figure 27–16c and d show. Patterns of stratification change at higher elevations. Trees get smaller, finally disappearing, so that on high mountain tops there is a treeless meadow, the *paramo*, which resembles the alpine meadows of northern regions in its abundant mosses and the low growth form of its shrubs.

Biomes

Up to this point we have stressed the continuity of communities and the changes in community structure along gradients. But there is one kind of large general classification of certain areas of the earth that does no violence to the continuum idea. We call such useful large classifications *biomes.* Biomes are recognized on the basis of the main plant components (forest, scrub, or grassland) as well as environmental factors (rain forest and seasonal forest, or tropical savanna and temperate grassland). Table 27–1 shows the distribution of biomes according to the classification scheme most often used.

This scheme is not universally accepted, however. For example, the noted tropical ecologist Leslie R. Holdridge of Costa Rica, feeling that the traditional schemes fail to describe the relation of mountain to lowland communities in the tropics, has proposed a new life-zone concept that recognizes many more biomes than we illustrate. The Holdridge life zone is a group of plant associations and their integrated animal components that are related by the effect of three climatic factors: heat, precipitation, and moisture. Figure 27–17 illustrates the Holdridge classification scheme.

Biomes merge gradually into one another. In other words, they are not truly discrete. Yet we can discuss them because their dominant vegetation occupies enormous areas and is itself distinctive enough to let us differentiate one biome from another. In this book, we will not go beyond generally recognized biomes. The brief descriptions in Table 27–1 introduce you to their nature.

Summary

Plants and animals occur in communities, complex assemblages whose composition is determined by many interacting factors. The amount of resources available, the equilibrium between colonization and extinction, the history of the region, habitat structure, climatic stability, habitat productivity, predation, and clines all influence the composition of communities. As a result of all these

Table 27–1 Principal Biomes

Biome	Dominant Plant Growth Form	Representative Organisms (Mostly Northern Hemisphere)
Aquatic Systems		
Open oceans	Plankton, floating algae	Diatoms, plankton, nekton, deep-sea fishes, and other animals
Estuaries and shores	Multicellular algae, grasses	Seaweeds, eelgrass, marsh grass, barnacles, mollusks, polychaetes, echinoderms, corals, anemones
Lakes and streams	Algae, mosses, higher plants	Plankton algae, filamentous algae, duckweed, water lilies, pondweed, water hyacinth, fishes, fresh-water crustaceans, clams and other mollusks, insects, amphibian larvae, water snakes
Swamps, marshes, bogs	Algae, rushes, etc.	Cattails, water plantains, pipeworts, rushes, sedges, sphagnum moss, tamarack, bald cypress, mangrove, fishes, alligators, salamanders, snakes, manatees, leeches
Forests		
Tropical rain forests	Trees, broad-leaved evergreen	Many species of evergreen, broad-leaved trees (unfamiliar to us), vines, epiphytes (orchids, bromeliads, ferns), vast diversity of insects, reptiles, amphibians, birds, and mammals
Tropical seasonal forests	Trees, evergreen and deciduous	Mahogany, rubber tree, papaya, coconut palm, monkeys, frogs, insects
Temperate rain forests	Trees, evergreen	Large coniferous species (Douglas fir, Sitka spruce, coast redwood, western hemlock, white cedar), frogs
Temperate deciduous forests	Trees, broad-leaved deciduous	Maples, beech, oak, hickory, basswood, chestnut, elm, sycamore, ash, salamanders, deer, mice, warblers
Temperate evergreen forests	Trees, needle-leaved	Pines, Douglas fir, spruce, fir, perching birds, tree squirrels
Boreal coniferous forests (taiga)	Trees, needle-leaved	Evergreen conifers (spruce, fir, pine), blueberry, oxalis, martens, grosbeaks
Reduced Forests—Scrubland		
Chaparral, etc.	Shrubs, sclerophyll evergreen	Live oak, deerbrush, manzanita, buckbrush, chamise, pocket mice, lizards
Thorn woodlands	Spinose trees, large shrubs	Acacia, large shrubs, reptiles, insects
Temperate woodlands	Small evergreen or deciduous trees, grass or shrubs	Piñon pine, juniper, evergreen oak, rattlesnakes, ground squirrels, sparrows
Grasslands		
Tropical savanna	Grass (and trees)	Tall grasses, thorny trees, sedges, antelopes, cheetahs, hyenas
Temperate grasslands	Grass	Bluestem, Indian grass, grama grass, buffalo grass, bluebunch wheat grass, bison, kangaroos
Tundras		
Arctic	Diverse small plants	Lichens, mosses, dwarf shrubs, grass, sedges, forbs, Arctic foxes, longspurs, ptarmigans
Alpine	Small herbs (grasslike)	Sedges, grasses, forbs, lichens, ptarmigans, flightless insects
Deserts		
Tropical warm	Shrubs, succulents	Spinose shrubs, tall cacti, euphorbias, geckos, snakes
Temperate warm	Shrubs, succulents	Creosote bush, ocotillo, cacti, Joshua tree, century plant, burr sage, kangaroo rats, kit foxes, roadrunners, sidewinder rattlesnakes, desert iguanas, solpugids (in USA)
Temperate cold	Shrubs	Sagebrush, saltbush, shadscale, winterfat, greasewood, jack rabbits (in USA)

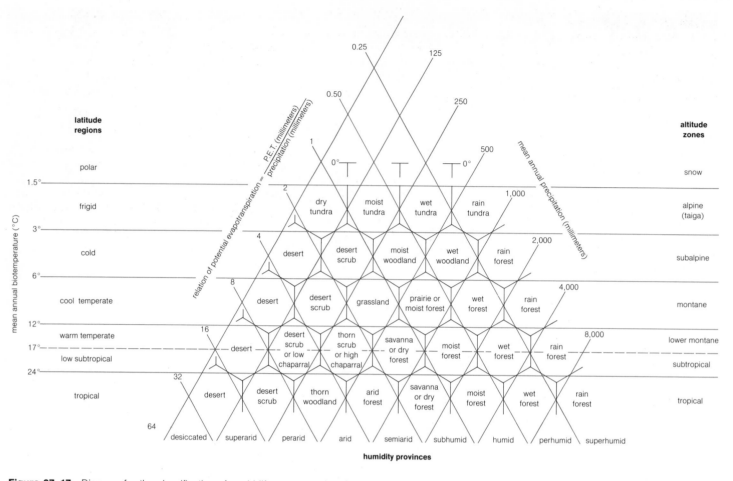

Figure 27–17 Diagram for the classification of world life zones or plant formations.

factors, there are particular community patterns. These are much easier to understand and map in aquatic communities than in terrestrial ones. Finally, biomes and their distribution were briefly discussed.

Supplement 27–1
A Hypothesis Rejected:
The Natural Distribution
of Vegetation

Graduate study can be a most exciting and rewarding time of your life. When your thought processes are stimulated by conflicting but logical views backed by brilliant

individuals, when you have free rein to develop your own ideas and test them objectively—discovering, perhaps, an exciting, unexpected answer—when you can combine all this with the physical stimulus of long data-collecting hikes under nature's canopy, then you can live with an intensity that you will seldom find again. Robert H. Whittaker, now at Cornell University in Ithaca, New York, tells about such a period in his life.

The University of Illinois at Urbana in 1946 was a stimulating environment when I came there as a graduate student in ecology. Victor E. Shelford in the zoology department, one of the major figures of early American ecology, represented the still strong influence of Frederic E. Clements. Clements had considered that natural communities were something like superorganisms and formed well-defined units called associations and formations, or biomes. In the botany department, Arthur G. Vestal subjected Clementsian ecology to a keen, analytic scrutiny—

and rejected it almost wholly in favor of conceptions related to those of Henry A. Gleason. Gleason had argued (in 1926) that plant species distribute themselves "individualistically"—that is, each according to its own genetics, physiological characteristics, and ways of relating to the environment (including the effects of other plants). Species, by this view, would not fit themselves into clearly defined, sharply bounded natural units. With Samuel C. Kendeigh, a distinguished ornithologist with ideas on communities developed from the Clementsian tradition, as my thesis advisor, and with Vestal as my second advisor, I worked in a cross fire of views on communities. Not only did the Clementsian and Gleasonian viewpoints conflict with each other, but each conflicted with a third viewpoint—the life zones of Charles H. Merriam, with which I had sympathy based on experience before coming to Urbana.

The topic of community units was thus subject to an exciting confusion. I began groping toward an understanding of my own and was led (with influences from both life-zone and biome concepts and from Charles Elton's book *Animal Ecology)* to frame a hypothesis. Because of their many interactions, species should tend to evolve toward coadapted groupings, each representing a favorable, balanced pattern of species interactions adapted to some range of environments. Between these favorable combinations should lie transitions of "community-level hybridization" with less balanced and more changeable mixtures of species populations. The favorable combinations should appear in some circumstances as life zones; biomes would in some cases correspond to the zones and would in others be groupings of a number of zones of similar vegetation structure. The distributional expression of the hypothesis could be tested. Unless Gleason was right, one could measure species population levels along a continuous environmental gradient and identify groups of coadapted species with parallel distributions, separated by transitions with irregular and rapidly changing population levels.

Kendeigh proposed that several of his graduate students do these measurements on animal communities in Great Smoky Mountains National Park in Tennessee and North Carolina. I welcomed the opportunity to test my hypothesis in connection with a study of insect communities in a superlative preserve of natural vegetation. In the summer of 1947 I sampled foliage insects in a number of different kinds of plant communities and also in a transect along the elevation gradient from deciduous forests up into spruce-fir forests. Fairly early in the summer it became apparent to me that, first, the latter test of my hypothesis would fail because of the irregularity of insect population levels and, second, the former study would lack effectiveness without knowledge of how the plant communities being sampled related to one another and to environments.

Hence I set out to study the vegetation of the Great Smoky Mountains. To test my hypothesis, I took field transects, a series of plant community samples along particular environmental gradients. But this was not enough, since I was confronted with a range of plant communities whose complexity I was observing in fascinated but baffled admiration. How to study, how to find order in this complexity? As a broader basis for vegetation study, I took 300 samples, largely at random, over as much of the surface of the mountain range as I could reach by extensive hiking. This sampling procedure was quite unorthodox at the time. I rejected the usual practice of subjectively choosing samples to represent particular associations, feeling that the natural units I was looking for should emerge from the interrelations of objectively chosen samples. I assumed also, but without clear plans at the time, that I could fit the many samples together into transects to test my hypothesis. My sample procedure was an intuitive gamble—somehow I could reduce the Smokies vegetation to a comprehensible order.

Back in my student cubicle at Urbana, I compiled my data on plant populations in the field transects, then grouped larger numbers of the 300 samples into elevation transects, and then discovered that I could use weighted averages of species composition of the samples to group them into transects of topographic moisture gradients. There emerged in all these transects a characteristic pattern: bell-shaped population distributions of species, with scattered centers and broad overlaps that together formed a population continuum. My hypothesized groups of coadapted species with parallel distributions were not there, and (with certain limited exceptions) the transitions I had been looking for were not in evidence as such, since the many kinds of forest communities intergraded continuously. I spent some time in reluctant astonishment at the overturning of my hypothesis, followed by delight in the feeling of having made a first-order discovery of far-reaching implications for ecological research and understanding.

I then tried plotting my 300 samples, and densities of plant populations in them, on a chart with elevation and the topographic moisture gradient as axes. The diversity of communities that had baffled me fell into place as a pattern of intergrading combinations of species corresponding to the pattern of environments. The vegetation was to be understood, not as a mosaic of associations, but as a coherent design, a subtly wrought tapestry of differently distributed species populations variously combining to form the mantle of plant communities covering the mountains. In my thesis, in 1948, I stated the conception of communities as forming complex and predominantly continuous patterns of populations. I called my approach to the study of such patterns *gradient analysis,* a technique for analyzing relations of gradients of environments, species populations, and community characteristics, to one another. The pattern concept led later to the study of vegetation of western mountain ranges and to inquiry into the evolutionary meaning of patterns and their relation to problems of species diversity.

The discovery should be placed in further context. The scientific stature of the men with whom I worked at Urbana, who both stimulated my thinking and encouraged my research when it departed from their ideas, must be mentioned. It was the ideas of Gleason, caustically rejected by many ecologists when proposed, that my work belatedly vindicated. I do not know why Gleason's ideas had gone untested for a generation, but the time was ripe for myself and others in 1947–1950. During that period, John T. Curtis and his students at the University of Wisconsin came independently to the ideas of species individuality and community continuity. We found at a meeting in 1950 that we were codiscoverers of a new approach to communities. Still later, we learned that Gleason's ideas had been independently advanced in the Soviet Union by L. G. Ramensky (1924 and thereafter).

In this account you may note a number of themes—the continuity through generations of scientific developments, the roles of readiness, of timing, and of independent discovery, the close interrelation of new theory and new method, the fortunate choice of object of study, the recognition of unresolved conflict of understanding, and the essential intuitive jump toward resolution—that are frequent elements in the advance of science.

28

Population Biology and Evolution

Change is a condition of life. Some change results from natural forces independent of organisms. In a forest pond, weather cycles determine water level and, to a large degree, water temperature. A drop in water temperature slows growth in many organisms and affects physiological processes. But organisms themselves modify their environment and change their conditions of life. Plant or animal wastes may degrade the environment, or animal wastes may provide nutrients for a plant, which in turn grows to become food for animals. Conditions constantly change, and adjustments are necessary if life is to continue.

As individuals, we can easily relate to change. We know about hunger and sleep, adolescence and aging. We also know much about reproduction, and we can see evidence in our own families of the operation of the laws of heredity. But variations must be expressed in groups, or populations, rather than in isolated individuals, if they are to have evolutionary significance. It is the success of individuals in leaving offspring that determines the genetic structure of future populations. Changes in this structure, the results of these recurring successes, constitute the process of evolution, one of the great unifying concepts of modern biology.

Populations

On the gravelly soils south of Puget Sound in west-central Washington, the nearly continuous lowland forest is broken by a series of prairies of varying size. The prairies may be natural, but their sizes probably are not. Native Americans were fond of the enlarged roots and bulbs of several plants that flourish on the prairies, especially the camas lily, and they may have enlarged some of the prairies through burning in order to encourage these plants. Today there is a mosaic of woods and prairies, with the prairies separated by woodland areas of varying size.

The prairies offer an excellent habitat for pocket gophers, *Thomomys talpoides* (Figure 28–1). These small mammals, about the size of rats, form stable local populations, and individuals cover a radius of no more than a few hundred meters in their whole lifetime. Thus the woodland strips, however small, form major barriers, and the population or populations occupying a given prairie are essentially permanent.

Individuals belonging to such closed local populations share a certain limited number of genes. Each individual has its own unique genotype, and all of the local genotypes taken together form what we call a **gene pool.**

If we could look at the genotypes of all the individuals in a population, we would find different forms, or *alleles,* of the same gene. Where only two alleles—say, A and a—are present in the population, an individual may have one of three genotypes: $AA, aa,$ or $Aa.$ The first two are homozygous for this gene; the third is heterozygous (see Chapter 7). If we counted all members of the population by genotype, we could express each kind as a percentage of the total. This would be the *genotype frequency.* From these data we could easily calculate the relative

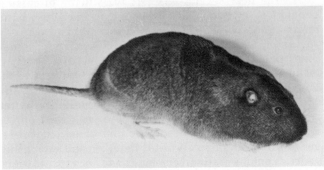

Figure 28–1 The pocket gopher, *Thomomys talpoides.*

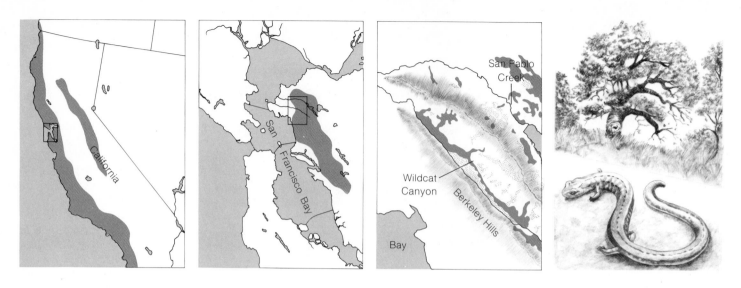

Figure 28–2 The range of the California slender salamander. Within the range indicated, the animals are actually found only in moist oak groves and small canyons.

proportions of the two alleles in the population. Biologists are often interested in this statistic, which is called the *gene frequency.* Each population of a given kind of organism has a unique array of gene frequencies for the many genes represented in the population. Population biologists define the gene pool as the frequencies of all the genes in the population.

Although most individuals in populations interact only with the local gene pool, occasionally animals will move across the barriers separating them from other populations. Among our pocket gophers, for example, young animals dispersing from the nest may breach the woodland barriers, or the woodland barriers may burn or be bulldozed away. When these wanderers enter a new prairie and a new population, they can successfully mate, even though they may differ in color or in some structural feature from most or all of their new neighbors. New alleles are thus introduced into populations.

In the next several chapters we will be discussing populations, and you will need to understand certain basic terms, concepts, and assumptions. When we say *population* we usually mean the local population, sometimes called a **deme.** We usually assume that any two members of opposite sex in a local population have equal probability of mating with each other and producing offspring. Probably no natural population meets this criterion, but some come close. When an individual moves from one population to another and breeds, it carries genes from one population to the other and establishes genetic communication between them. Local populations usually are not completely closed systems, but there are limits to communication between populations. The

pocket gophers can exchange genes with other local populations of pocket gophers, but not with meadow voles. All the populations that can communicate genetically in the manner described form a large, inclusive gene pool that we call the **species.** Each species usually is distinct from other species in features of structure and biology that reflect its genetic uniqueness.

Local populations, sharing a common gene pool, are relatively stable, both in area occupied and in genetic structure, but they are not unchanging. A community of potentially interbreeding individuals—a **local population**—is the fundamental evolutionary unit in which heritable change is established and passed on to generations. In fact, **evolution** may be defined as any change in the gene pool of a population over time.

Populations differ greatly in size, in area occupied, and in density. Eels *(Anguilla)* living in streams as far apart as New York and Georgia can be considered to belong to the same enormous population, since they gather together to breed in one place—the Sargasso Sea—and presumably exchange genes randomly. In contrast, several discrete populations of house mice *(Mus musculus)* can live in the same barn. Although the barn may be full of mice, each population contains only about 10 individuals, occupying its own corner, and individuals are in close, regular contact. Mountain lion *(Felis concolor)* populations, however, cover hundreds of square miles, and one mountain lion might go for days without seeing another.

Some species are so restricted that they consist of but a few populations or even a single population. The famous whooping crane *(Grus americana),* a relic of the

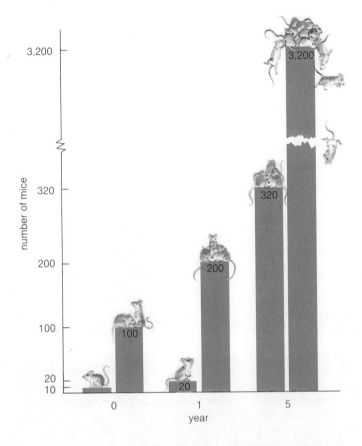

Figure 28–3 Effect of initial population size on population growth.

Pleistocene epoch, has had its numbers reduced by climatic change and, recently, by humans. It now breeds in a restricted area in northern Canada and migrates each winter to a wildlife refuge on the Texas coast. Other species seem never to have had very large ranges. The distinctive California shrub *Carpenteria,* a relative of mock-orange *(Philadelphus),* occurs only in an area of wind-shielded chaparral with a total range of about 12 by 17 miles in the foothills of the Sierra Nevada of California. Yet it is a large, vigorous plant that reaches a height of 16 feet.

Perhaps the most typical species pattern is that illustrated in Figure 28–2 for the California slender salamander, *Batrachoseps attenuatus.* The species is widespread and abundant in California, but in any local region it is not continuously distributed. As an amphibian, it cannot tolerate the extremes of heating and drying that commonly occur in California. The species occupies relatively moist oak groves and small canyons, avoiding the intervening areas of grassland scrub. We see here how species become divided into different local populations.

Populations also differ in persistence. The eel popula-

tion has been in existence for millions of years, but house mouse populations in barns are likely to be short-lived.

Population Growth

When we start looking at patterns of population growth, we find a bewildering diversity. Mice can reproduce several times in their first year of life, but elephants do not reproduce until they are 15 to 20 years old. In this respect, humans are more like elephants than like mice; but humans, like mice, can achieve high densities. Although some mice have wide distributions and high densities, others with similar reproductive habits are restricted in distribution and have low densities in local populations. Thus, the potential for growth is independent of the density or distribution a population ultimately achieves.

In most real populations, we do not find homogeneity in age, nor do we find equal chances of reproducing, having identical numbers of young, or surviving. Males may outnumber females, and we often find that some individuals are more likely to mate with one kind of individual of the opposite sex than with another. But we assume equality so that we can set up models to study. We even assume that all adults leave the same number of offspring, even though it is differing ability to leave offspring that is the central point in evolution, as we will see shortly.

Population growth at any given rate is proportional to the size of the original population. For example (Figure 28–3), take a mouse population with an initial size of 10, half of which are females. Each female has 4 offspring at a particular time each year, then both parents die. All offspring live to reproduce in the same manner as their parents, living but a single year. In year 0, the population size is 10. In year 1, it has increased to 20; in year 5, it reaches 320, having doubled each year. Now consider a population that starts at 100. If it grows the same way, in year 5 it will number 3,200. The females in both cases demonstrate the same **fecundity,** being equally productive of offspring (fecundity, in this case, is 4 per year and 4 per generation); and the two populations are growing at the same *rate;* but the second population is adding 10 times as many members each year.

Growth of Unlimited Populations Nature's statistics, of course, are never so neat. In reality, generations overlap, and some adults die before reproducing. Further, the time of reproduction, the proportion of mating pairs, and the numbers of offspring per female all vary. In a population as a whole, however, we can count the numbers of births and deaths during a particular period, and calculate *birth and death rates.* Then, simply from the difference between birth and death rates (Figure 28–4), we can calculate *rate of increase.*

The birth rate is calculated as the average number of offspring per population unit (often 1,000 people in the case of humans) per unit of time. About 14 babies are born each year for every 1,000 people in the United States. This means a birth rate of 14 per 1,000, or 1.4 per 100, or 1.4 percent per year. The death rate is the average number of deaths per unit of time.

The difference between the birth rate and the death rate, the population's growth rate or rate of increase, is often abbreviated r. Obviously, the value for r can be affected by anything that affects birth or death rates. Age structure is important, for populations consisting mainly of very old people may reproduce little; populations full of young adults may reproduce heavily; and populations full of children have a great potential for growth. Climate may also have important effects. Algal populations grow much faster in early summer, when temperature and nutrient levels are favorable, than in autumn, when nutrients are scarce, the temperature is lower, and ponds are drying. Fecundity is also important, and a small increase in the number of young produced per average female can dramatically increase a population's growth rate, unless the death rate also increases. Finally, a downward shift in age of females at time of first reproduction can also raise the growth rate sharply.

Change in the number of organisms in a population over extremely short periods of time is written as dN/dt, where N is the number of individuals in the population, t is the elapsed time, and d is the shorthand notation indicating instantaneous change. Now, the change in the number of organisms during some period of time equals the number of births minus the number of deaths times the number of organisms in the population. We have defined r as birth rate minus death rate, so we can construct the following very useful equation:

$$\frac{dN}{dt} = rN$$

The kind of growth described by this equation is called *exponential* (Figure 28–5), which means simply that populations grow by multiplication, not by addition. In other words, as we saw with the mice, $10 \times 2 \times 2 \times 2$. . . . And, assuming ideal conditions, all populations of all organisms would indeed grow in this rather frightening way. In fact, most natural populations are limited in their growth by external factors. We must distinguish between the theoretical rate of increase (r max), which is based on the intrinsic capacities of organisms to produce, and the realized growth rate, which is simply the birth rate minus the death rate.

Growth Potential of Populations Since all populations could grow by multiplication, population scientists use

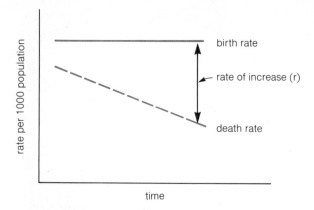

Figure 28–4 The relation of birth rate, death rate, and rate of increase in a population. In this example, the birth rate is constant, but the death rate decline results in a large rate of increase. Given a constant death rate, how could the same result be obtained?

the concept of *doubling time*—the length of time it takes a population to double in size. For the mice cited earlier, doubling time was one year. Even in humans, where age at first reproduction is one of the highest known among animals, the doubling time is shorter than you might think. A 1972 study led by Dennis L. Meadows of the Massachusetts Institute of Technology reported that doubling time for urban population growth in developing countries was 15 years, and doubling time for the whole world (based on $r = 0.021$ per year, or a growth rate of 2.1 percent per year) was only about 33 years. A 1977 United Nations report estimated a growth rate of 1.9 percent—which would mean that the present population of 4 billion would double to 8 billion by 2011. Whichever source you accept, the figures are frightening. What are the limits to our growth? (See Supplement 28–1 for a discussion of human population.)

Limits to Population Growth In spite of all this potential, we see that natural populations do not grow this way, at least not for very long. In fact, for most populations, realized values for r are near zero, or even below zero (a condition in which population size is decreasing). How does this happen?

Many factors in nature keep populations from growing exponentially. Perhaps we can see these limiting factors most easily by returning to the pocket gophers. If these populations increased at their maximal rate, the prairies would be teeming with gophers, yet they are not. Why?

Summers are dry in Washington. The grass dries, and roots of annuals shrivel. Food may get so scarce in late summer that only the hardiest young can survive. Further, hawks patrol the air by day, and owls, coyotes, and wea-

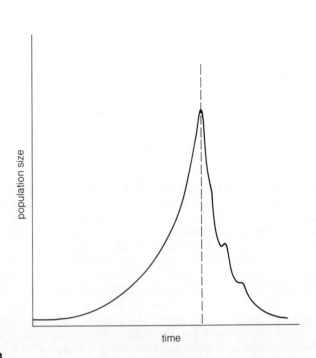

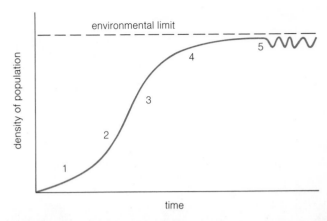

Figure 28–6 The S-shaped, or sigmoid, growth form of a population. The indicated points along the curve are (1) the establishment of positive acceleration phase; (2) the logarithmic increase phase; (3) the inflection point, where the growth rate begins to slow down; (4) the deceleration phase; (5) the maximum population size, or carrying capacity.

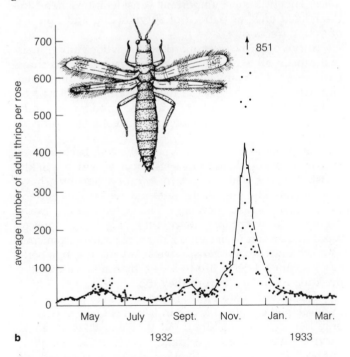

Figure 28–5 (**a**) A J-shaped population growth curve, produced when exponential growth is followed by a crash or die-back. The J portion of the curve, to the left of the broken line, reflects exponential population growth. (**b**) The J-shaped population curve as demonstrated by the insect *Thrips imaginis* on roses in South Australia.

sels take up the death watch by night. Winters are long, cool, and damp; these factors favor disease, which may empty an entire burrow system in a few days. Finally, most areas of appropriate soil are already occupied by adult gophers. The young gophers thus face unfavorable odds for maturing and reproducing.

In each of the isolated areas, the gophers have reached a fairly stable population size. The number of births and deaths is nearly equal over a given time period.

Environmental Resistance and Population Growth

Like the pocket gophers, most populations never grow as fast as they might *(r* max), even at the beginning of their growth. However, growth can still be fast. During early growth periods, some populations may, indeed, grow exponentially. But limiting factors are already coming into play. As we saw with the pocket gophers, resources (especially food), climate, disease, and predators are some of these factors. We can group them all under a single heading: **environmental resistance.** At some particular population density, environmental resistance becomes so great that population growth ceases.

In nature, populations rarely stop growing suddenly. Rather, environmental resistance gradually becomes stronger as populations grow denser. Of course, a high density for elephants is not necessarily a high density for mice, and each population has a unique relationship with the environment in this respect.

Growth of populations that occurs under conditions of steadily increasing environmental pressure usually takes the form of a sigmoid (S-shaped) curve (Figure 28–

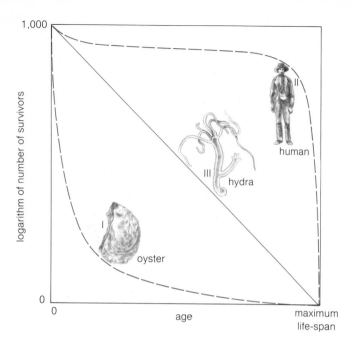

Figure 28–7 Three different types of survivorship curves, depicting the number of individuals still living at various times after birth. (I) Survivorship of a group of 1,000 oysters from the time of hatching of the larvae to the end of the adult life-span. (II) Survivorship of a human population. (III) Survivorship of a *Hydra* population.

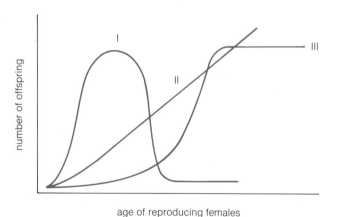

Figure 28–8 Patterns in population in terms of production of young. (I) Populations showing a period of postreproductive senescence, as in humans. (II) Reproductive capacity increases with age. This is true for many organisms as diverse as salamanders and trees. (III) Organism achieves maximum productivity late in life and remains maximally reproductive until death. This is true for many plants.

6)—slow at first, speeding up and becoming exponential, and then reaching a density at which environmental resistance becomes significant and growth no longer accelerates. Eventually, growth begins to slow down and continues to slow as environmental resistance increases, until it reaches a plateau. This plateau, or equilibrium density, is the *carrying capacity* of that population's environment. Natural populations may not follow the expected curve exactly. The plateau is not really stable, and the carrying capacity may fluctuate widely from year to year (see the waves in the curve after 5 in Figure 28–6); but the concept is useful to population scientists.

Modes of Reproduction and Population Growth A population that is reproducing itself in a nearly constant environment close to carrying capacity will achieve a virtually stable *age distribution*. That is, it will contain the same proportion of juveniles and adults year after year. This stability helps ecologists predict population growth, especially when a previously stable environment starts to change. For example, a population consisting only of old adults is in trouble. When a population consists primarily of adults past their reproductive prime, the prospects for growth are radically different than in an otherwise similar population consisting mainly of juveniles. The two most important considerations here are the related ones of pattern of survivorship and number of young (fecundity).

Survivorship (the proportion of newborn individuals that remain alive at a given age) patterns fall into three basic types: (1) many offspring and few but long-lived and very fecund adults; (2) few offspring, almost all of whom survive to be adults who also produce few offspring; and (3) survival rate that is about the same at any age.

An example of the first type is the cod. Each mature female produces millions of eggs, most of which quickly perish. Females who survive to maturity, however, have a comparatively low death rate and repeat the pattern by laying millions of new eggs. The extreme of this fairly common survivorship pattern is the "big-bang" type of reproduction found in salmon and some other organisms. The few individuals that survive larval and juvenile periods wait until late in life to breed, then make a massive breeding effort just before they die.

The survivorship pattern seen in human populations in developed countries is one in which couples may have only two or three children, but they expect them all to survive. Infant mortality is higher than juvenile mortality, but both are low relative to other species, and individuals surviving infancy have an excellent chance of achieving maturity and living the characteristic life-span of the population. This general pattern of survivorship appears in a less extreme form in many large vertebrates.

Examples of the third type are many invertebrates, such as *Hydra,* and temperate-zone songbirds that have survived their first winter. In such populations the death rate is constant, or nearly so, and the chances of survival for the young are about equal to those of the adults.

If we plot the number of fertilized eggs that survive to different ages in each of the three patterns, we get curves like those of Figure 28–7.

Fertility Patterns and Population Growth   Populations also have three characteristic patterns in the relation between fertility and age (Figure 28–8). Some organisms have a distinct reproductive period in their life-span, roughly corresponding to physical adulthood. Then comes a period of postreproductive **senescence,** during which the individual draws on resources but no longer directly contributes to the population by producing young (curve I). This pattern appears in humans but not in many other organisms. Perhaps the fact that young humans often learn from listening to their elders has led to selection for life past reproductive age in humans, although this is rare among organisms as a whole.

Alternatively, reproductive capacity may increase with age (curve II), and the older a female, the more valuable she is in terms of producing young. This pattern is characteristic of such diverse organisms as salamanders and cherry trees.

Curve III illustrates a condition in which the organism achieves maximum productivity late in life and remains maximally productive at each reproduction until death. The most extreme form of this condition is the "big-bang" type of reproduction mentioned earlier.

Factors that Regulate Population Growth   Growth-regulating factors in ecology are generally of two kinds: those that depend on the density of the population and those that do not. Suppose a population of hole-nesting birds lives in a woods. When all the holes are full of nests, the population has reached carrying capacity, even though there might be enough seeds or insects to feed many more birds. The tree holes, then, are what we call a *limiting resource,* braking population at a certain density. They are an example of a **density-dependent factor** that regulates population growth.

Food is the most obvious limiting resource. A shrinking food supply will slow down growth some time before a population reaches its environment's carrying capacity. If growth continues beyond that point, or if food gets still scarcer, competition for food will become intense, the death rate will soar, and the population will drop below capacity again. The food limit is always there, but its effect is not felt until populations reach a certain density, so food supply is considered density-dependent.

Large numbers of individuals may attract predators. For example, a population explosion among voles in an area often brings in such wide-ranging predators as hawks and coyotes. The death rate goes up, and the population size goes down—again, a regulating effect.

When densities become too great, individuals may simply leave the population, or emigrate. This, too, reduces population size.

Parasites and disease-causing organisms spread more easily in denser populations. Classic examples are fungal diseases in crop plants that do relatively little damage in the wild, but may completely devastate closely packed fields. Organisms may even die of poisoning by their own waste products, as fish do in overcrowded ponds—and as humans may one day do.

Density may also have positive results. Below a certain density, males and females have a low probability of meeting and mating. This is one of the worries of people who are trying to protect whales. Also, in social species, learning and communication are better at higher rather than lower densities. A large hive of bees, for example, finds food more efficiently than a small one. Of course, the negative effects of density keep these populations from growing forever.

Certain regulators of population are unrelated to density. These **density-independent factors** include violent storms, abnormal droughts, and other natural disasters. Gophers living by a stream may be drowned after a violent storm. An unexpected freeze may kill newly hatched quail. These deaths will happen whether there are 10 gophers or 10,000, whether the quail are sparse or above carrying capacity at that moment. The point is that chance may reduce population size or greatly impede population growth.

Genetics of Populations

Even a casual look around you in a restaurant or at a football game shows that people vary. They don't look the same or act the same. Similar variations appear in every population. Some pocket gophers, for example, are larger than others, some have darker hair than others, and some have longer ears. Some of these differences are the result of genetic differences, but others are not. For example, some of the larger gophers may simply be better fed than the smaller ones.

Much of what we see as variation in nature is the result of differing responses to environmental factors. Thus, among tide-pool fishes, the number of vertebrae depends on the temperature at which the fish developed rather than on genetic difference. The number of vertebrae changes from bay to bay along the Pacific Coast, increasing toward the north, and can be predicted by the water temperature. However, a certain portion of the observed variation in populations is inherited. Mutation is the ultimate source of this variation, but genetic recombination is the principal means of its distribution and spread in a population.

The important thing is to find which variations are genetically controlled—and thus inherited—and which are not. Assuming we can do this, we can learn where the genetic variation in populations comes from.

Mutation   A mutation is a change in a polypeptide produced by a given gene (Chapter 8). In order for a gene to produce a different polypeptide, the DNA molecule itself must change. At any given **locus** (or place) on the chromosome there are discrete limits to how much a gene can change. Thus, the number of different alleles of the same gene that can arise there is limited and is characteristic of that locus. The presence of two or more alleles (such as A and a) per locus is common (10 to 60 percent of all loci) within natural populations.

Obvious mutations are usually harmful. To get an idea why, imagine typing the letters b-r-a-i-n. Should you make a mistake on the last letter, you would produce either a nonword (24 chances in 25 in English) or the word *braid*. Even if you hit d and got *braid*, substituting it for *brain* would almost surely reduce the sentence to nonsense. Similarly, a new mutation introduced into a population of well-organized, efficiently operating organisms is more likely to lead to trouble than to increased efficiency.

However, some mutations are neutral, and rarely one is highly beneficial. Mutations for resistance to antibiotics in pathogenic bacteria are beneficial to them—although certainly not to us! Ultimately, mutations are the source of variation. How this variation is distributed within a population depends on other factors.

Recombination   Chromosomes are packages of genes that must travel together. However, the chromosomes can sort and recombine in several ways. All the recombinations occur during meiosis or at fertilization. The simplest way of sorting is *segregation,* that is, the separation of maternal and paternal chromosomes during meiosis. Another mechanism is *independent assortment,* whereby each pair of chromosomes segregates independently of every other pair. A final mechanism is *crossing over,* a process whereby parts of the maternal and paternal chromosomes are exchanged during meiosis. Together these processes (see Chapter 8 for more detailed explanations) result in a great deal of reorganization of the alleles. This reshuffling is what produces most of the variant individuals in a population, and evolution could not take place without them.

Chromosomal Changes   Structural changes in the chromosomes themselves occasionally occur (Chapter 11). These can be of considerable significance for several reasons. First, they can produce segments that do not line up properly with homologous chromosomes during meiosis. This will tend to insulate the changed portion (usually a duplicated, inverted, or reversed segment) from crossing over. The genes of this segment will tend to be inherited as a group, and only mutations of genes within the segment, or a further chromosomal modification, will change the situation. Finally, should part of a chromosome be dropped, the offspring may be defective, with both vitality and potential to reproduce lowered.

Changes in Chromosome Number   The chromosome number may change because of an error during cell division. Changes in the number of chromosomes can have profound effects on the variability of a population. Often such changes take the form of adding a whole set of chromosomes (triploidy) or a double set (tetraploidy). Among the higher plants, and rarely in animals, these superoffspring may be fertile. These cases are discussed in more detail in Chapter 29.

Fundamentals of Population Genetics

Consider first a single genetic locus on a particular chromosome. We will deal with diploid organisms, in which the chromosomes are paired. Thus, each individual will have two doses of this gene, one derived from each parent. If the gene has only two forms, allele A or allele a, individuals will be of three kinds: (1) those with allele A at both loci, AA, (2) those with allele A at one locus and allele a at the other, Aa, and (3) those with allele a at both loci, aa. Because these are diploid organisms, there are twice as many alleles as there are individuals.

Now assume that a biologist establishes a population of organisms containing this gene. Since the two parents used to start the population are homozygous $(AA$ and $aa)$, all of their offspring (the F_1 generation) are heterozygotes (Aa). The population then increases as later generations are produced. Except for the original cross, mating is random, and all the genotypes (AA, Aa, aa) have equal possibilities of survival and reproduction. What proportions of genotypes will be produced? When we discussed Mendelian inheritance, we learned that heterozygotes crossed to other heterozygotes produce young in a 1:2:1 ratio of $AA:Aa:aa$. The expected genotypic frequencies of the F_2 generation are thus 0.25 AA, 0.5 Aa, and 0.25 aa. If the environment does not favor one genotype over another, and if there are no mutations (or the mutations cancel out), this ratio should remain unchanged. We call it the *equilibrium frequency* for the gene being considered.

In a large, isolated, randomly breeding population, this is exactly what happens, generation after generation, as the English mathematician G. H. Hardy and the German physician G. Weinberg independently realized in 1908 from simple mathematical considerations. Under the conditions described, gene frequencies will remain stable from the start, and the genotype frequencies that are established by the first breeding will also be stable. However, biologists soon learned not to expect "Mendelian" frequencies (such as 3:1 and 1:2:1), because a wide variety of factors, including chance alone and, most importantly, the fact that most populations are founded by more than a single pair of individuals, lead to other frequencies.

The importance of the Hardy-Weinberg equilibrium concept to evolutionary biology is its value as a model. It tells us what *should* happen. When something else happens, and gene frequencies seem to be changing, we are alerted to ask why. Since the model is based on sound mathematical logic, we look to environmental, physical, or behavioral changes in organisms—and even chance— to explain the unexpected.

The Hardy-Weinberg equilibrium establishes a baseline for evolutionary biologists. It defines a condition of **genetic inertia.** Overcoming this inertia and introducing change in the genetic structure of populations is the fundamental process of evolution.

Since populations and their environments seldom meet the conditions necessary for a Hardy-Weinberg equilibrium, we often find gene frequencies shifting. Let us look at some of the many factors responsible for such change.

Mutation   Mutation alone can slowly change the frequencies of the alleles of a particular gene, if all other factors remain constant. If one allele *(A)* of a two-allele gene mutates to the other *(a)* at a certain rate per generation (and *a* does not mutate to *A* at all), the population will change in a predictable manner to contain ever higher percentages of *a* genes. In fact, it is only a matter of time (about a million generations from a start of all *A* alleles at a typical mutation rate of *A* to *a* of 0.000001 percent) until all the *A* alleles mutate and the population has only *a* alleles.

Usually, however, mutation in the opposite direction is going on at the same time. That is, some *a* genes will be mutating to *A*. If so, change will obviously be slower, and it can stop altogether if the two mutation rates cancel each other out. Then we would again have a condition that would look like a Hardy-Weinberg equilibrium, as long as other factors stayed constant.

Nonrandom Mating   We have assumed (and the Hardy-Weinberg equilibrium demands) that all individuals have an equal chance of mating with any individual of the opposite sex. But such random mating is rare in nature. Suppose, for example, that *AA* insects produced a secretion that *AA* individuals of the opposite sex found irresistible but that *Aa* and *aa* individuals ignored. Then, *AA*s might mate only with *AA*s, while the other genotypes showed no preference. This kind of **assortative mating,** where individuals consistently choose mates in response to a genetically determined feature, results in non-Hardy-Weinberg gene frequencies.

Disassortative mating can also occur. Here, individuals select mates who differ from them genetically. Both assortative and disassortative selection are examples of nonrandom mating, and either will change populations from Hardy-Weinberg expectations.

Gene Flow   One way that gene frequencies can change is for certain genotypes simply to leave a population more (or less) often than others. All young animals (and plants, in the form of seeds or spores) tend to *disperse,* or leave the immediate area of their birth (or parent plants). Young pocket gophers brave the forest barriers that separate them from other populations, and cocklebur seeds hook onto the fur of animals. Both take their genes with them.

Some evolutionary biologists have argued that dispersal or **gene flow** from one population to another has an important influence in maintaining genetic communication among different populations of a species. In that sense, gene flow slows the divergence of populations. However, empirical evidence on the amounts of gene flow occurring in nature is sorely lacking.

There is always the possibility that a dispersing individual might introduce an entirely new genetic combination into a recipient population. Here, gene flow is a potential source of novelty for a population. The novelty may often—even usually—be harmful rather than helpful. In such cases, it should quickly be lost. But when a beneficial change is introduced, gene flow can be a positive evolutionary factor. Even when the migrants do not differ in some clearly adaptive way from the recipient population, they may exert a significant force on the genetic structure of that population.

Random Genetic Drift   If you toss a coin 10 times you expect heads 5 times. But often you get 4 or 6 heads, and even 3 and 7 are not uncommon. If you toss a coin 1,000 times, however, you should get close to 500 heads, and numbers as low as 400 or as high as 600 would be most exceptional. Small deviations from expectation are simply a matter of chance, and they are unpredictable.

Such deviations are much more important for a small

series (say, 10) than for a large one (say, 1,000). Suppose that a local population contains only 12 sexually mature, diploid organisms that breed all at once, and then die. They produce 24 offspring, of which, on the average, 12 reach maturity. The organisms occur in two genetically determined color phases. Red is dominant *(AA* and *Aa* individuals), and white is recessive *(aa).* The frequency of *A* in the starting generation is 0.5, and by the second generation the population reaches a Hardy-Weinberg equilibrium (9 red and 3 white individuals).

Assume that the population is under equilibrium of two kinds. One is genetic, and we know that it is stable. The other is caused by predation, in which a color-blind predator takes prey randomly—usually in the proportion in which they occur. We expect 18 red young and 6 white young per generation. Suppose that, by chance, 19 red young and 5 white young are produced in one generation. If predation takes the usual 9 red and 3 white young, the population then contains 10 mature red but only 2 mature white individuals. If there are 4 *AA* and 6 *Aa*

Essay 28–1

Darwin and Wallace

Charles Darwin was born in 1809, the son of a well-to-do English physician. His father wanted him to enter medicine, but after a year at the University of Edinburgh medical school he quit and entered Cambridge University to study theology. He was no more satisfied with theology than with medicine, but at Cambridge he came into contact with well-known biologists and scientists working in natural history. One in particular, J. S. Henslow, a gentle professor of botany, took great interest in young Charles. Darwin got such good training in natural history that in 1831 he was named naturalist on the *H.M.S. Beagle,* a British naval vessel about to leave on a world-wide mapping expedition. Before embarking, he got the first volume of Charles Lyell's just-published *Principles of Geology,* with advice from Henslow to "read Lyell . . . but on no account accept his views." Henslow was disturbed with Lyell's view that vast amounts of time had made gradual changes in the earth possible, the record of which Lyell felt could be seen in the distribution and arrangement of rocks.

This voyage, which exposed Darwin to the astonishing variety and diversity of the biological and physical world, helped formulate many of his ideas on evolution. In the five years of the voyage, he saw tropical jungles, desolate plains, coral atolls, soaring mountains, and desert islands. His exposure to world geology converted him to Lyell's thoroughly modern viewpoint of gradual changes over immense time spans. On his return, he recorded his experiences in a book entitled *The Voyage of the Beagle.*

Darwin had no quarrel with special creation (the idea that species were created and that they remained essentially immutable throughout their existence) when he em-

barked on the *Beagle* and hence was not offended by Henslow's warning to read Lyell's geology but not to believe it. But he did believe it. Lyell's arguments that all present geological features could have been formed by the same slow and gradual processes taking place today were overwhelmingly persuasive. Besides, Darwin was seeing evidence backing up these arguments virtually everywhere the *Beagle* touched shore.

Furthermore, he observed that the birds on the Cape Verde Archipelago off the coast of Africa were similar to those on the African mainland, whereas the birds on the Galapagos off the coast of South America were similar to South American birds. Why, he wondered, weren't the same kinds of birds found on islands all over the world? Darwin further observed that species of finches on the various Galapagos Islands were different but still closely related. He reasoned that the differences had developed through gradual changes, over a long time, in response to the relative isolation of the birds on their individual islands. Perhaps species were not immutable after all.

When he returned to London in 1836, he was aflame with this idea and eager to study all the relevant facts. In his bachelor's quarters in London, he interviewed many people and read a great deal. "I am surprised at my industry," he recorded. "I soon perceived that selection was the keystone of man's success in making useful races of animals and plants. But how selection could be applied to organisms living in a state of nature remained for some time a mystery to me." In October 1838, about 15 months after he began his systematic study, he read Thomas Malthus' *Essay on the Principle of Population.* This gloomy treatise, written in 1798, showed that continual increase in population caused by birth rates higher than death rates would ultimately bring disaster. Food supplies, said Malthus, increase arithmetically, but populations increase red individuals, the frequency of A rises, from the initial 0.5 to 0.58. This population is no longer in Hardy-Weinberg equilibrium, because of chance alone. But if conditions remain unchanged, the population will reach a new equilibrium in the next generation—20 red and 4 white young, and 10 red and 2 white survivors.

What happens if the predator gets lucky and happens to get one or two unexpected white animals in some future generation? We can see that chance alone can profoundly affect small populations, often in a short period of time. The smaller the population, the more effect random factors have on its genetics. The change in gene frequency that is based on chance alone is known as **genetic drift.**

Population size is thus of great evolutionary significance, especially when it is small. And many natural populations *are* small, on the order of 10 to 100. Population size varies from generation to generation, and periodically most populations become small enough for random factors to be important. Thus, even in populations that are

geometrically. The ultimate result must be famine and a "struggle for existence." Darwin immediately saw that, in such a struggle, favorable variations should tend to be preserved. He had his theory of natural selection!

For many years, Darwin was torn by an inner conflict. He had a theory that he strongly felt was true and that he knew would be both fascinating and important to science and to the world at large. Yet he also knew what a shock the theory would be to his quiet Victorian world. He must have been afraid to publish his theory—and also afraid that he might die before he could publish.

During those years, Darwin was far from idle. Despite poor health, he wrote a detailed treatise on barnacles, presented a still-accepted theory on the origin of coral atolls, wrote books about volcanic islands and South America, and published dozens of papers on various subjects. Darwin's reputation was clearly established by 1858, when he received a paper from Alfred Wallace, advancing the same evolutionary theory conceived but unpublished by Darwin.

Wallace, 14 years younger than Darwin, had not immediately become a naturalist either, being first a surveyor and an architect. Later, he became interested in botany. Then, while teaching in an English preparatory school, he met the famous naturalist H. W. Bates, who persuaded him to become a beetle collector. Moreover, as Henslow did to Darwin, Bates urged Wallace to visit and collect in the tropics, and in 1848 went with Wallace on an expedition to the Amazon. Most of Wallace's collections from the Amazon were lost, but the information inside his head was safe from shipwreck.

In 1854 he traveled and collected in the Malay Archipelago—Wallace's Galapagos. Wallace found that he could draw a line dividing the archipelago into two parts and that the animals living on islands on opposite sides of the dividing line (now called the "Wallace line") differed widely in form. This was true even if the islands were close together.

He began thinking about the mechanism of evolution in 1855. But not until 1858, during a severe attack of fever, did he begin to think about Malthus, and "there suddenly flashed upon me the idea of survival of the fittest." Lacking Darwin's hesitation to present the theory, he wrote his paper in three days and sent it to Darwin. A shocked Darwin wrote: "If Wallace had my manuscript sketch written in 1842, he could not have made a better short abstract! Even his terms now stand as heads of my chapters."

This event finally forced Darwin to publish his book *The Origin of Species* in the summer of 1859. His influential position, the number and power of his friends, and Wallace's long absences from England put Darwin rather than Wallace at the center of the controversy set off by *The Origin of Species* and made Darwin the one remembered in most biology books.

Wallace himself respected Darwin's prior claim to the theory. He did not write a book on evolution until 1889 and then entitled it *Darwinism.* Wallace lived until 1913, to the ripe old age of 90. His masterwork is his *Geographical Distribution of Animals,* published in 1876.

Darwin continued to be productive throughout his long life. By the time he died in 1882, at the age of 73, he had published hundreds of papers and eight books. The subject matter of these studies is wonderfully diverse, covering earthworms, climbing vines, human evolution, growth responses in plants, mechanisms of pollination in flowers, geological subjects, and many, many others.

Both Darwin and Wallace would have held important places in the history of biology even if the theory of natural selection had never appeared. Yet since both were men of genius, it is fitting that they independently discovered one of the most important of all biological theories.

usually large, rare reductions in population size can produce genetic change. The smallest, not the largest, extreme is important when population size fluctuates.

Even if populations stay large, chance deviations from expectation can occur if the population exists for millions of years. Some biologists think that chance factors have been significant in long-term evolutionary trends, such as the evolution of macromolecules (especially proteins).

Natural Selection The most important mechanism for changing gene frequencies is natural selection. This is the process whereby certain genotypes become more frequent than Hardy-Weinberg equilibrium would predict, by being more successful in leaving offspring.

This idea is deceptively simple. Naturally, you may say, individuals having more offspring will leave more descendants. But you will have missed the point unless you ask *why* the bearers of certain genotypes are more successful reproductively. It is not just that they produce more young, but that they produce young that are biologically superior (in terms of their performance in the environment) to those produced by individuals with different genotypes. These biologically superior young may, in turn, be reproductively superior. If so, in time they will dominate the population.

In other words, natural selection is *qualitative* as well as quantitative. And it is the qualitative aspect that distinguishes natural selection from other processes that can lead to changes in gene frequencies.

Up to this time we have dealt with loci, alleles, genes, gene frequencies, and genotypes. When we discuss natural selection, however, we focus on *phenotypes* of *individuals.* When we speak of quality, we refer to the performance of organisms in the environment. The environment "selects" the genotypes that will survive, and it does so on the basis of phenotypes—the expressed, somatic (structural and functional) properties of individuals. These are largely the result of the genotype of the individual, but of the *whole* genotype and not simply one or two loci that we may wish to consider.

Darwin and Natural Selection

When we consider natural selection, we come to one of the world's greatest biologists: Charles Darwin. Darwin was an unlikely revolutionary, but his work and writings transformed our whole way of looking at ourselves and the world around us. The publication of the *Origin of Species* in 1859 is a milestone, not only in the history of biology, but in the history of human intellectual development (Essay 28–1, p. 286).

What was so revolutionary about Darwin's ideas?

Darwin proposed that species evolve through the process of natural selection. His hypothesis was based on two premises: (1) All species have the potential to overpopulate, but generally they do not; and (2) variations exist among the individuals of a given species.

It takes no genius to observe that organisms produce, on the average, far more offspring than can survive to reproduce themselves. Some fish deposit millions of eggs, and some toads produce over 10,000. A single dandelion plant can produce many thousand seeds. If all survived, we would quickly have seas filled with cod, and land covered with toads and dandelions. But this does not happen. Darwin correctly observed that the potential for geometric increase in numbers is seldom, if ever, realized. Not all offspring survive. Darwin's genius was to see that it is the environment that decides which ones survive. In other words, the environment *selects.*

One need only glance about in a crowd to observe that individuals vary. All mice may look alike to the casual observer, but an expert—or a mother mouse—can quickly spot differences in proportions, odor, coloration, and other characteristics. Darwin, too, made this observation and generalized it to all life. Further, he knew that there was a correlation between characteristics present in the parent and in the offspring.

In Darwin's day, it was well known that interactions between individuals of the same population in a community can be intense in mating, food-gathering, nest-building, and so on. Such interactions ordinarily are much more severe and costly on a day-to-day basis than are interactions with members of other populations living in the same community. Darwin argued that individuals faced a struggle for existence, and this led to the notion that became known as "survival of the fittest." Those individuals who survive to reproduce are naturally the ones that have the greatest influence on the direction of evolutionary change; and, in general, the more offspring they have, the more influential they are. Darwin said:

As many more individuals of each species are born than can possibly survive: and as, consequently, there is a frequently recurring struggle for existence, it follows that any being, if it vary however slightly in any manner profitable to itself, under the complex and sometimes varying conditions of life, will have a better chance of surviving and thus be *naturally selected.* From the strong principle of inheritance any selected variety will tend to propagate its new and modified form.

Clearly Darwin implied that individuals with a better chance of surviving also had a better chance of reproducing. We can simplify the statement now, and say that **natural selection** is the differential reproductive success among individuals in a population, no matter what the means of that success.

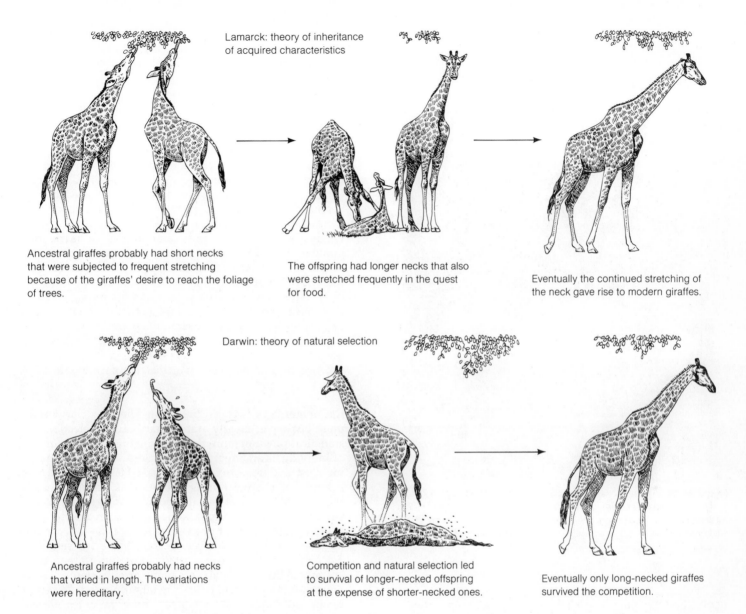

Lamarck: theory of inheritance of acquired characteristics

Ancestral giraffes probably had short necks that were subjected to frequent stretching because of the giraffes' desire to reach the foliage of trees.

The offspring had longer necks that also were stretched frequently in the quest for food.

Eventually the continued stretching of the neck gave rise to modern giraffes.

Darwin: theory of natural selection

Ancestral giraffes probably had necks that varied in length. The variations were hereditary.

Competition and natural selection led to survival of longer-necked offspring at the expense of shorter-necked ones.

Eventually only long-necked giraffes survived the competition.

Figure 28–9 A comparison of Lamarckian and Darwinian theories of evolution.

Earlier scientists, such as the famous French biologist Lamarck, had stated evolutionary hypotheses (Figure 28–9). However, these were all tinged with mysticism and inconsistencies. Darwin's theory was beautifully consistent with known biological facts, then and now.

Some Practical Examples of Selection Darwin was familiar with the principles of animal- and plant-breeding from his own observations as a pigeon breeder and from his readings. Thus, many of his examples of selection

deal with the kind of selection used by breeders, which he called **artificial selection.** Results of this kind of selection are impressive indeed. For example, all our breeds of dogs, from the St. Bernard to the chihuahua, are thought to have come from wolflike or coyotelike basic stocks. Over the centuries, owners interbred individuals with traits they considered desirable, and distinctive races appeared. Even today, new breeds of dogs are being developed, sometimes becoming stable, or true-breeding, in only five or six generations.

Other examples can be found in most domesticated

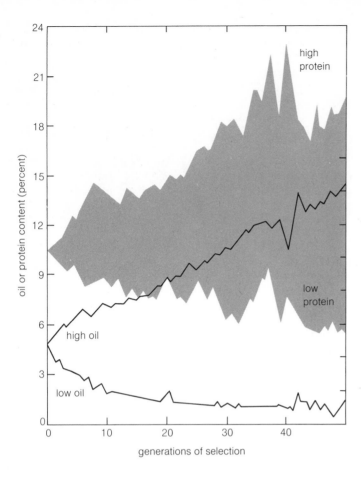

y-axis: oil or protein content (percent)

x-axis: generations of selection

Labels on graph: high protein, low protein, high oil, low oil

Figure 28–10 The response to artificial selection (over 50 generations) for genetic strains producing both increased and decreased protein and oil content of corn kernels, starting with an open-pollinated variety of field corn. Note that considerable divergence of the different strains occurs.

plants and animals, especially economically important species. Breeders have selected cattle for increased milk production alone (Holstein), for rich milk as well (Jersey), or for meat (Hereford).

The impact of controlled breeding programs (artificial selection) on human food plants has been so dramatic that we call it the "green revolution." Plants have been selected not only for increased yield, but also for quality of yield, for resistance to pests, for mechanical harvesting, and for adaptability to environmental variables such as moisture, temperature, and day length. Some of the most dramatic results have been achieved with corn, which has been under intense artificial selection in the United States since 1893. Breeders grew field corn for 50 generations, always selecting plants for high and low protein content and high and low oil content of the kernels. Figure

28–10 shows the results. The high-protein line doubled the amount present in the original stock, and the low-protein line halved it. The same was true for the oil. Since corn is a cross-pollinating species and highly heterozygous, it gave the results expected.

How close is the analogy between artificial and natural selection? The breeder can exercise conscious choice, but the environment cannot. Nonetheless, the environment has selected survivors to produce viable offspring as surely as if it had been conscious. Take, for example, the red scale insect that preys on citrus trees in California. In the early years of this century, growers used hydrocyanic gas to fumigate entire trees, and for a while it worked. But in 1914, certain insects survived fumigation. Gradually these survivors became numerous in the local population, and resistance began to appear in other areas of the state as well. This resistance was conferred by a mutant allele of a single gene. As fumigation continued, obviously more mutants than nonmutants survived to reproduce, and their numbers grew; that is, the frequency of the mutant allele increased in the populations. Individuals bearing the mutant genes apparently dispersed to neighboring populations, thus extending the area of resistance. Should we call this artificial or natural selection?

Refinements in Natural Selection Theory Darwin implied but never clearly stated that natural selection is a statistical phenomenon; that is, it involves large numbers of individuals and deals in populations. The great English biologist and statistician Sir Ronald Fisher showed that the rate of change during evolution is proportional to the genetic variability in the population, so long as this variability relates to reproductive success in a direct way. More variable populations evolve faster than less variable populations, for they offer greater choice for selection. Selection is not an all-or-nothing situation. Rather, there are probabilities that given genetic arrangements will be more or less successful in leaving offspring under a certain set of environmental variables. Those variables we can identify are usually called **selection pressures,** and success is ordinarily measured in terms of relative reproductive contributions of various genotypes.

One of the most troublesome terms that Darwin introduced is **fitness.** From the way Darwin and especially some of his early advocates spoke of "survival of the fittest," we tend to think of fitness in physical terms— the kind you get by doing your morning exercises or jogging 5 miles every day. But Darwinian fitness is an utterly different concept, with only one criterion: the rate at which an individual produces offspring who in turn survive to reproduce.

If an individual produces 4 offspring in its life-span, its *absolute fitness* is 4 per generation. If another produces 8 offspring in its life-span, its absolute fitness is 8 per generation. If we arbitrarily choose 8 as our standard, we can calculate *relative fitness* by dividing absolute fitness by the standard. Thus, the class of individuals with 4 offspring would have a relative fitness of 0.5. Those individuals producing 12 offspring would have a relative fitness of 1.5.

What, then, is survival of the fittest? Suppose a population consists of a mixture of phenotypes having different relative fitnesses. Assume that fitness is genetically controlled, at least in part. As long as the environment remains stable, the *average fitness* of the population will increase, for the population will become more and more dominated by the "fittest" phenotypes (those with highest relative or absolute fitness). This is another way of stating the concept of natural selection. It seems a simple idea, but it has remarkable predictive power. For example, if the average fitness of the population changes, and the environment has not changed, the proportions of the phenotypes must have changed. Those individuals with higher relative fitness are more frequent. Since the phenotypes have different genetic bases, proportions of genotypes have changed, and gene frequencies have changed. Evolution by natural selection has occurred.

It is important to see fitness in relative terms. Natural selection can occur whether populations are stable in size, growing, or even shrinking. If one genetic type is becoming *relatively* more numerous, natural selection is taking place.

The consequences of Darwinian fitness can be startling. A mentally deficient person who has 10 children has high Darwinian fitness, while Leonardo da Vinci, Beethoven, and Queen Elizabeth I of England, all childless, had a fitness of zero. But, as we shall see, fitness is not the only important factor in evolution.

Kinds of Natural Selection

Natural selection functions on many different levels and has been studied by thousands of scientists using diverse methods. Using the work of the past century, let us look at some of the best-known and most thoroughly understood kinds of natural selection.

Stabilizing Selection We ordinarily think of evolution only in terms of change, yet much of the evolutionary process seems to be of a *stabilizing,* or antichange, kind—maintaining a successful phenotype or a successful genetic structure in a population. For example, on February 1, 1898, a severe snow, rain, and sleet storm struck southern New England. H. C. Bumpus, a biology professor at Brown University in Rhode Island, noticed many house sparrows left stunned or even killed by the storm. Of

136 birds brought to Bumpus, about half died. Taking eight measurements of all 136 specimens, he found that the survivors were closer to the mean values of most measurements than were the victims. In other words, the environment had "trimmed" from the population individuals who differed from the average structure. This is an example of negative stabilizing selection, or **normalizing selection.** Individuals on the extremes of variability in the population are eliminated.

There is also a positive kind of stabilizing selection called **canalizing selection.** It favors the evolution of complex positive feedback systems that work during development to ensure the production of a functioning individual. In other words, the kinds of phenotypic deviants popularly called freaks are simply not produced, or rarely produced, and usually do not survive. For example, mammals born without the suckling instinct do not survive. Selection works in the normal way, with the fittest individuals being those that produce "standard" offspring, those that function best in the particular environment and, accordingly, are also best at reproducing. The genetic mechanism for producing a standard kind of offspring is often complex and, as a result, stable. Among mammals, even a minor disturbance in the overall genetic control of development may lead to spontaneous abortion.

Balancing Selection Variation in populations is often expressed as **genetic polymorphism** (from the Greek for "having various forms"), where discrete kinds of individuals occur without intermediates. In certain species of European moths, there are individuals with one, two, three, or four spots on the back wings. Each kind is called a *morph,* in reference to a characteristic structural trait. An obvious way for genetic polymorphism to be maintained in a population is for the morphs to have equal fitness. That is, if predators find two-spotted moths as easily as three-spotted ones, and females have no mating preference for any particular morph, spot number will have no reproductive advantage or disadvantage, and the population will reach some state of genetic equilibrium, keeping both two- or three-spotted individuals.

However, we sometimes find that morphs have different fitnesses, yet still they all persist. Selection that maintains such polymorphism is called *balancing selection.* We can illustrate balancing selection with a simple example. Suppose that a population of flies contains two morphs, one with a dark and one with a light body. Both breed throughout the year. The dark-colored flies can stand cold better because they absorb more solar energy, and during cold weather they have the higher fitness. The light-colored flies do better in the hot, sunny summer, when they can reflect the sun's rays. During warm periods, the light morph has the higher fitness. The fluctuating environment acts as a selection pressure that produces

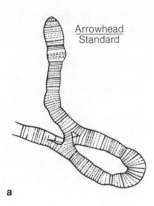

Figure 28–11 (a) An example of abnormal chromosome pairing in the salivary gland cells of *Drosophila*. Such configurations can easily be seen in the microscope and were used by Dobzhansky. (b) Seasonal changes in the frequencies of third chromosomes with two different inversions (ST and CH) in a *Drosophila* population at a given locality.

a

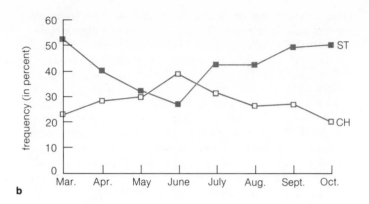

b

a dynamic balance, with the frequency of dark morphs rising in the winter and falling in the summer, predictably.

For 40 years the late geneticist Theodosius Dobzhansky studied natural selection in fruit flies, genus *Drosophila*. These little flies have only four pairs of chromosomes, and in the salivary glands of the larvae the chromosomes are gigantic, containing over 100 chromatin strands as big as normal chromosomes. These chromosomes are strongly marked by consistent banding patterns that provide useful landmarks. They have been mapped in detail, in a series of beautifully controlled breeding experiments, so that we have a good idea where the genes controlling diverse processes are located on the chromosomes. Further, each of the four chromosomes is uniquely banded, and changes in structure can be readily identified.

Among the chromosomal reorganizations that occur in *Drosophila,* the most useful for scientists are *inversions* (Figure 28–11a), in which a section of chromosome is turned end for end. During meiosis, the inverted portions of chromosomes fail to pair up with the uninverted or differently inverted portions of other chromosomes. Thus, the loci on the inverted section cannot cross over and will remain together as a cohesive unit. They can be directly observed in the salivary gland chromosomes. Dobzhansky took advantage of these facts to analyze selection in natural and laboratory populations. The two chromosome types in Figure 28–11a, Standard (ST) and Arrowhead (AR), produce three phenotypes: ST ST; ST AR; and AR AR. We can test these as if they were simple genetic homozygotes, heterozygotes, and homozygotes, respectively. In reality, many so-called genes may be groups of genes as tightly linked as these.

After many years of studying these and other chromosomal inversions in *Drosophila pseudoobscura,* Dobzhansky convincingly demonstrated many instances of balancing selection. Predictable seasonal changes in inversion frequencies can be demonstrated (Figure 28–11b). Further, within a species, there is a general pattern of

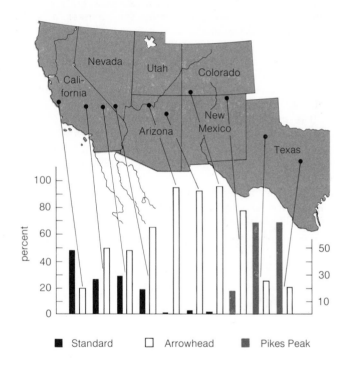

Figure 28–12 The change in frequencies of chromosomes in populations of *Drosophila pseudoobscura* in various localities of the southwestern United States.

balance. Some inversions are more frequent at low elevations than at higher ones (Figure 28–12). Others are more frequent in dry areas than in wet ones.

Still another kind of balance can be demonstrated with *Drosophila* inversion systems. Heterozygotes tend to have a higher fitness than homozygotes. This phenomenon, called *heterosis* (see Chapter 8), may well be impor-

From the experiments of Dr. H. B. D. Kettlewell

a   b

Figure 28–13 (a) Dark and light forms of the peppered moth on a tree trunk in an area unpolluted by soot. The light form is indicated by a bracket. (b) Dark and light peppered moths on a tree trunk blackened by pollution. The dark form is indicated by a bracket.

tant in all of nature, for it maintains genetic variability in a population. Even if heterozygotes mate only with other heterozygotes, about half their offspring will be homozygotes! Among experimental populations of *Drosophila pseudoobscura,* individuals that are heterozygotes in respect to a particular set of inversions generally have higher fitness than homozygotes. This should mean that a balance will be maintained, with good representation of homozygotes; and indeed such a balance is found in nature.

Directional Selection   In the most familiar kind of selection, one trait replaces another. This so-called directional selection is the kind that concerned Darwin and most biologists since Darwin. It forms the core of Darwin's evolutionary theory.

The most spectacular and best-documented study of directional selection involves the rather inconspicuous but now famous British peppered moth, *Biston betularia* (Figure 28–13). Long interested in natural history, the British have been recording information and saving biological materials for hundreds of years. Societies and individuals competed to see who could obtain the most extensive natural history collections. In nineteenth-century England, moths were collected as enthusiastically as stamps and coins.

About the time that Darwin began his work, naturalists were avidly seeking a rare black morph of the normally gray-and-white peppered moth. During the following century, the black morph became more and more common, especially around cities, and the bottom fell out of the black morph market! By about 1900, the black morph had almost replaced the normal morph in areas surrounding cities. In rural areas, the normal moth continued to predominate.

Biologists were soon asking why this should be so and suggesting an explanation. It was the age of the Industrial Revolution, and coal was the primary fuel. Coal smoke and soot spread over everything in the vicinity of such cities as Manchester and Birmingham. Even the pale gray lichens on the tree trunks where the gray-and-white moths had always been nearly invisible to predatory birds got noticeably darker. Now the black morphs, which had always been obvious and easy prey before, had a better chance of escaping the birds, and hence of surviving in industrialized areas, than did the normally colored morphs.

H. B. D. Kettlewell, an English biologist, recognized that the peppered moth offered an unusual opportunity to study natural selection experimentally. For his first experiment he raised large numbers of both morphs in the laboratory. Reasoning that fewer black moths should survive in rural settings (where tree trunks were not blackened by soot) and fewer normal moths in industrialized areas, he decided to release his moths (each marked with a dot of paint on its wing) in rural Dorset and in industrial Birmingham.

Nearly 800 marked moths, about half black and half normal, were released in Dorset. In 11 days Kettlewell recaptured 73 of them. This was a small, but statistically adequate, sample. Among these were 13.7 percent of the original release of normal moths, as opposed to only 4.7 percent of the blacks. When he repeated the experiment near Birmingham, he found 27.5 percent of the original blacks, as opposed to only 13 percent of the normals. Kettlewell thus showed that a black moth was 2.1 times as likely to survive in the blackened woods near Birmingham as a normal moth, but that a normal moth was 2.9 times as likely to last the 11 days in rural Dorset as a black moth.

For his second experiment, Kettlewell went into woodlands and placed moths of the two morphs on trees. Watching from a blind, he could see various species of birds feeding on the moths. Clearly, the birds more often chose the light-colored morphs on darkened tree trunks, although they ate some of both morphs.

The mechanism of natural selection in this example is obvious. The black moths are protectively colored in sooty woods and thus less subject to predation by birds than gray-and-white moths. They have a better chance to survive and reproduce, and, with the passage of time, they increase their frequency in smoke-polluted areas.

The peppered moths offer a convincing example of the operation of natural selection, but they are not alone. More than 60 other species in England and several species in the United States have also experienced *industrial melanism* (darkening induced by pollution) during the past 150 years. The many examples, taken together, produce pervasive evidence of the power of natural selective processes.

The example of industrial melanism is particularly instructive, for it shows that not all populations of a spe-cies necessarily evolve in the same direction. Rather, following a period of directional selection, different populations reach new gene frequencies, which then tend to be maintained by stabilizing selection. Natural selection is a dynamic process, and its various patterns change and interact through time.

Selection for Population Parameters In the final analysis, the most important kind of selection is that which changes the breeding biology of the population. In the last chapter, we saw that growth of populations is a product of the interaction of the intrinsic rate of increase and the carrying capacity of the environment, through time. Further, we know that birth and death rates determine the rate of increase. Now we shall look at selection that affects components of population growth.

Of course, all kinds of selection influence evolution by affecting birth and death rates. And in all but a few exceptional situations, selection is probably directed at individuals, through whom the population parameters change. But population growth models can be useful tools for studying evolution. Thus, instead of trying to correlate changes in frequencies of phenotypes with environmental change, one asks new kinds of questions. What kind of selection could raise the death rate of a population? What would be selected to raise the rate of increase or the carrying capacity of the environment?

We know from the population growth equation that rate of increase and carrying capacity are entirely separate concepts. Thus, we should expect to find that selection can apply separately to births, deaths, and carrying capacity. Our own experience bears out this expectation. Timber wolves are ideally suited for life in North America. When undisturbed, they leave more offspring than the average human. Yet humans, who do not reproduce until age 20 or more on the average, now have much higher densities in many areas of North America than wolves ever reached. The environment, as now altered, has a higher carrying capacity for humans than it ever had for wolves. Clearly there is a complex web of interaction between rates of increase and carrying capacity. Selection that leads to changes in these components can have dramatic, long-term effects.

In the next two chapters we will see how natural selection applies to speciation and evolutionary patterns.

Summary

Evolutionary changes are detected and expressed in populations of organisms, and it is important to understand the ways in which populations grow and change genetically. The interrelation of these factors is of crucial importance in evolutionary biology. A mutation (chromosomal modification) or other genetic change that occurs early in the growth phase of a population has a much greater chance of becoming widespread in the population than changes occurring when populations are near carrying capacity. Genetic changes can increase in frequency in populations by mutation pressure, by assymmetries in mating patterns, by differential movement of individuals into or out of the population under consideration, by sampling errors or other random factors, or by natural selection. Natural selection occurs when some individuals are more successful in leaving offspring than are others, as a result of qualitative (rather than accidental) differences in the way in which they function as biological organisms. Natural selection may result when individuals deviating from mean phenotypes are pruned, or not produced (stabilizing selection); when certain phenotypes are consistently more successful in leaving offspring than are others, and the population in time comes to be dominated by them (directional selection); or when populations change in time from one phenotype dominating to another dominating, and the reverse, as a result of shifts in the advantage of one relative to the other through time (balancing selection). Natural selection seems to be the only creative evolutionary process, in the sense of introducing advantageous novelties into populations.

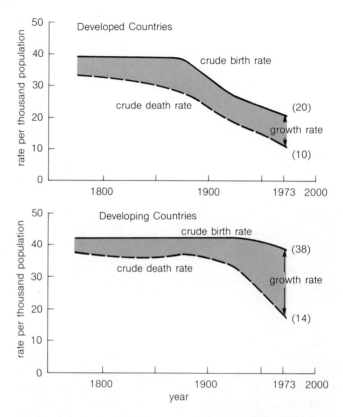

Figure 28–14 Approximate crude birth and death rates in developed and developing countries between 1770 and 1972. The difference between the two curves represents the growth rate or natural increase.

Supplement 28–1
The Human Population

In this chapter we have dealt with the general principles that control populations. One special case deserves greater attention—the human population. The world's population is now in an exponential phase of growth, increasing by a billion human beings in shorter and shorter time periods.

Yet average worldwide birth rates have stayed about the same for the past 200 years, with even a slight decline during the past 50 years (see the crude birth rate curves in Figure 28–14). Why, then, the present population explosion?

Factors Underlying the Human Population Explosion

Birth Rate and Death Rate To determine population growth or decline, you never look only at birth rate, but at the difference between the birth rate and the death rate—called the *rate of increase,* as we saw earlier. As long as the world's birth rate is higher than its death rate over a given time period, there will be net growth in the world population.

Global birth and death rates are now so far out of balance (Figure 28–14) that world population is increasing by 200,000 people each day, or 73 million each year. To get an idea what these numbers mean, suppose you decided to take only 1 second to say hello to each new person added during the past year. Working 24 hours a day, you would need 2.5 years to greet them, and during that time 183 million more people would have arrived, putting you almost 6.5 years behind in your greeting program.

How did this imbalance between birth and death rates come about? For several hundred years before 1940, the birth rate on this planet was normally only slightly above the death rate. But during the early 1940s, the powerful antibiotics penicillin and streptomycin and the insecticide DDT were introduced to control human disease. Since

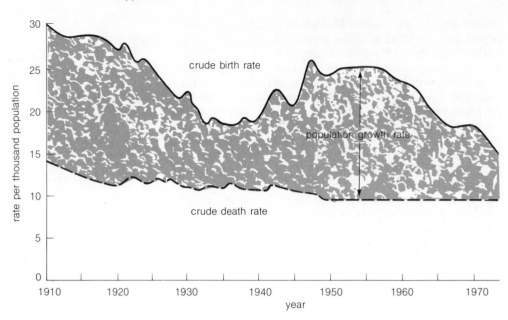

Figure 28–15 Crude birth and death rates in the United States between 1910 and 1973. (Figures from U.S. Department of Health, Education and Welfare, Public Health Service)

that time, the world has not been the same. The worldwide use of these chemicals and other modern public health and sanitation programs has drastically lowered the death rate (Figure 28–14). As the gap between the birth and death rates widened, the human population increased dramatically.

Although population is growing in almost all countries, the greatest differences between birth rate and death rate are in the developing countries (Figure 28–14). In 1973, some 74 percent of all births occurred in Asia, Africa, and Latin America. Over half these babies were born in Asia.

The crude birth rate in the United States has declined from a high of 26.6 per thousand in 1947 to around 15 to 17 per thousand in recent years (Figure 28–15). And although the crude death rate has remained almost constant at 9.1 to 9.7 deaths per thousand, the U.S. population continues to increase because we still have almost twice as many births as deaths each year, and we gain 400,000 people a year through immigration. At present rates, natural increase accounts for 80 percent and immigration for 20 percent of our yearly population growth. Our population is now about 214 million, almost twice as many as in 1921. At this rate, we should have 60 million more Americans by the year 2000. Yet the U.S. Census Bureau predicts 70 to 80 million as a medium projection. What factors have we neglected in our population predictions?

The Effect of Age Structure on Human Population Growth During the next 10 to 20 years, most couples

will probably have fewer children, but in spite of this the U.S. birth rate may rise. Why? Here we must consider a second factor in population dynamics, the *age structure* or *age distribution* of a population—the number or percentage of people at each age level. A major factor in determining future population growth is the number or percentage of women of childbearing age, arbitrarily set at 15 to 44, with the prime reproductive years from 20 to 29.

When the soldiers came home after World War II, the United States had a baby boom, with high birth rates and low death rates prevailing from 1947 to 1957 (Figure 28–15). Women born in 1947 entered their peak reproductive years in 1967 and stayed there until 1977; those born in 1957 entered this phase in 1977 and will stay there until 1987. The high birth rate of the postwar baby boom 15 to 25 years ago becomes the potential "mother boom" of the 1970s and 1980s. Sharp increases in population growth thus affect the future growth and structure of society for 20 to 40 years, as the population bulge moves through the entire generation. Because of this potentially explosive age structure, the United States will be sitting on a population time bomb through 1987.

We can make an age structure diagram for the world or for a given country by plotting the percentages of the total population in three age categories: preproductive (ages 0 to 14), reproductive (ages 15 to 44, with prime reproductive ages 20 to 29), and postproductive (ages 45 to 75), as shown in Figure 28–16. The percentage of males is shown to the left and the percentage of females to the right of the center line.

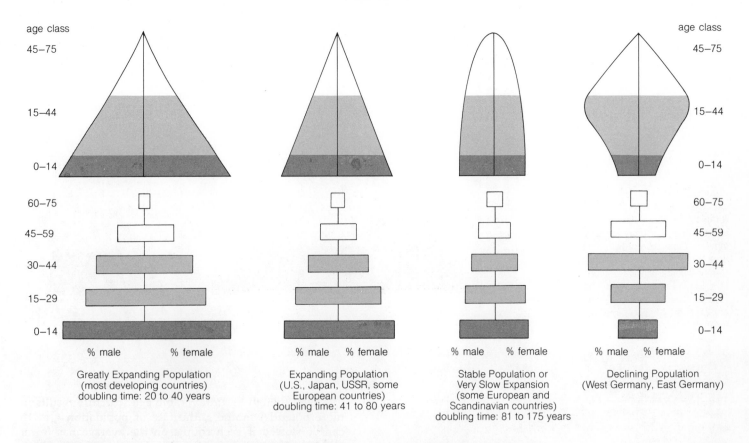

age class

45–75

15–44

0–14

60–75

45–59

30–44

15–29

0–14

% male   % female

Greatly Expanding Population
(most developing countries)
doubling time: 20 to 40 years

% male   % female

Expanding Population
(U.S., Japan, USSR, some
European countries)
doubling time: 41 to 80 years

% male   % female

Stable Population or
Very Slow Expansion
(some European and
Scandinavian countries)
doubling time: 81 to 175 years

age class

45–75

15–44

0–14

60–75

45–59

30–44

15–29

0–14

% male   % female

Declining Population
(West Germany, East Germany)

Figure 28–16 Major types of age structure diagrams for human populations. Dark portions represent preproductive periods (0–14); shaded portions represent reproductive years (15–44); and clear portions are postproductive years (45–75).

The shape of the age structure diagram is a key to whether a population might expand, decline, or remain stationary. A rapidly expanding population would have a pyramid shape, broadest at the base. Such a population would have many people already in the reproductive category and an even larger percentage of young ready to move into this category over the next 15 years. A declining population, by contrast, would have a small base, with more old or postproductive people than young people; and a stable population with zero growth rate would tend to be symmetrical.

Alarmingly, about 37 percent of the people now on this planet are under 15 years of age. In developing countries the average is around 42 percent, compared with around 28 percent in developed countries. These young people are the broad base of the world age structure and represent the even greater explosion of population to come.

With the general shapes of age structure diagrams in mind, let us look at two actual examples. Figure 28–

17 compares India and Sweden. Already at 600 million people, the second largest population in the world, India is growing by 2.5 percent each year. Doubling time is only 28 years. This rapid growth is expected to continue because of the broad-based age structure, in which 42 percent of the population is under 15 years of age. In contrast, Sweden, with a population of only 8.2 million, is close to zero population growth. In 1973, only 21 percent of its population was under age 15, and it is growing at a rate of only 0.3 percent. Long before its doubling time of 231 years, Sweden's population could easily reach a balance between births and deaths.

Fertility Rate and Human Population Growth We have seen the importance of two factors—rate of increase and age structure—in population dynamics. Now we must consider a third factor—*fertility rate,* or the average number of children a woman has between ages 15 and 44. This is an often unpredictable but important factor.

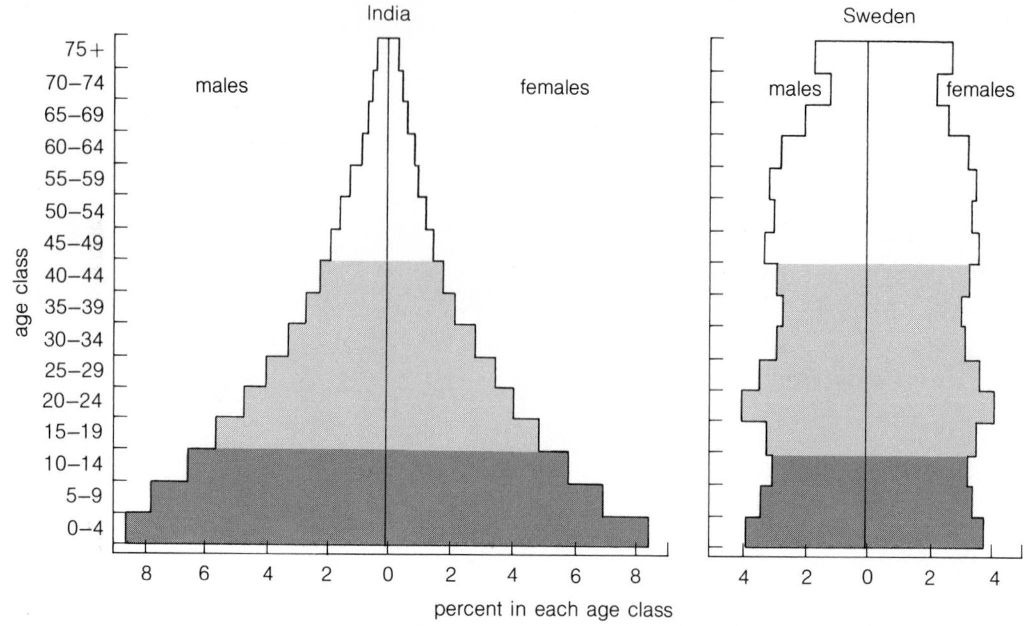

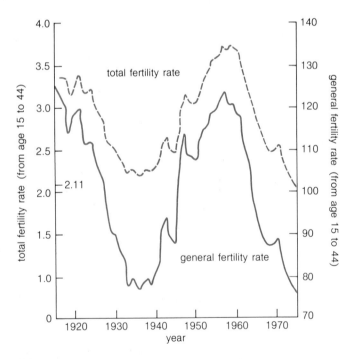

Figure 28–17 The 1970 age structure diagrams for India, with a rapidly expanding population, and Sweden, a country approaching zero population growth. (Figures for India from International Demographic Statistics Center, U.S. Bureau of the Census. Figures for Sweden from UN Population Division, working paper no. 11.)

Figure 28–18 Total fertility rate and general fertility rate for the United States between 1915 and 1973. Total fertility is the average number of children expected of a woman from age 15 to 44, based on current childbirth statistics. General fertility rate is the average number of births per 1,000 women in the age group. Replacement level is 2.11. (Source: Population Reference Bureau, *Population Profile*, March 1973)

Even though an unfavorable age structure results in more potential mothers, the rate of population growth can be slowed if each couple on the average has fewer children. If average family size decreases drastically, a population can even level off in 30 years. Similarly, a slight rise in the fertility rate can bring about a catastrophic increase in population.

The average number of children per family (the fertility rate) in the world today is 4.7 (about 5.7 in the developing countries and 2.6 in the developed countries). If the world's population is eventually to stop growing, this average would have to drop to 2.5 children per family (2.7 in the developing countries and 2.1 in the developed countries). It takes more than 2 children to replace each set of parents because some girls die before reaching childbearing age; developing countries have higher replacement figures because of their higher average mortality.

Even after this replacement level was reached and maintained throughout the world, population would still grow for 70 to 100 years, because of our youthful age structure. In the unlikely event that we reach and maintain a replacement level of 2.1 in developed countries and 2.7 in developing countries by the year 2000, for example, the world's population would not stop growing until about 2100, when it would level off at about 8.2 billion, or over twice our present size. If the developed world reaches replacement level by the year 2000 but the developing world not until 2045, world population would not stabilize until about 15.5 billion—four times our present size—around the year 2145.

Two things become startlingly clear from these figures. First, population momentum from a youthful age structure is probably the most powerful growth force in the world. Second, any plans to control population growth must be made and put into effect 70 to 100 years before the goal date and level. No other form of governmental or social activity requires this sort of advanced planning.

How close are we to zero population growth in the United States? Let's look briefly at U.S. fertility patterns (Figure 28–18). During the 1950s and most of the 1960s, the average U.S. family included around 3 children, resulting in our present potential "mother boom." In contrast, the number of children per U.S. family has fallen in recent years and by 1972 it fell slightly below the replacement level of 2.11.

Do we have zero population growth? This lowered fertility rate will have to be maintained until 1987 to overcome the effects of the large increase in potential mothers between 1970 and 1987. Some observers think our fertility rate will stay low. But even if it slowly drops to an all-time low of 1.8 and stays there, our population will still rise to about 251 million by the year 2000 and level off at about 265 million by the year 2020. We are a long way from ZPG.

Some concerned citizens are calling for an immediate halt to U.S. population growth, including immigration. Too rapid a population change could cause serious economic and social problems, however. For example, after 20 or 30 years, there might be too few working adults to support the larger population of retired and aged people. But it seems urgent that orderly zero population growth be attained in the next 40 to 60 years, with a national population goal of no more than 275 million Americans. To reach this goal, average fertility would have to be reduced to around 1.8 children per family or to around 2 children per family if immigration were sharply curtailed. This is not so harsh as it seems. *Average* is the key word. Not every family would have just 2 children. For example, economist Stephen Enke has shown that one of a large number of schemes that would lead to ZPG is: 50 percent of families with 2 children, 30 percent with 3 children, 10 percent with 1 child, 5 percent childless, and 5 percent with more than 3 children.

Average Marriage Age and Human Population Growth
A fourth factor in population dynamics is the average marriage age or, more precisely, the average age of the mother at the birth of her first child. If a society by custom, as in Ireland, or by coercion, as in China, raises the marriage age, the fertility rate will normally drop and the average generation time be lengthened. Older

brides tend to have fewer children. By raising the average marriage age to 24, the reproductive period is shrunk by nine years and the prime reproductive period is cut almost in half. This will almost certainly lead to fewer children per family and is probably one of the fastest and surest routes to zero population growth. For example, in 1951, British women marrying at age 19 had an average of 2.9 children; those marrying at 20 to 24 had an average of 2.2 children; and those marrying at age 25 to 29 had only about 1.8 children.

Some Implications of Population Growth

The world is rapidly polarizing into two major groups—rich, literate, and overfed; and poor, largely illiterate, and hungry. One of the most difficult problems of poor or developing nations is runaway population growth. Such growth is the biggest single factor keeping these countries in poverty. Poverty is not just an abstract economic number. It is the grief of parents, who themselves may die by age 35, watching perhaps three or four of their seven children die as infants from routine childhood diseases or infections.

India gets most of the publicity about population growth, but Latin America is the fastest growing region in the world—2.8 percent a year, doubling every 25 years. For instance, if 10,000 houses could be built every day for the next 10 years in Latin America, about 100 million people would still be ill housed.

At present, Brazil has 96 million inhabitants. At its current growth rate, it should surpass the United States in population before 2030. Although Brazil has made impressive gains in total gross national product, most of the gains have been eaten up by rapid population growth, so that per capita gross national product is only $420.

It is encouraging that governments in over 31 countries, including China and India, with almost 40 percent of the world's population, have instituted official policies to reduce population growth. However, these policies have met with varying degrees of success. India's birth rate has actually *increased* during the 20-odd years of the program. Another 31 countries officially support family-planning activities, but for nondemographic reasons. However, many governments in Africa and Latin America have policies for deliberately increasing, not decreasing, their populations even though these two continents have the highest population growth rates in the world. Some of these governments feel that only a rapidly growing population can give them enough producers and consumers for Western-style economic development and enough citizens for world prestige and power.

Some Alternative Futures for Humanity

We can project three major alternative futures for humanity based on our present rate of population growth. One is a population crash, with billions dying (perhaps 50 to 80 percent of the world population) through a combination of famine, disease, war, and widespread ecological disruption. The second is population stabilization. The third involves population stabilization followed by a slow decline to a more optimum level.

Let us look realistically at these three alternatives. Population could keep on growing as it is today until the crash comes, and continue to oscillate between runaway growth and crashes without ever reaching stability. Population could stabilize at the *maximum,* or upper limit—that is, at the carrying capacity of the earth. Living at this limit means survival for most, but a much lower standard of living for all. At least half the world's population already lives this way, just above starvation. And even this equilibrium level would probably be achieved only when billions of people had died after population momentum carried us past the maximum. Or population could stabilize at the *optimum* (ideal) level, where all could have a high-quality life and their fair share of the planet's resources.

What is the maximum number of people the earth can support? No one knows, but many scholars have attempted crude estimates. For example, the National Academy of Sciences' projection of 30 billion people by about 2075 at present growth rates could mean that everyone would be living on a grain diet without meat of any kind and at near starvation level. By contrast, agricultural economist Colin Clark sees a world of 45 billion people in which everyone has a good diet by present U.S. standards. To do this we would use nuclear power and mine much of the earth's crust to a depth of 1 mile. On a Japanese-type diet, Clark feels, the world could support 157 billion people.

The most important point is that asking how many people the earth can support is asking the wrong question. It is like asking how many cigarettes you can smoke before you get cancer. Instead we should be asking how many people the earth can support with a decent standard of living—adequate food, housing, employment, education, health care, freedom, and all the things that contribute to the quality of life. In other words, what is the optimum or best possible population?

But this, too, is an extremely elusive question. *Optimum* is a qualitative adjective based on much cultural bias. Values vary with individuals, cultures, and times. Furthermore, optimum figures would vary with climate, availability of vital resources, and the impact a given population has on the environment.   (by G. Tyler Miller, Jr.)

29

Patterns of Evolution
and Speciation

So far we have been examining how populations of organisms can change as selection acts on individuals, changing the frequency of genes in the gene pool. But how do populations of a single species give rise to separate species? To understand the answer to this question, we must first discuss what we mean by a species.

The definition of *species* is hard to pin down. It may be different for plants than for animals, and its meaning has changed over the years. The first definition of species was a common-sense one: If one group of animals or plants all looked very much alike, and looked different enough from all other groups, that group was considered a species. Or, to put it more formally, groups of organisms in which the individuals closely resemble one another but have clear-cut and significant differences in form from related groups are species. This approach, defining species according to their differences in form, is called the *morphological species concept.* In practical terms, most plant and animal species are still recognized this way.

Then, in this century, zoologists suggested that there is really only one way to tell whether any two groups are separate species or not, and that is to see whether they do or can interbreed. If so, they are members of a single species; if not, they are not. This is called the *biological species concept.* Biologists assume, of course, that differences in appearance reflect the *reproductive isolation* that defines biological species. The biological species approach works quite well with animals, but it is difficult to apply to plants.

Some biologists prefer a more abstract species concept which states that species are independently evolving units, with their own unique tendencies. This *evolutionary species concept* incorporates elements of the morphological and biological species concepts. This concept recognizes explicitly that it may be impossible to assign some populations to one species or another.

When interbreeding is the only species criterion, we find groups of organisms that look quite different classed as races of the same species (as in dogs, for example),

whereas other groups that look much alike are classed as separate species (as in some plants). **Races** are simply populations within a species that have changed through natural selection so that they differ in form from other populations of the species. They can still interbreed with these other populations or races.

To restate our original question, How does a population develop distinct races, and how does a race become a species? One obvious way is for parts of a population to live in different environments or to have what ecologists call different *spatial distributions* from other parts over a long period of time. Different environments produce different selection pressures and guide evolution in different directions. Let us look now at the various environmental factors that influence the formation of races and species.

Spatial Distribution
of Organisms

Nearly all species are made up of local populations separated by areas that are unoccupied or thinly populated by the species in question. These separate populations are all unique in some features. Virtually all have unique genetic structures, too, although the differences may be so subtle that only sophisticated analyses of proteins can detect them.

Often, one characteristic changes more or less steadily from one population to the next in one geographic or climatic direction. Such a character gradient is called a *cline.* For example, the number of vertebrae in certain fish populations increases regularly from south to north, as environmental temperature drops. But clines are not always so steady. There may be regular change in a character over some distance, but then a sudden greater change, or a change in a shorter distance may occur.

When analyzing clines, we must be sure they repre-

Classification of Organisms

The cornerstone of the classification scheme is the *species*. The problems in defining a species are many, but this unit is merely the smallest in a hierarchy that progresses through a logical series of steps to larger and larger groupings, or *taxa*. Thus, a number of closely related species make up a *genus* (pl., genera). The genus and species names, used together, identify a given species and are known as the *binomial*. Related genera are grouped together in a *family*, related families in an *order*, and related orders in a *class*. Classes, in turn, are put into a *phylum* (pl., phyla). Finally, at the highest level of all is the *kingdom*. Sometimes these main groups need to be subdivided, as in subphylum or superclass. Figure E29–1 shows how to classify a plant and an animal according to this system.

Note that the names used are Latin or Greek in origin.

Kingdom	Animalia	Plantae
Phylum	Chordata	Tracheophyta
Class	Reptilia	Angiospermae
Order	Squamata	Asterales
Family	Colubridae	Compositae
Genus	*Thamnophis*	*Xanthium*
Species	*elegans*	*strumarium*
Common name	garter snake	cocklebur

Figure E29–1 The classification of a plant and an animal.

This is done both for historical reasons (Latin was the only acceptable language for intellectual discourse for many centuries) and to avoid confusion among the various modern languages.

sent genetic differences. In plants especially, geographic variation may be a developmental or physiological response to some changing feature of the environment, such as temperature or type of soil, without any genetic change at all. For example, flowering plants of the genus *Achillea* (yarrow) grow at a wide variety of elevations in California. In the lowlands they are tall, but in high mountain meadows they are short. When seeds from the upland populations are grown in lowland gardens, they get nearly as tall as the lowland populations. When the differences do turn out to be genetic, however, biologists seek an evolutionary explanation.

Through the years, biologists have come to recognize several specific patterns of geographic variation. These consistent patterns, most intensively studied in birds and mammals, are called **ecogeographic rules** of variation. Although there are many exceptions to these rules, they do seem to apply in species after species. They are usually named for the scientists who first studied them in detail.

Bergmann's rule relates environmental temperature and body size. In many species of warm-blooded vertebrates, populations living in cool areas have larger bodies than those living in warm areas (Figure 29–1a). Warm-blooded vertebrates use heat produced by their own metabolic activities, and selection has apparently favored organisms that lose this heat relatively slowly in cold environments. Heat loss is directly associated with the relation between surface area and body volume. As organisms increase in mass, their surface area increases much more slowly than their body volume, and this is a favorable situation for cutting heat loss.

Bernard Rensch, a German evolutionary biologist, has shown that most birds and mammals in the Northern Hemisphere are larger in cooler than in warmer climates—8 percent larger for sedentary birds, and up to 40 percent larger for Old World mammals. Small warm-blooded animals have an unfavorable surface-to-volume ratio and thus are seldom found in very cold areas.

Allen's rule is nearly a corollary of Bergmann's rule: Warm-blooded vertebrates typically have shorter extremities (such as ears, wings, and legs) in cold areas than in warm areas. Jack rabbits or hares, for example, have shorter ears in North Dakota than in Texas, and song sparrows have shorter bills in Maine than in Florida (Figure 29–1b). Extremities lose heat more rapidly than the rest of the body, and short extremities lose less heat than long ones.

Gloger's rule, also mainly applicable to warm-blooded vertebrates, relates dark and light coloring to the moisture in the environment. There seems to be no evolutionary explanation for this rule: We find no clear relation between color and survival in environments of differing humidity. And indeed the rule has several exceptions. The desert forms of some beetles and mammals

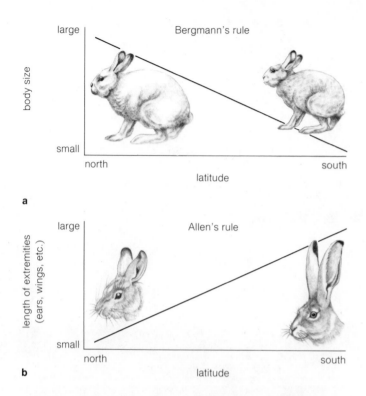

Figure 29–1 The ecogeographic rules of variation: (**a**) Bergmann's rule, relating change in body size to latitude and thus to temperature; (**b**) Allen's rule, relating length of extremities to latitude and thus to temperature.

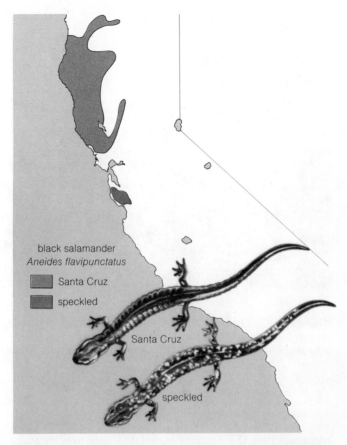

Figure 29–2 Two races of black salamander and their distribution in northern California. Uniformly black animals are found only south of San Francisco.

are darkened or even black. Conflicting selection pressures may be at work here. Perhaps radiative heat gain, increased by black coloration, helps some desert organisms. The nights are often cool, and organisms that can get started first in the morning (because the morning sun warms them faster) may have a competitive advantage. There is also some evidence that dark pigmentation slows down water loss and may protect tissues against radiation damage. But the real question in each case is whether these advantages outweigh the loss of protective coloration. If, for example, an organism's predator hunts mainly at night or by means other than sight, dark color might be almost entirely a plus for survival.

The many exceptions to ecological rules are to be expected. Any phenotype is a compromise among conflicting selection pressures, and clearly some pressures that are insignificant in one group must dominate in another. The rules are not laws, but statistical phenomena, giving us only a general, if important, idea of how selection works.

We can accept geographical variation in phenotypes and genotypes, then, as fact. But are these geographical variations enough to give rise to races and eventually to species?

Geographic Race Formation

In the coastal foothills of Mendocino County in northern California, the black salamander, *Aneides flavipunctatus,* is common in open woodlands, under logs and in rock piles. It is an attractive animal with shiny black skin spotted with white (Figure 29–2). To the south of San Francisco Bay, in the Santa Cruz Mountains, one also finds black salamanders. Here, however, the animals prefer stream sides, and they have no spots. These two populations are recognized as races. As it turns out, races are simply geographic segments of a species often named for their most conspicuous characteristic. Ideally, races are

Figure 29–3 The black volcanic soil of the Tularosa Basin of New Mexico contrasts with the adjacent brilliant white gypsum of the White Sands. This difference in color is reflected in the animals that live there. The matching of fur coloration with the background is especially striking in pocket mice *(Perognathus)*.

recognized on the basis of correlated variation in many features.

Ecological races result from specific responses to the environment. The Tularosa Basin of south central New Mexico provides a good example of geographic variation. The Tularosa Badlands are about 150 square miles of dark rock and soil, formed from a lava flow. Just to the south lie the famous White Sands, a relatively barren area of brilliant gypsum dunes. Most species of small mammals living in the area of dark lava have evolved local races with darkened coloration. Some of these same species have evolved local races with very light fur in the White Sands area (Figure 29–3). Similar trends are seen in local lizards and insects. Other areas of strong environmental contrast show similar examples.

Regional races develop in response to environmental conditions that characterize large areas, such as deserts, mountains, and grasslands. In this instance, the regional climates are broadly similar, and local areas differ from one another only in detail. For example, many plants and animals have generally uniform populations over the area of the Great Plains in Central North America. Usually these races have broad distributions, often hundreds or even thousands of square miles. By contrast, in the southwestern United States, where radical climatic changes can occur in a few miles, distributions are much narrower.

For instance, in southern California, many plants and animals have distinct desert and coastal races. Among reptiles alone there are distinctive regional races for the western banded gecko, the side-blotched lizard, the rosy boa, the glossy snake, and the gopher snake.

Historical races result when populations of a given species are separated by new geographical barriers. During the period of isolation the populations diverge, and even if the barriers are removed so that the populations once again come together and freely exchange genes, average differences between them persist.

Marginal races characteristically occur at the outermost fringes of a species' range. These are segments of a species consisting of geographically isolated populations that have diverged from the main mass of the populations. For example, zoologists recognize few races and little distinctive geographic variation among the widely distributed gray foxes of western North America. Yet the Channel Islands of Southern California have populations of gray foxes that differ in body size, behavior, and population characteristics from those on the adjacent mainland. Marginal isolates are much more likely to have distinctive features than are populations from the center of the species' distribution.

Pseudo-races are populations that differ from others of the same species in perhaps only a single feature. If

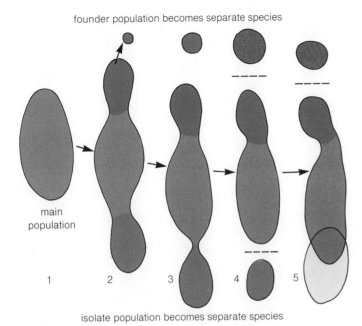

founder population becomes separate species

main population

1 2 3 4 5

isolate population becomes separate species

Figure 29–4 Forms of allopatric speciation. Events that lead to the production of different races and species, starting with a homogeneously similar group of populations.

this feature is spectacular and readily observed, the populations have likely been recognized as a race. We are now beginning to see that many so-called races are in reality not races at all, but simply populations with high frequencies of certain alleles.

Human races have been recognized both popularly and scientifically for centuries. Even the Bible distinguishes between races. However, biologists and anthropologists have long recognized that racial classification is based on but a few characteristics, and even for these there are many exceptions. Certain population groups do not fit neatly into any race. Others may display features supposedly characteristic of two or more races. When we study individual traits in human populations, we see a pattern of distribution that is characteristic of many widespread natural populations. In the blood groups, for example, the I^A allele is frequent in parts of Scandinavia, and well represented generally in Europe through Central Asia into China. But it is rarer in most of Africa and North America and nearly absent in natives of South America. The allele, however, transcends recognized racial boundaries.

Many biochemical and genetic traits have now been studied for human populations, and the variability within populations and the absence of any sharp boundaries to variation from population to population is impressive.

In fact, in a recent, detailed study of the genetic evidence for geographic variation of traits in human populations, Richard Lewontin of Harvard University found little correlated variation among the so-called races and challenged the basis for recognizing races in humans. It is increasingly clear that we have been influenced too strongly by a few superficial features in our attempts to describe geographic variation in humans. Human races may prove to be pseudo-races from a biological point of view.

Speciation

During the past century, biologists have analyzed many patterns of geographic variation and race formation with the goal of learning something about how new species arise. From these studies have come generally accepted models of the speciation process.

Allopatric Speciation Although Charles Darwin and even earlier biologists emphasized that geographic isolation preceded speciation, the modern geographic model of speciation is closely associated with Ernst Mayr, now at Harvard University. Each population of a species varies in a unique way from all others, and studies of geographic variation within species show that remote, isolated populations are often the most distinctive. Therefore it seems logical that barriers to dispersal—and thus to gene flow—might favor the development of unique combinations of genes in populations. Given enough time, and environments distinct enough that different genetic combinations are favored among the isolates, the isolate population will tend to diverge from the main population to the point where, even if the barrier is removed, genetic information cannot be exchanged (Figure 29–4). In other words, individuals from the isolate population cannot or will not mate with individuals of the opposite sex in the main population. When this happens, the isolate population is considered a separate species. Such **allopatric speciation** *(allo,* different; *patria,* habitat) is perhaps the simplest mechanism of speciation: the combination of geographic variation and a barrier to gene flow. An important variant of this model is what can happen at the extreme margin of the range of a species. A new population may be founded by a very few individuals (even a single pregnant female in extreme cases in animals). These "founders" carry only a sample of the total genetic variation in the source population. Further, since they are beyond the normal limits of the species, selection pressures are likely to be great. Under such circumstances, evolution may be rapid, and this is now thought to be an extremely important mode of speciation.

Many different kinds of evidence support the geo-

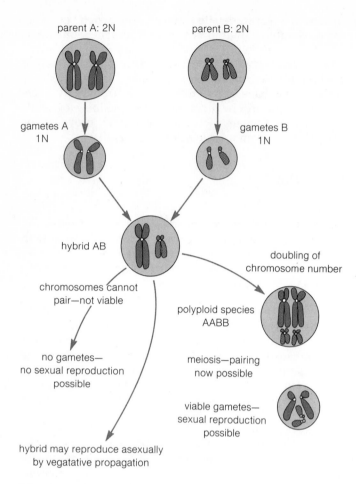

Figure 29–5 Chromosome behavior during the formation of a hybrid and a polyploid, which results from doubling the chromosome number of the hybrid. Note that the hybrid cannot produce viable gametes because the chromosomes lack homologous pairs; the polyploid can, because such chromosomes are now present.

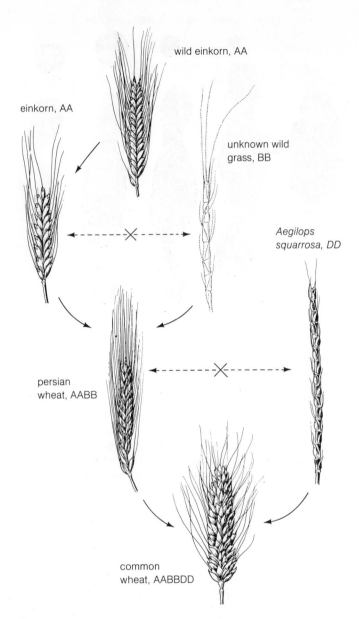

Figure 29–6 The evolution of new species by polyploidy, illustrated in the genetic history of wheat.

graphic model of speciation, in one or another of its modes. The strongest evidence is the almost universal phenomenon of geographic variation itself, which suggests that different populations of a given species are constantly adapting to the selective pressures prevalent in their communities.

Instantaneous Speciation Species sometimes seem to arise in a surprisingly short time—almost instantaneously. These occurrences are usually associated with the formation of *hybrids,* or individuals that result from matings between individuals with great genetic differences, usually members of different species. When such individuals do succeed in interbreeding or mating—contrary to the

usual idea of species—their offspring are usually infertile, as the chromosomes cannot pair at meiosis and no viable gametes are produced (Figure 29–5). However, in rare cases they can mate with the parent population and produce offspring that are in effect a new species, able to mate only with each other.

In hybrid plants, the entire complement of chromosomes may double, producing a condition called *polyploidy.* The offspring in this case are fertile, because individual chromosomes can continue to pair with their homologues during meiosis, rather than trying to pair with an ill-fitting "pseudo-homologue" derived from another parental stock (Figure 29–5). They are also true-breeding and isolated from the parental species because the offspring of a cross with either parent are sterile. In other words, they are a new species because their chromosomes cannot pair during meiosis with those of individuals outside the group.

Consider wheat, for example. This important crop was one of the first brought into cultivation; wheat grains dating from 7000 B.C. are known from eastern Iraq. Several wild wheat species with 14 chromosomes (two sets of 7, designated AA) are found growing in that area. Also growing there is a wild grass with 14 chromosomes different from A, as judged by their failure to pair with set AA, and thus designated B (two sets of 7, designated BB). Several of the common cultivated wheats have a chromosome number of 28, and careful analysis of their meiosis shows that these wheats are AABB—a new polyploid species. Another common cultivated wheat has 42 chromosomes (6 sets of 7) that can be designated AABBDD—the last chromosome set coming from another wild grass. These wheats arose, so far as we know, from hybridization without direct human intervention, but humans did act as agents of selection, encouraging new species that were most useful to them (Figure 29–6).

Many new species of plants have evolved through hybridization followed by chromosome doubling. The phenomenon is not so widespread among animals, but in recent years some interesting examples have been discovered among the lower vertebrates (fish, amphibians, and lizards). In a stable environment, hybrids might not survive. But when environments are changing, their new combinations of genes might chance to fit the new conditions better than either of their parents' genotypes, and thus they become established. With animals, as with plants, there is usually a correlation between the establishment of these unusual new species and some climatic or other drastic change, such as advancing desert, receding glaciers—or human beings deciding to invent agriculture.

Sympatric Speciation   Some scientists have argued that species can form without either geographic isolation or hybridization. Two populations, they say, could emerge from a single one in a single community without ever being geographically isolated. This is called **sympatric speciation** (*sym,* same; *patria,* habitat). In order for it to happen, however, there must be polymorphism (see Chapter 28) in the population, and morphs must mate only with like morphs for many generations. It is hard to imagine such conditions being met in nature. Indeed, the idea arose mainly because scientists were trying to explain *species swarms,* vast numbers of closely related

species living together in a single area. For example, in the Great Lakes of Africa, a single lake contains many closely related species of fresh-water fishes, and a geographic model to explain their speciation would have to be highly contrived.

Increasing attention is being given to a special kind of speciation allied to sympatric speciation. This form of speciation arises when chromosomal rearrangements occur, often at the edge of the range of a species or in some sparsely inhabited portion of the range. Individuals having a heterozygous condition of this rearrangement are often unfit—the condition may even be lethal—but individuals homozygous for the new arrangements survive well. Thus, if animals with the new arrangement interbreed, they quickly become separated from adjacent family groups with no change in habitat. The great Australian geneticist M. J. D. White has made detailed studies of this phenomenon, called by him *stasipatric speciation* (*stasis,* standing still; *patria,* habitat), in grasshoppers. Other researchers use the term parapatric speciation (*para,* beside; *patria,* habitat) for this phenomenon.

Isolating Mechanisms

What kind of genetic change is necessary for speciation? For many years it was thought that speciation involved a kind of genetic revolution, with extensive changes in the genetic structure of the diverging populations. Now we think that drastic genetic changes may not always accompany speciation. Theoretically, a single mutation could produce a feature that would isolate its bearers from others in the population. Moreover, different populations of a species may be almost as different from one another as they are from populations of other, closely related species. Both these facts suggest that no great genetic reorganization is necessarily associated with speciation. The crucial question is whether the change that has occurred, whatever its size, prevents fertile interbreeding. If it does, we call it an **isolating mechanism.** A large number of isolating mechanisms are known, some acting before fertilization takes place and some after.

Prezygotic Mechanisms   Prezygotic mechanisms are efficient, for they tend to keep members of different species from trying to mate, or, if they do mate, from producing a fertilized zygote. The best-studied isolating mechanisms are those that work before mating occurs, not surprisingly called *premating isolating mechanisms.*

Earlier we mentioned that each species seems to play a unique ecological role. That is, closely related species living in the same geographic area often prefer different habitats. For example, two closely related midwestern species of toads, *Bufo fowleri* and *Bufo americanus,* live in the same general area, but *B. fowleri* stays mainly in the plains, whereas *B. americanus* prefers woods. Thus, under ordinary conditions the two species do not often meet, and during the breeding season they move to different ponds. Habitat preferences, then, can serve as an *ecological isolating mechanism.*

Breeding seasons are another possible isolating mechanism. Tropical biological communities, for example, often contain many species of birds and frogs. Among these neighboring species, some breed at the beginning of the wet season and others at the end. Some even choose the height of the dry season, seemingly an unfavorable time, to start reproductive activity. This is a kind of temporal isolation, or isolation by time, with species that breed in the wet and the dry seasons, respectively, never receptive to courtship behavior at the same time. This kind of isolation is also common in flowering plants, where close relatives in the same general location may flower at different times.

When temporal isolation is not operating, some close relatives that live in the same community breed at the same time. How do they avoid exchanging genes? Such species are often *ethologically isolated* from one another. That is, the sexual behavior of one species is not at all attractive to members of another species. Basically, through the operation of natural selection, they have come to respond only to certain cues, so that although potential mates live close together and breed simultaneously, they do not mate. The kinds of cues vary greatly from group to group. Many species of birds use elaborate visual stimuli in their courtship behavior. Grebes, a group of aquatic birds, engage in a highly ritualized and elaborate display between the two sexes in the water. If the sequence is not exactly correct, one partner or the other breaks off the courtship. No other species can duplicate this behavior, which surely must be inherited.

Frogs, birds, and many insects use auditory cues or stimuli; other insects (Chapter 14) and some mammals use chemical stimuli; certain salamanders use mechanical stimuli—and all of these must be right, or breeding will not take place.

Should neither ecological (habitat preference, breeding time) nor ethological (courtship signals) isolating mechanisms function, several possibilities for premating isolation still remain. One of these is a *mechanical isolating mechanism.* Males of many organisms have a copulatory organ that fits the reproductive tract only of females of the same species. Simple physical size can also be a barrier. For example, one of the smallest toad species, the oak toad *Bufo quercicus,* occurs in the same communities with *Bufo terrestris,* the terrestrial toad, which is one of the largest. They are not known to try to interbreed, but if they did, a male *B. terrestris* would likely drown a female *B. quercicus.* Conversely, a male *B. quercicus*

could never mate with a female *B. terrestris,* for its arms are not long enough to grasp her.

In addition to these premating isolating mechanisms, there are isolating mechanisms that function after mating has occurred. These *postmating mechanisms* work by preventing the formation of the zygote. Among the fruit flies, genus *Drosphila,* for example, the sperm enter a hostile environment in the reproductive tract of a female of another species. The female's vagina swells enormously in an immunological reaction, blocking the passage of the sperm. The general term for such postmating but prezygotic mechanisms is *gametic incompatibility.* In plants, the incompatibility is expressed during growth of the pollen tube, which is either incomplete or otherwise faulty.

Postzygotic Mechanisms Postzygotic isolating mechanisms operate after mating has occurred, fertilization is accomplished, and hybrid zygotes are formed. At this point, and in later development, things go wrong for the hybrid. The hybrid zygote may be so weak and defective that it dies either as an embryo or as a juvenile. This is called *hybrid inviability.* In flowering plants, the endosperm often collapses and the hybrid embryo dies. Sometimes the zygote is vigorous, as in the mule, which is a hybrid of horses and donkeys, but the gonads either fail to develop or, if they do, they produce no gametes or mainly genetically defective ones. This phenomenon is called *hybrid sterility.*

Finally, even if hybrids manage to be both vigorous and fertile, premating isolating mechanisms will act against them. For example, if the hybrid has a combination of courtship behaviors derived from its two parent species, the combination works with neither species, and maybe not even with other similarly formed hybrids. This is a kind of *secondary hybrid isolation.* Clearly the deck is stacked against most hybrids, as far as evolutionary success is concerned.

In nature, often not just one but a whole complex of isolating mechanisms keeps species from hybridizing. For example, in two sympatric species, *Drosophila pseudoobscura* and *Drosophila persimilis,* biologists know at least six isolating mechanisms: (1) habitat preference *(D. persimilis* favors cooler and moister habitats), (2) shape of the male sexual organ, (3) difference in courtship behavior, (4) temporal isolation *(D. persimilis* is active early in the day and *D. pseudoobscura* in the evening), (5) sterility of hybrid males, and (6) weakness or sterility in backcrosses of hybrid females to either parental stock. Together, these factors keep the two species separate.

Evolution of Isolating Mechanisms There is a high premium on efficiency of mating in nature, but it is hard to imagine how natural selection could "improve" postmating isolating mechanisms. Premating mechanisms are in any case by far the more economical. Thus, only premating mechanisms are thought to be under the direct influence of natural selection. For one thing, individuals with the highest precision—those responding only to the correct courtship behavior at the optimal time of year—are the most successful in leaving offspring. As a result, the populations evolve increasingly in the direction of highly precise signals, environmental or behavioral, that keep different species separate. Theoretically, contact between two incompletely isolated species should reinforce poorly developed isolating mechanisms, the number of mismatches steadily dropping through time. Since hybrids are almost always deficient, both they and their parents will be less fit in the Darwinian sense. Thus, the "fitter" individuals who are "most careful" in mate selection will come to dominate the population.

Summary

Species are the "kinds" of organisms that are recognized as discrete groups of organisms by humans. We usually have no difficulty differentiating most of the species that occur at a single place, but characteristically species vary geographically. Thus, when we find a population of a species in a distant place we may not always immediately recognize it as the familiar one we know. Geographic variation most commonly results from adaptation of local populations to environmental selective pressures. These pressures differ from place to place, sometimes according to general phenomena of earth climatology (colder in the extreme north and south, for example), but sometimes to regional and even local conditions of soil, precipitation, amount of sunlight, and other such phenomena. When groups of populations inhabiting one area are morphologically differentiated from groups of populations inhabiting another area, we sometimes name the two groups as subspecies. So long as gene flow continues between the two groups they remain members of the same species, but should a strong barrier—such as a river for desert reptiles or a mountain range for lowland birds—intervene for a long period of time, the two groups may diverge to such a degree that isolating mechanisms evolve. Then we say that speciation has occurred. Should the barrier later disappear or be breeched by some migrants, the sister species may come to coexist. This geographic model of speciation is thought to be common, but it is not the only way that speciation can occur. In recent years we have come to understand that special mechanisms, such as chromosomal modifications, can produce speciation within a single limited region. Isolating mechanisms evolve coincidentally with speciation processes. The most efficient isolating mechanisms are those requiring minimal energy investments to operate. Thus premating mechanisms, in which potential mates meet but do not mate, are common, and it can be inferred that natural selection has reinforced incipient isolating mechanisms that arose by chance during periods of geographic separation of two populations in the process of speciation. Frequently postmating isolating mechanisms also are found, but these appear to arise by chance and not by direct selection.

30

Adaptation in Populations and Communities

Adaptation is a core concept in modern evolutionary biology, yet there is continuing confusion over its meaning. Let us get it straight from the beginning. **Adaptation** is the outcome of natural selection; specifically, the genetic modification of populations by natural selection.

Biologists often refer to particular features of organisms as adaptations. Birds' wings are called adaptations for flight, and horses' teeth adaptations for chewing grass in a special way. These usages imply that natural selection is responsible for the mechanical modification of structures to perform particular functions. It would be better to use the word **specialization** for such processes, and to describe structures modified to perform a particular function as *specialized.* However, biologists have so long used the word *adaptation* in both the general sense of change in a population and the more restricted sense of modification of particular structures that change is unlikely, and you should be alert whenever you see the term to determine which way the author is using it.

Natural selection is the mechanism whereby organisms evolve means of coping with "problems" in nature. These problems can be visualized in a variety of ways. For example, how should organisms be expected to change in order to run or swim or fly more efficiently? Engineering approaches may be useful in analyzing functional designs and making predictions for future change. In using them we must keep in mind the realm of possibilities, for all organisms are faced with design constraints, and any change must be a compromise. To a large degree, adaptations are the evolved solutions to the problems of nature.

Adaptation is basically a population concept that applies to the evolutionary status of populations in **communities,** which are associations of interacting populations in nature. We cannot explain industrial melanism in peppered moths (Chapter 28)—or in other species—without considering their community relationships, for all such organisms have predators that hunt them by sight. The peppered moths, once protectively colored, lost average fitness when they became conspicuous on lichens blackened by soot. In response to this new selection pressure in the environment, the population adapted, increasing its proportion of black members and thus its average fitness. Natural selection was visibly at work.

Are these moths as well adapted as they were before industrialization? We do not know how to answer such questions precisely, but we do know that the species has survived and still enjoys densities not obviously lower than those of the eighteenth century.

How can adaptation be measured? It cannot, at our current state of knowledge, for too many interactions must be considered. In measuring fitness, one need only determine the success of certain individuals in leaving offspring, in comparison with competing members of the population. But in communities, we must also take into account competitive interactions between members of different species, and predator-prey interactions. The only criterion for success in such situations is whether a population maintains its relative strength in the community. In other words, simply surviving implies that the population has adapted somewhat. Populations that become extinct must be less well adapted than those that survive, for their average fitness obviously declined to zero.

But does an increase in numbers always mean better adaptation? It may—or it may signal disaster from overpopulation. After a certain point, it becomes maladaptive to produce young, for resources are almost always limited. For example, a population of flour beetles *(Tribolium)* in a flour container will die off sooner if it increases its rate of reproduction. Even where the food resource is renewable, long-term survival of a population may demand adaptation that produces young no faster than adults die off. Is our rising human population a sign of better or worse adaptation?

The Community as the Arena of Evolution

Our familiar western Washington pocket gophers (Chapter 28) eat various plants and plant products. The gophers in turn provide food for predators, and their waste products provide a substrate for the growth of bacterial and fungal colonies. Each patch of prairie forms a *natural community*, defined as an assemblage of populations of plants, animals, bacteria, and fungi that live in an environment and interact with one another, together forming a distinctive living system with its own composition, structure, environmental relations, development, and function. The community is, in essence, the living portion of an ecosystem. There are many distinctive kinds of communities: piñon pine and juniper woodlands on desert mountain ranges, gallery forests following rivers into treeless steppes, grasslands, redwood forests, creosote bush deserts, and many more.

Communities   When pocket gopher populations adapt, they are doing so within the context of their communities. For example, the gophers' fur tends to be about the same color as the soil in which they burrow. Such adaptive coloration, found in many organisms, usually has to do with avoiding predation, and the pocket gophers are no exception. Thus, biotic interactions—in this case between variably colored rodents (the color being genetically determined) and visually hunting predators (perhaps hawks and coyotes)—are extremely important.

Adaptations also occur in response to positive selective forces. The prairies are rich in wildflowers, and many of these are pollinated by insects. The flowers provide food (in the form of nectar and pollen) for the insects, and flowers and insects can adapt jointly. The flowers that attract the most insects are the most successful in reproduction; the insects that find the most rewarding flowers are the best nourished, and thus presumably the fittest in terms of reproduction. This coadaptive process, often called *coevolution,* is a community phenomenon.

Communities are units of both ecology and evolution, and may be stable and long-lasting even though the populations within them change. As an evolutionary unit, a community is certainly far less well defined than a single population, but some general comparisons can be made. Within populations, frequency of alleles changes through time, and some alleles may be replaced. Populations in a community change over time, however, and may even be replaced by other, similar populations.

The evolution of populations has often been studied, but only recently has much attention been given to quantitative work on community structure and evolution. Of special interest are population interactions that result in adaptive responses. Ecologists are only beginning to learn how extensive these interactions can be.

Habitat   In order to understand population interactions in a community, we must first see how individual populations are related to their environment. The actual place where the members of a population live is called the **habitat.** Think of a place in nature where communities exist, such as the gopher prairies of western Washington. We soon see features within this habitat that vary from place to place and time to time: exposure to wind and sun, elevation, soil fertility and texture, and so forth. Each species in the community, like the gophers, can exist over a particular range of environmental variables in a place, and all of the species in the community (including gophers, coyotes, camas lilies, and pollinating insects) together cover an area that is their common habitat.

The habitat of a species can also be seen as the range of environments or communities over which a species occurs. Most gophers live in deep soil, but a few can manage in the gravelly glacial till so common in western Washington. The separate populations of a species respond to the particular selective pressures that characterize each habitat. The two aspects of the habitat to which populations respond in an evolutionary sense are the *abiotic* or physical aspects, and the *biotic,* living, or interactive aspects. Through the years, many biologists have studied details of specific organisms, particularly their responses to habitat variables. This kind of investigation, directed at the populations of a single species in relation to their environment, is called **autecology.** In contrast, the study of interactions of populations in a community is called **synecology.**

Ecological Niche   The biological role or "profession" of a species in a community is its *ecological niche.* This refers not, as you might think, to a particular *place* in the environment, but to the way a species makes its living. As many as six species of tits live in some English forest communities. They can live together in part because they hunt for food in different places (Figure 30–1). Some of the larger species eat large food and can perch only on sizable twigs. Smaller species eat smaller food, can move through denser foliage than larger ones, and can perch on tiny twiglets. All six species glean foliage in the same forest, but they have six different ways to glean foliage, six "professions," six ecological niches—which we might name foliage gleaner class 1, class 2, and so on. The resources of the community are great enough to support all six.

Each species in a community has a particular range of temperature and moisture tolerances, a particular choice of nutrients, and a special activity period during the year and even during a day. Most vertebrates have courting and egg-laying seasons, and each plant has a reproductive period. Within a community, there are variables of many sorts. For tits, food comes in many sizes. For gophers, soil comes in many textures. For salamanders, different places have different amounts of moisture. For plants, light is available in differing intensities.

The ecological niche of a species, like the personality

of an individual, is real, yet complex and hard to quantify. When a species becomes extinct, there is no vacant "space" left in the community. Usually, there is not even a vacant function for long. The extinction may have happened partly because another species was beginning to perform the same role. Suppose, for example, that a mutation among the foliage-gleaning tits allowed the class 3 gleaners to eat slightly larger food. Now more versatile in their "profession" and probably better nourished, they might produce more young. Gradually, they might encroach so much on the niche of class 4 tits that the latter species becomes extinct. Thus, parts of the abandoned niche are incorporated into the niches of other species. The multidimensional niche disappears, together with the species of which it was an attribute.

Is there such a thing as an empty niche? For example, does such a niche await the arrival of hoofed mammals, such as cattle or antelopes, in Australia, where kangaroos were the only large, grazing mammals? Not exactly. When cattle were brought to Australia, they did not enter a vacant niche, but the existing communities changed to accommodate them.

Community Interactions That Influence Adaptation

Theoretically, at least, two populations of a community can interact so that neither is affected. Other possibilities, however, are more likely. One population may be affected more than the other, or both populations may be affected to about the same degree, positively or negatively. By negative effect, we mean that population success or Darwinian fitness is reduced—fewer surviving young, repressed population growth, or some other factor related to long-term reproductive success of the population relative to others in the community. Positive effect, of course, is just the opposite.

The various kinds of interactions between two species are illustrated in Table 30–1. Some of these are fairly unimportant, such as *neutralism,* in which neither of the populations affects the other, and *amensalism,* in which one population affects another without any cost or benefit to itself (for example, a box turtle stepping on several members of a population of small, fragile flowering plants). Others, however, such as *competition, predation, parasitism,* and *coevolution* (which includes the various forms of symbiosis), are quite important, and we will look at each of these in more detail.

Competition Generally speaking, competition means the same thing in ecology as it means in families, when two children want the same chicken drumstick. Competi-

Figure 30–1 Various kinds of tits can live in the same community because they hunt for different types of food and have different requirements for perches and other physical attributes of the environment.

tion seems to be a universal feature of communities, and as such it has great actual and theoretical significance. But it is hard to study directly. From one point of view, it hardly seems to need discussing—surely some resources are limited in all communities, and there must be competition for them. Fishes in a pond, for example, fight over a bit of food and vigorously defend a carefully circumscribed area, their territory, against all comers. But what is the significance of this behavior? Can the effects be measured directly, or must we content ourselves with the outward signs of competition?

Competition is interaction between populations trying to use resources too scarce to support all competitors. A *resource* is some feature of the environment that individuals in a population need for one of the three general phenomena of life: growth, maintenance, and reproduction. Scarce resources are called *limiting resources,* for they can directly affect population growth and stability. Some resources are present in fixed amounts, and when that amount is allotted, the population can no longer grow. The resource has limited its size. For example, a given patch of forest contains only a certain number of

Table 30–1 Analysis of Two-Species Population Interactions

Type of Interaction*	Species† 1	Species† 2	General Nature of Interaction
1. Neutralism	0	0	Neither population affects the other
2. Competition: direct interference type	−	−	Direct inhibition of each species by the other
3. Competition: resource use type	−	−	Indirect inhibition when common resource is in short supply
4. Amensalism	−	0	Population 1 inhibited, 2 not affected
5. Parasitism	+	−	Population 1, the parasite, generally smaller than 2, the host
6. Predation	+	−	Population 1, the predator, generally larger than 2, the prey
7. Commensalism	+	0	Population 1, the commensal, benefits while 2, the host, is not affected
8. Protocooperation	+	+	Interaction favorable to both but not obligatory
9. Mutualism	+	+	Interaction favorable to both and obligatory

* Types 2 through 4 can be classed as "negative interactions," types 7 through 9 as "positive interaction," and 5 and 6 as both.
† The symbol 0 indicates no significant interaction; + indicates growth, survival, or other population attribute benefited (positive term added to growth equation); − indicates population growth or other attribute inhibited (negative term added to growth equation).

holes suitable for bird nests. As the number of birds in a given population of hole nesters increases, *intraspecific competition* (competition within a species) for the few remaining holes becomes intense. If two species of hole-nesting birds occur in the same community, *interspecific competition* (between species) may occur. The holes are limiting resources.

Some resources, such as food, are usually *renewable.* They are either produced at a predictable rate, or are available periodically. When they are scarce, either because the rate of production has dropped, or because populations have achieved high density, both intraspecific and interspecific competition may occur.

Early naturalists observed that there seemed to be only one kind of organism doing any given thing in a given community. They also saw that a particular set of plant and animal types tended to be repeated in communities occupying similar, though widely separated, environments. Joseph Grinnell of the University of California, Berkeley, was the first naturalist to formulate a clear theory to account for this phenomenon. He thought that each species of plant and animal had a unique relationship to the environment, and that, if a species was introduced into a community already occupied by an ecologically similar species (having that same relation to the environment), the two would eventually come into competition for some limiting resource, and one or the other would, in time, be eliminated. In other words, species whose niches overlap too far cannot coexist. This idea is now called the **competitive exclusion** principle.

G. F. Gause, a Soviet scientist, conducted classic studies in the 1930s that stimulated a good deal of further research on competition. He studied two species of protozoans, *Paramecium caudatum* and *Paramecium aurelia,* that were cultured in his laboratory. Each grew well by itself, but if both were placed in a culture tube where food input was controlled, only one species survived: *P. aurelia* (Figure 30–2). This finding led biologists to direct special attention to competition, in an effort to understand what was happening in nature. Why could closely related species not coexist in the laboratory? Naturalists already understood how species *can* coexist in nature. They understood that, for species to coexist in communities, they must be able to pursue different professions; that is, there must be as many niches as there are species.

We all know that species share certain resources in an environment. Plants share soil. Many species of rodents may eat the same species of grass, or its seeds. In certain South American tropical lowlands, over 100 species of generally similar frogs may live together in the same area, many eating the same species of insects. In the Great Lakes of Africa, hundreds of species of cichlid fish live in close proximity and share many resources. We must focus not on the whole environment but on scarce resources—*limiting resources.* The competitive exclusion principle states that two species cannot exist in the same community when both depend on the same limiting resources. But limitation may be a sometime thing. Resources may usually support both species, but 1 year in 10, or 30, the limiting resource may be so reduced that one species is excluded.

Gause's results stimulated others to study competition directly and experimentally. Thomas Park and his students at the University of Chicago grew flour beetles (*Tribolium*) on wheat flour medium. Two species, *T. confusum* and *T. castaneum*, have different growth curves when kept by themselves. When the species are mixed, the outcome varies from environment to environment. When the climate is warm and damp, *T. castaneum* invariably wins. When the climate is cool and dry, *T. confusum* always wins. But in intermediate climates, the results are less predictable. For example, in a warm and dry climate, *T. confusum* wins 87 percent of the time; but in a cool and damp climate, it wins only 69 percent of the time. However, in time, one or the other species is always eliminated, and the remaining one thus wins the competitive encounter.

In real communities, with their many species, the situation is almost certainly more complex than in the laboratory. Yet many examples suggest that species exclude other species by competing better. For example, certain salamander species in the southern Appalachian Mountains replace each other as altitude changes. The exact elevation where the replacement occurs may differ from place to place, depending on exposure and other factors, but the order of replacement is always the same. This suggests that one species competes better at higher elevations and lower temperatures, while the other competes better at lower elevations and higher temperatures.

Predation One living organism feeding on another—predation—is, of course, one of the most characteristic features of an ecosystem. The search for food and the avoidance of predators dominate the lives of most animals in natural communities. It is not surprising, therefore, that a great many structural and behavioral features of animals have evolved in response to predator-prey interactions. Plants, too, are much affected by predation and have evolved specializations in response to selection pressure associated with interactions with animals. A few plants, such as the sundews and pitcher plants, have evolved leaf specializations permitting them to become predators. The specializations we will discuss here are those associated with improving predatory efficiency or success, on the one hand, and those facilitating avoidance of predation, on the other.

Adaptations or specializations for improved predation fall generally into three categories: those for finding, catching, and devouring prey (Figure 30–3).

Carnivores have many specialized features for finding

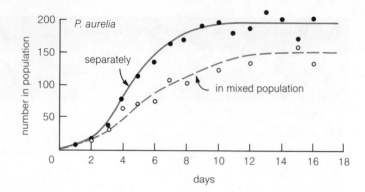
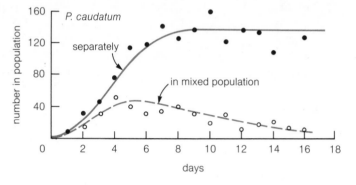

Figure 30–2 Growth in *Paramecium caudatum* and *P. aurelia* cultivated separately and in the mixed population.

prey. The great predatory birds, the hawks, falcons, and eagles, have remarkable flight power and sharp eyes that let them spot moving rodents from great heights. As we saw earlier, bats use a kind of sonar system, not only to avoid obstacles in flight, but to find prey.

Terrestrial herbivores ordinarily have no problem catching their prey, for it usually stands rooted before them. Carnivores, not so fortunate, have evolved a variety of prey-catching specializations. The enlarged canine teeth of carnivorous mammals are well-known specializations for capturing prey. However, different groups use them in different ways. Wolves and dogs use them to tear at exposed soft parts of potential prey; cats use them to crush the back of the brain case. Rattlesnakes have not only long piercing teeth, but also a potent toxin. However, the jaws holding these teeth are surprisingly weak, and one theory concerning the origin of snake venom is that selection has favored venom systems ensuring that the prey would be dead, and thus still, during the long swallowing process. A large, struggling mammal might break the fragile jaw bones. Frogs, chameleons, and many salamanders have long tongues they can shoot out to capture prey. Birds have a vast array of sizes and shapes of beaks, often used in a precise way for capturing a particular kind of food. Hunting, or raptorial, birds like

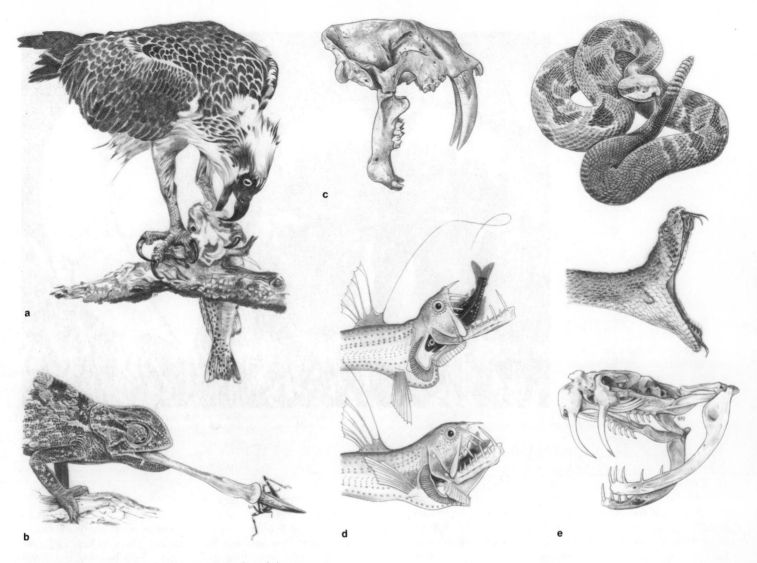

Figure 30–3 Adaptations of carnivores to catch and devour prey.

hawks and owls use long claws to grasp the captured prey.

Not all predators actively seek prey. Some set traps and passively wait for it. Spiders build complex webs for trapping insects. Pitcher plants have elongated, specialized leaves. As insects enter these leaves, they find themselves in a vertical trap, with hairlike structures that point down, stopping them from climbing back out. Eventually they fall to the fluid at the bottom and die, to be digested by bacteria and by the plant's enzymes.

Once food is found and held, it must be ingested, taken into the predator's body. Rattlesnakes' highly mobile skulls let them use tooth-bearing bones one at a time to draw food into the throat, where peristalsis can carry

it on down to the stomach. Since the lower jaws are held together only by a ligament, they can separate widely to accommodate a large food object. Insects show a wide variety of specialized mouthparts. Beetles have specialized appendages around their mouths that reduce food to fragments small enough to be swallowed. Butterflies, in contrast, have mouthparts elongated to form a tubelike structure that is inserted into a flower to retrieve nectar. Hummingbirds' tongues can also reach deep into flowers for nectar. Spiders liquefy their prey by squirting digestive jucies onto it while holding it near their mouths; then they need take only a fluid into their digestive systems.

Nearly every species of mammal has a highly distinc-

Figure 30–4 The dead-leaf butterfly demonstrates crypsis. With its wings closed, it appears to be a dead leaf; with them open, it looks like a butterfly.

tive pattern of tooth structure associated with its predominant prey. In carnivores, the teeth shear flesh; in rodents, they chew abrasive vegetation; and in shrews, they puncture insects. In some bats, they puncture mammalian flesh and provide an opening for blood, which is then swallowed.

We cannot always separate catching from ingestion. Baleen whales take vast quantities of water into their mouths and sieve the food out with their baleen ridges. Some turtles, fishes, and aquatic amphibians open their mouths wide at the instant of greatest throat expansion and ingest their prey along with all the water (and objects in it) near their heads.

A wide variety of behavioral specializations is associated with prey capture in carnivores. For example, wolves hunt in packs and take prey much larger than they are. Some predatory fishes swim together in schools, and when one attacks a prey item, often large and strong, the others quickly join it and overcome the prey. Many predators, especially those with well-developed central nervous systems, such as birds, develop **search images.** For example, a crow that feeds on herring gull eggs

has a search image of the characteristically spotted gull's eggs. It will pass over chicken eggs that are more easily seen and are suitable food, unless the chicken eggs are painted to look like gull's eggs. Search images permit predators to concentrate on special kinds of prey and, in general, to increase their own efficiency.

As many specializations are associated with avoiding predators as with predation. In general, potential prey have three choices: They can hide, run, or fight. Hiding often means running first. Consider a mouse dashing madly into a hole to hide from a cat, or a deer hiding from hunters in a thicket. Many animals have retreats they rush to in the face of danger.

There are alternative ways of hiding. Over many generations of natural selection, certain animals have come to look like parts of their environment. For example, walking stick insects look so much like twigs that they can hide on a branch in plain sight, simply by not moving. Leaf hoppers, small true bugs, often resemble plant spines. Even plants, such as the stone plants of the African deserts, may be nearly indistinguishable from objects in their environment. This method of hiding—by looking like

Figure 30–5 Countershading in a mako-shark. Light falling on a uniformly colored fish would throw a shadow on the underparts (top figure). A countershaded fish in all-round illumination would appear as in the middle figure. When sunlight falls on countershaded coloration, the fish becomes inconspicuous. The dark back merges with the light, but shadowed, underparts.

Figure 30–6 Disruptive coloration. The stripes of a zebra obscure it in its native habitat.

some object in the environment that does not interest potential predators—is called **crypsis** (Figure 30–4).

The examples just given are a particular kind of crypsis called *matching*. There are also other kinds. For example, many fishes that generally inhabit the upper levels of open waters are *countershaded*, their upper surfaces dark and their lower surfaces light. They are thus protected both from surface predators such as birds, which look down into relative darkness, and from submerged predators such as fish, porpoises, and crocodiles, which look up into light (Figure 30–5).

Yet another common kind of crypsis is illustrated by the color pattern of the zebra. Zebras seem conspicuously, even spectacularly, colored when you see them in zoos, but they blend into their background extremely well when running across the plains of Africa. This is called **disruptive coloration** and is typical of many organisms (Figure 30–6).

Running is a common defense. Caribou try to outrun wolves in North America, and zebras run from lions in Africa. The adaptations of hoofed mammals (such as horses, antelopes, and deer) for long stride and speed

(long legs, shoulders that are functionally part of limbs, running on the "fingernails" and "toenails" of but one or two digits) are the result of selection for speed to outrun predators. The leaping of frogs is but a kind of running, involving a series of structural specializations—greatly shortened bodies, huge heads, and long legs—that have severely constrained their way of life. Flight in birds may also have evolved in response to predation pressure.

When we think of fighting, we think of the battles between wolves and caribou, lion and antelope, and other such examples. But fighting can be more subtle than this. Many plants and animals have structural features that act as defenses. We are familiar with the thorns of roses, the spines of cacti and porcupines, and the armor of turtles and armadillos, all evolved as a result of predation pressure. Defenses may also be chemical. Poison oak and poison ivy are inedible to many organisms and cannot even be touched without harm by some—particularly humans! Many plants contain substances that make them poor food for any animal.

Occasionally one particular species of animal adapts its metabolism to the chemical deterrent of a species of plant. Few animals eat milkweed, which produces a highly distasteful substance that is poisonous to many animals. Yet an insect known as the milkweed bug lives exclusively on milkweed. The insect is distasteful as a result of its diet and has no known predators. The monarch butterfly also has few predators because of its noxious taste; its larvae, too, feed on milkweed.

Many animals produce chemical substances that make them either unpalatable or poisonous. Several species of fishes have poisonous substances in their flesh, and most amphibians produce distasteful skin secretions. Yet European hedgehogs, already protected by sharp and numerous spines, seize toads, and after killing them, spread their saliva, now richly supplied with toad venom, over their spines. Many insects produce noxious fluids that are used in defense. Tenebrionid beetles are often called "stink bugs" because of a secretion of their abdomens. This does not provide universal protection, however, for some insect-eating mice of North America *(Onychomys)* grab them and bury the abdomen in the ground before eating the rest of the beetle.

Many species that have distasteful or poisonous substances have also evolved some system of advertising their distastefulness to potential predators. Selection has favored not those animals that get eaten and kill the predator, but those that get tasted and rejected and thus survive—and that have some structural or behavioral feature the predator can easily "remember" and thenceforth avoid. One such memorable feature is bright color. The poison-arrow frogs *(Dendrobates* and *Phyllobates)* produce a highly toxic substance in their skin long used by the indigenous humans of South America to poison their arrows. Most of these frogs advertise their danger by their bright color and are left alone by most predators.

Once a species is established among predators as noxious, selection favors not only it, but morphs of other species that happen to look like it. For instance, several apparently tasty or only mildly noxious species of butterfly have come to look like, or **mimic,** the bad-tasting monarch butterfly (Figure 30–7). In this kind of mimicry, called **Batesian mimicry,** the **model,** such as the monarch, is toxic, distasteful, or dangerous, and the mimic is less harmful. In **Müllerian mimicry,** several harmful species resemble each other, thus reinforcing the advertising of their common dangerousness, to their mutual advantage.

When the predator is the kind that develops a strong search image, rare phenotypes in the prey population have a great survival advantage. In the case of the crows and the herring gull's eggs mentioned earlier, were there gulls whose eggs were not spotted, these eggs might have a great survival advantage. In such instances, a special kind of frequency-dependent balancing selection will favor variability in the population. The stronger the predation pressure (that is, the more accurate the search image), and the more dominant the most frequent morph in the prey population, the greater the adaptive value of a rare morph. The population then evolves in the direction of one rare morph until the predator changes its search image, and then the pattern repeats.

Snails of the European genus *Cepaea,* preyed upon by birds that hunt by sight, have an extensive and complex system of color polymorphism. The frequencies of

Figure 30–7 Mimicry between butterflies. The foul-tasting monarch butterfly (**a**) is mimicked by the edible viceroy (**b**).

the morphs change constantly as one rare morph after another gains advantage and increases in frequency until it becomes common and predators "switch" to it.

No one is surprised to learn that predators can limit the population density of their prey. What might surprise you is how closely the two populations, predator and prey, can be linked. In fact, the two sometimes seem to oscillate together. For example, Hudson's Bay Company data extending back to the early 1800s show regular 10-year oscillations in the population density of snowshoe hares. Populations build to a peak, then decline abruptly (Figure 30–8). This decline is followed by a sharp decline in the population density of the hare's main predator, the lynx. Similar oscillations appear among Arctic lemmings and their predators, owls and jaegers, and in many groups of herbivores and their predators.

If predators can reduce population densities among

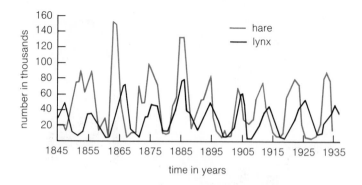

Figure 30–8 Changes in the abundance of the lynx and the snowshoe hare, as indicated by the number of pelts received by the Hudson's Bay Company. This is a classic case of cyclic oscillation in population density.

their prey, obviously they can exert great selection pressure. Whenever a population is stressed, the genotypes are tested anew, and certain genotypes will probably have greater adaptive value than others. We have examined some of the diverse responses of prey populations to predators. The complex relationship between predator and prey extends over scores of generations and is a major factor in the evolution of what we call specializations.

Parasitism **Parasitism** is an extreme form of interaction, one that produces a great range of specializations, both in the predator (parasite) and in the prey (host). The parasite benefits and the host is harmed; but, in contrast to normal predation, the parasite-host relationship is prolonged. There are plant parasites (such as mistletoe) and vast numbers and kinds of animal parasites, varying from external, temporary parasites (fleas) to internal, permanent forms (tapeworms).

Coadaptation Ecologists—and biologists in general—have usually dwelt on the negative aspects of natural selection and community evolution, especially predation and competition. Recently, however, there has been a great rebirth of interest in positive interactions, largely as a result of studies done in complex tropical communities. Such positive interactions are increasingly being grouped under the term **coadaptation.**

Nearly everyone has seen bees and butterflies visit flowers as a part of their daily ritual. The immediate conclusion is that the insects are getting food, and that is correct. But they are also being exploited by the plants, for they are carrying pollen from one flower to another. Often plants have flowers that either cannot or, because

Figure 30–9 Some orchid flowers, like the one shown here, mimic the female of the insect species that pollinates them. The male tries to copulate with the flower; pollination—and frustration—result as he goes from flower to flower.

of some structural specialization, do not fertilize either themselves or other flowers on the same plant. This protection against self-fertilization ensures variability in the population, but it is possible only where pollen is transported with the cooperation of various kinds of animals. Plants that somehow dispersed their pollen must have left a greater variety of offspring than those that self-pollinated and thus were inbred. Some of these variants were reproductively superior to any offspring of self-pollinating plants, and the whole population may eventually have shifted to cross-pollination.

Insects are the most commonly exploited agents of pollen dispersal, paid for their work with pollen or nectar. Some flowers, notably orchids, have bizarre specializa-

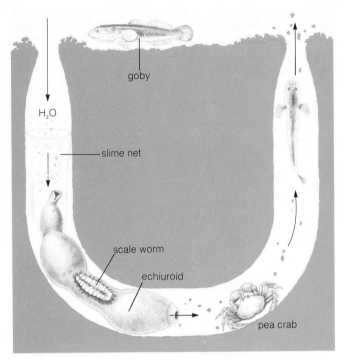

Figure 30–10 *Urechis*, the fat innkeeper, and the commensals that live in and around its burrow.

Figure 30–11 The shark sucker *Remora rembra*.

tions for attracting the insect. In several cases, the flowers mimic the pollinating insect so perfectly that the insect may even try to copulate with them, pollinating them in the process (Figure 30–9). This is perhaps the ultimate in exploitation. Birds and bats are also used as pollinators, and flowers have adaptations for each.

Biologists have learned much of what they know about coadaptation from these diverse pollination systems, but in recent years they have discovered many other examples, especially in tropical communities. One of the most interesting is the relationship between certain species of ants and bull's-horn acacia trees in the dry forests of Mexico and Central America. This acacia has large thorns along its branches. Queen ants bore into the soft interior of the large thorns and establish colonies. The ants protect the trees from herbaceous insects, which they vigorously attack. In return, the trees offer "housing" and have even evolved means of feeding the ants. Specialized nectaries at the bases of leaves produce a rich secretion, and the tips of the leaflets may bear specialized nodules that the ants use as food. Further, unlike most of the trees in these communities, the acacias keep their leaves throughout the dry season, supplying food to the ants the year around. Most species of ants are diurnal, but this one is active around the clock, providing continuous

protection. This interaction would be perfect for both species, except that a few insects mimic the ants so well that the ants do not recognize them as different, and the mimics attack the trees.

Symbiosis Coadaptation generally refers to long-sustained interactions between two or more species, interactions that may or may not be required for their existence. One special category of coadaptation is **symbiosis.** Literally, this word is neutral, meaning simply "living together," but it is generally interpreted in a positive sense: Either both species benefit, or at least one benefits and the other is not harmed. Symbiosis is usually restricted to two closely interacting species.

One kind of symbiosis is **commensalism,** an interaction with positive effects for one participant and no effects on the other. For example, small crabs and small polychaete worms often share the tubelike burrows of *Urechis,* a worm of the Pacific Coast of North America (Figure 30–10), who does not seem to benefit from their presence.

In **protocooperation,** both participants benefit, at least occasionally, but the interaction is not essential for either species. For example, remoras, or shark suckers, attach themselves to sharks with a specialized clinging organ and feed on the debris from the sharks' feeding attacks (Figure 30–11). Remoras, for their part, may help the sharks locate potential prey.

A final type of symbiosis is **mutualism,** where neither species can exist without the other. A flagellate protozoan dwelling in the digestive tract of termites digests the cellulose in the wood the termites eat. The protozoan, of course, could not bite off wood for itself and would die if exposed to the dry outside air; the termite would starve to death without the protozoan.

When termites molt, they shed the entire lining of the digestive tract along with the exoskeleton. This could leave the protozoans high and dry and the insect incapable of digesting its only food. But the protozoans respond to the release of molting hormone in the insect by encysting, forming a water-resistant structure inside of which they can remain dormant for long periods. Then the termite either eats part of the molted tissue, where the protozoans survive in their cysts, or touches the mouthparts or licks the anus of another adult, thus regaining its essential symbiont. Young termites cannot feed on wood until they have infected themselves with protozoans by ingesting fecal deposits of adults in the population.

Lichens demonstrate a special kind of symbiosis that defies precise classification. A lichen is a combination of a fungus and a one-celled (usually) green alga. Both can be cultured in the laboratory, but in combination they form a unique kind of life. The organism resembles neither its algal nor its fungal component, but is a special integration. By carrying out photosynthesis, the alga provides the association with food; the fungus provides a structure to house the alga, anchors it in place, and helps to keep it from drying out.

Sociality

The organization of populations is a direct result of their interactions in communities—interactions that may have persisted for thousands of years, producing features of organization that appear to be characteristic of the entire group. We should always seek the evolutionary basis of a particular organizational pattern. Sociality is scarcely meaningful when applied to plants, for they are, by and large, rooted in place and cannot associate consciously with others of their species. However, they can produce chemicals that keep other plants from growing too near, so perhaps they can be said to demonstrate antisociality.

Among vertebrates and more complex invertebrates, individuals within a given population spend their entire lives in a definite area called the *home range*. This is the area of normal activity, where they feed, hide, sleep, mate, and rear their young. Sometimes they defend part, but usually not all, of the home range from intruders, whether members of their own or a different species. The defended part is called a *territory*. There are advantages to holding a territory. The food supply is protected, and competition with individuals of the same species is reduced, as in fighting.

Many vertebrates reveal a distinct chain, or network, of dominant-subordinate relationships within populations. This is the familiar pecking order, observed in any barnyard chicken flock. Often just one or a few of the more dominant males in such a *social hierarchy* do most or all of the breeding. Clearly, selection has strongly favored the traits that make those particular individuals dominant, and they will have more than their share of influence on the future genetic constitution of the population.

Recently a new field called *sociobiology* has developed. We cannot go into much detail here, but this field is concerned with social behavior, interactions among individuals of the same species. These are, of course, rich and diverse, ranging from mere chance encounters to such complex societies as ant colonies and human cities. The general biological principles governing social behavior have been studied at least since the time of Darwin, but during the past decade there has been a great rebirth of interest and what appears to be a new synthesis. In the simplest possible terms, all social interactions can be interpreted as being of benefit, direct or indirect, to the individual organisms, in agreement with the general principles of natural selection.

Life Histories

Adaptation of populations often involves some aspect of population growth patterns or density. We see many different ways of life in nature, and obviously they are all successful to some degree, for the populations still survive. To be sure, some kinds of life histories, those of house mice, for example, are more successful in our modern world than are others, such as that of the California condor. But in the recent past, nearly all proved to be good "strategies," that is, adapted ways of life that led to something approaching a steady state in the population.

David Lack spent a lifetime studying the adaptive responses of birds. Because of the importance of reproductive success in natural selection, he focused primarily on life-history (birth to death) parameters. Lack observed, for example, that it is sometimes better (in terms of number of offspring that live to reproduce) for birds to have fewer young but spend more time with them, than merely to keep increasing the clutch size. Observing adult swifts with nestlings, he found they made 16.0 visits per hour to single nestlings, 13.2 visits to clutches of two, and 8.5 visits to clutches of three. Lack hypothesized that the parent birds can probably give high-quality care to one or two young, but all young might be undernourished if there were three. And of course the parents might be poor reproductive competitors with but a single young.

Thus, most female swifts lay two eggs. Clutch size, then, has been determined by natural selection so that the number of eggs laid corresponds with the largest number of young for which the parents can, on average, provide enough food.

Theoretically, the swifts in this example could increase their evolutionary success in two extreme modes. They could produce young that leave the nest sooner and thus need less parental care. This might tip the balance in favor of producing more young, and a genotype producing three young would be favored over one producing two. In time the population would shift in the direction of three young per clutch. During the shift, the population might grow fast if mortality rates did not change, for what has been changed is the intrinsic growth rate *(r)* (see Chapter 28).

On the other hand, parents with but a single offspring might be able to give it so much attention that it would be not only well fed but also almost completely protected from predators while in the nest. It might be so strong that it would be much more likely to survive to reproduce than birds from two- and three-young nests.

It is the old question of quantity versus quality. Allocating resources to reproduction might be favored under certain circumstances—for example, in species that are colonizing new environments. Under such circumstances, those genotypes will be favored that find the new environments or habitats quickly and reproduce rapidly to monopolize available resources, before potential competing species can invade the new habitat. A good example of this life-history mode is a dandelion, which flowers quickly, produces vast numbers of seeds, and can reproduce vegetatively as well.

However, when conditions are crowded or resources are scarce, a species is in great danger of overeating its food, or of overextending in some other way. It is then that selection favors nonreproductive improvements, such as more efficiency in finding, processing, or even metabolizing food; longer generation time, diminishing the effects of periodic droughts, unexpected disasters, and the like; and generally improved competitive ability and maintenance efficiency. For example, in contrast to the dandelion is the avocado, a tropical forest tree. It takes many years to reproduce and produces relatively little fruit. Each fruit has but a single seed. However, the seed has a large energy supply, which can sustain the seedling for a long time in the dark forest floor. This mode is characteristic of populations that are near the carrying capacity of the environment, where specializations can enable organisms and populations to increase their efficiency and hence contribute to their long-range reproductive success without necessarily increasing the frequency of reproduction or the number of young produced. But this mode can be maladaptive if the population is always maintained below the carrying capacity by density-independent factors.

Mice represent the first mode and elephants the second, relative to each other. But some species of mice are more like elephants and others less. Even in a single population, the two strategies might be expressed in different genotypes. In fact, there may be more than two strategies. For example, perhaps organisms that "hedge their bets" will be favored. These might produce a wide variety of offspring, some of which will be optimally suited to the environments they encounter in the future.

Summary

When natural selection operates on a population, the population functions more effectively in the community of which it is a part than would be the case if the population had not changed. The progressive change in populations under the influence of natural selection is a continuous process, called adaptation. The precise direction that adaptation will take seems to be predictable in only general ways, for there are many alternative routes that may be followed. This dynamic process takes place in the context of biological communities, and parts of the community interact intimately so that they coadapt through time. The interactions of populations in communities are intricate and varied. They result from the almost infinite interconnections of the "professions," or ecological niches, of the individual populations of the species. The ultimate adaptation is the life history pattern of each population. Since natural selection operates through differential reproductive success, the way in which phenotypes and populations reproduce is of crucial importance. This is an active area of research, but we do not yet have a complete theory of life history evolution that allows us to make reliable predictions. There are many effective ways to reproduce; exactly which one is best under any given set of environmental variables cannot be predicted, and this poses a worthy challenge for evolutionary biologists.

31

Species in Time and Space

Most of us are fascinated by the "terrible lizards" we call dinosaurs. Their size alone commands attention. The thought that these awesome creatures are now extinct, while we survive, fills us with a false sense of longevity—false because the dinosaurs' time on earth lasted hundreds of millions of years, and ours to date is less than 10 million. Our introduction to dinosaurs, whether through books, fossils, or actual bones in place at the Dinosaur National Monument in Utah, may have been the first time we were impressed by the vast scope of events in the past of our planet.

Dinosaurs, of course, are only one of the more fascinating parts of the earth's biological past. We know that many organisms came before the dinosaurs and many more came after. In this chapter, we will examine briefly some past evolutionary events, and we will see how these events have shaped the present distribution of life on earth. We will consider evolution in the larger groups of plants and animals.

Evolution and the Geologic Past

Life probably arose on earth over 4 million years ago (in Chapter 32 we will discuss how life may have arisen). For more than 2.5 billion years, life was represented by one-celled organisms, but these were just the beginning of life as we know it. To continue our discussion of the evolution of life, we must look at the later history of the earth. This history is divided into various time spans, the largest of which are called *eras* (Table 31–1). These are divided into *periods,* which may then be divided into *epochs.* We are living in the *Quaternary* period of the *Cenozoic* (meaning "recent life") era. The era before that was the *Mesozoic* ("middle life"), and the one before that was the *Paleozoic* ("old life"). The earliest period of the

Paleozoic is the *Cambrian;* time earlier than this is called *Precambrian.* The eras vary greatly in duration, the longest being the Precambrian.

By the end of the Precambrian era, there were well-established lines of bacteria, blue-green algae, and eukaryotic algae. Single-celled and filamentous organisms were present, and the modern cell had evolved.

The pace of development and the diversity of both plants and animals increase as we move through geologic time to the present. With the beginning of the Paleozoic era, life began to get more complex. The Cambrian through the Silurian periods were a time of life in the oceans and seas (Figure 31–1a). The algae diversified and became more complex in organization and structure. Marine invertebrates evolved and became the largest and most numerous animals. Indeed, all the major invertebrate phyla known today were present in the Cambrian seas. The climate was warm and mild, a condition that continued through much of the Paleozoic era.

Between the end of the Silurian and the middle of the Devonian period of the Paleozoic, plants became adapted to life on land (Figure 31–1b). The earliest were simple vascular plants, but later more elaborate species evolved into the treelike horsetails and club mosses of the great coal swamps of the Carboniferous period (Figure 31–1c). Ferns and early gymnosperms also appeared in considerable numbers during the middle Paleozoic. The shallow seas were alive with fish. A few reptiles adapted to the land during the Carboniferous, but insects were the first animals to become numerous on land during this period.

As the Paleozoic ended, the warm, mild climate cooled and dried, and the seas shrank. The coal forests were reduced, and with them the great tree horsetails and club mosses. The reptiles began to diversify, and the first mammal-like reptiles appeared.

The Mesozoic era is more familiar, since it is the time of the great dinosaurs (Figure 31–1d). The gymnosperms, now the prominent plant species, showed a great diversi-

Era	Period	Epoch	Beginning (Millions of Years Ago)	Events
Table 31–1 Geological Eras, Periods, and Epochs				
Cenozoic				
	Quaternary			
		Holocene		Historic time
		Pleistocene	2	Ice ages; humans appear
	Tertiary			
		Pliocene	5 ⎫	Apelike ancestors of humans
		Miocene	23 ⎭	appear
		Oligocene	38 ⎫	Origins of most modern mammals
		Eocene	54 ⎬	
		Paleocene	65 ⎭	
Mesozoic				
	Cretaceous		136	Flowering plants appear; dinosaurs become extinct
	Jurassic		190	Conifers, mammals, and birds first appear; dinosaurs dominant
	Triassic		225	Mammal-like reptiles appear
Paleozoic				
	Permian		280	Origins of most modern orders of insects
	Carboniferous		345	Origins of reptiles; amphibians dominant
	Devonian		395	Bony fishes dominant; first seed plants
	Silurian		430	First land plants
	Ordovician		500	First vertebrates
	Cambrian		570	Origins of most invertebrates

fication of types. The climate in what is now the temperate zone was fairly mild. In the final period (Cretaceous) of the Mesozoic era (Figure 31–1e), the flowering plants became more numerous, both in species and in individuals, and they spread to all types of environments. This increase in flowering plants took place while the dinosaurs were becoming extinct and the early mammals were appearing.

The Cenozoic era began with a mild climate, and the subtropical rain forest extended over most of the earth (Figure 31–1e). As the era went on, the climate became more temperate, and plants and animals that had been living together began to be distributed on the basis of their ability to withstand cooler conditions. Giant mammals appeared during the Cenozoic, only to become extinct not long after. During the mid-Cenozoic, the great development of grasses and similar flowering plants, as contrasted with the earlier woody species, occurred. Many present-day animals also evolved during this period. At the end of the Cenozoic, humans appeared on the scene.

This, then, is the general sweep of the evolution of life on earth—an event of such magnitude that it is almost impossible to comprehend. One thing we know: The changes, large and small, were all based on adaptations. Let's look at the question of adaptations on this grander scale, starting with the horse.

a

b

c

d

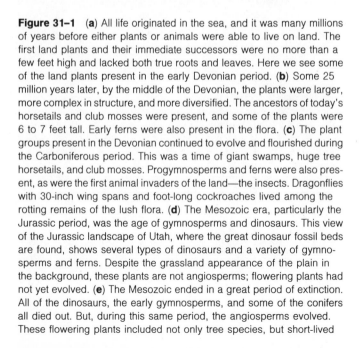

e

Figure 31–1 (**a**) All life originated in the sea, and it was many millions of years before either plants or animals were able to live on land. The first land plants and their immediate successors were no more than a few feet high and lacked both true roots and leaves. Here we see some of the land plants present in the early Devonian period. (**b**) Some 25 million years later, by the middle of the Devonian, the plants were larger, more complex in structure, and more diversified. The ancestors of today's horsetails and club mosses were present, and some of the plants were 6 to 7 feet tall. Early ferns were also present in the flora. (**c**) The plant groups present in the Devonian continued to evolve and flourished during the Carboniferous period. This was a time of giant swamps, huge tree horsetails, and club mosses. Progymnosperms and ferns were also present, as were the first animal invaders of the land—the insects. Dragonflies with 30-inch wing spans and foot-long cockroaches lived among the rotting remains of the lush flora. (**d**) The Mesozoic era, particularly the Jurassic period, was the age of gymnosperms and dinosaurs. This view of the Jurassic landscape of Utah, where the great dinosaur fossil beds are found, shows several types of dinosaurs and a variety of gymnosperms and ferns. Despite the grassland appearance of the plain in the background, these plants are not angiosperms; flowering plants had not yet evolved. (**e**) The Mesozoic ended in a great period of extinction. All of the dinosaurs, the early gymnosperms, and some of the conifers all died out. But, during this same period, the angiosperms evolved. These flowering plants included not only tree species, but short-lived

annuals such as grasses. This meadow from the Tertiary period of the Cenozoic era in Middle Europe looks almost like one we could walk in today. Of course, we would not find palms at this latitude today, and the animals (including a mastodon—an antlerless type of antelope) are surely not what we would expect to see.

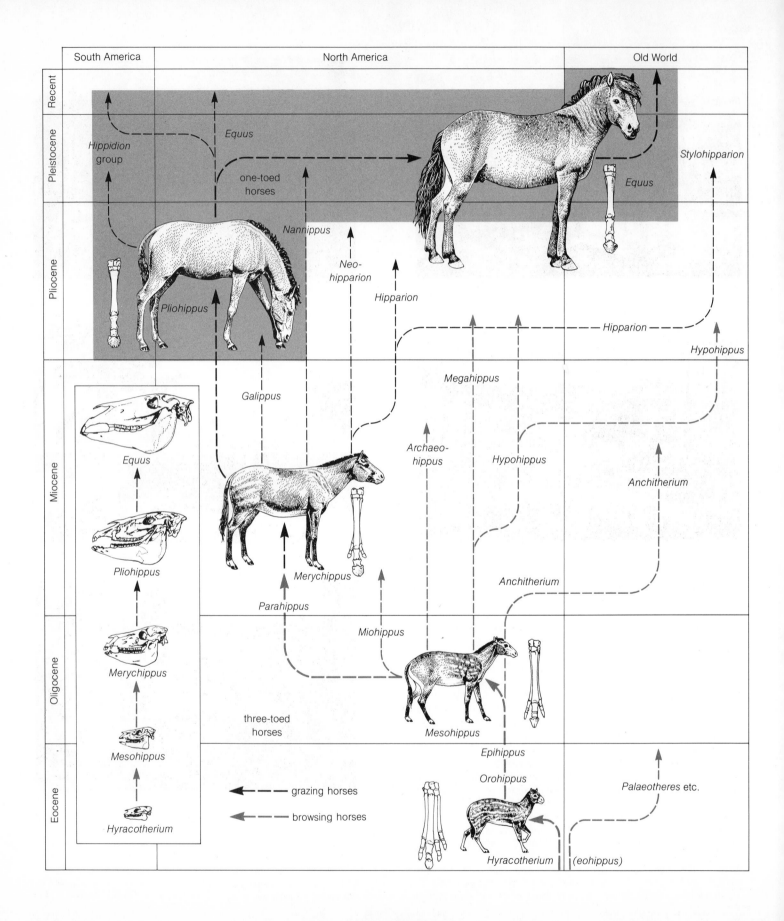

Adaptation and Evolution

Some 60 million years ago lived the ancestor to the horse—barely a foot tall, with four toes, and teeth adapted to feeding on tender leaves and shoots. Over the years, through many distinct species, we can follow changes in the horse through the fossil record to the present species—some 5 feet tall at the shoulder, having but a single toe, its teeth adapted to grazing on hard, gritty grasses (Figure 31–2). Because of the completeness of its fossil record, the horse is often used as an example of the changes in a **lineage,** an ancestor-descendant sequence extending through time. We lack such complete records for most plants and animals, but we have enough to trace evolution on a larger scale than considered so far in this text.

When we look beyond the events occurring within a species to groups of species as they are distributed in space and time, we find that adaptation operates in many different ways. In all this diversity, however, we can distinguish two basic categories of adaptation: *general adaptations,* which fundamentally reorganize an organism's way of life and are characteristic of whole groups or lineages of organisms; and *special adaptations,* which fit an organism or population more closely to its immediate environment. In the case of horses, the elongate leg bones and general arrangement of the toes are general adaptations, but one rather than three toes is a special adaptation relating to the open plains as opposed to forest habitat (more appropriate for speed, less versatile than the three-toed foot). The history of life has been an interplay between general and special adaptation, tempered by large doses of extinction. From this interplay arise the patterns of evolution.

Special adaptations are usually associated with structural or ecological detail, with features of organisms or populations that adapt them to particular details of their immediate environment. For example, as we noted earlier, each population of pocket gophers differs slightly from every other in coat color. In areas of light-colored soil, paler coat colors are fitter, in that predators kill more of the contrasting dark individuals. Thus, the population evolves in the lighter-colored direction.

Most of the features we call *specializations* fall into the category of special adaptations also. For instance, birds' beaks are general adaptations, but the particular forms they take in a community of interacting and competing species result from special adaptation. Up to now we have dealt largely with the origin of special adaptations. Here we look more closely at general adaptations.

General Adaptations General adaptations often involve the internal organization or physiological efficiency of a group of organisms. For example, the high, nearly constant internal temperature of mammals involves many body structures and physiological processes. It is such a fundamental change from the reptiles, which do not maintain constant internal temperature, that it not merely adapts one particular mammal to one environment, but enables mammals in general to live in environments that would otherwise be closed to them. Typically, a general adaptation involves the complex interaction of different systems in organisms. Maintaining a high, stable body temperature, for example, involves heat-generating (increased basal metabolic rate, increased production of thyroxine), regulating (hypothalamic temperature control), insulating (subcutaneous fat, hair), and many other systems.

Other examples of general adaptations are flight in birds, flowering in plants, the characteristic body plan in insects, the jaws of higher vertebrates, and the vertebrate head itself with its brain, sense organs, and feeding mechanisms.

General adaptations often result in further diversification and adaptations, especially if they are followed, as they often are, by *adaptive radiation.* In this process, a group of organisms expands into a new environment through modifications of the general adaptation. Thus, in the case of the general adaptations establishing mammals, later adaptations gave rise to mammals that could survive by hunting (such as lions, tigers, and wolves), swimming (including whales, porpoises, and sea lions), flying (bats), and eating plants (deer, mice, rhinoceroses, antelope, and so forth) (Figure 31–3). Adaptive radiation can also take place on a less grand scale. The general adaptations that established the basic characteristics of the rodents—sharp incisor teeth with continuous growth, short hair, eyes suited for night vision, and a host of other changes—led to adaptive radiation that gave rise to rats, mice, beavers, squirrels, prairie dogs, porcupines, and South America's capybara, the largest of all rodents.

Adaptive Radiation and Evolutionary Patterns
Adaptive radiations typically occur within certain bounds set by the gene pool. Bird species, for example, can diverge far from one another in their choice of food and in modifications of bill and tongue. Hawks and owls have bills specialized for tearing flesh, finches for crushing seeds, hummingbirds for drinking nectar, and ducks for sieving small vegetation from water. But, early in their history,

Figure 31–2 The lineages of horse evolution. Redrawn from Horses by G. G. Simpson. Copyright 1951 by Oxford University Press, Inc. Courtesy of G. G. Simpson and Oxford University Press.

birds lost the genetic capacity to produce grinding teeth, and no process that we can imagine could produce grinding teeth like a rodent's in a grass-eating bird. Not only have birds lost the capacity to produce teeth, but their jaws have so changed that rodentlike jaw movements could almost surely not evolve.

Birds are thus evolving in a particular **adaptive zone,** which is the way of life of a group of organisms, looked at from an evolutionary point of view. It is an abstraction that includes both the idea of constraints, or limits, and the idea of radiation and diversification within these limits. Not only do all the new species of a particular adaptive radiation resemble their common ancestor in some general adaptation, but they also tend to have many other features in common with each other and with their ancestor. The concept of an adaptive zone is most helpful when applied in an analytical framework. It directs attention to several crucial aspects of the way of life of organisms in a lineage, including patterns of resource use, resistance to predation, and life history. The concept is generally most useful at high levels of the taxonomic hierarchy (order and class), but it can be applied almost anywhere above the species level. Thus, one can define an adaptive zone for the class Aves (birds), which would include reference to evolution of (1) the bill structure as an element determining how the group uses food resources, (2) feathers as structures modifying the space resource pattern and relating to resistance to predation (flying enables birds to escape many predators), (3) warm-bloodedness as a physiological feature relating to the fundamental framework of the way of life, and (4) large, yolky eggs as a fixed life-history feature.

We can also speak of the adaptive zone of more limited groups, for example, of the family Geospizidae, the Darwin's finches of the Galapagos Islands (Figure 31–4). Here we cannot define the zone so precisely as before. We know, at least, that the ancestor was a seed-eating ground finch. We also know that its jaw and beak structure was not so specialized that it could not change any further, although there are limits to such change. Even in color there appears to be a rather narrow range of possibility. However, although the ranges of potential diversification for various features are narrow, there has been a significant adaptive radiation. Most of the radiation is based on ecological segregation and specialization to particular kinds of food within particular habitats. For example, there are tree species and ground species. Within the tree species, there are seed eaters and insect eaters. Among the insect eaters are big-billed forms that eat large, relatively heavy-bodied insects, and smaller-billed forms that eat small, relatively weak-bodied insects.

But even among the small-billed, insect-eating, tree-dwelling finches there is room for diversification, and we find species that work bark, that glean the foliage, and, in one extreme case, a species that behaves as a wood-

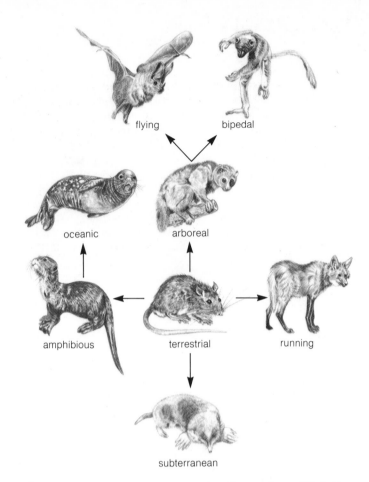

Figure 31–3 Adaptive radiation in modes of locomotion and life habits in mammals. (Precise relationships between various forms are not intended.)

pecker does, probing for insects (Table 31–2). Here we see the limits of the adaptive zone. The finches' bill, jaw, and tongue are less modifiable than the same structures in the woodpeckers' ancestors. The finches' short tongue is not really suitable for probing. Instead, behavior has been specially adapted, and the finches have evolved the ability to use a cactus spine as a tool to probe for insect larvae in holes in trees.

When we try to reconstruct the history of a small and rather recently derived group such as the Darwin's finches, which have no fossil record, we must rely on studies of living species. The *comparative method* of analysis—generating hypotheses on the basis of one set of observations and testing them by increasing the sample to more species—then becomes useful. For example, we might study the jaw and beak in great detail and hypothesize that features common to four or five different species

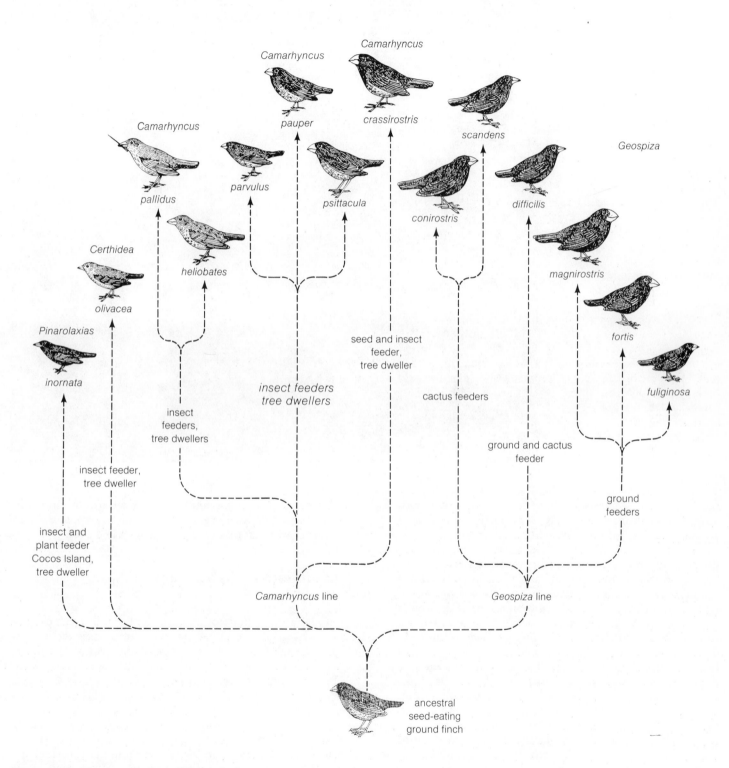

Figure 31–4 Adaptive radiation in Darwin's finches.

The following labels appear in the figure:

Camarhyncus
pauper

Camarhyncus
crassirostris

Camarhyncus
pallidus

parvulus

psittacula

scandens

Geospiza

conirostris

difficilis

Certhidea
olivacea

heliobates

magnirostris

Pinarolaxias
inornata

fortis

fuliginosa

seed and insect
feeder,
tree dweller

*insect feeders
tree dwellers*

cactus feeders

insect
feeders,
tree dwellers

ground and cactus
feeder

insect feeder,
tree dweller

ground
feeders

insect and
plant feeder
Cocos Island,
tree dweller

Camarhyncus line

Geospiza line

ancestral
seed-eating
ground finch

Table 31–2 Adaptation in Darwin's Finches

Bill Shape	Species	Bill Type	Feeding Habit
	Certhidia olivacea	Probing bill	Insect eater (trees)
	Camarhynchus pallidus	Probing bill	Insect eater; uses twig or cactus spine to probe insects from crevices
	Camarhynchus heliobates	Grasping bill	Insect eater (trees)
	Camarhynchus crossirostris	Crushing bill	Cactus seed eater
	Geospiza magnirostris	Crushing bill	Seed and nut eater (ground)

must also have been present in a common ancestor. Examining more and more species, we can finally hypothesize an ancestral jaw and beak condition.

But the comparative method must be applied cautiously and only when detailed information is available. Similar special adaptations can appear in distantly related or unrelated lineages, a phenomenon called *convergent evolution.* For example, many vertebrates other than birds can fly or glide. Some frogs have skin flaps and large, fully webbed feet that form gliding surfaces (Figure 31–5). A few lizards have developed abdominal "wings," with long, collapsible ribs to support them. Others have free flaps of skin around the abdomen and limbs that open as the lizard falls. Clearly these are not like the wings of birds, which are built around limb bones and muscles. Bat wings are far closer, with many of the same engineering principles; indeed, bats were once thought to be birds. But feathers in birds and smooth skin in bats are fundamental differences. Wings evolved convergently in bats and in birds from forelimbs used to get around on earth. Just because both have wings we must not assume that they evolved from a common winged ancestor.

Evolution of Higher Taxa

In higher taxa, perhaps at the order or class level, extinctions have been so numerous that comparative techniques give only a fragmentary picture of the group. The limits of the group can be defined, and subgroups recognized, but for anything further we must turn to the fossil record. Although incomplete, usually only the record can establish intermediate stages in the evolutionary history of a lineage, between its origin and its present condition. Using the fossil record, we compare the structure of earlier fossils with later fossils and with known living forms. Gradually we build a picture of the limits of a lineage.

Figure 31–5 Adaptations for gliding in frogs and lizards are specializations for locomotion in the air that are unrelated to the wings of birds, bats, or insects.

We can then begin to ask about the lineage's origin and its entrance into a particular adaptive zone. We look for the first appearance of a *key character*—a structural feature or physiological element that more or less commits its bearer to a particular new way of life. This way of life then becomes the adaptive zone of the newly evolved lineage.

Such a structural change as the rearrangement of the wrist and ankle bones of ancestral hoofed mammals is in one sense a special adaptation, but it is really general, for it can provide access to a new adaptive zone associated with fleetness of foot as a specialization for avoiding predators. Physiological features, hard to trace in the fossil record, often involve general adaptations. For example, biologists have long debated just where mammals diverged from reptiles. The most basic difference between a mammal and a reptile is physiological—the matter of temperature regulation, with mammals being homeothermic and reptiles poikilothermic. Mammals gain energy for their activity by aerobic mechanisms, relying on a metabolism that keeps up with demand or runs only a bit behind. Reptiles, by contrast, rely on anaerobic mechanisms. They build up enormous lactic acid debts during activity and tire quickly. Warm-bloodedness involves a great many features not found in the fossil record (including hair, sweat glands, and a particular heart and

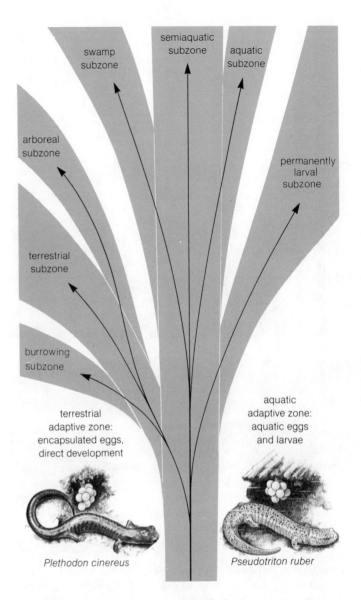

Figure 31–6 Evolution in the lungless salamanders illustrating adaptive thresholds.

lung structure). Thus, indirect approaches must be used. Reptiles are traced as far forward as possible, and mammals as far back. Finally there must be a decision. It is, of necessity, often based on rather technical grounds. Underlying any decision is the assumption that a physiological adaptation accompanied whatever structural features are chosen as the criteria for classification. The physiological change was warm-bloodedness, and it opened an utterly new way of life to what were, at that time, highly specialized reptiles.

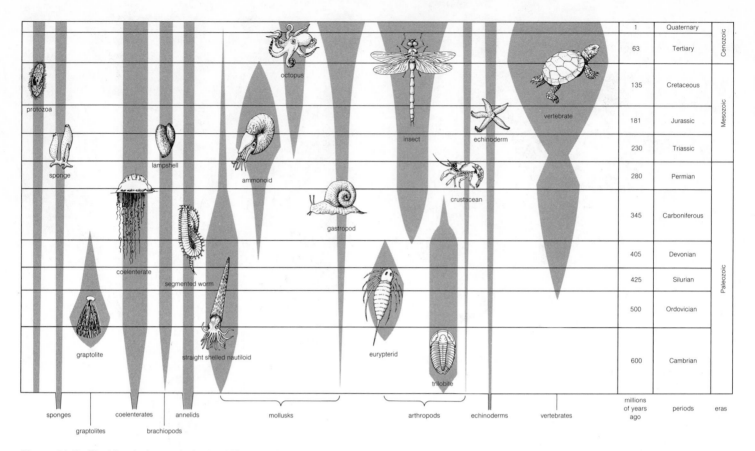

Figure 31–7 The historical record of animal life.

Trends in Lineages When one lineage diverges from another, a whole way of life changes. The change is not instantaneous, and evolutionary biologists often focus on the transition areas and transition period between adaptive zones, called *adaptive thresholds* (Figure 31–6). In terms of evolutionary time, however, adaptive thresholds are probably brief. When two lineages first begin to diverge, the threshold separating them is narrow. However, if truly key characters have opened a new adaptive zone, the new zone will probably develop rapidly, and the adaptive threshold will broaden just as rapidly.

Visualize a carving tool making a U-shaped groove in a piece of wood. One groove cuts completely across the piece of wood. It represents the ancestral adaptive zone, extending through time. A second groove starts, going off at an angle from the first. This represents a newly evolved adaptive zone. At its starting point, it is relatively shallow, and is separated from the original groove only by a narrow piece of uncut wood. This narrow area represents the adaptive threshold. The second groove deepens and moves away from the first, representing further development of adaptations that both widen and deepen the new adaptive zone. The small gap that first separated the two ways of life, the adaptive threshold, grows wider and wider as time passes, and the members of the two zones become increasingly isolated in an evolutionary sense. It is now impossible to pass from one zone to another, and the two separate taxa are firmly launched, evolving in different directions.

Within a given lineage, then, evolution tends to run in "grooves." Groups evolve within a system of constraints and along lines of least resistance. Ducks' bills change subtly from one species to the next, but they do not become hawks' or hummingbirds' bills. This phenomenon, which has been called *evolutionary canalization,* seems to characterize most major groups.

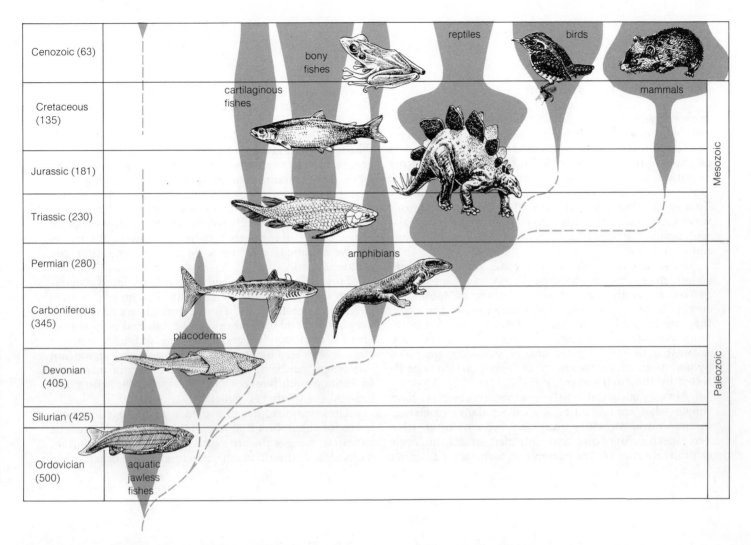

Figure 31–8 The historical record of vertebrates. The width of the shapes represents relative dominance; for example, reptiles were dominant in the Permian and more or less codominant with bony fishes in the Cretaceous. Age is shown in millions of years.

The Flow of Species in Time

From the fossil record, we know that groups replace each other as time passes. Amphibians and reptiles once dominated the earth. Eventually, birds and mammals expanded at about the same time, and reptiles and amphibians largely disappeared. The few surviving lineages (such as frogs and snakes) were usually fairly small and specialized in some unique way (Figures 31–7 and 31–8). Insects, birds, and mammals have prevailed now for about 70 million years. Among the plants, there has also been a

general pattern of origin, radiation, dominance, disappearance, and apparent replacement by another group.

The forests of Paleozoic times were rich in species of plants and animals. If all these species were alive today, we would recognize only a few of them, and the organization of communities must have been very different. However, the ecological roles remain more or less the same—just as a play stays much the same even though different actors perform in it.

During the last 150 million years, there have been great changes in the mammals. But the major lineages seem to have been established about 70 million years

On the Flow of Evolution

In recent years, a great debate has raged between biologists who consider natural selection the central process in evolution and those who do not. The debate centers on the role played by so-called neutral mutations in the evolution of lineages. Many mutations occur, say one group, that neither help nor harm their bearers' fitness. That is, they have no outward effect on the organism's functioning, and thus the survival or frequency of the mutation in the population cannot be influenced by environmental selective pressures.

For example, in the genetic code, several different codons may code for the same amino acid. One does it as well as another and thus is as likely as another to survive in the gene pool, all else being equal. Similarly, they argue, the high levels of heterozygosity and polymorphism seen in proteins in natural populations also show that there is a great deal of variation, produced by mutation, whose frequency in the population is unaffected by the environment.

Also, in one protein after another, biochemists have shown what appears to be a great regularity of change in amino acid structure over time. Proteins often have very specific functions, and only part of the molecule can tolerate change. For example, cytochrome c tolerates relatively little change, for it is a small molecule with a very specific function in oxidative metabolism. In contrast, fibrinogen, a nonspecific protein with a general role in blood-clotting, can tolerate great changes. The paleontological record gives us a pretty good idea when various groups of organisms diverged from a common ancestral stock. The structure of a given protein—hemoglobin, for example—can be compared in living members of two groups under study, and the number of differences in the order of amino acids recorded. This method reveals, among other things, that there are fewer differences among mammals than between any mammal and any reptile or bird.

From this process has come the still-disputed generalization that the amino acid structure of a protein evolves at a nearly constant rate. Thus, cytochrome c has evolved very slowly, and differences can be detected only at high levels of taxonomic organization—say, at the phylum and class levels. Hemoglobin evolves faster, and significant differences can be detected between orders of mammals. In fibrinogen, differences can be detected between families and even between genera.

These various lines of evidence, taken together, have led to the suggestion that a great, invisible flow of genetic variation changes the structure of biologically important molecules without much affecting their functioning.

---

ago, and there have been relatively few utterly new kinds since then. It is almost as though some sort of long-range equilibrium limits the kind and number of species that can exist. Periodically, new mammalian groups have appeared, differentiated, and then apparently stabilized. On occasion, for example with bats, the appearance of a new group opens up an entirely new adaptive zone, and the total number of species soars. But in general, as new groups have appeared, old groups have disappeared.

In fact, there seems to be a constant extinction rate within a given adaptive zone (see Essay 31–1). If species are becoming extinct at some constant rate, and yet the lineage maintains itself through long periods of time, then species within that adaptive zone must be appearing at about the same rate, although great bursts of speciation could alternate with relatively stagnant periods. The number of species balances itself in some such manner. Equilibrium in time is an important consideration in analyzing change in lineages.

The Distribution of Life on Earth

So far in this chapter, we have examined how groups of organisms diversify through time. Now we will consider how this diversification has led to the present spatial distribution of species.

Some individual species occur in a great variety of habitats; others may be restricted to a single, isolated one. But even widespread species are more common in some habitats than in others and are entirely missing from certain areas. No species is uniformly distributed over the whole world. We must therefore ask what limits the distributions of organisms.

Each species has a particular distribution, or **range.** Usually we think of an organism's *geographical range,* the general area in which it lives. For example, our familiar pocket gophers occupy a particular map range of the northwestern United States. However, each species also

Changes in function certainly occur by adaptive processes—that is, by natural selection. But the apparently constant rates of evolution of different molecules are determined by chance events accumulated over vast periods of time. Genetic drift (Chapter 28) is most effective when populations are small. In larger populations, we can see the drift, according to theory, only over vast periods of time. Of course, such periods have been involved in the history of life, so we must incorporate time into our thinking.

Regardless of how we explain the apparent regularity of changes at the molecular level, this does not help much in explaining the rate at which adaptive change in a lineage occurs. We know, for example, that selective pressures operating on a particular locus can be great. The rapid spread of industrial melanism is a good example. How can this kind of change be studied over vast periods of time, when all that we have to study are an often fragmentary fossil record and the living species themselves?

Adaptive change can be studied by examining extinction rates, which may then be used as guides for the study of adaptive evolution. Extinction rates can be calculated for several groups from the fossil record. From such data we can get some idea of the general evolutionary "health" of a given group. In theory, lineages with more or less average extinction rates are evolving in balance with their changing environments and are maintaining a "healthy" status as they pass through time. Other lineages are experiencing rapid extinction rates, presumably because of an adaptive spurt. Perhaps they are early in a phase of adaptive radiation, or perhaps their lineage is just evolving into a new adaptive zone. Still others apparently are becoming extinct slowly. Perhaps these are relatively stagnant groups, and in looking at examples (for instance, the lungfish) we see they are generally forms that are out of the mainstream. They have a well-adapted way of life and are not under pressure to change it.

Recently Leigh Van Valen, an evolutionary biologist at the University of Chicago, has discovered that, within any given adaptive zone, following its initial establishment phase, extinction rates appear to be constant throughout the zone's existence, changing only when a lineage as a whole, and the adaptive zone with it, becomes extinct. Throughout the history of a lineage, adaptive change occurs in an unpredictable manner. Periods of rapid adaptive change in lineages are likely to be episodic and extremely short-lived. They occur when a lineage is crossing an adaptive threshold and establishing a new adaptive zone. It is not surprising, therefore, that complete records of transition from one adaptive zone to another are so rare in the fossil record.

has a discrete *ecological range.* If we magnify part of the map range enough so that we can identify the ranges of individual demes, and then even further so that we can spot the home ranges of individuals, we can use such data to identify the ecological range. For example, we can determine what proportion of the total species range lies in forest, in grassland, in desert, and so on. The pocket gophers' primary range is in grassland areas, and within the grasslands the gophers are restricted to areas of relatively deep soil.

The third component of a species' distribution is its *geological range,* or its range back through time. Pocket gophers go back only a few million years, to the Pliocene epoch. Although pocket gophers have always had about the same range, this is not the case for some other kinds of organisms. Redwood forests once covered much of the western United States, for example. During the Pleistocene epoch, the musk ox, now restricted to far northern areas, ranged as far south as Texas.

The geographical, ecological, and geological ranges of a species are the products of limiting factors in the environment. Every species has a certain tolerance for each of the many factors or conditions in its environment, and any environmental condition or combination of conditions that exceeds this tolerance becomes a limiting factor. For example, near the northern boundary of a species' range (in the Northern Hemisphere), any one of several factors may set the limit. For an animal, perhaps it is the lowest temperature reached, or the number of days with freezing temperatures. For a plant, it may be the minimum winter or maximum summer day length.

As one moves from east to west across North America (short of the Sierras and Cascades), the average annual rainfall gradually decreases. This pattern seems to be a major element in limiting the distribution of many species of plants and animals. However, we cannot always find a single factor that limits the ranges of organisms; it is often a delicately balanced combination of factors. In the

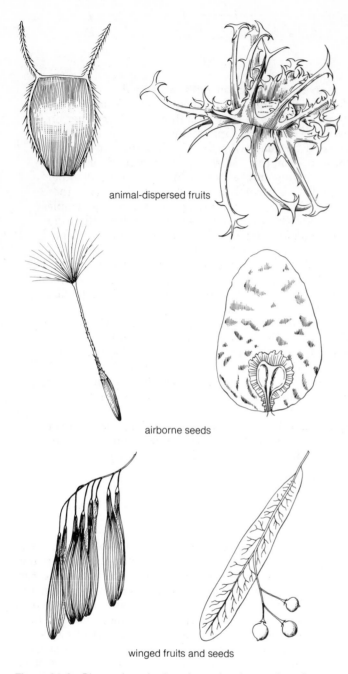

animal-dispersed fruits

airborne seeds

winged fruits and seeds

Figure 31–9 Dispersal mechanisms in seeds: wings, spines, floats.

the northern end of the range and inadequate rainfall the southern end.

In addition to features related to physiological tolerance, other factors limit the distribution of species. Some of these may be strictly physical barriers, such as rivers, mountains, deserts, oceans, and forests, depending on the species' means of dispersal. To a gopher, the ocean is a physical barrier and a dense forest is a *biotic barrier* (a barrier presented by other organisms). To the far north, a climatic barrier prevails, in that temperature cycles have created permanently frozen subsoil *(permafrost)*. During much of the year, the soil above the permafrost zone is also solidly frozen, so gophers and other burrowing vertebrates cannot survive.

Physical and climatic barriers have been generally understood for many centuries, but we continue to discover subtle, specific factors that act as barriers even though they may occur as seldom as once per human generation. During the winter of 1974, for the first time in the memory of most residents, a severe storm produced a heavy, wet snowfall in central coastal California. The heavy snow had devastating effects on trees that do not shed their leaves, such as live oaks, madrones, and laurels, and many ancient trees were destroyed. Some of the heaviest snow fell at elevations of about 2,000 feet—the approximate upper limits of distribution in these species. It appears that even rare climatic events can determine range boundaries.

Competition and predation are often considered part of biotic barriers. If we lay out strips of land (transects) of a certain width, running from high mountains to lowlands, and study their populations carefully, we find a layering of species from one elevation to the next. For example, if two insect-eating birds occur on a mountainside at 3,000 to 5,000 and 5,000 to 7,000 feet, respectively, we may find that they live in nearly identical ways. They may forage in the same manner, eat the same kinds of food, and nest in similar situations. Thus, they would compete for the same resources if they lived in the same range. However, one is slightly more tolerant of cold temperatures, and the other somewhat more tolerant of warm temperatures. Over many generations, the two species divide the total space and thus also the resources, reaching a kind of equilibrium. Similar phenomena occur along latitudinal transects and also along transects that show a temperature or moisture gradient at the same general elevation.

American West, for instance, species may be kept out by the high temperature and low soil moisture during the summer, which together have a far greater effect than either by itself. Further, the environments change from place to place, so that lack of winter daylight may limit

Dispersal of Species *Dispersal* can be defined as spread from a point of origin. It involves individuals—often in the form of *propagules* (seed, pollen), immatures (larva), or juveniles—moving permanently into or out of a population. Animal migration, in which individuals move from one area to another but eventually return, is not dispersal.

The point during the life cycle at which dispersal occurs is characteristic for each species and is usually restricted to that part of the cycle. The inherent power of movement (either by individuals or propagules) is called **vagility.** On the average, birds have greater vagility than amphibians, and large running mammals such as antelopes have greater vagility than small burrowing mammals such as gophers. Trees with large, heavy fruit and seeds have lower vagility than trees such as cottonwoods, with light seeds bearing fine, threadlike processes that are caught by the wind.

Plants have evolved a great variety of dispersal mechanisms (Figure 31–9). Plant pollen may travel inches or miles in the wind or on the bodies of bees, flies, moths, bats, or birds. Some seeds are airborne; others are equipped with spines or hooks that catch in fur or feathers and are then carried for some distance. Still other seeds are surrounded by fleshy fruit. When birds or mammals eat the fruit, they carry the seeds away from the parent plant, where they eventually pass from the animals and germinate. Seeds of some tropical palms float for great distances across the sea in waterproof outer coatings. The legendary "Coco de Mer" is the giant palm fruit that drifted onto the shore of India and Africa for many years before its source area in the Seychelles Islands was discovered.

Animals also have a wide variety of dispersal mechanisms. Many microscopic organisms form spores that are resistant to harsh physical conditions and serve as effective dispersal agents. Even fairly complex but tiny animals may be transported in mud caked on the feet of aquatic birds.

Winged insects and birds have the most obvious mechanisms for dispersal. Many insects devote the entire adult stage to dispersal and reproduction. In certain marine organisms, that adult stage is sedentary, entirely confined to a tube in the substrate. Here the free-floating larvae are the dispersal stage, moved about by the water masses in which they float. Spiders have no larval stages, but they pass through a succession of growth stages. Some young spiders spin a long thread that catches in the wind like a balloon, carrying the spider for great distances.

In general, both animals and plants have far greater dispersal powers than are commonly recognized. We have seen that the geographical ranges of individual species are determined by the barriers to dispersal. Often, however, distributional patterns can be understood most readily in terms of historical factors. Such historical factors are even more important for understanding the distributions of whole faunas and floras.

Biologists have long studied distributional patterns by examining many species at the same time. They can do this because some kinds of barriers are truly general and seem to have prevailed for millions of years. For example, Australia is today entirely bounded by the sea, a highly effective barrier for most land-dwelling plants and animals. Certain areas that have such boundaries have evolved distinctive faunas and floras. Not surprisingly, these areas correspond rather closely with continents, but the match is not perfect.

About 100 years ago, Alfred Wallace defined six zoogeographic realms, based on the distribution of animals (Figure 31–10). Each realm has a distinctive fauna, characterized by the presence of species and groups either absent or poorly represented elsewhere. These units generally have stood the test of time, and the distribution of many plants corresponds fairly well with them. However, Wallace had difficulty in precisely defining the borders of areas, and later scientists have shifted them in one direction or another. Wallace personally studied the border between the Oriental and Australian regions for many years, and he finally drew the boundary (Wallace's Line) between two islands in Indonesia (Figure 31–11). That boundary has long been debated by biologists.

Australia has perhaps the most distinctive fauna of any biogeographical region. The marsupials have diversified widely in Australia, producing many unique organisms. In addition, the monotremes (spiny anteaters, platypus) occur only in this region, as do many bird groups (including the emu and the cassowary). This distinctive fauna and the almost equally distinctive flora are taken as signs of long isolation.

But how was this to be explained? The islands of the Indonesian archipelago seem close enough to be a source of a new colonizing species, separated from Australia's north coast by only 50 miles of ocean. Could such a short expanse of water have kept primitive mammals from leaving Australia, and others from coming in, throughout the many millions of years of mammalian evolution?

The puzzle was solved only recently, when we learned that the great continental masses have been drifting over the face of the earth for millions of years. According to the now accepted theory of plate tectonics and continental drift, as recently as 200 million years ago the earth had only one gigantic land mass, which geologists call Pangaea, containing all our present continents, and one even more gigantic ocean called Panthalassa (Figure 31–12).

By about 180 million years ago, Pangaea had broken into a northern mass called Laurasia (modern North America, Europe, and Asia) and a southern mass called Gondwanaland (modern South America, Africa, Antarctica, and Australia). Before Australia broke loose from Gondwanaland and moved slowly northward, and before the Indonesian archipelago left Laurasia and moved slowly southward, these two regions (Australian and Oriental) were much farther apart than they are now. During the long history of mammals, these regions have drifted nearer and nearer to each other. Now there is an ever increasing interchange between the two faunas, those or-

Figure 31–10 The six continental faunal regions.

1	Mandan Indian	7	tiger salamander	13	motmot
2	chinook salmon	8	yellowthroat (warbler)	14	tamandua (anteater)
3	porcupine	9	turkey	15	arrow-poison frog
4	kangaroo rat	10	musk ox	16	capuchin monkey
5	pronghorn antelope	11	Auca Indians	17	piranha
6	bison	12	black jaguar*	18	guanaca

19	water opossum	25	lapwing
20	agouti	26	hedgehog
21	rhea	27	white stork*
22	Atlantic salmon*	28	saiga antelope
23	Lapp reindeer herder	29	jerboa
24	reindeer*	30	yak

* Forms not strictly confined to these regions.

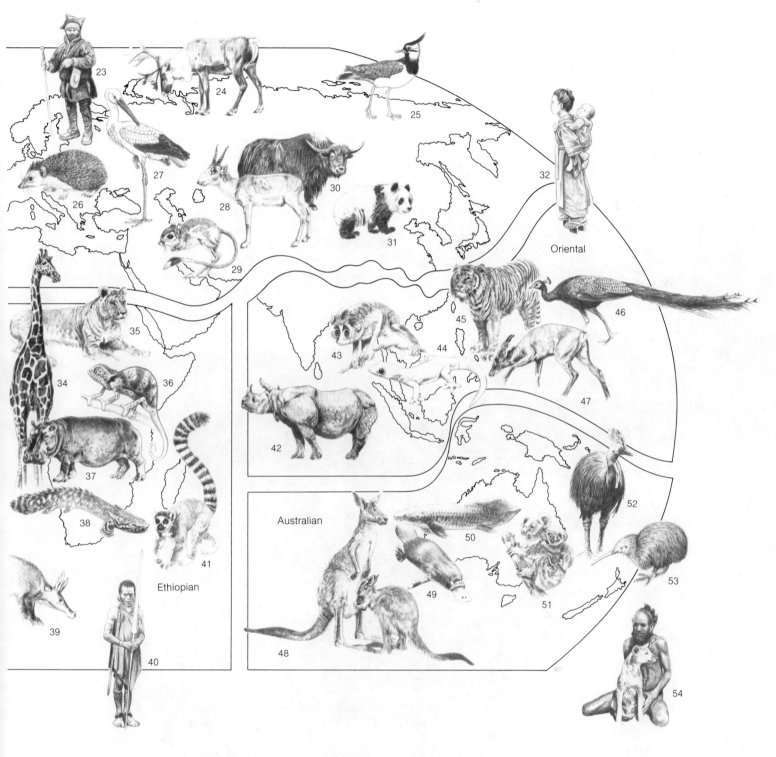

31	greater panda	37	hippopotamus	43	slender loris	49	platypus
32	Japanese woman and child	38	bichir *(Polypterus)*	44	gecko*	50	Australian lungfish
33	vulturine guinea fowl	39	aardvark	45	tiger*	51	koala
34	giraffe	40	Kikuyu warrior	46	peafowl	52	cassowary
35	lion*	41	lemur	47	muntjac	53	kiwi
36	chameleon*	42	Indian rhinoceros	48	kangaroos	54	Australian aborigine and dingo

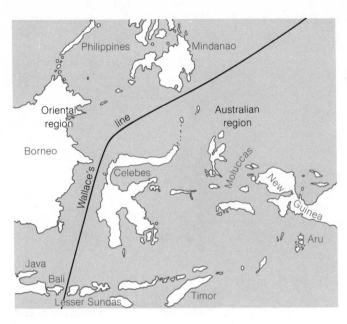

Figure 31-11 Wallace's Line, in the Indonesian archipelago, marks the boundary between two strikingly different faunas.

ganisms with superior dispersal powers making the most impressive inroads.

It is interesting that Wallace's Line, determined from biological evidence long ago, lies near the point where the two plates are colliding.

Biogeographic realms are useful conceptual tools, even though their boundaries cannot always be precisely defined. North and South America have markedly different faunas and floras, and we know that an oceanic barrier separated the two continents until a few million years ago. In fact, when we examine all available information, we see that all the recognizable biogeographic realms were surrounded by formidable barriers for long periods of time.

However, the use of the realms should not be pushed too far. For many years biologists were so intent on discovering the "exact" borders of realms that they tended to gloss over many interesting distributional patterns. For example, lungfishes are fresh-water forms occurring in Australia, but also in South America and Africa. The tongueless frogs (which include the clawed frogs and the Surinam toad) occur both in Africa and South America. Pangolins occur both in Africa and in southeast Asia,

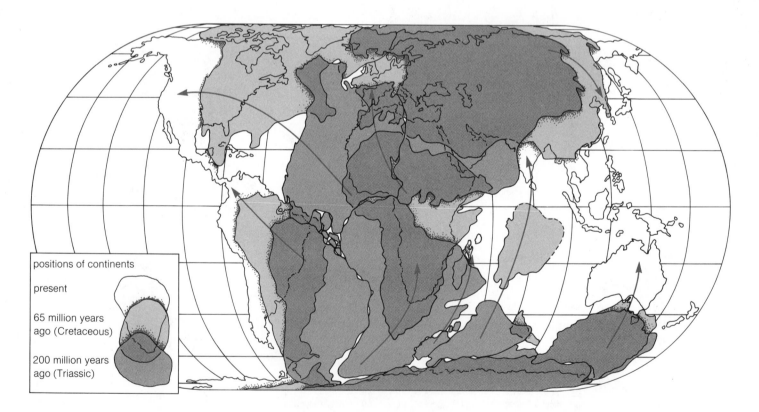

Figure 31-12 Summary of the paths of various continents as they have drifted with time.

as do rhinoceroses. Lemurs occur on Madagascar and in India and southeast Asia. How can these and other seemingly inexplicable patterns be understood? In the same way as the uniqueness of the Australian fauna—by understanding the pattern of continental drift.

Lungfishes, for example, were probably common in Gondwanaland before it broke up, and they survive in three separate segments of this ancient land—Australia, South America, and Africa.

Jack Fooden, of Chicago State University, has made some interesting suggestions about mammalian history in the light of drifting continents. Suppose that 200 million years ago, when Pangaea was the world continent, the world fauna of mammals consisted only of monotremes and marsupials. As Pangaea first began to break up, the Australia-Antarctica unit retained this faunal element; but elsewhere, including South America, Africa, and other sections of Gondwanaland, the monotremes became extinct, possibly in the face of the newly evolved eutherians, the familiar modern placental mammals. The eutherians either evolved or gained prominence on the remainder of Pangaea following the separation of Australia-Antarctica.

About 135 million years ago, South America became separated from the world continent, leaving what is now Africa, and for the succeeding 125 to 130 million years it was an isolated island continent. On the larger land mass left behind, the eutherians prevailed, successfully competing against and replacing the marsupials. Here, then, the familiar groups of mammals evolved—the rodents, primates, artiodactyls (sheep, antelope, hippopotamus, pig, and so on), perissodactyls (such as the horse, tapir, and rhinoceros), carnivores, and rabbits.

In isolated Australia, whose fauna was originally the same as Pangaea's, the monotremes continued (even today the platypus and spiny anteater survive), and the marsupials underwent a remarkable independent adaptive radiation. In North America, only the generalized and adaptable opossum represents the marsupials, but in Australia there are kangaroos instead of pronghorns or bison, wombats rather than peccaries, flying phalangers rather than flying squirrels, phascogales rather than mice, Tasmanian wolves rather than coyotes, marsupial moles rather than moles, and catlike marsupials rather than wildcats. The marsupials have diversified in the same general way as placental mammals in the rest of the world.

In South America, too, there was an independent adaptive radiation of mammals. Marsupials became highly diversified and differentiated, although less so than in Australia. The primitive edentates (anteaters, armadillos, and sloths) diversified greatly (as did other primitive eutherian groups that are now entirely extinct). But some 5 to 10 million years ago, South America became joined by a land bridge to North America. Even before that, some highly mobile elements of the competitively superior world fauna had managed to reach the continent. Thus, little remains today of the once highly distinctive South American fauna of mammals.

Madagascar became separated from Africa less than 135 million years ago. Since then it has been isolated, and evolution has pursued a unique course, unaffected by the world fauna. By the time of this separation, both monotremes and marsupials had become extinct on the world continent. Only five groups of mammals are represented in prehuman Madagascar: insectivores, primates, carnivores, rodents, and aardvarks, all relatively ancient groups. On Madagascar these groups, particularly the carnivores and primates, have undergone independent evolutions. Groups missing from Australia, Madagascar, and South America but present elsewhere include rabbits and their relatives, horses and their relatives, cattle, antelope, deer, and so forth.

India is an especially interesting case. It was an island continent for a long time. Among its surviving primitive animals are lemurs and a group of burrowing snakes, the xenopeltids, probably representing a stage of evolution about 180 million years old. Today, India's fauna contains remnants of the period of isolation living together with invaders from the world continent that have successfully occupied most of the subcontinent.

The southern continents, once joined in Gondwanaland, show much evidence of ancient connection in their faunas. We have mentioned the lungfish already. Giant birds such as the emu (Australia), the extinct moa (New Zealand), rheas (South America), and ostriches (Africa) also occur only in the south, and are likely derived from an ancestral stock that originated when Gondwanaland was a unit. Leptodactyloid (sometimes called southern) frogs occur in Australia, South America, and South Africa. Southern beech *(Nothofagus)* forests are found in Australia, New Zealand, and neighboring islands, and in South America, and there are beech fossil remains in Antarctica and South Africa. Some closely related fossil reptiles *(Lystrosaurus)* have been found in South America, Antarctica, South Africa, and India.

Continental drift has had profound effects on the distribution of plant and animal life, as well as upon our theory of historical biogeography. The well-documented modern theory of plate tectonics provides a superb new framework for studying and analyzing both the dependence and the independence of the earth's faunas and floras over many millions of years.

Summary

Over the past several hundred million years there has been a dynamic progression of plants and animals over the face of the earth. Nearly all the major lineages (phyla) are represented in the fossil-rich rocks of the Cambrian. Since that time there has been a complex interplay among the lineages. Some lineages have replaced others in time, such as the mammals replacing the ruling reptiles. Other lineages have replaced each other in space at one moment in time; thus the placental mammals common in most of the world were replaced by the marsupial and monotreme mammals in the Australian area. Paleontology deals with the temporal changes and ecology with the spatial changes, while the field of historical biogeography attempts integration. Temporal changes have involved a complex interplay between general and special adaptations in lineages, with the former being important in the establishment of new ways of life and the latter in the adaptive radiations that lead to exploitation of subcategories of each way of living. In the past biogeography used only habitat relations of organisms in trying to reconstruct history. Habitats were known to have been subject to change, but the focus was on particular organisms and their abilities to adapt or disperse or both. Organisms were thought to disperse around a framework of fixed continents, in relation to ever-changing climatic conditions. The discoveries of the past two decades have demonstrated that continents and parts of continents have shifted almost continuously through time. Plate tectonics theory accordingly offers an additional approach in explaining the present distribution of many lineages of plants and animals.

Suggested Readings

Books

Calder, N. 1973. The Life Game: Evolution and the New Biology. Viking Press, New York.

Dobzhansky, T., F. J. Ayala, G. L. Stebbins, and J. W. Valentine. 1977. Evolution. W. H. Freeman, San Francisco.

Ehrlich, P. R., A. H. Ehrlich, and J. P. Holdren. 1977. Ecoscience: Population, Resources, Environment. W. H. Freeman, San Francisco.

Emmel, T. C. 1976. Population Biology. Harper and Row, New York.

Endler, J. A. 1977. Geographic Variation, Speciation, and Clines. Monographs in Population Biology. Princeton University Press, Princeton, N.J.

Gould, S. J. 1977. Ever Since Darwin. W. W. Norton, New York.

Krebs, C. J. 1978. Ecology: The Experimental Analysis of Distribution and Abundance, 2nd ed. Harper and Row, New York.

Lerner, I. M., and W. J. Libby. 1976. Heredity, Evolution and Society, 2nd ed. W. H. Freeman, San Francisco.

MacArthur, R. H. 1972. Geographical Ecology. Harper and Row, New York.

Miller, G. T., Jr. 1971. Energetics, Kinetics, and Life: An Ecological Approach. Wadsworth Publishing Co., Belmont, Calif.

Moorehead, A. 1969. Darwin and the *Beagle*. Harper and Row, New York.

Pianka, E. R. 1978. Evolutionary Ecology, 2nd ed. Harper and Row, New York.

Ricklefs, R. E. 1976. The Economy of Nature. Chiron Press, Portland, Ore.

Rosenzweig, M. L. 1974. And Replenish the Earth: The Evolution, Consequences, and Prevention of Overpopulation. Harper and Row, New York.

Savage, J. M. 1977. Evolution, 3rd ed. Holt, Rinehart and Winston, New York.

Smith, R. L. 1974. Ecology and Field Biology, 2nd ed. Harper and Row, New York.

White, M. J. D. 1978. Modes of Speciation. W. H. Freeman, San Francisco.

Wilson, E. O. 1975. Sociobiology. Harvard University Press, Cambridge, Mass.

Articles

Bush, G. L. 1975. Modes of animal speciation. Annual Reviews of Ecology and Systematics 6:339–364.

Scientific American editors. 1978. Evolution (entire issue). Scientific American 239(3).

SEVEN

The Origin and Diversity of Life

32

The Evolution of Life

Cellular life originated between 3 billion and 4 billion years ago. The earlier date, 4 billion years ago, is the probable age of the earth itself. At this time, our planet condensed out of primordial matter and began its long transition into the environment we know today. By the more recent date, 3 billion years ago, chemical evolution of the earth's surface and organic evolution of life among the materials of the surface reached a state remotely resembling that of modern times. The climate and topography were favorable for cellular life, and life itself had reached the cellular level of organization.

Few fossil records remain to tell us of the characteristics of cellular life 3 billion years ago; there is just enough information to indicate that the earliest cells were probably bacterialike prokaryotes (Figure 32–1). Nothing at all exists to inform us about the earlier period, between 4 billion and 3 billion years ago, when the transition from inanimate to living matter took place. No intermediate forms survive, and there is no readable fossil record of the period. We are left with hypothesis, speculation, and conjecture about the events of this time.

It is not necessary to assume that life appeared spontaneously on earth through the chance aggregation of nonliving matter. The first life may have appeared through divine intervention, as many religions teach. Alternatively, life may have been carried to earth on particles from outer space or on machines delivered here by civilizations based elsewhere in the universe. Divine intervention is not within the domain of science, and speculation or revelation about the origin of life by this route is the concern of theologians rather than biologists. Assuming that life was introduced from somewhere else in the universe merely begs the question, because we are then left with the difficulty of reconstructing the origin of life on some other planet. Therefore, from the scientific point of view, basing hypotheses on the assumption that life on earth originated here through inanimate chemical processes is the only productive and reasonable approach to follow and test.

A number of working hypotheses based on this assumption have been successfully advanced about some of the stages in the first evolution of life. A few of these hypotheses, concerning the spontaneous appearance of complex organic molecules in the primeval earth and their aggregation into groups able to carry out some of the processes of life, have even been successfully tested in the laboratory.

Discussion of the evolutionary origins of life requires a somewhat special definition of "life." Probably, the first spark of life appeared in collections of molecules much simpler than even the simplest cells we know today. Present-day cells at a minimum have a boundary membrane separating the cell interior from the external environment; one or more nucleic acid coding molecules; a translation system, including the various RNAs and ribosomes, capable of converting the coded information into biological molecules; and a metabolic system providing the energy to carry out the cell's activities. Because these systems are so complex, it is hardly conceivable that life appeared at this level without being preceded by less highly organized forms. Presumably, then, the transition from nonliving matter to the cell was gradual, and no sudden event caused cellular life in all its complexity to appear at one instant.

The problem, then, is to decide what minimum level of organization and activity is required for a collection of interacting molecules to be considered alive. Some claim that very simple groups of molecules are alive if they can use an energy source to carry out a single, continuous energy-requiring reaction. Others insist that life at a minimum must include the ability to grow, reproduce in kind, and react to environmental changes by internal adjustments in chemical activity.

Because the latter, more complex requirements are closer to our present-day concept of life, this level will be used to define precellular life in this chapter. Therefore, collections of molecules will be considered to have the first spark of life if they (1) use either light or chemical energy to drive internal reactions requiring energy for completion, (2) increase in mass by controlled synthesis, and (3) reproduce into additional collections of matter of the same kind. This ability to reproduce in kind is considered to carry with it a requirement for an information-coding system and a system for translating the coded information into finished molecules. At this minimum level, life is relatively complex, but still much simpler than the simplest cells we know of.

Stages in the Evolution of Life

Although probably a continuous process, the first evolution of life can be broken for convenience into four successive stages. The first stage was the formation of the earth and its primordial atmosphere. This stage provided the basic raw material for evolution in the form of inorganic molecules and set up the conditions under which they could interact. In the second stage, complex organic molecules were produced by inanimate forces such as lightning or ultraviolet radiation acting on the inorganic chemicals in the environment. This hypothetical stage has been duplicated in the laboratory, with results indicating how this step probably took place. In the third stage, the newly produced organic molecules collected into clusters capable of chemical interaction with the environment. Experiments testing ideas dealing with this stage have also met with some success. In the fourth stage, some of these clusters were able to convert the energy of complex molecules absorbed from the environment into useful chemical energy; some of this energy was used to synthesize other complex molecules such as proteins and nucleic acids. Gradually, in this stage, the coding function of the nucleic acids became established, and the relationship between the sequences of nucleotides in nucleic acids and amino acids in proteins became fixed into the genetic code. This development provided direction to synthesis and reproduction in the aggregates of molecules. At this level of organization, life appeared in the aggregates. Evolution then changed from chemical to organic, with natural selection of favorable mutations in the coding system as the basis for further evolutionary development and change. Although the development of the coding relationship between nucleic acids and proteins is the most difficult part of the entire process of evolution to duplicate experimentally, it has recently been partially reconstructed and tested with some success.

To study the evolution of life by the scientific method, we must assume that all the processes occurring in these successive stages were inanimate and took place by chance. Given similar conditions and sufficient time, it is possible that a similar process could occur again and

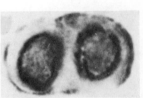

William J. Schopf

Figure 32–1 Some of the truly dramatic discoveries of modern paleobotany were the Precambrian microfossils. Here we show some of the fossils of blue-green algae and eukaryotic algae that have been recovered from the Bitter Springs Formation in central Australia. These are from the late Precambrian and are believed to be some 900 million years old. Recognizable microfossils are also known from the early Precambrian—around 3.1 billion years ago.

that life has evolved or is evolving at other locations in the universe.

First Stage: The Origins of the Earth and the Primordial Atmosphere The contemporary view of the earth's origins is that the sun and planets of our solar system, and all the stars and other bodies of the universe, condensed out of cosmic clouds of gas and dust particles. The composition of these clouds, which still occur in the universe (Figure 32–2), has been studied by analyzing the light transmitted or reflected by them. Most of the matter of the clouds is hydrogen gas at extremely low concentrations; lesser amounts of helium and neon are also present. Other elements and compounds—including metallic iron and nickel, the silicates, oxides, sulfides, and carbides of these and other metals, inorganic carbon compounds, ammonia, and frozen water—are also present in the clouds as solid particles (see Table 32–1).

According to the condensation hypothesis, stars, suns, and planetary systems are continually forming from clouds of gas and dust and disintegrating into dust again. Our own solar system condensed from one of these large

Figure 32–2 A cloud of gas and dust (the Horsehead nebula in Orion) in space some 1,300 light years from the earth.

clouds of dust. Most of the cloud condensed rapidly around a single center, causing high pressure and heat to develop in the interior and setting off a thermonuclear release of energy; this release created the sun. The rest of the dust and gas cloud presumably condensed around the sun as concentric rings of material, which gradually aggregated into the planets. Cool in temperature at first, the condensing planets heated because of gravitation and increasing internal pressure. Although never reaching the temperature of the sun, the heat inside the condensing earth was high enough to melt and stratify the collected materials. The lighter planetary materials, such as silicates and carbides, rose to the surface, and as the sphere cooled, they solidified into the surface crust.

The original atmosphere of the cooling earth probably contained large quantities of hydrogen and water vapor originating from the cosmic cloud. As the atmosphere and surface cooled, much of the water vapor condensed into droplets and rained down on the dust particles and rocks of the crust. Eventually, after years of torrential rains, the water collected into the rivers, lakes, and seas of the primordial earth.

The remaining water vapor and hydrogen of the at-mosphere reacted with carbides, nitrides, and sulfides in the exposed crust to release gaseous methane and ammonia into the primordial atmosphere. Carbon dioxide may also have been expelled from the interior of the earth, along with other gases, by erupting volcanoes. Free oxygen is not believed to have been present in the atmosphere in significant amounts; had it been present, it would have reacted quickly with particles and rocks of the crust to form oxides. The absence of oxygen and the presence of hydrogen, methane, ammonia, and water vapor gave the primordial atmosphere a nonoxidizing rather than an oxidizing quality. This nonoxidizing quality was fundamental to the next stage of evolution: the appearance of complex organic molecules through the action of various energy inputs on the inorganic matter of the crust and atmosphere.

Second Stage: The Spontaneous Production of Organic Molecules on the Primordial Earth The hydrogen, methane, ammonia, and water vapor of the primordial atmosphere, and the same gases dissolved in bodies of

Table 32-1 Atoms, Molecules, or Chemical Groups Detected in Cosmic Clouds or Outer Space

Atom, Molecule, or Radical	Symbol
Hydrogen atom	H
Hydroxyl radical	OH^-
Ammonia	NH_3
Water	H_2O
Formaldehyde	HCHO
Carbon monoxide	CO
Cyanogen radical	CN^-
Hydrogen cyanide	HCN
Cyanoacetylene	HC_2CN
Methyl alcohol	CH_3OH
Formic acid	HCOOH
Carbon monosulfide	CS
Formamide	$HCONH_2$
Silicon oxide	SiO
Carbonyl sulfide	OCS
Acetonitrile	CH_3CN
Isocyanic acid	HNCO
Hydrogen isocyanide	HNC
Methylacetylene	CH_3C_2H
Acetaldehyde	CH_3CHO
Thioformaldehyde	HCHS
Hydrogen sulfide	H_2S
Methylene imine	H_2CNH

Adapted from S. W. Fox, *Molecular and Cellular Biochemistry* 3:129 (1974); courtesy of S. W. Fox and *MCB*.

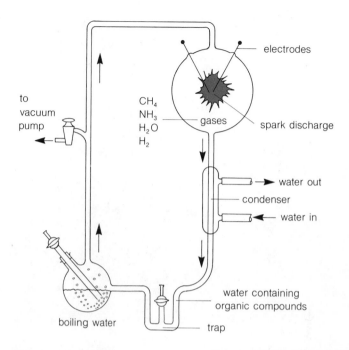

Figure 32-3 The Miller apparatus for demonstrating that organic molecules can be produced spontaneously in a primitive atmosphere.

water, were exposed to continual inputs of energy from a number of sources. One source, then as now, was sunlight. Besides the visible light from the sun, greater quantities of ultraviolet light than at present reached the lower atmosphere and acted on surface chemicals and waters. (At present, most of the ultraviolet light approaching the earth is absorbed by oxygen and ozone in the outer atmosphere and never reaches lower levels or the surface.) Another energy source was provided by heat from absorbed light and volcanic activity. Electrical discharges during the violent rainstorms of the period also supplied energy to the atmosphere and surface of the earth.

In 1924, the Soviet biochemist A. I. Oparin proposed that these energy sources, acting on the inorganic matter of the earth, could cause complex organic molecules to form. The same idea was advanced independently a few years later by an English geneticist, J. B. S. Haldane. Both Oparin and Haldane reasoned that a great variety of organic chemicals would be produced and would accumulate, because oxidation and decay by microorganisms, the chief routes by which organic matter is broken down in the present-day environment, were absent from the primordial earth. Haldane thought that the concentration

of these organic substances might even reach the consistency of a "hot, dilute soup."

Oparin and Haldane's ideas have been tested in the laboratory. The first significant test of their hypothesis was carried out in 1953 at the University of Chicago by S. L. Miller, who put together an apparatus designed to test the effect of electrical discharges on the gases believed to have been present (Figure 32–3). In the apparatus, water vapor, methane, ammonia, and hydrogen flowed continuously through a chamber subjected to repeated sparking from electrodes. Below the chamber, the water vapor and any organic chemicals produced were cooled, condensed, and trapped at a low point in the tubing. Operating the apparatus for one week produced a surprising variety of organic chemicals, including urea, several amino acids, lactic acid, formic acid, and acetic acid.

Since Miller's experiments, many other organic chemicals have been spontaneously produced by similar techniques. Adding hydrogen sulfide to the gases in the apparatus produced sulfur-containing organic compounds. Electrical discharges in mixtures of gases containing hydrogen cyanide and ammonia have yielded many other compounds, including most of the naturally occurring

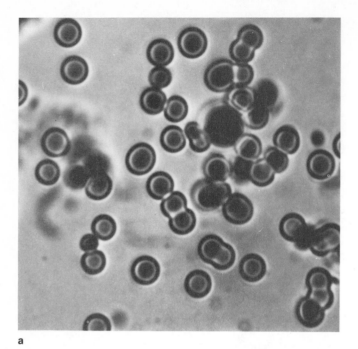

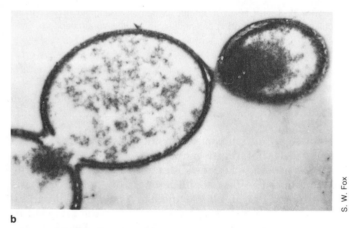

S. W. Fox

Figure 32–4 Fox's microspheres. (**a**) Light micrograph of a collection of microspheres made by cooling a heated solution of proteinoids (×1,000). (**b**) Electron micrograph of microspheres fragmenting in a process superficially similar to cell division (×11,000).

amino acids. (Both hydrogen sulfide and cyanide are found in interstellar matter and were probably components of the earth when it was first formed.) Mixing inorganic compounds such as hydrogen cyanide, formaldehyde, and ammonia and exposing them to various energy sources yielded quantities of deoxyribose and ribose sugars, the nitrogenous bases guanine, adenine, and uracil, and the nucleoside adenosine. Adding phosphorus produced the nucleotides AMP, ADP, and ATP. Other ex-

periments produced many carbohydrates and lipids. Thus, the building blocks of proteins, nucleic acids, and polysaccharides are all likely to have appeared spontaneously during this stage. A key feature in these experiments is the absence of oxygen in the "atmosphere"; if oxygen is added to the mixture of gases used, very few or no organic molecules are produced.

Even proteinlike chains of amino acids have been synthesized in laboratory experiments approximating the primitive environment. S. W. Fox of the University of Miami produced proteinlike molecules by heating dried mixtures of amino acids to 160 to 210°C for several hours. Fox has termed these proteinlike molecules *proteinoids.* Fox and others have also formed nucleic acids by heating mixtures of nucleotides and phosphates to about 60°C and holding them at this temperature for some time.

These experiments establish that a wide variety of complex organic substances could have arisen spontaneously over the millions of years between the formation of the earth and the first appearance of living matter. Accumulation of these compounds, which probably included all the major molecules found in living organisms, provided the raw materials for the third stage in the evolution of life, the collection of organic matter into aggregates capable of interacting with the environment.

Third Stage: The Spontaneous Collection of Organic Molecules into Functional Aggregates As the organic substances in the environment became increasingly concentrated, one or more types of molecules clumped into aggregates. As a preliminary step, they may have been concentrated somewhat by evaporation of water from lakes, inland seas, and tidal flats. These concentrated solutions of organic matter may then have followed one or more of several possible routes of aggregation into discrete, functional units. Some of these routes have been shown to be possible under conditions like those of the primitive environments.

Oparin proposed that a mechanism known as *coacervate formation* was important in the aggregation process. Coacervates form when protein molecules in solution separate into discrete, more concentrated droplets as a result of attraction between charged and polar groups on the surfaces of different polypeptide chains. If many charged or polar groups face the surfaces of the protein droplet, the surrounding water molecules tend to form an ordered film or boundary layer several molecules thick around the coacervate. The film of water molecules gives the boundary between the droplet and the surrounding medium many of the properties of a membrane. As a result, coacervates may concentrate molecules from the surrounding medium into their interior, or they may shrink or swell in response to changes in inside and outside concentrations, as do living cells.

A similar process of droplet formation has been studied by S. W. Fox and his colleagues. They showed that if solutions of proteinoids are heated in water and then allowed to cool, small spherical particles containing the proteinoids separate out of solution. These small particles (Figure 32–4), termed *microspheres* by Fox, are similar to coacervates in activity and can take in various substances from the surrounding medium. Some of the microspheres in Fox's experiments acted as catalysts; they were able to speed the rate of various biochemical reactions (see Table 32–2), including the hydrolysis of ATP to ADP, with release of free energy. Other microspheres were observed to bud, split, or fragment in a process superficially similar to cell division (see Figure 32–4b).

Other mechanisms of aggregation may also have been important. J. D. Bernal of the University of London has reasoned that molecules may have collected into aggregates on particles of clay in mud around tidal flats and river mouths. Beds of clay mixed with complex organic molecules might have reached extensive sizes and provided large sun-warmed areas in which life might slowly have evolved. Some clays containing metallic compounds also act as catalysts; they can enhance the breakdown and conversion of organic molecules into other substances and thus can carry out limited chemical interactions.

Another possible route of aggregation is the formation of films or particles by lipid molecules. When placed in water, lipid molecules spontaneously form layers two molecules thick called *bilayers,* which have many of the properties of cell membranes. Often, lipid bilayers suspended in water round up into closed vesicles, consisting of a saclike continuous "membrane" enclosing a central space. Other organic molecules, including proteinlike polypeptides, nucleotides, or nucleic acids could have been absorbed into the lipid bilayers or trapped in the space inside the closed vesicles.

Supposedly, one or more of these aggregation processes took place repeatedly in the primitive environment. Sooner or later, within some of the aggregates of protein, nucleic acid, lipid, and carbohydrate molecules, life appeared.

Fourth Stage: The Development of Life in the Primitive Aggregates For our purposes, life has been defined as the stage at which the primitive aggregates could (1) use an energy source to drive energy-requiring reactions to completion, (2) increase in mass by controlled synthesis, and (3) reproduce into additional aggregates like themselves. These steps toward life have proved more difficult to reconstruct and test in the laboratory than the earlier stages. Nevertheless, some ideas about them have been developed, and a few have even been tested experimentally.

Energy to drive reactions "uphill" was probably first

Table 32–2 Catalytic and Related Activities of Microspheres

Type of Reaction	Substrate Broken Down by Microspheres
Hydrolysis	ATP
	Nitrophenyl acetate
	Nitrophenyl phosphate
Decarboxylation	Glucaronic acid
	Pyruvic acid
	Oxaloacetic acid
Amination	Glutamic acid
Reduction–oxidation	H_2O_2
Synthesis (with added ATP)	Nucleic acids
	Peptides

Adapted from S. W. Fox, *Molecular and Cellular Biochemistry* 3:129 (1974); courtesy of S. W. Fox and *MCB*.

obtained by breaking down molecules absorbed from the surrounding medium, including ATP and the other high-energy nucleotides CTP, GTP, UTP, and TTP. Some of the reaction mechanisms might also have removed hydrogens and electrons from other organic molecules, with the release of usable energy. All of these energy-yielding reactions were probably speeded by the natural catalytic activity of aggregates such as microspheres or coacervates.

Transition from the relatively slight catalytic activity of the aggregates to specific enzymes must have taken place gradually. Scientists have developed some ideas about the possible steps in this process by studying the catalytic properties of substances like iron in pure form and in various compounds. Metallic iron can act as a catalyst to increase the breakdown of hydrogen peroxide (H_2O_2) to water and oxygen. Combining iron into iron oxide further increases the rate of the reaction. If iron is bound into a complex organic structure known as a *protoporphyrin ring* (see Figure 32–5), its catalytic ability is increased about a thousand times. Although complex, this ring structure could have been synthesized spontaneously in the primitive environment. Finally, if combined with the correct protein, the catalytic properties of iron are increased millions of times. Thus, catalysts probably first evolved from small inorganic and organic molecules and later increased in activity by combining with more complex organic molecules and polypeptides. With the development of directed protein synthesis, the catalytic complex became more specific, and the first enzymes appeared.

The energy released by these reactions was coupled to other reactions requiring energy. If the reactions included combining amino acids into protein or nucleotides

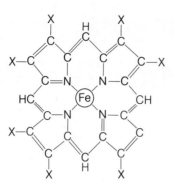

Figure 32–5 A protoporphyrin ring. Different chemical groups may be bound to the ring at the positions marked with an X.

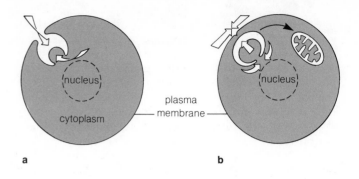

Figure 32–6 A hypothetical route by which mitochondria may have formed, through inpocketings of the plasma membrane of a prokaryote.

into nucleic acids, the aggregates could increase in mass. Upon reaching a large enough size, they could fragment into one or more smaller pieces. By this route, the first functional aggregates probably approximated growth and reproduction.

Fox has recently demonstrated that his microspheres are capable of carrying out these first steps toward life. After mixing complex microspheres containing both proteinoids and nucleic acid chains (Fox calls these complex microspheres *protoribosomes)* with ATP and amino acids in solution, Fox was able to detect the synthesis of new peptide chains by the microspheres. If ATP was omitted from the solution around the microspheres, no polypeptides were produced, indicating that the protoribosomes were using ATP as an energy source. Some of the new peptides made were released into the medium, and some remained tightly bound to the protoribosomes.

Sooner or later in the progress toward life, an information system arose to direct synthesis in the aggregates along controlled pathways. The first parts of the information system may have been adapted from molecules that were originally part of the energy-yielding reactions. The various nucleotides, particularly ATP and GTP, are important in transferring energy from one reaction to another and probably first acted in this capacity. Occasionally, under the proper conditions, nucleotides probably combined into nucleic acids. Some of these nucleotides may have favored the absorption of particular amino acids into the aggregates. As a result, the sequence of amino acids in any protein made would become less random, and the first vestiges of directed synthesis might have appeared. With time, enough specificity developed to lay down the beginnings of the coding system.

Fox has also been able to demonstrate that his complex nucleic acid–protein microspheres are able to carry out the first steps in this process; that is, some correlation can be detected between the type of nucleic acid included in his microspheres and the type of amino acid absorbed.

For example, including a poly-A nucleic acid (a nucleic acid containing only the base adenine) in the microspheres favored the absorption of the amino acid lysine; including poly-C favored the absorption of proline; and poly-U favored the absorption of phenylalanine. These correlations are of more than passing interest, because AAA is a nucleic acid code word for lysine, CCC is a code word for proline, and UUU is a code word for phenylalanine!

Once the coding system appeared and the relationships between the nucleic acid code words and the specified amino acids became fixed, the way was open for the transition from inanimate chemical evolution to organic evolution. Mutations in the sequence of the coding nucleic acids would then cause changes in the proteins of the aggregates. Some of these changes would be favorable and would increase the ability of the aggregate to compete for organic molecules in the medium. Other favorable mutations would allow the aggregates to manufacture some of their required organic substances for themselves. By this stage, the aggregate would probably have been separated from its environment by a membrane, either arising during the initial aggregate formation or developing as an adaptation that improved the ability of the aggregate to survive and compete. At this stage, these first living beings would have assumed the important characteristics of cellular life, including a coding system, a system capable of translating the coded information into specific proteins, and systems coupling energy-yielding to energy-requiring reactions, all surrounded by a limiting surface membrane.

While the functions of life were developing, the aggregates were dependent on absorbed organic molecules for their energy. The most efficient energy-yielding reactions would probably involve breakdown of ATP or removal of electrons and hydrogen ions from absorbed fuel molecules. Because there was no free oxygen in the environment, the aggregates probably used inorganic compounds such as sulfates or nitrates as final electron accep-

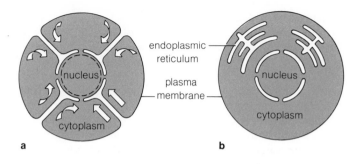

Figure 32–7 The hypothetical origin of eukaryotic cells through inpocketings of the plasma membrane. Fragments destined to form endoplasmic reticulum could have originated through the same route.

tors for these reactions, and would thus be anaerobic. As the anaerobically respiring aggregates became more widely distributed, the supply of organic fuel substances dwindled until the lakes and seas more closely resembled those of today. The dwindling fuel supply would certainly have led to extinction of the new life except for the appearance of photosynthesis among the early cells. As this capacity appeared, some of the primitive cells used light instead of absorbed organic substances as their energy source and were liberated from dependence on the supply of fuel molecules from the environment.

Photosynthesis in most contemporary organisms consists of two basic processes. In the first, water is split into oxygen and hydrogen ions. Two electrons are released from each water molecule as part of this reaction. In the second reaction, the electrons removed from water are increased in energy by the absorption of visible light. The electrons, raised in chemical potential by this mechanism, then power the chemical work of the cell. Thus, in photosynthesis, the energy of the electrons entering chemical reactions comes from sunlight rather than from complex organic molecules.

The initial reaction, the splitting of water into hydrogen ions, oxygen, and electrons, probably took place spontaneously in the primitive environment, because ultraviolet light has enough energy to split water molecules directly. Without the filtering ozone layer in the outer atmosphere, greater quantities of ultraviolet light probably reached the surface of the primitive earth than do today to carry out this reaction. This spontaneous splitting of water reduced the requirement for specific catalysts and provided a supply of low-energy electrons in the primitive aggregates.

Absorption of light energy to raise electrons to higher energy levels probably also took place spontaneously if pigmented organic molecules, which are capable of absorbing light, were included on the surface or interior of the aggregates. As with other activities of life, water-

splitting and light absorption gradually became regulated as enzymes appeared in the developing aggregates.

With the development of photosynthesis, large numbers of primitive organisms began living on the energy of light rather than the energy of absorbed organic molecules. As these organisms increased in number, greater and greater quantities of oxygen were released to the environment as a by-product of photosynthesis. This caused a basic change in the atmosphere and led to the balance of gases present in our environment today. With the appearance of oxygen, some primitive cells became adapted to the use of oxygen as a final electron acceptor, and aerobes appeared on the earth.

As a result of these developments, the complexities of cellular organization had probably advanced to levels resembling present-day prokaryotes. Among these primitive cells were photosynthesizers capable of living with only a source of inorganic chemicals (autotrophs). Other cells (heterotrophs) still required an energy source in the form of absorbed or ingested organic chemicals. Some of these organisms adapted to the use of oxygen as the final electron acceptor and were thus aerobic, and some remained anaerobic. These four types of prokaryotelike cells—aerobic autotrophs, anaerobic autotrophs, aerobic heterotrophs, and anaerobic heterotrophs—then evolved and possibly interacted to produce the first eukaryotic cells.

Eukaryotic cells undoubtedly developed from one or more of the different kinds of prokaryotelike cells that first became abundant some 3 billion years ago. This process, which included evolution of new structural features and organelles such as mitochondria, chloroplasts, and the nuclear envelope, may have taken place by one or more of several routes. One possibility is that many of the membranous structures characteristic of eukaryotes may have arisen through inpocketings of the surface of the plasma membrane of a prokaryotic cell (Figure 32–6). In prokaryotes, the enzymes and biochemical activities associated with electron transport and ATP synthesis in respiration and photosynthesis are linked to the plasma membrane. Possibly, through inpocketings of the plasma membrane (Figure 32–6a), portions of the plasma membrane containing these enzymes and biochemical activities extended into the cytoplasm and pinched off (Figure 32–6b), giving rise to membranous organelles that gradually evolved into mitochondria and chloroplasts. Mitochondria undoubtedly arose first, because these organelles are common to all eukaryotic cells. Later, some early eukaryotic cells with mitochondria formed chloroplasts and gave rise to cell lines leading to the plants.

The nuclear envelope membranes may have developed by the same process, through inpocketings of the plasma membrane that extended inward and gradually surrounded the nucleus (Figure 32–7). Similar inpocketings may also have given rise to the membranes of the rough and smooth endoplasmic reticulum.

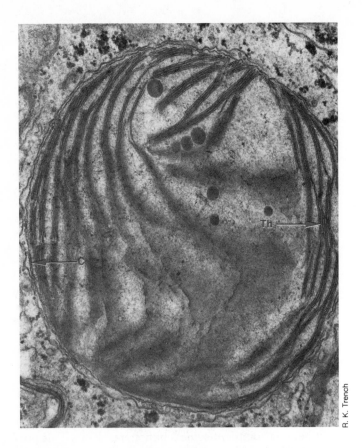

Figure 32-8 An active chloroplast in the cytoplasm of a digestive cell of the snail *Elysia*. C = chloroplast boundary membranes; Th = thylakoid membranes.

R. K. Trench

A different hypothesis about the possible origins of eukaryotic organelles is based on the obvious resemblances among mitochondria, chloroplasts, and contemporary free-living prokaryotes. According to this idea, advanced in its most recent and complete form by Lynn Margulis of Boston University, the major cytoplasmic structures of eukaryotes arose from ancient prokaryotes that became established as permanent residents in the cytoplasm of other prokaryotes. Mutations increasing the interdependence between the parasitic prokaryotes and their host cells led to the appearance of chloroplasts and mitochondria, and thus to the first ancestral eukaryotes.

According to this idea, the prokaryotes giving rise to mitochondria and chloroplasts probably entered the cytoplasm of their host simply by being eaten. Although this seems at first to be an unlikely process, there are present-day examples of this route of entry of one cell into another.

Margulis proposes that, in the development of mitochondria, prokaryotes evolved to the point where photosynthesis was common and oxygen was present in the atmosphere. But many types of prokaryotes were still unable to carry out photosynthesis and were required to capture and ingest other prokaryotes as a source of fuel molecules. Among these heterotrophic organisms were both aerobes, capable of using oxygen as a final electron acceptor, and anaerobes, still limited to the use of intermediate substances as final electron acceptors.

This hypothesis postulates that mitochondria began to evolve when anaerobic heterotrophs captured and ate aerobic heterotrophs in large numbers. Some of the aerobes, instead of being digested, persisted intact in the cytoplasm of the capturing cells and continued to respire aerobically. As a result, the cytoplasm of the host anaerobe, limited to the use of intermediate electron acceptors, became the place of residence of an aerobe capable of the much more efficient transfer of electrons to oxygen. Supposedly, some of the chemical energy from the parasitic aerobe diffused into the surrounding cytoplasm of the host anaerobe. The association would have been advantageous for the ingested aerobe as well if the host cell was highly successful at capturing and ingesting organic fuel substances. Mutations would have then appeared, increasing the interdependence of the host and ingested cells. Among these postulated mutations would be the loss of many motile and synthetic functions of the ingested aerobe, because these activities could be supplied by the host. Theoretically, the ingested aerobe then became more and more specialized as an energy-converting organelle and eventually completed its transition into the mitochondrion. Finally, the hypothesis proposes that all that remains today of the nucleus of the ancestral parasites that gave rise to the mitochondrion are small circles of DNA that code for a few limited and essential functions that evidently cannot be supplied by the "host."

Margulis proposes that chloroplasts appeared later, by a similar route. In this case, some of the feeding cells already containing mitochondria ingested autotrophic prokaryotes capable of photosynthesis. Instead of being digested, some of these autotrophs persisted in the cytoplasm of the ingesting cells and supplied their hosts with the ability to use light as a source of energy. As this relationship became established, the ingested autotrophic prokaryotes were transformed into chloroplasts. These primitive eukaryotes, containing both mitochondria and chloroplasts, founded the cell lines leading to the plants.

Some modern examples of recently established relationships between ingested and host cells provide evidence that this route of origin could actually have been followed. Many kinds of animal cells, including representatives of eight major phyla, pick up algal cells or chloroplasts and retain them as beneficial visitors. One of the most interesting cases, first described by Robert K. Trench at Oxford University, involves a marine snail and chloroplasts derived from an alga the snail uses as food. The snails hatch without chloroplasts, but as soon as the larvae begin to feed on algae, chloroplasts from the algae are absorbed directly into cells of the larvae without being digested. These chloroplasts survive in the gut cells of the snail (Figure 32–8) and continue to carry out photosynthesis in their new location. Trench has shown by the use of radioactive labels that molecules synthesized by the ingested chloroplasts diffuse into the cytoplasm of the host and are used as a source of fuel substances by the snail.

As these membranous organelles and structures developed in the cytoplasm of the primitive eukaryotes, other major adaptations appeared that completed the conversion of eukaryotic cellular life to contemporary forms. These adaptations were increases in the complexity of chromosomes and the development of microtubules to regulate cell division and motility. Later evolution led to the aggregation of cells into colonies and eventually to the specialization of cells and the appearance of many-celled organisms of greater and greater complexity, including our own form of life.

Summary

Life appeared on earth some 4 billion years ago. The exact processes involved in the formation of life are unknown, but it is possible to construct a series of logical hypotheses that conform to known physical-chemical laws. These hypotheses are based on the idea that the atmosphere of the early earth was different in composition from the earth's present atmosphere. Evidence indicates that it was rich in hydrogen, methane, ammonia, and water vapor. Energy in the form of sunlight, ultraviolet radiation, and lightning was abundant. It is possible to recreate in the laboratory the conditions believed to be present on the early earth. When this is done, a great variety of organic compounds can be produced, including amino acids, urea, nitrogen bases, and various sugars. Once such organic compounds are present, the next probable step is the formation of coacervates. These are small drops of protein molecules that can become surrounded by a lipid bilayer. In the next proposed stage, energy-yielding reactions evolved and the energy used to control the mass of the aggregate. Eventually mechanisms would be formed whereby the aggregates would be able to reproduce additional aggregates like themselves. This step would probably involve an information system and ultimately the formation of DNA. Only later, is it hypothesized, photosynthesis evolved and with it the presence of oxygen in the atmosphere. The presence of oxygen drastically alters the course of evolution and makes possible the much more elaborate energy-yielding mechanism found in cells. The first cells were prokaryotic in nature with eukaryotic cells evolving later.

33

Kingdom Monera and Viruses

From the simple beginning of single cells, life evolved into the rich myriad of forms that surround us today. Beginning with this chapter, we will explore the various groups of organisms that share the planet with us. But before we can do this we must ask, How do we separate this vast array into groups we can meaningfully discuss?

Over the years, biologists have devised various systems to arrange organisms in kingdoms. The simplest and oldest of these systems recognizes but two kingdoms: Plantae (plants) and Animalia (animals). However, the cellular characteristics of the bacteria and the blue-green algae are so different from those of other organisms that most biologists now think they should be placed in their own kingdom: the Monera. Many biologists further feel that the protozoa (single-celled animals) and many of the single-celled algae have more in common with one another than they do with either higher plants or animals, and place these together in a kingdom known as the Protista. The most recent proposal places the fungi in a kingdom by themselves. Thus, the final breakdown is five kingdoms: Monera (bacteria and blue-green algae), Protista (protozoa and some single-celled algae), Fungi (slime molds and true fungi), Plantae (most algae and higher plants), and Animalia (multicellular animals) (Figure 33-1).

Which system is right? All—and none! This is because all the systems are artificial. Organisms have evolved in response to environmental pressures and genetic laws, not for the convenience of biologists who wish to study and classify them. Clearly, some system is necessary for studying organisms, but the need of the biologist doing the study may determine what system is best in a particular case. As we gain more and more information about organisms, our ideas change; and the general tendency appears to be toward a more specialized system, as evidenced by the recent five-kingdom proposal. In this book, we have decided to follow the most recent proposal and have grouped all organisms into five kingdoms: Monera, Protista, Fungi, Plantae, and Animalia.

One final note: Traditionally, botanists have used the term *division* and zoologists *phylum* for the same taxonomic group. In the interests of simplicity, we will use phylum throughout all the kingdoms.

Kingdom Monera

All members of the kingdom Monera are prokaryotes, and the prokaryotes—bacteria, blue-green algae, and possibly mycoplasms—are the simplest of all independently living organisms. They are also among the deadliest to both plants and animals, including humans. Between 1348 and 1350, the Black Death, caused by a bacterium, reduced the population of Europe from 80 million to 60 million. Currently, the bacterial diseases syphilis and gonorrhea are at epidemic levels in many U.S. cities. Bacteria also cause many plant diseases that destroy millions of dollars' worth of crops each year. At a time of increased population pressure on the world food supply, such losses must be measured in more than dollars.

At the same time, prokaryotic organisms are essential for life on earth. Bacteria and blue-green algae can convert atmospheric nitrogen into nitrate and ammonia that can be used by other organisms. Bacteria are essential in breaking down organic matter so that the elements are available to be used again. Bacteria are also important as industrial sources of many organic compounds that we use every day.

In Chapter 3, we discussed the major characteristics of prokaryotic cells. Compared to eukaryotic cells, they are simple in organization, lacking membrane-bound nuclei and other membrane-bound organelles. Their chromosomes are simple strands of DNA arranged in a circle and lacking associated histones. They are surrounded by cell walls, but these walls are different in composition from those of eukaryotic cells. For all their seeming sim-

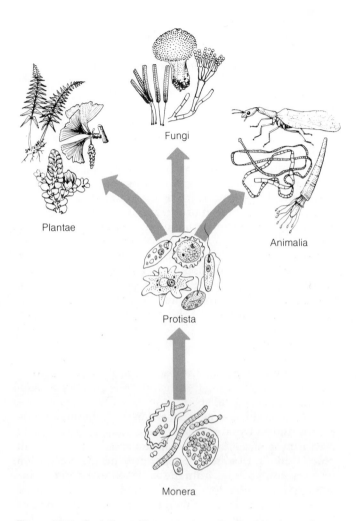

Figure 33–1 Evolutionary lineages among the five kingdoms of organisms.

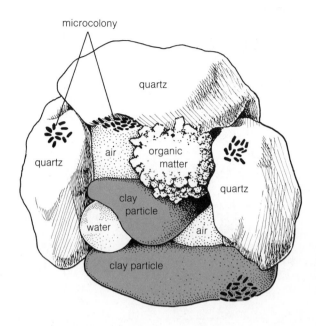

Figure 33–2 A section through a soil crumb showing the distribution of soil bacteria. Most soil bacteria are not found as free individuals in the soil solution but as microcolonies attached to soil particles. A soil particle is a complex and diverse environment composed of mineral and organic components, with soil microorganisms localized within it.

plicity, however, they are remarkably successful organisms.

In our survey of the prokaryotes, we have included the viruses. As viruses are not cells, they are not prokaryotes—indeed, many question whether they can be classified as living. Outside the living cell, they are inert chemicals; inside the proper type of cell, they can reproduce through the machinery of the cell they have infected. We covered viral genetics in Chapter 9 and here will cover their general biology.

To begin our survey of the prokaryotes, let's consider the bacteria.

The Bacteria

Where do bacteria live? Everywhere! They are found not only as disease-causing organisms in plants and animals but also in soil, ponds, lakes, streams, hot springs, gasoline storage tanks, food, snow, sewage, stomachs and intestines, concrete, and stone.

Every gram of soil contains about 100 million living bacteria. A section of a piece of soil less than 25 millimeters long (Figure 33–2) is a complex environment for a multitude of bacteria.

Bacteria also abound in all types of water—from birdbaths to the open ocean. They live on the bottom, floating in the water, or attached to all manner of plants and animals. The bacteria found in the lower depths of the ocean live at temperatures of 4 to 10°C and under pressures ranging as high as 50 to 100 atmospheres. Certain so-called thermophilic bacteria can grow at high temperatures, 55 to 70°C; the extreme is a filamentous bacterium called *Flexibacterium* that flourishes in hot springs at temperatures up to 90°C.

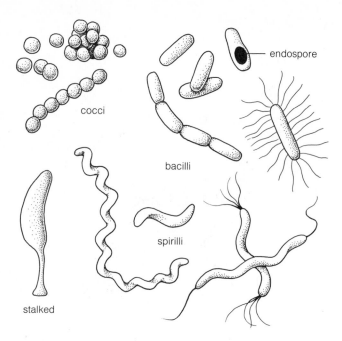

cocci

endospore

bacilli

spirilli

stalked

Figure 33–3 Various shapes of bacterial cells.

On the whole, bacteria do not tolerate acidity; hence the age-old popularity of pickling foods in dilute acid to preserve them (see Supplement 33–1). Some bacteria, however, not only tolerate high acidity, but actually produce sulfuric acid.

The primary basis for the universal distribution of bacteria is their versatility in metabolism. They can live in aerobic and anaerobic conditions and can use just about anything as an energy source. Unfortunately for us, there are a few things they cannot use. One of the most important roles bacteria play in the ecosystem is breaking down organic matter so that elements bound within it can be liberated and recycled. There are few natural compounds that bacteria cannot attack, but various plastics have so far resisted bacterial attack, thus posing major disposal problems.

Bacterial Structure Bacteria range in size from 0.5 to 5 micrometers, with most between 1 and 5 micrometers. Traditionally, bacteria have been grouped into three categories: *rods,* or *bacilli* (*Escherichia coli,* which is small and relatively simple in shape, is an example); *spheres,* or *cocci;* and *spirals,* or *spirilli* (Figure 33–3). But this is too limited a view of bacterial form. Bacterial shapes include filaments similar to fungi as well as multicellular chains and stalked forms.

Why such various shapes? They must have adaptive ecological significance, but what can this be? First, all

nutrients enter the bacterium by diffusing through the cell wall. The surface area of the wall will thus determine the efficiency of uptake. In other words, rod-shaped bacteria will be more efficient than spherical bacteria in absorbing nutrients. Interestingly, most bacteria found in the water of lakes and oceans, where concentrations of organic materials are low, are either filamentous or rod-shaped, and spherical bacteria are more common in the nutrient-rich sediments of these bodies of water.

Nutrition is not the only factor involved in bacterial shape; resistance to drying is another. Spherical cells are usually more resistant to drying than elongated ones, possibly because of more uniform pressures on the cell wall during drying. This explanation is supported by the fact that spherical bacteria are the most common type isolated from the air.

But what about the spiral and stalked bacteria? The spiral shape has been related to the fact that these bacteria are usually motile, and the helical pattern of motion resulting from their shape helps them to penetrate the debris and viscous media in which they live. Stalked bacteria, on the other hand, are epiphytes, and the stalk anchors them to the substrate. In both stalked and filamentous forms, only one end of the bacterium is attached to the substrate, thus allowing the organism the greatest possible area for absorbing nutrients.

The typical bacterium, as seen in the light microscope, is surrounded by a rigid wall that may be surrounded in turn by a slime layer of varying thickness. The wall, which defines the gross morphology of the bacterium, differs in composition from the walls of most other organisms. Its major components are *peptidoglycans* (or mucopeptides), compounds consisting of amino acids and sugar derivatives (Figure 33–4). The sugars form a polysaccharide-based network connected by polypeptide (amino acid) cross-linkages. Penicillin kills bacteria—acts as an antibiotic—partly by blocking the formation of these cross-linkages. Also present in the cell walls of some bacteria are lipopolysaccharides, which occur as an outer layer on top of the peptidoglycans.

Projecting through the wall in a pattern that depends on the species may be one or more *flagella* (Figure 33–5), 3 to 12 micrometers long. Bacterial flagella, unlike those of higher organisms, are only 10 to 20 nanometers in diameter and consist of a single fibril composed of a single protein, flagellin. This protein belongs to the same group of proteins as the contractile proteins of animals. Also present are smaller hairlike *pili* (Figure 33–5). The flagella give the bacteria considerable mobility. Some nonflagellated bacteria appear to move by an unexplained gliding motion.

In the section of *E. coli* seen in Figure 33–6, you can distinguish the cell wall and the plasma membrane. The center of the cell is occupied by an irregular mass of fibrous material containing the DNA. This is called

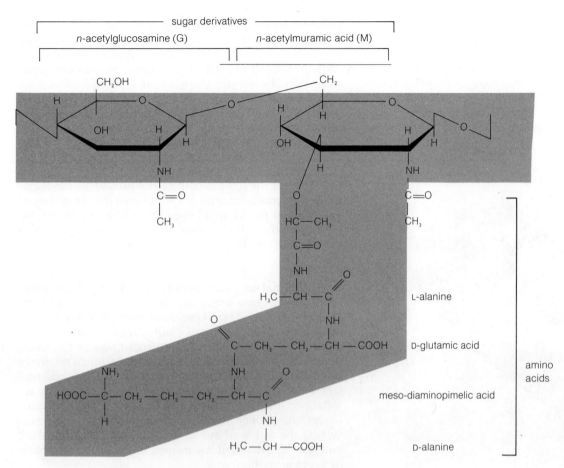

sugar derivatives

n-acetylglucosamine (G)   n-acetylmuramic acid (M)

L-alanine

D-glutamic acid

meso-diaminopimelic acid

amino acids

D-alanine

peptidoglycan—the basic repeating unit

Figure 33–4   The basic chemical structure of the bacterial cell wall. This basic peptidoglycan unit varies in different bacteria, and other compounds may be present within the framework shown.

the *nucleoplasm,* and it lacks a limiting membrane. The only regular membranous structure found in bacteria is an array of concentric membranes associated with the plasma membrane and known as the *mesosome* (Figure 33–7). The cytoplasm is devoid of all membrane-bound organelles. Ribosomes are present, as are accumulations of polyphosphates, which are a reserve supply of phosphorus. Even in much more complex bacteria, little more is found than assorted vacuoles, some filled with gas that permits the bacterium to float, and occasional membranous sacs (Figure 33–8).

Cell Division in Bacteria   In bacteria, all the hereditary information is coded into a single DNA molecule that exists as a closed circle with no free ends. At least at one point, this circle is believed to be attached to the plasma membrane. Replication and division of the circle are believed to take place on the molecule, as shown in

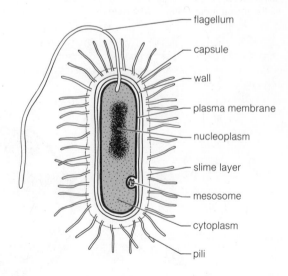

flagellum

capsule

wall

plasma membrane

nucleoplasm

slime layer

mesosome

cytoplasm

pili

Figure 33–5   Structure of a generalized bacterium.

Figure 33–9. Replication starts at one point on the molecule and proceeds in both directions from this point, producing two forks that gradually advance around the circle (Figure 33–9b). As the replication forks complete their circuit around the DNA molecule, the duplicated circles become completely separate (c). Presumably, the membrane attachment point is duplicated at the same time, with the result that both circles are now attached to the plasma membrane at two closely spaced points. The membrane begins to grow between the attachment points (d), separating and pushing the two DNA circles to opposite ends of the cell (e). Thus, the plasma membrane rather than a spindle divides the replicated chromosomes.

The cytoplasm is then compartmentalized into daughter cells by furrowing. An indentation in the plasma membrane appears around the midpoint of the cell between the separated DNA circles (Figure 33–9f). This indentation or furrow deepens, eventually pinching off the cytoplasm into two completely separate halves. The cell wall follows the inward path of the furrow; as the membranes separate, the growing wall becomes complete and extends as a partition between the daughter cells (g). Inward growth of the plasma membrane and cell wall of a bacterium can be seen in Figure 33–10.

Spore Formation   One reason some bacteria can withstand adverse environmental conditions is that they form *spores* (Figure 33–11). These are resting cells not directly comparable to the spores of plants. In bacterial spore formation, part of the cytoplasm, often most of the living material in the cell, becomes concentrated into a relatively small mass that is then surrounded by a thick cell wall laid down within the original wall. Spores are highly resistant to drastic extremes of temperature and moisture. Under the proper environmental conditions, the spore swells; the old wall splits; and a single vegetative cell emerges. The new cell has its own cell wall and is capable of rapid growth.

Metabolism in Bacteria   Glucose forms the staple diet of bacteria, but they can and will take on almost anything. Many bacteria secrete enzymes that break down proteins,

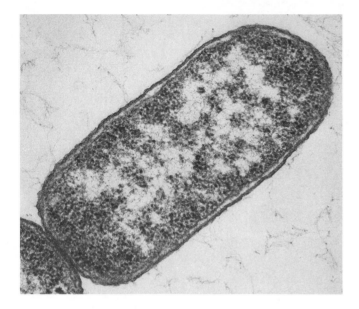

Figure 33–6   *Escherichia coli* consists of a wall and a plasma membrane that surrounds a region lacking membranes. The center of the cell is occupied by the nucleoplasm, which contains the DNA. Ribosomes fill the region around the nucleoplasm.

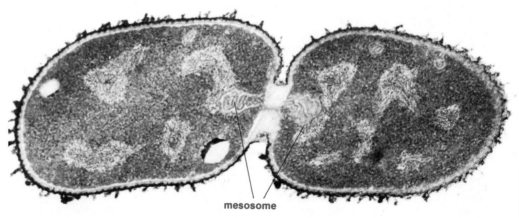

Figure 33–7   The relationship of the mesosome to the formation of the cell wall during bacterial division is clearly shown in this electron microscope photograph.

mesosome

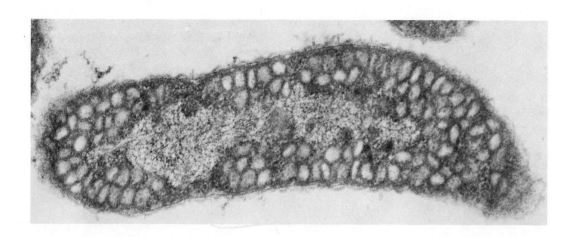

Figure 33–8 Photosynthetic bacteria, such as the *Rhodospirillum rubrium* shown here, contain photosynthetic membranes in the form of membranous sacs.

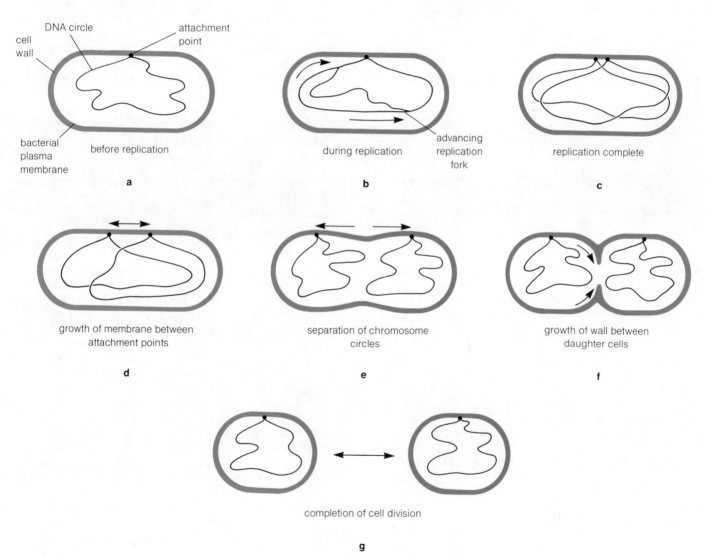

DNA circle

attachment point

cell wall

bacterial plasma membrane

before replication

a

during replication

advancing replication fork

b

replication complete

c

growth of membrane between attachment points

d

separation of chromosome circles

e

growth of wall between daughter cells

f

completion of cell division

g

Figure 33–9 Cell division in a bacterial cell.

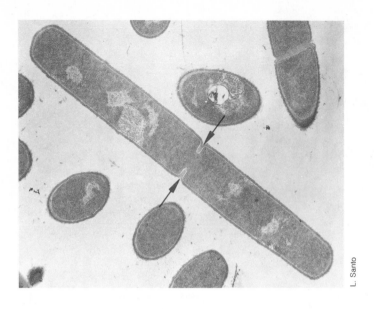

Figure 33–10 Division of the cytoplasm in *Bacillus ceres*. Inward growth of a new wall (arrows) is separating the daughter cells.

L. Santo

fats, and other large molecules into molecules they can use. In their metabolism, they do not necessarily carry the oxidative reactions to CO_2 and H_2O, as discussed earlier. The acetic acid bacteria, for example, start with alcohol produced by yeasts and convert it to acetic acid, releasing a significant amount of energy in the process. This, of course, is fine for the bacteria but not so good for the winemaker, who wants a good vintage, not vinegar. Milk provides the sugar for other bacteria, which convert the lactose to lactic acid, again releasing energy. The lactic acid sours the milk and makes it undrinkable for most people. Yet it is harmless and is necessary for the production of butter, buttermilk, and yogurt (Essay 33–1).

Some *chemosynthetic* bacteria have more unusual ways of obtaining energy. These species use oxidation-reduction reactions in which high-energy electrons are moved to lower energy levels. The energy liberated is used to synthesize organic matter in the cell directly from carbon dioxide. The bacteria use such starting compounds as ammonia, ferrous iron, sodium nitrate, and hydrogen sulfide.

Essay 33–1

Bacteria and Yogurt

One of the oldest and most nutritious of all foods is yogurt, enjoyed for centuries by peoples of the Near East, Turkey, and the Balkans. More recently, millions in western Europe and the United States have developed a taste for this natural food. Yogurt is essentially soured milk. Specifically, two bacteria are normally found in yogurt: the common *Lactobacillus casei*, which converts the lactose in the milk to lactic acid, thus making the environment more acid and unfavorable for the growth of pathogenic bacteria; and *Streptococcus thermophilus*, which adds a characteristic creamy flavor to the finished product.

Yogurt is easy to make with a minimum of equipment, and we now present one standard recipe, recognizing that there are almost as many ways to make yogurt as there are people who eat it.

Bring one quart of milk (to which may be added ½ cup of cream or 2 T powdered milk) to a boil in a glass, porcelain, or ceramic container (enameled iron pots can be used only if they are not chipped or cracked, since the presence of metals will inhibit the growth of the bacteria). Pour the milk into a nonmetal container (a glass or earthenware bowl with a cover is best) and cool to lukewarm. Dissolve 1 T to ½ cup of yogurt (commercial unflavored yogurt may be used) in a little warm milk to act as the starter. Before adding the starter, test the temperature of the milk with your wrist to make sure it is no more than lukewarm. A thermometer is, of course, more accurate and should register 110 to 125°F when the starter is added. Mix in the starter, cover the bowl with a lid, and wrap a heavy towel over it. The bacteria will generate enough heat that no more is needed; if you wish to speed up the process, place the milk in a warm place (in the oven with only the oven light as a source of heat or in a box warmed with a light bulb). Within 3 to 12 hours, depending on the amount of heat, the mixture will become as thick as heavy cream or custard. Now, put it in the refrigerator to set for several hours before eating. In the immortal words of Julia Child, "Bon appétit!"

The chlorophylls of photosynthetic bacteria, although related to plant chlorophylls, absorb solar radiation best in the near-infrared portion of the spectrum. This means that bacteria can carry on photosynthesis in what we would call darkness. They also photosynthesize in the absence of oxygen and release no oxygen. Just as in plants, the photosynthetic pigments in bacteria include carotenoid pigments and are located in membranes. However, these membranes are not organized into chloroplasts, but form simple structures similar to vacuoles (Figure 33–8).

An example of the type of photosynthetic reaction found in bacteria is that of the purple sulfur bacteria, in which carbon dioxide and hydrogen sulfide react to produce glucose and sulfur. This reaction is analogous to the kind of photosynthesis that involves water and releases oxygen. Other bacteria use other hydrogen donors (for example, organic compounds and molecular hydrogen) in place of H_2S.

Blue-Green Algae

Closely related to the bacteria are the second group of living prokaryotes, the blue-green algae, or Cyanophyta. The ultrastructure of a blue-green algal cell is shown in Figure 33–12. Although it lacks cell organelles, the cell is far from simple. The most conspicuous features are the parallel rows of membranes and masses of ribosomes.

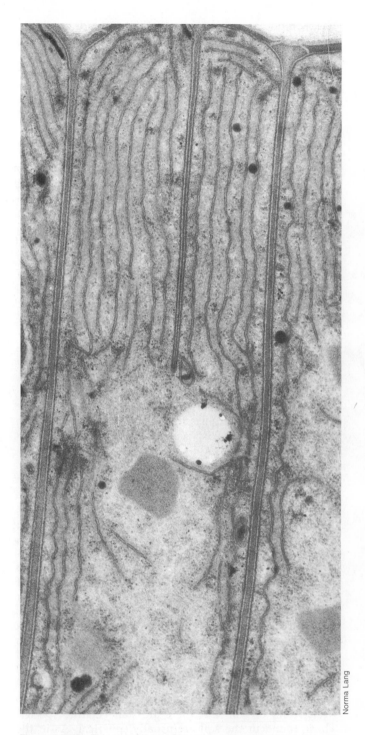

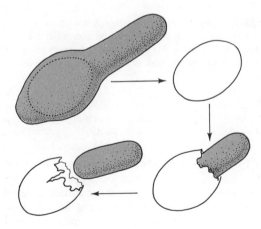

Figure 33–11 Bacterial spores. The light oval region in the bacteria indicates the presence of a spore. It forms within the original cell wall and is released when the cell wall breaks down. The spore wall then breaks down when the spore germinates, releasing a new cell.

Figure 33–12 This is a common blue-green alga, *Oscillatoria*, as seen with the electron microscope. It is a filamentous alga in which the cells are arranged in files. Thylakoid membranes run parallel to the end walls; the lighter areas contain the DNA; and the crystal-like shapes are phosphate deposits. The cells are dividing. The wall is beginning to grow from the existing wall inward. It will eventually reach a similar wall growing from the opposite side, thus dividing the cell.

The membranes are photosynthetic thylakoids similar to those found within the chloroplasts of eukaryotic green cells. Each thylakoid is essentially a flattened sac composed of two membranes. The thylakoids contain chlorophyll *a* and carotenoids. A major feature of the blue-green algae is the presence of the *phycobiliproteins,* the red and blue pigments that together with chlorophyll give the algae their characteristic color, which ranges from black to blue-green to rusty red. The phycobiliproteins are located in granules on the surface of the thylakoid.

Some blue-green algae, like some bacteria, can fix atmospheric nitrogen. In rice paddies, for example, blue-green algae are important in maintaining soil fertility. Fixation occurs in the thylakoids. Photosynthesis and nitrogen fixation are closely related in the blue-green algae, with respiration probably also taking place in the thylakoids.

The cells contain a nucleoplasm, similar to that of bacterial cells, within which are located strands of DNA. The numerous ribosomes are smaller than the cytoplasmic ribosomes of plants, about the same size as those of bacteria. Cell walls in the blue-green algae are similar in composition and structure to bacterial cell walls and quite unlike those of other algae and of higher plants. They contain mucopeptides and pectic substances but little, if any, cellulose.

Many blue-green algae contain gas vacuoles (Figure 33–13). These are important in maintaining buoyancy in floating species. The gas vacuoles, usually cylindrical, can be formed quickly by the cell.

Blue-green algae vary in form, but all lack flagella. Many are single-celled organisms, but many others form multicellular colonies. Some of these colonies are squares or cubes of cells; others are loosely organized globules. Still others form simple filaments, as in *Oscillatoria,* which derives its name from its slow, oscillating motion, easily observed under a microscope. In most of these filamentous blue-green algae, the cells are arranged on top of one another, looking much like a stack of coins but surrounded by a gelatinous sheath instead of a paper wrapper.

The filamentous species have mechanisms for fragmenting the file of cells, each fragment then growing into a new filament. Asexual spores similar to those found in bacteria also develop in some blue-green algae. Sexual reproduction is unknown.

Where do blue-green algae live? Most are found in fresh water or soil, but there are some marine species. Those found in the soil are usually spherical. As in the bacteria, this shape, together with the thick gelatinous sheath, appears to be important in the cells' ability to withstand drying. How well blue-green algae can withstand not only drying, but extremes of cold, is shown by the Antarctic species. These cells are exposed to severely desiccating conditions as a result of low atmos-

Figure 33–13 Gas vacuoles in a blue-green alga, as seen with the electron microscope. The vacuoles can be seen both in cross section and in a longitudinal view. The preparation is freeze-etched and reveals the structure of the membranes of the vacuoles. ×200,000.

pheric humidity, and temperatures may drop as low as −88°C.

Blue-green algae are important inhabitants of many lakes and ponds. Algal growth in a lake is limited by the available nutrients, particularly nitrogen and phosphorus. When lakes are low in these compounds, algal growth is low. But gradually, as a result of natural processes, these nutrients accumulate. We may increase the rate of accumulation by dumping material into the lake.

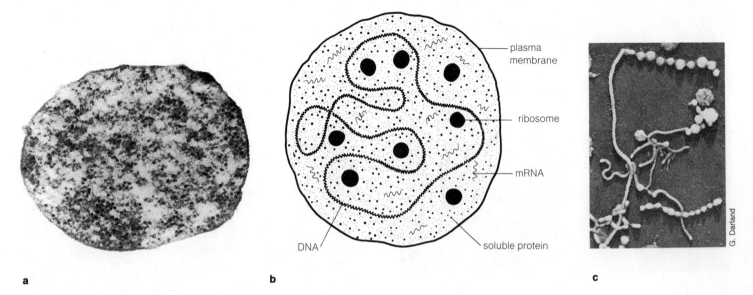

a   b   c

Figure 33–14 Mycoplasms are among the least complicated of all living things. Electron microscope studies indicate that they consist of little more than a plasma membrane, ribosomes, enzymes, and nucleic acids (**a** and **b**). Mycoplasms may have a wide variety of shapes, which appear to change quite readily. In the electron microscope photograph of metal-shadowed mycoplasm (**c**), the diversity of the shapes is clearly shown. In this technique, solutions of the organism to be observed are sprayed on plastic-coated grids. Then the particles are lightly covered, or "shadowed," with a metal so that their structure is brought out under the electron microscope.

Sewage, even fully processed, adds calcium, sulfur, nitrogen, and organic carbon. Most important, it adds phosphates, primarily from laundry detergents. Another rich source of phosphorus is fertilizers applied to the surrounding farmlands. Industrial wastes also play an important role in water pollution, even when they are not noxious. When the nutrient level becomes high enough, other factors come into play. Summer's high temperatures and long days produce conditions favorable to algal growth, and the higher than usual phosphate levels result in massive *blooms* of algae (usually blue-green). Such enormous concentrations of algae may be disastrous.

Some blue-green algae, such as *Microcystis,* produce toxic compounds that can kill birds and mammals that drink the water in which the algae are growing. With less drastic consequences, masses of algae may clog filtration systems and impart a disagreeable taste to drinking water. The decomposing algae may give off extremely offensive odors. Both the bloom and its decay rob the water of oxygen, killing not only animal plankton but fish as well. Finally, as winter comes, the dead masses of algae sink to the bottom, where bacteria decompose them. This depletes the oxygen under the ice and may result in massive fish kills.

Mycoplasms

Even simpler than the bacteria, and considerably smaller (0.1 to 0.25 micrometer in diameter), are a group of organisms known as *mycoplasms.* (They are also known as pleuropneumonialike organisms, or PPLOs, after the disease some of them cause in animals.) Their internal structures are almost nonexistent (Figure 33–14): They have a plasma membrane but no cell wall, a few ribosomes, and a granular matrix. However, they do contain a wide range of proteins and lipids, RNA, and DNA.

At first, all mycoplasms were believed to be parasites, causing disease in cattle, chickens, humans, and vascular plants. Then, in 1970, a mycoplasm was discovered in a burning coal refuse pile at the Friar Tuck mine in southwestern Indiana—a strange habitat indeed. The mycoplasm found there was both thermophilic and acidophilic, with an optimum pH for growth of 1 to 2. The organism was aptly named *Thermoplasma acidophila.*

Little is known about mycoplasms, but they have been tentatively classified among the prokaryotes. Many more will almost certainly be discovered in future years, since microbiologists now know how to culture them.

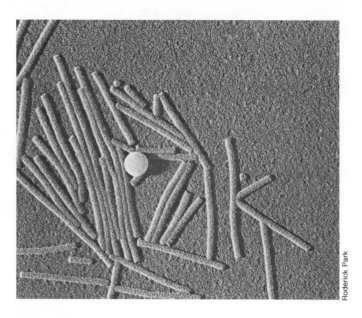

Figure 33-15 Isolated TMV can be seen with the electron microscope after metal-shadowing. TMV treated in this way appears as tiny rods. The sphere (88 nanometers in diameter) seen in this photograph is polystyrene and is used to calibrate the size of the virus particles.

Viruses

In the 1880s, Holland was a center for the production of fine cigars, but the Dutch farmers were faced with a serious problem. A disease had appeared in the fields that resulted in brown spots in the tobacco leaves, making them unsuitable as cigar wrappers. By 1887, it was known that the disease was contagious and that the sap from diseased plants would cause the disease in healthy ones. The Dutch microbiologist Martinus Beijerinck first attempted to demonstrate the presence of bacteria, but could find none. He then discovered that the juice pressed from diseased plants remained contagious even after passing through a porcelain filter fine enough to hold back all bacteria. Beijerinck next showed that a small amount could be used to infect a plant, and that this multiplied in the plant to the point where the plant could infect many others. He also showed that the causal agent was completely unable to replicate in plant juices alone. Heating extracts of the causal agent to 90 to 100°C inactivated it. At the end of his investigations, he concluded that the infective agent was not cellular, but instead was a soluble contagious agent.

We now call the contagious agent in this case tobacco mosaic virus (TMV). It and the virus that causes foot-and-mouth disease were the first viruses to be character-

ized. TMV was also the first virus to be crystallized. In 1935, Wendell Stanley of the University of California, Berkeley, purified TMV and was able to crystallize it. Crystallized TMV prepared by Stanley in 1935 is still infective today. The crystallization of TMV opened a new era in the study of viruses. The viral particles were finally seen when the electron microscope became available (Figure 33–15).

There has been a great deal of research on the nature and function of viruses. Much of the early interest was based on their importance as disease-causing agents in both animals and plants. Polio, influenza, smallpox, measles, mumps, and rabies are caused by viruses in humans. Viruses are equally important in plants and cause great crop loss every year. But viruses are also of great interest for what they can reveal about the nature of the gene and gene expression. Research on viruses helped form the foundation of the modern field of molecular biology (see Chapter 9).

The Nature of Viruses Every virus contains either DNA or RNA (a few rare ones contain both) surrounded by a protein coat. The protein coat protects the nucleic acid from enzymes in the environment of the host that might break it down and acts as a mechanism for the identification of the correct host cell. By showing a specific affinity for the surfaces of certain kinds of host cells, the protein coat permits the virus to attach itself only to appropriate host cells and not to other kinds of cells within which the virus could not reproduce. Thus, the protein determines the host range of the virus.

Viruses range in size from polio virus, about 10 nanometers in diameter, to some that are about 0.5 micrometer in diameter. Most viruses are spherical, but some are bricklike, rodlike, cubic, or irregular in shape. Some are surrounded by a plasma membrane, but this membrane is produced by the host cell when the virus is released and does not function as a plasma membrane in the sense of those surrounding cells.

The truly unique feature of viruses is that they exist in two distinct states. Outside a living cell, they are essentially inert chemicals that, as in the case of Stanley's TMV crystals, remain potentially active for years in a bottle on the shelf. Once inside a living cell, they are capable of taking over the metabolic machinery of the cell they are in, converting it to make viral protein and nucleic acid.

Viruses are known that infect just about all known groups of organisms, but the distribution among hosts is decidedly uneven. Viral diseases of flowering plants are very common, but they are rare in gymnosperms, bryophytes, ferns, algae, and fungi. Almost all vertebrates are susceptible to viral infection, but among the invertebrates there are few viral diseases. The exception is the insects, in which viral infections are common. Viruses also infect both the bacteria and the blue-green algae.

We discussed the reproduction and genetics of viruses in Chapter 9 and will not repeat that information here. Instead, let's look at some common viral diseases in animals and plants.

Viral Diseases Individual polio viral particles are very small—about 10 nanometers in diameter—and, when highly purified, form distinctive crystals. Polio is an RNA virus, and each particle is composed of 25 percent RNA and 75 percent protein. A single infected cell may produce as many as 10,000 virus particles before it dies and breaks open.

Polio virus replicates primarily in the cells of the intestinal tract of humans, producing only mild symptoms. The virus may in rare cases invade the central nervous system and cause the paralytic disease known as poliomyelitis, or infantile paralysis. Thus, although infection with the virus is very common, paralytic polio is relatively rare.

The virus particles are excreted in the feces in large numbers and, through contamination of food and water with fecal material, the disease is passed from individual to individual. Where sanitation is poor, most infants are infected early, when they still possess immunity transmitted to them by their mothers. They build up an active immunity that may be lifelong. Under these conditions, paralysis is rare and found primarily in infants. In more advanced countries with better sanitary conditions, infant infection is rarer. Infection occurs later in life, when the individual has no immunity. Such infections are more likely to lead to the paralytic disease, and this form is more common in developed countries in older children and young adults.

Increased knowledge of the virus together with the development of cell culture techniques made it possible to grow the virus in large numbers. This in turn led to the development of two vaccines. The first was the Salk vaccine, in which the virus particles were inactivated with formaldehyde and used to induce immunity. In the second type of vaccine, mutant strains of the virus were isolated that induced immunity but were not themselves pathogenic. Such strains are called *attenuated* and are used to combat a number of diseases. The introduction of these vaccines led to a dramatic drop in the number of cases of polio in the United States. The problem now is to make sure that children continue to be inoculated against the disease.

One of the worst viral diseases in humans was smallpox. The course of history was repeatedly changed as ruler and commoner alike were stricken with this deadly disease. Smallpox was far more important than firearms in the defeat of the Indians by the European colonists.

Yet today we have eliminated the disease worldwide. The difference in the past 200 years is the development of a vaccine that induces an active immunity to the disease. Let's first look at the virus itself and then at the vaccine and its history.

The smallpox virus is large, some 0.2 to 0.25 micrometer, and relatively complex. It is a DNA virus and, as usual, the nucleic acid is surrounded by a protein coat. In addition, however, the virus particle is surrounded by a protein envelope produced by the host cell.

The smallpox virus is found only in humans and monkeys. It spreads through the air and enters the host through the upper respiratory tract. It grows first in the mucous membranes, but soon enters the lymph and the bloodstream, from which it passes to the heart, liver, spleen, and kidney. During the initial stages, fever, headache, and backache occur. A rash appears on the skin. Soon after, the temperature falls to normal and vesicles teeming with the virus form crusts on the skin. These develop into the characteristic pockmarks that permanently disfigure the skin. In severe cases, death occurs, usually due to hemorrhaging.

Recovery from smallpox infection confers complete immunity. For several centuries it was known that material from smallpox pustules could induce immunity in individuals inoculated with it. This practice was widely employed and was generally successful—although occasionally severe, rather than mild, cases occurred, resulting in death.

The breakthrough came in 1798, when Edward Jenner, an English physician, noted the similarity of the pox diseases in cows and humans. Moreover, he noted that milkmaids and others who worked around cows were often immune to smallpox. He reasoned that artificial inoculation with cowpox material might confer immunity to smallpox. His experiments confirmed this, and he introduced the process of vaccination (from *vacca,* "cow"). Cowpox virus is similar to smallpox virus but causes only a mild, self-limited disease in humans. This simple procedure confers immunity to smallpox and is a most effective control of the disease. By systematically tracking down all known cases of the disease and inoculating against it throughout the world, we have come close to totally eradicating smallpox.

Although most of us have been vaccinated for smallpox, few indeed have ever had or even seen the disease in the United States. This is not true, however, for influenza. Most of us have been stricken, not once, but repeatedly, by the "flu." The fever, sore throat, and cough that frequently mark the presence of the virus are unpleasant, but rarely more than that.

The influenza virus is round or oval and 80 to 120 nanometers in diameter. It is a rather complicated virus, surrounded by a membrane and containing RNA, protein, carbohydrate, and lipid.

Human influenza virus exists in nature only in humans. It is capable of causing worldwide epidemics. One of the best-studied of these was the "Asiatic" flu epidemic of 1957. It first appeared in the interior of China in late February 1957, and by early April had been brought to Hong Kong by refugees. It spread from Hong Kong to San Diego, California, on navy ships; another navy ship had carried it to Newport, Rhode Island, by May. Other outbreaks occurred in various parts of the United States, and by the last two weeks of October some 22 million bed cases had been reported. From then on there was a progressive decline.

Summary

The kingdom Monera consists of the bacterica and the blue-green algae. All of the Monera are prokaryotic organisms. Bacteria are particularly ubiquitous in distribution. They are mainly saprophtyes or parasites, although a few are capable of a special type of photosynthesis and some are chemosynthetic. They are both very useful and very harmful to humans. The blue-green algae all carry on photosynthesis of the type found in plants. Some species of both the bacteria and the blue-green algae are capable of fixing atmospheric nitrogen into organic forms. Viruses are discussed in this chapter although they are not in the kingdom Monera. Viruses are packets of nucleic acid, either DNA or RNA, that once inside a living cell are capable of causing the machinery of that cell to reproduce the virus. Many viruses are parasites on plants and animals; many human diseases are caused by viruses.

Supplement 33–1
Bacteria and Food

Bacteria pose an immediate threat to us when the wrong kind occurs in large numbers in our food. Unfortunately, bacteria like many of the same foods we do. Let's look at some examples and some of the ways we have of preventing bacterial decay of food.

Consider an egg. This remarkably designed package of foodstuffs comes in a case, the shell, that allows gas exchange but is waterproof and impervious to bacteria, if handled correctly. Once the egg is opened, however, the contents provide an almost ideal culture medium for the growth of bacteria. If the outside of the egg is not clean, or the cook's hands have not been washed well, or the bowl is slightly dirty, bacteria can be immediately introduced into it. No harm is done if the egg or the food it is in is cooked; high heat will kill virtually all the bacteria. But if the egg is used in a sauce, dressing, or custard, where it is either left raw or only gently heated, the bacteria will multiply rapidly, particularly if the mixture is left at room temperature or warmer. Ten staphylococci bacteria can become 600,000 overnight, and the food eaten the next day could cause a severe case of food-poisoning. Simply refrigerating the food can reduce the bacterial growth to a safe point.

Simple bacteria growth in food does not necessarily cause food-poisoning or gastroenteritis. In our egg example, we were careful to select a staphylococcus bacterium that causes food-poisoning. Many bacteria spoil food, but few cause food-poisoning. Americans on the whole tolerate less food spoilage than most peoples. At the other extreme are the Eskimos, who relish a delicacy known as "titmuck." This is fish buried to allow bacterial fermentation to occur and is a semiliquid, foul-smelling substance that even the Eskimos' dogs refuse to eat.

Food spoilage can easily be detected by smell, appearance, and taste—but poisoned food cannot. There are two distinct types of food-poisoning. One is caused by bacteria that produce toxins. The most famous of these is the toxin produced by *Clostridium botulinum,* which causes botulism. This toxin is so potent that 1×10^{-10} grams can kill a mouse. The bacteria in this case can form spores resistant to high temperatures and can grow in anaerobic conditions of essentially neutral pH. Another toxin-producing group of bacteria are the staphylococci, but they are less often fatal.

The second type of food-poisoning is caused by the presence of the bacteria themselves. This group includes *Salmonella* food-poisoning, which has been increasing steadily and receiving much publicity in recent years—so much so that Senator Magnuson of Washington introduced a bill in the Senate to have the name of the bacteria changed because he felt it was adversely affecting the sale of salmon!

The rod-shaped *Salmonella* bacteria are natural inhabitants of the colon (bowel) of humans and other animals. They are passed out with feces and are introduced into food from soiled hands or contact with animal excreta. This is the only source, and it is incredible that in a highly industrialized, affluent society this type of food-poisoning should exist at all, let alone be increasing at the rate it is.

Food preservation has been called a race between people and the lower forms of life to see who gets the food supply first. Our efforts in the race have been largely devoted to slowing down the bacteria. One of the most effective means is to keep food cold, so that bacterial growth is drastically reduced. Drying achieves the same end, since bacteria cannot grow without water. Most bacteria also grow slowly or not at all in acid media; so pickling in vinegar, which is dilute acetic acid, is also a good way to keep food. Strong salt and sugar solutions also inhibit bacterial growth. All of these are ancient methods. (Even freeze-drying was first developed by the Incas and is still used in the same form by the natives of the high mountains of Bolivia.)

Modern civilization has added many refinements to these procedures. We have developed mass means of freezing, cooling, drying, and canning (which is essentially heat preservation) food and of storing and shipping it. Modern food technology also relies on the addition of chemicals that retard the growth of bacteria. This is hardly a new procedure, but it has been receiving more attention in recent years. Such additives have been called poisons; in truth, many are. Sodium nitrate, for example,

has been used since the Middle Ages and, when used in small amounts, causes changes in food, usually meat, that inhibit bacterial growth; yet used in high concentrations it can be fatal. Many such additives do little to preserve the food, but are there to ensure a particular color, texture, or manufacturing character. These are clearly less defensible but are often of great economic importance. The revulsion that many people feel toward food additives and agricultural chemicals in general has led in recent years to the establishment of natural food stores. In general, the food runs to the less perishable types and usually requires the development of different tastes.

In the end, the race metaphor seems apt. Without a doubt, bacteria and insects (with fungi running a strong third) are our strongest competitors for domination of the ecosystem, and it is by no means clear who will win.

34

Kingdom Protista

The kingdom Protista is a collection of fascinating single-celled or colonial microscopic organisms. First seen by humans in the late 1600s, when Anton van Leeuwenhoek looked at a drop of pond water under one of the earliest microscopes, they continue to amaze and delight us. Some members of the Protista are important in maintaining the productivity of the oceans, but others cause some of the worst diseases in animals and humans.

One of the most recently recognized kingdoms, the various phyla of the Protista had earlier been placed in the plant and animal kingdoms. Many biologists now feel that the phyla have more in common with each other than with the phyla of other kingdoms. Yet the Protista is still a collection of rather diverse organisms, and the various evolutionary relationships are not clear. Many of the common names reflect the older associations with other kingdoms.

All Protista are eukaryotic organisms. Many species contain chloroplasts and photosynthetic pigments similar to those of plants. Other species are heterotrophic and actively hunt and ingest prey. This major difference in nutrition divides the Protista into two groups: the single-celled algae and the protozoa. The algae carry out photosynthesis using chlorophyll *a* and other pigments. The protozoa are nonphotosynthetic and live by ingesting organic matter or absorbing it from their surroundings. Almost all protozoa are motile in at least some stage of their life cycle.

Let us look first at the algal phyla and then at the protozoa.

Phylum Euglenophyta: Euglena

The members of the small phylum Euglenophyta (some 450 species) are unusual one-celled organisms (Figure 34–1). Some species contain chloroplasts and have rigid exteriors and flagella. Other species lack chloroplasts, have a flexible exterior, move in a creeping or gliding motion (lacking flagella), and ingest food through a gullet. Some species are truly autotrophic, others are heterotrophic, and still others may be either, depending on the environment.

Phylum Chrysophyta: Golden Algae and Diatoms

Chrysophyta is a much larger phylum of algae than the Euglenophyta. It consists of about 5,800 species divided into two classes: Chrysophyceae, or golden algae, with 300 species; and Bacillariophyceae, or diatoms, with over 5,500 species.

The golden algae, found mainly in fresh water, are often naked protoplasts equipped with flagella. In contrast, the diatoms have rigid walls containing silica (SiO_2)—as much as 95 percent of the wall—in addition to cellulose. The walls of individual diatoms have such various and beautiful markings and sculpturings that they are used to test the optical efficiency of microscopes. Diatoms in mass, however, are anything but beautiful: They appear as a yellowish brown scum or slime on rocks and other objects in rivers, lakes, and intertidal regions of the ocean.

The walls of diatoms (Figure 34–2) are extremely resistant to decay. As a result, these walls accumulate on the ocean floor. Raised above sea level by geologic activity, such deposits form diatomaceous earth, which is commercially mined and used in polishes, air and water filters, and window cleaners, among other products. The number of diatoms involved in these deposits can only be imagined—more than 6 million shells can be found in 1 cubic millimeter, and the deposits may be 200 meters or more deep.

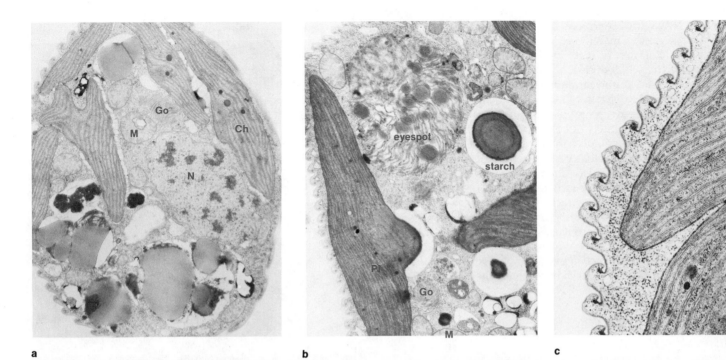

Figure 34–1 (**a**) Many features of the Euglenophyta can be seen in this low-power view of a species in the genus *Colacium*. There is a single nucleus (N), surrounded by chloroplasts (Ch) and mitochondria (M). A large Golgi complex (Go) is also present near the nucleus. The opposite end of the cell is filled with food storage material and a few mitochondria. The flagellum and associated structures are not visible in this section. (**b**) The swirling mass of membranes and disks in the upper left corner of this photo is the eyespot, which appers red under the light microscope. The white ring below and to the right of the eyespot is starch surrounding a small portion of the chloroplast. The relation of the starch to the chloroplast can be seen better in the plastid (Pl) running down the lefthand side. The starch is present in the cytoplasm in the form of a dome over a special region of the chloroplast. (**c**) This higher-power view of the edge of the cell shows the nature of the surface and the structure of the chloroplast. No external wall is present, and the beautiful, wavelike pattern of the cell surface is associated with an area of dense material (in each trough) and a couple of microtubules (in each crest). Within the chloroplast, the thylakoids are stacked in groups of two to five, with three being the most common. The chloroplasts also contain large numbers of ribosomes.

Phylum Xanthophyta: Yellow-Green Algae

Until fairly recently the 400 mostly fresh-water species of yellow-green algae, Xanthophyta, had been considered part of the Chrysophyta, but because of differences in pigments and other characteristics, many biologists have placed them in a phylum by themselves. The most common forms are bulbous or threadlike and are found in ponds as plankton. Others are epiphytes, and a few others are motile, having two flagella. The cells of the motile species are often so small that they are overlooked.

Phylum Pyrrophyta: Dinoflagellates

There are some 1,000 species of Pyrrophyta (dinoflagellates and cryptomonads) in 125 fresh-water and marine genera. Most genera are one-celled motile flagellates. They can be autotrophs or heterotrophs. The cells have two flagella; in dinoflagellates one flagellum is ribbonlike, encircling the cell, and the second runs backward in a longitudinal groove (Figure 34–3).

The cells have one nucleus. Dinoflagellate chromosomes are most unusual for eukaryotic cells: They consist only of DNA, with no protein (Figure 34–4). Thus, they resemble the genetic material of prokaryotic cells and appear to have a similar ultrastructure. No centromeres are present, and the DNA divides directly, as in prokaryotic cells. A well-defined nuclear envelope is present, but mitosis is most unusual. Early in division, the spherical nucleus is invaded by deep cytoplasmic invaginations that become essentially tunnels of cytoplasm through the nucleus. The nuclear envelope remains intact, and the tunnels, which cross the nucleus parallel to the division axis, contain 8 to 20 microtubules. The nucleus now elongates and becomes dumbbell-shaped. The chromosomes, which have duplicated earlier, are attached to the nuclear envelope and separate, not necessarily in unison, as the nucleus elongates. At all times, the nuclear envelope remains intact. Eventually, the nucleus pinches in two, and separate daughter nuclei are formed. This is a most remarkable— and probably primitive—division mechanism.

Phylum Protozoa

Protozoa are found in fresh water, in the sea, and as parasites in many animals, including humans. Although a few are colonial and display some degree of specialization among cells of the colony, most live as single cells. In spite of this unicellular mode of life, many of the protozoa are complex and have specialized internal organelles not found in the cells of other organisms. The ciliate protozoans (the Ciliophora) have internal organelles so complex that several authorities see in them an evolutionary trend away from cellularity, toward the development of complex organization without separation of interior functions into multiple, specialized cells.

The protozoa are all small, but they vary in size from about the diameter of bacteria to specks just visible to the naked eye. Most are between 100 and 300 micrometers in diameter.

The organisms potentially classifiable as protozoa are so varied that no completely satisfactory system of classification has been designed. For example, many single-celled forms so closely resemble the algae that protozoologists and botanists have been fighting over them for at least a century.

The protozoa are usually divided into three major subphyla. The first, Sarcomastigophora, includes protozoa that move with flagella, *pseudopodia* (footlike, lobed extensions of the cell surface), or both. A second major subphylum, the Sporozoa, contains only parasites with complex life cycles, typically forming spores at some stage. The Sporozoa have no motile organelles except for the flagellated reproductive cells of a few species. Both the Sarcomastigophora and the Sporozoa have a single nucleus. Members of the third subphylum, the Cilio-

Figure 34–2 The wall of a diatom as seen in the scanning electron microscope.

phora, have two kinds of nuclei in each cell, the *micronucleus* and the *macronucleus*. In at least one stage of their life cycle, all of the Ciliophora have many short appendages called *cilia* (singular, *cilium*). The Ciliophora are among the most structurally complex and the most behaviorally interesting of the protozoa. Almost any sample of pond or stream water will contain one or more species of Ciliophora, almost all of which are free-living.

Subphylum Sarcomastigophora The subphylum Sarcomastigophora is divided into two major superclasses, the Mastigophora and the Sarcodina. The Mastigophora have one or more flagella and usually a semirigid outer covering, the *pellicle,* that gives the cell difinite shape. This superclass includes both chloroplast-bearing organisms that carry out photosynthesis, and others without chloroplasts that feed by ingesting organic material or capturing prey.

The several human parasites among the Mastigophora have caused incalculable misery. The most important of

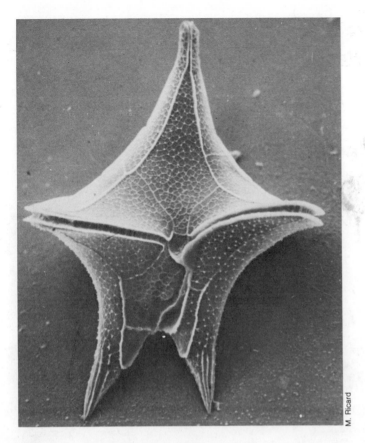

Figure 34-3 The dinoflagellate *Peridinium*.

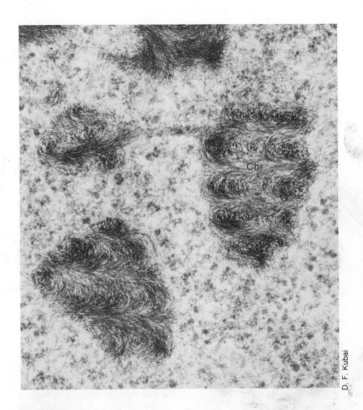

Figure 34-4 Nuclear chromatin bodies (Chr) of the dinoflagellate *Gyrodinium*.

the parasitic Mastigophora are members of the genus *Trypanosoma* (one species of which is shown in Figure 34–5), which in humans cause African sleeping sickness and Chagas' disease.

The sleeping sickness trypanosomes spend most of their life cycle as parasites of the blood, brain, and spinal fluids of humans and other vertebrates. The parasite is transmitted by a small biting insect, the tse-tse fly, common in many parts of Africa. The fly bites an infected individual and takes in trypanosomes with the blood. The parasites multiply in the fly's stomach, eventually migrating to the salivary glands, from which they are injected along with the salivary fluids when the fly bites its next victim. In the host, the trypanosomes attack the lymphatic glands and central nervous system, causing fever, loss of weight, and extreme lethargy. Efforts to control the disease, which is most prevalent in equatorial Africa, are concentrated primarily in attempts to eradicate the tse-tse fly. Isolating victims doesn't work because domestic and wild animals serve as a reservoir of trypanosomes that can readily be transmitted to people.

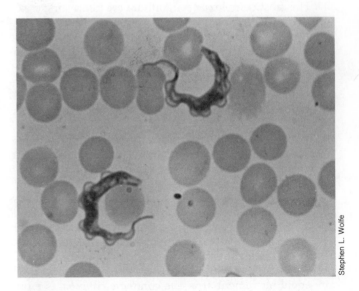

Figure 34-5 *Trypanosoma equiperdum*, a blood parasite of horses and donkeys.

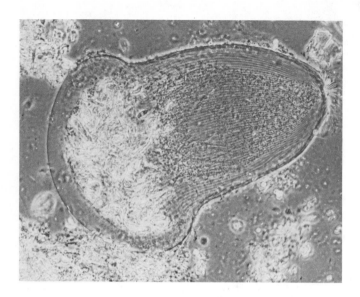

Figure 34–6 *Trichonympha agilis*, a protozoan that lives symbiotically in the gut of termites.

A less dangerous but more prevalent mastigophoran that is also parasitic on humans is *Trichomonas vaginalis,* which inhabits the reproductive tracts of both men and women. Infection by *Trichomonas vaginalis,* normally transmitted by sexual intercourse, often produces no symptoms in men, but causes inflammation, itching, and vaginal discharge in women. Infection by *Trichomonas* is easily controlled by drugs.

The most complex Mastigophora are interesting creatures that live as benevolent parasites or symbionts in the intestinal tracts of termites and roaches. These protozoans, the class Hypermastigida (Figure 34–6), are often completely covered with flagella and contain complex internal organelles. The Hypermastigida, which include many diverse body forms, may make up from 20 percent to as much as 50 percent of the total body weight of a termite.

The Sarcodina, in contrast to the Mastigophora, have no pellicle, no defined body form, no chloroplasts, and no flagella except in the reproductive forms of a few species. A single type of nucleus is present in the cytoplasm. The most familiar example of this superclass, and perhaps the most familiar protozoan, is *Amoeba proteus* (Figure 34–7). *Amoeba proteus* changes its body shape constantly as it glides or flows through its environment. A long pseudopod is extended, and the mass of the organism simply flows into the projection. This type of movement, called amoeboid movement after the common *Amoeba,* is more or less characteristic of all the Sarcodina.

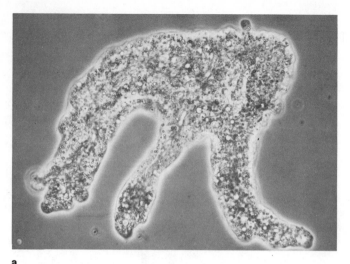

a

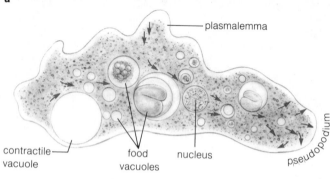

b

Figure 34–7 (**a**) A light micrograph of *Amoeba proteus* about to extend a pseudopod in order to move. (**b**) A diagram of *Amoeba proteus* showing its component parts. The arrow indicates its direction of movement.

Pseudopodia also act in the capture of prey and organic food matter by flowing around a food particle until it is completely enclosed by the cytoplasm in a membrane-bound food vacuole. After the food is digested by enzymes secreted into the vacuole, any undigested waste materials are excreted to the outside by a reversal of the ingestion mechanism: The food vacuole migrates to the surface of the amoeba, fuses with the plasma membrane, and is expelled. *Amoeba proteus* and other fresh-water Sarcodina also contain a *contractile vacuole* that eliminates excess water from the cytoplasm by a similar mechanism.

The other members of the Sarcodina include the Foraminiferida (from *foramen,* "hole," or "opening"; Figure

Figure 34–8 Examples of foraminiferans.

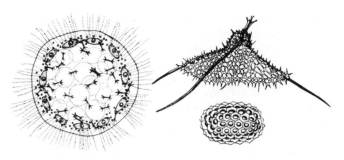

Figure 34–9 Examples of radiolarians.

34–8), which have chambered shells shaped like snail shells, usually containing calcium carbonate. These protozoans are among the most numerous living organisms in the seas. The siliceous-shelled or chitinous-shelled Radiolaria (Figure 34–9) also reach large numbers in both fresh and salt water. Radiolarian pseudopodia are often long, raylike processes that extend outward in all directions through openings in the shell. The shells of dead organisms form deposits that can be used to determine the geologic age of the formations in which they are found.

Although most of the Sarcodina are free-living, some are parasitic in humans and other animals. *Entamoeba hystolytica* is a parasite in the intestine, liver, and other organs of humans that causes amoebic dysentery, a debilitating disease affecting as many as 50 percent of the people in tropical and subtropical parts of the world. This amoeba causes ulceration and bleeding of the intestinal lining and other organs and characteristically severe dysentery. During infection, the amoebae regularly form cysts, rounded cells with a protective exterior coat, that are eliminated with the feces and form the main route of infection for new hosts. Outbreaks of amoebic dysentery are no longer common in the United States and are not likely to occur unless drinking-water supplies are contaminated with raw sewage in accidents or natural disasters.

Subphylum Sporozoa   The Sporozoa are all parasites of animals. These protozoans lack flagella and pseudopodia, except for the flagellated male gametes of some species. Many of the Sporozoa have complex life cycles involving more than one host organism. Malaria is caused by species of the sporozoan genus *Plasmodium* that live alternately in humans and in the mosquito *Anopheles*. As the name suggests, the Sporozoa typically form resistant spores at some time in their life cycle.

The sporozoans that cause malaria have probably had a greater effect on human life, society, and history than any other pathogenic microorganism. The life cycle of these parasites, typical in complexity for the sporozoans, is shown in Figure 34–10. Malaria is characterized by general debilitation, chills, and fever. It is sometimes fatal. In the most common form, caused by *Plasmodium ovale,* the *Plasmodium* cells within the bloodstream multiply in synchrony in infected cells and are released together at 48-hour intervals to infect other blood cells. At these intervals, the victim experiences severe chills and fever, probably caused by accumulated toxins released along with the *Plasmodium* by the disintegrating red blood cells.

Four different *Plasmodium* species infect humans and cause forms of malaria with slightly different symptoms. All forms can be recognized by the periodic nature of the chills and fever. Malaria is essentially incurable, although the attacks of chills and fever may be prevented or moderated by drugs. Some 10 percent of the world's population is presently infected with malaria, and the disease is estimated to cause between 1 and 2 million deaths each year. Control of the disease consists primarily of eliminating the *Anopheles* mosquito and administering drugs such as quinine, which will prevent the symptoms from recurring if taken regularly.

Subphylum Ciliophora   All the members of the varied and complex subphylum Ciliophora have numerous cilia in at least one stage of their life cycle. In contrast to other protozoan subphyla—and indeed to all other living organisms—the Ciliophora have two types of nuclei, a *macronucleus* and a *micronucleus*. The macronucleus functions throughout the cell cycle as the source of the messenger, transfer, and ribosomal RNA necessary for cellular synthesis of proteins. The micronucleus evidently functions only during reproduction (mitosis and meiosis), when genetic information is replicated and passed on to daughter cells. Almost all the Ciliophora are free-living and exist by eating other organisms or organic matter;

none are photosynthetic. The Ciliophora have a rigid pellicle and exhibit a definite body form.

Paramecium caudatum (Figure 34–11) is typical of the Ciliophora. The body is slipper-shaped, with the narrowest part of the "slipper" forming the anterior end of the cell. The entire cellular surface is covered with about 25,000 cilia that beat in unison and propel the organism through water in a spiral path. At one side, an *oral groove,* lined with cilia, leads to a mouth opening. Within the cytoplasm are a large, conspicuous macronucleus and a much smaller micronucleus. Two contractile vacuoles are present, each with smaller canals radiating into the surrounding cytoplasm. One is located near the anterior and one near the posterior end of the cell. Numerous food vacuoles are distributed throughout the cytoplasm.

Studies of reproduction in *Paramecium* and other Ciliophora have revealed the origins and functions of the micronucleus and the macronucleus. In asexual reproduction (Figure 34–12), both nuclei divide after DNA replication and are passed on to daughter cells. The micronucleus divides by a typical mitotic mechanism, but the macronucleus divides simply by pinching into two parts after DNA replication is complete. No spindle is formed in this pro-

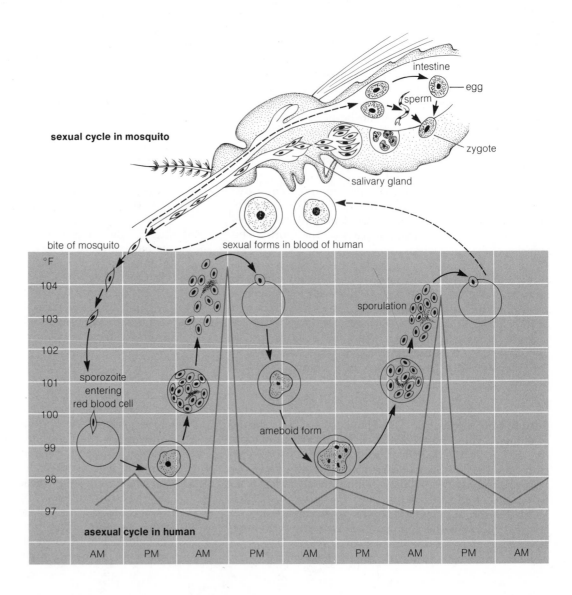

Figure 34–10 The life cycle of the malarial parasite *Plasmodium.* The phases of the cycle in the mosquito are shown in the diagram of the mosquito. The phases in the human bloodstream are shown superimposed on the temperature chart of a malaria patient. The bursting of millions of red blood cells, liberating spores and black granules, coincides with the regular periods of high fever.

cess, called *amitosis,* and no chromosomes are recognizable within the macronucleus or daughter macronuclei at any stage of asexual cell division. During the interphase following asexual reproduction, the micronucleus remains inert except when DNA is replicated. All the RNA synthesis necessary for cellular activities takes place in the macronucleus.

Sexual reproduction (Figure 34–13) in *Paramecium* occurs by a process known as **conjugation.** As two conjugating *Paramecia* come together, the macronucleus in each cell disintegrates and disappears into the surrounding cytoplasm. The micronucleus enters meiosis, producing four nuclei in each of the conjugating cells. Two of these break down, leaving two gamete nuclei in each cell. As these nuclear events are taking place, the protoplasm of the conjugating cells is joined by a cytoplasmic bridge that forms near the opposed mouth regions. One daughter micronucleus from each of the conjugants acts as a male gamete and crosses the bridge, fusing with the "female" gamete nucleus of the opposite cell. As a result of these nuclear fusions, each of the cells has a zygote nucleus at the close of conjugation. The cells then separate, and the zygote nucleus enters a series of mitotic divisions. Some of the products of this division sequence remain as micronuclei, and some are transformed into macronuclei. Eventually, after several divisions of the cytoplasm, daughter cells are produced that each have one micronucleus and one macronucleus. As a result of conjugation, meiosis, and fertilization, these daughter cells and their asexual descendants contain new combinations of the alleles derived from the parent cells entering conjugation.

The development of macronuclei from micronuclei following conjugation has been the subject of intensive research in recent years. This work has shown that, as the micronucleus is transformed into a macronucleus, chromosomal DNA is replicated many times. As a result,

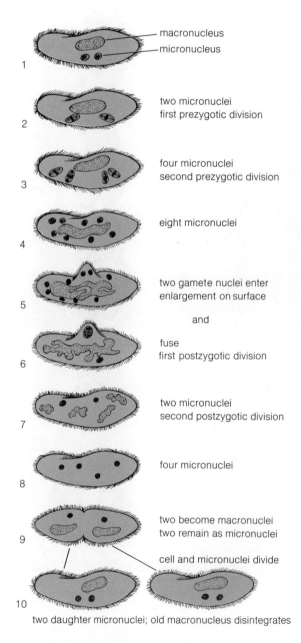

1 macronucleus
micronucleus

2 two micronuclei
first prezygotic division

3 four micronuclei
second prezygotic division

4 eight micronuclei

5 two gamete nuclei enter enlargement on surface

and

6 fuse
first postzygotic division

7 two micronuclei
second postzygotic division

8 four micronuclei

9 two become macronuclei
two remain as micronuclei

cell and micronuclei divide

10 two daughter micronuclei; old macronucleus disintegrates

Figure 34–12 Asexual reproduction in *Paramecium,* showing the changes in the macronucleus and the micronuclei.

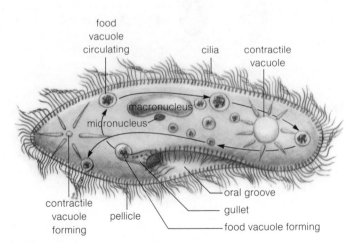

food vacuole circulating
cilia
contractile vacuole
macronucleus
micronucleus
oral groove
gullet
contractile vacuole forming
pellicle
food vacuole forming

Figure 34–11 The structure of *Paramecium caudatum,* a fresh-water ciliate.

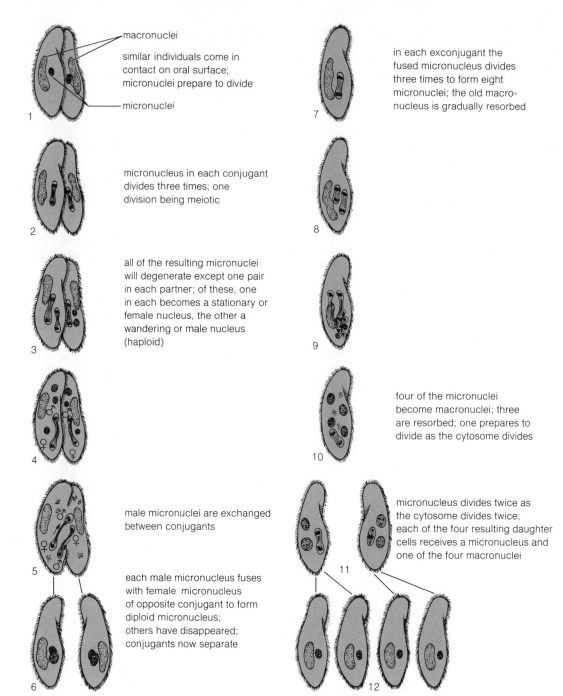

macronuclei

1. similar individuals come in contact on oral surface; micronuclei prepare to divide

micronuclei

2. micronucleus in each conjugant divides three times; one division being meiotic

3. all of the resulting micronuclei will degenerate except one pair in each partner; of these, one in each becomes a stationary or female nucleus, the other a wandering or male nucleus (haploid)

5. male micronuclei are exchanged between conjugants

6. each male micronucleus fuses with female micronucleus of opposite conjugant to form diploid micronucleus; others have disappeared; conjugants now separate

7. in each exconjugant the fused micronucleus divides three times to form eight micronuclei; the old macronucleus is gradually resorbed

10. four of the micronuclei become macronuclei; three are resorbed; one prepares to divide as the cytosome divides

11/12. micronucleus divides twice as the cytosome divides twice; each of the four resulting daughter cells receives a micronucleus and one of the four macronuclei

Figure 34–13 Sexual reproduction (conjugation) in *Paramecium*.

the mature macronucleus contains many copies of genes in the micronucleus from which it is derived. Analysis of the DNA of mature macronuclei reveals that only some of the genes of the micronucleus are replicated and preserved in the macronucleus; others are eliminated during macronuclear development. Thus, the macronucleus contains only partial genetic information, presumably only that required for maintaining cell functions during interphase. Each cell retains an intact micronucleus that does not undergo transformation into a macronucleus and conserves the complete genetic information of the individual.

Summary

The kingdom Protista consists of a collection of phyla of single-celled or colonial eukaryotic organisms. Two of the phyla, the Chrysophyta and Xanthophyta, have only autotrophic organisms; the remaining phyla have both green and nongreen species. Thus, the Protista sit at the base of the evolutionary lineages leading to the fungi, plants, and animals. Many of the Protista are beneficial to humans, but many are harmful, including some that are deadly parasites.

35

Kingdom Fungi

The spring rains cease, and a few days later the mushrooms spring up in a ring as if by magic. This is a common event, but how do we account for it? And how do we know whether the mushrooms are edible? The answers are simple: On moonlit nights the fairies dance in a ring in the fields, and where they dance, mushrooms spring up. When the dance is finished, they rest on the mushrooms. But the toads also come out and sit on the mushrooms. Those chosen by the fairies are edible, but those used by the toads (and therefore called toadstools) are poisonous.

This explanation is an old folktale, but barely over a hundred years ago we knew so little about the Fungi that many people accepted this type of reasoning. Even today, many people rely on equally old-fashioned ways of telling which mushrooms are edible—and many a mushroom collector dies as a result. People still attribute mystical properties to mushrooms and other Fungi, although they know perfectly well that Fungi are living organisms not unlike plants and animals.

All Fungi are eukaryotic organisms. They all lack photosynthetic pigments and do not carry on photosynthesis. They live by absorbing organic matter from their surroundings. Frequently, their surroundings are dead matter, such as fallen leaves or dead insects, but many Fungi live off of living organisms and may be deadly parasites.

The Fungi are divided into two phyla: the slime molds, or Myxomycophyta, which generally lack cell walls; and the Eumycophyta, which have true cell walls. The Myxomycophyta are divided into the true slime molds and the cellular slime molds on the basis of structure and modes of growth and reproduction. The Eumycophyta are divided into classes, primarily on the basis of their mode of reproduction, but also on the basis of differences in such features as the composition of their walls and the structure of their motile cells. These various relationships are shown in Figure 35–1.

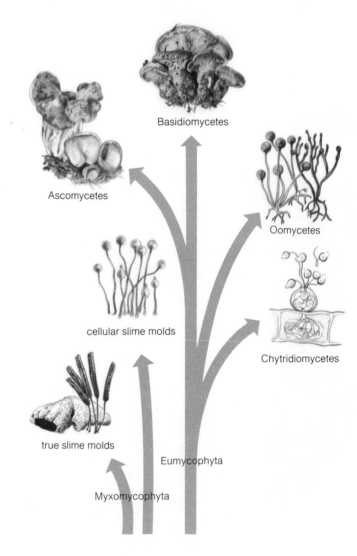

Figure 35–1 Evolutionary lineages in the kingdom Fungi.

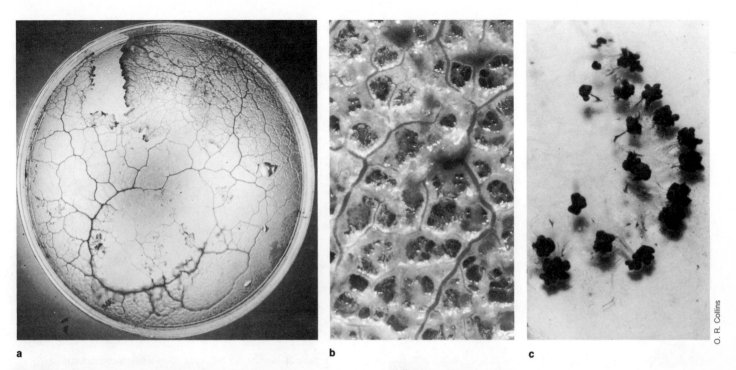

Figure 35–2 (a) The plasmodium of a slime mold, *Physarum*, as it crawls over the surface of agar. (b) A higher magnification of a plasmodium showing the detailed structure of the moving mass. (c) The fruiting bodies of *Physarum*.

Phylum Myxomycophyta: Slime Molds

The Myxomycophyta, or slime molds, lack cell walls except in their fruiting bodies. They exist as multinucleate masses or as uninucleate naked cells in moist soil, wood, decaying vegetation, or dung, and they ingest solid food particles; but their fruiting bodies and spores are similar to those of the Fungi (Figure 35–2). Thus, they sometimes resemble animals and sometimes fungi. One of the founders of mycology, the German botanist Heinrich de Barry (1831–1888), considered slime molds to be animals. Some modern scientists agree with him and have even placed them in the Protozoa. Yet most investigators who have worked intensively with them believe they are Fungi and place them in a separate phylum, as we have done here.

True Slime Molds A prime example of the class Myxomycetes is the common slime mold *Physarum polycephalum* (Figure 35–3). This yellow, netlike fungus can often be seen beneath the bark of rotting tree trunks. If some of the naked protoplasm or *plasmodium* network is kept moist, it can be taken back to the laboratory, grown on moist towels, and fed oatmeal. The plasmodium will engulf pieces of the oatmeal just as it does bacteria, wood

particles, and other organic matter in nature. This method of assimilating food is identical to the phagocytosis of amoebae. Liquids are taken into the plasmodium by pinocytosis. The plasmodium moves slowly over the substrate in response to gradients of moisture and dissolved nutrients.

When conditions for vegetative growth become unfavorable (for example, dry and cold), the plasmodium may produce a resistant structure, the *sclerotium,* which has a hard outer cover and may live for seveal years. Similar environmental conditions, together with decreasing food supply and other unknown factors, may stimulate the plasmodium to form spores instead of a sclerotium.

When sporulation is about to take place, the plasmodium creeps to an exposed location and forms several masses of protoplasm that become the sporangia. Each sporangium consists of a stalk and a spherical or rounded tip in which the spores form (Figure 35–3). Meiosis occurs in the spores, which are surrounded by a rigid wall.

Upon germination, the spores of *Physarum polycephalum* form *myxamoebae,* which are similar to true amoebae and move only by amoeboid motion. In other species, *swarm cells* are formed. These cells may either swim with flagella or move by amoeboid motion. Both myxamoebae and swarm cells can feed, divide, and form resistant cells. They also function as gametes. In *Physarum,* two myxamoebae fuse to form a zygote (Figure 35–3). The zygote

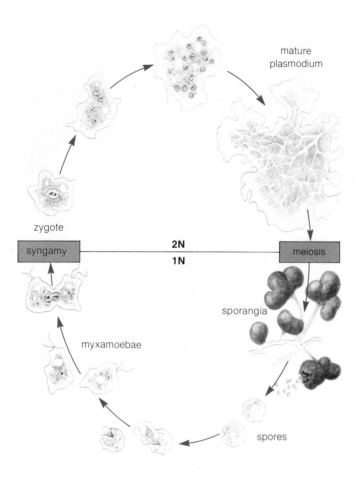

Figure 35–3   The life history of *Physarum polycephalum*.

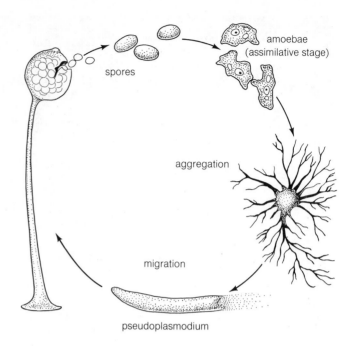

Figure 35–4   The life history of *Dictyostelium discoideum*.

nucleus divides mitotically in a matter of hours, and the daughter nuclei continue to divide (usually synchronously), forming a plasmodium. A plasmodium may form from a single zygote or from several zygotes; most commonly, small plasmodia may coalesce.

The sporangia of the true slime molds are small, beautiful objects in a range of forms. Many consist of a filamentous stalk supporting tiny spherical, cylindrical, or feathery heads. In others, the stalks are missing, and the sporangia consist of groups of globular or cylindrical bodies.

Cellular Slime Molds   Not all slime molds have true plasmodia. The ones that do not are called the cellular slime molds. The best understood of these is *Dictyostelium discoideum*, which has proved to be an excellent organism for the study of various developmental processes (Figure 35–4).

The vegetative stage of *Dictyostelium* is not a plasmo-

dium but the free-living myxamoebae. These independent cells feed and divide until the population density reaches a certain point. Then some of the cells begin to secrete a hormonelike substance called *acrasin*, which is actually cyclic adenosine monophosphate (cAMP), formed from ATP. Acrasin attracts nearby amoebae to congregate, and as the amoebae come together, more of them produce acrasin, so that amoebae are attracted over an ever widening area. As the cells come together, they form a conical aggregation called a *pseudoplasmodium*. Unlike the cells in true plasmodia, the cells in pseudoplasmodia retain their individuality, and feeding does not occur.

In *Dictyostelium*, the pseudoplasmodium, which is known as a slug, migrates. After a period of migration, marked by a characteristic trail of slime, the slug stops and rises upward. A sporocarp consisting of a stalk, a basal disk, and a bean in which the spores are produced is formed by a remarkable series of interactions of the individual cells. The cells that formed the tip of the slug form a stalk and secrete a cellulose sheath that is strong enough to support the rest of the slug. This pulls itself up the stalk by adding new cells to the stalk tip. The cells that remain after the stalk is complete form a globular mass at the apex of the stalk. These form the spores, each of which is surrounded by a thin cellulose wall. Experiments have shown that, although each cell is independent in the slug and all are morphologically similar,

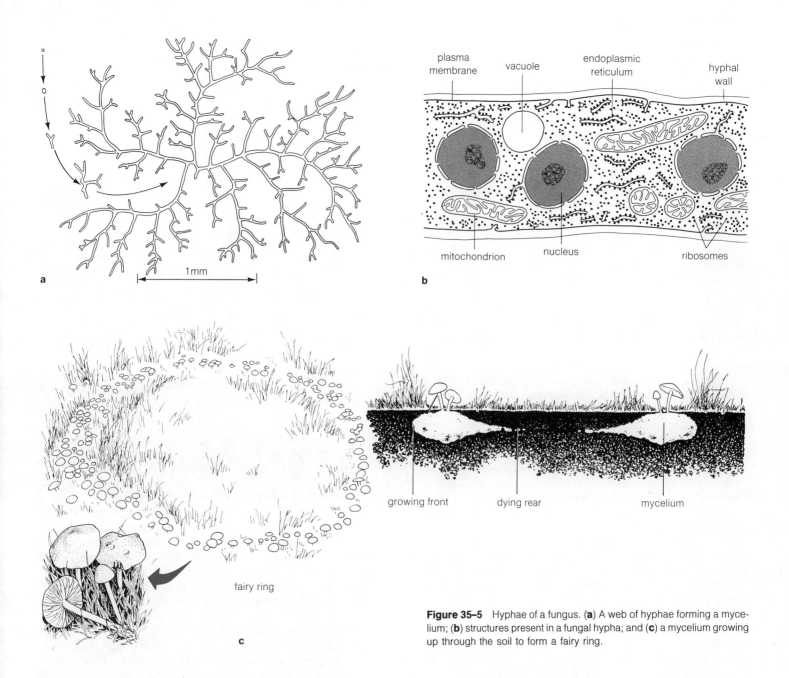

plasma membrane   vacuole   endoplasmic reticulum   hyphal wall

mitochondrion   nucleus   ribosomes

a

b

growing front   dying rear   mycelium

fairy ring

c

Figure 35–5 Hyphae of a fungus. (**a**) A web of hyphae forming a mycelium; (**b**) structures present in a fungal hypha; and (**c**) a mycelium growing up through the soil to form a fairy ring.

the position of a cell in the slug determines its position in the fruiting body. Thus, the cells in the slug have some means of interacting so that they retain their correct position in the differentiated slug.

Upon germination, the spores form myxamoebae. Where—and whether—sexual reproduction takes place in the cellular slime molds is not known. The evolutionary relation of the cellular to the plasmodial slime molds is also unknown.

Phylum Eumycophyta: True Fungi

The Eumycophyta are subdivided into a number of classes on the basis of mode of reproduction, cellular structure, number, type, and position of the flagella (which are similar to those discussed for the algae), and composition of the cell wall. The most important classes are the Chytridiomycetes, the Oomycetes, the Zygomycetes, the Ascomycetes, the Basidiomycetes, and the Deuteromycetes (or Fungi Imperfecti).

The basic body unit of the Eumycophyta is a microscopic filament called the *hypha* (plural *hyphae,* from *hypha,* "web"). These filaments (Figure 35–5), although small in diameter, may be many millimeters or even centimeters long. Hyphae branch and are interwoven in masses that form the body structure of the fungus, the **mycelium** (from *mykes,* "fungus"). There is little differentiation or specialization of the mycelial filaments. Typically, the main mycelium is embedded directly in matter being absorbed as food; only the reproductive structures are exposed. This is the basis of the fairy rings of mushrooms: The mycelium is present in the lawn, but the presence of the fungus is not detected until the fruiting bodies, the mushrooms, are formed (Figure 35–5).

Class Chytridiomycetes: Water Molds The Chytridiomycetes are water molds characterized by motile cells with a single flagellum and by cell walls composed predominantly of chitin.

Three major cell-wall compounds are found in fungal walls: *cellulose, chitin,* and *glucan* (Figure 35–6). Cellulose, discussed earlier, is found in walls throughout the plant kingdom. Chitin is much like starch, but the basic unit is a glucose amine, which contains nitrogen. Glucan is also a polysaccharide and, like cellulose, has glucose as the basic repeating unit. The way in which the glucose molecules are joined together is different than in cellulose, resulting in a distinct compound.

Most of the Chytridiomycetes (chytrids, for short) are extremely small single cells, although some are filamentous. Many chytrids are parasites of filamentous and unicellular algae, as well as other fungi—including even other chytrids. One is *Rhizophydium planktonicum,* which parasitizes a planktonic diatom *(Asterionella formosa)* (Figure 35–7).

The unwalled motile zoospores, swimming by means of their posterior whiplash flagella, attach themselves to the diatom cell. Each settled zoospore forms a wall and grows a fine extension, or *haustorium,* into the host cell. Several zoospores may attach themselves to the same diatom, which they rapidly kill by draining it of its nutrients. The attached spherical zoospore enlarges, and a number of new unwalled, or naked, zoospores are produced. The new zoospores are released when part of the original wall dissolves. This chytrid, common in lakes, is specific about its host and attacks only one species of diatom.

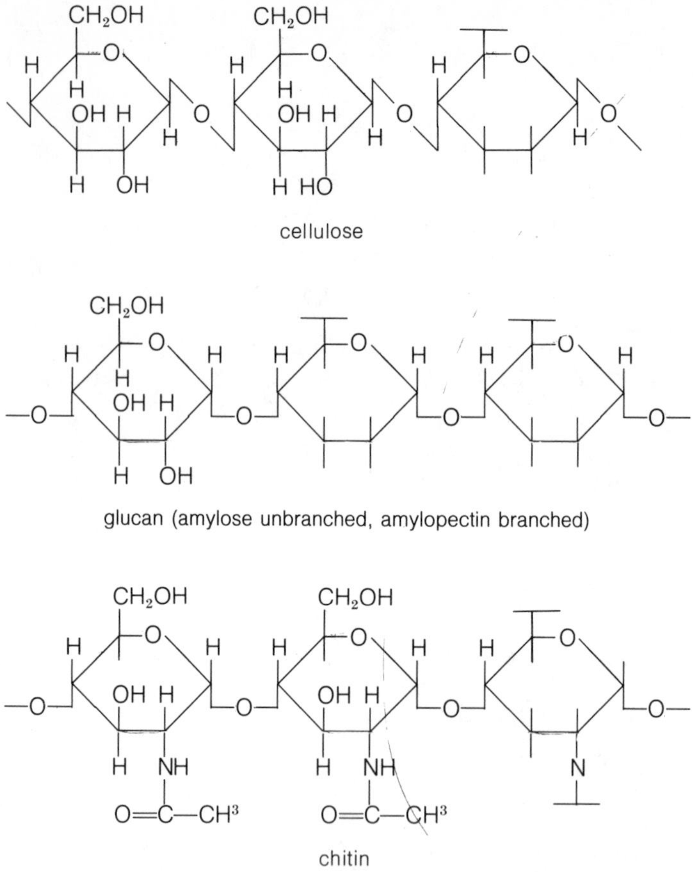

Class Oomycetes: Mildews The Oomycetes are another structurally varied class of Fungi, containing both saprophytic and parasitic species and both aquatic and

cellulose

glucan (amylose unbranched, amylopectin branched)

chitin

Figure 35–6 Compounds found in the cell walls of Fungi.

terrestrial forms. One of the major characteristics of the class is the presence of zoospores lacking cell walls and possessing two flagella, both anterior. Another significant feature is the presence of cellulose and the absence of chitin in the wall of the mature cells.

The effective and deadly parasites among the Oomycetes include *Phytophthora infestans,* the late blight of potato, and *Plasmopara viticola,* the downy mildew of grapes that has caused tremendous losses in vineyards. *Peronospora parasitica* causes downy mildew on a wide range of hosts and is a pest commonly found on plants in home gardens. *Albugo candida* is the white rust on cabbage and related plants.

Plants become infected with downy mildew when airborne spores land on the young leaves and germinate in a drop of dew or rain. Upon germination, the spore may give rise to a biflagellate zoospore, which forms a wall and then in turn germinates to form a sporangium. Alternatively, the spore may directly form a hypha. The hypha grows along the surface of the leaf and into it through a stoma. Once inside, the hypha forms a myce-

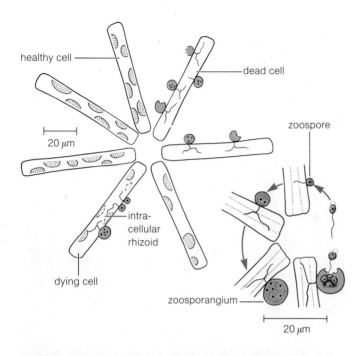

Figure 35-7 *Rhizophydium planktonicum,* a chytrid that is a parasite on diatoms.

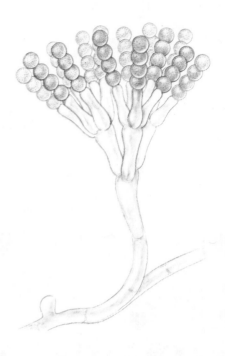

Figure 35-8 Conidia spores.

lium of unbranched hyphae that are multinucleate and lack cross walls. Large, branched haustoria are put forth into the host cells to absorb nutrients for use by the fungus. The more virulent mildews rapidly kill the host cells; the milder forms allow the host plant to survive.

After the period of vegetative growth, the fungus develops straight hyphae that emerge from the stomata and grow outward. The ends of these hyphae branch several times, and the tip of each branch enlarges to form a spore. These are called *conidia spores,* and the structures that produce them are *conidiophores* (Figure 35–8). The masses of conidiophores on the diseased plant give it the appearance we recognize as mildewed.

Conidia spores are a most effective means of asexual reproduction and can rapidly carry a disease through a crop. Sexual reproduction is also found in the powdery mildews. The male and female gametangia are produced within the larger intercellular spaces of the host. The thick-walled zygote is the resting, resistant stage and is released only after the decay of the host plant. When the zygote germinates, it may either divide and give rise directly to zoospores or first produce a short filament, and then form zoospores.

Class Zygomycetes: Black Bread Molds The best-known member of Zygomycetes, is the black bread mold

Rhizopus stolonifer. Like all the Zygomyetes, it is a terrestrial fungus completely lacking flagellated cells. The hyphae have cross walls, and chitin is a predominant component of the cell wall.

Rhizopus can be found on almost any piece of moldy bread or soft fruit (Figure 35–9). There it is a saprophyte— the hyphae secrete enzymes that help break down the substrate so that it can be absorbed by the fungus. But *Rhizopus* can also be found causing soft rot of sweet potatoes and strawberries. Here, the secreted enzymes kill the cell and partly dissolve the wall in advance of the growing hyphae. No haustoria are formed; the hyphae simply grow throughout the rotten tissue. Strictly speaking, *Rhizopus* in this case is still a saprophyte, but it kills the tissue it is using.

The structure of *Rhizopus* is more complex than it first appears. One part of the hypha grows rapidly over the surface of the substrate. At some places, hyphae extend into the substrate. Opposite these, hyphae differentiate sporangia that give rise to numerous dark-colored spores. These spores are light enough to float in the air for long periods and great distances.

Sexual reproduction occurs when the hyphae of two different strains (called plus and minus) grow together. Short lateral branches develop and come together; then their ends are cut off by cross walls to form multinucleate gametangia. These fuse as the wall dissolves, allowing

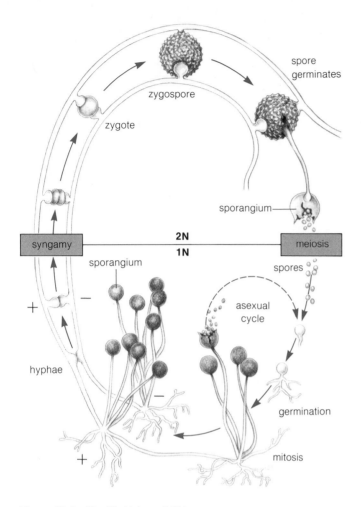

Figure 35–9 The life history of *Rhizopus*.

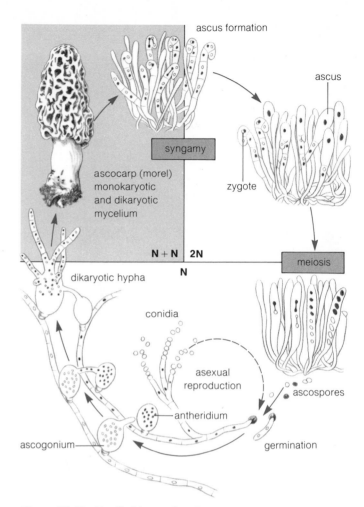

Figure 35–10 The life history of an Ascomycete.

the gametangia from the plus and minus strains to meet. The multinucleate protoplasts fuse, and one nucleus from the plus strain and one from the minus strain pair and fuse, giving rise to a 2N zygote. This is repeated many times, so that the resulting structure is a multinucleate zygote. It develops a thick, warty wall and becomes the dormant stage of the organism, capable of withstanding adverse environmental conditions. When the multinucleate zygote germinates, the nuclei first undergo meiosis; then a single sporangiophore is formed and produces haploid spores.

Class Ascomycetes: Penicillium, Yeasts, and Other Sac Fungi The yeasts responsible for fermentation are members of the Ascomycetes, a great class of Fungi comprising some 30,000 species and containing forms as diverse as

Penicillium, the deadly ergot, and various wood-rotting fungi. Despite all this diversity, however, the Ascomycetes are a long-recognized natural class, united by several similarities. None of the Ascomycetes have motile cells, and all have walls containing chitin and insoluble glucans. In most cases (the yeasts are an exception), the hyphae are subdivided by cross walls that have a central pore. This allows cytoplasm to move from one cell to the next; not only cell organelles such as mitochondria but also nuclei have been seen to migrate from cell to cell. The resulting cells may have one nucleus or many. The sexual spores are produced on special structures called *asci* (singular, *ascus*).

In most Ascomycetes (Figure 35–10), sexual reproduction is based on the germination of different kinds of spores that look identical but actually represent the different genetic strains, usually designated plus or minus. Under the proper conditions, a branch of one hypha enlarges,

Figure 35-11 An ascocarp of *Sarcoscypha coccinea*, called scarlet cup because the inside is a deep red.

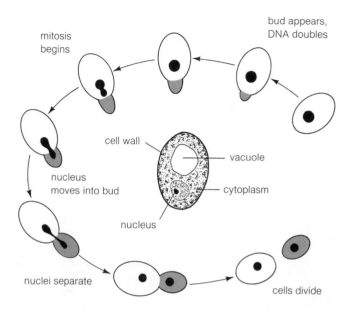

Figure 35-12 A yeast cell reproduces asexually by developing a small bud.

and many nuclei accumulate in the swollen tip, called an *ascogonium.* The tip of the second hypha enlarges only slightly, but again many nuclei are concentrated in it. The smaller tip, the *antheridium,* fuses with the larger, and the nuclei of the plus and minus strains mingle. From this enlarged, swollen hypha many new hyphae form. As these hyphae grow, new cells are constantly being formed. The cells closest to the fused hyphal mass have several nuclei. As the tips of the hyphae grow farther and farther away, the cells become increasingly binucleate—always in pairs of plus and minus nuclei. Each of these contains two haploid nuclei, so that the cells function as if they were diploid. A cell that contains two paired nuclei with one from each part is called *dikaryotic.* Eventually, the two nuclei of the subterminal cell or ascus fuse, forming a single diploid nucleus, the zygote nucleus. Almost immediately, this nucleus undergoes meiosis, and the resulting four nuclei divide mitotically, forming eight haploid nuclei. At the same time, the ascus is elongating; eventually the eight nuclei will line up in a neat row. A wall is formed around each nucleus so that eight haploid *ascospores* are formed. The hyphae producing the ascospores are surrounded by many hyphae that do not produce spores and that are usually *monokaryotic.* This entire hyphal mass composes a fruiting body or *ascocarp* (Figure 35-11).

Sexual reproduction is not common in *Penicillium,* but it is common in such Ascomycetes as *Neurospora* (pink bread mold).

One of the better known of the Ascomycetes is the yeasts. Many claims have been made for yeasts: They are one of the most important of all domesticated organisms and they were the first to be domesticated. *Saccharomyces cerevisiae* (baker's or brewer's yeast) is grown around the world, in complex modern bakeries and breweries and in earthen pots containing dough or mash. The Egyptians and other ancient or primitive peoples used yeasts without understanding that they were dealing with a microorganism. Anton van Leeuwenhoek was the first to see yeasts (in 1680), but he did not recognize that they were alive. In the 1830s, various biologists discovered that yeasts were living organisms, but it was Louis Pasteur who first established (in 1857) their importance in fermentation.

Yeasts are single-celled and simply organized, but the electron microscope shows them to be eukaryotic cells. A yeast reproduces asexually by producing a small protuberance or bud on the side of the mother cell. The nucleus divides as the bud is forming, and one daughter nucleus moves into the bud (Figure 35-12).

The *Penicillia* fungi are famous as the source of the first effective antibiotic, penicillin. Other species of *Penicillium* are also important to humans: *Penicillium roquefortii* and *Penicillium camemberti* are responsible for much of the flavor, texture, and color of Roquefort and Camembert cheese. The veins of blue that characterize Roquefort or blue cheese are due to the spores produced by the fungus.

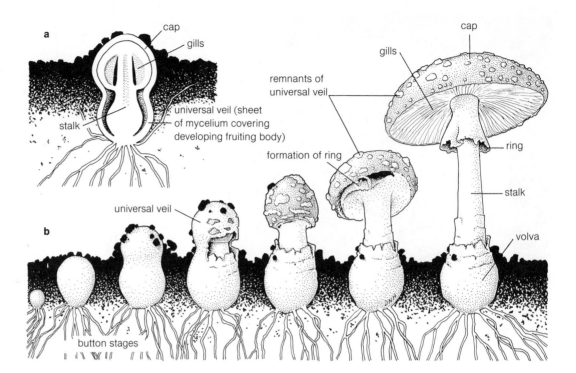

Figure 35–13 (**a**) The development and structure of the fruiting body in Basidiomycetes. (**b**) The stages of development are shown for a species with a universal veil, but this structure is not present in all species.

Class Basidiomycetes: Mushrooms and Rusts The Basidiomycetes contain many of the Fungi of which we are commonly aware. There are some 13,000 species of Basidiomycetes in a vast array of forms. These include the common mushrooms, the wood-rotting shelf fungi, as well as the black-stem wheat rust *(Puccina graminis)* that causes enormous losses in wheat crops each year.

Let's look at the life cycle in the most familiar of the Basidiomycetes, the common mushroom, starting with the germination of some of the millions of spores produced by a single mushroom. The spores that land on moist soil will germinate rapidly and form mycelia consisting of branched hyphae that contain cross walls with pores in them. The hyphae fuse with one another, forming a three-dimensional network. The cells in the hyphae of newly germinated spores are uninucleate, but soon two hyphae fuse. The cytoplasm, but not the nuclei, fuses to form binucleate, or dikaryotic, cells.

The formation of the dikaryotic cells is really the start of a process of sexual reproduction that will not end until the fruiting body (the mushroom that we eat) is formed and the nuclei fuse. The structure and development of the fruiting body are shown in Figure 35–13. In a mushroom, gills line the lower surface of the cap, where the spores, called *basidiospores,* are formed.

Along the surface of the gills are many hyphal ends,

the terminal cells of which enlarge, forming the *basidium.* The two nuclei present in the basidium fuse, forming true diploid nuclei (see Figure 35–14). These nuclei, representing the zygote stage, immediately undergo meiosis. The cells themselves continue to enlarge, and the four nuclei that are produced by the meiotic division in each cell migrate to the outer edges of the cells. Walls are then formed around the nuclei. The spores are discharged a distance some 10 to 20 times their diameter. This gets them free of the hyphae and into the space between the gills, where they drop away from the cap. Once below the cap, the spores can travel for miles on air currents.

Class Deuteromycetes: Fungi Imperfecti The 10,000 species of Deuteromycetes are called *imperfect* because they do not reproduce sexually, and the sexual phase is considered the perfect stage. In effect, the grouping is a taxonomic dustbin into which fungi are dumped until they are studied more thoroughly. If a sexual stage of a member is observed, it is placed in the appropriate phylum. Many mycologists believe that most of the Fungi Imperfecti are really Ascomycetes, if their vegetative characteristics are the criteria.

The Fungi Imperfecti include *Trichophyton interdigitale,* the cause of athlete's foot, as well as several other

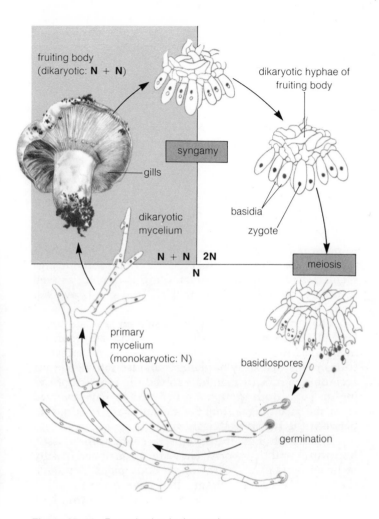

Figure 35–14   Reproduction in the mushrooms.

Labels in figure: fruiting body (dikaryotic: **N** + **N**); dikaryotic hyphae of fruiting body; gills; syngamy; basidia; zygote; dikaryotic mycelium; N + N | 2N; N; meiosis; primary mycelium (monokaryotic: N); basidiospores; germination

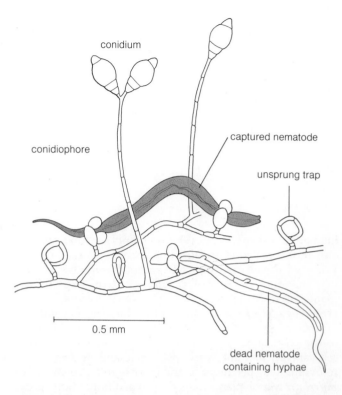

Figure 35–15   *Dactylella* is a fungus that actually traps nematodes and then grows into their bodies.

Labels in figure: conidium; conidiophore; captured nematode; unsprung trap; 0.5 mm; dead nematode containing hyphae

human parasites. Cat and dog ringworms are also Fungi Imperfecti, as are head and skin ringworms of humans.

On the brighter side, one of the Fungi Imperfecti traps the nematodes or eelworms that are major pests of the roots of orchard trees and many crops. *Dactylella bembicodes* captures nematodes with a trap consisting of a three-celled ring attached to the mycelium by a short cell acting as a stalk (Figure 35–15). The nematode enters the ring, triggering the cells, which rapidly increase in diameter. The harder the nematode struggles, the tighter it is held. Hyphae then grow into the nematode, using its body as a food source. *Dactylella* can grow in the soil without nematodes to prey upon, but if nematodes appear, or if the fungus is placed in water in which nematodes have been growing, it will rapidly form traps. In other words, the prey itself induces the formation of the trap. Researchers had hoped *Dactylella* would prove an effective biological control for nematodes, but thus far it has not.

Summary

The kingdom Fungi contains organisms both beneficial and harmful to humans. All fungi are eukaryotic, heterotrophic organisms. They are all either saprophytes or parasites. The fungi are divided into two phyla: the slime molds, or Myxomycophyta; and the true fungi, or Eumycophyta. The slime molds lack true cell walls in the vegetative state but may form them during reproduction. The true fungi all have cell walls and vary greatly in form, habitat, chemical composition, and mode of reproduction. There are few places where Fungi will not grow and few things that some species cannot use as food. The Fungi are particularly important in the breakdown of dead organic matter and, thus, are critical to the recycling of minerals.

36

Kingdom Plantae

Like our ancestors, we depend on plants for food, shelter, and clothing. Plants are also important to us esthetically: We enjoy their form, color, and scents. It is hardly surprising, therefore, to find that humans have long studied plants, often going to extraordinary lengths to collect and grow them. One of the first plant-hunting expeditions of ancient Egypt is magnificently recorded on the walls of the great temple at Karnak: Queen Hatshepsut's great sea caravan to Punt (perhaps Somaliland or Abyssinia) to obtain live specimens of the incense tree *(Commiphora myrrha)*. Many plant-collecting expeditions, both great and small, have followed over the years, and great collections have been made of both living and preserved specimens. Work has continued in the laboratory, and data on the microscopic and chemical composition of plants have accumulated. We now have a good—although by no means complete—idea of the plant kingdom and the major relationships within it.

An Overview of Plants

There are estimated to be some 300,000 living species of plants on earth at the present time. Logically classifying such an assemblage is a formidable task, and it is not surprising that various biologists have arrived at different ways to do it. We have picked one scheme that makes sense to us (Table 36–1 and Figure 36–1), but we hasten to add that the data can be evaluated in several ways. The way you choose determines whether a scheme will look right or wrong to you. Organisms do not care how they are classified; only biologists care.

The algae are placed in three phyla differing in the accessory pigments involved in photosynthesis, the composition of their cell walls, the nature of their stored food, and their mode of reproduction. The algae are basically aquatic, although some species are found in the soil and on trees, and are adapted to living in an abundance of water. We will discuss the algae later, but here we want

to point out that, of all the algae, only the Chlorophyta (green algae) have the same types of photosynthetic pigments found in all land plants. On this ground, as well as others, botanists believe that all land plants evolved from the green algae.

The remainder of the plant kingdom is made up of land plants. (Some species now grow in aquatic environments, but they appear to have evolved from terrestrial forms.) This vast array of plants is divided into two phyla according to how its members solved the problem of living on land. Both groups are believed to have evolved from the green algae, but the evolutionary relationship between the two is not clear.

The first phylum, the Bryophyta, contains the mosses, hornworts, and liverworts. They are all small and usually inconspicuous plants with simple body plans. Although they lack the vascular tissues that have made the other phylum of land plants so successful, they are true land plants.

The second phylum of land plants, the Tracheophyta, contains the plants we are most familiar with—essentially all the plants we see around us every day. The members of this great phylum all have vascular tissue, both xylem and phloem. The group is broken down into four subphyla, three of which (Psilopsida, Lycopsida, and Sphenopsida) have few living species and one of which (Pteropsida) has a great many. Although Psilopsida, Lycopsida, and Sphenopsida contain plants that are not important in today's world, their ancestors were dominant species in the geologic past.

The subphylum Pteropsida contains plants that are anything but inconspicuous. True, some are small, like the tiny desert flowers or the minute floating ferns, but this subphylum also includes the giant redwood and the mighty oak. The unifying feature of this group of plants is the presence of true leaves, stems, and roots, and all share common features of growth and development.

The largest classes of the Pteropsida are the Filicineae (ferns), the Coniferophyta (conifers), and the Angiospermae (flowering plants). The major criterion for separating

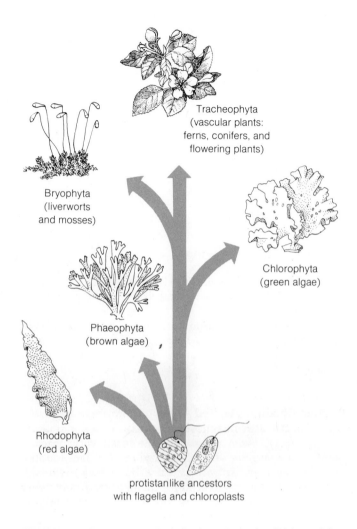

Tracheophyta
(vascular plants:
ferns, conifers, and
flowering plants)

Bryophyta
(liverworts
and mosses)

Chlorophyta
(green algae)

Phaeophyta
(brown algae)

Rhodophyta
(red algae)

protistanlike ancestors
with flagella and chloroplasts

Figure 36–1 Evolutionary lineages among the major divisions of the kingdom Plantae.

Table 36–1 The Plants	
Group Name	Common Name
Phylum: Chlorophyta	Green algae
Phylum: Phaeophyta	Brown algae
Phylum: Rhodophyta	Red algae
Phylum: Bryophyta	
Class: Musci	Mosses
Class: Hepaticae	Liverworts
Class: Anthocerotae	Hornworts
Phylum: Tracheophyta	
Subphylum: Psilopsida	Psilopsids
Subphylum: Lycopsida	Lycopsids
Subphylum: Sphenopsida	Horsetails
Subphylum: Pteropsida	
Class: Filicineae	Ferns
Class: Cycadophyta	Cycads
Class: Coniferophyta	Conifers
Class: Ginkgophyta	Ginkgo
Class: Gnetophyta	Gnetum
Class: Angiospermae	Flowering plants
Subclass: Dicotyledoneae	
Subclass: Monocotyledoneae	

the classes is the way they reproduce and the nature of the sporophyte and gametophyte. Finally, the Angiospermae are divided into two subclasses, the Dicotyledoneae and the Monocotyledoneae. The dicots and the monocots, as they are usually called, differ in a wide range of features, including the arrangement of the vascular tissue in the stem, the basic leaf pattern, the number of flower parts, the number of storage leaves in the seed, and other features as well.

Algae

The various phyla of algae constitute a vast array of plants growing in all types of aquatic and some terrestrial envi-

ronments. They range in size from minute one-celled species to some of the longest plants known. On the whole, they are relatively simple in structure, with many species consisting of single cells, filaments of cells, or colonies. Marine algae are generally more complex than fresh-water species, but even the most elaborate, the giant kelps, are simpler than the land plants.

Because of the several unifying features shared by all algal phyla, it is not surprising that the algae were once classed together in one large group. Typically, algae contain chlorophyll *a* (as do the blue-green algae but not the photosynthetic bacteria) and produce oxygen as a by-product of photosynthesis (photosynthetic bacteria produce sulfur). The algae also have one-celled reproductive structures, with eggs and sperm not surrounded by layers of sterile protective cells. Vascular tissue, composed of cells specialized for carrying water and foodstuffs within the plant, is missing from the algae, except for the kelps (in which only a phloemlike tissue is present).

The marked differences among the various phyla are based on biochemical, morphological, and cytological criteria. The most striking biochemical difference, pigmentation, gives rise to the common names of the algae: green algae, brown algae, and red algae.

All algae have chlorophyll *a,* but many also have a second chlorophyll slightly different in structure. Also present are fat-soluble pigments, the yellow to orange

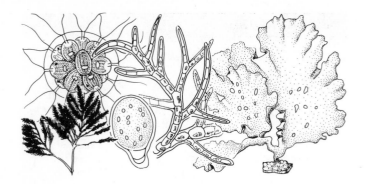

Figure 36–2 Phylum Chlorophyta (green algae).

Figure 36–3 Phylum Phaeophyta (brown algae).

carotenoids (carotenes and xanthophylls), and water-soluble pigments, the blue or red phycobiliproteins (phycocyanin and phycoerythrin). Various modifications of these pigments provide major criteria for classifying the algae.

Phylum Chlorophyta: Green Algae With over 7,000 species, the green algae are the largest of the algal phyla and show a tremendous range in size and shape. The fresh-water green algae are single-celled, colonial, filamentous, or (rarely) leafy. The marine green algae are often larger than the fresh-water ones. They have the same general forms as fresh-water green algae but are more often leafy.

Some of the diversity in form of the Chlorophyta is shown in Figure 36–2. Reproduction in the Chlorophyta was covered in Chapter 10.

Phylum Phaeophyta: Brown Algae The Phaeophyta contain some 1,500 species, all of which (except for those in three genera) are marine (Figure 36–3). Brown algal cells generally have only one relatively large nucleus. The chromosomes contain DNA and histones and are similar to those found in higher plants. The nuclear membrane persists during mitosis, except at the poles, where spindle microtubules originate near paired centrioles.

The cytoplasm contains numerous small vacuoles. The chloroplasts are usually disk-shaped, but plates, stars, and spiral, ribbon-shaped chloroplasts are also found. Laminarin, the food-storage product, is deposited outside the chloroplast and is different from starch.

No brown algae are motile, although motile cells are produced. These generally have two flagella.

The great range in form among the brown algae is based on relatively simple differences in patterns of devel-

opment. The basic pattern is cells arranged in a filament. The more elaborate brown algae consist of complex arrangements of such filaments, or files, of cells.

The brown algae contain the largest and vegetatively most complex marine algae: the giant kelps *Nereocystis* and *Macrocystis.* A life history of a brown alga is given in Chapter 10.

Phylum Rhodophyta: Red Algae There are nearly 4,000 species of red algae in almost 600 genera, almost all marine (Figure 36–4). They are, to say the least, very unusual plants. They are the only algal phylum that lacks motile cells of any sort. They may once have had them and then lost the capacity to produce them, or they may never have evolved them in the first place. As yet, no true centrioles have been observed in the Rhodophyta.

The cells in the simpler red algae are uninucleate, but in the more advanced forms, they are generally multinucleate. In the simpler red algae there may be only one chloroplast. The more complex red algae generally have numerous small, disk-shaped chloroplasts.

The Rhodophyta possess an accessory red pigment in their chloroplasts that absorbs blue light. As light penetrates water, red light is absorbed first, so that in deeper water only blue light is available. In the red algae, the red accessory pigment absorbs blue light and passes the energy on to chlorophyll, which is far less effective at these depths. Thus, the red pigment allows the Rhodophyta to survive at depths many other algae cannot tolerate.

Without doubt, some of the most beautiful plants in the world are found in the Rhodophyta. Many have exquisite red colors in a wide range of hues, and delicate, lacelike forms. However, not all red algae are handsome. Some are single cells or simple filaments, and others are flat sheets of cells.

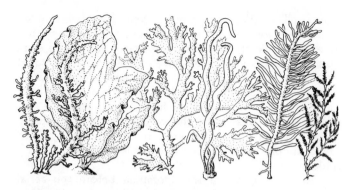

Figure 36–4 Phylum Rhodophyta (red algae).

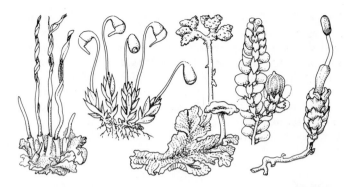

Figure 36–5 Phylum Bryophyta (mosses and liverworts).

Phylum Bryophyta: Mosses, Liverworts, and Hornworts

Everyone is familiar with the mosses growing along the banks of streams, between the bricks of a walk, or on the sides of walls, houses, and flower pots. Their world-wide distribution, tiny leafy shoots, beautiful green color, and softness to the touch have excited interest for centuries. Fewer people are familiar with the liverworts. These may be leafy or flatter in shape, sometimes forming mats of dark green over the soil, particularly in forests and under benches in greenhouses. Still fewer people are aware of hornworts, although they occur worldwide. In fact, awareness of all bryophytes directly parallels their numbers: There are 14,500 species of mosses, 9,000 of liverworts, and 100 of hornworts (Figure 36–5).

Most bryophytes are either terrestrial or epiphytic; only one is truly aquatic. They are generally quite luxuriant in wet, humid regions but are also found widely in northern and arctic areas. The rain forests of the temperate and tropical latitudes are where bryophytes flourish. In the rain forests of the Olympic Peninsula in the American Northwest, the trunk and branches of every tree are covered with bryophytes and lichens—plants that are often found together.

When we see a mat of mosses, liverworts, or hornworts, we are looking at a gametophyte generation. All the cells are haploid, a condition familiar enough in green algae but decidedly uncommon among land plants. In the bryophytes, the gametophyte or haploid (1N) plant is the dominant independent generation; the sporophyte or diploid (2N) plant is small and generally partially dependent on the gametophyte. In vascular plants, the reverse is generally true: The sporophyte (2N) generation is dominant and the gametophyte (1N) generation dependent.

The life history of a moss is presented in Chapter 10. In general, the pattern of reproduction represented by the mosses is typical of liverworts and hornworts as well.

Bryophytes are simply constructed compared with vascular plants, but even the simplest is far more complex than the type of green alga thought to be the progenitor of land plants. A cuticle is often present, and so are pores or stomata. All bryophytes lack the true stems, leaves, and roots that exist in vascular plants, but the functional equivalents are present. Rhizoids with one or many cells take the place of roots, and in many bryophytes there is a main axis bearing small lateral projections involved in photosynthesis.

Phylum Tracheophyta: Vascular Plants

Some 400 million years ago, the shallow marshes on the edges of the ancient seas were filled with mats of slender, erect green plants. The first plants to dominate the land included the Psilopsida (psilopsids), Lycopsida (club mosses), and the Sphenopsida (horsetails). All these members of the early Tracheophyta have features in common (Figure 36–6). All are true land plants with stems containing well-differentiated xylem and phloem. The sporophyte is the dominant, independent generation, with the gametophyte only rarely seen and often subterranean. In the sporophyte, the stem is the principal organ; both roots and leaves are poorly developed. The leaves are missing in some of these plants and scalelike in others. All have motile sperm and true embryos.

There are also differences among the various groups—differences in vegetative structure and methods of reproduction. To understand these ancient and interesting plants, let us examine each class separately.

Subphylum Psilopsida: Psilopsids The psilopsids are represented by two genera: *Psilotum,* with 2 or 3 species, and *Tmesipteris,* with 8 to 10 species. These are extremely simple vascular plants. The psilopsid most commonly seen in botany classes is *Psilotum nudum,* since it grows well in greenhouses. This plant, which is native to the tropics and subtropics, consists of a stem that is erect and branched above ground, but underground becomes a rhizome (Figure 36–7). Photosynthesis occurs in the green cells of the above-ground stem. There are pointed emergences on the stem, but these are not leaves (nor does the bulk of photosynthesis occur here). The stem has an epidermis with well-developed stomata and guard cells. The parenchyma cells of the cortex immediately below the epidermis are rich in chloroplasts and have large intercellular spaces that interconnect with the stomata. Below this layer, the cortex consists of closely packed parenchyma. The center of the stem is occupied by a simple stele composed of xylem surrounded by phloem.

Meiosis occurs in the sporangia before the aerial spores are produced (Figure 36–8). All the spores are the same size. The gametophytes, which are formed on the germination of the spores, lack chlorophyll and are subterranean. Antheridia and archegonia are produced on the same gametophyte; the sperm have several flagella.

Subphylum Lycopsida: Lycopods Like the psilopsids, the lycopsids are ancient plants dating back to the Lower Devonian. Today, only five genera survive, and all are herbaceous. In the Upper Devonian and Carboniferous periods, there were large numbers of lycopsids, many of them trees up to 30 or 35 meters tall. The smaller herbaceous lycopsids were also present at the time, and the

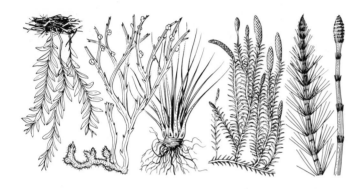

Figure 36–6 Subphyla Psilopsida, Lycopsida, and Sphenopsida (psilopsids, lycopsids, and horsetails).

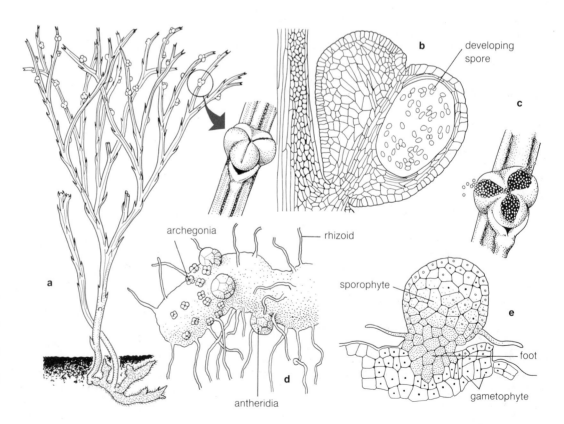

Figure 36–7 The details of the life history of *Psilotum.* The haploid spores are produced following meiosis of the cells in the lateral sporangia (**a** and **b**). These are released when the sporangia open (**c**) and germinate fo form the subterranean gametophyte (**d**). Antheridia and archegonia form on the gametophyte, and after fertilization the new sporophyte begins its development on the gametophyte (**e**).

fossils of some of these are similar to present-day species.

All the lycopsids have a basic pattern of dichotomous branching (in which the two arms are more or less equal), and all have some type of true root and true leaves (Figure 36–8). The sporangia may be grouped together into cones, or they may be spread out along the stem.

The species of the genus *Lycopodium* growing in the

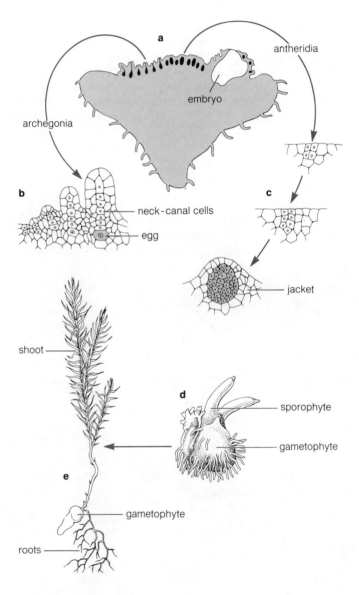

Figure 36–8 Life history of a lycopsid. A longitudinal section of the gametophyte (**a**) shows the position of the antheridia, archegonia, and one embryo. The stages in the development of an archegonium (**b**) and an antheridium (**c**) are also shown. The further growth of the embryo into a mature sporophyte is shown (**d** and **e**).

temperate zones are generally erect, shrubby plants. Probably the most familiar is the one commonly known as ground pine, found in northern temperate forests.

Subphylum Sphenopsida: Horsetails   The evolutionary history of the Sphenopsida closely parallels that of the Lycopsida. The earliest sphenopsid (horsetail) fossils are found in the Devonian, and during the Carboniferous these plants grew worldwide. It was also during the Carboniferous that they reached their greatest diversity in shape and size, ranging from small, herbaceous species (some similar to those existing today) to large treelike species 15 to 25 meters high. During the Permian the larger species began to disappear, and by the end of the Triassic only a few herbaceous species were left. Today's one surviving genus, *Equisetum,* may be one of the oldest living vascular plant genera, essentially unchanged since the Carboniferous period.

The appearance of the sphenopsids is characterized by the presence of whorls of small leaves or branches (Figure 36–9). This gives the stem a jointed look. The stem is the main photosynthetic unit of the plant. In some species, what appear to be long needlelike leaves are actually whorls of thin branches.

The stem is marked by ridges of tissue composed of cells with thick walls rich in silica. The cells make the plant useful for scouring pots (as American Indians and early European settlers knew), but can injure livestock that eat them.

Subphylum Pteropsida: Ferns   The ferns are part of the subphylum Pteropsida, which also includes the conifers and flowering plants, among others. The ferns themselves are in the class Filicineae (Figure 36–10).

The ferns are a large, successful group of plants but have never dominated the world's flora. The Carboniferous period was once called "the age of ferns" because of the large number of fernlike fossils dated to that period. Later investigations have shown that many of these fossils were actually lycopsids, sphenopsids, and early gymnosperms (including the seed ferns). This is not to say that ferns were not abundant—they were; but they were not, as they are not now, large conspicuous plants that dominated the flora. The ferns of the Carboniferous grew under the large tree lycopsids, sphenopsids, and gymnosperms, as they now grow under the redwoods and other forest trees. And then as now, they successfully filled these particular environmental niches.

Like the lycopsids and sphenopsids, the ferns evolved in the Devonian, flourished during the Upper Devonian and Carboniferous, and declined during the Permian. But from there on, the ferns show a different pattern. Some primitive ferns persisted, but new fern species and families evolved during the Mesozoic and Cenozoic and are apparently continuing to evolve today, particularly on tropical islands.

As a result of this long history and evolutionary adaptability, the ferns show a wide range of characters, some resembling those of lower vascular plants and others more advanced. Many ferns have underground stems or rhizomes and, in general, lack elaborately developed root systems. All have vascular systems with little or no secondary growth. All reproduce by spores and have independent gametophytes that produce sperm dependent on free water to reach the egg.

The leaves of ferns vary enormously in size and shape. The familiar fern leaf is a large, compound leaf or *frond* (Figure 36–11). It consists of a stemlike stipe and a blade, which may be divided into pinnae that may, in turn, be divided in a number of ways. When such compound leaves are formed, often underground, the stipe and pin-

nae are tightly coiled. Most of us have seen the coiled frond, called a "fiddlehead," uncoiling gracefully in the spring (Figure 36–12), as the cells on the lower side elongate faster than those on the upper side. In many ferns the leaves are annual, but several immature leaves, which will develop in future years, may already be developing underground.

The stems of most ferns are either underground (rhizomes) or short erect structures near ground level. In a few species, however, the erect stems may be barrel-like or quite tall, as in the tree ferns.

The rhizomes of ferns play an important role in the spread of the plant. Growing through the soil, they are relatively immune to drying out or being eaten by animals. By branching, they can rapidly spread and provide stiff competition to grasses and other plants. New Zealand farmers and ranchers have spent generations trying to eradicate the bracken fern, since it crowds out the grass

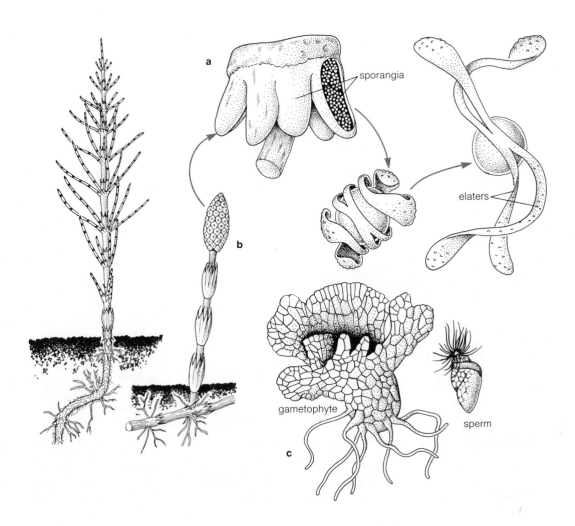

Figure 36–9 Details of the gametophyte and sperm in *Equisetum*. The sporangia are clustered on scales (**a**) that are grouped into cones (**b**). The spores themselves have long extensions of their outer walls (the elaters) that aid in their dispersal (**c**).

on open hillsides, ruining the pasture. Trying to dig out
or grub up the rhizomes merely spreads the plants, and
trying to force the sheep to overgraze the ferns usually
leads to sick sheep. The most successful method involves
cutting the fronds back repeatedly so that the starch reserves in the stem will be depleted and the plant will
starve. Some ferns have rhizomes so tough that they cannot be pulled out by hand, and wire cutters must be
used to sever them.

Ferns have small, inconspicuous roots arising from
the stem near the leaves. These are adventitious roots;
they originate from the pericycle of the stem, forcing

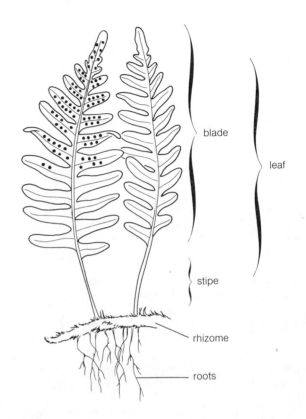

Figure 36–11   A large, compound fern frond.

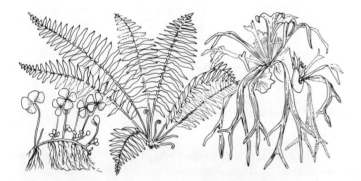

Figure 36–10   Class Filicineae (ferns).

Figure 36–12   Fiddleheads.

their way through the tissues of the stem as they develop. A few ferns—some of the water ferns and the delicate, filmy ferns of the tropics—have lost their roots in the course of evolution and are now rootless.

Progymnosperms   This extinct group of ancient plants has only recently been recognized. Paleobotanist Charles Beck of the University of Michigan showed that fossil leaves thought to belong to a fern were actually attached to a gymnosperm-type trunk. The progymnosperms have now been established as important plants in the late Devonian and Carboniferous forests, and this knowledge has added a new dimension to our understanding of the past.

Seed Plants   All pteropsids except ferns are seed plants. The seed as a vehicle of reproduction appears to have evolved independently several times. It is a self-contained unit capable of dispersing and establishing the next generation with the maximum probability of success. It contains not only a small edition of the sporophyte, ready to grow, but a built-in supply of food and a protective outer covering to minimize the impact of adverse environmental conditions.

The seed ferns are an example of seed-bearing plants that have become extinct. The gymnosperms and angiosperms, however, are currently the dominant plants in our environment.

The gymnosperms and angiosperms have other adaptive features as important as seeds for life on land. Free water is no longer required for fertilization because of the evolution of pollen and pollen tubes. There are great advances in the structure of the vegetative sporophyte and the reduction of the gametophyte to a structure nurtured and protected by the massive sporophyte. We can appreciate the great diversity of plants with such adaptations by examining the gymnosperms and the angiosperms.

Subphylum Pteropsida: Gymnosperms   The term *gymnosperm* is an ancient one, first used by Theophrastus around 300 B.C.: "Some seeds again are enclosed in a pod, Some in a husk, Some in a vessel, and some are completely naked." *Gymnosperm* in Greek means "naked seed."

All gymnosperms are trees or shrubs. Moreover, as their name implies, gymnosperm seeds are uniformly born uncovered on the plant, not contained within an ovary as in flowering plants. (Looks alone can be misleading here, since the seeds of pines, which are gymnosperms, appear similar to the seeds of many flowering plants.) Yet gymnosperms vary so greatly in form and in the details of their reproduction that many botanists no longer

feel they are a natural group and have separated them into the following four classes: Cycadophyta, Ginkgophyta, Coniferophyta, and Gnetophyta.

Reproduction in pines, treated in Chapter 10, can be taken as the general pattern in the gymnosperms, with the understanding that there are marked differences in details in the various groups.

When the dinosaurs ruled the earth during the Triassic and Cretaceous periods of the Mesozoic era, the Cycadophyta, or cycads, grew worldwide, and their young leaves and seeds most likely provided food for some dinosaurs. The cycads and the dinosaurs reached their maximum development at the same time, and although the cycads did not become totally extinct, both decreased in importance in the Cenozoic era. The living cycads number only 9 genera and roughly 100 species, all confined to the subtropics and tropics.

The cycads are striking plants (Figure 36–13). They have large, attractive, palmlike leaves that may be 1 to 2 meters long and stems that are usually barrel-shaped but may, in some species, reach a height of 10 to 20 meters. Cycads grow extremely slowly: One plant, some 2 meters tall, was calculated to be at least 1,000 years old!

The reproductive cycle in the cycads closely follows the general pattern discussed earlier for gymnosperms. The male and female cones are borne on separate plants. Except for one genus, *Cycas,* cycad cones are tightly organized (Figure 36–14). The female cone in some species is tremendous, weighing 25 to 35 kilograms. The male cones are smaller but still respectable in size.

The Coniferophyta, or conifers, are found in the vast forests of pine, spruce, fir, cedar, and other conifers that stretch for miles across western North America, eastern and central China, parts of Australia and New Zealand, and in the great northern boreal forests that encircle the globe. Smaller stands of conifers are found in most parts of the world. Individual conifers (shrubs and trees) are widely planted as ornamentals, and ancient individual trees were once honored with shrines in China. The conifers are a truly magnificent group of plants (Figure 36–15). Their evolutionary history is longer than that of the cycads, but whereas the cycads are essentially living fossils, the conifers comprise some 50 living genera and some 500 species. This is only about one-thirteenth the number of species represented by the ferns or one-five-hundredth the number of angiosperms. Nevertheless, conifers are very important in many present ecosystems and have been important since the late Carboniferous period, 300 million years ago.

All the leaves, even the very small ones, have well-developed vascular tissue (Figure 36–16), and the stems all possess a prominent vascular system. Secondary growth, which is found throughout the group, is often extensive. As a rule, the xylem is anatomically uniform,

Figure 36-13 Cycads.

Figure 36-14 The pollen and seed cones of a cycad. The male cone on the left is mature, and the pollen has been released. But the seed cone or female cone on the right is still immature.

consisting of tracheids and parenchyma cells. Cones are the most common reproductive structure, but these may be highly modified from the typical form and may even be fleshy, as in juniper berries. In all cases, the female cones are larger and longer-lived than the male. In general, the two types of cones are produced on different trees, and all conifers are wind-pollinated.

The Ginkgophyta, or ginkgos, are living fossils—relic plants that may have survived because of people's interest in their majesty and beauty. The fan-shaped leaves have been found as fossils in rocks from the Triassic and Jurassic periods of the Mesozoic era, when there were several ginkgo species distributed worldwide. There is, for example, a national park in the state of Washington that is an ancient petrified ginkgo forest. These gradually declined in number and range. If any ginkgo trees still grow in the wild today, they are found only in the mountains of southeastern China. The single surviving species, *Ginkgo biloba,* has been called the oldest living species of seed plant.

A mature *Ginkgo biloba* is a tall, stately tree with

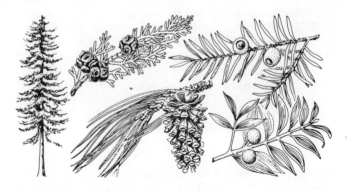

Figure 36-15 Class Coniferophyta (conifers).

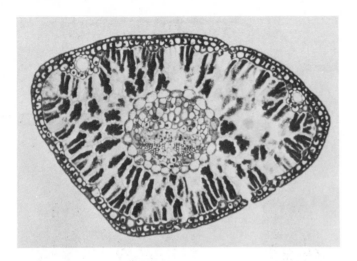

Figure 36–16 The needles of the conifers are complex, tightly organized structures, as seen in this photomicrograph of a cross section of the needle of *Cedrus deodora*.

Figure 36–17 Class Angiospermae (flowering plants).

several gracefully curving branch stems uniting in a thick, basal trunk. The beautiful fan-shaped leaves are borne on distinct, short shoots and are shed each year. The male and female reproductive structures are always produced on separate trees. These structures differ markedly in external form from those of the cycads and conifers, and the female structures have a most unpleasant smell at maturity (hence, only male trees are usually planted along streets). In the female, the ovules are found at the top of a stalk and are essentially naked. The reproductive cycle follows that for gymnosperms in general. Large multiflagellate sperm are produced.

The Gnetophyta, or gnetales, are a strange group of gymnosperms composed of only three genera: *Ephedra, Welwitschia,* and *Gnetum,* each in its own family, its own order, and its own class.

Ephedra has some 35 species, all of them either shrubs or small trees. Most are desert plants with green stems and reduced, scalelike leaves. *Gnetum,* in contrast, is a tropical forest genus with some 30 species. The *Gnetum* species have broad leaves and are climbing vines or trees. *Welwitschia* inhabits a single desert area (the Kaokoveld) in South Africa. Over the desert surface trail the long, strap-shaped leaves continually produced by the top of the long underground stem. The only visible part of the stem is a woody, concave disk that produces both the leaves and reproductive structures.

Subphylum Pteropsida: Angiosperms Consider for a moment a world without flowering plants. The first thing we would miss would be the bright colors. We would be surrounded by a sea of greens—a wide range of greens—but still only green. Missing would be the reds, yellows, blues, oranges, violets, and all the other hues found in flowers and many fruits. But the color of angiosperms is not only in their flowers and fruits. In fact, the greatest display of color is in the autumn, when whole trees and forests are a blaze of red, yellow, and orange. Only ginkgo trees and two or three other gymnosperms have leaves that change color in the fall.

A little later, we would notice that the seasons would be much less sharply marked to our senses. Spring would not be heralded by early flowers and sprouting seeds, nor would summer feel so full without the waving fields of grass and flowers. Autumn would be less the harbinger of winter, lacking the change of leaves, and winter, although just as cold, would surely seem less bleak without the bare-limbed trees.

The flowering plants, or *angiosperms,* affect our smell and taste, as well as our sight. This is no mere accident of evolution but an important facet of the life of angiosperms. When we come to the flowering plants, we come to plants that have truly evolved not only for life on land but for life on land filled with animals.

The colors and aromas of flowers, which so delight us, attract insects and other animals necessary for pollination. The colors and aromas of fruits entice us and a wide range of other animals to eat them and spread the seed. The spines and hooks on seeds would be useless without fur on animals. The intercalary meristems of grass, which provide us with a summer of both green lawn and lawn-mowing, allow the plants to survive grazing by cattle, deer, or buffalo. In addition, angiosperms

Table 36–2 Characters Distinguishing Dicots From Monocots

Dicotyledoneae	Monocotyledoneae
Two cotyledons or seed leaves in the embryo	One cotyledon or seed leaf in the embryo
Principal veins of the leaves branching out from the midrib or from the base of it, not parallel, forming a distinct network	Principal veins of the leaves parallel to each other
Flower parts arranged in twos, fours, or usually fives	Flower parts arranged in threes
Root system characterized by a taproot, that is, a large primary root with branch roots growing from it	Root system fibrous—that is, without a principal or taproot, the evident roots being of about equal size
Stem with the vascular bundles in a single cylinder ("ring" as seen in cross section)	Stem with the vascular bundles scattered apparently irregularly through pith tissue
The cambium adding a new cylindroidal layer ("ring") of wood each year or each growing season	Stem and root without a cambium, not increasing in girth by the formation of annual cylindroidal layers of wood

are the only vascular plants to trap and digest animals.

There are some 275,000 species of angiosperms, grouped into several hundred families. This largest group of plants comes in every conceivable size and shape: There are angiosperm trees that challenge the great conifers in size; there are tiny, floating aquatic angiosperms that are little more than drifting leaves; there are great climbing vines that can kill large trees; there are the beautiful and delicate plants that bloom in the spring; there are shrubs of many sizes and shapes; and there are the miles of grasses that cover our prairies (Figure 36–17). Not only do angiosperms assume almost every shape and size, but they grow in almost all climates. The cacti and certain other angiosperms are adapted to living in all but the most arid deserts. The tropical rain forests are a fantastic collection of angiosperms, including thousands of species of orchids, many growing as epiphytes on flowering trees and vines. Other angiosperms grow near the tops of all but the highest mountains and to the edge of the great polar ice packs. Some have adapted to life in the intertidal zones, and many inhabit marshes and the edges of lakes. Temperate, arctic, tropical, wet, or dry—these are all environments successfully invaded by angiosperms.

Angiosperms can be divided into two broad groups or subclasses: the Dicotyledoneae and the Monocotyledoneae. The names refer to the number of cotyledons (storage seed leaves) present in the seed. But the differences between dicots and monocots are much more profound than this (Table 36–2). The Dicotyledoneae contain more than 200,000 species, compared with fewer than 50,000 species of Monocotyledoneae.

The morphology of the flower and the general life cycle of the angiosperms were covered in Chapter 10.

Summary

The kingdom Plantae contains those plants with which we are most familiar. This vast collection contains some of the smallest and largest organisms to be found on earth. There are three algal phyla: the green algae, the brown algae, and the red algae. While the brown algae contain many species with elaborate body plans, it is from the structurally simpler green algae that the land plants probably arose. The mosses, liverworts, and hornworts are all land plants, but are considered less advanced than the other land plants because they lack a well developed vascular system for the transport of water and organic materials. The phylum Tracheophyta contains the vascular land plants, those plants that are most conspicuous in our surroundings. The psilopsids, lycopsids, and horsetails were once the major plants on land, but now they are relatively inconspicuous parts of the flora. Ferns are highly successful vascular plants that reproduce by spores. The most successful of all land vascular plants are the seed-bearing plants: the gymnosperms and the angiosperms. The gymnosperms or cone-producing plants include the conifers and other needle-leaved trees. The flowering plants or angiosperms are the most recent and most successful of all land plants. Their great diversity has assured a place for them in almost all land ecosystems.

37

Kingdom Animalia

Since we are animals, we are apt to think of all animals in terms related to our own structure and behavior. Yet even a cursory study of the animals sharing our planet with us reveals that most are, in fact, markedly different from us. Most have no backbones, and many have external rather than internal skeletons. Few have hair. The largest number of species of animals are insects, and a great many animals spend most of their lives as firmly "rooted" as plants.

As we did with the plants, we are going to present one scheme for classifying the major animal groups that makes sense to us (Table 37–1).

We will start this chapter with a brief overview of the entire animal kingdom, emphasizing the major trends in animal evolution, leaving the details to later in the chapter. These trends emphasize increasing complexity of organization to achieve efficiency of function. Evolution does not proceed in a straight line from "lower" forms to "higher," but occurs with selection and adaptive radiation within each major group, as emphasized in Chapter 30. Our overview of animals begins with the simplest and evolutionarily oldest and proceeds through more complex and, often, more recently evolved forms. A hierarchy of organization is emphasized, as well as an evolutionary scale (Figure 37–1).

An Overview of Animals

The animal kingdom is composed of many-celled forms. The evolution of multicellular organisms has meant an increase in efficiency. How multicellular forms were first established is not clear; they may have arisen differently in different lines. What is clear is that there is a tremendous diversity in the form, function, and behavior of multicellular animals.

We will examine several features of structure and their development in various major groups of animals in order to assess their relationships and possible evolutionary history. We will consider cellular organization—how the cells are organized to provide division of labor through tissues and organs. We will consider the kinds of symmetry of body organization and such developmental features as cleavage patterns, patterns of organ and body-space formation, and types of larvae. The structures of digestive, respiratory, circulatory, reproductive, locomotor, and nervous systems also provide clues to probable relationships among animal groups.

The phylum Porifera, the sponges, includes the simplest of all animals. They are little more than elaborate collections of loosely organized cells.

The phylum Coelenterata contains sea anemones, corals, jellyfishes, and hydra. These are organized, *radially symmetrical* animals—like a spoked wheel lying flat on the ground, they have a top and bottom, but no right or left side. Coelenterates have true tissues, and the center of the animal is occupied by a digestive cavity with a single opening. Members of the phylum Ctenophora also have this plan of organization.

In the phylum Platyhelminthes, or flatworms, we find animals that are *bilaterally symmetrical* (they have not only a top and bottom, but a left and a right side). Smaller than many of the coelenterates, the flatworms are considerably more complex: They have three embryonic cell layers, muscles, and excretory system, and a rudimentary centralized nervous system. The digestive tract of the flatworms still has but one opening; but in the phylum Rhynchocoela, or ribbon worms, an anus has been added and digestion becomes a one-way street.

In the phylum Annelida, or segmented worms (which includes the earthworm), we see all these trends continued and a new feature added. This is a body plan organized on the basis of a series of segments. The great advantage of this modular construction lies in the control of motion, as well as the specialization of the various segments into specific functions.

The phylum Mollusca—the clams, snails, and squid—

Table 37-1 The Animals

Group Name	Common Name	Group Name	Common Name
Phylum: Porifera	Sponges	Order: Decapoda	
Phylum: Coelenterata		Anomura	Hermit crabs
Class: Hydrozoa	Hydra	Brachyura	True crabs
Class: Scyphozoa	Jellyfishes	Order: Euphausiacea	Krill
Class: Anthozoa	Sea anemones, corals	Order: Stomatopoda	Mantis shrimp
Phylum: Ctenophora	Comb jellies	Class: Arachnida	Spiders, scorpions, ticks, and mites
Phylum: Platyhelminthes	Flatworms		
Phylum: Rhynchocoela	Ribbon worms	Class: Chilopoda	Centipedes
Phylum: Aschelminthes	Roundworms and rotifers	Class: Diplopoda	Millipedes
Phylum: Phoronida		Class: Insecta	Insects (some orders listed below)
Phylum: Ectoprocta	Bryozoans		
Phylum: Brachiopoda	Lamp shells	Order: Odonata	Dragonflies
Phylum: Annelida	Segmented worms	Order: Orthoptera	Grasshoppers and cockroaches
Phylum: Mollusca		Order: Isoptera	Termites
Class: Monoplacophora	Segmented mollusks	Order: Anoplura ⎫	Lice
Class: Amphineura	Chitons	Order: Mallophaga ⎭	
Class: Pelecypoda	Clams, mussels, oysters, and scallops	Order: Hemiptera	True bugs
		Order: Lepidoptera	Moths and butterflies
Class: Scaphopoda	Tooth shells	Order: Diptera	True flies
Class: Gastropoda	Snails and slugs	Order: Siphonaptera	Fleas
Class: Cephalopoda	Squids and octopuses	Order: Coleoptera	Beetles
Phylum: Arthropoda		Order: Hymenoptera	Bees, wasps, and ants
Class: Crustacea		Phylum: Echinodermata	Starfishes, sea urchins
Subclass: Copepoda	Microscopic crustacea	Phylum: Urochordata	Sea squirts
Subclass: Cirripeda	Barnacles	Phylum: Cephalochordata	Lancelets
Subclass: Ostrocoda	Fairy shrimps, clam shrimps, water fleas	Phylum: Chordata	Vertebrates
		Class: Cyclostomata	Jawless fishes
Subclass: Malacostraca		Class: Chondrichthyes	Sharks, skates, rays
Order: Isopoda	Pill bugs, wood lice	Class: Osteichthyes	Bony fishes
Order: Amphipoda	Beach hoppers, scuds, well shrimp	Class: Amphibia	Frogs, toads, salamanders
		Class: Reptilia	Turtles, snakes, lizards, alligators
Order: Decapoda	Shrimps, lobsters, crabs, crayfish	Class: Aves	Birds
Macrura	Shrimps, lobsters	Class: Mammalia	Mammals

includes bilaterally symmetrical animals complete with a circulatory system including a heart, kidneys for excretion, a well-developed nervous system, a complete digestive tract, and usually gills for respiration.

The phylum Arthropoda contains what must be among the most successful of all animals: the insects, spiders, scorpions, crabs, shrimps, pill bugs, and many others. This phylum shows remarkable adaptations to life both in the water and on land. All have an exoskeleton that provides remarkable protection, jointed legs that increase the efficiency of movement, jaws, antennae, and compound eyes. Many have wings and other specialized adaptations that fit them to particular habitats. Particularly diverse are the insects that successfully challenge the vertebrates for dominion of the earth.

The phylum Echinodermata—starfishes, sea urchins, and so forth—are radially symmetrical and have a unique type of digestion (they essentially force their stomach, turned inside out, into their prey and digest it within its own shell) that makes an anus unnecessary. One should not be misled, however, because this reduction is an evolutionary specialization, and the Echinodermata are advanced and highly successful animals.

When we reach the phylum Chordata, we come to the phylum in which our own species is found. We share with other vertebrates such characteristics as an epithelial body covering of epidermis and dermis, an internal jointed skeleton with vertebrae, a closed circulatory system with a heart, and a functionally differentiated brain and endocrine organs.

There are seven living classes of vertebrates. The most primitive are the jawless fishes (Cyclostomata), whose

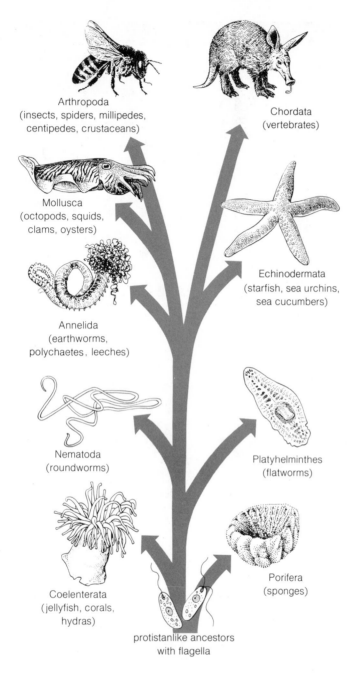

Arthropoda
(insects, spiders, millipedes,
centipedes, crustaceans)

Chordata
(vertebrates)

Mollusca
(octopods, squids,
clams, oysters)

Echinodermata
(starfish, sea urchins,
sea cucumbers)

Annelida
(earthworms,
polychaetes, leeches)

Nematoda
(roundworms)

Platyhelminthes
(flatworms)

Coelenterata
(jellyfish, corals,
hydras)

Porifera
(sponges)

protistanlike ancestors
with flagella

Figure 37–1 Evolutionary lineages and relationships among the major phyla of the animal kingdom.

ancestors were swimming in the oceans some 400 million years ago. The class Chondrichthyes contains the sharks and rays. The evolution of jaws was an experiment of the Placodermi, a now extinct group, some of whose members are thought to be the ancestors of sharks. Jaws are but one of the many features that mark the sharks

and their relatives as more advanced than the jawless fishes. Sharks have a cartilaginous skeleton and were long thought more primitive than the class Osteichthyes, or bony fishes. They are now regarded as a specialized evolutionary offshoot. The bony fishes make up the great bulk of today's fishes, animals superbly adapted to life in water.

The next evolutionary thrust in the vertebrates came with the invasion of the land. As we saw with the vascular plants, aquatic organisms must change in many different ways to live successfully on land. In the case of the vertebrates, this transition seems to have been made in a series of stages reflected in the class Amphibia. Such amphibians as frogs, toads, and salamanders may live most of their life on land and have the structural and physiological adaptations for such life, but most must return to water to lay their eggs. Water must also be available for the fishlike larval young. The reptiles (class Reptilia) have cut this tie with an aquatic environment: They have watertight eggs and young that are themselves true land animals. The tremendous success of the dinosaurs shows how successful their adaptation was. The birds (class Aves), which evolved from reptiles, show still other adaptations to terrestrial living, including feathers and wings.

More recent in the evolution of the vertebrates is the class Mammalia, or mammals, animals familiar to us all. Hair, the birth of live young, the production of milk in mammary glands, and many other features mark the mammals as land animals, even when they have returned to the aquatic environments from which their ancestors evolved.

In this brief overview, we have given but the barest of details and neglected whole phyla. Now let's look at the animals in somewhat more detail.

Phylum Porifera: Sponges

Sponges are multicellular organisms of many shapes (Figure 37–2). They have loosely organized outer and inner tissue layers. The sponges represent an increase not only in size, but in specialization and efficiency, over unicellular organisms. Single cells can get only so big, since they must have enough surface area to allow diffusion to accommodate the entire cell. As the volume of the cell increases, its surface area relative to that volume decreases, so that soon diffusion cannot service the total cell volume. Multicellularity may have arisen in two ways: (1) Cells increased in size, nuclei divided, and cell membranes grew to partition them and their cell contents, giving rise to a mass of cells; or (2) the members of colonies of unicellular organisms lost their autonomy, remained aggregated, and effected a division of labor so that certain cells did the feeding, others the reproduction, and so on for the entire mass.

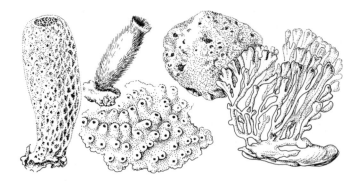

Figure 37-2 Phylum Porifera (sponges).

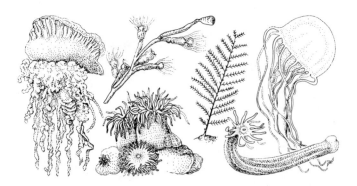

Figure 37-3 Phylum Coelenterata (jellyfish, corals, sea anemones, hydras).

A sponge is a mass of cells supported by mineralized, needle-like, often branched structures or by fibers and penetrated by many channels. The cells are specialized so that certain cells of the mass ingest food particles from water flowing through the channels and nourish the entire organism, throwing water *out* the main channel, or "mouth"; other cells secrete the support structures; still others produce eggs and sperm; and yet others may produce asexual reproductive units called *gemmules.* Sponges can also reproduce by budding or branching. Sponges lack a coordinating system and have other cellular characteristics suggesting that they arose from a different unicellular group than other multicellular organisms.

Sponges are not restricted to warm or shallow seas; some occur in cold waters and at great depths, and a few have invaded fresh water. With the advent of synthetic sponges, some of the commercial value of natural sponges was lost, but sponges without spicules are still harvested to be sold. Calcareous sponges aggregate on pilings and may foul boat channels. Since dissociated sponge cells readily reaggregate (see Chapter 13), sponges have become useful tools in the study of cellular development.

The Radiate Phyla

Phylum Coelenterata (Cnidaria): Anemones, Corals, Jellyfishes, Hydra, and Others The coelenterates are highly successful and diverse—anemones, corals, jellyfishes (including the Portuguese man o' war), and hydras (Figure 37–3). They share a common body plan that exhibits radial symmetry. They have a mouth as entrance to the digestive cavity, and most have tentacles around the mouth with stinging cells, or *nematocytes,* for food-gathering and defense. They have a more complex organization than sponges: The body wall has two or even three distinct layers, or tissues. With this "tissue level of organization," cellular specialization has proceeded, including the appearance of a network of "nerve" cells that provides coordination.

Coelenterates are further characterized by *polymorphism,* the presence of more than one kind of body form in members of the same species. One form is the *polyp* or *hydroid,* attached to the substrate at one end and with its mouth, encircled by tentacles, at the other (Figure 37–4). Polyps may be individual or colonial. The other form is the free-swimming *medusa,* which has marginal tentacles and its mouth projecting from the underside of its bell-shaped body (Figure 37–4). Polymorphism is thought to be a balance of form to cope with environmental conditions. The general radial body plan and tissue organization are found in both forms. Reproduction is usually asexual in polyps (giving rise to more polyps or to medusas) and sexual in medusas. Colonial hydroids have an efficient strategy. Some individuals in the colony are specialized to gather food, others to reproduce. In effect, individuals in the colony act as organs, efficiently dividing the work of the colony.

The three major classes of coelenterates are characterized by features of the tentacles, sex cells, digestive tract, and life cycle. Certain coelenterates, the sea anemones, enter fascinating symbiotic relationships in which tiny shrimp and fish live unharmed among the anemones' tentacles and "clean" them. Many other coelenterates, called corals, secrete a calcareous shell and have a symbiotic relationship with unicellular algae called zooxanthellae, which promote calcification and supply oxygen to the polyps.

Coelenterates have a long and diverse fossil record, beginning in the Precambrian. Some coelenterates (hydras) are now found in fresh-water habitats, although most species are marine. Medusas float in currents on the open ocean. Polypoid forms, because of their bottom-dwelling existence, are usually found in shallow waters,

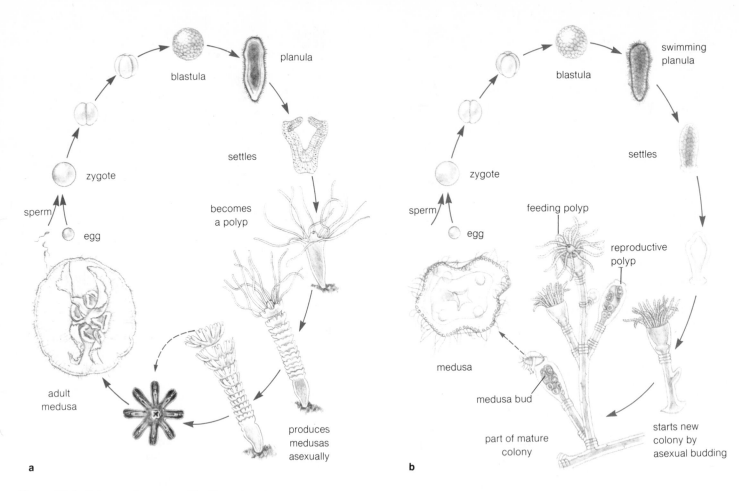

Figure 37-4 Polyps and medusas. The life cycles of (**a**) *Aurelia* and (**b**) *Obelia*.

planula

blastula

settles

zygote

sperm

egg

becomes
a polyp

adult
medusa

produces
medusas
asexually

a

swimming
planula

blastula

settles

zygote

sperm

egg

feeding polyp

reproductive
polyp

medusa

medusa bud

part of mature
colony

starts new
colony by
asexual budding

b

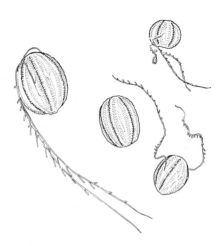

Figure 37-5 Sea gooseberries, a variety of Ctenophora.

including tide pools, although they may occur in the depths as well. Corals, particularly, are found in shallow, warm, subtropical or tropical waters, where their skeletons provide the structures for entire communities of plants and animals.

Phylum Ctenophora: Comb Jellies The ctenophores, or comb jellies (Figure 37–5), resemble coelenterates in many ways—tissue level of organization and lack of organ systems, radial symmetry, and digestive cavity extending from a mouth. They look like medusae, except that they move by means of eight rows of comblike structures (from which the phylum and common names were born) and have adhesive organs on their tentacles for capturing prey, rather than nematocytes. Most forms have a pair of long tentacles and may have numerous shorter tentacles. Their prey include tiny animals, larvae, and ova of other aquatic

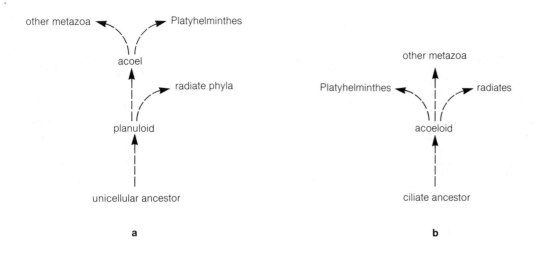

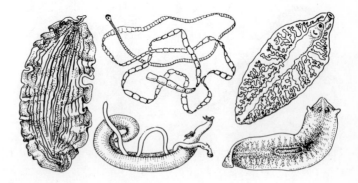

Figure 37-6 (**a**) The theory that multicellular animals arose from a unicellular ancestor via a planula; (**b**) the line of origin if multicellular forms arose from a ciliate via a hypothetical noncoelomate adult form (an acoel).

animals. The comb jellies reproduce sexually and are hermaphrodites (that is, gonads of both sexes are present in each individual).

All the comb jellies are marine, and most are pelagic, floating near the water's surface with the currents. They are among the most beautiful of marine animals. Delicate and translucent, they are iridescent by day and bioluminescent by night, making the currents in which they are found unusually attractive.

Phylum Platyhelminthes: Flatworms

The flatworms are of great evolutionary interest because they "introduce" a number of features that are of adaptive significance for all higher groups. Many scientists consider a simple flatworm to be the ancestral stock from which more advanced phyla are derived. Flatworms have bilateral symmetry, three embryonic cell layers, musculature, an excretory system, and a rudimentary centralized nervous system. However, circulatory, respiratory, and skeletal systems are not developed. Flatworms exhibit a variety of lifestyles: They may be free-living, crawling about on mucky lake or river bottoms or rocks, or living in or on the soil; they may be ectoparasites, such as the species that lives on the shell of the horseshoe crab; or they may be endoparasites with complex life cycles in which successive stages parasitize different hosts.

There are two major theories of the origin of multicellular animals. One theory derives them from a multinucleate ciliate protozoan and the other from a *planuloid*—a multicellular, larvalike organism that came from an unknown single-celled ancestor. The ciliate theory says that

Figure 37-7 Phylum Platyhelminthes (flatworms).

three major groups arose from the hypothetical ancestor: (1) the flatworms, (2) the coelenterates and ctenophores, and (3) all the other animal phyla. The planuloid theory says that the coelenterates and ctenophores arose from the planuloid organism, then another ancestral form arose, and it in turn gave rise to two metazoan lines, the flatworms and all the other phyla (compare these lines of origin as diagrammed in Figure 37-6).

The three classes of platyhelminths—Turbellaria (free-living flatworms), Trematoda (parasitic flatworms), and Cestoda (tapeworms)—are shown in Figure 37-7. Most members of the flatworm class Turbellaria have a centrally located ventral mouth that scoops food into the intestine and that also serves as the exit for undigested material. Since they have no anus, their digestive tracts are called "incomplete." Turbellarians can reproduce both asexually, by regenerating from fragments, and sexually. Most are hermaphrodites, but they cross-fertilize via a

Figure 37–8 The derivation of mesoderm and the coelom in various lineages. (**a**) In forms without coeloms, mesoderm buds from cells to form blocks of tissue. (**b**) In protostomes, mesoderm arises as a solid ingrowth of cells from a single cell, and the coelom forms as a hollow in the mesoderm block. (**c**) In deuterostomes, the mesoderm arises from saclike outpocketings of the gut wall, and these pockets give rise to the coelom.

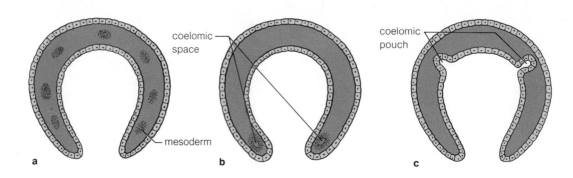

coelomic space

mesoderm

coelomic pouch

a   b   c

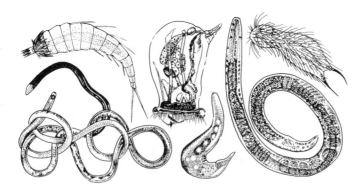

Figure 37–9 Phylum Aschelminthes (round worms and rotifers).

copulatory mechanism. Most turbellarians are small and transparent, but some are up to 50 centimeters long, and some are brown, black, or brightly colored in warm marine waters. *Dugesia,* a small, brownish black fresh-water turbellarian, is well known to biologists as a subject for regeneration experiments, and the nature of the platyhelminth nervous system is such that biologists consider it the "lowest" form useful in memory and learning experiments. Turbellarians usually live in cool permanent waters and fasten to the undersides of plants and rocks.

The Trematoda are a class of ecto- and endoparasites, living primarily in vertebrates. Equipped with large adhesive organs or suckers, these flatworms or flukes ingest blood, mucus, and tissue of the host. The life cycles may be simple, with a single host, or complex, with intermediate hosts.

The class Cestoda includes a small group of nonsegmented worms that parasitize primitive fishes, and the large group of true tapeworms that parasitize many animals, especially vertebrates. They look segmented because they have many, sometimes hundreds, of repeating body units (proglottids), each containing excretory and both male and female reproductive apparatuses. The adult tapeworm has a head with hooks and suckers for attachment and has many "segments." Some grow to a length of 8 meters. Adults are intestinal parasites of vertebrate animals, including humans; larvae may find intermediate hosts among invertebrates and vertebrates. Cestodes are highly specialized parasites that lack a digestive system; they absorb nutrients directly from the fluids surrounding them in the body of the host. Sexual reproduction may take place in the same proglottid, between proglottids of one worm, or between those of two different worms. Survival is well assured, for by these mechanisms huge numbers of larvae can be produced.

Phylum Rhynchocoela: Ribbon Worms

The rhynchocoels, ribbon or nemertine worms, have a complete digestive system (an anus is present) and a projectable proboscis that is used for food capture, burrowing, and protection. This proboscis gives nemertines more sophisticated means of dealing with the environment than flatworms have. The nemertines are free-living, usually on muddy or sandy marine bottoms beneath rocks and shells.

The Protostome-Deuterostome Divergence

We have seen that a simple flatwormlike ancestor is thought to have given rise to many, if not all, phyla of animals. The phyla more advanced than cnidarians and flatworms seem to represent two assemblages or lines of evolution, based on several embryological characteristics. In one assemblage, which includes the annelids, arthropods, and mollusks, the embryonic blastopore becomes the mouth, or opening to the gut. Hence that group is called the Protostomia ("first mouth"). In the other assemblage, the echinoderms and chordates, the blastopore becomes the anus and a new mouth forms. These phyla are the Deuterostomia ("second mouth").

Other differences distinguishing the two groups include those of early cleavage. In protostomes, cleavage is *determinate* (cell fates are partly fixed; see Chapter 13) and *spiral* (early cleavages are oblique to the vertical axis). The cleavage of deuterostomes is *indeterminate* (early cells can be separated and still develop into complete organisms) and *radial* (cell division is parallel or at right angles to the axis). The larval types resulting from these two patterns of cleavage also distinguish the two groups.

Further, in protostomes, mesodermal tissue arises as a solid ingrowth of cells from a single cell. Deuterostome mesoderm arises from saclike outpocketings of the gut wall (Figure 37-8).

Another difference between the two assemblages is in the formation of the coelom, if one is present. The **coelom** is a cavity between the digestive tract and the body wall that is completely lined by mesoderm. Not all protostomes have a true coelom, but in those that do, it forms as a split in the initially solid block of developing mesoderm. The deuterostome coelom is derived from the cavity or pocket of the outpocketings that form the mesoderm, as diagrammed in Figure 37-8.

Six major differences, then, exist between protostomes and deuterostomes: the fate of the blastopore, the cleavage pattern, the determinateness of its cells, the type of larva, the origin of the mesoderm, and the formation of the coelom. We shall see the significance of these features as we discuss the phyla in each group.

The Pseudocoelomate Protostomes

Phylum Aschelminthes: Rotifers and Round Worms
The five classes composing the aschelminths do not form a "natural" group, for each arose separately, although they are all protostomes (Figure 37-9). The groups share several characteristics. They have a complete digestive tract (an anus is present as well as a mouth), and they are pseudocoelomate (the space between the gut and the body wall exists from early development and lacks a mesodermal lining tissue layer). They are usually small and cylindrical, have a cuticularized epidermis, and have discrete nervous, muscular, and excretory systems. The most common of the Aschelminthes are the rotifers and the round worms.

The Rotifera live mostly in fresh-water lakes, ponds, streams, and puddles, although there are a few marine species. They look segmented, but are not truly so, and they have "wheel organs"—bands of beating cilia that create water currents around the mouth to bring in food. Many have a chewing pharynx equipped with crushing plates, and most forms have adhesive organs on the posterior part of the body that are used to grip the substrate. Rotifers eat organic debris.

The most common of the round worms are nematodes that live everywhere! The round worms are unsegmented and usually tiny. There are probably hundreds of thousands of species, and they live in every available habitat—some are free-living, some are parasites, some live in root nodules, soil, seeds, blood, and other restricted habitats. Having longitudinal muscle fibers only, they move with a whiplike motion, and they lack circulatory and respiratory structures.

Many, perhaps most, nematodes are parasites, and the parasitic forms may be of more ecological significance than the free-living ones. A commonly studied nematode is *Ascaris*, a parasite primarily of pigs, but also of humans. Young pigs ingest larvae from feces and dirt in the pig yard or on the mother. The larvae mature in the young pigs' intestines. Adult worms copulate, and the female may then lay 200,000 eggs per day of the 27 million she has in her body at one time. Deposited in the host's intestine, the eggs reenter the barnyard via the pig's feces. They undergo a developmental period and are often then ingested. In the pig who eats them, they hatch, burrow through the veins and lymph vessels into heart and lungs, and finally, at a larger size, they erupt through the lung membranes, move back to the intestine via the coughing of the host into its trachea, and then are swallowed into the esophagus and stomach, where they mature.

The Lophophorate Phyla

The Bryozoans and Company The phyla Phoronida, Ectoprocta, and Brachiopoda are truly coelomate animals that have a lophophore—a foldlike extension of the coelomic wall that bears tentacles (Figure 37-10). The lophophore, surrounding the mouth, is used to generate water currents to bring food particles into it.

Phoronids are exclusively marine and most commonly live just below the water's surface in mud flats. They secrete tubes around their bodies to which sand grains adhere, and the tubes may form masses attached to the substrate. Muscular, digestive, nervous, and excretory systems are well developed. Phoronids have a closed circulatory system but lack a heart. They reproduce both sexually and by fission.

The ectoprocts, or bryozoans (Figure 37-10), are

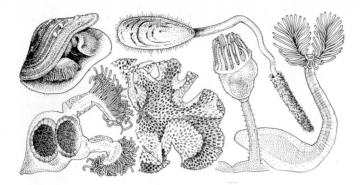

Figure 37–10 The lophophorate phyla (phoronids and bryozoans, or ectoprocts).

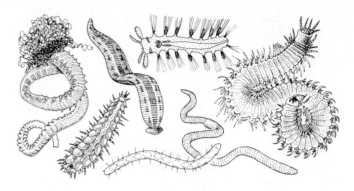

Figure 37–11 Phylum Annelida (earthworms, leeches, and polychaetes).

widely distributed in marine and fresh water. They usually live in colonies attached to rocks, vegetation, or debris. They secrete an exoskeleton around their bodies, which may become calcified. Polymorphism is highly evolved, and members of a colony are modified variously for feeding, cleaning, reproduction, and other maintenance tasks.

Brachiopods (or lampshells) have a long fossil history. One living genus, *Lingula,* has existed for 400 million years. The lampshells were once abundant, but few species remain. They are marine, attached forms that live from low tide level into the deeps.

The Coelomate Protostomes

The phyla Annelida, Mollusca, and Arthropoda are allied by sharing protostome characteristics and a true coelom.

Phylum Annelida: Segmented Worms The phylum Annelida includes animals that have refined the wormlike body plan (Figure 37–11). The annelid body plan is based on a series of true segments that repeat the organization of each unit, a plan called *segmentation,* or *metamerism.* One great advantage of metamerism (also found in arthropods and chordates) is the promotion of controlled movement. Muscular movement compresses the coelomic fluid differentially in each segment, lengthening, contracting, or curving it.

There are three classes of annelids. Oligochaeta are primarily terrestrial or fresh-water forms and include the "typical" earthworms. Hirudinea, or leeches, primarily prey on fresh-water and terrestrial animals, although some species are marine, parasitizing sharks and rays. Polychaeta are mostly marine.

Most oligochaetes have tiny chitinous setae or rods in each segment that are used in locomotion. Their well-developed coelom is divided by *septa* (connective tissue segmental boundaries). Figure 37–12 diagrams the structure of a body segment. Each worm may have one hundred to several hundred segments. The mouth is located on the first segment. The major digestive, circulatory, and reproductive organs are concentrated in the anterior region of the animal, although each posterior segment has a pair of excretory organs and a nerve ganglion with lateral nerves. Earthworms and other oligochaetes are present in soils that are well aired, moist, and full of organic matter. They are not often found in acid or dry soils. They avoid light and therefore are nocturnal. They feed on small particles of vegetation and other organic debris. Earthworms are of long-term importance to the standing crop of food for many organisms, for they turn over great amounts of soil as they burrow. This allows air and water to penetrate and "folds" organic material into the soil. Other oligochaetes live in fresh water in lakes or streams. Some, such as *Tubifex,* can live in polluted water. *Tubifex* feeds on bottom debris and aids in purifying polluted bodies of water.

The class Hirudinea, or leeches, includes segmented worms that have terminal suckers for movement and attachment. They usually have 34 segments. They are scavengers, active predators, or parasites. Perhaps most notorious are the blood-sucking leeches that prey on humans and other warm-blooded vertebrates. Attaching to the skin with their suckers, they pierce the skin and suck out the blood with their muscular pharynx. The blood is stored in the crop, a "pocket" of the digestive tract, and is kept from coagulating by a salivary enzyme secreted by the leech. The nutrients from the stored blood are then broken down and absorbed over several months. Leeches live primarily in fresh water, but some occur in moist places on land and others as marine fish parasites.

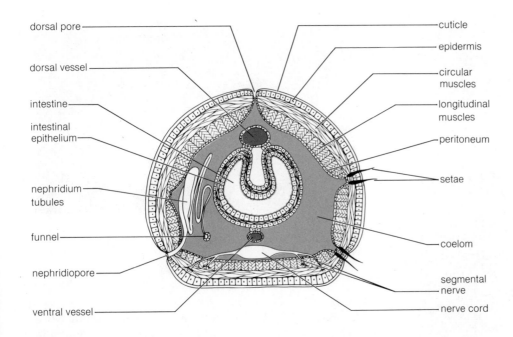

dorsal pore
dorsal vessel
intestine
intestinal epithelium
nephridium tubules
funnel
nephridiopore
ventral vessel

cuticle
epidermis
circular muscles
longitudinal muscles
peritoneum
setae
coelom
segmental nerve
nerve cord

Figure 37–12 Diagrammatic cross section of an earthworm.

Usually these animals do little harm to their prey or hosts, but if their numbers are great, they may seriously damage or even kill the animals they infest.

Polychaetes have a distinct head with eyes and tentacles. Respiratory, excretory, nervous, and circulatory systems are present and well developed. Each segment has a pair of lateral expansions used in locomotion, reproduction (gamete discharge), respiration (via gills), and sensing the environment. There are two main groups of polychaetes, those that move about actively, and those that live in tubes and expose only their heads.

Phylum Mollusca: Clams, Snails, Octopuses, and Squids The phylum Mollusca (Figure 37–13) is large and varied, allied to the annelids by the segmentation found in one genus of mollusk and by certain embryological features. The diversity of the group is well known to shell collectors, for nearly all the shells that wash ashore are those of mollusks. The animal that produces the shell is bilaterally symmetrical and usually has a developed head region, a ventral muscular foot for locomotion, and a complete digestive tract (with a mouth containing a rasping organ, the *radula*). It has a circulatory system with a heart, a well-developed nervous system, kidneys for excretion, and usually gills for respiration. The body is contained in a fleshy "mantle" that usually secretes a calcareous shell.

Most molluscan species are marine and live on or

Figure 37–13 Phylum Mollusca (clams, oysters, squids, octopuses, snails).

under rocks, sand, or mud, although land snails and slugs are familiar mollusks as well. They are found in all depths of the ocean. There are seven classes of mollusks, discussed briefly here.

The Monoplacophora are primitive mollusks with a long fossil record. They have only one living representative, *Neopalina,* which retains the segmental characteristics of annelids, lost in all other mollusks.

All of the Amphineura, or chitons, are marine. They have a broad, flat, muscular organ, the foot, with which they crawl about and attach firmly to the solid substrate. Their shell is dorsal, and usually of eight plates, although

Figure 37–14 Various pelecypod and snail shells.

Figure 37–15 A nudibranch.

some species have only one or two plates or none at all. Most chitons scrape minute algae and diatoms from rock surfaces to obtain food, but some are detritus feeders. Their shells are often covered with algae, hydroids, or bryozoans.

The Aplacophora, or solenogasters, are tiny wormlike marine mollusks. They lack a shell but are covered by a spiny cuticle.

The Pelecypoda—clams, mussels, oysters, and scallops—are characterized by a double shell of right and left halves. Each pelecypod species has a different shell shape, set of muscle scars, or color, and most of the 11,000 living and 15,000 fossil species can be identified by the shell alone (Figure 37–14). They have a strong muscular foot usually used for digging. It may also be used to fasten the animal to its substrate, although some forms secrete fibers to anchor themselves in place. Respiration occurs as water is pumped in through one "siphon," over the gills, and out an excurrent siphon. The same water current also brings in food particles which are trapped in sheets of mucus over the gills and then transported

by the current, along with the mucus, to the mouth. Most pelecypods are marine, but fresh-water clams abound in some lakes and streams. Pelecypods are important in the marine food web. They are the food items of many forms, from walruses to starfish. One form, the shipworm (really a clam), does millions of dollars of damage each year to wooden ships, pilings, and other structures.

The Scaphopoda, or tooth shells, live in sand or mud below water level, at all depths. Their shell is a tapered tube open at both ends. The foot is extended through the larger opening to dig in the sand. Scaphopods lack hearts and gills. They feed on small plants and animals living in the sand.

Gastropoda include snails, slugs, abalone, limpets, cowries, nudibranchs, and many other forms. They typically have a distinct head with eyes and tentacles. Their bodies are twisted into a spiral shell that can be used to distinguish the various species (except for the nudibranchs, whose shells are very reduced or lost). Shelled marine mollusks extract calcium from seawater to build their shells, and the shells are modified according to the lifestyle of the animal. The snail has a thick, protective shell; the limpet shell is flattened and fitted to provide less resistance to wave action; the abalone has a flattened shape that allows its broad foot to exert great suction to hold onto a rock. The foot is also used to glide along the substrate. Gastropods scrape the algal film from rocks or cut up leaves for food. Some snails are scavengers that feed on dead flesh (performing a valuable cleaning function), and others prey on clams, barnacles, or other gastropods. Birds, fish, and humans also eat gastropods.

The terrestrial snails and slugs of the subclass Pulmonata lack gills, and the lining of the mantle cavity constitutes an air-breathing lung. They are bilaterally symmetrical, but often all of the body except the head and foot

Figure 37–16 A 20-kilogram Pacific octopus. The more than 200 sucking disks are used for locomotion, food capture, and manipulation. Each disk reportedly exerts 4 ounces of pressure.

is coiled 180° into the shell. Primarily nocturnal, they move by muscular "waves" in the foot as they proceed along a mucus track secreted by the pedal gland below their mouth. They have a well-developed nervous system with numerous ganglia and sensory organs, and a pair of retractile tentacles, the posterior ones bearing complex eyes and olfactory organs. Both male and female reproductive organs are present in each individual, and a pair copulates as each inserts its penis into the other. Each then lays eggs that develop in damp places. Development is direct (there is no larval stage), and miniature adult snails are hatched.

The shell-less gastropods are of interest because of their beauty. The nudibranchs and tectibranchs (Figure 37–15) are often brightly colored and patterned. The swimming flaps extending from the foot and the tentacles or fingerlike projections of the body used for respiration all contribute to the unusual appeal of these bright, gliding animals. The "sea hares" from this group, especially the genus *Aplysia*, are often used as experimental animals in studies of neurology.

The Cephalopoda include the familiar octopus and squid, and the nautilus. Most have well-developed heads and eyes. The foot is divided into 8 or 10 arms that have suckers for locomotion and prey-grabbing (Figure 37–16).

Only *Nautilus* has a large external shell; in other genera, the shell is internal, usually reduced, or even lost. Inside the mantle are the gills, and a mouth with biting mouthparts is below the "head." These animals can "swim" with their arms or jet along by taking water into the mantle cavity and forcibly ejecting it through the siphon. All octopuses and squids prey on other animals, including fish, worms, other mollusks, and crustaceans. In several species, poison "glands" in the mouth secrete a substance that kills the prey. The octopus then tears the prey apart and eats it. In turn, cephalopods are preyed upon by moray eels, some pelagic fish, and humans. Most octopuses live in crevices in rocks, in caves, under rocks, and on the ocean bottom in the deeps; squids live near the water's surface in plankton beds.

Reproduction in mollusks involves the laying of eggs, either before or after fertilization, and the hatching of larvae, which swim about freely, often near the water's surface. The nearly adult form settles to the bottom and completes metamorphosis. Some octopuses exhibit maternal behavior. They guard their eggs and brood them, refusing to eat during the brooding period. The mothers clean the eggs to prevent fungal growth and fight off all approaching animals. With this meticulous care, nearly all of the 45,000 eggs per female will hatch!

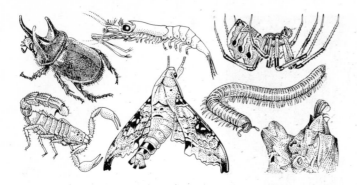

Figure 37–17 Phylum Arthropoda (insects, lobsters, crabs, spiders, millipedes, and centipedes).

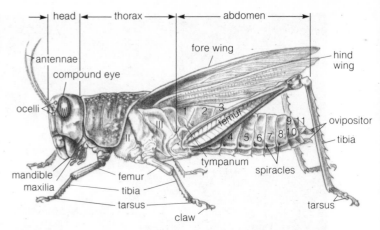

Figure 37–18 External segmentation in an insect. The appendages such as antennae, legs, and wings carried by different segments differ in structure.

Phylum Arthropoda: Crabs, Barnacles, Shrimp, Spiders, and Insects

The amazingly successful arthropods (jointed-legged animals) include three-quarters of the known animals—over 900,000 species (Figure 37–17). They live in water, on land, and in the air. Arthropods respire through gills, lungs, or tracheae and have an open circulatory system. They all have a hard exoskeleton, jointed legs, a body of several segments or units, jaws, and antennae. All of these structures reflect the segmentation seen in arthropods. Jaws, antennae, legs, and swimming structures are paired appendages, each of which is derived from a particular segment. Muscles and nerves effect coordinated movement for the several functions of these appendages. Figure 37–18 is a diagrammatic representation of a "typical" arthropod.

Arthropod classes are characterized by the pattern of fusion of their body segments and the numbers of pairs of legs they have. Let us first consider members of the class Crustacea—the crabs, shrimp, crayfish, and barnacles. The subclass Copepoda contains many species of minute animals. They are important in the food web, for they eat diatoms in plankton and are in turn eaten by larger animals. Living in fresh, brackish, or salt water, they are a rich and abundant food source. Many copepods are commensals of mollusks and coelenterates, and others are parasites on the skin or gills of fishes and whales.

Barnacles are members of the subclass Cirripeda. Although they look little like copepods, their embryology is similar, and a structurally similar larva hatches from the egg. All barnacle adults are sessile (fixed to the substrate) (Figure 37–19) and can attach to almost any substrate—including rocks, pilings, boats, and the shells of other animals. The larvae have a cement gland on the first antenna that produces an adhesive substance. They settle and complete metamorphosis. A barnacle feeds by extending its cirri (homologues of the thoracic appendages of other crustaceans) through the opening of the divisions

Figure 37–19 Stalked barnacles, with the cirri exposed.

of its shell. The cirri sweep the water, trap small particles of food, and stuff them into the mouth. Starfish, worms, and snails feed on barnacles, and occasionally fish and birds do, too. Several kinds of barnacles are parasitic on crabs or shrimp; they are often reduced and lack a digestive tract. The nonparasitic barnacles often have specific habitat preferences, forming great clumps in strips defined by the depths of the tide.

The subclass Ostracoda includes tiny crustaceans that are enclosed in bivalved shells. They use their antennae for swimming. Many species are planktonic, and some live on ocean or lake bottoms.

Figure 37–20 An American lobster *(Homarus)*, which ranges along the east coast of North America.

Bodega Marine Lab/Steve Renick

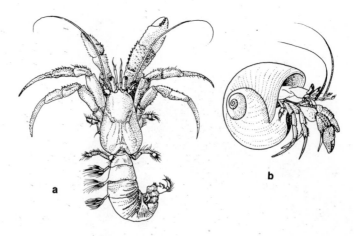

Figure 37–21 A hermit crab *(Pagurus):* (**a**) removed from snail shell; (**b**) as it is usually seen within a snail shell.

Members of the subclass Branchiopoda live primarily in fresh water. Included in the subclass are fairy shrimps, clam shrimps, and water fleas. *Daphnia pulex,* the water flea, is often examined in biology labs. *Daphnia* reproduce often, and females carry large numbers of eggs inside their bivalved shells. They are an important source of food in many fresh-water systems.

The subclass Malacostraca contains several orders. Isopoda are flattened and have seven pairs of legs of equal size. Some (including pill bugs and wood lice) live on land and some in fresh and salt water. They are scavengers, but you know that the flesh they eat doesn't have

to be dead if you've had your toes bitten by them. Some isopods are commensals; others are parasites on the gills and mouths of fishes.

The Decapoda include all the crabs, hermit crabs, lobsters, crayfish, and shrimps. They all have five pairs of legs, and their bodies have two regions, a head-thorax and an abdomen. There are three groups of decapods: the large-tailed forms, such as shrimp and lobster (Figure 37–20); hermit crabs; and the true crabs. The large-tailed forms include all sorts of species that humans eat, and the commercial shrimp and lobster catches are million-dollar enterprises. Some shrimp occur in great schools and can be netted easily.

Members of the second group have shorter abdomens and include burrowing shrimp, hermit crabs, stone crabs, and sand crabs. The burrowing shrimp are constantly maintaining their burrows, shifting huge quantities of sand out of the opening. The burrow may house many kinds of commensals—goby fish, pea crabs, copepods, small clams, and scale worms. Hermit crabs (Figure 37–21) have soft asymmetrical abdomens that they insert into the snail shells in which they live. The legs on their right sides are reduced because of the whorl of the snail shell. As a hermit crab grows, it must find larger shells in which to live, and it will fight other hermit crabs to get the right-sized shell. Many hermit crabs are scavengers, picking up bits of food as they scurry about, and some are detritus feeders that fan up mud with their claws and strain detritus from it.

The true crabs have a large carapace, or dorsal shell, and the abdomen is a small flap tucked under it. Their first pair of legs always has pincers. Some crabs eat detritus or plankton, but most are scavengers or predators. Many crabs seem to have a "masking instinct"—they

Figure 37–22 Horseshoe crab *(Limulus)*.

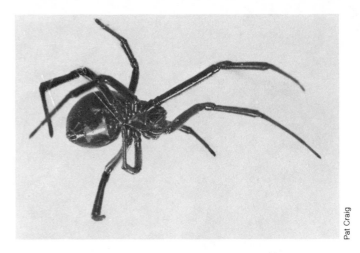

Figure 37–23 A black widow spider.

cover their exoskeletons with seaweed, anemones, tunicates, or debris, much of which continues growing and further disguises the animal.

The order Euphausiacea comprises few species but millions of individuals. They are tiny pelagic animals characterized by legs that have two parts of equal length. They are the "krill" eaten in vast quantities by whales.

Members of two other major arthropod classes are marine. The Xiphosura, or horseshoe crabs, are of interest because of their long fossil record. They have survived for hundreds of millions of years. These animals have an arched, horseshoe-shaped carapace over their cephalothorax (head-thorax) and a wide unsegmented abdomen (Figure 37–22). They eat worms and other small animals and live in shallow waters on the east coasts of North America and Asia. The Pycnogonida, or sea spiders, are usually tiny animals with small bodies and long legs. They often live on algae and coelenterates, or in the mantle cavity of mollusks. With their sucking mouthparts, they eat the tissue of the organisms on which they live.

Members of the class Arachnida are usually terrestrial and include spiders (Figure 37–23), scorpions, ticks, and mites. They have four pairs of legs and two pairs of mouthparts. The spider's body consists of a cephalothorax that has eight simple eyes anteriorly and bears the mouthparts and legs. It is covered by a hard cuticle. Spiders are free-living predators that usually eat insects. Some spiders hunt and seize their prey; others trap it in webs. The web is composed of proteinaceous threads secreted by special abdominal glands; the threads harden when they leave the body and are used variously as drag lines, multiple radiating threads, and as complex webs. (They are also used by humans as cross hairs in telescopes.) Nests and shelters are also constructed of these threads.

Spiders live in almost every sort of terrestrial habitat, from deserts to high mountains.

Arachnids are of some importance to humans. Most scorpions subdue prey by injecting a venom, and the venom of some species has a serious effect on us. Spiders have a reputation for lethal bites, but, in fact, few are toxic. Even the bites of the most dangerous spiders in the United States—the brown recluse and the female black widow—are fatal less than 5 percent of the time. Ticks and some mites are parasites and feed on blood. Not only can they cause anemia in their hosts if their numbers are great, but they may carry disease. For example, Rocky Mountain spotted fever is transmitted by ticks. Other forms, especially mites, suck plant juices and can cause extensive crop damage.

Chilopoda and Diplopoda are arthropod classes of centipedes and millipedes, respectively (Figure 37–24). They have 15 to 200 body segments; their heads bear antennae and chewing mandibles. Most segments of millipedes are double, with two nerve ganglia, two breathing tubes, and two pairs of legs. Both groups live in warm humid places under rocks and logs. Centipedes are active at night, pursuing small insects and small invertebrates. They kill their prey with poison inserted by a modified poison claw and then chew it with their mandibles. Millipedes are slow-moving animals that eat primarily dead plant material.

The class Insecta is one of evolution's greatest achievements. Insects are the most abundant of all land animals, in numbers of both individuals and species (nearly 900,000 of the 900,000+ species of arthropods known). Some can fly, and they can live in dry environments, because their chitinous external covering resists desiccation and protects the viscera against injury, and

Figure 37–24 A centipede, with one pair of legs per segment.

Figure 37–25 Order Orthoptera (crickets, praying mantises, walking sticks, locusts, grasshoppers, and cockroaches).

Figure 37–26 Order Lepidoptera (moths and butterflies).

also because they have a tracheal air-breathing system. They live in fresh and brackish (but not salt) water, in and on the soil, in and on plants and animals; from the tropics to the Arctic; and from sea level to 20,000 feet. As a group, they can use a tremendous variety of food sources—some parasitize plants or animals; some scavenge on dead material, and others eat parts of plants (leaves, seeds, fruits, nectar, roots, stems, and sap). The diets of individual species, however, may be highly specialized.

A typical insect has a distinct head, thorax, and abdomen, one pair of antennae, three pairs of legs, two pairs (or one or none) of wings, and mouthparts that are modified to correspond to the feeding mode.

Some 25 orders of living insects are currently recognized; let's look at the major groups among them. Four orders of tiny insects living primarily in soil and litter lack wings and lack metamorphosis. Members of the other 21 orders, all winged, are divided into two groups based on their developmental patterns. Eleven orders, including the Odonata (dragonflies), Orthoptera (grasshoppers and cockroaches), Isoptera (termites), Anoplura and Mallophaga (lice), and Hemiptera (true bugs), have a gradual metamorphosis in which the young are *nymphs* whose wings grow externally. The remaining 10 orders, including the Lepidoptera (moths and butterflies), Diptera (true flies), Siphonaptera (fleas), Coleoptera (beetles), and Hymenoptera (bees, wasps, and ants), have a complex metamorphosis in which young are larvae and wings are grown internally during a pupal stage.

In order to appreciate the diversity of insect adaptation, let us contrast the habits of selected members of some of the major orders. The order Orthoptera includes crickets, praying mantises, walking sticks, grasshoppers, locusts, and cockroaches (Figure 37–25). Several of these have strong jumping legs. Nearly all use their chewing mouthparts to eat leaves; however, the cockroaches eat almost everything. Plagues of locusts, in which fields are stripped of growing crops by hordes of migrating animals, have been recorded since antiquity.

The Lepidoptera (moths and butterflies) have distinctive habits (Figure 37–26). Often brightly colored, some are active by day and some by night. Their larvae (caterpillars) have chewing mouthparts and may do great damage to fruit and leaves of crops and trees. The adults, however, have sucking mouthparts formed of fused "jaws" and eat plant juices, including nectar. Many of these forms have highly specific diets—certain kinds of larvae, such as the tomato worm, can decimate particular hosts; the larvae of the silkworm moth feed only on mulberry leaves, and mulberry trees may be grown solely to support the valuable silk-producing larvae. Certain insect adults also feed from specific hosts and have incidental but significant effect as pollinators.

The Diptera, or true two-winged flies, include mosquitoes, gnats, syrphid flies (which mimic bees), houseflies, and fruit flies, as well as others (Figure 37–27). Most flies have piercing-sucking mouthparts, as anyone who has been bitten by a mosquito will attest. Several kinds

Figure 37–27 Order Diptera (mosquitoes, gnats, house flies, fruit flies).

Figure 37–28 Order Coleoptera (beetles).

Figure 37–29 Order Hymenoptera (wasps, bees, and ants).

of flies are disease carriers, such as the mosquitoes that transmit malaria and yellow fever and the flies that carry the agents for sleeping sickness. Not all flies are merely damaging, however. The tiny fruit flies *(Drosophila)* have served as laboratory animals for many of the basic studies on genetics and evolution. Like many insects, they reproduce quickly and in large numbers; their small size and simple diet make them excellent laboratory animals.

The largest order of insects, the Coleoptera, or beetles, includes some 300,000 species. They have chewing mouthparts and are excellent predators and scavengers. Beetles have their forewings modified as thick leathery structures that cover the flying wings, thorax, and abdomen (Figure 37–28). Some beetles burrow in soil, some live in caves, some dive under water, others live on the water's surface, and many live on land in litter and on plants. Both adults and larvae may damage crops and trees, for they include borers, weevils, and other harmful forms. Several kinds of beetles are beneficial, however; the ladybird beetles feed on aphids and scale insects, other beetles prey on various insect larvae as well, thereby constituting an insect control mechanism. Glowworms and fireflies are beetles and, if not beneficial, are certainly tolerated for their entertainment value. Flour beetles are the subject of much attention from population biologists.

The order Hymenoptera, including wasps, bees, and ants (Figure 37–29), exhibits some interesting habits. They have chewing or chewing-lapping mouthparts; larvae are often parasitic on larvae of other insects. Most species are solitary, but others live in colonies and have a complex social structure, as described in Chapter 23. The development of social structure, in which the population is separated into reproducers, food getters, nurses, guards, and workers, is of substantial evolutionary significance.

The class Insecta, in summary, constitutes a living laboratory in which the reproductive, developmental, physiological, and other adaptations seen in all animal life can be studied—within the constraints imposed by insect life and structure. Insects therefore provide a vehicle for the study of evolution and adaptation, even as they compete with one vertebrate species, *Homo sapiens,* for dominance of the biosphere.

The Deuterostome Phyla

Several phyla comprise the deuterostome lineage (Figure 37–30). They are diverse in structure, function, and habitat, but share fundamental features of development.

Phylum Echinodermata: Starfishes, Sea Urchins, and Friends The phylum Echinodermata includes the classes Asteroidea, the starfishes; Echinoidea, sea urchins and sand dollars; Ophiuroidea, brittle stars; Holothuroidea, sea cucumbers; and Crinoidea, sea lilies (Figure 37–31).

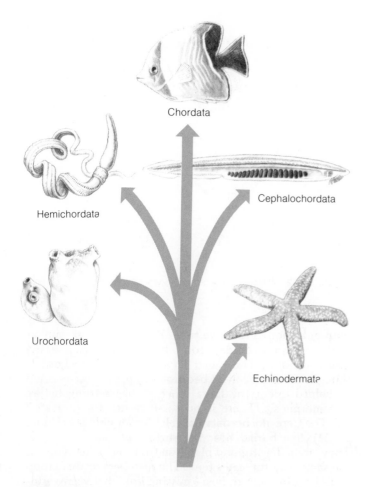

Figure 37–30 Deuterostome relationships.

Labels on figure: Chordata, Hemichordata, Cephalochordata, Urochordata, Echinodermata

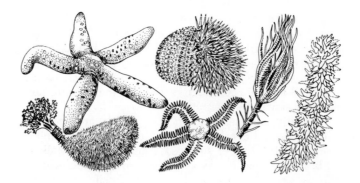

Figure 37–31 Phylum Echinodermata (starfish, sea urchins, sea cucumbers).

Echinoderms are radially symmetrical with a central mouth and have a calcareous skeleton of plates or spicules. All are marine, and few are parasitic. All are bottom dwellers and are found from the deeps of the oceans to tide pools. Their deuterostome mode of development allies them more closely to chordates than to the annelid-mollusk-arthropod lineage.

The Asteroidea include the familiar starfishes. Most have 5 arms (Figure 37–31), although one species has 4 and one has 40. The mouth is on the under surface, the anus on the upper. A starfish moves around by forcing structures called tube feet out of grooves on its arms. These are forced in and out as part of the water vascular system that controls them. Tube feet on the ends of the arms are sensory and help to locate food; some tube feet are respiratory. Starfish are carnivores and feed on clams, mussels, oysters, and other mollusks. "Crown of thorns" starfishes are having a destructive effect on Australian coral reefs through their predation. The starfish wraps itself around its prey, forces open the shell with the sucking disks on the ends of its tube feet, and extrudes its stomach around the soft body. Digestive juices dissolve the flesh, and the starfish ingests the resulting fluid. Sexes are usually separate in starfish, and eggs and sperm are spawned into the water. The bilaterally symmetrical larvae are free-living and planktonic for two weeks to two months before metamorphosis.

The echinoids have a body pattern similar to starfish, but without "arms." Sea urchins have spines that hinge to the skeleton (Figure 37–31). The spines may be used in locomotion, but tube feet are the primary means of movement. Some species live on rocky shores, others on sandy or muddy ocean bottoms. Sand dollars are flattened, and some forms live standing on end, oriented crosswise to the direction of the tidal flow. When the tide goes out, the dollar flops down and covers itself with sand. Echinoids use their spines to propel small food particles to their mouths.

The ophiuroids, serpent stars or brittle stars, have long arms around a small central disk. The arms are the primary means of locomotion, and the tube feet are primarily sensory and respiratory. These fragile animals are secretive and live under seaweed and rocks and on the deep ocean floor. They live primarily on detritus. An interesting defense mechanism of brittle stars is that of breaking off an arm if disturbed. Since they can regenerate structures in a short time, leaving a predator holding an arm while the rest of the star escapes is not much of a sacrifice.

The sea cucumbers, or holothurians, are soft-bodied, elongate echinoderms (Figure 37–31). The mouth is at the anterior end of the body, the anus posterior. Five body regions are homologous with starfish arms, and tube feet are present. Holothurians live on rocks, sand, or deep mud. They have tentacles that create water currents and trap small food items such as diatoms and detritus. Sea cucumbers have a large space at the end of their digestive tract. Water currents are brought into this, and oxygen is distributed to the body from here via the so-called respiratory tree—an unusual respiratory mechanism. This space in some species also houses commensals, including crabs and fish! In reproductive biology, the holothurians resemble starfish.

Sea lilies and sea feathers, or crinoids, are deep-sea animals with the echinoderm rayed body plan and a mouth on their upper surfaces. Anchored by a stalk to the bottom, especially as younger animals, they feed on detritus or small organisms. Many fossil species are known, and these animals were relatively abundant in the Paleozoic sea. Fewer species are alive today, and, unfortunately, little is known of their way of life.

The Protochordata   Three phyla of animals that have a notochord, or gill slits, and/or a dorsal hollow nerve cord compose the superphylum Protochordata. These characters are shared with vertebrates, but protochordates lack a vertebral column and other "truly chordate" features. (Some authors consider Protochordata a phylum and the three groups in it subphyla.) Members of the phylum Hemichordata, the acorn worms (Figure 37–32), live in mud flats or on the ocean bottom. They have a long proboscis, once assumed to contain a notochord. Some biologists now doubt that the proboscis has a notochord, so the phylum may not actually belong in the protochordates. In any case, in most hemichordates, there are paired pharyngeal gill slits, and the collar nerve cord is dorsal and hollow. The embryos and larvae of hemichordates are very much like those of starfish, suggesting a relationship to echinoderms as well as to chordates.

The tunicates, or sea squirts (phylum Urochordata), include many sessile forms that typically attach to rocks, floats, boats, and pilings. Their name is derived from the horny "tunic" that covers them, which itself forms a substrate for many animals. Sponges, bryozoans, mussels, tubeworms, and other tunicates all fasten to individual animals. They have two siphons through which they take in and expel water (Figure 37–33a). Tunicates have a large cavity, the branchial basket, with many gill slits that strain food (plankton or detritus) from the water. Some forms are colonial and share a common water system. They form masses of brown, white, yellow, red, or transparent animals that cover their substrate and may foul piers and ships' hulls.

It is the tunicate larva that has the other chordate characters. Eggs hatch into "tadpoles" with long tails, gill slits, a notochord, and a dorsal nerve tube (Figure 37–33b). The tadpoles swim freely for a time, then attach to the substrate and metamorphose to the sessile adult form that shows little resemblance to a vertebrate, lacking

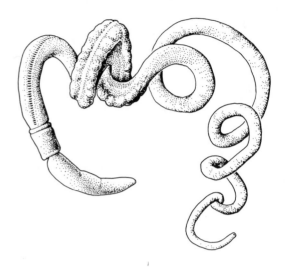

Figure 37–32   *Saccoglossus*, an acorn worm.

both notochord and nerve tube. The larva may be similar to the ancestral chordate form, but is not, of course, the direct ancestor. The tailed larva with its notochord is thought to have evolved because it improved locomotion or helped orient the larva when it was settling before metamorphosis. There is no fossil record for tunicates.

The Cephalochordata, lancelets or amphioxus (Figure 37–34), live buried near the surface of sandy beaches. They strain detritus and plankton from the water in much the same way tunicates feed. The members of this group were long thought to be a "missing link" between vertebrates and invertebrates because they have the characteristic gill slits, notochord, and tubular nerve cord, as well as a body form superficially resembling that of the jawless fishes. The notochord is well developed and extends the length of the body, as does the dorsal nerve cord. Since these features, and gill slits, are present in both larvae and adults, a close relationship with vertebrates is suggested.

Phylum Chordata: The Classes of Vertebrates   Probably because of our close kinship with them, we humans often find ourselves interested in and even sympathetic to other vertebrates (Figure 37–35). We share with them such characteristics as an epithelial body covering of epidermis and dermis, an internal jointed skeleton with axial vertebrae, a closed circulatory system with a heart, and a functionally differentiated brain and endocrine organs, as well as the developmental characters common to all chordates. (See Table 37–2 for a summary of their variations.)

There are seven classes of living vertebrates: Cyclo-

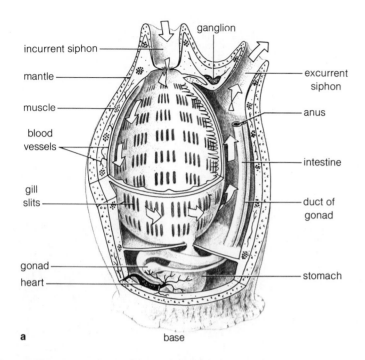

incurrent siphon
mantle
muscle
blood vessels
gill slits
gonad
heart

ganglion
excurrent siphon
anus
intestine
duct of gonad
stomach

a
base

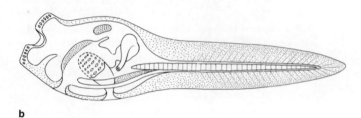

b

Figure 37–33 A tunicate. (**a**) An animal that is totally unlike the vertebrates in adult characteristics proves to have a notochord, a primitive vertebrate structure, during embryonic development. (**b**) The tunicate larva.

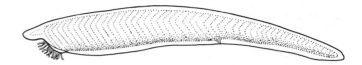

Figure 37–34 Amphioxus *(Branchiostoma).*

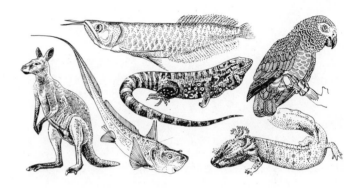

Figure 37–35 Phylum Chordata (vertebrates).

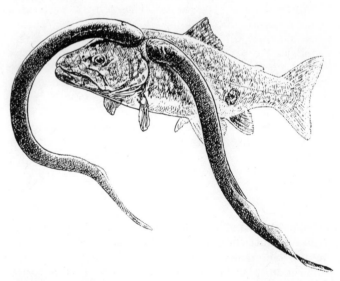

Figure 37–36 A sea lamprey attached to a lake trout, which also shows the scars of the attacks of other lampreys.

stomata, the jawless fishes (lampreys and hagfish); Chondrichthyes, the cartilaginous fishes (sharks, skates, and rays); Osteichthyes, the bony fishes; Amphibia (frogs, toads, salamanders, and caecilians); Reptilia (tuatara, turtles, snakes, lizards, and alligators); Aves, the birds; and Mammalia, the mammals. Vertebrates, with arthropods, dominate land, air, and marine habitats. These several groups have common ancestors at various points in evolutionary time, and each has undergone an extensive adaptive radiation.

Class Cyclostomata: Jawless Fishes The first vertebrates appear in the fossil record in Ordovician times,

some 400 million years ago. These were ostracoderms, armored jawless fishes now extinct, whose thick, plated armor has been preserved. There must have been earlier vertebrates, but we assume they had soft bodies not easily preserved as fossils. A great variety of armored jawless

Table 37–2 Vertebrate Classes and Their Characteristics

Class	Number of Living Species	Skeleton	Locomotion	Respiration	Circulation
Cyclostomata	50	Cartilaginous skull, visceral arches, notochord; vertebrae rudimentary; no jaws	Swims; tail compressed, median fins	5–16 pairs gills; one nostril	2-chambered heart; 6 pr. aortic arches; round, nucleated red blood cells
Chondrichthyes	650	Cartilaginous cranium with sensory capsules; notochord; vertebrae complete; pectoral and pelvic girdles	Swims; median and paired fins well developed	5–7 pairs gills; 2 nostrils	2-chambered heart; 6 pr. aortic arches; oval, nucleated red blood cells
Osteichthyes	30,000	Usually bone, some cartilage, fins supported by fin rays	Swims; lateral body movements; fins for balance and directional change	Gills covered by operculum; air bladder	2-chambered heart; 4 pr. aortic arches; oval, nucleated red blood cells
Amphibia	2,600	Bone, 5–185 vertebrae; limb and girdle modifications	0–2 pr. limbs, swim, walk, crawl, jump, or burrow	Lungs, gills in larvae and some adults; skin	3-chambered heart; 1–3 pr. aortic arches; oval, usually nucleate red blood cells
Reptilia	6,500	Bone, limbs, vertebrae modified	0–2 pr. limbs; swim, walk, run, climb, crawl, burrow	Lungs	Incompletely 4-chambered heart; 1 pr. aortic arches; oval, nucleated red blood cells
Aves	8,600	Bone, limbs modified for flight	Flight, walking, swimming	Lungs	4-chambered heart; right aortic arch; oval, nucleated red blood cells, internal temperature regulation
Mammalia	4,060	Bone, vertebrae regionally modified	Limbs modified for walking, flying, running, swimming, burrowing, climbing	Lungs	4-chambered heart; round enucleate red blood cells. Internal temperature regulation

fishes evolved and flourished in the seas, but they disappeared almost completely as more highly evolved fishes became abundant.

The only jawless fishes living today are relic forms, the hagfish and lampreys. These most primitive of living vertebrates are called the cyclostomes. They have long slender bodies equipped with dorsal and ventral fins; they lack jaws, scales, and paired lateral fins. Their brains are differentiated, and eyes, ears, and olfactory systems are functional sensory mechanisms. Although they lack jaws, they are effective feeders. Most lampreys are parasitic, with suctorial mouths for attachment to the fish they parasitize. The mouth forms a funnel and is equipped with horny teeth. The tongue, which also has teeth on

it, is used as a rasp to penetrate the circulatory system of the host, so that blood can be sucked (Figure 37–36). Hagfish have a protrusible biting mouth and eat invertebrates that live on ocean bottoms, as well as dead and dying animals that sink to the bottom. Both have survived by becoming specialized kinds of predators.

Class Chondrichthyes: Cartilaginous Fishes Fish life was transformed by the evolution of jaws and teeth. The anterior gill arches found in cyclostomes developed into upper and lower jaws, and the gill arch muscles evolved to open, shut, and maneuver the jaw elements. The jaw apparatus provides a stronger and better-controlled bite,

External Body Covering	Feeding	Nervous System	Reproduction	Habitat
Skin with mucous glands	Mouth suctorial or biting	Brain—8–10 cranial nerves; extensive nervous system; sense organs	Sexual; external fertilization; direct development or larval stage	Marine and fresh-water (lampreys), marine (hagfish)
Skin with mucous glands and placoid scales	Biting jaws, teeth; predaceous	More complex brain; 10 pr. cranial nerves; eyes, smell well developed	Internal fertilization; oviparous or ovo- or viviparous; larva or direct development	Marine; very few fresh-water
Skin with mucous glands and dermal scales	Mouth terminal; teeth well developed	More complex but as above	Usually external fertilization; usually oviparous; larva	Marine and fresh-water
Skin with many mucous and serous glands	Mouth with wide gape; suck, bite, flip tongue to catch prey	Increasingly complex brain; eyes, ears, smell well developed in most	External or internal fertilization; most oviparous; larva	Fresh-water, terrestrial
Skin with few glands; epidermal scales	Teeth or beak used to grab or rip food	Complex brain; 12 pr. cranial nerves; well-developed sense organs	Internal fertilization; oviparous to viviparous; amniote egg, no larva	Terrestrial primarily; some fresh-water, marine
Skin with feathers; oil glands	Beak modified for sucking, piercing, cracking	Complex brain; 12 pr. cranial nerves, eyes, ears well developed	Oviparous; no larva; amniote egg	Terrestrial
Skin with glands, hair	Mouth, teeth modified according to diet	Complex brain; expanded cerebrum and cerebellum	Internal fertilization; viviparous except monotremes; amniote egg	Terrestrial; some marine, some fresh-water

allowing fish to evolve many different ways of feeding. Toothed jaws enabled fish to become effective predators on other animals and on each other.

At about the same time that fish acquired jaws, they acquired lateral fins. With these modified skin folds, they could swim much more efficiently, turning and trimming more precisely and moving faster through the water. This new versatility and efficiency in both feeding and locomotion made fishes far more adaptable than before, and they branched out into various ways of life, especially during the Devonian period. In both salt and fresh water, fish with a variety of body forms and feeding habits abounded. Particularly successful were members of the class Placodermi—armored jawed fishes. Many different placoderms appear in the Devonian fossil record, but few survived that period and none live today.

Another group that apparently evolved during the Devonian is still successful today, however. This is the Chondrichthyes, the sharks, skates, and rays (Figure 37–37). The Chondrichthyes are cartilaginous fishes that have jaws, open gill slits, and pectoral, dorsal, anal, and tail fins. An important feature is that the skin is covered with tiny scales rather than armor, and the mouth contains enamel-covered teeth; both scales and teeth are of the same embryonic origin. The skates and rays have flattened bodies and generally live on the ocean bottoms, especially along the continental shelf. Several species eat detritus and small invertebrates, but some eat larger animals that

Figure 37–37 Class Chondrichthyes (sharks and rays).

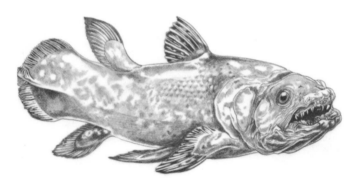

Figure 37–38 Coelacanth.

they uncover in the bottom sand. Sharks, which have a more "fishlike" body form, are active swimmers from the moderate deeps to the surface of the ocean. Most are predators that often feed on schools of fishes; some eat cephalopods and crustaceans; and some large species will attack seals or porpoises. Some forms of both groups, however, are exceptions to the general rule. The manta rays, for example, swim near the surface and eat planktonic animals; and some bottom-dwelling sharks have flattened bodies and forage for their food.

The main enemies of rays are sharks, many of which are active predators. The largest species of sharks, however, are not predators. The basking and whale sharks have no seizing teeth, but are equipped with long gill rakers that strain huge volumes of plankton from which they obtain food.

Chondrichthyans are of some economic importance because of their positions in food chains. Also, in spite of the wrecked nets, stolen fishes, and general harassment they cause, we make considerable use of sharks and rays. They are a food source; their skins are processed for sharkskin leather; and their livers are the primary source of oil containing vitamin A. Their speed, size, and maneuverability (sharks) and efficiency in exploiting the ocean bottoms (rays) have enabled them to compete successfully with the other major group of fishes that evolved during the Devonian—the bony fishes, or Osteichthyes.

Class Osteichthyes: Bony Fishes The Osteichthyes have developed in two distinct directions, represented by the lobe-finned and the ray-finned forms. The lobe-finned fishes lived prehistorically, as now, primarily in fresh, usually shallow, water. They have lungs, allowing them to use oxygen from air obtained at the water's surface, and *choanae,* or passages for transporting that air. Their fleshy fin lobes were probably adapted for support and maneuverability in shallow water. There have been three groups of lobe-finned fishes in evolutionary time, and members of two are alive today. One of these groups is represented by only a single known living species, the coelacanth *(Latimeria chalumnae),* which is found in deep waters off Madagascar and in the Indian Ocean (Figure 37–38). This "living fossil" has revealed a great deal to biologists investigating the evolution of vertebrates.

The other group of lobe-finned fishes survives in fresh-water streams in Africa, Australia, and South America. These are the lungfish, whose distribution implies that they are relics of a once greater range. These fish are adapted for life in streams that may dry up periodically. Besides being air-breathing, they burrow in the mud of the stream bottom and form a mucous cocoon in which they lower their metabolism and survive the dry period.

The ray-finned fishes have a broad adaptive zone and inhabit a great variety of marine and fresh-water situations. They are represented by a great diversity of feeding and locomotor types and have undergone an extensive adaptive radiation in the seas and in fresh water (Figure 37–39). They range from finless eels to blind cave fish to huge tunas to velvet black deep-sea forms.

All bony fishes are characterized by several features—among them, of course, a skeleton composed primarily of bone. Many are covered by dermal scales, and the gills are in a chamber covered by a bony operculum and open to the outside through a single opening on each side. Bony fishes are thought to have evolved in fresh water. Lungs may have developed in response to oxygen deficits in certain fresh-water habitats. Marine bony fishes maintain their fluids at a distinctly hypotonic level

Figure 37–39 Class Osteichthyes (the bony fishes).

Figure 37–40 Class Amphibia (frogs, toads, and salamanders).

relative to the environment, whereas fresh-water bony fishes are hypertonic to the surrounding medium. Salt-water forms maintain osmotic balance by drinking the salt water and using special mechanisms to eliminate large quantities of salts; fresh-water fishes constantly urinate.

Ray-finned fishes live in almost every available aquatic habitat—fresh, brackish, and salt water; various kinds of bottoms; from the water's surface to depths of over 30,000 feet in the ocean and in lakes at 15,000 feet in altitude; from the tropics to the Arctic. Some range widely among microhabitats; others are restricted in distribution and can survive only in a limited habitat.

Class Amphibia: Frogs, Toads, and Salamanders
Members of the class Amphibia (Figure 37–40), which arose from the lunged lobe-finned fishes, are the "first" vertebrates to spend a substantial part of their lifetimes out of water. Since the lobe-finned fishes were already equipped to breathe air, the major adaptation giving rise to amphibians was the modification of the lobe fin into a walking structure. The struts and rays of the fin gave rise to the limb. The primitive amphibian is thought to have been a small animal with a large head, a glandular skin, external scales, four limbs, a long, fleshy tail, and a three-chambered heart.

Living on land was a major adaptive shift. The habitat of Permian amphibians was probably a swamp, but the ancestral amphibians faced two major problems: how not to dry up, and how to get about on land. The first problem was solved by staying in moist habitats, retaining scaly skins that reduce water loss from their bodies, and having

mucous glands in the skin that secrete a moist covering. The second problem, locomotion, was solved in part when fins became limbs. However, water had given a great deal of body support that air could not provide. The earliest amphibians dragged their bodies, propelling themselves by thrusts of their limbs. They moved on land with a wavelike motion, much as fishes do in water.

Amphibians had an extensive radiation as amphibious predators. Some were large, bulky animals; others were small and agile. Still others were elongate, sometimes limbless burrowers. Many of these forms could not compete in the age of reptiles, but three orders of amphibians survive today. These are the Urodela, newts and salamanders, which still superficially resemble the presumed ancestral form in having long tails and four limbs of nearly equal length; Anura, the frogs and toads, which have elongate hind legs for jumping, swimming, or digging, and lack a tail; and the Gymnophiona, or caecilians, which are blind, limbless elongate burrowers that live exclusively in the tropics.

Several skin modifications equip modern amphibians to live on land. These include numerous mucous glands to keep the skin moist and poison glands that discourage or incapacitate predators. The skin has many blood vessels in some forms and is used for respiration in such forms as lungless salamanders. The mouth lining is also used for respiration. Only some of the caecilians have scales.

Amphibians range throughout the world. All caecilians are tropical; most burrow in loose, damp soil; and one South American family lives in fresh-water streams and rivers. Frogs live in the tropics, where they are diverse. They also range to high altitudes in the Andes

Figure 37–41 Class Reptilia (turtles, snakes, lizards, and alligators).

primarily on land. Not all live in hot, dry places. Many inhabit plains and forests, both temperate and tropical. A problem faced by all reptiles, and particularly by those living in hot places, is overheating. Reptiles cannot regulate their internal body temperatures as fully as birds and mammals can. Like fish and amphibians, reptiles are ectotherms, and their body temperatures are near those of the environment. They must seek food at a time of day when they will not overheat and must retreat to cooler shelter at peak heat periods. Conversely, they must emerge and allow themselves to warm up before they can move about at optimum levels of energy. (Ectothermy and its metabolic correlates are discussed in detail in Chapter 19.)

The mode of development and reproduction in reptiles is a significant evolutionary step beyond the amphibian condition. Reptiles copulate, usually after courtship; fertilization is internal. Most forms lay eggs, although in some species eggs are retained in the body of the female and living young are born.

Let us examine the structural features and habits of some living reptiles. Turtles' bodies are encased in a bony shell. This constrains movement but protects against predators or desiccation, for head and limbs can usually be drawn into the shell, leaving only a hard surface exposed. Turtles live in diverse habitats, both aquatic and terrestrial, in temperate and tropical areas. Some aquatic forms, including all the marine ones, have limbs modified for swimming. Tortoises are highly terrestrial and live in such sparse environments as the deserts of the southwestern United States (gopher turtles) and the Galapagos and Aldabra islands (giant tortoises). Turtles eat a variety of foods. Some are herbivores eating algae or land vegetation; several species, especially large aquatic ones, prey on invertebrates, fish, and young water birds.

Crocodiles and alligators have many modifications for life in water. They have short stout legs and webbed hind feet, long laterally compressed tails, and flattened heads. Their skins are thick and their backs are covered by a double row of bony scales. Their hearts are completely four-chambered, unlike those of other reptiles, and they have a hard palate that allows inspired air to be conducted to the pharynx via the internal nares (nasal openings) even though the mouth may be open under water. These modifications provide for efficiency of oxygenation. Alligators and crocodiles live in tropical and subtropical humid areas in Asia, Africa, Australia, and North and South America. They prey upon animals and eat anything from insects to medium-sized mammals—including humans—although fish are the dietary staple.

Lizards vary in size and form, from limbless, wormlike burrowers to the giant Komodo dragon, to the "Jesus Cristos" lizard that can run on the water's surface, to the algae-eating marine iguana of the Galapagos. Most lizards have paired limbs and a tail and are covered by

(some 16,000 feet) and well into the Arctic. Some frogs are aquatic, some are burrowers, some are terrestrial from deserts to forests, and some live exclusively on vegetation. Most families of salamanders live in temperate climate zones, but one family has had an extensive radiation in the New World tropics. The most primitive salamanders are aquatic, and a few species in more advanced families never complete metamorphosis—they remain aquatic, retain gills and other larval features, but become reproductively mature. Amphibians have to avoid desiccation, so they cope with seasonal change by burrowing during extreme cold or dryness, seeking water, or sealing themselves in a mucous envelope that hardens on contact with air.

Class Reptilia: Turtles, Snakes, Lizards, and Alligators
The class Reptilia shows the effects of several early evolutionary divergences (Figure 37–41). During the Mesozoic, the historic age of reptiles, the 14 diverse orders of reptiles were the dominant land animals. By the end of the Mesozoic, only 4 reptile lines remained. These 4 survive today: turtles and tortoises (order Chelonia), alligators and crocodiles (order Crocodilia), lizards and snakes (order Squamata), and tuatara (order Rhynchocephalia). These are the first vertebrates well adapted for life in dry places. They have nearly glandless skins covered with epidermal scales; their legs, when present, are structures for rapid, efficient locomotion. Each digit ends in a claw that makes locomotion even more efficient. They all have lungs, and their hearts approach the four-chambered condition that keeps oxygenated and deoxygenated blood from mixing.

Reptiles live in a great diversity of environments,

small, usually overlapping, scales. Their jaws have teeth, and they variously seize prey and take bites from it, gnaw vegetation, or use their tongues to catch insect prey, as do chameleons. Lizards live in the tropics, from lowlands to high altitudes (16,000 feet), in litter, near streams, and on vegetation. They also range throughout temperate zones toward the Arctic and live in both rich and poor habitats. Many species of lizards live in both New World and Old World deserts and have morphological and physiological adaptations to that habitat.

Snakes apparently arose from the same stock as lizards, but diverged to become subterranean, limbless burrowers with reduced eyes, which have lost their function through selection. Some snakes retain vestiges of a pelvic girdle, but most show no trace of limbs. As snakes diversified, they came to occupy a variety of habitats and developed a fully functional eye only slightly different from that of lizards. Some snakes still burrow or live in leaf litter, but the majority crawl on the surface of land, although they may hide under rocks or vegetation. Still others live almost exclusively in trees, especially in the tropics; and one family, the sea snakes, lives its entire life in warm marine waters. Several snakes, such as the water moccasin, live in fresh-water habitats as well. Snakes, like lizards, are covered with small scales, and most snakes have ventral scales modified as transverse plates that aid in locomotion. All snakes are predators, eating insects and other invertebrates, small vertebrates, and sometimes even larger vertebrates, including mammals.

The venomous reptiles particularly fascinate us. Only two species of lizard, the beaded lizard and the Gila monster, are poisonous; they occur in the southwestern United States and in Mexico. Their venom apparatus is poorly developed, and they rarely attack sizable mammals. Among the snakes are many venomous species found on most major land masses. The venom, produced in modified salivary glands, may be either conducted to a modified "injection" fang or released into the mouth where it mixes with saliva. The fangs can be either at the front or at the back of the jaw. The venom usually is a mixture of agents attacking nerves (neurotoxins) and blood (hemotoxins) in various ways. The venom is used to subdue small prey so they can be ingested whole and is used defensively against larger animals. The 19 species of venomous snakes in the United States kill many stock animals, but few people, each year.

Class Aves: Birds Birds, members of the class Aves, are well known to us because of their numbers, their distinctive colors, their songs, and their aerial and diurnal habits (Figure 37–42). Arising from the same reptilian stock that gave rise to crocodilians, they have adapted significantly to their aerial way of life. They are the only

Figure 37–42 Class Aves (the birds).

feathered vertebrates, they fly (with few exceptions), and they internally regulate their body temperatures.

A fossil creature has been found that is nearly intermediate between reptiles and birds, called *Archaeopteryx*. It has the long bony tail and toothed jaws of a reptile, but the well-developed wings and feathers of a bird.

Feathers are specialized structures that insulate birds against loss of body heat and function in flight as well. The front pair of limbs is modified to form wings, although not all birds fly, and the back pair is variously adapted for walking, perching, scratching, seizing prey, or swimming. The skeleton is modified for flight: The sternum (breast bone) is enlarged for the attachment of wing muscles, and many of the bones have air spaces within them, reducing their weight. The jaws lack teeth, but the horny bill is modified for obtaining various kinds of food. Birds have a four-chambered heart and an efficient circulatory system in which high blood pressures are maintained. They have a high rate of metabolism, which is necessary not only to meet the energy requirements for flight, but to maintain a constant body temperature. The efficient circulatory system facilitates both flight and temperature regulation by maintaining high oxygen levels for the muscles and organs of the body. Another adaptation for maintaining high energy levels involves the food that birds choose. It is often either high in sugar (nectar) or high in fat (nuts and seeds), both good sources of quick energy. Birds do not store fat as an energy reserve.

Birds have a *syrinx,* or voice box, and many communicate by complex calls or songs in mating, maintaining territories, warning, and other situations. Birds reproduce by copulating; fertilization is internal. All birds lay eggs,

Figure 37–43 Class Mammalia (the mammals).

usually with a hard shell reinforced by calcium carbonate. Some birds lay their eggs on bare ground; others make hollows; many construct elaborate nests. Some hatchlings, such as those of most songbirds, require a long period of parental care; others, such as chickens, are able to fend

for themselves in a very few days. The parental care, nest-building, and sociality of birds are evolutionary advances that help ensure survival.

Birds are distributed widely, from the Artic to the Antarctic, from sea level to above 20,000 feet, on sea coasts and islands, in forests, grasslands, and even in urban areas. There are nearly 9,000 species of birds, all with distinctive habits. Most species have fairly specific nesting sites. Some birds stay in an area all year, but many that nest in temperate regions retreat from cold by overwintering in subtropical or tropical areas, returning in the spring to the breeding sites. These migrations often cover thousands of miles over land and water.

The food of birds is diverse. Many species, as we noted, use high-energy seeds; others eat fruits. Some are carnivorous, and many eat invertebrates, especially insects. Fresh-water and marine birds often feed on fish; hawks and owls prey upon reptiles and small birds and mammals. Some birds eat carrion (dead animals), and the hummingbird feeds on nectar.

Class Mammalia: Mammals Members of the class Mammalia appeared at the end of the Mesozoic. The 30-odd orders of mammals alive today are the dominant group of land vertebrates (Figure 37–43). They are allied by being endothermic (warm-blooded), having hair to provide insulation, and having mammary glands so that females can nourish the young after they are born. Let us look at representatives of these several orders, both to see the diversity of animals in the class of which we are members, and to give us a basis for considering mammalian adaptations.

1. Monotremata. Egg-laying mammals. Duckbill platypuses and spiny anteaters. Australian region.

2. Marsupialia. Pouched mammals. Opossums, Tasmanian devils, koalas, wombats, kangaroos. Australia, where they are the dominant mammals; opossums in the New World.

3. Insectivora. Shrews, moles, hedgehogs. Europe, Asia, Africa, North and South America.

4. Dermoptera. Flying lemurs. Southeast Asia.

5. Chiroptera. Bats. Worldwide.

6. Primates. Lemurs, tarsiers, monkeys, apes, humans. Primarily subtropical and tropical Asia, Africa, South America; humans, worldwide.

7. Edentata. Anteaters, armadillos, and sloths. Subtropical and tropical New World.

8. Pholidota. Pangolins. Africa, southeast Asia.

9. Lagomorpha. Pikas and rabbits. North America, Europe, Africa, Asia; introduced in Australia.

10. Rodentia. Squirrels, gophers, mice, porcupines, guinea pigs. Worldwide.

11. Cetacea. Whales, dolphins, porpoises. Worldwide oceanic, and some fresh-water areas of Asia and South America.

12. Carnivora. Dogs, cats, bears, raccoons, foxes, weasels, otters, skunks, mongooses, hyenas, seals, walruses, sea lions. Worldwide.

13. Tubulidentata. Aardvarks. Africa.

14. Proboscidea. Elephants. Africa, southeast Asia.

15. Hyracoidea. Hyrax or coneys. Africa and Syria.

16. Sirenia. Manatees and dugongs. Worldwide in subtropical and tropical seas and rivers.

17. Perissodactyla. Horses, zebras, tapirs, rhinoceroses. Eurasia, tropical America, Africa.

18. Artiodactyla. Pigs, hippopotami, camels, deer, giraffes, antelopes, cattle, sheep, musk oxen, water buffalo. Worldwide except Australia.

From this cursory review of the kinds of mammals and their general distributions, we can see a tremendous range of adaptations. Whales, dolphins, and seals have many limb and feeding modifications for their marine life; bats have modified their forelimbs for flight; moles have adapted theirs for burrowing; and monkeys have modified theirs for an arboreal existence.

Mammals eat a great variety of foods, and their teeth are modified according to food type. Rabbits, mice, and beavers have long, sharp incisors for cutting vegetation; dogs and cats have long canines for capturing prey and bladelike cheek teeth for ripping meat; we omnivorous humans have sharp incisors for biting, canines for tearing, and molars for crushing and grinding. These are among the several adaptations for efficient food-processing. When food becomes temporarily or permanently scarce, mammals may migrate to a new area, store food as a reserve, or hibernate or aestivate, reducing their metabolic activity and staying inactive during seasons of scarcity.

Mammals, like birds, are endotherms that internally regulate their body temperatures. This makes their physiology generally more efficient and also allows them to exploit a number of habitats that ectotherms could not tolerate. Several of the mammalian characteristics already mentioned work together in endothermy; for example, hair is good insulation against heat loss. The several adaptations, morphological and physiological, that provide efficient food-processing help maintain the proper metabolic level. Modifications of the digestive tract, including teeth, facilitate the effective breakdown of food. The highly efficient circulatory system transports those food molecules, and oxygen, to cells that need them for homeostasis. Endothermy and homeostasis are significant factors in mammals' biological success.

Reproductive specializations in mammals are many. Males have penises; fertilization is internal. Monotremes lay eggs and incubate them; they then nurse and otherwise care for the hatchlings. Marsupials retain fertilized eggs in the uterus, where the modified membranes of the eggs are pressed close to the uterine wall to form a primitive kind of placenta. When the tiny, immature young are born, they crawl into the marsupium, or pouch, where they nurse and grow, emerging when they can begin to fend for themselves. All other mammals retain developing young in the uterus for a relatively long time and have an efficient placenta that facilitates nutrition and gas exchange. When young are born, they are nursed for a time. Mammals have highly developed systems of parental care and training of young. They build nests and shelters where they nurse and protect the young, and in many species both male and female actively train the young in getting food, defending themselves, and interacting socially. Thus, the subclasses of mammals show a trend toward more parental care, more efficient placentas, longer gestation time, and more efficient milk production.

Many species are markedly social. Deer, horses, and whales move in herds; thousands of bats may share a cave; gophers and prairie dogs construct whole underground "cities"; wolves travel in packs, baboons in troops. There may be a definite social structure to these aggregations. Dominance hierarchies in which each individual has an ordered place are seen in many species, including lions, wolves, and baboons. Although members of many species seem to mate randomly, males and females of some species form stable pairs that may last for a season in order to raise the young, or perhaps for a lifetime. Some systems involve a dominant male who has a "harem" of females that he defends against other males. Members of other species may be solitary, aggregating only for breeding.

Our relationship to other mammals is a long-standing one. Cave paintings thousands of years old show humans hunting large mammals for food; we continue to eat various mammals, both wild and domesticated. We also study the biology and psychology of mammalian species, especially rats and monkeys, in order to gain a better understanding of ourselves.

Summary

The Kingdom Animalia includes multicellular organisms that have great diversity of structure, function, and habitat. A number of general evolutionary trends are seen in members of the kingdom. The evolution of multicellularity provides an increase in efficiency through division of labor among organized cellular units (which may become tissues and organs). More primitive groups of animals are aquatic; terrestriality, including flight and burrowing, seem to be derived features.

The symmetry of body plan reflects evolutionary trends as well. Coelenterates are radially symmetrical; flatworms and many more advanced groups are bilaterally symmetrical. The acquisition of complex organ systems also characterizes animal evolution. Coelenterates have tissues, a nerve net, and a digestive system with only one opening. Flatworms have an excretory system and the beginnings of a centralized nervous system. Annelid worms are segmentally organized, with organs from the several systems of the body present in each segment in a coordinated manner. Mollusks, related to annelids by some features of embryology, also have a circulatory system with a heart, a well-developed nervous system, and a respiratory system. The arthropods, also related to annelids, include great numbers of species that inhabit a great variety of situations in water and on land. They all have a protective exoskeleton and jointed appendages. Echinoderms are evolutionarily specialized marine animals, more closely related to chordates than to other invertebrates by embryological characteristics. The chordates have an internal skeleton, a dorsal nervous system with a differentiated brain, endocrine organs, and a closed circulatory system with a heart. The seven classes of living chordates include the Cyclostomata (jawless fishes, such as lampreys), the Chondrichthyes (sharks, rays, chimaeras), Osteichthyes (bony fishes), Amphibia (frogs, salamanders, caecilians), Reptilia (lizards, snakes, turtles, and alligators), Aves (birds), and Mammalia (platypus, kangaroo, bat, whale, elephant, mouse, human). The evolutionary trends in complexity, efficiency, and habitat diversity are reflected in these subgroups as well.

Suggested Readings

Books

Barnes, R. D. 1968. Invertebrate Zoology, 2nd ed. Saunders, Philadelphia.

Bold, H. C. 1964. The Plant Kingdom, 2nd ed. Prentice-Hall, Englewood Cliffs, N.J.

Borror, D. J., and D. M. Delong. 1971. An Introduction to the Study of Insects, 3rd. ed. Holt, Rinehart and Winston, New York.

Calvin, M. 1969. Chemical Evolution. Oxford University Press, Oxford.

Dillon, L. S. 1967. Animal Variety. Wm. C. Brown Co., Dubuque, Iowa.

Hanson, E. D. 1964. Animal Diversity, 2nd ed. Prentice-Hall, Englewood Cliffs, N.J.

Harris, R. M. 1969. Plant Diversity. Wm. C. Brown Co., Dubuque, Iowa.

Hyman, L. H. 1940–1967. The Invertebrates. McGraw-Hill, New York, 6 vols.

Jensen, W., and F. Salisbury. 1972. Botany. Wadsworth, Belmont, Calif.

Kudo, R. R. 1966. Protozoology, 5th ed. C. C Thomas, Springfield, Ill.

Romer, A. S. 1970. The Vertebrate Body, 4th ed. Saunders, Philadelphia.

Scagel, R. F., R. J. Bandoni, G. E. Rouse, W. B. Schofield, J. R. Stein, and T. M. C. Taylor. 1965. An Evolutionary Survey of the Plant Kingdom. Wadsworth, Belmont, Calif.

Storer, T. I., R. L. Usinger, R. L. Stebbins, and J. T. Nybakken. 1972. General Zoology. McGraw-Hill, New York.

Young, J. Z. 1962. The Life of Vertebrates, 2nd ed. Oxford University Press, New York.

Articles

Adler, Julius. 1976. The sensing of chemicals by bacteria. Scientific American 234(4):40–47.

Alldredge, Alice. 1976. Appendicularians. Scientific American 235(1):94–102.

Banks, Harlan P. 1975. Early vascular land plants: proof and conjecture. BioScience 25(11):730–737.

Barghorn, E. S. 1971. The oldest living fossils. Scientific American 224(5):30–42.

Berg, Howard C. 1975. How bacteria swim. Scientific American 233(2):36–44.

Caldwell, Roy L., and Hugh Dingle. 1976. Stomatopods. Scientific American 234(1):80–89.

Calvin, M. 1956. Chemical evolution and the origin of life. American Scientist 44:248–263.

Faulkner, Douglas. 1977. A living fossil, the nautilus, glides through the ages. Smithsonian 8:76.

Feder, H. M. 1972. Escape responses in marine invertebrates. Scientific American 227(1):92–100.

Fox, S. W. 1960. How did life begin? Science 132(3421):200–208.

Glassner, M. F. 1961. Pre-Cambrian animals. Scientific American 204(3):72–78.

Hohn, E. O. 1969. The phalarope. Scientific American 220(6):104–111.

Jensen, D. 1966. The hagfish. Scientific American 214(2):82–90.

Johansen, K. 1968. Air breathing fishes. Scientific American 219(4):102–111.

Kooyman, G. L. 1969. The Weddell seal. Scientific American 221(2):100–106.

Lapan, Elliot A., and Harold J. Morowitz. 1972. The Mesozoa. Scientific American 227(6):94–101.

Margulis, L. 1971. The origin of plant and animal cells. American Scientist 59:230–235.

Melnick, Joseph L., Gordon R. Dreesman, and F. Blaine Hollinger. 1977. Viral hepatitis. Scientific American 237(1):44–52.

Rothschild, M. 1965. Fleas. Scientific American 213(6):44–53.

Spector, Deborah H., and David Baltimore. 1975. The molecular biology of poliovirus. Scientific American 232(5):24–31.

Yonge, C. M. 1975. Giant clams. Scientific American 232(4):96–105.

Zentmyer, George A. 1976. Phytophthora—Plant destroyer. BioScience 26:686–689.

Glossary

A

absorption spectrum The amount of light absorbed by a pigment at various wavelengths.

actin A contractile protein in muscle cells and other cells where movement is found.

action spectrum The effectiveness of a process, such as photosynthesis, at various wavelengths of light.

activation energy The minimum energy required to bring about a reaction.

active transport The movement of materials into and out of the cell by the cell's expenditure of energy.

adaptation A genetically determined characteristic that enhances the ability of an organism to cope with its environment.

adaptive zone A particular environmental opportunity requiring similar adaptations for diverse species. Species in different adaptive zones usually differ by major characteristics of form and function.

adenosine triphosphate (ATP) A nucleotide used as a short-term energy storage and transfer compound in cells.

adrenal glands In humans, two glands that sit atop the kidneys and produce the hormones epinephrine and norepinephrine. They also produce male and female sex hormones.

allantois An extraembryonic saclike extension of the hindgut of amniotes, serving in excretion and respiration.

allele One of an array of forms of a given gene.

allopatric speciation Formation of two or more species from a preexisting one as a result of disruption of a formerly continuous range and adaptation in isolation to the restricted range remnants.

amino acids A class of organic compounds based on a structural plan of a central carbon atom with an amino group (NH_2) on one end and a carboxyl (COOH) on the other.

anaphase The time during mitosis and meiosis when the chromosomes move to the poles. In meiosis, the homologous chromosomes move to opposite poles of the spindle.

animal pole The upper, less yolky pole of an egg.

antheridia Structures in plants in which sperm are produced.

antibodies Proteins produced by leucocytes in response to the presence of antigens.

antigen A foreign protein or substance that enters the body.

apical (or primary) meristems Regions at the apex of shoots and roots where cell divisions occur.

archegonia Structures in plants in which eggs are produced.

arteries Blood vessels that carry blood away from the heart.

artificial selection Change in the genetic constitution of a breeding line of organisms as a result of mating patterns selected by humans.

assortative mating The choosing of mates that are genetically similar.

atom The structural unit on which the chemical characteristics of life are based.

atomic nucleus The center of the atom, consisting of protons and neutrons.

atomic number The number of protons in an atomic nucleus; the number that identifies that atom.

autecology The study of organisms in relation to their physical environment.

autonomic nervous system That portion of the nervous system that reflexively controls bodily functions generally considered outside conscious control.

autosomes The chromosomes of a cell, except for the sex chromosomes.

autotroph An organism that can manufacture its own food. Green plants and photosynthetic bacteria are autotrophs.

axons Short branches of the cytoplasm of neurons that generally lead nerve impulses away from the cell body.

B

balancing selection Selection that maintains polymorphism in a population.

Batesian mimicry The similarity between edible and noxious organisms by which the edible ones gain some protection by being less heavily preyed upon.

bile salts Compounds that dissolve the fat droplets of food.

blastocoel The cavity in the blastula.

blastocyst The hollow sphere of early mammalian development.

blastodisk The mass of dividing cells atop a yolky egg that will give rise to the embryo, as in a chicken.

blastopore The opening into the gastrocoel.

blastula An early embryo, the cells of which are commonly arranged in the form of a hollow sphere.

C

callus A mass of undifferentiated plant cells that has developed from a single cell or collection of cells.

Calvin cycle or C_3 cycle One of the carbon dioxide fixation pathways usually associated with photosynthesis. The first product of the reactions is a three-carbon sugar.

canalizing selection A kind of natural selection in which individuals producing offspring with a more or less standard and uniform morphology or physiology are favored.

capillaries Tiny blood vessels that connect arteries and veins.

carbohydrate One of a series of compounds that contain carbon, hydrogen, and oxygen in the approximate ratio of 1:2:1.

cardiac muscle Multinucleate, striated muscle found in the heart.

carnivores Organisms, usually animals, that eat animals.

carotenes Pigmented lipid molecules always found associated with chlorophyll in chloroplasts but also found in high concentrations in chromoplasts.

carrying capacity The number of individuals of a population that the environment can sustain.

catalyst A substance that increases the rate of a reaction without appearing in the overall net equation of the reaction.

cell The basic unit of life. All organisms are composed of cells, and all cells arise by the division of existing cells.

cell wall The rigid or semirigid outer covering of plant cells.

centrioles A pair of small, barrel-shaped structures in the cytoplasm just outside the nuclear envelope of animal cells and some plants that acts as the spindle organizing center.

centromere The point of attachment of the spindle fibers to the chromosome.

cerebellum In the vertebrate brain, the center for involuntary muscular coordination. It is most highly developed in birds and mammals.

cerebrum In humans, the part of the brain that coordinates sensory stimuli and directs motor responses; it is the site of conscious mental processes. In lower vertebrates, such as the fishes and amphibians, the cerebrum is concerned primarily with smell and is not essential to survival.

character displacement The idea that, if two species are in competition for some limited resource, they will diverge in features associated with utilization of the resource to the point where competition is reduced.

chemical bonds Those forces that hold atoms together in molecules. See specific types.

chiasmata Points on homologous chromosomes where there has been an exchange of chromatid segments.

chlorophyll A green pigment found in plants that absorbs the light energy used in photosynthesis.

chloroplast A chlorophyll-containing cell organelle in which photosynthesis occurs.

chorion The outer membrane surrounding the embryo.

chromatid The basic organizational unit of the chromosome; two chromatids united in the region of the centromere compose a chromosome.

chromatin An association of DNA and protein found in the nucleus of eukaryotic cells.

chromoplasts Plastids found in plants that contain high concentrations of nonphotosynthetic pigments.

chromosome One of a group of structures found in the nucleus of a cell, composed of DNA and protein; the genes are located on the chromosomes.

cilia Shortened flagella usually present in large numbers. Found in protozoa and various cells in higher animals—for example, the cells lining the human respiratory tract.

circadian rhythm A change in an organism that occurs daily.

citric acid cycle The cycle that results in the complete breakdown of pyruvic acid to CO_2 and H_2O with the release of energy.

cleavage The successive divisions of the zygote to form the multicellular blastula.

climax community The end point of ecological succession; the particular mix of plants and animals that represents a near stable condition, regenerating itself through time after disturbances.

coadaptation The result of two or more populations responding to selection pressures to their mutual advantage.

coelom A cavity bounded by mesodermal epithelium.

collenchyma Long-celled parenchyma with thickened walls, particularly in the corners. Found in young, rapidly growing organs.

commensalism An interaction between populations in which members of one are benefited, but members of the other are neither benefited nor harmed.

community An association of interacting populations, usually living in one restricted area.

competition Use or defense of a resource by an individual or population so that use of the resource by others is restricted.

competitive exclusion The exclusion of members of one species if certain resources required by potential competitors are limited in quantity. One species is usually superior in its ability to utilize the limited resource.

compound A pure substance composed of two or more elements in fixed proportions.

compound eye The type of eye found in the arthropods, composed of many separate optical units called ommatidia.

cones One of the two types of sensory cells found in the eye (the second are rods). Cones are the less sensitive of the two, but are responsible for sharpness and color.

conjugation (bacterial) The process in which a connection is formed between two bacterial cells and a segment of DNA is passed from one cell to the other.

cork cambium The layer of dividing cells that forms the bark or outer covering of a plant.

cornea A transparent portion of the spherical shell of the eye.

cortex The outer portion of an organ or part. In vascular plants, a tissue located between the epidermis and the stele.

covalent bonds Chemical bonds based on the sharing of electrons by two or more atoms.

crypsis Any feature of an organism that reduces its visibility to potential predators or prey.

cuticle The waxy covering on the surface of land plants. Also, the chitinous exoskeleton of insects.

cytochromes Compounds containing ring structures with iron in the center, primarily involved in electron transport.

cytokinesis The division of the cytoplasm of the cell, usually following closely after mitosis or meiosis.

cytokinins A class of plant growth hormones that is important in the control of cell division.

cytology The science of the cell.

cytoplasm That portion of the cell not occupied by the nucleus or vacuole in a eukaryotic cell or the nucleoid in a prokaryotic one.

D

dark reactions A series of reactions in which CO_2 is incorporated into organic compounds at the expense of high-energy compounds; usually considered the second half of photosynthesis.

deme Potentially interbreeding individuals that live and interact with each other; a local population.

denaturation The disorganization of a protein molecule by changes in its structure, frequently through the disruption of hydrogen bonds.

dendrites The tapered, highly branched extensions of the neuron that carry nerve impulses toward the cell body.

denitrifying bacteria Bacteria that convert nitrate to gaseous nitrogen.

density-dependent factors Influence of environmental conditions on population that varies with the number of individuals present.

density-independent factors Influence of environmental conditions on the population that is independent of the number of individuals present.

diastole That part of the heartbeat that takes place when the ventricles are relaxed and filling with blood from the atria.

differentiation Process by which cells develop specific characteristics or patterns.

diffusion The movement of suspended or dissolved particles from a more concentrated to a less concentrated region as a result of the random movement of the individual particles.

diploid A cell or organism with double the basic chromosome complement.

disassortative mating The choosing of mates that are genetically dissimilar.

disruptive coloration A color pattern that has the effect of distracting another organism and making the individual less visible by making it blend in with the background.

dormancy In plants, a period where growth does not occur, as in seeds or perennials.

double bond A covalent bond in which two pairs of electrons are shared by two atoms.

double fertilization A process unique to flowering plants in which the pollen tube discharges two sperm; one fuses with the egg to form the zygote, and the second unites with the two polar nuclei to give rise to the endosperm.

E

ecdysis The process of molting, or the discarding of an old, smaller cuticle or exoskeleton and the formation of a new, larger one.

ecocline Gradual change in a character along some ecological or geographical gradient.

ecogeographic rules Generalizations concerning natural selection as it affects many species over broad geographic and ecological ranges.

ecosystem The interacting parts of the biotic and physical environment.

ectoderm The outermost of the three primary germ layers.

ectotherms Animals that take up heat from the environment.

electron A small, negatively charged particle that orbits the atomic nucleus.

element A pure substance consisting of distinctive atoms.

embryo An immature stage of an organism.

endocrine glands Ductless glands that liberate their products directly into the blood.

endocytosis The mechanisms by which substances are taken up by the cell that involve membrane invaginations and fusion.

endoderm The innermost of the three primary germ layers.

endodermis The innermost layer of the cortex of the vascular plant, specialized to help regulate the flow of materials to the xylem.

endoplasmic reticulum (ER) A tubular membrane system found in eukaryotic cells to which ribosomes are frequently attached.

endosperm A tissue formed by the union of a sperm nucleus with the two polar nuclei during double fertilization in flowering plants. Usually triploid, it is essential to the development of the embryo in seed plants.

endotherms Animals that use heat from their own metabolism to regulate their body temperature.

entropy The randomness or disorder of a system.

environmental resistance The sum of factors that cause growth of populations to slow.

enzyme An organic catalyst. All enzymes are proteins.

epidermis The outer epithelial portion of the skin or the outer layer of cells of the primary plant body.

epithelium Cellular layer covering a free surface.

erythrocytes Red blood cells, the oxygen-carrying cells of the blood.

estrogen Hormone produced primarily by the ovarian follicle.

estrus Periods of hormonally regulated sexual excitement in the female.

ethology The study of the inborn behavior of animals under field conditions. Includes the study of the evolution of behavior and its adaptive significance.

eukaryotic cell A cell whose internal structure is based on a series of membrane-bound organelles. This type of cell is found in all organisms except the bacteria and blue-green algae.

Eustachian tube A duct that connects the middle ear to the pharynx and allows for the equalization of pressure in the middle ear.

evolution Gradual change in a species resulting from changes in frequencies of alleles in a gene pool.

exocrine glands Glands that release their products to the exterior or into a body cavity through a duct or tube. Examples are the mucous, sweat, digestive, and salivary glands.

exocytosis The process by which cells excrete materials through the fusion of the membrane of the vesicle in which the material is present with the plasma membrane.

exoskeleton A relatively hard exterior covering that acts as a skeleton. In arthropods, the exoskeleton consists of a cuticle to which the muscles are attached from inside.

F

fecundity The number of young produced by a given female during a stated period.

fertilization Union of egg and sperm resulting in the formation of the zygote.

fitness The relative advantage of one individual over another in terms of the production of offspring.

flagella (sing., flagellum) Hairlike extensions of the cell surface with a core of microtubules capable of whiplike movement, found in motile animals and in plant cells.

flame cells The excretory cells of the flatworms.

fluorescence The emission of a photon by a molecule that has been excited and then falls to the ground state.

free energy Usable energy in a chemical system.

G

gall bladder An organ that temporarily stores large amounts of bile.

gametophyte That generation of an organism that consists of haploid cells.

gastrin A hormone secreted by the cells of the stomach that circulates in the blood back to the stomach wall, where it promotes the production of gastric juices.

gastrocoel The cavity within the gastrula.

gastrula The embryo in the stage of development following the blastula, consisting of a sac with a double wall.

gene A length of DNA in the chromosome containing the information required for a unit of function in the cell; the fundamental unit of heredity in the cell.

gene flow Movement of genes from one population to another as a result of movement of individuals.

gene pool All the genes in a population.

genetic drift Changes in the frequency of alleles from generation to generation as a result of chance deviations from expectations.

genetic inertia The tendency for frequencies of alleles to remain stable.

genetic polymorphism The presence of two or more alleles of a given gene in a population.

genome The total complex of genes in an organism.

genotype The genetic constitution of an organism.

geotropism The growth response of plants to gravity.

gibberellins A class of plant growth substances primarily involved in the control of cell elongation.

glycolysis The breakdown of organic molecules into short two- or three-carbon segments that are the immediate fuels for oxidative respiration.

Golgi complex A cell organelle consisting of a collection of flattened vesicles appressed to one another. The site of assembly of substances to be secreted by the cell.

Golgi tendon organs Receptors associated with tendons that register changes in the tension of the muscle.

gonad A reproductive gland, ovary or testis.

guard cells Highly specialized epidermal cells that regulate the size of the stomata.

H

habitat Actual place where members of a population live.

haploid The basic chromosome number of an individual or species.

Hatch-Slack pathway or C_4 pathway An alternative to the Calvin cycle for the initial fixation of CO_2 in photosynthesis that results in the formation of four-carbon compounds.

Haversian system The structural unit of bone consisting of repeating, roughly cylindrical units composed of a central canal enclosed within concentric layers of calcium phosphate.

hemoglobin The iron-containing protein in the blood that carries oxygen.

herbivores Animals that eat plants.

heterotrophs Organisms that cannot manufacture their own food and live by eating or digesting autotrophs. All animals and nongreen plants are heterotrophs.

heterozygote The genetic condition in which the two alleles of a gene in a diploid individual are different.

homeotherms Animals that maintain a constant body temperature despite environmental temperature fluctuations.

homologous chromosomes A pair of identical chromosomes found in a diploid cell that will pair at meiosis.

homology Similarity in structure and origin of two different organs or parts in different species.

homozygote The genetic condition in which both alleles of a gene in a diploid individual are the same.

hormone An organic molecule produced in one part of the body but affecting some other, separate part.

hybridization Mating and production of offspring by two dissimilar individuals, usually members of different species.

hydrogen bond An intermolecular bond based on the unequal sharing of electrons in covalent linkages.

hyphae The filaments that constitute the vegative stage of most fungi.

hypothalamus The brain center responsible for internal homeostasis; it monitors and controls body temperature, water balance, blood pressure, and carbohydrate and fat metabolism.

I

imprinting A permanent change of behavior in an animal resulting from a stimulus or group of stimuli.

independent assortment Segregation of one chromosome pair occurring independently of that of all other pairs.

indoleacetic acid (IAA) One of a group of plant growth hormones known as auxins that control a large number of processes in the plant, particularly cellular elongation.

induction Production of one part under the influence of another during development.

instinct Generally a behavioral pattern that appears to be inborn and usually not modified by learning.

insulin A hormone produced by the pancreas that regulates carbohydrate metabolism in cells.

interphase That period in the life of the cell when it is not dividing and during which replication of cell parts takes place.

interstitial Relating to space or structure within parts.

ion An atom or molecule that has lost or gained one or more electrons.

ionic bond A chemical bond based on the losing or gaining of an electron.

isolating mechanism Any genetically controlled feature of populations that restricts interbreeding with members of other populations.

isotopes Different forms of an element, all with the same number of protons but varying numbers of neutrons.

J

joint receptors Complex nerve terminals located at joints that can sense movement within the joint.

K

karyotype The collective morphology of the chromosomes of an organism.

L

larva Developmental stage that differs from the embryo in being able to secure its own nourishment, and from the adult in lacking maturity, especially sexual.

lateral (or secondary) meristems Regions of cell division located along the axis of the plant that give rise to tissue that results in increase in diameter of the plant.

learning The modification of behavior by experience.

learning theory The concept that at birth the brain is a blank slate and that behavior is a result of learning based on rewards and punishments.

leucocytes A class of cells found in the blood that is the basis of the immune system; also known as the white blood cells.

ligaments Bands of tough connective tissue that span the joint and bind its elements together.

light reactions Those reactions of photosynthesis in which chlorophyll absorbs light and energy-rich compounds, including ATP, are produced.

lineage An ancestor-descendant sequence extending through time.

linkage The phenomenon of genes located on the same chromosome moving as a unit during reproduction.

lipids A diverse group of organic compounds that are greasy to the touch and insoluble in water but soluble in organic solvents.

local population All the individuals of a species that live together and interbreed in a place.

locus The place on a chromosome where a given gene is found.

lysogeny The incorporation of viral DNA into the chromosomes of the host, where it may remain inactive for long periods of time before finally becoming active, destroying the cell.

lysosomes Single membrane-bound cell organelles containing enzymes capable of breaking down most biological molecules.

M

Malpighian tubules Sacs or pouches branching out from the gut of insects that act as excretory organs.

medulla oblongata That portion of the hindbrain of a vertebrate that connects the spinal cord to the remainder of the brain. It regulates such involuntary functions as breathing, swallowing, vomiting, and vasoconstriction and is an important part of the autonomic nervous system.

megaspores Spores formed by meiosis that give rise to the megagametophyte that contains the egg.

meiosis The division of a diploid nucleus that results in the formation of four nuclei each with half the chromosome number of the original nucleus.

meristem Localized regions in plants where cell divisions take place.

mesenchyme Embryonic connective tissue.

mesoderm The middle of the three primary cell layers of the embryo; gives rise to skeleton, muscular, blood vascular, and other body systems.

messenger RNA An RNA made from template DNA that moves to the ribosomes, where it codes the sequence of amino acids in a protein.

metamorphosis Change of shape, especially from one larval stage to another or from larval to adult.

metaphase The stage in mitosis during which the chromosomes line up at the equatorial plate and spindle attachment occurs. In mieosis I, the tetrads line up rather than individual chromosomes.

microfilaments Fibers of contractile protein 5–7 nanometers in diameter.

microspores Spores formed by meiosis that will give rise to male gametes.

microtubules Long, hollow cylinders, 18–27 nanometers in diameter, involved in chromosome movement, in cytoplasmic streaming, and as cytoskeletal elements.

middle lamella The portion of the wall common to two plant cells. The middle lamella holds the cells together and is composed of pectic substance.

mimic A relatively harmless organism that comes to resemble a relatively harmful one through the action of natural selection.

mitochondrion The cell organelle involved in the oxidation of food molecules in the cell.

mitosis The division of the nucleus of a cell that results in the formation of two nuclei with the same genetic composition.

model The species that is the basis for natural selection for mimicry.

molecule A structural unit consisting of two or more like or unlike atoms.

morphogenesis The origin and development of form.

Müllerian mimicry The case where two equally unpalatable species evolve to resemble one another and both gain increased protection from predators.

muscle spindle receptors Nerves that measure the shortening, or contraction, of a muscle.

mutagen A chemical or physical agent that can cause mutations.

mutation A change in a gene to produce a new allele.

mutualism An interaction between populations in which both benefit from the association.

mycelium A mass of fungal hyphae.

myofibrils The contractile elements of muscle fibers.

myoglobin An oxygen-binding, heme-containing protein found in muscle.

myosin A contractile protein found in muscle.

N

natural selection Nonrandom reproduction leading to a shift in the frequency or average distribution of alleles in a population.

nephridium The excretory organ of the earthworm.

nephrons The functional unit of the human kidney. There are some 1 million in each kidney.

nerves Discrete bundles of axons and dendrites traveling through the tissues.

neurohormones Hormones secreted by nerve cells into the bloodstream.

neurons Nerve cells, the functional unit of the nervous system. Each neuron contains a nucleus and cytoplasm plus long cytoplasmic branches leading away from the cell body.

neurosecretory cells Nerve cells that are specialized for secreting neurohormones into the bloodstream.

neurotransmitter A chemical released by a nerve that travels across a synapse to the nerve or muscle it affects.

neurulation Process of establishment of the neural plate and its formation of neural folds and the neural tube.

neutron An atomic particle that has mass but no charge.

niche The precise relationship of an organism and its total environment, especially in relation to its utilization of resources.

nicotinamide adenine dinucleotide (NAD) An important electron acceptor in cellular respiration.

nicotinamide adenine dinucleotide phosphate (NADP) A nucleotide that acts as an electron receptor and a source of energy in the oxidized state in reactions in the cell.

nitrifying bacteria Bacteria involved in the conversion of ammonia to nitrate.

nitrogen fixers Bacteria and blue-green algae capable of incorporating atmospheric nitrogen into organic nitrogen compounds.

nodes The places at which leaves are attached to the stem.

nondisjunction The failure of two homologous chromosomes to separate during meiosis.

normalizing selection A kind of natural selection in which individuals near the mean value of a particular feature under consideration are favored.

notochord A fibrocellular rod constituting the primitive skeletal axis.

nuclear envelope The double membrane that surrounds the nucleus and is continuous with the endoplasmic recticulum. The nuclear membrane is pierced with pores.

nucleic acids A group of organic compounds based on polymers of nucleotides.

nucleoid The DNA-containing region of the prokaryotic cell.

nucleolus (pl., nucleoli) A body composed of RNA and protein found in the nucleus of the cell.

nucleotides The structural units of all nucleic acids. They consist of a nitrogen-containing base, a five-carbon sugar, and a phosphate group. They may also have more than one phosphate group and be energy-transferring compounds (see adenosine triphosphate).

nucleus A membrane-bound cell organelle that contains the majority of the genetic information of the cell.

O

omnivores Animals that eat both plants and animals.

oncogene The hypothesis that viruses, particularly lysogenic viruses, cause cancer.

oocyte The primitive ovum in the ovary.

oogenesis The process of formation and development of the ovum.

oogonia In animals, the primitive ova from which the oocytes are developed. In certain fungi and algae, a unicellular female sex organ that contains one or several eggs.

optic nerve The collection of nerve fibers that connect with the receptors of the eye and run to the brain.

orbitals The most probable locations of the electrons around the atomic nucleus.

organogenesis The formation of organs.

osmoconformers Animals whose body fluids have the same osmotic concentration as the medium in which they live.

osmoregulation The maintenance in animals of a constant osmotic concentration of body fluids despite changes in osmotic concentration of the surrounding medium.

osmosis The movement of water across a semipermeable membrane in response to a free energy gradient.

oviparity Egg-laying.

ovule Structure in the flower that contains the egg.

oxidation The removal of electrons from an atom or molecule.

P

Pacinian corpuscles The receptors for touch and vibration located in the underlayers of the skin and muscles.

parasitism Members of one species existing at the expense of another species, but usually without causing the death of any individual parasitized.

parasympathetic nervous system One of the two divisions of the autonomic nervous system; it inhibits the sympathetic response. It contains the cholinergic fibers, and the transmitter substance is acetylcholine.

parathyroids In humans, four pea-sized glands located on the rear surface of the thyroid that produce parathyroid hormone (PTH).

parenchyma cell The basic cell type found in plants.

parthenogenesis Development without fertilization.

pepsinogen The enzyme produced in the stomach and converted to its active form, pepsin, in the stomach. Pepsin is a protein-digesting enzyme.

pericycle A layer of plant cells that retains the ability to divide and is the source of secondary roots.

peristalsis Wavelike contractions of circular muscles in the walls of the passageway of the esophagus.

petals The second ring of parts in a flower, frequently highly colored.

petiole A leaf stalk.

phagocytosis The engulfing of food organisms or particles by the invagination of the plasma membrane of the cell.

phenotype The physical appearance of an organism.

pheromone A chemical used as a means of communication between individuals of the same species.

phloem The tissue that transports solutions of sugars and other organic materials within the vascular plant.

phosphoglyceric acid (PGA) The first stable product in the Calvin cycle.

phospholipids Lipids that contain a phosphate group.

photon The basic particle of light. Photons have various amounts of energy depending on the color of the light.

photoperiodism The response of organisms to changes in day length. Flowering in many plants and the beginning of migration in birds are examples.

photosynthesis A series of reactions in which light energy is used to produce organic compounds, most commonly carbohydrates.

photosynthetic unit A group of pigment molecules that act as a unit in the absorption of light.

phototropism The growth response of plants to light.

phragmoplast A layer of microtubules and vesicles that forms between the daughter nuclei of plant cells at the end of mitosis and marks the beginning of the new cell wall.

piloerection The raising of feathers or hairs to form an insulating layer of air in response to cold.

pineal gland The "third eye" of lizards that secretes the hormone melatonin. Present in mammals, but with an unknown function.

pinocytosis The process by which molecules become attached to the plasma membrane of the cell, which then invaginates and forms a vesicle.

pituitary (hypophysis) A small gland suspended by a stalk from a part of the brain. It produces a number of neurohormones and hormones that regulate the function of other endocrine glands.

placenta The organ of physiological communication between mother and fetus.

plasma membrane The outer membrane of the cell.

plasmodesmata Cytoplasmic connections between plant cells.

plasmolysis The shrinking away of the cytoplasm of a plant cell from the wall as a result of decreased osmotic pressure.

plastids A group of cell organelles involved in the synthesis and storage of food molecules by the cell.

poikilotherms Animals whose body temperature is the same as that of the environment.

polymers Large molecules consisting of repeating units.

predation Members of one species feeding on members of another. Usually restricted to animal interactions.

primary wall That cell wall surrounding the young dividing and elongating cells. This wall is composed of pectic substances, hemicellulose, and cellulose.

progesterone A hormone of the corpus luteum that helps prepare the uterus for reception of the embryo.

prokaryotic cell A cell without membrane-bound cell organelles; only the bacteria and blue-green algae have prokaryotic cells.

pronucleus The nucleus of the spermatozoon or the ovum prior to fertilization.

prophase The first stage of mitosis or meiosis when the chromosomes condense and become clearly visible. In meiosis, the stage during which homologous chromosomes pair and crossing-over takes place.

proprioceptors Internal receptors that monitor the position and status of the various parts of the body.

proteins A class of large organic molecules consisting of one or more long chains of amino acids.

proton A positively charged particle in the atomic nucleus.

R

r The intrinsic rate of increase; the rate at which a population with limitless resources will grow.

race A geographic segment of a species, identifiable by several to many independently evolving characteristics.

range The geographic occurrence of a species.

recombination The exchange of portions of the chromatids of homologous chromosomes during meiosis.

reduction The acceptance of electrons by a substance.

reflex arcs Neuronal circuits linking receptor and effector.

regeneration Ability to reproduce a part that has been lost or to reorganize dedifferentiated cells.

resource axis The total spectrum of availability of a substance or object required for the maintenance, growth, and reproduction of species in a community.

respiration Cellular respiration: Those chemical reactions within the cell that release energy through the oxidation of carbon bonds. External respiration or ventilation: Those mechanisms by which organisms provide oxygen to and remove carbon dioxide from their tissues.

retina The rear surface of the eye that contains the sensory receptors.

reverse transcription The formation of DNA from RNA through the action of an RNA-dependent DNA polymerase.

rhodopsin Visual purple, the pigment present in the rods and cones of the eye.

ribosomes Small spherical bodies found in the cytoplasm composed of RNA and protein. They function as the site of protein synthesis in the cytoplasm.

rods One of the two types of receptors found in the eye (the second are the cones). The rods are the more sensitive of the two. Both types contain rhodopsin as the light-absorbing pigment.

root hairs Extensions of the epidermal cells of the root that are important in the uptake of water and minerals from the soil.

ruminants Animals, such as cattle, sheep, and deer, that have four-chambered stomachs. Newly ingested food is held in the first stomach, or rumen, which harbors cellulose-digesting bacteria and protozoa.

S

sclerenchyma A specialized support cell found in plants.

search image A behavioral mechanism by which individuals focus on certain prey types and ignore others, generally increasing their feeding efficiency by eliminating distractions.

secretin A hormone released by the intestine wall into the bloodstream that promotes the secretion of bile from the liver.

segregation Separation of maternal and paternal chromosomes during meiosis.

selection pressure A measure of the relative disadvantage of one kind of organism relative to another in a population.

senescence Growing old, especially living past reproductive age.

sepals The outermost ring of parts of a flower, frequently green.

sex chromosomes Morphologically distinct chromosomes that determine the sex of the individual in those species in which they are found.

skeletal muscles The most abundant muscles of the body, they are striated in appearance and under direct voluntary control.

smooth muscles Found in the intestines, uterus, stomach, and so on, smooth muscle fibers are interlaced into sheets that contract slowly; their contraction is usually involuntary.

sori Clusters of sporangia on the leaves of ferns.

specialization Modification by natural selection for performance of some restricted function.

species Biologically unique groups of individuals, living together or not, but capable of sharing the same gene pool.

spermatid The rudimentary spermatazoon derived from the division of the spermatocyte.

spermatocyte A cell resulting from the division of the spermatogonium, which in turn by division forms the spermatid.

spermatogenesis The formation of the spermatozoa.

spermatogonia The primitive sperm cells giving rise, by division, to the spermatocytes.

sporangia Structures in plants in which spores are produced.

sporophyte That generation consisting of diploid cells and producing spores.

stamens That portion of the flower in which the pollen is produced.

statoliths Particles postulated to be responsive to the force of gravity in plant cells and to be the detectors in geotropism; also the structure involved in balance in crustaceans.

stigma The portion of the flower on which the pollen lands and germinates.

stomata Openings in the surface of leaves through which gases diffuse.

survivorship The proportion of individuals born at the same time that are living at some later time.

symbiosis The close, long-lasting, and often necessary interaction of members of one species with those of another.

sympathetic nervous system One of the two divisions of the autonomic nervous system, working in opposition to the parasympathetic. The transmitter substance is norepinephrine.

synapse The junction between an axon terminus and another cell. Synapses may occur between two nerve cells or between a nerve cell and a muscle cell.

synecology The relationships of populations to biotic factors in their environment.

systole That part of the heartbeat that results from the contraction of the ventricles.

T

taxis The movement of an animal in response to a stimulus, such as light, temperature, or gravity.

teleology In science, the assumption that an organism purposefully pursues an action when there is no objective evidence for this.

telophase The stage of mitosis and meiosis during which the new nuclei are formed by the appearance of a new nuclear envelope.

tendons Inelastic connective tissues that attach the muscles to the bone.

testosterone One of the male hormones (androgens) secreted by the testes.

tetrad The paired homologous chromosomes.

thalamus A portion of the forebrain of vertebrates which in humans is completely concealed by the cerebrum. It is the center where incoming sensory information is integrated.

thermoperiodism A phenomenon in plants in which there are specific and different day and night temperatures for maximum growth.

thymus Endocrine gland derived from the embryonic pharyngeal pouches; involved in establishment and regulation of the immune response.

thyroid glands A pair of glands, fused in humans, that produce the hormone thyroxine.

thyroid-stimulating hormone (TSH) A hormone released by the pituitary that triggers the release of thyroxine by the thyroid.

thyroxine A hormone produced by the thyroid that regulates cellular metabolism.

totipotent Able to reproduce a whole organism from a single cell or small group of cells.

trachea A rigid tube that leads down toward the lungs.

transduction Transfer of bacterial DNA from one cell to another through the intervention of a virus.

transfer RNA Low-molecular-weight molecules of RNA that transfer specific amino acids to the ribosomes, where these amino acids are incorporated into protein.

transformation The process in which fragments of DNA released from one bacterium are absorbed by healthy bacteria from the medium.

translocation Movement of carbohydrates and other organic molecules through the plant.

transpiration Water loss to evaporation in the aerial parts of the plant.

trophoblast The outer wall of the mammalian blastocyst, concerned with nutrition of the embryo within.

tropism The growth of a plant or plant part in response to some environmental factor, such as gravity or light.

turgor pressure The pressure that results within a plant cell as a result of the osmotic pressure of the cytoplasm and vacuole pressing against the wall.

twitch A single muscle contraction. Each twitch is divided into three phases: the latent period, the contraction phase, and the relaxation phase.

tympanic membrane The eardrum, a membrane stretching across the auditory canal of the ear and separating the outer from the middle ear.

U

urea An organic form of nitrogen, CH_4ON_2, into which the waste products of protein metabolism in many animals are converted for excretion.

V

vacuoles Saclike structures, bounded by a single membrane, containing dilute solutions of a variety of substances. Particularly characteristic of plant cells.

vagility The ability of an organism to transport itself from one place to another.

vagus nerve The largest single nerve of the parasympathetic system.

vascular cambium The lateral meristem that gives rise to the secondary xylem and phloem.

vegetal pole The end of an animal egg containing the yolk.

veins Blood vessels that carry the blood back to the heart.

villi Surface projections of the cells lining the small intestine.

virus A particle consisting of a segment of nucleic acid (either DNA or RNA) coated by protein, capable of infecting a cell and converting the machinery of the cell to make more virus particles.

vitamins Organic compounds that are essential in small amounts for a variety of biological processes. Many vitamins are coenzymes.

vitelline Relating to the yolk of an egg.

vitreous humor The transparent fluid that fills the chamber of the eye.

viviparity The condition of giving birth to living young.

X

xylem The water-conducting tissue of the vascular plants.

Answers to Genetics Problems

Chapter 7

1. a. Monohybrid cross (involves only one trait)

b. #7 = Tt; #4 = TT or Tt

c. #6 must be heterozygous to bear 1/2 homozygous recessive offspring.

d. Autosomal recessive trait

e. Law of dominance

f. 50 percent

2. a. Aa: *A, a; AaBb: AB, Ab, aB, ab; AaBbCc: ABC, ABc, Abc, aBC, aBc, abc, AbC, abC; AaBbCcDd: ABCD, ABCd, ABcd, Abcd, AbCD, AbcD, ABcD, AbCd, aBCD, aBCd, aBcd, abcd, abCD, abcD, aBcD, abCd*

b. Yes. Formula = 2^n, where n = number of traits

c. *AABb: AB, Ab; AaBBCc: ABC, ABc, aBC, aBc; AABBCc: ABC, ABc; AABBCC: ABC*

d. The formula is appropriate only when both allele types are present (when the parent is heterozygous).

3. No, Because in the roundworm all the traits are carried on the one chromosome. Thus, all the traits are linked and do not follow Mendel's ratios for independently assorting genes. If Mendel had studied more pea traits he would have observed some non-Mendelian ratios, however.

4. The breeder should perform a test cross: B __ × bb. If the mother is homozygous dominant then all F_1 phenotypes will be brown (100 percent heterozygous).

5. a. 25 percent

b. 1/2 (50 percent)

c. $1/2 \times 1/2 = 1/4$

d. $1/2 \times 1/2 \times 1/2 \times 1/2 \times 1/2 \times 1/2 = 1/64$

6. He inherited the trait from his mother. Color blindness is a sex-linked trait and he inherited his Y from his father and the X chromosome with the color-blind gene from his mother.

7. a. *AABbCc:* 4 (*ABC, AbC, ABc, Abc*)

b. The same number, 4

8. ACBD

9. Synapsis (at crossing over):

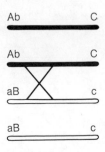

At the end of meiosis I:

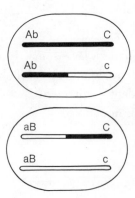

At the end of meiosis II:

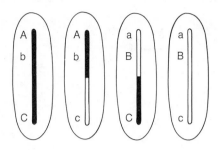

10. a. Man = $X^C Y$; woman = $X^C X^c$ (C = normal color vision; c = color-blind)

b. 1/4 (1/2 that it will be a boy times 1/2 that he will be color-blind)

c. 75 percent

11. a. 50 percent

b. No, because the autosomal dominant condition produces homozygous dominant and heterozygous conditions both having the disease.

c. No, if the gene is present is will show itself somewhere.

12.

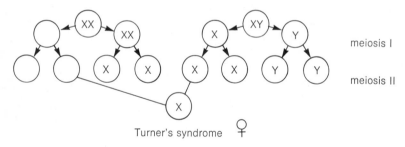

meiosis I

meiosis II

Turner's syndrome ♀

13. a. Both are sex-linked recessive traits.

b. Map distance = $\dfrac{23 + 27}{75 + 74 + 23 + 27} = \dfrac{50}{199} = 25$

14. 9:3:3:1

15. a. Bristle shape is autosomal recessive.

b. Eye color is dominant sex-linked.

c. Independent assortment

d. They are not linked.

Chapter 8

1. The parents are $I^A i$ and $I^B i$, so an $I^B i$ child could be theirs and the information does not support the man's case.

2. B > C > Y

3. The color ratios of the offspring will be 1 black:2 gray:1 white

4. a. Complete dominance

b. None

c. 3/16

5. 50 percent wild, 50 percent vestigial

b. Environmental effect on gene expression: Vestigial flies grow almost normal wings when exposed to higher temperatures. The genotypes were as expected, but the environment affected the phenotype of the vestigal flies.

6. Hybrid vigor: Whether the large beaned plant is self-pollinated (Aa) or is pollinated by a homozygous dominant or recessive for bean size, the result will still be 50 percent Aa (hybrids) and 50 percent homozygotes.

7. Advise her that she probably need not worry about the exposure to her hand since mutations in somatic cells are not passed on to offspring. She need only worry if there was a possibility that her ovaries were exposed to a high dosage.

8. Check the calf's pedigree. If there have been no dwarfs in his family for a number of generations then it probably is a mutation; if his grandparents include dwarfs then it is probably due to chance matings. To check for environmental modification compare his growth with other calves who have been living under the same conditions and eating the same food.

9. A mutation in gene H will be more deleterious than one in gene F because with no enzyme H it will never be able to synthesize substance A; if enzyme F is unavailable but substance C or B may be available from other sources, substance A may still be produced.

10. a. Transition

b. Transversion

c. Chromosomal mutation

Index

tendons, **371**
terminal cisternae, 366, *367*
termites, 277, 404, 520–521, 572
territory, 521
tertiary consumers, 443, *444*
testcross, defined, 125
testes, 205, 208, 239, 240, *241*, 250–252
testosterone, 207, **208**, 247, 250, 251, *252*
tetanus, 350, 367
tetrad, **113**, 115
tetraploidy, 484
tetrose, *29*, 30
T-even bacteriophages, 155–156, *158*
thalamus, **257**, 352, **357**
thermodynamics, 23–24
thermoperiodism, **330**–331
Thermoplasma acidophila, 563
thermoregulation, 331–342, 527, 531
thermotaxis, 410
thiamine, 76, 275
Thomomys talpoides. See pocket gophers
thoracic duct, 311
Thorpe, W. H., 419
three-carbon cycle, **67**–68, *69*
threonine, *36*, 274
threshold stimulus, 382
thrombin, 313
thromboplastins, 313
thylakoids, *62*, 63, 66, 561–562, *569*
thymine, 40
thymus gland, 244, 251, **259**, 314
thyroid glands, 247, 251, **252**–253, 334
thryoid releasing factor, 252, *253*
thyroid-stimulating hormone (TSH), 249, 251, **252**–253, 255, 334
thyroid-stimulating hormone releasing factor (TRF), 257
thyrotropin releasing factor, 249
thyroxine, 251, **252**, *253*
  and body temperature, 249
  in metamorphosis, 231, 252–253, 263
  in thermoregulation, 334
tide pools, 461–462, *466–467*, 468, 483
Tinbergen, Niko, 409, 414, 417
tissue
  fate of, 224–226
  origin of, 221, 240–244
tissue culture, 44, 187, 189, 223
tits, 511, 512
Tmesipteris, 592
toads, 417, 507–508, 540
tobacco, *186*, 187, 350
tobacco mosaic virus (TMV), 564
tongue, 389, 514
torpor, 335, 430
totipotency, **188**–191, 229
toxins, 563, 566
trachea, **259**, *290*, 291
tracheae, 289
tracheids, 191, *192*, 197
tracheoles, 289
Tracheophyta, 588, 591–599
tranquilizers, 350, 359
transcription, of nucleotides, 88–89, 158
transducer, defined, 381
transduction, **155**, *157*, 159–160
transfer RNA, 83, **88**–89
  in amino acid activation, 90–92
  in polypeptide assembly, 93–94
  types of, 89–90
transformation, *154–155*
transition, *140*, 141
translocation, 298, **299**–301
transmitter substances, 347, 350
transpiration, **288**–289, 298–299
transpiration-tension-cohesion hypothesis, 298–299
transport
  of CO_2, 287, *310*, 311
  in humans, 302–308
  in invertebrates, *300*, 301–302
  in plants, 297–301
  in vertebrates, 302
transverse system (T-system), 366, *367*
transversion, *140*, 141
tree frog, 223, 460
tree rings, 195

Trematoda, *605*, 606
Trench, Robert K., 552–553
trial-and-error learning, 417, 419
Tribolium, 510, 514, 616
triceps muscle, 371–372, *373*
Trichomonas vaginalis, 572
Trichophyton interdigitale, 586
triglycerides, 33–34
triiodothyronine, 251, 252
triose, *29*, 30
triploidy, 484
Triticale, 274
Tritonia, 355
trophic levels, 445–446
trophoblast, **236**
tropisms, *184*, **185**, 409–410
tropomyosin, 364, 365
troponin, 364, 365, 366
Truman, James W., 427–428
Tryon, R. C., 416
Trypanosoma, 571
tryptophan, *36*, 266, 274
tse-tse fly, 571
tube feet, 369, 373, 617
Tubifex, 608
tubule, 325–326, *327*
Tubulidentata, *539*, 627
tubulin, 108
Tucker, Vance, 430
tunicates, 618, *619*
Turbellaria, 605–606
turbinate bones, *290*, 291
turgor pressure, **57**, 318, 378
Turner, Bruce, 336
Turner's syndrome, 135
twitch, **367**–368
tympanic membrane, **386**
tympanic organs, 384–385
tyrosine, *36*, 254

ulcers, 282
ultraviolet light, 142, 152, 156, 547, 551
umbilical cord, 237, *238*
upwelling, 461–462
uracil, 40
urea, **231**, 320, 322, 323
Urechis, 520
ureter, 325, *326*
urethra, 240, 325, *326*
uric acid, 322, 323
Urochordata, 618, *619*
Urodela, 623. *See also* salamanders
Urtica, 194–195
uterus, 235–237

vaccination, 565
vacuoles, **51**
  in bacteria, 557
  in blue-green algae, 562
  contractile, 323–324, 572
  in epidermis, 193
  food, 277
  in *Paramecium caudatum*, 574, *575*
  in prokaryotes, 49
vagility, **537**
vagus nerve, 282, 307–308, **359**, 361
valine, *36*, 274
valves, in human circulatory system, 303, *304*, 305, 306
van Helmont, Jan, 267
Van Valen, Leigh, 535
vascular cambium, **183**–184, *194*, 195
vascular tissue, in stem, *196*, 197
vasopressin, 249, 251, 255, *256*, 326–327
vegetal pole, **203**, 220
vegetarians, 285–286, 446
veinlets, 197
veins, **303**, 305
  of leaves, 197
  human, 248, 303, 311, *312*
venereal disease, 212, 554
ventilation, 287, 292–293, *294*
ventral light reaction, 410
ventricle, 302
venules, 303, *305*

vertebrates. *See* Chordata
Vesalius, 297
vesicles, 49, 59, 97–99, 111–112, 323–324
vessels, in xylem, 191, *192*
Vestal, Arthur G., 474–475
vibration receptors, 383–388
villi, 236–237, **283**
Virchow, Rudolf, 45
viruses, **44**, 155, 157–158, 555, 564–566
visceral organs, development of, 239
vision, 387, 390–395, 514
vitelline membrane, **203**
vitreous humor, **391**
viviparity, **229**
vitamin A, 64, 275, 286, 390
vitamin B, 266, 275, 283, 285
vitamin C, 275
vitamin D, 253, 275, 286
vitamin E, 232
vitamin K, 275, 283, 286
vitamins, 49, **275**–276, 286
von Frisch, Karl, 404–407, 409, 428–429
von Mering, Johann, 253–254
vultures, 376

Wald, George, 392–393
walking, 373
Wallace, Alfred, 487, 537
Wallace's line, 487, 537, *540*
Ware, Katherine, 284
water
  in animal reproduction, 198
  bacteria in, 555
  as barrier to colonization, 459, 460
  cohesiveness of, 298, 299
  in cosmic clouds, 545, 547
  and gamete transfer, 200–201
  larval development in, 201
  necessity of, 317
  properties of, 20–21, 22, 26, 56
  and seed germination, 181
  transport of, in plants, 297–299
water balance
  in animals, 320–323, 623
  in plants, 318–320
water boatman, 410
water cycle, *450*, 451–452
water flea. *See Daphnia*
water molds, 582, *583*
water strider, 383–384
water table, 451–452
Watson, James D., 7, 86–87
weasel, 374
Weinberg, G., 485
Weisenberg, R., 108
Wells, Patrick, 406–407
Welwitschia, 598
Wenner, Adrian, 406–407
Went, Frits, 3–4, 330–331
whales, 277, 375, 388, 399, 516
wheat, 274, *506*, 507, 586
whippoorwills, 335
White, M. J. D., 507
white blood cells, 59
Whittaker, Robert H., 474–476
whooping crane, 478–479
Willingham, Mark, 55
Wilson, Donald, 354
Wilson's snipe, 397
wings, 376–377, 530, *531*
Witt, Peter, 423
Wollman, E. L., 154
wolves, 399, 401–402, 516
woodchucks, 430
woodcock, 401
wood thrush, 399
Woodward, John, 267

Xanthophyta, 569
xenopeltids, 541
Xenopus, 163, 205
Xiphosura, 381, 614
X rays, as mutagens, 139, 142, 152
xylem, **183–184**, 191–192, 195, 197
  in plant nutrition, 274
  water transport in, 298–299

yeasts, 74, 152, 584, 585
yellow-green algae, 569
yogurt, 560
yolk, *202*, 203, *216*, 217–219
yolk platelets, 203
Young, Thomas, 393
yuccas, 289

zebras, 517
zeitgeber, 426
zero population growth, 498–499
zinc, 14, 267, 268, 276
Z-line, 363–364, *365*, 366, *367*
zooplankton, 443, 472
zoospores, 166–167, 582
zooxanthellae, 468, 603
Zygomycetes, 583–584
zygote, formation of, 117
zygotene, 113, *114*, *116*

To the owner of this book:

We hope that you have enjoyed *Biology* as much as we enjoyed writing it. We'd like to know as much about your experiences with the book as you care to offer. Only through your comments and the comments of others can we learn how to make *Biology* a better book for future readers.

School _____ Your Instructor's Name _____

1. What did you like *most* about *Biology?* _____

_____

2. What did you like *least* about the book? _____

_____

3. Were all of the chapters of the book assigned for you to read? _____

(If not, list the numbers of ones that weren't.) _____

_____

4. How interesting and informative were the essays and supplements for you? _____

_____

_____

5. Would you like to see the book continue to include essays and supplements in future editions? Why or why not?

_____

_____

_____

6. Were there any unclear illustrations? (If so, list the figure numbers.) _____

_____

How would you compare the graphics in *Biology* to those in other college textbooks you have read? _____

_____

7. Were there any unclear chapters? (If yes, list the chapter numbers.) _____

_____

_____

8. In the space below or in a separate letter, please let me know what other comments about the book you'd like to make. (For example, were any chapters *or* concepts particularly difficult?) I'd be delighted to hear from you!

Optional:

Your Name _____ Date _____

May Wadsworth quote you, either in promotion for *Biology* or in future publishing ventures?

Yes _____ No _____

Sincerely,

William A. Jensen

_____

FOLD HERE

CUT PAGE OUT

_____

FOLD HERE

FIRST CLASS
PERMIT NO. 34
BELMONT, CA

BUSINESS REPLY MAIL
No Postage Necessary if Mailed in United States

Dr. William A. Jensen

Wadsworth Publishing Company
10 Davis Drive
Belmont, CA 94002